INTRODUCTORY APPLIED PHYSICS

INTRODUCTORY APPLIED PHYSICS

FOURTH EDITION

NORMAN C. HARRIS

Professor Emeritus of Higher Education
The Center for the Study of Higher Education
The University of Michigan

EDWIN M. HEMMERLING

Chairman Emeritus, Division of Science,
Mathematics, Engineering and Technology
Bakersfield College

Gregg Division/McGraw-Hill Book Company

New York	*Madrid*
Atlanta	*Mexico*
Dallas	*Montreal*
St. Louis	*New Delhi*
San Francisco	*Panama*
Auckland	*Paris*
Bogotá	*São Paulo*
Düsseldorf	*Singapore*
Johannesburg	*Sydney*
London	*Tokyo*
	Toronto

Library of Congress Cataloging in Publication Data

Harris, Norman C
  Introductory applied physics.

  Includes index.
  1. Physics. I. Hemmerling, Edwin M., joint author.
II. Title.
QC21.2.H37   1979   621   79-19475
ISBN 0-07-026816-9

Introductory Applied Physics, Fourth Edition

1 2 3 4 5 6 7 8 9 0 D O D O 7 7 6 5 4 3 2 1 0 9

Contents

Preface

In keeping with the central challenge of the 1980s, the unifying theme of this Fourth Edition of *Introductory Applied Physics* is *energy and matter*. In addition to preserving the energy theme throughout the usual topical subdivisions of physics, much new material on energy resource development has been added. The chapters and sections on electronics, thermodynamics, and wave motion have been extensively revised and updated. A completely new chapter on quantum physics, containing an introduction to high-energy (particle) physics, is included. The chapter on nuclear energy has been extensively revised and updated; and introductory discussions of solar energy, geothermal energy, wind energy, and energy from ocean thermal currents are new features.

The Fourth Edition puts increased emphasis on SI-Metric units. The authors have purposely placed the book well ahead of the present status of metrication in the United States, anticipating nearly complete metrication in a decade or two. Users in other countries, already on the SI-Metric system, will welcome this feature. British (engineering) units are retained in some sections of the book however, since these units still represent the "state of the art" in many fields in which technicians and semiprofessionals now work.

College-level technical education has gone through a significant evolution since 1955, when the first edition of *Introductory Applied Physics* was published. Engineering technicians rapidly demonstrated their worth during the 1950s and 1960s, and other segments of the economy were quick to adopt the idea of semiprofessional support personnel. Agricultural firms found that the production, processing, and marketing of their products was significantly improved by the work of technicians. Medicine added a score or more of "paraprofessional" job titles prior to 1975. As computers and electronic devices revolutionized the business world, semiprofessionals have become key members of its work force. Even in the realm of public and human services, the use of paraprofessionals has increased at a rapid rate during the 1970s.

As a consequence of these trends, four-year colleges, community and technical colleges, and area vocational-technical schools now provide hundreds of different curriculums for the career education and specialized training of technicians, paraprofessionals, and semiprofessionals in all fields of the economy. A basic course in applied physics is recommended or required in most of these programs. This edition of *Introductory Applied Physics* has been especially prepared for use in such courses.

The book still retains a strong emphasis on the needs of engineering, science, and industrial technicians; but the revision meets the needs and interests of students from programs in agriculture, business, allied health, and public services as well.

All the features which made the earlier editions popular with students and faculty alike have been retained. The book *really applies* physics to the job requirements of paraprofessionals and technicians. The scores of photo-illustrations and diagrams clearly point up the relationship of physics to industry and engineering; and also to agriculture, business, and allied health. The mathematics requirement consists only of plane geometry, elementary algebra, and simple trigonometry. A review of essential mathematics is provided in the first chapter. There is also an important section on dimensional analysis early in the book. Illustrative problems are numerous throughout the text, and these are completely worked out for the student, showing all the steps involved and emphasizing dimensional analysis.

Problems for out-of-class assignments are included at the end of every chapter, and these are graded in approximate order of difficulty. Many completely new problems have been prepared for this edition. Answers to odd-numbered problems are given in the Appendix, and all the end-of-chapter problems are completely solved in an *Instructor's Manual*, which is available to faculty members using the book as a class text.

The *Instructor's Manual* contains, for each of the book's thirty-four chapters, such features as:
1. Suggestions for organizing a basic applied physics course, including lectures, discussions, demonstrations, and laboratory.
2. A sample set of instructional objectives.
3. Sample quizzes, including short-answer type items and numerical problems.
4. Complete, step-by-step solutions to every problem in the book.

In addition, the first section of the manual contains many suggestions specifically intended to be of help to faculty members teaching applied physics for the first time to students enrolled in various career-oriented programs.

The authors deeply appreciate the helpful suggestions offered by the many professors and instructors who have used prior editions of the text with their students. All suggestions received were carefully evaluated and many of them have been incorporated in the revision.

Special acknowledgment is made of the assistance provided by Professor Dwight D. Hinds of the Washtenaw Community College in Michigan, who served as a consultant for the revision of portions of the section on Modern Physics.

Sincere appreciation is also due to Professor Tom Sandin, Physics Department, North Carolina Agricultural and Technical University, Greensboro, North Carolina, who carefully reviewed the entire manuscript and made many valuable comments.

Professor Ramon H. Palmer, Physics Department, Williamsport Area Community College (Pennsylvania), and Professor D. W. Silvey, Physics Department, Bakersfield College (California) both made many valuable contributions, and their interest is recognized and appreciated.

Finally, the authors' appreciation is here acknowledged for the courtesy and helpfulness of the many manufacturers, corporate offices, hospitals, agricultural firms, and government agencies that provided information and hundreds of photographs and charts, from which the illustrations for the Fourth Edition were selected.

Norman C. Harris
Edwin M. Hemmerling

Definitions, Conversion Factors, Equivalents, and Physical Constants

(Most values are given to four significant figures in these tables. However, three significant figures are sufficient for most applied physics problems)

Length (Distance)

1 inch = 2.540×10^{-2} m
   = 2.540 cm

1 foot = 0.3048 m
   = 30.48 cm
   = 3048 mm

1 yard = 0.9144 m
   = 91.44 cm
   = 3 ft

1 nautical mile = 6076 ft

1 statute mile = 5280 ft
    = 1760 yd
    = 1609 m
    = 1.609 km

1 light/year = 9.463×10^{15} m
    = 5.880×10^{12} statute miles

1 meter = 100 cm
   = 1000 mm
   = 10^{-3} km
   = 1.094 yd
   = 3.281 ft
   = 39.37 in.

1 kilometer = 1000 m
    = 0.6214 mile
    = 1094 yd
    = 3281 ft

1 angstrom (Å) = 10^{-10} m = 10^{-8} cm

1 microinch (μin.) = 10^{-6} in.

1 micron (μ) = 10^{-6} m

1 millimeter = 10^{-3} m

Area

1 in.^{2} = 6.452 cm^2

1 ft^2 = 929.0 cm^2

1 m^2 = 10.76 ft^2

1 acre = 43,560 ft^2 = 0.4049 hectare

1 hectare = 10,000 m^2 = 2.471 acres

1 sq mile = 640 acres

Volume

1 cubic meter = 10^6 cm^3
     = 10^3 liters
     = 35.31 ft^3
     = 264.2 gal

1 cubic foot = 1728 in.^{3}
    = 7.481 gal
    = 0.02832 m^3
    = 28.32 liters

1 liter = 1000 cm^3
   = 10^{-3} m^3
   = 0.2642 gal
   = 1.056 qt

1 quart = 946.5 cm^3

1 gallon = 231.0 in.^{3}
    = 3.785 liters
    = volume of 8.337 lb water at 60°F

Angle

1 degree (°) = $\frac{1}{360}$ the angular measure
     around a point (revolution)
    = 60 minutes of arc
    = 3600 seconds of arc

1 minute of arc = 60 seconds of arc

1 radian = 57.30°
    = 0.1592 rev

1 revolution = 360°
     = 2π rad

Speed

1 m/sec = 3.60 km/hr
    = 3.281 ft/sec

1 km/hr = 0.6214 mi/hr

1 ft/sec = 0.3048 m/sec
    = 30.48 cm/sec

1 mi/hr = 1.467 ft/sec
   = 1.609 km/hr

30 mi/hr = 44 ft/sec

1 knot = 1 nautical mile/hr = 1.152 statute mile/hr

Mass

1 kilogram = 10^3 gm
= 10^6 mg
= 0.0685 slug

1 slug = 14.59 kg

1 international atomic
mass unit (amu) = 1.6605×10^{-27} kg

Time

1 year = 365.2 days

1 day = 24 hr
= 86,400 sec

1 min = 60 sec

Frequency: 1 hertz (Hz) = 1 cycle/sec = $\sec^{-1}$

Force

1 newton (N) = 0.1020 kg force
= 102 gm force
= 0.225 lb force

1 pound force (wt) = 16 oz force
= 7000 grains (gr)
= 4.448 newtons
= 453.6 gm force
= 0.4536 kg force

1 ton force = 2000 lb force (wt)

1 gram force = 0.00981 newton (N)

1 kilogram force (wt) = 2.205 lb force
= 9.807 newtons

1000 kg force = 1 metric ton

1 metric ton = 1.103 English tons

Density of "Standard" Substances

Mass density of mercury, ρ = 13,600 kg/m^3
= 13.6 gm/cm^3

Mass density of water (4°C), ρ = 1000 kg/m^3
= 1 gm/cm^3
= 1 kg/liter

Weight density of water (4°C), D = 62.4 lb/ft^3

Mass density of dry air (STP), ρ = 1.293 kg/m^3
= 1.293×10^{-3} gm/cm^3

Weight density of dry air (STP), D = 0.0807 lb/ft^3

Pressure

1 newton/m^2 = 1 pascal (Pa)
= 1.450×10^{-4} lb/in.^{2}
= 9.869×10^{-6} atm
= 7.501×10^{-4} cm Hg
= 10^{-5} bar
= 0.00750 torr

1 lb/in.^{2} = 144 lb/ft^2
= 6.895×10^3 newtons/m^2
= 5.171 cm Hg
= 27.68 in. water

1 atmosphere (atm) = 76 cm Hg = 760 mm Hg
= 1.013×10^5 newtons/m^2 (Pa)
= 1.013 bar
= 1013 millibars
= 14.70 lb/in.^{2}
= 29.95 in. Hg
= 33.90 ft water

1 bar = 10^5 Pa

1 millibar = 10^2 Pa

1 torr = 1 mm Hg
= $\frac{1}{760}$ atm = 133.3 Pa

Mechanical Energy

1 joule (J) = 1 newton-meter (N-m)
= 1 watt-sec
= 0.7376 ft-lb
= 2.778×10^{-7} kW-hr
= 6.242×10^{18} electron volts (eV)

1 foot-pound (ft-lb) = 1.356 joules (J)
= 3.766×10^{-7} kW-hr

1 kilowatt-hour (kW-hr) = 3.600×10^6 J
= 2.655×10^6 ft-lb

1 horsepower-hour (hp-hr) = 1.980×10^6 ft-lb
= 0.7457 kW-hr

1 electron-volt (eV) = 1.602×10^{-19} J

Heat Energy

1 kilocalorie (kcal) = 1000 cal
= 3.970 Btu
= 2.61×10^{22} eV

1 British thermal unit (Btu) = 252 cal
= 0.252 kcal

1 therm = 10^5 Btu

1 ton of refrigeration = 12,000 Btu/hr

1 Btu/lb = 0.555 kcal/kg

Power

1 watt = 1 joule/sec

1 kilowatt (kW) = 1000 watts
= 737.6 ft-lb/sec
= 1.341 hp

1 megawatt (MW) = 10^6 watts

1 horsepower (hp) = 550 ft-lb/sec
= 33,000 ft-lb/min
= 745.7 watts
= 0.7457 kW

Electricity and Magnetism

electric charge Q

1 coulomb (coul) = 6.242×10^{18} electron charges (e)

1 electron charge (e) = 1.602×10^{-19} coul

1 amp-hr = 3600 coul

1 coul = 2.788×10^{-4} amp-hr

1 faraday = 96,487 coul

magnetic flux, ϕ

1 weber (wb) = 10^8 maxwells
= 10^4 gauss-m^2
= 1 tesla-m^2
= 1 volt-sec
= 1 N-m/A

magnetic flux density, or magnetic induction, B

1 tesla (T) = 1 weber/m^2 = 1 newton/(amp-m)
= 10^4 gauss

1 gauss = 10^{-4} weber/m^2

magnetic field intensity, H

1 amp-turn/m = 1 N/wb = 1.257×10^{-2} gauss

1 gauss = 79.55 amp-turn/m

electrical relationships

1 amp = 1 coul/sec

1 coul = 1 amp-sec
= 1 farad-volt

1 ohm = 1 volt/amp

1 farad(F) = 1 coul/volt

1 microfarad (μF) = 10^{-6} farad

1 volt = 1 joule/coul

1 henry = 1 volt-sec/amp
= 1 ohm-sec
= 10^6 microhenrys (μH)
= 10^3 millihenries (mH)

1 joule = 1 volt-coul
= 1 weber-amp

1 eV = 1.602×10^{-19} joule

1 MeV = 1.602×10^{-13} joule

Conversion Factors for Heat Energy and Mechanical Energy

1 Btu = 778.0 ft-lb (Joule's constant
J, English system)
= 1055 joules
= 2.930×10^{-4} kW-hr

1 therm = 10^5 Btu = 29.3 kW-hr

1 kcal = 4.186×10^3 joules (Joule's
constant J, mks system)
= 3.086×10^3 ft-lb
= 1.163×10^{-3} kW-hr

1 calorie (cal) = 4.186 joules

1 ft-lb = 1.285×10^{-3} Btu
= 3.239×10^{-4} kcal

1 joule = 2.389×10^{-4} kcal
= 9.480×10^{-4} Btu
= 2.778×10^{-7} kW-hr

1 kilowatt (kW) = 0.2389 kcal/sec
= 3.413×10^3 Btu/hr

1 horsepower (hp) = 0.1781 kcal/sec
= 2.545×10^3 Btu/hr

1 Btu/hr = 0.2160 ft-lb/sec
= 3.928×10^{-4} hp
= 6.998×10^{-5} kcal/sec
= 2.930×10^{-4} kW

1 kilowatt hour (kW-hr) = 1.341 hp-hr
= 3.600×10^6 joules
= 3413 Btu
= 860.6 kcal

Mass-Energy Equivalency Factors

$$1 \text{ amu} = 931.5 \text{ MeV}$$

$$1 \text{ electron (rest) mass} = 0.5110 \text{ MeV} = 8.187 \times 10^{-14} \text{ joule}$$

$$1 \text{ kg} = 5.61 \times 10^{29} \text{ MeV} = 8.99 \times 10^{16} \text{ joules}$$
$$= 24.98 \times 10^9 \text{ kW-hr}$$

$$1 \text{ lb} = 2.55 \times 10^{29} \text{ MeV} = 11.33 \times 10^9 \text{ kW-hr}$$
$$= 15.19 \times 10^9 \text{ hp-hr}$$

If 1 kg of mass could be *entirely* converted to energy at a controlled rate in one 24-hr day, the power being generated would be slightly in excess of 1 billion kilowatts!

Important Physical Constants

Factor	Symbol	Accepted Value
Acceleration of gravity (earth, sea level, equator)	g	9.807 m/sec^2; 32.17 ft/sec^2
Universal gravitational constant	G	$6.670 \times 10^{-11} \text{ newtons-m}^2/\text{kg}^2$; $6.670 \times 10^{-11} \text{ m}^3/\text{kg-sec}^2$; $3.44 \times 10^{-8} \text{ lb-ft}^2/\text{slug}^2$
Standard atmospheric pressure	p_{atm}	$1.013 \times 10^5 \text{ newtons/m}^2$ (Pa); 14.7 lb/in.^2; 76 cm Hg; 29.95 in. Hg; 33.9 ft water
Mechanical equivalent of heat (Joule's constant)	J	4186 joules/kcal; 778.0 ft-lb/Btu
Absolute zero	0 K	$-273.16°C$
	0°R	$-459.7°F$
Stefan-Boltzmann constant	σ	$1.36 \times 10^{-11} \text{ kcal/(sec)(m}^2)(\text{K}^4)$
Planck's constant	h	$6.626 \times 10^{-34} \text{ J-sec}$
Ice point of water, at 1 atm		0°C; 32°F
Boiling point of water, at 1 atm		100°C; 212°F
Speed of light in vacuum	c	$2.998 \times 10^8 \text{ m/sec}$; 186,300 mi/sec
Speed of sound in air (0°C, 32°F)	v	331.4 m/sec; 1087 ft/sec
Charge on the electron	e	$1.602 \times 10^{-19} \text{ coul}$
Charge-to-mass ratio of electron	e/m	$1.759 \times 10^{11} \text{ coul/kg}$
Permittivity of free space	ε_0	$8.854 \times 10^{-12} \text{ coul}^2/(\text{newton-m}^2)$
Permeability of free space	μ_0	$4\pi \times 10^{-7} \text{ T-m/A}$
Rest mass of neutral hydrogen atom	m_h	$1.6734 \times 10^{-27} \text{ kg}$; 1.0078 amu
Diameter of hydrogen atom		$1.058 \times 10^{-10} \text{ m}$
International atomic mass unit ($_6C^{12} = 12$ amu)	amu	$1.6605 \times 10^{-27} \text{ kg}$

Rest masses of selected particles ($_6C^{12} = 12$ amu)

Particle	Rest mass amu	Rest mass kg
Electron	5.486×10^{-4}	9.109×10^{-31}
Proton	1.00728	1.6725×10^{-27}
Neutron	1.00867	1.6748×10^{-27}
Alpha-particle	4.00151	6.6441×10^{-27}

Some Earth Data

Factor	Symbol	Accepted Value
Acceleration of gravity (at sea level)	g	9.807 m/sec^2; 32.17 ft/sec^2
Atmospheric pressure (at sea level)	p_{atm}	14.7 lb/in.^{2}; 1.013×10^5 newtons/m^2 (Pa); 760 mm Hg; 29.95 in. Hg; 760 torrs; 33.9 ft water
Density of dry air (0°C; 32°F; 1 atm)	D_{air}	1.293 gm/liter; 0.0807 lb/ft^3;
Density (mean) of earth		5520 kg/m^3
Magnetic induction of earth's field	B_{earth}	5.7×10^{-5} weber/m^2
Mass of earth	m_{earth}	5.983×10^{24} kg; 6.594×10^{21} tons
Mass of the sun	m_{sun}	1.99×10^{30} kg
Mean orbital speed of earth around the sun	v_{earth}	2.977×10^4 m/sec; 18.50 mi/sec; 66,700 mi/hr
Mean angular velocity of daily rotation	ω_{earth}	7.29×10^{-5} rad/sec
Mean surface speed of daily rotation		0.464 km/sec; 0.288 mi/sec; 1037 mi/hr
Mean distance from center of sun		149.5×10^6 km; 92.9×10^6 miles
Mean distance, earth to moon		3.85×10^8 m = 239,000 miles
Radius of earth, equatorial		6378 km; 3964 miles
Polar		6357 km; 3950 miles
Mean		6370 km; 3960 miles
Solar constant of radiation (power received per unit of area of earth's surface perpendicular to sun's rays)		0.323 kcal/(sec)(m^2) = 19.4 kcal/(min)(m^2) 1.35 kw/m^2; 7.16 Btu/(min)(ft^2) 4.296×10^2 Btu/(hr)(ft^2)

PART 1 Introduction, Measurement, and Graphic Methods

Optical flats utilize the wavelength of light to check measurements to an accuracy of a few millionths of an inch. (DoALL Company.)

1 Introduction and Mathematics Review

There was a time not very long ago when scientists dealt with theory and principles, engineers turned theory into practice, and manual workers produced the goods and services that maintained the economy. These three groups have now been joined by a fourth—an occupational group called *technicians* and *technologists*. Scientific and engineering theory and practice have become very complex and the machines and instruments of industry and business are so sophisticated that millions of new professional and semiprofessional workers with technical training are now essential in the nation's work force.

New developments in chemistry have created entirely new industries like petrochemicals, plastics, and synthetics. Research in biology has almost revolutionized agriculture and medicine. And physics and engineering research have produced the computer, microwave communications, space exploration, automated industrial production, and nuclear energy. All of these new industries require thousands of technicians and technologists.

Besides these newer developments of physics research, there are many ordinary everyday applications that demand highly trained technicians. Examples include the automobile, the jet airplane, air conditioning and refrigeration, electronic communications, biomedical equipment, electric light and power, and the highly productive machines in business, industry, and agriculture. Technicians and technologists are involved in manufacturing and construction; in science and engineering research and development; in energy resources development; in business and finance; in medicine and allied health fields; in agriculture; and in public and human services.

In introductory physics you will learn a great deal about the science which is basic to technology and about the applications of physics to the career field you may be considering. If you have already decided on a career as a technician or technologist, this book will provide basic knowledge for the study of your career field. It will also serve as a valuable reference for you later when you are working in industry, business, medicine, or agriculture. If you are still uncertain about a career choice, or if you enrolled in college physics as a basic science requirement for some other field, you will find much of interest and value in this book. No person in modern society is truly educated without an understanding of basic physics. As a citizen you cannot afford to be ignorant about physics; and as a future technician or technologist your job will depend directly on your knowledge and application of physics principles.

WHAT IS PHYSICS?

1.1 Physics Is the Science of Matter and Energy

Matter is the stuff of which everything you see and touch is made. *Energy* is the ability to do work. Advanced technical societies such as ours depend on high-quality materials of all kinds—metals, plastics, fuels, synthetics, wood products,

Fig. 1.1 Molten iron from a blast furnace is poured into an open-hearth furnace where steel will be made. The open-hearth furnace will operate at a temperature of about 2900° F. Tremendous amounts of heat energy are used to produce steel, one of the basic materials of industry. (Bethlehem Steel Corporation.)

chemicals, food and fiber—and they use vast quantities of energy in refining, shaping, and fabricating these materials. Physics is the study of both matter and energy, and of the factors which make industry and advanced technology possible.

The future demands that we develop not only better ways of using resources but better ways of conserving them. The list of critical materials—those in a severely depleted state—grows longer each year. It is now increasingly evident that petroleum and natural gas, our main sources of energy for half a century, are on the way to exhaustion. Many experts on petroleum reserves agree that at present and future predicted rates of use, the world supply will be nearly gone by the end of this century. The United States has an ample supply of coal reserves, but the problems of mining coal, transporting it, and burning it within acceptable environmental and pollution standards are yet to be solved.

The search for, and development of, new sources of energy is already a major worldwide concern, and the science of physics is basic to the development of all potential energy sources. This book will give you an understanding of how physics relates to the development, use, and conservation of energy resources. In summary, the twin themes of the entire book are *matter* and *energy*.

1.2 Matter Exists in Three Forms

For convenience in study, matter is ordinarily classified as existing in one of three *states*—as a *solid*, a *liquid*, or a *gas.**

Solids and their properties are of great importance industrially, and one family of solids, the metals, constitutes the very foundation of our

---

* In "modern physics" some scientists suggest a fourth state—the *plasma state*—involving gaseous mixtures of electrified atomic particles at temperatures reaching millions of degrees.

Fig. 1.2 Both heat energy and hydraulic energy can be converted to electric energy. (a) The rotor of a giant steam turbine undergoing final inspection. When assembled in the turbine, it will spin at 3600 rev/min and will generate 500,000 kW of electric power. (Westinghouse Electric Corporation.) (b) The Noxon Rapids Hydroelectric Project on the Clark Fork River in western Montana. The electric power output of this project is also 500,000 kW. (The Washington Water Power Company.)

(a)

(b)

modern machinery and construction industries. Many new metals have been developed in recent years for use in devices and machines for space exploration. Certain alloys of magnesium, titanium, and nickel are examples of new "exotic" metals.

Liquids and their properties, such as pressure, surface tension, viscosity, and specific gravity, will be studied. The applications of hydrostatics and hydraulics to many industries will be explained. Two-thirds of the earth's surface is covered by the oceans, and we are only beginning to realize their importance as a natural resource for food, minerals, chemicals, and, perhaps eventually, energy.

Gases are highly important in today's industrial economy. Their use in welding and as anesthetics is well known. Gases expanding as a result of heat serve as the working substance in heat engines—steam and internal combustion engines, jet engines, and rocket motors. The properties of several gases make them suitable for use as refrigerant vapors in refrigerating and air-

conditioning systems. Aside from industrial uses, the very atmosphere we breathe is a mixture of gases which exerts a considerable pressure at the earth's surface. We have taken the atmosphere for granted in the past, believing that an inexhaustible supply of air was free for the breathing. Current problems of air pollution around the world are now serious enough, however, to cause scientists, engineers, medical doctors, and the general public to become increasingly concerned about the thin envelope of air which supports life on this planet.

Solids, liquids, and gases as forms of matter will be discussed in detail in later chapters.

1.3 Different Forms of Energy

Energy shows up in many forms. *Potential energy* is energy due to position, like that possessed by water at the top of a dam or by a pile-driver hammer at the top of its travel; or energy due to an unusual arrangement of matter, as in a coiled spring or in compressed air. *Kinetic energy* is energy of motion, for example that of a rushing stream, a pile-driver hammer as it falls, a speeding automobile, or a rocket exhaust.

Before an object or substance can possess potential energy, *work* must be done on it. For example, in winding up a spring or in compressing air, the coiled spring and the compressed air possess potential energy. By the use of a suitable *machine*, this potential energy can, with some losses, be converted back into work. For the purposes of physics the definition of a machine is *a device to convert energy into work.*

Some examples of forms of energy in common industrial use are: *electric energy, heat energy, chemical energy, hydraulic energy,* and *nuclear energy.*

Energy Sources Petroleum, natural gas, coal, and oil shale are known as *fossil fuels,* since they were produced and laid down in the earth in past geological ages from the remains of animals and plants. (The word "fossil" is from a Latin root which means "to dig up.") With fossil fuels, except coal, getting dangerously scarce and very costly, great interest and much research and exploratory activity are focused on developing energy from sources which are "renewable" or inexhaustible. Examples of these sources are: *solar energy, wind energy, geothermal energy* (heat from the interior of the earth), and *tidal energy.* Nuclear energy is undergoing intensive development also. There are indications that large amounts of energy could be obtained from agricultural projects (using the sun as the energy source and selected plants as energy converters), and also from the recycling of garbage and waste.

The study of energy involves the concepts of *mass, force,* and *motion,* and leads to an understanding of *heat, work,* and *power.* These terms, and many more, will all be defined and discussed in detail in later chapters.

1.4 Heat Energy and Mechanical Work

The study of heat and temperature is a vital part of technical physics. Heat engines run most of the nation's transportation system and generate a major share of our electricity, and heat energy warms our homes and factories. Reversed heat engines perform refrigeration and air-conditioning jobs, and heat is a basic factor in the manufacture of steel and in most other metallurgical processes. Many of the machines of industry and nearly all of those in agriculture are powered by heat engines, and the electricity that powers the rest of them is, in large part, generated in steam plants which convert heat energy to electric energy. The exact relationship between heat and mechanical work will be explained in detail in a later chapter.

1.5 Energy and Wave Motion

Wave motion is an interesting study, and we shall encounter it as we discuss the fields of *sound* and *acoustics* and again in *light* and *optics.* The communications and entertainment industries, including telephony, sound movies, phonograph and tape recordings, radio, and television, require a knowledge of how people hear and how hearing in rooms can be improved by the application of acoustical materials.

The motion-picture industry, camera manufacturing, optical-instrument manufacturing, and the preparation and fitting of eyeglasses for the millions who have subnormal vision—all depend on the basic principles of light and optics.

1.6 Electric Energy is the Foundation of Technology

What is electricity? Despite our near total dependence on electricity, it is little understood by the majority of people. Electricity is *energy* in the form of *electron flow,* or the flow of electric charge. It will be the purpose of several chapters of this book to explain in detail the basic principles and common industrial practices in the production and use of electric energy. We shall develop electrical theory at some length, then devote considerable space to lessons in electric circuits, motor and generator principles, construction and use of electrical instruments, and methods of generating and distributing electric energy. Alternating-current circuits will be explained, and the principles of commercial power distribution will be outlined. Single-phase and three-phase circuits will be discussed, as will motors operating in these circuits. Electrical communications and electronics will be treated and many of their industrial applications explored. Transistors, computers, oscilloscopes, and electronics in medicine will be covered in considerable detail.

1.7 "Modern Physics" and the Future

The study of physics is sometimes divided into two main parts: *classical physics*, comprising the body of knowledge outlined above; and *modern physics*, which includes atomic and subatomic phenomena, radioactivity, x-rays, relativity, quantum (or "particle") physics, the photoelectric effect, atomic fission, and thermonuclear reactions.

This book will close with a section on selected topics from "modern physics." There we will point out how these discoveries are being used, and attempt a look into the future to see how well they may eventually be adapted as tools of industry, commerce, agriculture, and medicine.

The remainder of this introductory chapter concerns itself with a review of some basic mathematics essential to an understanding of physics.

MATHEMATICS REVIEW

1.8 Mathematics Is a Basic Tool of the Technician

Ability in computation often determines the rate at which technicians advance to positions of greater responsibility. The purpose of the following sections is to review some elementary principles of decimal notation and to illustrate some concepts of algebra, geometry, and trigonometry essential to technical calculations. Construction and interpretation of graphs will be explained also, since they are used frequently in technical work. Even if you feel well qualified in mathematics, this brief review should be carefully studied. If you find these mathematical concepts extremely difficult, it would be wise to enroll at once in a technical mathematics course.

1.9 Decimal Notation and Exponents

Consider the number 10,385.6491. Spreading it out across the page and numbering the digits both ways from the decimal point affords a quick review of decimal terminology.

ten-thousands	thousands	hundreds	tens	units		tenths	hundredths	thousandths	ten-thousandths
1	0	3	8	5	•	6	4	9	1

A convenient way of handling decimals and calculations with them is to deal with exponents. An *exponent* is a small number printed to the right and above another number. This small superscript tells how many times that other number is to be taken as a factor. Thus, in the expression $5 \times 5 \times 5 \times 5$, five is taken as a factor four times. We would express this as 5^4. Here are two more examples:

$$4^2 = 4 \times 4 = 16$$
$$2^5 = 2 \times 2 \times 2 \times 2 \times 2 = 32$$

An *exponent indicates the power to which a given number is to be raised.* We read 10^4 as "ten to the fourth power."

The table at the top of the next page lists some powers of 10 and their decimal values.

Exponential notation frequently simplifies numerical calculations, especially if very large or very small numbers are involved. The method is also known as *scientific notation*, since it is the method used by scientists for numerical calculations. When numbers are written in scientific notation, they are often written as numbers between 1 and 10 times 10 to a power.

Some Powers of 10 and Their Decimal Values

$10^6 = 1,000,000$	$10^{-1} = 0.1,$	read as	"one-tenth"
$10^5 = 100,000$	$10^{-2} = 0.01,$	read as	"one-hundredth"
$10^4 = 10,000$	$10^{-3} = 0.001,$	read as	"one-thousandth"
$10^3 = 1000$	$10^{-4} = 0.0001,$	read as	"one ten-thousandth"
$10^2 = 100$	$10^{-5} = 0.00001,$	read as	"one-hundred-thousandth"
$10^1 = 10$	$10^{-6} = 0.000001,$	read as	"one-millionth"
$10^0 = 1$			

Analysis of the above table reveals that increasing the power of 10 by 1 unit is equivalent to moving the decimal point one place to the right; and conversely, decreasing the power of 10 by 1 unit is equivalent to moving the decimal point one place to the left. Consequently, the number 18,750,000 can be expressed by any of the following:

$$18,750 \quad \times 10^3$$
$$18.75 \quad \times 10^6$$
$$1.875 \times 10^7$$

In like manner the decimal fraction 0.0000785 can be correctly expressed in scientific notation as

$$7.85 \times 10^{-5} \quad \text{or as} \quad 78.5 \times 10^{-6}$$

You have probably observed from these examples that the value of the exponent or power of 10 is equal to the number of places the decimal point is moved—to the right for increasing values of positive exponents, and to the left for increasing values of negative exponents. Here is another example:

$3.48 \times 10^0 = 3.48$
decimal point not moved at all
$3.48 \times 10^1 = 34.8$
decimal point moved right 1 place
$3.48 \times 10^2 = 348$
decimal point moved right 2 places
$3.48 \times 10^3 = 3480$
decimal point moved right 3 places
$3.48 \times 10^{-1} = 0.348$
decimal point moved left 1 place
$3.48 \times 10^{-2} = 0.0348$
decimal point moved left 2 places

$3.48 \times 10^{-3} = 0.00348$
decimal point moved left 3 places

To repeat, the exponent of the 10 tells how many places, right or left, to move the decimal point. And, conversely, when you move a decimal point right or left for your own convenience in calculations, you must keep track of this movement by supplying a 10 to the proper power. For example, if you want to express 348,500 as a number between 1 and 10 times 10 to a power, note that the decimal point will have to be moved 5 places to the left (348500.), so the result is 3.48×10^5. Or, if you want to express 0.0000348 as a number between 1 and 10 times 10 to a power, move the decimal point 5 places to the right (0.0000348), and write the result as 3.48×10^{-5}.

Some examples follow. Remember that in scientific notation, we usually express all numbers as numbers between 1 and 10 times 10 to a power. Thus,

$$3460 = 3.460 \times 10^3$$
$$0.0018 = 1.8 \times 10^{-3}$$
$$5,284,000 = 5.284 \times 10^6$$

Setting up problems for computer, pocket-calculator, or slide-rule operations is simplified by using scientific notation.

Illustrative Problem 1.1 Compute the value of

$$\frac{648.5 \times 0.000397}{0.0016}$$

Solution Express each factor as follows, using powers of 10:

$$\frac{6.485 \times 10^2 \times 3.97 \times 10^{-4}}{1.6 \times 10^{-3}}$$

Divide 10^{-3} into 10^{-4}, leaving 10^{-1}, which, when multiplied by the 10^2 in the numerator, gives 10^1. We now have

$$\frac{6.485 \times 3.97 \times 10^1}{1.6} = 160.9 \qquad ans.$$

Note again how this method is especially applicable to calculations with pocket calculators and slide rules.

ALGEBRA REVIEW

It is assumed that students of applied physics have completed at least a basic course in algebra. Those whose ability in mathematics is limited should, of course, be enrolled in a mathematics course concurrently and will be acquiring the mathematical skills needed for the study of physics. The brief review presented here merely relates some elementary concepts of algebra to the solution of simple physics problems. Since technicians often work with formulas, our discussion begins with several such examples.

1.10 Algebra Applied to Formulas

A formula is a mathematical statement of equality in which letter symbols are combined with numbers to express a physical relation in a more convenient form for problem solving than is possible by the use of words. As an example, take the verbal statement, "Distance traveled is equal to the average velocity multiplied by the time." By letting s stand for distance, v for velocity, and t for time, we can write the simple formula

$$s = vt \qquad (1.1)$$

Another rather familiar example may be drawn from a problem in finance. A certain sum of money, called the *principal*, draws simple *interest* at a certain rate for a stipulated *time*. The interest which accrues in a given time is equal to the product of the principal, rate, and time. The principal and the interest are usually expressed in dollars, the time in years, and the rate in percent per year. As a formula, letting i stand for the accrued interest, P for the principal, r for the rate, and t for the time, we find that the interest accrued is

$$i = Prt \qquad (1.2)$$

As a final example, a formula from the field of electricity will be chosen. It is known that the amount of current in a wire, measured in units called *amperes*, is equal to the electrical pressure, measured in units called *volts*, divided by the resistance of the wire, measured in units called *ohms*. As a word equation, then

$$\text{Amperes} = \frac{\text{volts}}{\text{ohms}}$$

Let I stand for current intensity, V for voltage, and R for resistance. The formula then becomes

$$I = \frac{V}{R} \qquad (1.3)$$

Solving Problems with Formulas   In formulas the various letters and symbols can be thought of as *variables*, the value of any one depending on or *varying with* the values assigned to the others. For example, in Eq. (1.1) above, the value of s depends on the values assigned to v and t. This formula as written is said to be solved *explicitly* for s. Should s and t be given and the value of v desired, the formula must first be solved explicitly for v. Then the known values can be substituted for s and t and the numerical value of v can be determined. Solving Eq. (1.1) for v requires the use of one of the basic axioms of algebra, which states that *when equals are divided by equals, the results are equal*. The process involves dividing both sides of the equation (or formula) by t:

$$\frac{s}{t} = \frac{v\cancel{t}}{\cancel{t}}$$

The t's on the right side cancel out, leaving

$$\frac{s}{t} = v \qquad \text{or} \qquad v = \frac{s}{t}$$

If it were desired to solve Eq. (1.3) explicitly for R, we would use another axiom, namely, that when equals are multiplied by equals, the results are equal. First multiply both sides by R, obtaining

$$RI = \frac{V\cancel{R}}{\cancel{R}}$$

Then divide both sides by I,

$$\frac{R\cancel{I}}{\cancel{I}} = \frac{V}{I}$$

and obtain the formula solved explicitly for R, as

$$R = \frac{V}{I}$$

Illustrative Problem 1.2 A sum of money earns $812.50 interest when loaned out for 5 years at $6\frac{1}{2}$ percent per annum simple interest. What was the principal?

Solution Solve Eq. (1.2) explicitly for P by dividing both sides by rt:

$$\frac{i}{rt} = \frac{P\cancel{rt}}{\cancel{rt}}$$

or

$$P = \frac{i}{rt}$$

Substituting the given values,

$$P = \frac{\$812.50}{0.065 \times 5} = \$2500.00 \qquad ans.$$

Illustrative Problem 1.3 If a waffle iron has an effective electrical resistance of 8.6 ohms while operating on a household voltage of 120 volts, what current, in amperes, does it draw?

Solution Note that Eq. (1.3) is already in a form to calculate the current, I. Substitute the given values for V and R, obtaining

$$I = \frac{V}{R} = \frac{120 \text{ volts}}{8.6 \text{ ohms}} = 14 \text{ amp.} \qquad ans.$$

NOTE: *Answers to all illustrative problems have been obtained by a pocket electronic calculator. This is true also of the answers to the odd-numbered problems at the back of the book.*

See Section 2.12, p. 33, for an explanation of how many significant digits should be retained in answers to problems.

1.11 Algebraic Symbols

Algebra replaces the numerical units of arithmetic with letters which can represent *any* arithmetic numbers. Letters of the alphabet are used as symbols to represent sets of numbers. The letters are called *variables*. In some cases, where physics relationships are involved, a letter will also be assigned to represent a *constant* value—a value which remains unchanged over time—for example, g, the symbol for the acceleration of gravity.

The four fundamental processes, indicated by the symbols + (plus), − (minus), × (times), and ÷ (divided by), have the same meaning and application in algebra that they have in arithmetic. If x and y represent two numbers, we would write their sum as

$$x + y$$

If y is to be subtracted from x, the difference is written

$$x - y$$

To indicate the product of numbers represented by x and y we write

$$(x)(y) \text{ or simply } xy$$

or sometimes

$$x \cdot y$$

Sometimes the "times sign" (×) is used as an indicator for multiplication. The quotient of x divided by y is generally written as

$$\frac{x}{y} \qquad \text{not as} \qquad x \div y$$

1.12 Equations

An *algebraic expression* is a combination of variables and constants, together with certain indicated operational symbols such as addition, subtraction, multiplication, or division. Some algebraic expressions also include operations like taking roots, raising to powers, and trigonometric relationships.

An *algebraic equation* is a statement that two algebraic expressions are equal or that they are equivalent to the same number. We will use two kinds of equations in this book, illustrated by the two following types:

Type 1 examples: $3x + 4x = 7x$; $s = vt$

Type 2 examples: $3x + 4x = 35$; $x^2 + 3x = 4$

Equations of the first type are called *identities*. Equations of the second type are called *conditional equations*.

The statement

$$3x + 4x = 7x$$

is true for *any* value that might be assigned to x and is an *identity*.

The equation relating distance, velocity, and time units is true for all sets of consistent units, and is also an identity. Thus, if we know $v = 60$ miles per hour (mi/hr) and $t = 3$ hours (hr), we use

$$s = vt$$

to determine that

$$s = \left(60 \frac{\text{mi}}{\text{hr}}\right)(3 \text{ hr}) = 180 \text{ mi} \qquad ans.$$

Equations like $s = vt$, relating physical quantities in predictable relationships, are called *formulas* (Sec. 1.10).

In contrast to these identities, the conditional equation

$$3x + 4x = 35$$

is true if, and only if, the value of x is 5. The number 5 is called the *root* of the equation, and the expression is an equation *only on the condition* that x is assigned a value of 5.

In like manner, the equation

$$x^2 + 3x = 4$$

is true if, and only if, $x = 1$ or -4.

1.13   Solving Equations

The value or values for a letter symbol in an equation which make it true, or "satisfy" it, are called the *solution* to the equation. The process of finding these values is called *solving the equation*.

In solving equations use is often made of a number of "self-evident" truths or agreed-upon rules called *axioms*. One of these is:

Any operation performed on one side of the equals sign of an equation must also be performed on the other side.

Adding the same quantity to both sides of an equation or subtracting the same quantity from both sides does not alter the equation, and it will still have the same roots. In like manner, multiplying or dividing both sides of an equation by the same quantity will not change the validity or the roots of the equation. Operations such as raising to the same power or taking the same root of both sides of an equation also will not change the truth of the equation.

Illustrative Problem 1.4 Solve the equation $2x - 9 = 21$.

Solution   Add 9 to both sides of the equation:

$$2x - 9 + 9 = 21 + 9$$

$$2x = 30$$

Now divide both sides by 2:

$$\frac{2x}{2} = \frac{30}{2} \qquad and \qquad x = 15 \qquad ans.$$

To verify that 15 is truly the root of the equation, substitute it back in the original equation. Thus

$$2x - 9 = 21$$

$$2(15) - 9 = 21$$

$$30 - 9 = 21$$

$$21 = 21$$

Illustrative Problem 1.5 Solve the equation

$$\frac{5x + 10}{2} = 3(31 - x)$$

Solution   First, multiply both sides by 2:

$$2\left(\frac{5x + 10}{2}\right) = 2[3(31 - x)]$$

$$5x + 10 = 6(31 - x)$$

Now remove the parentheses by multiplying each term inside them by 6.

$$5x + 10 = 186 - 6x$$

Next add $6x - 10$ to both sides. The result is

$$5x + 10 + 6x - 10 = 186 - 6x + 6x - 10$$

Collect like terms:

$$11x = 176$$

Finally, divide both sides by 11.

$$\frac{11x}{11} = \frac{176}{11}$$

$$x = 16 \qquad\qquad ans.$$

1.14   Transposing

It will be noted in the last two illustrative problems that when the proper term is added to or subtracted from both sides of an equation, terms disappear from one side of the equation and reappear on the other side with signs changed. This property leads to a short cut in solving some equations. The process is called *transposing* and it can be expressed by the following rule:

> *If any term in an equation is moved from one side to the other with its sign changed, an equivalent equation will result.*

The key word here is "term." A term is a number or symbol separated from another by a $+$ or a $-$ sign. For example, in the expression $7x + 3$, $7x$ and 3 are *terms* but 7 alone is not a term and neither is x alone.

Illustrative Problem 1.6   Solve the equation

$$7x - 6 = 3x + 14$$

Solution   Transpose both the -6 and the $3x$ terms. (Remember to change signs.)

$$7x - 6 = 3x + 14$$

$$7x - 3x = 14 + 6$$

$$4x = 20$$

$$x = 5 \qquad\qquad ans.$$

1.15   Laws of Exponents

The solution to many problems in science and technology involves multiplying a number by itself many times. A convenient bit of symbolism is used to indicate this process. The symbol or indicator is the *exponent*, discussed in Sec. 1.9 as it applies to the *base* 10. However, any number can serve as a *base* and the superscript or exponent indicates how many times the base is taken as a factor in repeated multiplication.

The numbers $2^2, 2^3, 2^4, \ldots$ are called powers of 2. Thus 2^5 is read "two to the fifth power" or "the fifth power of 2."

The following laws of exponents dealing with the product and quotient of two powers will be found useful in solving problems, both those in this book and the ones technicians will encounter on the job.

Some Laws of Exponents

1. The exponent of the product of powers with the same base is the sum of the exponents of the factors. In general,

$$(x^m) \cdot (x^n) = x^{m+n}$$

Some examples will illustrate:

$$(x^4) \cdot (x^7) = x^{4+7} = x^{11} \qquad 2x^3(4x^6) = 8x^9$$

2. The exponent of the quotient of two powers with the same base is the difference of the exponents of the powers. In general,

$$\frac{x^m}{x^n} = x^{m-n} \qquad \text{(if } x \text{ does not equal zero)}$$

Some examples are:

$$\frac{x^8}{x^2} = x^6 \qquad \frac{10x^2}{2x^6} = 5x^{-4}$$

3. The exponent of a power of a power is the product of the exponents of the powers. In general,

$$(x^m)^n = x^{mn} \qquad \text{and} \qquad (x^m y^n)^k = x^{mk} y^{nk}$$

Examples:

$$(x^2)^3 = x^6 \qquad (5x^2 y^3)^4 = 625x^8 y^{12}$$

4. *Any quantity* raised to the *zero power* is equal to 1. Examples:

$$\frac{12x^3}{x^3} = 12x^{3-3} = 12x^0 = 12$$

$$(5x)^0 = 1 \qquad \text{(by definition)}$$

5. The roots of an equation will not change if both sides are raised to the same power.

Illustrative Problem 1.7 Solve the equation (formula)

$$T = 2\pi\sqrt{m/k}$$

for *m*.

Solution In order to eliminate the square root sign, square both sides of the equation, with the result

$$T^2 = 4\pi^2 \left(\frac{m}{k}\right)$$

Then multiply both sides by *k*

$$kT^2 = 4\pi^2 m \qquad \text{or} \qquad 4\pi^2 m = kT^2$$

Finally, divide both sides by $4\pi^2$ to get

$$m = \frac{kT^2}{4\pi^2} \qquad\qquad ans.$$

GEOMETRY REVIEW

Certain plane and solid geometrical figures are often referred to in the study of physics. The areas and perimeters of plane figures enter into many physics problems and so do volumes and surface areas of solid figures. Technicians and technologists in all fields frequently become involved with problems on the job which require the ability to work with geometrical relationships. A brief review of a few simple geometrical figures follows.

1.16 Plane Geometrical Figures

We are not concerned here with formal propositions or with theorems or proofs. Formulas will be used to express relationships for problem solving. The rectangle will be reviewed first.

The Rectangle Figure 1.3 shows a general rectangle. Note that all four angles are right angles (90°). Opposite sides of rectangles are equal in length, and consequently only two sides need to be designated. The length is lettered *l*, and the width, *w*, as shown. The area of a rectangle is simply the product of length and width. As a formula

$$A_{\text{(rectangle)}} = lw \qquad (1.4)$$

Areas of all geometrical figures are expressed in square units, as square feet (ft^2), square inches (in.^2), square meters (m^2), and so on.

A *square* is a rectangle with all four sides equal, that is $l = w$. The area of a square is (see Fig. 1.4)

$$A_{\text{(square)}} = l \cdot l = l^2 \qquad (1.5)$$

The General Triangle Figure 1.5 shows the typical general triangle and its nomenclature. Note that no two sides are equal and that no two angles are equal. The angles are designated by capital letters and the sides opposite them by corresponding lowercase letters. The side *AC*, on which the triangle appears to rest, is called the *base* and is lettered *b*. The perpendicular distance from the vertex *B* to side *AC* is called the *altitude*, lettered *h*. The area of a triangle, regardless of its shape, is equal to one-half the product of the base and the altitude, or

$$A = \frac{bh}{2} \qquad (1.6)$$

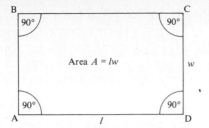

Fig. 1.3 The general rectangle, length *l*, width *w*.

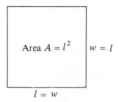

Fig. 1.4 The square. A rectangle with all four sides equal, *l* = *w*.

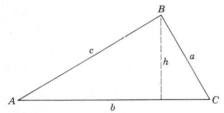

Fig. 1.5 The general triangle—no equal sides and no equal angles.

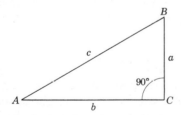

Fig. 1.6 The right triangle. Angle *C* is 90°.

The Right Triangle Figure 1.6 shows a right triangle. Angle *C* is a 90-degree angle or *right angle*. The side *c* opposite the right angle is called the *hypotenuse*. Sides *a* and *b* are called *legs*.

If one leg *b* is taken as the base, the other leg *a* becomes the altitude *h*, since it is perpendicular to *b* by definition. The area of a right triangle is also

$$A = \frac{bh}{2} \qquad (1.7)$$

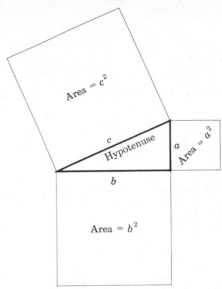

Fig. 1.7 Geometric representation of the theorem of Pythagoras. The square on the hypotenuse is equal to the sum of the squares on the other two sides: $c^2 = a^2 + b^2$.

The method illustrated here of designating the sides and angles of triangles should be carefully noted, since it is a standard labeling system.

Finding the Hypotenuse of a Right Triangle; Pythagorean Theorem The hypotenuse *c* of a right triangle can be found if the legs *a* and *b* are known. Pythagoras, a Greek mathematician and philosopher of the sixth century B.C., is credited with the discovery that in right triangles, *the square of the hypotenuse is equal to the sum of the squares of the other two sides.* As a formula

$$c^2 = a^2 + b^2 \qquad (1.8)$$

or

$$c = \sqrt{a^2 + b^2} \qquad (1.8')$$

Geometrically, the relationship is shown in Fig. 1.7, where the area of the larger square, constructed on the hypotenuse as a side, is equal to the sum of the two smaller areas, each constructed on one of the legs as a side.

The Circle The simple geometry of the circle can be studied with the aid of Fig. 1.8. A circle is defined as a closed curve in a plane, all points of which are equidistant from a point called the

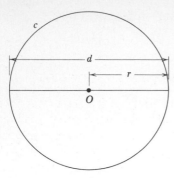

Fig. 1.8 The circle. The diameter is twice the radius: $d = 2r$. The circumference $c = \pi d = 2\pi r$. The area $A = \pi r^2 = \pi d^2/4$.

Fig. 1.9 The sphere. Every point on a spherical surface is equidistant from the center C. This distance is the radius r.

center O. The distance from the center O to the circle is the radius r. The *diameter d* is a line through the center of the circle with end points on the circle. The diameter has a length twice that of the *radius r*. The distance c around a circle is called the *circumference*. Greek mathematicians long ago discovered the relationship between the circumference of a circle and its diameter. Regardless of the size of a circle, the ratio of the circumference to the diameter is always a constant. This constant value was given the name of a letter of the Greek alphabet, π, pronounced "pi." As a formula,

$$\frac{c}{d} = \pi$$

or $$c = \pi d = 2\pi r \qquad (1.9)$$

The value of π cannot be expressed as an exact

decimal fraction. Worked out to four decimal places, its value is 3.1416. For most computations in applied physics we may use the value 3.14 without introducing a significant error in the results.

Area of a Circle It is often necessary to find areas of circular regions. The area of a circle is given by

$$A = \pi r^2 = \frac{\pi d^2}{4} \qquad (1.10)$$

1.17 Solid Geometrical Figures

Of the many possible geometrical solids, only three will be briefly studied here. These three, which occur with some frequency in the application of physics to industrial and technical problems, are the *sphere*, the *cylinder*, and the *cone*.

The Sphere A sphere (Fig. 1.9) is a solid figure every point of whose surface is equidistant from a fixed point within called the *center*. The distance from the center to the surface is called the *radius*. The volume of a sphere is computed from the formula

$$V = \tfrac{4}{3}\pi r^3 \qquad (1.11)$$

and the surface area from

$$S = 4\pi r^2 \qquad (1.12)$$

Right Circular Cylinder Figure 1.10 illustrates a right circular cylinder. A cross section of such a cylinder is a circle of radius r. The length of the cylinder is h. Its volume is given by the formula

$$V = \pi r^2 h \qquad (1.13)$$

The *lateral surface* area can be calculated from

$$S = 2\pi r h \qquad (1.14)$$

and the total surface area, including top and bottom, from

$$A_T = 2\pi r(h + r) \qquad (1.15)$$

Right Circular Cone A right circular cone is a solid figure generated by a right triangle as it rotates about one of its legs as an axis. In Fig.

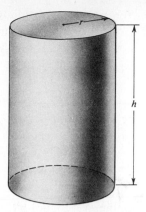

Fig. 1.10 A right circular cylinder.

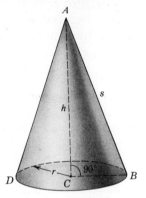

Fig. 1.11 A right circular cone.

1.11 right triangle *ABC* has been rotated around *AC* as an axis to form the cone. A line from the *vertex A* perpendicular to the base passes through the center *C* of the base circle. This line (*h* in the diagram) is called the *altitude* of the cone. The distance *AB*, labeled *s*, is called the *slant height* of the cone. It is, of course, the hypotenuse of the right triangle *ABC*.

The volume of a cone is equal to one-third the area of the base times the altitude. As a formula,

$$V = \frac{\pi r^2 h}{3} \qquad (1.16)$$

The lateral surface area is given by

$$S = \pi r s \qquad (1.17)$$

and the total surface area by

$$A_T = \pi r(s + r) \qquad (1.18)$$

TRIGONOMETRY REVIEW

Certain problems of the right triangle are more readily solved by trigonometry than by any other method. The word *trigonometry* literally means "measuring three-sided figures." A brief introduction to trigonometry and the solution of a sample problem will serve to illustrate applications of trigonometry to technical physics.

1.18 Basic Trigonometric Relations

The right triangle of Fig. 1.12 has vertices *A*, *B*, and *C*; angles α (alpha), β (beta), and γ (gamma); and sides *a*, *b*, and *c*. This notation is conventional in trigonometry and should be memorized. The right angle is always at *C*, or in other words, $\gamma = 90°$.

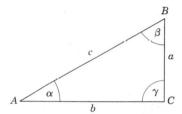

Fig. 1.12 A right triangle, labeled with the conventional symbols of trigonometry.

The fundamental principle of trigonometry lies in expressing the angles in terms of the ratio of two sides. The following definitions of the *trigonometric functions* are given; of the six, *sine*, *cosine*, and *tangent* are of most importance for technical physics and the ratios for these should be memorized:

$$\text{Sine } \alpha = \frac{\text{opposite side}}{\text{hypotenuse}}$$

$$\text{Cosine } \alpha = \frac{\text{adjacent side}}{\text{hypotenuse}}$$

$$\text{Tangent } \alpha = \frac{\text{opposite side}}{\text{adjacent side}}$$

$$\text{Secant } \alpha = \frac{\text{hypotenuse}}{\text{adjacent side}}$$

$$\text{Cosecant } \alpha = \frac{\text{hypotenuse}}{\text{opposite side}}$$

$$\text{Cotangent } \alpha = \frac{\text{adjacent side}}{\text{opposite side}}$$

Using the approved abbreviations and substituting the corresponding letters for the sides, the *six trigonometric functions* of angle α become

$$\sin \alpha = \frac{a}{c} \qquad \sec \alpha = \frac{c}{b}$$

$$\cos \alpha = \frac{b}{c} \qquad \csc \alpha = \frac{c}{a}$$

$$\tan \alpha = \frac{a}{b} \qquad \cot \alpha = \frac{b}{a}$$

A similar set of trigonometric functions could be written for angle β.

Inverse Functions   Just as we can write

$$\sin \alpha = \frac{a}{c} = 0.185 \text{ for example,}$$

we can also write

$$\sin^{-1} 0.185 = \alpha,$$

meaning "α is the angle whose sine is 0.185." Slide rules and electronic calculators have scales and keys, respectively, to deal with "inverse" functions.

Since sides can be measured, the ratios representing the trigonometric functions of an angle can be evaluated as decimal numbers and used in numerical computations. Mathematicians have computed the values of the six trigonometric functions for all angles from 0 to 90°, so that reference may be made to these trigonometric tables for the solution of numerical problems. Appendix III contains such a table in steps of whole degrees, accurate to three decimal places, for *sines*, *cosines*, and *tangents*.

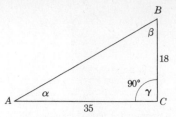

Fig. 1.13   A right triangle with two sides known. The hypotenuse and angles α and β are to be found.

1.19   Algebra and Trigonometry

The methods of algebra are used to solve expressions involving trigonometry. Figure 1.13 shows a right triangle with the two sides AC and BC known. Find the third side and the angles α and β.

Find angle α.

$$\tan \alpha = \frac{\text{opposite side}}{\text{adjacent side}} = \frac{18}{35} = 0.514$$

Thus α is the angle whose tangent is 0.514. It is common practice to express such a relation by either of the following:

$$\alpha = \tan^{-1} 0.514$$

or $\qquad \alpha = \arctan 0.514$

both of which mean "α is the angle whose tangent is 0.514." From the table of Appendix III, to the nearest whole degree,

$$\alpha = \tan^{-1} 0.514 = 27° \qquad \textit{ans.}$$

Find angle β. The sum of the angles of any triangle is 180°. Since $\gamma = 90°$ and $\alpha = 27°$,

$$\beta = 180 - (90 + 27)$$

$$= 180 - 117 = 63° \qquad \textit{ans.}$$

Find hypotenuse AB. This could be done using the pythagorean theorem, that is,

$$AB = \sqrt{\overline{AC}^2 + \overline{BC}^2}$$

but it is more easily done using the cosine function, as follows:

$$\cos \alpha = \frac{\text{adjacent side}}{\text{hypotenuse}} = \frac{AC}{AB}$$

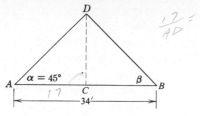

Fig. 1.14 Diagram for Illustrative Problem 1.8.

Substituting known values,

$$\cos 27° = \frac{35}{AB}$$

This is an algebraic equation and is solved by regular methods of algebra. Divide both sides by cos 27° and multiply both sides by AB to obtain

$$AB = \frac{35}{\cos 27°} = \frac{35}{0.891} = 39.3 \qquad ans.$$

Illustrative Problem 1.8 A house is being constructed, and the plans call for a simple gable roof to be pitched at 45°. If the width of the house is 34 ft, what length are the rafters, assuming no overhang?

Solution The diagram of Fig. 1.14 shows the facts of the problem. $AB = 34$ ft, and angle $\alpha =$ angle $\beta = 45°$. Draw DC perpendicular to AB. The midpoint of AB is C, and therefore $AC = 17$ ft. In right triangle ACD, the hypotenuse AD represents a rafter.

$$\cos 45° = \frac{AC}{AD} = \frac{17 \text{ ft}}{AD}$$

$$AD = \frac{17 \text{ ft}}{\cos 45°}$$

$$= \frac{17 \text{ ft}}{0.707} = 24 \text{ ft} \qquad ans.$$

Trigonometry is of course not limited to the solution of right triangles, but for the time being we shall not need any more advanced analysis. Ordinarily only the first three functions defined above (sine, cosine, tangent) will be involved in technical physics problems.

THE USE OF GRAPHS

Just as a picture is often more meaningful than word explanations, so graphs are an effective aid to the interpretation of data and of technical principles and practices.

1.20 Construction and Interpretation of Graphs

A graph is a type of picture or diagram that shows the relationship existing between two or more quantities. Newspapers and magazines frequently use graphs to interpret economic, business, health, and cost-of-living data. As a simple form of graph, consider Fig. 1.15, which plots electric energy production from a steam plant against time for the 12 months of the year. This is a "line graph."

A short study of the graph should give a clear picture of the production record of this plant for the given year. Peak months and slumps are clearly shown. Many hundreds of words would be necessary to present this production record as clearly and as simply as the graph does.

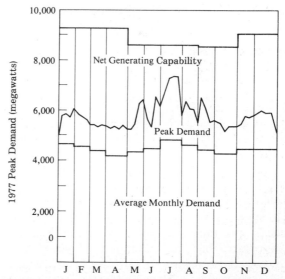

Fig. 1.15 Net generating capability, peak demand, and average monthly demand for electric power for a recent year, as reported by a large public utility. Note the amount of information that can be easily portrayed in graphi form. (Detroit Edison Company.)

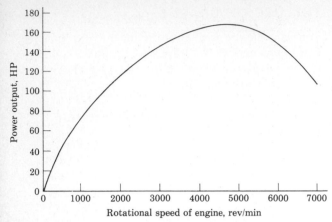

Fig. 1.16 Power-output vs. engine-speed curve (smoothed) for an auto engine rated at 165 hp.

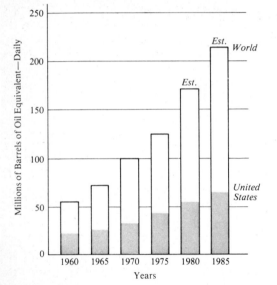

Fig. 1.17 World and United States daily energy demand 1960–1985. An example of a bar graph. (Exxon Corporation.)

As a second example, look at Fig. 1.16, which plots the maximum horsepower output of an auto engine against the speed in revolutions per minute (rpm) at which it is tested. This graph has been "smoothed" so that a flowing curve results. It portrays results which should be typical of all engines of the same model when run under identical conditions. Note that this engine "peaks" at 5000 rpm and that power output drops off sharply thereafter. Such a graph would be constructed by the test engineer from data obtained during actual test runs on a test instrument called a *dynamometer*. What is this engine's maximum power output at 3000 rpm?

The two graphs just considered are both *line graphs*, a type which is well suited for presenting technical data. Other forms of graphs are *bar graphs*, *circle graphs*, and *pictographs*, which are especially suited to the portrayal and interpretation of business data and economic statistics, as well as medical and health data.

Figure 1.17 portrays, by means of a bar graph, the world and United States energy demand from 1960 to 1985, in millions of barrels of oil daily.

In general, graphs of physical data are constructed by plotting the two quantities against each other along two axes—one horizontal, the other vertical. *The quantity whose variation is of the most immediate interest is usually plotted along the vertical axis*, since its "ups and downs" are thus made more evident. Every graph should have a clear, well-phrased title, and each axis must be carefully labeled so there will be no doubt about the units or the quantities being plotted.

We shall have occasion to use graphs or "curves" many times in the presentation of physical principles and data throughout this book. Laboratory work frequently calls for the preparation or interpretation of graphs.

HOW TO STUDY PHYSICS

The brief mathematics review presented in the foregoing sections is intended to emphasize the *minimum* essentials for the successful study of technical physics. The mathematics requirements in this book are not rigorous, but skill in the processes outlined above will make the solution of applied physics problems a challenge to be met rather than a mystery to be avoided. In addition, all students should become skilled in slide-rule or pocket-calculator operations during the early weeks of the course.

In the following paragraphs we shall summarize a few other factors important to the study of physics.

1.21   Reading Comprehension

Reading scientific and technical material is not to be compared with reading fiction, where the plot can be appreciated by rapid scanning with only casual attention to detail. In technical material *every word* is essential to the meaning. Because words alone cannot always make the meaning clear enough, graphs, charts, drawings, and photographs are included. They are to be used with the reading material so that a clear understanding of the principle or application will emerge. Word meanings are important. Liberal use should be made of a good dictionary, and reading should be accompanied by thinking. If you are presently a poor reader, go to your college's reading clinic for diagnostic testing and enroll in a reading improvement program if possible. The *summary* found at the end of each chapter of this book will help you to build a technical vocabulary and to recall some of the essential concepts in the chapter.

1.22   Lectures and Individually Paced Learning

Much of what you will learn will be from class discussion and demonstrated lectures in which your instructor brings you selected lessons from his accumulated experience in science, engineering, and technology. You cannot hope to remember adequately the content of these lectures unless you write down the significant portions of each day's lesson. Keep a well-organized lecture notebook. The very act of writing something down helps you remember it, and the notebook preserves the material in a ready form for review. If your college or school has a Learning Center, be sure to make good use of the films, slides, charts, tapes, and programmed materials available there for individually paced learning in physics.

1.23   Laboratory Experiments and Field Trips

Basic principles of physics will be rediscovered by you as you perform laboratory experiments planned by your instructor. You will have the privilege of using valuable precision equipment in a laboratory similar to that maintained by many industrial corporations for testing and improving their products. Make the laboratory hours count for real learning. Turn in a report on each experiment—one you would be proud to submit to your employer in a real job situation.

Field trips to local industries, businesses, or hospitals serve two purposes: (1) They show how the principles of technical physics are being used out on the job, and (2) they acquaint you with certain segments of the local economy in a way that might determine your future occupation. Get the most out of field trips. Listen to explanations being made by the engineers, managers, or foremen in charge. Ask questions of these people. Accumulate the information you will need to write a concise but complete report of the field trip. Try to observe and report on the applications of applied physics that will later concern you in your career.

1.24   Problems Assignments

The ability to solve problems is a distinguishing feature of technicians and technologists. In almost every chapter, you will find many illustrative problems completely solved for you. At the ends of the chapters there are many typical problems from physics as well as selected problems from industry, business, and other fields. Work out and hand in at each session the problems your instructor assigns. Faithful performance on all problems assignments is the *only way to real understanding.* Develop habits of neat arrangement and proper organization in problem solving. Follow the pattern used in the **Illustrative Problems** in this book unless your instructor prefers some other format.

Very early, develop skill with either a pocket calculator or a slide rule. Solving physics problems by longhand methods takes too much time and compounds the chances for error.

SUMMARY

Modern industry, business, agriculture, and medicine would be impossible without the contributions made by the science of physics.

Physics is the study of *matter* and *energy*.

Physics is basic to the *development of new energy resources* and is also directly related to the *conservation of energy*.

Matter exists in three forms or states—*solid*, *liquid*, and *gas*. *Plasmas* are sometimes considered a fourth state of matter.

Energy is the ability to do work.

Kinetic energy is energy of *motion*.

Potential energy is energy of *position*.

A *machine* is a device to convert *energy* into work.

Energy is readily changed from one form to another in many industrial processes.

Some sources of energy are: *fossil fuels, solar energy, wind energy, geothermal energy, tidal energy,* and *nuclear energy.*

Mathematics is a basic tool of industry. Mathematical skills frequently needed are those of algebra, geometry, trigonometry, scientific notation, and interpretation of graphs.

A well-rounded approach to the study of technical physics will include textbook study, lecture note-taking, individual study in the Learning Laboratory, laboratory work, field trips, and problem solving.

QUESTIONS AND EXERCISES

1. Identify at least six substances that normally are (*a*) solids, (*b*) liquids, (*c*) gases.

2. In your chosen career, or in the one you are thinking seriously about (tell what it is), in what ways do you think you will be involved with (*a*) electrical energy? (*b*) heat energy? (*c*) chemical energy? (*d*) solar energy? (*e*) mechanical energy and machines? (*f*) nuclear energy? (*g*) other applications of physics? (List them.)

3. Write a one-page paper on the current arguments for and against the all-out development of nuclear energy.

4. Why are mathematical formulas a more convenient method of solving problems than the use of a statement in words?

5. Select an example of a varying relationship from any field you like (industry, business, health, agriculture, etc.) and prepare a graph (broken-line, curve, or bar-type) showing how one factor varies with some other selected variable. Obtain actual data and cite your sources. Be sure to include all the features of a good graph.

PROBLEMS

1. Express the following in decimal form: (*a*) six ten-thousandths, (*b*) three and one-half thousandths, (*c*) five-millionths, (*d*) seventy-five thousandths.

2. A cylindrical shaft is $3\frac{7}{16}$ in. in diameter. Express this in decimal form accurate to three places.

3. Express the following in scientific (or exponential) notation, as numbers between 1 and 10 times 10 to a power. (*Example:* $6785 = 6.785 \times 10^3$.) (*a*) 22,520, (*b*) 25,850,000, (*c*) 629, (*d*) 0.00629, (*e*) 0.0000785, (*f*) 0.084

4. Compute the numerical value of the following, expressing the result as a number between 1 and 10 times 10 to a power:

(*a*) $\dfrac{5.421 \times 728 \times 0.0036}{0.092}$

(*b*) $\dfrac{17.85 \times 440 \times 0.685}{27.6 \times 0.321}$

5. Solve the following algebraic equations (formulas) for the factors indicated:

(*a*) $V = IR$ for I (*b*) $\dfrac{F}{W} = \dfrac{L}{h}$ for W

(*c*) $s = \frac{1}{2}gt^2$ for t (*d*) $F = \dfrac{mv^2}{r}$ for m

6. From Eq. (1.1), if a rifle bullet impacts the target 400 yards (yd) away 0.55 seconds (sec) after the gun was fired, what was its average velocity in feet per second (ft/sec)?

7. A sum of $8450 was placed in a savings bank for 9 months at simple interest. The amount of interest earned was $367.50. Find the *annual interest rate.*

8. Find the area of a triangular plot of ground whose frontage along the street is 185 ft and whose depth (altitude of the triangle) is 140 ft.

9. An acre is 43,560 square feet (ft^2). How many acres are there in a rectangular plot of ground 925 ft long and 480 ft wide?

10. Find the hypotenuse of a right triangle whose sides are 8 ft and 5 ft. Use the theorem of Pythagoras.

11. A pulley has a circumference of 48 in. What is its radius to the nearest 0.01 in.?

12. A round concrete pier has a diameter of 2.5 ft. What is its circumference?

13. A sewage-treatment plant has a water aerator system with circular ponds 80 ft in diameter. What is

the surface area of the water exposed to the air in each pond?

14. The area of a square is 625 in.^{2}. What is the area of the largest circle that can be inscribed in the square?

15. One cubic foot (ft^3) contains 7.48 gallons (gal). Calculate the number of gallons in a cylindrical oil tank 20 ft high whose diameter is 35 ft.

16. Find the volume and the lateral surface area of a right circular cone whose height is 18 in. and whose base has a radius of 6 in.

17. A spherical tank for liquid oxygen storage is 8 ft in diameter. What is its capacity in cubic feet?

18. The hypotenuse of a right triangle is 36 in., and its angle with the base is 42°. Side a is 24.1 in. How long is the base? Solve first, using trigonometric functions (cosine or tangent function), and then using Pythagoras' theorem. Note the relative convenience of the trigonometric method.

19. A conveyor track with rollers is 35 ft long. One end is on a dock, and the other end is secured to the deck of a ship tied up at the dock. If the ship's deck is 16 ft above the dock, find (*a*) the angle which the conveyor track makes with the dock and (*b*) the horizontal distance from the base of the conveyor to the ship's side.

20. A communications antenna tower is 350 ft high. It is to be steadied by guy wires 400 ft long. What angle will these cables make with the ground?

21. From a point 200 ft from the base of a tree, the angle of elevation of the treetop is observed to be 30°. How tall is the tree?

22. A mountain road climbs steeply a 30 percent grade, which means it rises 30 ft for every 100 ft of horizontal distance. Find the angle of the incline in degrees from the horizontal.

23. The angle at which a steep mountain road climbs is measured by a transit and found to be 25°. What percent grade is this?

24. A circular building is to have a roof in the shape of a cone. The diameter of the building is 80 ft, and the roof overhang is negligible. The apex of the conical roof is 20 ft above the top of the circular building wall. Find the area of the roof.

25. If one gallon of a certain brand of aluminum paint covers 250 ft^2 of surface, how many gallons will be required to paint a spherical refinery tank 20 ft in diameter?

26. A loading chute delivers cartons from a ware-

house to a freight-car door 25 ft below and 18 ft out from the wall. How long is the chute?

27. Find the number of square feet of steel plate required to make a cylindrical tank 30 ft high and 20 ft in diameter. Include bottom and top of tank.

28. How many square feet of steel plate are required to construct a spherical refinery tank to hold 40,000 gal? (7.48 gal = 1 ft^3.)

29. Prepare a speed curve for a ship from the data given below. Use graph paper. Plot ship's speed in knots on the vertical axis and revolutions per minute (rpm) of the screw on the horizontal axis. "Smooth" the curve and write an interpretation including everything the graph tells you about the ship's performance.

Speed of Screw, rpm	Ship's Speed, knots
0	0
50	10
100	17
150	22
200	26
250	27.5
300	28.5
350	29.3
400	30
500	31

30. The following data were obtained on road tests of fuel consumption at different speeds in a new model automobile. Plot a "smoothed" line graph of the data. Choose carefully the variable you plot on the vertical axis. From 40 to 100 mi/hr what relationship seems to exist between fuel consumption and speed?

Speed, mi/hr	Fuel Consumption, gal/mi
10	0.046
20	0.043
30	0.039
40	0.037
50	0.041
60	0.049
70	0.058
80	0.069
90	0.098
100	0.145

2 Precision Measurement

THE IMPORTANCE OF PRECISION MEASUREMENT

Physics, without measurement, would be more of a philosophy than a science. Whether one deals with *macrophysics* (the earth, moon, and planets—great distances and enormous quantities of matter) or with *microphysics* (the realm of fog particles, crystals, molecules, atomic nuclei, or electrons), the necessity to measure, and *to measure with extreme precision*, is ever-present. Planning the journeys and predicting the exact arrival times and landing sites of spacecraft and space shuttles are examples of the accurate measurements required in macrophysics projects. In the world of microphysics, determining the mass of the atom, the charge on the electron, or the pressure inside a living cell are examples of the extremely small-scale measurements made by scientists.

In industry, precision measurements accurate to 1/100,000 in. are necessary in some cases; 1/10,000 in. is routine in the aviation and space industry; and accuracy to one or two-thousandths of an inch is common practice in the automotive and heavy machinery field.

Mass production in industry requires complete interchangeability of parts. For example, in the automobile, appliance, and aircraft industries, component parts may be manufactured by dozens of subcontractors, and the finished product may come off the assembly line in a plant thousands of miles away from the parts plants. Weeks or months later, these parts will be put

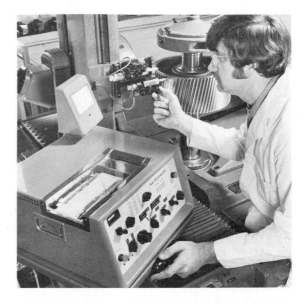

Fig. 2.1 Precision measurement in industry. Technician checks helical gear with electronically actuated "macrograph." Irregularities of only a few hundred-thousandths of an inch can be detected. (The Falk Corp.)

together by assembly-line workers on a tight time schedule where every part must "fit." The parts must be manufactured to the close *tolerances* specified by the design engineer and stipulated on drawings and in specifications prepared by technical draftsmen. For example, a valve stem might have a design diameter of 0.3125 in., with a tolerance of plus or minus ($\pm$) 0.0003 in. The *allowable tolerance*, then, is plus or minus 3 ten-thousandths of an inch. All stems measuring between 0.3122 in.

and 0.3128 in. are acceptable. Those with diameters beyond these limits are rejected.

Accurate measurements are also important to business and finance, agriculture, medicine and health, and to many everyday activities such as travel, sports and recreation, cooking and baking, and communications and entertainment.

2.1 Fundamental Units of Measurement

All measurements are actually relative in the sense that they are *comparisons with some standard unit of measurement*. Although we use many different units of measurement, there are only three *fundamental units* needed for the study of mechanics—*length*, *mass*, and *time*. Other well-known measures are called *derived units* to distinguish them from fundamental units. For example, the familiar unit of *area*, e.g., the square foot, is a product of a length times a length. Another derived unit is the unit of *velocity*, e.g., the foot per second, which is obtained by dividing a length unit by a time unit. Another is the unit of *density*, e.g., the kilogram per cubic meter, derived by dividing a mass unit by the cube of a length unit. In this manner all derived units in mechanics can be expressed in terms of length L, mass M, and time T.

The examples of derived units of measurement just given are mostly those from the system familiar to most of us in the United States—the *English system* of measurement. Most other nations of the world use a different system called the *metric system*. The United States is currently in the process of beginning a changeover to the metric system. The Secretary of Commerce recommended in 1971 that the changeover be completed by 1981. In 1975 the Metric Conversion Act was passed by the Congress and signed by the President, calling for voluntary conversion to the metric system, but not setting any specific date for the changeover. Adopting the metric system (called *metrication*) in the United States is a somewhat controversial issue because first, staggering costs are involved (estimates range into billions of dollars), and second, because people are reluctant to change from a system they are comfortable with to a new system which is foreign to them. Although it is unlikely that the suggested deadline of 1981 will be met, it seems probable that in the relatively near future, the United States will become a metric nation.

The metric system has been the universally used system in science for a long time. Engineering and technology make greater use of it each year, and instruction in the metric system is now a part of the curriculum in all public schools. This book will emphasize the metric system, but will give ample attention to the English system also, in discussions of both theory and practical applications. See Appendix I for a list of metric-system units and symbols.

The next several sections explain the metric and the English systems of measurement in detail and discuss some of the methods and instruments used in measuring length, mass, and time.

SYSTEMS AND UNITS OF MEASUREMENT

2.2 Measurement of Length

The Metric System The unit of length in the metric system is the *meter* (spelled *metre* in Great Britain and many other English-speaking countries). The standard meter was, until 1960, defined as the distance between two fine lines ruled on a platinum-iridium bar stored at the International Bureau of Weights and Measures in Sèvres, France. However, in October 1960, the International Committee on Weights and Measures held a General Conference in Paris and out of that meeting came an international agreement that the world's new standard of length would be the wavelength of the orange-red line of krypton-86. This agreement replaced an arbitrary standard with a constant of nature which can be reproduced and observed in any well-equipped laboratory anywhere in the world. The *standard meter* is now defined as 1,650,763.73 wavelengths of the orange-red line of krypton-86.

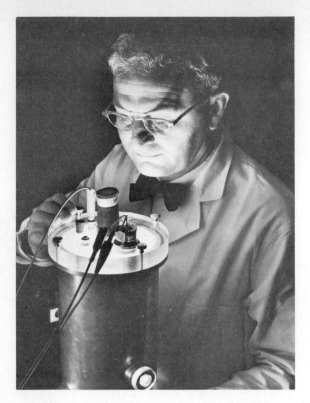

Fig. 2.2 The krypton 86 lamp at the National Bureau of Standards. The wavelength of the orange-red light emitted by Kr[86] is the new international standard of length. (National Bureau of Standards.)

The same General Conference of 1960 gave the name *International System of Units* to the metric system. The *meter* (length), the *kilogram* (mass), and the *second* (time), along with three other basic units which we shall define later in the book, make up the new International System. It is now commonly known as *SI-metric*, SI being the initials of the first two words of the French, *Système International d'Unités.*

The metric system is completely decimalized. Subdivisions of the meter include the *decimeter* (dm, 0.1 meter), the *centimeter* (cm, 0.01 m), the *millimeter* (mm, 0.001 m), the *micrometer* (μ, 0.000,001 m), and the *nanometer* (n, 0.000,000,001 m). For greater lengths the *kilometer* (km, 1000 m) is used.

If you have never seen a meter stick or a centimeter ruler, you can visualize the meter as being about $3\frac{1}{2}$ in. longer than a yard. The inch is slightly longer than 2.5 cm (see Fig. 2.3). Table 2.1 presents the more commonly used metric and English units, and Table 2.2 (p. 26) gives metric-to-English and English-to-metric equivalents.

The English System The standard unit of length in the English system is the *yard*. The legal definition of the yard was first established in 1878, and it defined the yard as the distance between two ruled lines on a bronze bar at the British Exchequer when the metal temperature is 62°F. A later definition (1963) defines the yard as 0.9144 meter. In the United States, although the yard is also a standard unit of measure, the *foot* ($\frac{1}{3}$ yd) is a more commonly used unit. Other length units are the *inch* ($\frac{1}{12}$ ft) and the *statute mile* (5280 ft). In navigation, the *fathom* (6 ft) and the *nautical mile* (6080 ft) are used. The nautical mile is defined as the distance along the surface of the earth at the equator or along any great circle subtended by one minute of arc.

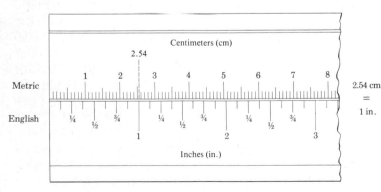

Fig. 2.3 Comparison of metric and English scales of measurement. One inch = 2.54 centimeters. (One meter = 39.37 inches, or about 3.4 inches more than a yard.)

Table 2.1 Units of measurement

Units of Length		Units of Mass		Units of Time	
SI-Metric System					
1 meter (m)	= the standard	1 kilogram (kg)	= the standard	1 second (sec)	= the standard
1 decimeter	= 0.1 m	1 gram (gm)	= 0.001 kg	1 minute (min)	= 60 sec
1 centimeter (cm)	= 0.01 m	1 centigram (cg)	= 0.01 gm	1 hour (hr)	= 3600 sec
1 millimeter (mm)	= 0.001 m	1 milligram (mg)	= 0.001 gm	1 day	= 24 hr
1 micrometer* (μm)	= 0.000001 m				= 86,400 sec
1 nanometer	= 0.000,000,001 m				
1 kilometer (km)	= 1000 m				
1 megameter (Mm)	= 10^6 m				
1 gigameter (Gm)	= 10^9 m				
English System					
1 yard (yd)	= the standard	1 pound (lb) (used in commerce and			
1 foot (ft)	= $\frac{1}{3}$ yd	industry—actually a unit of weight)			
1 inch (in.)	= $\frac{1}{12}$ ft	1 ounce (oz)	= $\frac{1}{16}$ lb		
1 microinch (μin.)	= 0.000001 in.	1 grain (gr)	= $\frac{1}{7000}$ lb	Same as in	
1 statute mile	= 1760 yd	1 ton	= 2000 lb	SI-Metric System	
	= 5280 ft	1 slug	= the engineering		
1 nautical mile	= 6080 ft		standard — to be		
	= 1 min of arc at		defined later		
	earth's equator				
1 fathom	= 6 ft				

* Formerly *micron* in the United States.

The inch is subdivided into smaller units in two different ways. Common practice, where precision is not required, makes use of the $\frac{1}{2}$, $\frac{1}{4}$, $\frac{1}{8}$, $\frac{1}{16}$, $\frac{1}{32}$, $\frac{1}{64}$ system of subdividing. Precision measuring instruments are, however, calibrated to read in *thousandths inch* and *ten-thousandths inch*. Some test-laboratory instruments can even measure accurately to one-millionth (0.000001) of an inch. This unit is called the *microinch* (abbreviated μin.). Standards of measurement in this country are legally fixed by the Congress, and the National Bureau of Standards is charged with the responsibility of maintaining accurate originals of all units of measurement.

2.3 Measurement of Mass

Mass is defined as the *quantity of matter contained in a body* and is to be distinguished from *weight*,

which is measured by *the pull of gravity on a body*. Numerically, mass and weight are ordinarily equal for industrial purposes, but *the two words do not mean the same thing*. A more precise distinction will be made in Chap. 4.

In the metric system, the standard unit of mass is the *kilogram* (kg) with a mass equal to that of a cube 10 cm on a side, of pure water at a temperature of 4°C, by original definition. After the kilogram was standardized, however, accurate determinations showed that 1 kg of water occupies 1,000.027 cm^3. This volume, 1000 cubic centimeters (cm^3), is called a *liter*. Subdivisions of the kilogram are the *gram* (gm, $\frac{1}{1000}$ kg), the *centigram* (cg, $\frac{1}{100}$ gm), and the *milligram* (mg, $\frac{1}{1000}$ gm).

Weight measurements in the English system are stated in *pounds* as the basic unit. Subdivisions of the pound are the *ounce* ($\frac{1}{16}$ lb) and the

Table 2.2 Metric and English equivalents of length, mass, and volume

Metric-to-English		*English-to-Metric*	
Length			
1 meter (m)	= 39.37 in.	1 yard (yd)	= 0.9144 m
	= 3.28 ft	1 foot (ft)	= 30.48 cm
1 centimeter (cm)	= 0.394 in.	1 inch (in.)	= 2.54 cm
1 millimeter (mm)	= 0.0394 in.		= 25.4 mm
1 micrometer (μm)	= 0.0000394 in.	1 microinch (μin.)	= 0.0254 micrometer (μ)
1 kilometer (km)	= 0.6214 mi	1 stat mile (mi)	= 1.609 km
	= 3280 ft		
Mass			
1 kilogram (kg)	= 2.205 lb*	1 pound (lb)*	= 454 gm
1 gram (gm)	= 0.0353 oz*		= 0.454 kg
1 milligram (mg)	= 0.0154 gr*	1 ounce (oz)	= 28.35 gm
1 kilogram	= 0.0685 slugs	1 grain (gr)	= 0.065 gm
			= 65 mg
		1 ton	= 907 kg
		1 slug (sl)	= 14.6 kg
Volume			
1 cu. meter (m^3)	= 1.31 yd^3	1 cu yd (yd^3)	= 0.764 m^3
	= 35.4 ft^3		= 27 ft^3
1000 cm^3	= 1 liter	1 cu ft (ft^3)	= 28.3 liter
	= 1.057 qt		= 7.48 gallon (gal)
1 cu. cm.	= 0.001 liter	1 quart (qt)	= 0.9434 liter
	= 1 milliliter (ml)	1 gallon (gal)	= 3.785 liter
1 liter	= 61.01 in.^{3}	1 cu in. (in.^{3})	= 16.39 cm^3 (ml)
	= 1.057 qt		

Miscellaneous
1 acre = 43,560 ft^2 = 4047 m^2 = 0.405 hectare
1 hectare = 2.47 acres
1 square mile = 640 acres = 2.59 square kilometers (km^2)
1 bushel (dry) = 2219 in.^{3} = 0.0352 m^3 = 35.2 liter

* In science and engineering, the pound, the ounce, and the grain are not used as units of *mass* and are regarded as *weights* or *forces*. However, industry, commerce, and agriculture use them as mass units as well as weight units.

grain (gr, $\frac{1}{7000}$ lb). Larger units are the *hundredweight* (cwt) and the *ton* (2000 lb). *Masses* in the English system are also still expressed in pounds, in industry and in commerce. The *engineering unit* of mass, however, is the *slug*, to be defined later in Chap. 4. In physics, the use of the pound as a mass unit is declining, but we will use it for that purpose occasionally in mechanics and heat, since industry does.

2.4 Measurement of Time

Both the metric and English systems of measurement employ the same time unit, the *second*,

which is defined as 1/31,556,925.9747 of the tropical year 1900. Because of irregularities in orbital movements, all years are not of exactly the same duration, and slight variations in the earth's rotation cause the length of the day to vary slightly. However, for most purposes, the second can be defined as

$$\frac{1}{86,400} = \left(\frac{1}{24 \text{ hr/day}} \times \frac{1}{60 \text{ min/hr}} \times \frac{1}{60 \text{ sec/min}}\right)$$

of the mean solar day.

A much more recent time standard is based on the spinning motion of electrons in atoms. In

Fig. 2.4 An atomic clock in the Boulder, Colorado, laboratory of the National Bureau of Standards. It is actuated by the electromagnetic oscillations in the atom of cesium-133. Its margin for error is estimated to be no more than 1 sec in 30,000 yr.

Fig. 2.5 A machinist's steel rule for English-system use, combined with a bevel protractor. (Lufkin Rule Co.)

In like manner we often refer to the English system as the *foot-pound-second* (*fps*) system. In engineering, a somewhat different system of English units, the foot-slug-second system, is used, which we shall introduce later.

SOME METHODS OF MEASUREMENT

2.5 Measuring Instruments

The basic length-measuring instrument of the mechanic or machinist is the steel rule. Figure 2.5 shows a machinist's steel rule combined with a bevel protractor, graduated in thirty-seconds and sixty-fourths of an inch. Rules are frequently combined with straightedges to make squares (if the angle is 90°) or to measure any given angle. Some rules are marked off in decimal fractions, that is in tenths and hundredths of an inch.

Rules of similar design are graduated (i.e. marked off) in the metric system and these will be seen more and more frequently as metrication is adopted in manufacturing industries.

Surveyors use long steel tapes marked off in meters in the SI-metric system and in feet in the English system. Interestingly, the English-system surveyor's tape is graduated in tenths and hundredths of feet, not in inches.

The best accuracy possible with an ordinary steel rule is of the order of 0.01 in., and this is far too crude for most machine work. For accurate work, the machinist makes use of the *micrometer caliper*. (Note that this is the same spelling as the word for 0.000,001 m given in Table 2.1. The pronunciation is different, however. The word used as a caliper is pronounced *micrómeter*, and

1967 the International Committee on Weights and Measures adopted a new definition of the second, making 1 sec equal to the duration of 9,192,631,770 periods of vibration of the outermost electron of the atom of cesium-133 (see Fig. 2.4).

Since scientists and technicians are nearly always interested in the time rate of change of physical phenomena we shall have frequent need to use time units, principally the *second* (and the *microsecond*, µsec, 0.000,001 sec), and the *hour*.

Since the standard units of the metric system are the meter, the kilogram, and the second, it is often referred to as the *meter-kilogram-second* (*mks*) system. More often recently, it is termed the *SI-metric* system, as mentioned earlier. Older references might be seen to the *centimeter-gram-second* (*cgs*) system, a designation used a few decades ago when the centimeter and the gram were looked upon as being basic units for scientific purposes. This designation is rarely seen today.

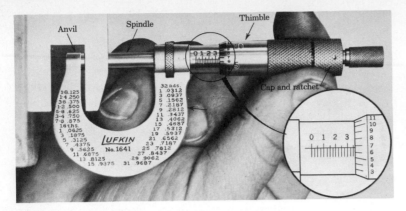

Fig. 2.6 A standard micrometer caliper for English-system measurements. Note the names of the parts. The machinist calls this instrument a "mike." The reading here is 0.357 in. (Lufkin Rule Co.)

the unit of length is *micrometer*). The word *micrometer* literally means *small (or fine) measurement*.

Figure 2.6 shows a standard micrometer caliper capable of reading to 0.001 in. The various parts are labeled to facilitate explaining its construction and operation. The entire instrument is rugged in construction to ensure uniformly accurate measurements. The work to be measured is inserted between the measuring faces. The *thimble* is then turned clockwise; this turn advances the *spindle* toward the *anvil*. The spindle is threaded with 40 threads to the inch. The *pitch* of the screw is then $\frac{1}{40}$ in. = 0.025 in. Consequently the main scale (on the *hub*) is graduated in divisions of 0.025 in., and one full turn of the thimble advances it and the spindle one main scale division. The thimble scale has 25 equal divisions, so that measurements to the nearest 0.001 in. can be read directly. The ratchet is to be used as the measuring faces close on the work to avoid the possibility of damage to the instrument. The reading, as shown in the figure, is 0.357 in. to the nearest thousandth. To get this value, read on the main scale 0.3 in. plus 0.05 in. = 0.35 in. Add to this the 0.007 from the thimble scale to obtain the reading of 0.357 in. Some micrometer calipers have scales and screw pitch so designed that direct readings to 0.0001 in. can be taken.

Micrometer calipers are also commonly used for metric system measurements. Figure 2.7 shows such an instrument.

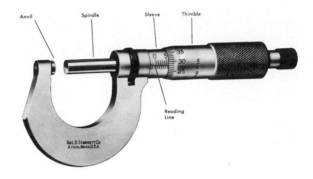

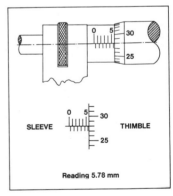

Reading 5.78 mm

EXAMPLE

Referring to the picture and drawing:

The 5 mm sleeve graduation is visible	5.00 mm
One additional 0.5 mm line is visible on the sleeve	0.50 mm
Line 28 on the thimble coincides with the reading line on the sleeve, so 28 × 0.01 mm	= 0.28 mm
The micrometer reading is	5.78 mm

Fig. 2.7 A metric micrometer capable of reading to 0.01 mm. The instrument illustrated shows a reading of 5.78 mm. (The L.S. Starrett Co.)

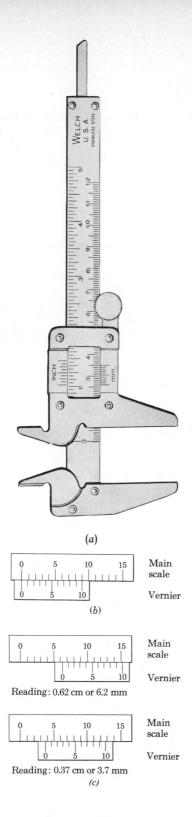

(a)

(b)

Reading: 0.62 cm or 6.2 mm

Reading: 0.37 cm or 3.7 mm

(c)

The micrometer principle is used in many other measuring instruments, for example the depth "mike" for use in measuring the depth of cuts or holes in machined parts.

The *vernier caliper* is another instrument capable of considerable accuracy. Figure 2.8*a* shows a student-type vernier caliper for metric-system measurements. The *main scale* is marked off in centimeters, each of which is divided into 10 parts, or millimeters. The *vernier scale*, which is movable, contains 10 divisions also, but each one is nine-tenths as long as the smallest main-scale division. Therefore the 10 divisions of the vernier scale have the same length as the 9 divisions of the main scale (see Fig. 2.8*b*). Each vernier division is then $\frac{1}{10}$ mm shorter than a main-scale division. This difference is called the *least count* of the instrument.

In taking a measurement, the jaws are brought against the work. Examination of the instrument reveals that the distance between the zeros of the main and vernier scales is equal to the distance between the jaws for any setting. The main scale is therefore read at the zero of the vernier scale. Referring to Fig. 2.8*c* (*top*), read from the main scale 0.6; count over on the vernier scale two divisions and find that point where the two scales coincide. The reading is 0.62 cm. Similarly, Fig. 2.8*c* (*bottom*) reads 0.37 cm. For both micrometer and vernier calipers, the zero should always be carefully checked before beginning to take measurements. An English system vernier caliper is shown in Fig. 2.9*a*, and a scale-reading exercise is given in Fig. 2.9*b*. Can you verify the reading as 0.883 in.?

Many measurements are of such a nature that they cannot conveniently be determined by either micrometer calipers or vernier calipers. For example, depths of certain machine cuts and grooves would be difficult if not impossible to measure with such instruments. Furthermore,

Fig. 2.8 (*a*) Student-type metric vernier caliper. (Welch Scientific Co.) (*b*) Vernier scales. Note that 10 divisions of the vernier scale span exactly 9 divisions of the main scale. (*c*) Exercise in reading vernier scales. The readings are: top, 0.62 cm; bottom, 0.37 cm.

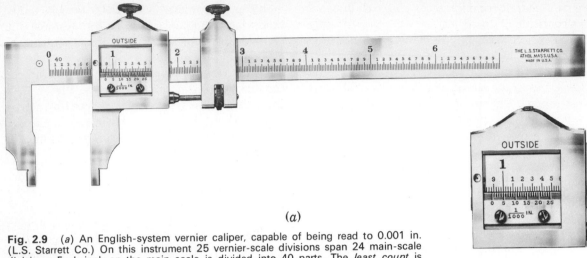

(a)

(b)

Fig. 2.9 (*a*) An English-system vernier caliper, capable of being read to 0.001 in. (L.S. Starrett Co.) On this instrument 25 vernier-scale divisions span 24 main-scale divisions. Each inch on the main scale is divided into 40 parts. The *least count* is $1/25 \times 1/40 = 1/1000$. (*b*) English-system vernier-caliper scale setting. The reading here is 0.883 in.

Fig. 2.10 A dial indicator and a set of gauge blocks. A bushing is being checked by the dial indicator, which is in turn calibrated by the gauge blocks. (DoALL Company.)

where continual checking of a dimension is necessary, as on every tenth part produced in a production run, a speedier method is required, but no sacrifice in accuracy can be permitted. The *dial indicator*, shown in Fig. 2.10, is a specialized measuring device which can be assembled in a variety of ways to measure accurately and speedily dimensions difficult to determine otherwise. Measurements on a dial indicator are taken with respect to some fixed point or reference plane

such as a lathe bed or a surface plate, and this arbitrary *zero* can be reset by the operator at any time. A variation of the dial indicator is the *comparator gauge*, which automatically registers how much a certain part is over or under the required dimension.

2.6 Calibrating Instruments

Dial indicators, micrometer and vernier calipers, and comparators can and do get out of adjustment. Accurate technical work demands frequent comparison of these instruments against standards whose accuracy is unquestioned. This process is called *calibration*. Modern, well-equipped shops make use of a *surface plate* and *gauge blocks* for periodic calibration of all shop instruments. The surface plate is a heavy block of high-grade steel or of stone with a smooth surface ground as flat as possible.

One typical full set of gauge blocks consists of 81 blocks, with which it is possible to obtain more than 100,000 practical combinations of dimensions. Figure 2.10 shows a complete set for a limited range of measurements, and also one application of the blocks.

Gauge blocks are made of case-hardened steel, carefully machined, finely ground and lapped, and with a surface finish so supersmooth that when "squeegeed" together they cling as if they were magnets. (This molecular attraction is called *cohesion* and is explained in Chap. 9.) Gauge blocks are manufactured to specified tolerances. Accuracy to 0.000001 in. is not uncommon. They are calibrated by using the wavelength of a standard light source. They must be used at a fixed standard temperature to avoid errors due to expansion. Gauge blocks can be assembled in various jigs and with many accessories to perform a variety of precision measurement tasks.

Optical and Electronic Methods The most accurate measurements of which man is capable are made by using principles involving the wavelength of light. For example, the orange-red line of krypton-86 has a wavelength of only a little more than one-half of a millionth of a meter (more precisely 0.6058 millionths of a meter) in vacuum. Since the krypton-86 radiation can be produced in any properly-equipped laboratory (see Fig. 2.2), its use for standardizing measurements is very important. Well-adjusted helium-neon gas *lasers* are also showing promise of great precision, with convenience unmatched by the krypton-86 lamp.

Optical flats are round disks of highly polished glass with surfaces made as flat as is humanly possible by the most advanced scientific and technical methods. They are used to standardize or *calibrate* gauge blocks and other precision measuring instruments, using the principles of light-wave interference.

2.7 Measuring Longer Distances

Until about forty years ago, only two reasonably accurate methods of measuring long distances were available—actual measurements by surveyors' tapes and associated mathematical calculations; and the use of astronomical observations and calculations (navigation). Developments in radio ranging (*radar*) and sound ranging (*sonar*) in recent decades have made accurate measurement of long distances commonplace. The theory and technique of radar and sonar measurements will be explained in later chapters. Also, later sections will discuss the measurement of fluid flow, temperature, mechanical energy, heat energy, electron flow, magnetism, sound intensity, light intensity, and atomic and nuclear radiation.

2.8 Measurement of Mass

Masses are measured by *balances* of various designs. The *beam balance* of Fig. 2.11a uses the principle of the lever to compare an unknown mass (left pan) with a known mass in the pan at the right end of the beam. The *platform balance* of Fig. 2.11b also makes use of levers, and compares an unknown mass to a known mass. The "known mass" in both cases is made up of selected small masses from a laboratory set. Actually, even though *masses* are being compared, the *force of gravity* is involved in the process. The assumption is that since the known mass and the unknown mass are both in essentially the same location on an equal-arm balance with respect to the earth and its gravitational field, the action of gravity on both is the same and therefore when the forces of gravity on both are equal (that is, when the instrument is "balanced"), the masses are equal.

A different kind of balance, called an *inertial balance*, is not dependent on gravitational forces, and it measures mass by utilizing one of the

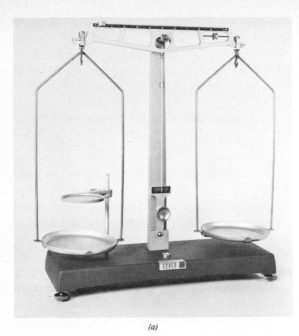

(a)

(b)

Fig. 2.11 Metric balances. (*a*) Beam balance. (*b*) Platform balance.

essential properties of mass, namely *inertia*. The relationships among mass, inertia, and force will be the subject of a later chapter. It is not necessary that you have a complete understanding of mass, force, and inertia at this point.

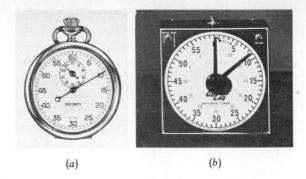

(a)   *(b)*

Fig. 2.12 (*a*) Stop-watch with one-fifth second precision. (*b*) Electrically-operated stop clock for student use in the laboratory. (Central Scientific Company.)

2.9 Time Measurement

The standard instruments for measuring time are clocks and watches. *Stop watches* and *stop clocks*, the latter electrically or electronically operated, are standard physics laboratory items (see Fig. 2.12). Timing devices utilizing the natural frequency of vibration of crystals such as quartz are commonly used in science and industry and in some watches for personal use. Extremely accurate "clocks" are based on the electromagnetic vibrations of certain atoms (see Fig. 2.4). Time periods of longer duration such as the *mean solar day* and the *year* are determined by the earth's rotation on its axis and its revolution around the sun, as measured by astronomical methods.

ACCURACY IN MEASUREMENTS AND COMPUTATIONS

2.10 Exact and Approximate Numbers

Some numbers are *exact*, like the number of wheels on an automobile (four) or the amount of money in your bank account on a given day ($865.22). But many numbers are the result of measurements, and these are *approximate numbers*, since no measurement is ever absolutely accurate. Most of the numbers we deal with in physics and technology are approximate numbers.

Certain other numbers like π (3.1416) and $\sqrt{3}$ (1.732), though not the direct results of measurements, are also approximate numbers, since the results never "come out even" no matter how far the calculation of their values is carried out.

2.11 The Decimal-Place Approach

In the normal processes of multiplication and division the products and quotients obtained frequently have an apparent accuracy indicated by several places of decimals which is not always justified by the data used in the problem.

Suppose it were required to compute the area of a circle whose diameter could be measured accurately to only two decimal places as 2.45 in. The area

$$A = \frac{\pi D^2}{4} = \frac{3.1416 \times (2.45 \text{ in.})^2}{4}$$

$$= 4.714383 \text{ in.}^2 \quad \text{to six decimal places}$$

Six-place accuracy for the computed result is of course misleading, since accuracy to only two decimal places was inherent in the measured data. The answer obtained for the area should be rounded off to 4.71 in.^{2}.

As a general rule, *do not retain decimal places which imply greater accuracy than the original measured data justify.* In most industrial calculations, π may be rounded off to 3.14.

2.12 Significant Figures

Engineers and scientists treat the problem of accuracy in computations from the viewpoint of *significant figures* or *significant digits*. The number expressing the magnitude of a measurement should be recorded with only as many digits as properly indicate the accuracy of the measurement. Only those digits are *significant digits*. When computations are subsequently made using numbers recorded from such measurements, the significant digits expressed in the final result of the computation should be limited to the number in the original data. For example, if the outside diameter of a steel shaft is measured with a micrometer caliper, and 2.42 cm can be read directly off the main and thimble scales, with a third *estimated* digit of 7 from the thimble scale, the reading would be 2.427 cm. Thus, we have three certain digits and one *doubtful* (but reasonable) digit. Only one doubtful digit is retained in recording the results of a measurement.

In computations with measured data or numbers known to be approximate, the following guidelines should be followed:

1. In adding and subtracting, do not include columns beyond the first one that contains a doubtful digit.

2. In multiplying and dividing, retain no greater number of significant digits in the final result than were in the *least* accurate of the approximate numbers involved in the calculation. One doubtful digit can be retained in the answer if desired.

3. Nonzero digits should always be regarded as being significant.

4. Zeros between nonzero digits (as 1.406) are significant.

5. Zeros before nonzero digits serve as placeholders, and are not significant digits. (*Example:* 0.00365.)

6. Zeros at the end of a *whole number* are usually not significant. (*Example:* A 75,000-horsepower (hp) rating on a marine engine could easily be in error by plus or minus 500 hp. The stated rating is probably accurate to only two significant figures).

7. Zeros at the end of decimal fractions are intended to be significant. (*Example:* A recorded measurement of 2.450 cm means that the zero is significant. If it were not significant the result should have been recorded as 2.45).

Some additional examples:

58,000	Probably correct to only two significant digits
0.0058	Correct to two significant digits
3.7825	Correct to five significant digits
4060	Correct to three significant digits
0.4060	Correct to four significant digits

0.004060   Also correct to four significant digits. (The two zeros immediately to the right of the decimal point serve as place-holders, not as significant digits.)

Illustrative Problem 2.1   A jet plane takes off and maintains a 25-degree angle of climb at a steady speed of 515 mi/hr. How far from the takeoff point is it (horizontal distance as measured along the ground) after 3.00 min?

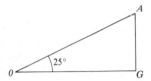

Fig. 2.13   Diagram for Illustrative Problem 2.1.

Solution   Diagram the problem as shown in Fig. 2.13. Point O is at takeoff. OA is the distance traveled through the air, and OG is the horizontal distance traveled. Note that all the given data are *measured data* (approximate numbers).

First, determine the air-line distance traveled.

$$OA = 515 \frac{mi}{hr} \times \frac{3.00 \ min}{60 \ min/hr} = 25.75 \ mi$$

retaining four digits *temporarily*.

Now use the cosine function (p. 15), and write

$OG = OA \cos 25$

   $= 25.75 \times 0.906$ (from Appendix III)

   $= 23.3295 \ mi$ if all digits are retained

The data given, however, justify only three significant figures (60 min/hr is an exact number, not a measurement). The value of cosine 25° (0.906) is accurate to three significant figures, but the 25° itself is a measurement, so we cannot place any real confidence in the third digit (6). It (the 6) is a very *doubtful* digit.

Therefore the answer should be written to only three significant figures, and is

23.3 mi   *ans.*

In this case, due to the uncertainty of measurement of the 25° angle of climb, the third digit is "doubtful."

It should be noted that although the decimal-place method and the significant-figure method are concerned with the same idea, they differ in their approach. The idea behind both is to prevent false or misleading degrees of accuracy from appearing in the results of computations, when the measured data which make up the problem have only limited accuracy. You should be familiar with both methods and use whichever your instructor prefers. In general, we will retain *one doubtful digit* in the answers to problems throughout the book.

2.13   Computing Devices

Either a pocket electronic calculator or a slide rule is a satisfactory device for solving the problems in this book. Longhand calculations are not recommended on two counts: (1) the opportunities for error are too great, and (2) the method is so burdensome and time-consuming that the principles of physics being studied tend to fade into the background.

Electronic calculators commonly yield answers containing up to 10 significant figures. That many digits are almost never justified in engineering and technical work. Since most input data are measured and are therefore approximate values, the seemingly very precise answers yielded by such calculators are really a gross overstatement. You should use the guidelines given above to determine how many significant digits to retain from electronic-calculator operations.

Slide rules (10-inch models) can be read to three significant figures (with a fourth figure estimated) at the left end of the C and D scales, but to only two significant figures (with a third estimated) at the right end of these scales.

QUESTIONS AND EXERCISES

1. In July of 1971 the Secretary of Commerce recommended that the United States should change over to the international metric system carefully and deliberately, with full conversion to be attained by 1981. Later, in 1975, the Congress passed the Metric Conversion Act. Do some library research and write a one- or two-page paper summarizing the present status of metrication in the United States.

2. What is meant by *fundamental units*? *derived units*? Give examples of three derived units and show how they are derived from fundamental units.

3. Why are machining operations which require a high degree of precision carried out in dust-free and temperature-controlled spaces?

4. Explain in detail how gauge blocks, comparators, dial indicators, and micrometers might be used in "tooling up" for the manufacture and inspection of a new line of electric motors for household appliances. Remember that subcontracting, interchangeability of parts, and assembly-line methods will all be involved.

5. Explain, with some examples, the difference between the "decimal-place method" and the "significant-digit method" of indicating the accuracy of a calculated result to a numerical problem.

6. In the career you have chosen or are considering (state what it is): (*a*) What kinds of measuring instruments will you be using either frequently or occasionally? (*b*) What is the normal degree of precision that will be expected (that is, to what fraction of a meter or an inch)? (*c*) If calculations result from these measurements, how many significant figures would typically be retained in the answer?

PROBLEMS

1. Convert the following: (*a*) 20 mm to inches, (*b*) 45 cm to inches, (*c*) 0.785 in. to centimeters, (*d*) 5.15 μin. to micrometers, (*e*) 60 nautical miles to kilometers, (*f*) 160 acres to hectares.

2. Express the following in decimal form: (*a*) $\frac{1}{1000}$ in., (*b*) $\frac{1}{10,000}$ in., (*c*) 1 μin., (*d*) 1 micrometer (μm), (*e*) $\frac{5}{8}$ in., (*f*) $\frac{22}{7}$.

3. The legal speed limit on U.S. freeways is 55 mi/hr. What would the speedometer of a European car read in km/hr at this speed?

4. An airplane is said to fly at Mach 1 when it attains the speed of sound. At or near sea level the speed of sound is about 330 m/sec. Express Mach 1 at sea level in miles per hour.

5. The 16-lb shot used in track and field meets has a mass of how many kilograms?

6. One ounce is the equivalent of how many milligrams?

7. Drugs are sometimes weighed in *grains* in the English system of measurement. Express the weight of a 5-gr aspirin tablet in milligrams of mass. If a nurse misinterpreted the labels and administered a 5-gm dose when a 5-gr dose was intended, what multiple of the correct dosage would this be?

8. A specification for a part manufactured in the United States reads, "allowable tolerance: +0.007; −0.003 in." Express these figures in micrometers.

9. A cylindrical shaft has the following measurements: Diameter, 0.964 in., length, 3.405 in. Calculate its volume in cubic inches, retaining only that number of significant digits justified by the measurements given (see Eq. 1.13).

10. Upon inspecting the lengths of a lot of rollers for roller bearings, an inspector finds that they range from 1.995 to 2.003 cm. The specification reads: 2, +0.001, −0.006 cm. Can he OK the entire lot? Why?

11. A metric system micrometer has two threads per millimeter along the spindle. There are 50 divisions around the thimble scale. What is the reading from the scales as shown in Fig. 2.14?

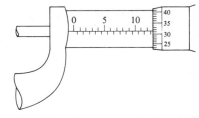

Fig. 2.14 Diagram for Problem 2.11.

12. An English-system micrometer caliper has 50 threads to the inch along the spindle. There are 200 divisions around the thimble scale. To what accuracy can the instrument be read? (See Fig. 2.6.)

13. A chemical plant ships acids in carboys (large glass bottles in wooden crates) containing 50 liters. How many gallons is this?

14. A motorcycle of a certain type and size has a mileage rating of 55 mi/gal of gasoline. How many kilometers should it travel on a liter of the same fuel?

15. A bushel of potatoes is weighed and found to contain 56 lb. How many kilograms weight (kilos) is this?

16. How many liters will a spherical tank contain if its diameter is 5.3 m? (See Eq. 1.11.)

17. Calculate the total volumetric piston displacement of an eight-cylinder auto engine if the cylinder diameter (bore) is $3\frac{7}{16}$ in. and the length of the piston stroke is $4\frac{1}{4}$ in. Assume each cylinder is a right circular cylinder and use Eq. (1.13).

18. What is the *great-circle* distance in nautical miles between the equator and a point at 38° 42′ N latitude?

19. The velocity of light is 2.998×10^8 m/sec. How long (in seconds) does it take light to travel from the moon to the earth, a distance of 384,500 km?

3 Vectors and Graphic Methods

In an earlier section the importance of portraying quantities and ideas graphically was pointed out. All too often in everyday life the *method of solving* a problem is so complicated that an understanding of the problem itself is erased. This is true in physics also. Many problems of length, time, direction, velocity, force, and momentum are more readily solved—and the problem itself better understood—by graphic methods than by the methods of algebra and trigonometry. Both *fundamental units* and *derived units* (see p. 23) lend themselves to graphic methods.

In particular, problems dealing with travel on land and sea and in the air; with velocity and acceleration; and with forces acting on, and stresses within structures can be more easily understood and solved by using graphic methods than by using analytic methods. In such problems much use is made of arrows drawn to a definite length and in a stipulated direction. These directed arrows are called *vectors*.

3.1 Vector And Scalar Quantities

Arrows used in diagrams and graphical solutions are handy devices for representing clearly the relationships of a physical problem. If the arrow is carefully drawn, its direction can indicate the direction of a motion or the line of application of a force. The *arrowhead* may also indicate the endpoint of the travel or the *point of application* of the force. And finally, the arrow itself may be drawn to an exact scale whose length will represent the magnitude, or size, of the distance, velocity, or force which it represents. Such arrows, drawn to careful scale, are *vectors*.

A *vector quantity* is a quantity which possesses the attributes of both magnitude and direction. Examples of vector quantities are force, displacement, velocity, acceleration, momentum, electric current, and magnetism.

A *scalar quantity* possesses magnitude only and has no attribute of direction. Examples of scalar quantities are such things as the volume of the city water tank, auto production for the year, your annual earnings, the number of cattle on your farm, or the number of beds in the local hospital.

The use of vectors in science and engineering is so important that courses and seminars lasting for a semester or a full year are offered at graduate and undergraduate levels under the name *vector analysis*. Our treatment here will be only an introduction to the topic, and of all the possible uses of vectors, we will deal in this chapter only with their application to the solution of problems involving changes of position (displacements), forces, and velocities. Some other applications will be presented in later chapters. In any one vector diagram all vectors must be of the same kind, that is *force* vectors, or *displacement* vectors, or *velocity* vectors.

VECTORS IN DISPLACEMENT PROBLEMS

3.2 Displacement or Change of Position

The simplest kind of problem readily solved by vector methods is one of displacement or change of position.

If, when you and a friend part on a street corner, he tells you he is going to walk 12 blocks, you will have no idea where he will be at the end of his walk. After all, he could walk any combination of streets and directions he chooses. He might even arrive back at the starting point. His "12 blocks" is a *distance*, not a displacement. Distance alone is not a vector quantity, since no direction is specified.

If, on the other hand, he says he will walk seven blocks *north* and then five blocks *east*, then you will know exactly where he will be at the end of the 12-block walk. This time he has described a *displacement*, or change of position, from where you are now, to a specified point on the city's grid of streets. His *net displacement* can be readily determined by using vectors as follows. Sketch the physical relationships of the problem as shown in Fig. 3.1. Then, using graph paper, plot the vector for north travel (**AB**), and the vector for east travel (**BC**). The vector representing the net change of position (displacement) is **AC**, the *resultant*, (**R**). The diagram itself is called a *vector diagram*, and the vectors are indicated in boldface type.

To determine the magnitude and direction of the resultant vector **AC**, place a compass or dividers with the points on *A* and *C*. Hold the point at *A* and swing the other point in an arc from *C* down to intersect the base line *AD* at *E*. Read the magnitude of *AE* = *AC* = 8.6 blocks, the actual displacement. **AC**, the *resultant*, is 8.6 blocks long and has a definite direction which can be determined by actual measurement of angle *BAC* with a protractor. The result is 35.5 degrees (east of north).

As a check on the vector solution we recall Pythagoras' theorem (Eq. 1.8) and solve analytically, as follows.

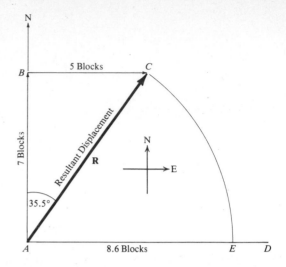

Fig. 3.1 Vector diagram for a displacement problem. Walking 12 blocks in a specified manner results in a measurable displacement.

$$\overline{AB}^2 + \overline{BC}^2 = \overline{AC}^2$$

Substituting,
$$7^2 + 5^2 = R^2$$
$$74 = R^2$$
$$R = \sqrt{74}$$
$$= 8.6 \text{ blocks}$$

To check the value of the direction angle *BAC*, note (see page 16) that

$$\tan \angle BAC = \tfrac{5}{7} = 0.7143.$$

From Appendix III (or from a pocket computer or slide rule), $\tan^{-1} 0.7143 = 35.5$ degrees. Graphic solutions by vectors generally give results of limited precision, since lengths and widths of lines and measurements of angles with protractors cannot be very accurate. To improve the accuracy of vector solutions, make your vector diagrams as large as possible.

Use of Vectors in Navigation A ship departs from port *A* (see Fig. 3.2) and steams due north for 650 nautical miles to point *B*, then turns northwest (45° to the left of due north) and steams 400 miles to *C*. Here it changes course

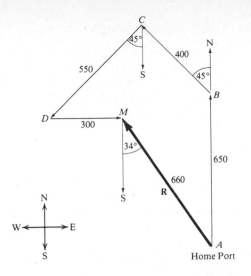

Fig. 3.2 Vector diagram for a navigation problem at sea.

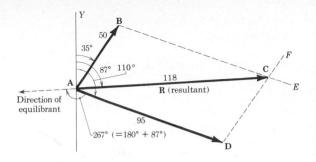

Fig. 3.3 The parallelogram of forces. Vectors applied to the solution of a force problem.

again and steams southwest for 550 miles to point *D*, then steams 300 miles due east to point *M*, where it receives a radio message to return to home port at *A*. How far away and in what direction is home port?

First sketch the problem, labeling all displacements (magnitude and direction) as shown in the diagram. Then plot the vector diagram on graph paper very carefully, choosing a scale which will nearly fill the sheet. The vector **R** represents the resultant displacement from home port. Careful measurement and conversion to your scale should find that **R** represents about 660 nautical miles. Measure angle *SMA* with a protractor to determine the direction of home port. It is about 34 degrees, so the course to home port is 34 degrees east of south. Later, a method of referring all directions in air and sea navigation to true north will be explained.

VECTORS APPLIED TO FORCE PROBLEMS

3.3 The Parallelogram of Forces

Two forces, *AB* and *AD*, act at a common point *A* as shown in Fig. 3.3. Forces that act at a common

point are said to be *concurrent*. The vector **AB** represents a force of 50 lb and is plotted 5 units long, each unit representing 10 lb. Vector **AD** represents a force of 95 lb and is therefore 9.5 units long. **AB** acts at 35° with the vertical and **AD** at 110° with the vertical. What are the magnitude and direction of the *net effect* (*resultant*) of these two forces? The resultant is found as follows. From *B* draw *BE* parallel to *AD*, and from *D* draw *DF* parallel to *AB*. (Use a set of parallel rulers.) These lines intersect at *C* and with the two original vectors, form the parallelogram *ABCD*. The diagonal *AC* of this parallelogram represents, to scale, the resultant **R** of the two given forces. Measuring **AC** and the angle *YAC* gives **R** as a force of 118 lb acting at an angle of 87° with the vertical. You should actually plot this exercise on graph paper and check the result given above.

3.4 Equilibrium of Concurrent Forces

If, instead of finding the net effect, or resultant, of the two forces *AB* and *AD*, it is desired to determine a single force which will balance or cancel them out, we may reason as follows. If vector **AC** is the *net effect* of the two forces, then a single force equal in magnitude and *opposite in direction* to *AC* will cancel the net effect of the two forces. This canceling force is of course just **AC** turned around, or **CA**. Such a force is called an *equilibrant*. Its magnitude is the same as that of vector **AC**, but its direction is 267°, 180° farther around clockwise from the vertical. (See Fig. 3.3.)

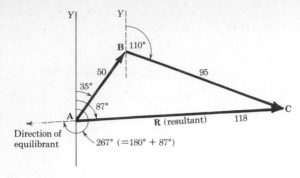

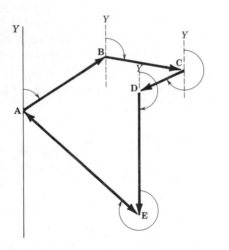

Fig. 3.4 The vector triangle. Determination of resultant (**R**) and equilibrant.

Fig. 3.5 The vector polygon. When a system of coplanar concurrent forces is in equilibrium, the vector polygon will close.

An *equilibrant* is a single force of such magnitude and acting in such a direction as to cancel out the net effect of a force system, or in technical terms, to put the force system into *equilibrium*. When a force system is in equilibrium, there is no net *unbalanced* force and consequently no acceleration of the body on which the force system acts will occur. Conversely, a body subjected to the action of a force system which is not in equilibrium will undergo an acceleration proportional to the magnitude of the resultant of the forces and in the same direction as the resultant. *The equilibrant is always equal in magnitude to, but opposite in direction from, the resultant.*

3.5 The Vector Triangle

Another, and sometimes more convenient, method of solving problems of concurrent forces is by means of the *vector triangle*. Take the same two forces as were used in Fig. 3.3. Draw **AB** as before representing the 50-lb force at 35° with the vertical (see Fig. 3.4). From the arrowhead end of this first vector, draw the vector **BC** representing the 95-lb force at 110° from the vertical. Now note carefully that if the triangle is closed by drawing the vector **AC**, we get **R**, the *resultant*, which, when measured, gives 118 lb at an angle of 87° with the vertical, the same answer obtained from the parallelogram method. If, on the other hand, we choose to close the triangle by drawing the vector **CA**, we get the *equilibrant*, which, as before, is 118 lb at 267° from the vertical.

3.6 The Vector Polygon

Sometimes more than two forces act at a point and it is desired to find the resultant or the equilibrant. The method is merely an extension of the vector triangle method. Figure 3.5 shows four concurrent forces represented by the plotted vectors **AB**, **BC**, **CD**, and **DE**. The resultant is **AE**, *drawn from the starting point to the arrowhead end of the last vector*. The *equilibrant* is obtained by drawing **EA** *from the end of the last vector and "going home" to the starting point*.

3.7 Resolution of Vectors into Rectangular Components

The solution of certain problems in mechanics is made easier by determining the so-called *rectangular components* of a vector. A *component* of a vector is another vector which indicates the net effect of the original vector in a specified direction. For example, in Fig. 3.6, the force vector **F**, which makes an angle θ (Greek *theta*) with the horizontal (x) axis, has components along the x and y axes, $\mathbf{F}_x$ and $\mathbf{F}_y$, respectively. $\mathbf{F}_x$, the component along the x axis, is obtained graphically by merely dropping the perpendicular PA from the end of the original vector to the x axis. $\mathbf{F}_y$ is obtained in a similar manner by constructing PB perpendicular to the y axis.

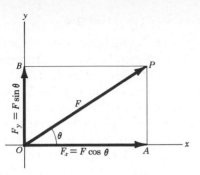

Fig. 3.6 Diagram showing the resolution of a force vector into rectangular (vertical and horizontal) components.

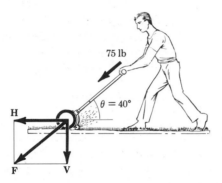

Fig. 3.7 The "lawnmower problem." Force acting at an angle to the desired direction of motion. The force vector **F** must be resolved into a horizontal and a vertical component.

Analytically, the values of F_x and F_y are readily obtained from the simple trigonometry of the right triangle, as

$$F_x = F \cos \theta \quad \text{and} \quad F_y = F \sin \theta \quad (3.1)$$

Many force problems are readily solved by resolving a known force into its components in specified directions. Two examples will be given, one rather simple, the other somewhat more complex.

The Lawnmower Problem In Fig. 3.7 a man is pushing on the lawnmower handle with a force of 75 lb when the handle is inclined at an angle of 40° with the horizontal. The pushing force can be resolved into its horizontal and vertical components either graphically, as shown in the diagram

by the **H** and **V** vectors, or analytically, using Eq. (3.1).

$$H = 75 \cos 40° = 57.5 \text{ lb}$$

$$V = 75 \sin 40° = 48.2 \text{ lb}$$

In this case the useful component is, of course, the horizontal one.

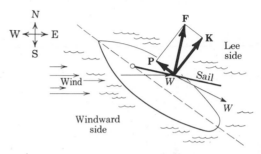

Fig. 3.8 The "sailboat problem." Vector diagram of forces acting on a sail when sailing almost into the wind.

The Sailboat Problem People from ancient times to the present day have wondered how it is possible for a sailboat to sail more or less against, or into, the wind. This is basically a problem in resolution of forces. Figure 3.8 shows the vector relationships of the problem. The wind is from the west, and the boat is headed in a generally northwest direction. With the sail set at the proper angle, the wind strikes it and is deflected at the same angle, as shown by the arrows labeled *W*. The resultant force on the sail indicated by **F**, is normal (perpendicular) to the sail's surface. The *useful component* of **F** is that component in the desired direction of travel, represented by vector **P**. The other component, athwart the boat, shown by vector **K**, is the force that causes the boat to heel over. This force is counteracted by the force against a centerboard, or keel, by the water beneath the hull, by the rudder, or by the boat's occupants themselves, who lean far out over the windward side of a light sailboat during a heavy blow. With careful adjustment of sails and rudder it is possible for small sailboats to make good progress upwind when sailing only 20° "by the wind," or 20° from the "wind's eye."

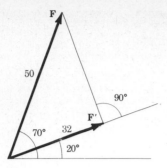

Fig. 3.9 Component of a force along a specified direction.

In summary, to find the component of any given force in any desired direction, merely drop a perpendicular from the end of the given force vector to the desired line of action. If a 50-lb force is acting at an angle of 70° with the horizontal (Fig. 3.9), and its component along a line making an angle of 20° with the horizontal is desired, the solution is obtained graphically as 32 lb. The desired force vector $\mathbf{F'}$ could also be obtained analytically from

$$\mathbf{F'} = F \cos (70 - 20) = 50 \cos 50$$

$$= 50 \times 0.643$$

$$= 32 \text{ lb}$$

3.8 Vector Addition; Coplanar Concurrent Forces

When several forces act on a rigid body at a point, their resultant can be found *graphically* by the vector polygon method (Sec. 3.6). The *method of components*, however, offers the best *analytical* approach to the solution of such problems.

In Fig. 3.10 a force system of four forces all in the same plane (coplanar), and concurrent at point O, is shown. Forces are in pounds. To find the resultant of such a force system analytically, take the following steps in order:

1. Resolve each force into x and y components as shown in Fig. 3.10.

2. Add (*algebraically*) all the x components to give the x component of the resultant (ΣF_x), and

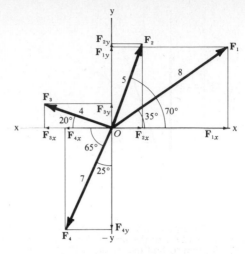

Fig. 3.10 Space diagram of a coplanar force system, concurrent at the origin of an x—y set of axes.

all the y components to give the y component of the resultant (ΣF_y).

3. ΣF_x and ΣF_y are then added *vectorially* to obtain the final resultant $\mathbf{R}$ (Fig. 3.11).

Components above the x axis or to the right of the y axis are considered positive; those to the left of the y axis or below the x axis are considered negative. The work for the exercise of Fig. 3.10 is summarized as follows:

x components

$$
\begin{aligned}
\mathbf{F}_{1_x} &= 8 \cos 35° &&= +6.55 \text{ lb}\\
\mathbf{F}_{2_x} &= 5 \cos 70° &&= +1.71\\
\mathbf{F}_{3_x} &= -4 \cos 20° &&= -3.76\\
\mathbf{F}_{4_x} &= -7 \cos 65° &&= -2.96
\end{aligned}
$$

$$\Sigma F_x = \qquad\qquad +1.54 \text{ lb}$$

y components

$$
\begin{aligned}
\mathbf{F}_{1_y} &= 8 \sin 35° &&= +4.59 \text{ lb}\\
\mathbf{F}_{2_y} &= 5 \sin 70° &&= +4.70\\
\mathbf{F}_{3_y} &= 4 \sin 20° &&= +1.37\\
\mathbf{F}_{4_y} &= -7 \sin 65° &&= -6.33
\end{aligned}
$$

$$\Sigma F_y = \qquad\qquad +4.33 \text{ lb}$$

These resultant components can now be combined vectorially as in Fig. 3.11. $\mathbf{R}$ is

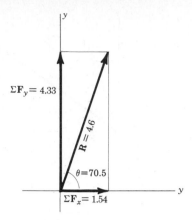

Fig. 3.11 The algebraic sum of *x*-components (ΣF_x) and *y*-components (ΣF_y) of the forces of Fig. 3.10, plotted in a vector diagram to find the resultant **R**.

determined from the plot as 4.6 lb, and the angle θ is measured with a protractor and found to be 70.5°.

Analytically, **R** can also be determined from the theorem of Pythagoras as

$$\mathbf{R} = \sqrt{(\Sigma F_x)^2 + (\Sigma F_y)^2}$$
$$= \sqrt{(1.54)^2 + (4.33)^2} = 4.59 \text{ lb} \qquad ans.$$

The angle θ can be calculated from the trigonometric relation

$$\tan \theta = \frac{4.33}{1.54} = 2.81$$

$$\theta = 70.4° \qquad ans.$$

In general,

$$\Sigma F_x = \Sigma \mathbf{F} \cos \theta$$

$$\Sigma F_y = \Sigma \mathbf{F} \sin \theta$$

$$\mathbf{R} = \sqrt{(\Sigma F_x)^2 + (\Sigma F_y)^2} \qquad (3.2)$$

$$\theta = \tan^{-1} \frac{\Sigma F_y}{\Sigma F_x} \qquad (3.3)$$

where the Greek letter *sigma* (Σ) is read "the algebraic sum of."

Fig. 3.12 Every truss, girder, pier, and cable of a bridge must be in equilibrium. (Reynolds Metals Co.)

3.9 Vectors Applied to Structural Problems

Buildings and other land-based structures are often designed to support great loads or withstand tremendous forces, without undue vibration or bending, or sway beyond established safe maximum values. The piers, girders, and beams of a multistory building, for example, are subject to tremendous stresses from the weight of the building and from the weight of the machinery, materials, and people it contains. Another example is provided by the piers, trusses, and cables of a bridge (see Fig. 3.12) which must support the weight of the bridge itself together with the traffic crossing it. Since forces tend to cause motion, the problem of the engineer and builder is to distribute the forces and stresses in the members of a structure in such a way that they cancel one another out, leaving each and every member, i.e., truss, girder, beam, cable, joist, in *equilibrium* under the action of whatever forces may be acting on it.

The section of physics that deals specifically with equilibrium conditions and the problems of forces in structures is called *statics*. A complete treatment of statics is found in textbooks of mechanical engineering; we shall discuss here only a few elementary examples that illustrate the principles. Some use of the principle of the lever will be required, and vector methods will be used.

3.10 Equilibrium Due to Concurrent Forces

Many construction problems involve the equilibrium (Sec. 3.4) of pins, joints, or unions at which three or more forces act at a common point. A common example of this type of force system is the ordinary derrick or hoisting crane. Figure 3.13a illustrates a hoisting crane in use on a ship, and Fig. 3.13b shows a diagram of the typical arrangement in which the mast MG is steadied by guy wire MT while the load w is lifted by the cable passing over the pulley at B. The boom AB is hinged at A and supported by cable MB. In the following analysis, the weights of structural parts (boom, pulleys, cables) are *assumed* negligible, compared to the load w.

In solving such problems we *isolate* one point at a time. To *isolate* a point means to consider, *for the time being*, only those forces acting at that point and to omit temporarily all other conditions of the problem. Isolating point B, we see that three forces act on it, namely, the *tension* in BL, which is 3000 lb vertically downward, the *tension* in BM, and the *compression* in BA, whose directions are known but whose magnitudes are unknown. Of course there are forces acting in MT and MG, but these are not considered while *isolating* point B. Vector methods offer a ready solution in determining the magnitudes of the tension in BM and the compression in BA.

(a)

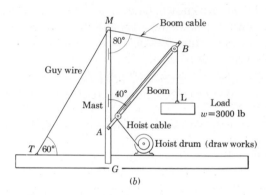

(b)

Fig. 3.13 Equilibrium of coplanar concurrent forces. (*a*) A hoisting crane in action. (Official U.S. Navy photograph.) (*b*) Diagram of the elements of the hoisting-crane problem.

Refer to the diagrams of Fig. 3.14. The *space diagram* (Fig. 3.14a) shows the three forces acting on the point B. **w** is plotted to scale, but only the *directions* of C (compression) and T (tension) are known; so they are shown with a wavy section to indicate that their lengths are unknown. Constructing such a space diagram is an extremely important step in the analysis of equilibrium of concurrent forces. *Space diagrams* need not be drawn to accurate scale, since their purpose is to portray the general conditions of the problem rather than to provide measures for obtaining the final solution. They guide the drawing of the *vector diagram*.

To draw the *vector diagram* in Fig. 3.14b start with the vector **w**, since *both its direction and*

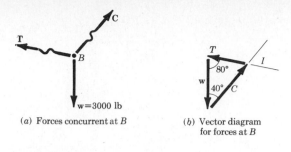

(a) Forces concurrent at B   (b) Vector diagram
                                   for forces at B

Fig. 3.14 Diagram of the vector solution of the hoisting-crane problem of Fig. 3.13. (a) Forces concurrent at B. (b) The vector triangle.

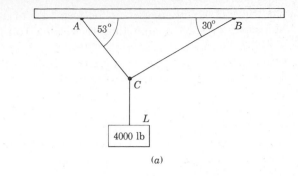

(a)

Fig. 3.15 Diagrams for Illustrative Problem 3.1. (a) Three coplanar, concurrent forces in equilibrium. (b) The space diagram. (c) The vector diagram.

magnitude are known. Draw it as shown vertically downward, just as the force acts on point B, to a scale representing 3000 lb. Now from the head of this vector draw a line in the direction of the compression force in the boom. This is the line C of the diagram, at 40° with the vertical and parallel to BC of the space diagram. Since the point B is in equilibrium, the three forces acting at B must cancel each other; i.e., they must have no net resultant. Therefore, when the vector representing the cable tension is added to the vector diagram, the vector triangle *must close*. The cable MB makes an angle of 80° with the vertical; so draw the line of vector **T** *back* toward the line of vector **C** at an angle of 80° with the vertical until **T** and **C** intersect at I. The vector triangle is now complete, and scaling values off the diagram gives C, the compression in the boom, as 3400 lb and T, the tension in the boom cable, as 2200 lb. All vector diagrams must be drawn to accurate scale, preferably on graph paper. Do this yourself with this exercise and verify the results given here.

Now that the tension in MB is known, the solution for the tension in the guy wire MT and the compression in the mast MG can be worked out in a like manner by isolating the point M, drawing a space diagram for it, and solving the vector triangle for the forces acting there. (The exercise just indicated should be performed by the student at this time. For checking results, the answers are: GM, 4150 lb compression, and MT, 4400 lb tension).

Illustrative Problem 3.1 A weight of 4000 lb is being supported by two cables under the conditions illustrated in Fig. 3.15a. Find the tension in each cable.

Solution Three forces, tension in CL, tension in BC, and tension in AC, are all concurrent at C. The forces are coplanar, and since C is not being accelerated, the system is in equilibrium; i.e., the resultant of the three forces is zero.

Isolate point C and draw a *space diagram* as shown in Fig. 3.15b. Both the magnitude and the direction of **L** (load) are known, but only the *directions* of A and B are known, and they are shown with wavy segments to indicate that the magnitudes are unknown.

Now draw the *vector diagram* carefully (the entire problem should be set up on graph paper) to scale, as shown in Fig. 3.15c. The vector **L** will be drawn 4000 units long to your chosen scale, vertically downward. **B** is drawn from the arrowhead end of **L** at an angle of 30° with the horizontal, and **A** is drawn *backward* from the

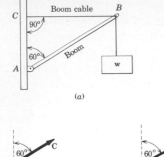

Fig. 3.16 Diagrams for Illustrative Problem 3.2. (*a*) Sketch of boom, cable, and load. (*b*) The space diagram. (*c*) The vector diagram.

"feathered" end of the **L** vector, at an angle 53° below the horizontal, until it intersects the line of **B** at *I*. Scaling off the vectors **B** and **A** gives the following results:

$$A = \text{tension in } AC = 3490 \text{ lb}$$

$$B = \text{tension in } BC = 2420 \text{ lb} \qquad ans.$$

Illustrative Problem 3.2 If the maximum allowable compression in the boom *AB* of Fig. 3.16*a* is 5000 lb, what is the maximum load *w* which can be handled? At this load what is the tension in the horizontal cable *BC*? Neglect weight of boom.

Solution Isolate point *B* and draw the space diagram as shown in Fig. 3.16*b*. The compression in the boom, **C**, is known in both magnitude and direction, but **w** and **T** are known in direction only, so they are shown with wavy sections.

Now draw the vector diagram (Fig. 3.16*c*). First, draw vector **C** to scale for 5000 units at 60° with the vertical. Then draw the line of vector **w** vertically downward from the arrowhead end of **C**, length indefinite. Finally, close the vector triangle with the line of vector **T** drawn horizontally from the starting point of vector **C** until the line of **w** is

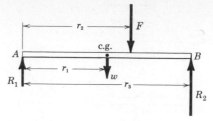

Fig. 3.17 Equilibrium of parallel forces.

cut. From the vector diagram as now completed, scale off the desired results:

w = load at maximum allowable boom compression = 2500 lb

T = tension in horizontal cable
$BM = 4330 \text{ lb}$ *ans.*

3.11 Equilibrium of Coplanar Parallel Forces

Up to this point we have been dealing with concurrent forces—several forces acting at a common point. What if the forces are parallel and act on a body at different points?

Consider the beam *AB* of Fig. 3.17 under the action of the parallel forces shown. *w* is the weight of the beam itself acting at a point which we call the *center of gravity* (c.g.). The center of gravity of an object can be generally described as *that point at which the entire weight of an object can be considered to be concentrated*. An object will balance if hung from, or supported under, its center of gravity. The center of gravity of a homogeneous body is at its geometric center. In Fig. 3.17, *F* is a force acting down on the beam, a force perhaps due to a heavy piece of machinery installed over the beam, and R_1 and R_2 are the reaction forces in the end piers which support the beam. All these forces act vertically, and since there are no forces acting in any other direction, if motion does occur it will have to be either vertical motion—the whole beam moving up or down—or a turning (rotation) of the beam about some point in its plane as a center.

Considering the first of these two possibilities, vertical motion either up or down can be

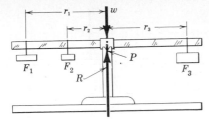

Fig. 3.18 Conditions for equilibrium of parallel forces.

ruled out if the sum of F and w equals the sum of R_1 and R_2. If this is true, all the vertical effects of the forces cancel out and therefore no acceleration of the beam up or down will take place.

Condition I for *equilibrium of parallel forces* may be stated: *The sum of the forces in one direction must be equal to the sum of the forces in the opposite direction.* Assigning one direction as positive $(+)$ and the other negative $(-)$, we obtain the following statement for condition I:

The algebraic sum of the forces acting on a body in equilibrium must be equal to zero.

Or, mathematically,

$$\Sigma F = 0 \qquad (3.4)$$

Condition I for equilibrium might be satisfied, with no tendency for vertical motion of the beam, but the distribution of the forces F, w, R_1, and R_2 might easily be such that there will be a net turning effect (torque) to rotate the beam. Such rotations must, of course, be prevented if equilibrium is to be attained.

Condition II for equilibrium states *that there must be no net turning effect about any axis, either within the body or outside it.*

To carry this analysis a bit further, refer to Fig. 3.18, which shows a meterstick mounted on a support at P, in equilibrium under the action of five forces F_1, F_2, F_3, w (its weight, acting downward through P, its center of gravity) and R, the reaction at the support. Since the meterstick is not being accelerated vertically up or down, condition I for equilibrium is satisfied; that is,

$$\Sigma F = R - F_1 - F_2 - F_3 - w = 0$$

What is necessary now to satisfy condition II, i.e., to prevent rotation? Obviously the net turning effect *clockwise* around point P must be equal to the net turning effect *counterclockwise*. *Each force times the perpendicular distance to the pivot F* contributes a turning effect or *turning moment*. In general, the *moment of a force* is the product obtained by multiplying the force itself by the perpendicular distance from the line of action of the force to the chosen center of rotation. This is the principle of the common *lever*, which we will deal with in detail in Chap. 6. *Moments* are classified *clockwise* or *counterclockwise* according as the effect of the force would be to cause clockwise or counterclockwise rotation. In Fig. 3.18 we note that the moment $F_3 r_3$ is a clockwise one and that $F_1 r_1$ and $F_2 r_2$ are counterclockwise moments. R and w have no moment at all, since they pass through the pivot point and their moment (lever) arms around P are zero.

Condition II for equilibrium of a body under the action of parallel forces can now be stated:

The sum of clockwise moments about any point in the plane of the forces must equal the sum of counterclockwise moments about the same point.

Letting the letter M stand for moments, we have

$$F_1 r_1 = M_1$$
$$F_2 r_2 = M_2$$
$$F_3 r_3 = M_3$$

Designating counterclockwise moments as positive (which is conventional) and clockwise moments as negative, we can write a shorter statement for condition II:

The algebraic sum of moments about any point must be equal to zero.

Mathematically, for Fig. 3.18,

$$M_1 + M_2 - M_3 = 0$$

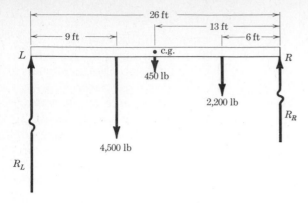

Fig. 3.19 Equilibrium of a loaded I-beam.

or, in general,

$$\Sigma M = 0 \qquad (3.5)$$

It is emphasized that there must be no net turning effect (moment) about *any point*, either within the body or anywhere else in the plane of the forces.

Returning now to the beam of Fig. 3.17, take moments about A (any other point would do as well, but A is convenient) and write the equation for the second condition of equilibrium:

$$R_2 r_3 - F r_2 - w r_1 = 0$$

Note that R_1 produces no moment when A is the moment center. The following problem will illustrate the foregoing principles.

Illustrative Problem 3.3 A steel I-beam (Fig. 3.19) weighs 450 lb and spans a distance of 26 ft. It is supported at each end by pillars. The beam supports a 2200-lb punch press 6 ft from the right end and a 4500-lb lathe 9 ft from the left end. Find the reaction forces in the supporting pillars.

Solution First diagram the problem as shown, using vectors to show the forces involved. The lengths of R_L and R_R are unknown, and this is indicated in the diagram by the wavy lines.

Applying condition I, we obtain

$$\Sigma F = 0 \quad (3.4)$$

$$R_L + R_R - 4500 - 450 - 2200 = 0$$

or

$$R_L + R_R = 7150 \text{ lb}$$

This equation for condition I cannot be solved for either reaction, since both are unknown.

Applying condition II, take moments about any point, say L, and solve for R_R,

$$\Sigma M_L = 0$$

$$26R_R - 2200 \times 20 - 450 \times 13 - 4500 \times 9 = 0$$

$$26R_R = 90{,}350$$

$$R_R = \frac{90{,}350}{26} = 3475 \text{ lb} \qquad ans.$$

which is the reaction in the right-hand pillar. Substituting this value in the equation for condition I above and solving for R_L gives

$$R_L = 7150 - 3475 = 3675 \text{ lb} \qquad ans.$$

the reaction in the left pillar.

VECTORS APPLIED TO VELOCITY PROBLEMS

Since velocity possesses the attributes of both magnitude and direction, it is a vector quantity. Therefore, velocities can be combined vectorially to yield a resultant velocity, and a given velocity can be resolved into components in specified directions.

3.12 The Resultant of Two Velocities. Composition of Vectors

If you have ever thrown a ball or a rock off a high cliff you noticed that in addition to the horizontal velocity you gave it, the force of gravity immediately began to give it a vertical velocity as well. In Fig. 3.20 the ball is thrown from cliff PC at a horizontal velocity $v_h = 20$ m/sec. Neglecting air resistance we can assume that the horizontal velocity is constant at this value. The ball begins immediately to pick up vertical velocity, v_v, which is calculated to be 45 m/sec on impact at X. What is the *resultant velocity* (magnitude and direction) just before impact?

The vector diagram for the problem is shown in Fig. 3.20*b*. The process is known as

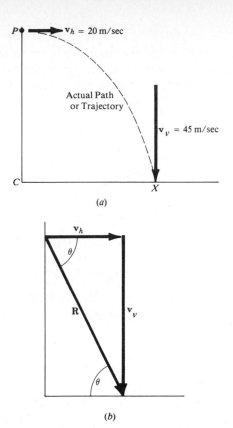

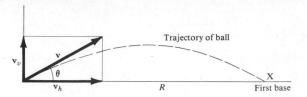

Fig. 3.21 Diagram showing a velocity vector resolved into horizontal and vertical components.

Fig. 3.20 Horizontal velocity is unaffected by increasing vertical velocity. (*a*) Space diagram. (*b*) Vector diagram.

vector addition ($\mathbf{v}_h + \mathbf{v}_v = \mathbf{R}$). You should plot the vector diagram carefully on graph paper to check the following results:

$$\mathbf{R} = 49.2 \text{ m/sec}$$
$$\theta = \tan^{-1} \tfrac{45}{20} = \tan^{-1} 2.25 = 66°$$

3.13 Resolution of Velocity Vectors

Throwing baseballs off cliffs is not a frequent occurrence, but throwing them around the baseball diamond is. If an infielder fields a ground ball and throws to first base for the out, he must project the ball both toward first base (horizontal velocity) and slightly upward so that its drop due to gravity will be compensated. The diagram of Fig. 3.21 shows the factors involved (neglecting air resistance). The actual velocity of the ball leaving the third baseman's fingers is indicated by $\mathbf{v}$. Note that it has a horizontal component, $\mathbf{v}_h$,

and a vertical component, which at the instant of release, is labeled $\mathbf{v}_v$. The vertical velocity is affected by gravity, reducing to zero at the top of the ball's flight or trajectory. Directed downward after that, it increases to a maximum as the ball reaches the first baseman's mitt.

The composition and resolution of velocity vectors has important military applications in gunnery, aircraft operations, rocket flight, and ship movements. In later chapters we will discuss these in detail and develop *mathematical* methods of solving practical problems.

3.14 Relative Velocity

A hungry passenger walking aft to the dining saloon along the deck of a ship which is steaming on a westerly course at 15 knots (1 knot is 1 nautical mi/hr) walks due east with a velocity of 5 ft/sec *with respect to the ship*. What is the passenger's velocity with respect to the earth? Vectors may be used for a ready solution. In Fig. 3.22, let vector **SG** represent ship's velocity with respect to ground, and vector **PS** represent passenger's velocity with respect to ship. Since **PS** is oppositely directed with respect to **SG**, the resultant or net velocity of the passenger with respect to ground is shown by the vector **PG**. Vectorially, this vector difference is written

$$\mathbf{SG} - \mathbf{PS} = \mathbf{PG}$$

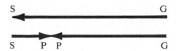

Fig. 3.22 Vector diagram for the analysis of the passenger's velocity relative to that of the ship.

To obtain consistent units, change ship's velocity to ft/sec. From Table 2.1, p. 25, note that a nautical mile is 6080 ft.

Ship's velocity

$$SG = \frac{15(\text{naut. mi/hr}) \times 6080(\text{ft/naut. mi})}{3600 \text{ sec/hr}}$$

$= 25.3$ ft/sec west, with respect to the earth.

Consequently

$$PG = 25.3 \text{ ft/sec} - 5 \text{ ft/sec}$$

$= 20.3$ ft/sec, west, the passenger's velocity with respect to the ground.

All motion is in fact relative, for although we commonly consider the earth to be stationary and refer all motions to it, actually the earth is in motion in a rather complex manner. It makes a complete rotation on its polar axis once every 24 hr. Each $365\frac{1}{4}$ days it makes an elliptical journey around the sun, a distance of some 585 million miles. Other planets in our solar system have their own equally complex rotations and orbits around the sun. There is evidence also that the entire solar system is moving through the Milky Way galaxy toward the constellation Hercules at a prodigious speed. And finally, our galaxy seems to move through space *away* from other galaxies. Consequently the launchings of man-made satellites, interplanetary rockets, and other space vehicles are fraught with exceedingly complex problems in relative velocity.

Motion of a Body in a Moving Medium Ship captains charting their courses for various ports desire to execute a motion with respect to the earth. However, their ships move *through* water, a medium that because of tides, storms, or ocean currents, moves appreciably with respect to the land. Aircraft fly from point to point on earth but must make their way through the air, a medium that is continually moving with respect to the earth's surface. Relative-velocity problems like these are readily solved by the use of vector methods, as follows.

First draw the vectors *separately*, freehand, *not assembled into a vector diagram* and not to exact scale, but with *approximate directions and relative lengths* indicated. Label each vector with a letter *at the arrowhead end* which stands for the *moving object or medium*, and a letter *at the tail* (feathered) end which indicates *the object or medium with respect to which it is moving*. Then, from these, construct the vector diagram accurately to scale, starting with a vector whose magnitude and direction are both known. Two examples illustrating the method are given. All directions in sea and air navigation are given in three digits, measured clockwise from due north—0° to 360°.

Illustrative Problem 3.4 A pilot flies on a compass heading of 045° at an indicated airspeed of 140 mi/hr. During the flight a wind is blowing at 40 mi/hr due south. What is the resultant velocity of the plane with respect to the ground?

Solution First sketch the vectors and label them as shown in Fig. 3.23a. Then draw the vector diagram Fig. 3.23b accurately on graph paper, starting with the wind vector **AG**, and *putting like letters together*, that is, A to A. Close the vector triangle by drawing vector **PG**, which gives the plane's speed with respect to ground. The student should check the answer as 116 mi/hr, actual course 059°.

Illustrative Problem 3.5 A pilot whose private plane cruises at 350 mi/hr (indicated airspeed) leaves San Francisco bound for St. Louis, 1700 miles due east. He is advised that a 50-mi/hr wind blowing southeast (i.e. 135°) will be encountered throughout the entire flight. (*a*) What compass heading should he use? (*b*) What is the plane's speed over the ground? (*c*) How long will it take to reach St. Louis?

Solution Sketch and label the vectors as in Fig. 3.24a. The wavy section in vector **PG** means that its *magnitude* is unknown. Showing vector **PA** dotted signifies that its *direction* is unknown.

Now construct the *vector diagram* as shown in Fig. 3.24b. Start with the wind vector **AG**,

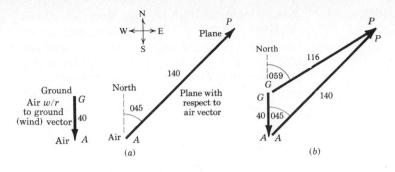

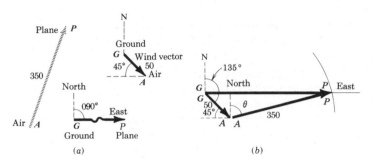

Fig. 3.23 Vector solution of a relative-velocity problem in aircraft piloting. (*a*) Preliminary sketch for identifying and labeling vectors (not to scale). Note that vectors are labeled with a letter at the arrowhead end which indicates the *moving* object or medium, and a letter at the "feathered" end indicating the object or medium *with respect to which* the motion takes place. (*b*) Vector diagram to solve for plane's speed with respect to ground (vector **PG**) (Illustrative Problem 3.4).

Fig. 3.24 Vector solution of a relative-velocity problem (Illustrative Problem 3.5). (*a*) Preliminary sketch for identifying and labeling vectors. (*b*) Vector diagram to solve for the plane's speed with respect to the ground (vector **PG**) and the compass heading necessary to make good the required course.

since both its magnitude and direction are known. Then draw the *line of action* of vector **PG**, *putting G to G*, and extending the *P* end out *indefinitely*. Now put the *A* end of vector **PA** at the *A* end of vector **AG** and swing an arc of radius 350 units, the *magnitude* of vector **PA**. The point at which this arc cuts the line of vector **PG** locates the third vertex of the vector triangle.

From **PG** scale off the magnitude of the plane's velocity with respect to the ground as 385 mi/hr. With a protractor, measure angle θ as 085°, the compass *heading* which the pilot must use to make good a *course* of 090° for the flight. The time required for the flight is:

$$t = \frac{s}{v} = \frac{1700 \text{ miles}}{385 \text{ mi/hr}} = 4.43 \text{ hr} \qquad ans.$$

Although *vector principles* are used in the solution of complex problems of relative motion—in gunnery, navigation, and aerospace operations—in actual practice these problems are usually solved by modern computers and associated electro-optical equipment.

SUMMARY

A *vector* is an arrow drawn to scale whose length indicates the magnitude of a quantity and whose direction indicates the direction of the quantity.

Vector quantities possess both magnitude and direction; *scalar quantities* possess magnitude only.

Vectors may be used to represent such quantities as forces, displacements, and velocities. Some other uses for vectors will be shown in later chapters.

The *resultant* of two vectors can be found by either the *parallelogram method* or the *vector-triangle method*.

Some vector problems are best solved by resolving the vectors into x and y components. $\Sigma F_x = \Sigma F \cos \theta$, $\Sigma F_y = \Sigma F \sin \theta$, where θ is the angle with the x axis. Analytically, $R = \sqrt{(\Sigma F_x)^2 + (\Sigma F_y)^2}$;

$$\theta = \tan^{-1}\left(\frac{\Sigma F_y}{\Sigma F_x}\right)$$

Structural problems, involving the principles of *statics*, are most easily solved by vector methods.

Condition I for equilibrium of a body acted on by parallel forces is as follows: *The algebraic sum of all* forces acting on the body must be equal to zero: $\Sigma F = 0$.

Condition II for equilibrium of such a body is: *The algebraic sum of moments about any point in the plane must be equal to zero:* $\Sigma M = 0$.

Concurrent force systems, which involve two or more forces acting at (toward or away from) a common point, are best solved by vector-polygon methods.

In solving any complex problem in statics, one should simplify the problem by *isolating* certain points or bodies in turn. A vector solution is determined for the body or point thus isolated, and results from this partial solution are used to obtain vector solutions for other parts of the problem.

Relative-velocity problems are readily solved by vector methods. Extreme care must be taken in labeling the vectors before constructing the vector diagram.

QUESTIONS AND EXERCISES

1. Which of the following are vector quantities and which are scalar quantities?

    The population of your city
    Mass of the earth
    Density of lead
    Time for the earth to revolve around the sun
    Force of a golf club hitting the ball
    Weight of a bushel of corn
    Capacity of your car's gas tank
    Velocity of an airliner flying overhead
    Weight of a sailboat's keel
    A walk from your office to the bank
    A push on your car's brake pedal

2. If a set of forces is known to be in equilibrium, what can be said of any one force with respect to the combined effect of all the others?

3. If three coplanar forces act at a point and the point is in equilibrium, why does the vector plot result in a closed triangle?

4. When a baseball or rock is thrown straight up, it is momentarily at rest at the very top of its flight. Is it in equilibrium at that point? Explain.

5. If a force in the positive "x-direction" of a pair of coordinate axes is plotted on a vector diagram with an equal force in the positive "y-direction," what is the direction of the resultant?

6. A balloon is 500 m above the ground and is stationary in still air. It would appear to be in equilibrium

at the moment. Is it? If so, what forces are equally balanced?

7. In riding a department store escalator from one floor to another you decide to walk up the steps as they are moving up the incline. What effect does this have on your velocity? your displacement?

8. When a set of parallel vertical forces acts on a body, will the body always be in equilibrium if the sum of the upward forces equals the sum of the downward forces? Explain.

9. From various sources, technical and nontechnical, obtain several definitions of *center of gravity*. List them, giving the source of each, and discuss how the concept of center of gravity is related to equilibrium.

10. If a beam is in equilibrium under the action of three coplanar but nonparallel forces, show that the *lines of action* of these forces must intersect in a single point. Assuming a beam of finite weight, where will the center of gravity be with respect to the single-point intersection?

PROBLEMS

NOTE: *Draw careful diagrams for all problems.*

1. An auto is driven 20 km due east, then 8 km due north, then 15 km along a road running northwest, then 16 km due west. Use the vector polygon method and graph paper to find its displacement (distance and direction) from the starting point.

2. Use the vector-parallelogram method to find the resultant of these two concurrent forces: one of 125 lb acting vertically upward, and one of 65 lb acting at an angle 20° below the horizontal to the right.

3. An airplane takes off and flies for 3 hr on course 045° at 350 mi/hr. The pilot then changes course to 110° and flies at 450 mi/hr for 2.5 hr. How far is he from the starting point? (Plot all courses clockwise from due north, which is 000°).

4. A 70-m displacement is at an angle of 35° with the positive x-axis. Find its x and y components graphically and then trigonometrically.

5. A velocity vector has a magnitude of 200 m/sec in a compass direction of 055°. Find its north and east components.

6. Find the horizontal and vertical components of a 500-lb force which acts upward and to the right, making an angle of 50° with the horizontal. Plot it accurately to scale, and then check your answers trigonometrically.

7. Three members of a bridge truss meet at the point A, as shown in the diagram of Fig. 3.25. AB and AC are under tensional stresses of 2200 lb each. What must be the stress in AD for equilibrium? (Solve by the vector-triangle method.)

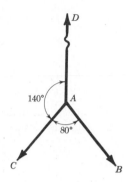

Fig. 3.25 Concurrent forces in a bridge truss (Problem 3.7).

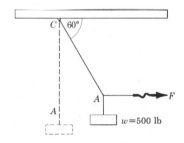

Fig. 3.26 Diagram for Problem 3.8.

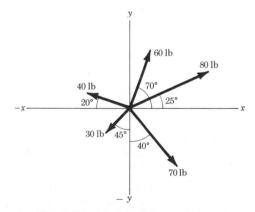

Fig. 3.27 Diagram for Problem 3.9.

8. What horizontal force F must be applied to pull the supporting rope AC aside until the angle at the ceiling is 60° as shown in the diagram of Fig. 3.26?

9. Find the resultant of the force system shown in the diagram of Fig. 3.27 by the method of components as presented in Sec. 3.8.

10. The following four forces are in the same plane and are concurrent: 500 lb at 035°, 350 lb at 130°, 450 lb at 220°, and 285 lb at 260°. What are the direction and magnitude of the single force which will put this system in equilibrium? (Use the vector-polygon method.)

11. A block A is held on an inclined plane, as shown in Fig. 3.28, by the cord P. If the weight of the block is 5 lb, find the normal force N with which the plane reacts against the block, and the tension in cord P. Neglect friction.

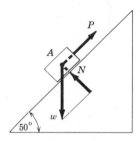

Fig. 3.28 Forces resulting in equilibrium of a block on an inclined plane (Problem 3.11).

12. Find the compression in the boom AB and the tensions in the guy wire DC and boom cable BC of the crane shown in the diagram of Fig. 3.29. Neglect the weight of the boom.

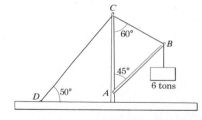

Fig. 3.29 Diagram of a hoisting crane (Problem 3.12).

13. Given the data shown in the diagram of Fig. 3.30 (next page), find the tensions in ropes AB and CD and also the value of angle θ when $\alpha = 45°$.

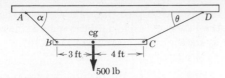

Fig. 3.30 Diagram for Problem 3.13.

14. A small bridge, supported by piers at each end, is being crossed by a 25-ton truck and semitrailer. If the bridge is 150 ft long and the center of mass of the truck and trailer is 50 ft from the left end, calculate the amount of the load carried by each pier.

15. The system of parallel forces shown in the diagram of Fig. 3.31 is in equilibrium. Find the value of the force F and its distance x from the 600-lb force. The markings on the beam are in feet. Neglect the weight of the beam.

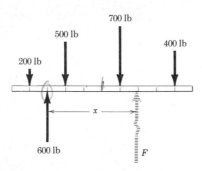

Fig. 3.31 Equilibrium of parallel forces. (Diagram for Problem 3.15.)

16. A river current flows south at 4 mph in a channel which is 0.5 mi wide. A swimmer who can swim at a steady rate of 1.5 mph in still water dives in and *heads* straight across the channel. (*a*) How far downstream is he when he reaches the far shore? (*b*) How far is he from his starting point? (*c*) How many minutes have elapsed since he left the starting point?

17. A ball is thrown from a moving auto with a velocity of 20 m/sec with respect to the auto, in a direction perpendicular to the auto's velocity, which is 80 km/hr due east. Find the velocity (speed and direction) of the ball with respect to the ground. Neglect gravitational factors.

18. A ferryboat crosses a north-south channel 500 yd wide in which a strong current runs due north at 5 knots. The ferry is capable of 10 knots (nautical miles per hour) in still water. What must be its compass heading in order to reach a dock directly across from its departure point?

19. A destroyer is headed due west at 30 knots. A 40-knot wind is blowing from the southeast. (*a*) Find the velocity of the wind with respect to the ship. (*b*) What angle does the flag make with the ship's course as it streams from the flagstaff in the relative wind?

20. A pilot leaves airport A to fly to airport B, a distance of 450 mi due south. He sets the plane's nose due south and flies at an indicated airspeed of 225 mph. Unknown to him, a 40-mph wind from the northwest was blowing throughout the flight. At the expiration of 2 hr he expects to see airport B, but despite excellent visibility it is nowhere in sight. How far and in what direction is he from airport B?

21. A jet transport plane will fly at 550 mph indicated airspeed, from airport A to airport B, 2400 mi distant. The true bearing (direction) of B from A is 072°. A 70-mph wind blowing due south is to be encountered throughout the entire flight. Find (*a*) the compass heading which must be maintained to stay on course to airport B and (*b*) the time required for the flight.

22. The sketch of Fig. 3.32 shows the cross section of a roof which weighs 800 lb per foot of length along the dimension AC. T is a heavy water-cooling tower (weight 5 tons) located as shown. Calculate the load borne by each of the supporting walls A and B.

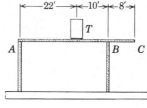

Fig. 3.32

PART 2 Mechanics

Giant machines must be driven by giant gears. This, the largest gear drive ever built, will provide an output of 33,800 horsepower at 514 revolutions per minute to drive an electric generator. The input energy is furnished at a speed of 60 revolutions per minute by a hydraulic turbine on the Arkansas river project. (The Falk Corporation, subsidiary of the Sunstrand Corporation.)

4 Force and Motion

Matter and energy are the twin themes of this book. Later chapters will take up the structure of matter in considerable detail, including molecular and atomic structure and the particles which are now thought to make up the atomic nucleus. At this point, however, we are ready to begin the systematic development of ideas about energy. Basic principles and practical applications of *mechanical energy* will be covered in this section of the book. Later sections will be devoted to heat energy, light and sound, electrical energy, and atomic and nuclear energy.

An understanding of mechanical energy depends on an understanding of *force* and *motion*, so we must begin with them.

The word *force* is used to convey many different meanings in everyday language. Political "forces" are said to decide elections; military "force" is threatened when diplomacy breaks down; store owners are "forced" into bankruptcy when business is bad; and the "forces of evil" threaten the good society. In physics, however, we are not concerned with these less precise meanings, but with the purely scientific and mechanical aspects of force, with *force* defined as *a push or a pull which tends to cause a change in motion.*

A force may start something moving or stop something which is already moving. Or, it may merely speed up (*accelerate*) or slow down (*decelerate*) a moving object. The essential thing to remember about an *unbalanced* force is that it invariably *causes a change in motion.*

All the machines of industry and the instruments and tools of science and engineering, agriculture, business, and medicine depend on the action and interaction of forces. And so does the utilization of the energy that we have now, as well as the development of energy sources for the future.

FORCE

A single observation of a tornado, a thunderstorm, an earthquake, or a pounding surf at the seashore makes a person appreciate the phrase, "the forces of nature." Winds, waves, lightning, tides, and ocean currents are all examples of natural forces which we would like to be able to control for the fantastic amounts of energy they produce. Other well-known forces, such as those from expanding steam, hot combustion gases, and electromagnetism are readily controlled, and they are the primary sources of energy for advanced technological societies.

4.1 The Force of Gravitation

One of the basic forces of nature is the force of gravity. We grow up with it from infancy and live with it all of our lives. It is ever-present, a help sometimes and a hindrance at other times, and it is among the least understood of all scientific phenomena. We can use it, measure it, deplore it sometimes, and define it; but we cannot explain it or eliminate it. The *laws* of gravity are known but not the *cause* of gravity.

Sir Isaac Newton (1642–1727), the British natural philosopher and mathematical genius, studied gravitation over many years and formulated many of the so-called laws of gravitation. He found that gravitation is *universal*, i.e. it applies to all bodies throughout the known universe; and that it is *mutual* in the sense that as the earth attracts the moon, so does the moon attract the earth. These general laws of gravitation will be presented in a later chapter (see Chap. 7). For the present we are concerned only with the earth's gravity and with the fact that the earth's gravity causes a predictable change in motion as it acts on bodies, unless there is a counteracting force of some kind.

4.2 The Force of Gravity Is the Measure of Weight

In the discussion of measurement (page 25), *weight* was defined as *the pull of earth's gravity on a body*. If you *weigh* 175 lb, this merely means that you and the earth attract each other with a *force* of 175 lb. The units in which forces and weights are measured need careful study, however, since past practices, from many different countries, still affect current usage.

In countries using the metric system (and the United States will probably soon be one) the approved *force* unit is now the *newton* (defined later in this chapter). The older unit, "kilogram of force" is no longer used in scientific and technical work. Only a few years ago, prior to the adoption of the entire SI-metric system, the kilogram was used both as a unit of mass and as a unit of force, or weight. In fact, everyday (nonscientific) use in metric countries still finds the kilogram as a unit of weight, as in the housewife's purchase of "a kilo of potatoes." In this book, when we use the word "kilogram" or its abbreviation, "kg," we will mean *mass*, not *force*. If, in some cases, a kilogram of force or weight is to be indicated, we will use the designations "kilogram of force," or the abbreviation "kg-*f*."

In the English (engineering) system the *pound* is the unit for both *weight* and *force* in everyday use as well as in technical work, but there is still lack of standardization when it comes to the unit of *mass*. Industry and commerce still use the pound as a mass unit, while engineering practice uses a unit called a *slug*, which will be defined and explained later in this chapter.

Despite the fact that the same words are sometimes used to describe forces and masses in the English system (and even to some extent in metric countries, for commercial transactions), it is strongly emphasized that *mass and weight do not mean the same thing*. *Mass* denotes the quantity of matter in a body, and the mass of a body is a constant no matter where it is measured—on the earth's surface, on the moon, or in a rocket ship on the way to Mars. *Weight* is merely the word we use to describe *the force of gravitational attraction between a body and the earth*. The moon's gravity is much less than that of the earth, since its mass is much smaller, and your 175-lb weight on earth would be only about one-sixth that, or 29 lb, on

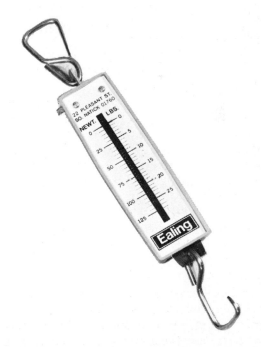

Fig. 4.1 Spring balances, or scales, measure gravitational force, or *weight*. They do not measure *mass*. (The Ealing Corp.)

the moon. An astronaut whose weight is 180 lb on earth becomes "weightless" when his space vehicle travels beyond the earth's effective gravitational field.

Mass and weight are related on the surface of the earth by the constant g, the acceleration of earth's gravity, which will be explained in a later section of this chapter.

Weights are commonly measured by *spring balances*, calibrated to read in pounds and ounces (English system); and in kilograms-force (kilos) and newtons (metric system). The *newton* (a unit of force) will be defined later in this chapter. Figure 4.1 shows a typical spring balance or pair of scales.

Mass is measured by the methods and apparatus described in Sec. 2.8, p. 31.

MOTION

4.3   The Concept of Motion

Motion may be defined as a continuous *change in position*. A change in position is called a *displacement*. Straight-line motion is said to be *rectilinear*, and motion along a curved path is called *curvilinear* (see Fig. 4.2).

As an airplane flies in level flight, as an automobile moves along a road, or as a piston moves back and forth in the cylinder of an engine, these bodies move along parallel lines. Motion of this kind is said to be pure *translation*, a linear motion. Motion which, like that of a turbine rotor or a centrifuge, involves the revolution of the object about a stationary axis, is called pure *rotation*. Complex motions may involve both translation and rotation, of course, but for the present we shall stay with simple forms of motion, considering translation in this chapter and rotation in a later chapter.

4.4   Velocity Is the Rate of Motion

The *velocity* of a body is *the distance it moves in a specified direction in a unit of time*. When the motion is such that equal distances are covered in each succeeding unit of time, the velocity is said to be *uniform*. Rectilinear motion at uniform velocity is the simplest kind of motion. *Speed* is also defined as *the rate of motion*, and in many types of problems *speed* and *velocity* can be used interchangeably. It is conventional to use the term *velocity* when both *rate* and *direction* of motion are specified and to use the term *speed* when direction is not known or is not important in the problem. *Speed* is the magnitude factor of velocity. Velocity is a *vector* quantity; speed is a *scalar* quantity.

Velocity or *speed* may be expressed in miles per hour (mi/hr), feet per second (ft/sec), meters per second (m/sec), kilometers per hour (km/hr), centimeters per second (cm/sec), etc. A fast sprinter can move at 30 ft/sec, an express train at 90 mi/hr, and an antiaircraft projectile at 950 m/sec.

Constant or Uniform Velocity   In any motion problem, it should be apparent that three variables—*distance*, *velocity*, and *time*—are involved. The defining equation for rectilinear motion at *uniform velocity (speed)* is

$$\bar{v} = \frac{s}{t} \qquad (4.1)$$

where $\bar{v}$ = uniform velocity
     s = distance covered
     t = time required

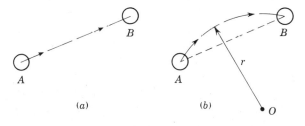

(a)   (b)

Fig. 4.2   Straight-line motion and curvilinear motion. In both sketches the sphere has moved from point A to point B. The displacement is AB in both cases, but the path distance is different. The motion in (a) is rectilinear (in a straight line), and that in (b) is curvilinear. The curve in (b) is an arc of a circle with radius r and center O.

Average velocity   A body traversing a finite distance may have a variable rather than a uniform velocity or speed. By suitable analysis, however, an *average* velocity for the motion can be determined. If average velocity is known for a motion problem, Eq. (4.1) can also be used, with $\bar{v}$ as the average velocity. This velocity formula, like any algebraic equation, can be solved for any one of the variables in terms of the others. The other two forms of the equation are

$$t = \frac{s}{\bar{v}} \tag{4.1'}$$

$$s = \bar{v}t \tag{4.1''}$$

Illustrative Problem 4.1   A jet plane flew from La Guardia Field, New York City, to Los Angeles International Airport in an elapsed time of 4 hr 12 min. The airline distance is 2885 mi. What was the average speed in miles per hour?

Solution   Since distance and time are known, and average speed is desired, we make use of Eq. (4.1), $\bar{v} = s/t$. Substituting known values,

$$\bar{v} = \frac{2885 \text{ mi}}{4.20 \text{ hr}} = 687 \text{ mi/hr} \qquad ans.$$

Representing uniform velocity graphically

Suppose a runner in an 800-m race intends to pace himself throughout the race at a uniform or constant speed and wants you to "clock" him. Markers are on the track at every 100 m, and as he passes each marker you note the elapsed time as recorded by an electronic timer. You might obtain data like those in Table 4.1.

From these data, the graph of Fig. 4.3 can be plotted, with the time scale on the horizontal axis and the distance scale on the vertical axis. Each plotted point (small circles) represents a position of the runner at the time indicated. Note that although some points fall slightly off the

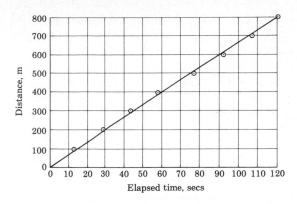

Fig. 4.3   "Smoothed" graph of the data from Table 4.1. A distance vs. time plot of a 400-m run.

"straight line," the runner did achieve his goal of uniform speed quite well.

When any two quantities are plotted on a graph of this kind (horizontal and vertical axes intersecting at the zero point or origin, of both axes), and the resulting plot is a straight line through the origin, the two quantities have a constant relationship to each other, or we say that they are *directly proportional* to each other. In this case, both from a visual inspection of the graph and from Eq. (4.1), the constant of proportionality is $\bar{v}$, the uniform velocity. Taking the race as a whole,

$$\bar{v} = \frac{s}{t} = \frac{800 \text{ m}}{120 \text{ sec}} = 6.67 \text{ m/sec}$$

4.5   Accelerated Motion

Objects do not ordinarily move with uniform velocity. Driving an automobile, for example, is a series of stops, starts, and turns with both speed and direction changing frequently. Moving parts in machinery are constantly changing direction and frequently changing speed under varying

Table 4.1   Time-distance measurements: 800-m race

Distance, m	0	100	200	300	400	500	600	700	800
Elapsed Time, sec	0	14.5	29.2	43.0	58.5	76.5	91.5	107.0	120.0

conditions of load and production demand. Motion in which velocity is changing is called *accelerated motion.*

Acceleration is *positive* if the moving object is gaining speed, *negative* if speed is decreasing. Just as uniform velocity is the simplest form of motion, so uniform acceleration is the simplest type of accelerated motion.

Acceleration is defined as *change in velocity per unit time* or *the time rate of change of velocity.* If an object is moving in a straight line with an initial velocity v_1 at a given instant, and t sec later its velocity has increased to v_2, the average acceleration is expressed by the formula

$$a = \frac{v_2 - v_1}{t} \qquad (4.2)$$

Illustrative Problem 4.2 Sales literature for a modern car claims that it will accelerate from 10 to 75 mi/hr in 12 sec. Find the average acceleration of which it is capable.

Solution The initial velocity v_1 is 10 mi/hr. The final velocity v_2 is 75 mi/hr.

The time required t is 12 sec. Using Eq. (4.2),

$$a = \frac{v_2 - v_1}{t}$$

and substituting,

$$a = \frac{75 \text{ mi/hr} - 10 \text{ mi/hr}}{12 \text{ sec}}$$

$$= \frac{65 \text{ mi/hr}}{12 \text{ sec}} = 5.4 \frac{\text{mi/hr}}{\text{sec}} \qquad ans.$$

Note carefully the units in which acceleration is expressed. Since 65 mi/hr equals 95.3 ft/sec (see Note below), the acceleration of the car referred to could be written as

$$a = \frac{95.3 \text{ ft/sec}}{12 \text{ sec}} = 7.9 \frac{\text{ft/sec}}{\text{sec}} \qquad ans.$$

NOTE: A convenient ratio of miles per hour to feet per second is

$$30 \frac{\text{mi}}{\text{hr}} = 30 \frac{\text{mi}}{\text{hr}} \left(\frac{5280 \text{ ft}}{1 \text{ mi}} \right) \left(\frac{1 \text{ hr}}{3600 \text{ sec}} \right) = 44 \frac{\text{ft}}{\text{sec}}$$

The student should memorize this convenient relation now.

"Feet per second per second" is the most common form of expressing acceleration in the English system. It is usually written as ft/sec². If this unit seems confusing at first, remember that acceleration is *change in velocity per unit time.* Velocity already involves time $(v = s/t)$, and hence the time unit appears twice in the expression for acceleration.

The standard unit for acceleration in the metric system is the meter per second per second, written m/sec². The unit cm/sec² is sometimes used for smaller accelerations.

The following problem illustrates the use of metric units in an acceleration situation.

Illustrative Problem 4.3 The Bullet Train of the New Tokaido line in Japan can accelerate from 30 to 210 km/hr in 32 sec. Find the acceleration (assumed uniform) in meters per second per second (m/sec²).

Solution From Eq. (4.2)

$$a = \frac{v_2 - v_1}{t} = \frac{210,000 \text{ m/hr} - 30,000 \text{ m/hr}}{32 \text{ sec}}$$

$$= \frac{5625 \text{ m/hr}}{1 \text{ sec}}$$

$$= \frac{5625 \text{ m/hr}}{1 \text{ sec}} \times \frac{1 \text{ hr}}{3600 \text{ sec}}$$

$$= 1.56 \text{ m/sec}^2 \qquad ans.$$

4.6 Velocity and Accelerated Motion

Multiplying both sides of Eq. (4.2) by t results in

$$v_2 - v_1 = at$$

But $v_2 - v_1$ equals the net gain in velocity in time t sec. Let

$$v_2 - v_1 = v$$

Then $\qquad\qquad\qquad v = at \qquad (4.3)$

or *change in velocity* equals *acceleration* multiplied by *time.*

Illustrative Problem 4.4 The brakes of a certain car are capable of giving the car a negative acceleration (*deceleration*) of 25 ft/sec^2. (*a*) How long a time (seconds) does it take for this car to come to a stop from 50 mi/hr? (*b*) How far does the car travel while being stopped?

Solution
(*a*) Using Eq. (4.3),

$$v = at$$

Solve for t, obtaining

$$t = \frac{v}{a}$$

It is now necessary to change the velocity as given in miles per hour, into feet per second. Remembering that 30 mi/hr = 44 ft/sec,

$$50 \text{ mi/hr} = (50 \times \tfrac{44}{30}) \text{ ft/sec} = 73.3 \text{ ft/sec}$$

Now substituting the values for v and a in $t = v/a$,

$$t = \frac{73.3 \text{ ft/sec}}{25 \text{ ft/sec}^2} = 2.93 \text{ sec} \qquad ans.$$

(*b*) The first step is to find the *average velocity* during the braking period. Assuming uniform deceleration, the average of the two velocities is merely their sum divided by 2:

$$\text{average velocity} \quad \bar{v} = \frac{v_2 + v_1}{2} \qquad (4.4)$$

Applied to the present problem,

$$\bar{v} = \frac{73.3 \text{ ft/sec} + 0 \text{ ft/sec}}{2} = 36.7 \text{ ft/sec}$$

Now, making use of Eq. (4.1''), $s = \bar{v}t$, and substituting values just determined,

$$s = 36.7 \text{ ft/sec} \times 2.93 \text{ sec} = 108 \text{ ft} \qquad ans.$$

4.7 Distance and Velocity Related to Uniformly Accelerated Motion

Relation of Distance to Time There are many motions in which acceleration is not uniform. The mathematical treatment of these cases is beyond the scope of this book, and the discussions to follow concern themselves with uniform or constant acceleration. It is convenient to have a single formula from which to calculate distance directly in accelerated-motion problems in which the time and the acceleration are known.

First the average velocity of the moving body during the interval must be expressed. If a body starts from rest, that is $v_1 = 0$, and is accelerated uniformly, the average speed is equal to one-half the final speed v_2:

$$\bar{v} = \text{average speed} = \tfrac{1}{2}v_2$$

From Eq. (4.3) we note that the final speed depends on the duration of the acceleration ($v = at$). Average speed, then, is $\tfrac{1}{2}v_2 = \tfrac{1}{2}at$. But *distance* equals *average speed* multiplied by *time*, and therefore the distance traveled in time t by a moving body starting from rest, whose uniform acceleration is a is

$$s = \tfrac{1}{2}at \times t$$

or $\qquad\qquad s = \tfrac{1}{2}at^2 \qquad\qquad (4.5)$

Relation of Velocity to Distance and Acceleration Suppose a rocket starts from rest and accelerates uniformly with a known acceleration. How can its velocity be determined at any specified distance from the starting point? Note that the problem involves a, v, and s.

From Eq. (4.3) we can write

$$t = \frac{v}{a}$$

and from Eq. (4.5), substituting for t,

$$s = \tfrac{1}{2}at^2 = \tfrac{1}{2}a \times \frac{v^2}{a^2} = \frac{v^2}{2a}$$

from which $\qquad v^2 = 2as \qquad\qquad (4.6)$

or $\qquad\qquad v = \sqrt{2as}$

A rocket with a uniform acceleration of 7.0 m/sec^2, for example, would, after traveling 1000 m, have a velocity

$$v = \sqrt{2 \times 7.0 \frac{m}{sec^2} \times 1000 \text{ m}} = 118 \text{ m/sec}$$

We now have a group of formulas for uniform motion and uniformly accelerated motion starting from rest, which are collected here for ready reference:

Uniform motion: $s = vt$ from (4.1″)

Accelerated motion: $v = at$ (4.3)

$$s = \tfrac{1}{2}at^2 \qquad (4.5)$$

$$v^2 = 2as \qquad (4.6)$$

4.8 Generalized Equations of Accelerated Motion

It should be noted that the three formulas for accelerated motion listed above apply only when the accelerating object is assumed to *start from rest* (if increasing speed) or to *come to rest* (if decreasing speed). Accelerations are *positive* if speed is increasing, *negative* if speed is decreasing.

When an object has an initial velocity v_1 before the acceleration begins, we may reason that

Final velocity = initial velocity
+ change in velocity

or $\qquad v_2 = v_1 + at \qquad (4.3')$

The distance traveled by a body which is moving initially with velocity v_1 and is then accelerated for a time t is the distance which would have been covered had the velocity remained unchanged, plus the distance resulting from the acceleration a acting for t sec. As an equation, this becomes

$$s = v_1 t + \tfrac{1}{2}at^2 \qquad (4.5')$$

Figure 4.4 illustrates graphically the relationships involved in both Eqs. (4.3′) and (4.5′).

An expression relating distance traveled to initial and final velocities may be derived as follows, the distance moved by such a body being the *average velocity* multiplied by *time*:

$$s = \left(\frac{v_2 + v_1}{2}\right)t$$

from which $\qquad v_2 + v_1 = \dfrac{2s}{t}$

But, from Eq. (4.3′) $\quad v_2 - v_1 = at$

Multiplying these two equations,

$$(v_2 + v_1)(v_2 - v_1) = \frac{2s}{t} \times at$$

from which $\qquad v_2^2 - v_1^2 = 2as$

or $\qquad v_2^2 = v_1^2 + 2as \qquad (4.6')$

Care must be taken to assign the proper algebraic sign, $(+)$ or $(-)$, to the values of a in all equations for accelerated motion.

These generalized equations for uniform and accelerated motion are collected here for your ready reference:

Uniform motion: $\quad s = vt \qquad$ from (4.1″)

Accelerated motion: $v_2 = v_1 + at \qquad (4.3')$

$$s = v_1 t + \tfrac{1}{2}at^2 \qquad (4.5')$$

$$v_2^2 = v_1^2 + 2as \qquad (4.6')$$

Illustrative Problem 4.5 The velocity-time graph of Fig. 4.5 was plotted by observing the movements of an automobile in traffic. Study the graph carefully and see what it tells you.

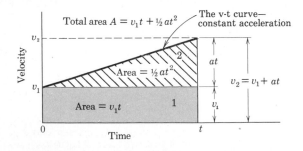

Fig. 4.4 Graph showing distance traversed by a body moving with initial velocity v_1, which is then accelerated for a time t. The shaded areas under the v-t "curve" represent the distance covered due to the initial velocity (area 1) and that due to the acceleration (area 2). The total area of the trapezoid represents the entire distance traveled in time t, or $s_{\text{tot}} = v_1 t + \tfrac{1}{2}at^2$.

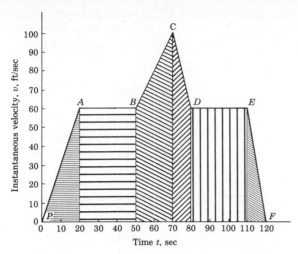

Fig. 4.5 Graph of instantaneous velocity vs. time for an automobile in traffic.

Solution First, you should note that the auto starts from rest ($v_1 = 0$) and accelerates uniformly (How do you know that the acceleration is uniform?) for 20 sec, at which time its velocity has increased to 60 ft/sec (about 41 mi/hr). Then, at point A on the graph, it proceeds at constant velocity (zero acceleration) for 30 sec at 60 ft/sec, moving with the traffic. At point B it accelerates for 20 sec (perhaps to pass another car), attaining a maximum speed of 90 ft/sec (about 61.4 mi/hr) at point C. It then slows down (negative, uniform acceleration) for 10 sec to its former cruising speed of 60 ft/sec (point D). It moves at this speed for 30 sec and then, at point E, the auto undergoes uniform negative acceleration coming to a complete stop in 10 sec.

The numerical values for the accelerations represented by the PA, BC, CD, and EF parts of the graph can be determined from the equation $v_2 - v_1 = at$. Solve this equation for a, obtaining

$$a = \frac{v_2 - v_1}{t}$$

The uniform acceleration for any one portion of the curve is merely the ratio of the change in velocity for that portion to the time required for that part of the motion.

For the motion represented by PA,

$$a_{PA} = \frac{60 \text{ ft/sec}}{20 \text{ sec}} = 3 \text{ ft/sec}^2$$

In like manner, you should now calculate the accelerations a_{BC}, a_{CD}, and a_{EF}.

Finally, what can be learned from the graph about the distances traveled? We know that for *constant velocity*, distance traveled equals uniform velocity multiplied by time, or $s = vt$. Therefore, for the AB portion of the travel, the distance traversed is

$$s_{AB} = v_{AB}t_{AB} = 60 \text{ ft/sec} \times (50 - 20) \text{ sec}$$
$$= 1800 \text{ ft}$$

As was pointed out in Fig. 4.4, this is equivalent numerically to the shaded area under the AB portion of the graph.

For the PA part of the travel, where the car is uniformly accelerated, recall that $s = \frac{1}{2}at^2$, and that $a_{PA} = 3$ ft/sec². Therefore,

$$s_{PA} = \frac{3 \text{ ft/sec}^2 \times (20 \text{ sec})^2}{2}$$
$$= 600 \text{ ft}$$

A quick look at Fig. 4.5 shows that this is numerically equal to the shaded area under the PA part of the graph.

Now, as an exercise, calculate the distances traveled in the remaining portions of the travel and sum up everything for the total distance traveled. Do you find the result to be 6700 ft?

4.9 Some Guidelines for Solving Uniform Motion Problems

Here are a few steps that will help you set up and solve problems involving uniform (constant) velocity and uniform acceleration.

1. Read the problem very carefully and make a labeled, free-hand sketch of the given relationships, whenever possible.

2. Identify the factors involved in the problem: Distance (s), velocity (v), and acceleration (a). Be sure to distinguish between average velocity, $\bar{v}$, initial velocity, v_1 and final velocity v_2.

3. Look out for positive or negative values of distances, velocities, and accelerations.

4. From the generalized equations of motion collected above, choose the one(s) that are needed and apply them *in the proper order*. Here is where a simple sketch of the problem relationships will assist you.

5. Substitute known values in the equations and solve them.

Illustrative Problem 4.6 Your speedometer reads 20 mi/hr as you press the accelerator to the floor. If your car is capable of an acceleration of 7.5 ft/sec², what should your speedometer read after exactly 6.0 sec?

Solution v_1, a, and t are given, and v_2 is desired, so select Eq. (4.3′). The value of a is positive $(+)$, since the car is accelerating:

$$v_2 = v_1 + at$$

Now since $20 \times \frac{44}{30} = 29.3$ (see Sec 4.5, p. 60),

$$20 \text{ mi/hr} = 29.3 \text{ ft/sec}$$

Substituting,

$$v_2 = 29.3 \text{ ft/sec} + (7.5 \text{ ft/sec}^2 \times 6.0 \text{ sec})$$
$$= 74.3 \text{ ft/sec}$$

or $\quad v_2 = 74.3 \text{ ft/sec} \times \dfrac{30 \text{ mi/hr}}{44 \text{ ft/sec}}$

$$= 51 \text{ mi/hr} \qquad\qquad ans.$$

Illustrative Problem 4.7 A jet plane is flying at 500 km/hr when the pilot cuts in the afterburner. The additional thrust gives an acceleration of 6 m/sec². If the afterburner is left on for 30 sec, how far (km) will the aircraft travel during the acceleration period?

Solution v_1, a, and t are given and s is desired. The use of Eq. (4.5′) is therefore indicated. Consistent units must first be obtained.
    Now,

$$500 \text{ km/hr} = 139 \text{ m/sec}$$

From Eq. (4.5′),

$$s = v_1 t + \tfrac{1}{2}at^2$$

Substituting gives

$$s = 139 \text{ m/sec} \times 30 \text{ sec}$$
$$+ \frac{6 \text{ m/sec}^2 \times (30 \text{ sec})^2}{2} \quad (a \text{ is positive})$$
$$= 4170 \text{ m} + 2700 \text{ m}$$
$$= 6870 \text{ m}$$
$$= 6.87 \text{ km} \qquad\qquad ans.$$

4.10 The Accleration of Gravity— Free Fall

The laws of uniformly accelerated motion apply to freely falling bodies, and the force of gravity which causes the acceleration is a constant at any given locality and becomes a known quantity in the solution of all free-fall problems.

Galileo (1564–1642) found that all bodies acted upon by gravity *fall with the same acceleration* if air resistance is neglected. It is true that the earth pulls with twice the force on a 100-kg body that it does on a 50-kg body, but the 100-kg body has twice as much mass which has to be set in motion. Its resistance to a change in motion (*inertia*) is twice as great, and therefore the accelerating force per unit mass is the same for both bodies.

The acceleration due to the force of gravity is approximately 9.81 m/sec² (981 cm/sec²) at sea level at the equator. This value is called g, the *acceleration of gravity*. In the English system the value of g is approximately 32.2 ft/sec². Slight variations exist due to altitude and latitude, but they are too small to affect the practical problems of technical physics. Air resistance definitely affects the rate of fall, and its effect must be allowed for in all real problems such as those encountered in predicting the flight of projectiles and rockets and the trajectory of bombs, as well as in the design of parachutes. Objects "falling" toward the earth from outer space encounter air resistance so great that the heat generated usually consumes them before they hit the earth's surface. The shooting stars (*meteors*) seen frequently in the August sky are actually "falling bodies" from outer space burning themselves up in the heat generated by atmospheric friction. Reentry of satellites and spacecraft into the earth's atmosphere is one of the very difficult problems which had to be solved before manned flights in space were possible. Heat shields capable of withstanding very high temperatures produced by atmospheric friction had to be perfected, and retrorockets had to be designed to slow down the returning spacecraft.

A high-speed photograph of a falling steel ball is shown in Fig. 4.6. The multiflash photograph was taken by a camera equipped with a stroboscopic light which emitted flashes at $\frac{1}{30}$-sec intervals. Observe the increased distance traversed in each succeeding time interval. Air resistance of such a ball is quite small at low velocities.

Figure 4.7 is a diagram of an apparatus used

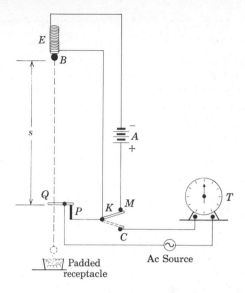

Fig. 4.7 Diagram of apparatus to measure distance and time of fall of a freely falling body (air resistance assumed negligible).

in the laboratory to obtain data on freely falling bodies. E is an electromagnet which holds a small steel ball B when the magnet is energized by battery A. K is a two-position switch. At M it closes a dc circuit, energizing the magnet E. When snapped from M to C, it breaks the magnet circuit and starts the stop clock T. Ball B thus starts falling, pulled by the force of gravity, at the same instant that the stop clock starts operating.

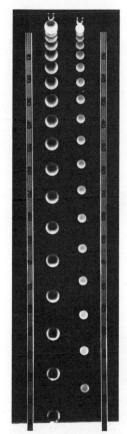

Fig. 4.6 High-speed multiflash photograph of (left) a falling billiard ball, and (right) a pingpong ball released at the same time. Note that air resistance has a much greater effect on the lighter pingpong ball. The flash interval is $\frac{1}{30}$ sec. (From *PSSC Physics*, D.C. Heath and Co., Boston, 1965.)

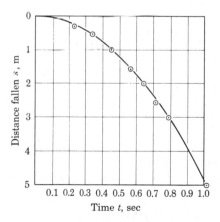

Fig. 4.8 Graph of data from a falling-body experiment. Distance fallen is plotted against time of fall for a freely falling body whose mass is large compared to the force of air resistance it encounters.

Table 4.2 Data for a freely falling body

Elapsed Time, sec	Distance Fallen, s		Instantaneous Velocity	
	m	ft	m/sec	ft/sec
0	0	0	0	0
1 $(s = \frac{1}{2}gt^2)$	4.9	16	9.8 (approx)	32 (approx)
2	19.6	64	19.6	64
3	44.1	144	29.4	96
4	78.4	256	39.2	128
5	122.5	400	49.0	160

NOTE: $g = 9.8$ m/sec^2 = 32 ft/sec^2 approx. More exactly $g = 32.2$ ft/sec^2, or 9.81 m/sec^2

When the falling ball hits trigger switch Q, the clock circuit is deenergized as the points at P snap apart, stopping the clock.

The magnet E can be positioned at any measured height s above Q. The clock can be read accurately to the nearest 0.01 sec, and estimates to the nearest 0.001 sec are fairly reliable if several timed trials from the same height are averaged. Figure 4.8 shows the curve obtained when s is plotted against the time of fall t, for a typical set of experimental data from such an apparatus.

Note that the smoothed curve of Fig. 4.8 has the shape of a parabola, conforming to the mathematical form of Eq. (4.5). Substituting g for a in Eq. (4.5) gives the distance fallen in any time t of a body starting from rest, as $s = \frac{1}{2}gt^2$. When any two variables (in this case, s and t) are related so that one is proportional to the square of the other, the resulting curve is a parabola.

Table 4.2 shows the distances fallen and the velocities attained at the end of each second by a body in free fall from rest over a total time period of 5 sec (air resistance neglected). Values are for the earth's surface.

Collected Equations for Free Fall By merely replacing a by g in Eqs. (4.3), (4.5), and (4.6), we obtain a set of formulas for freely falling bodies for the special case where $v_1 = 0$, that is, when the fall starts from rest.

$$v = gt \qquad (4.7)$$

$$s = \tfrac{1}{2}gt^2 \qquad (4.8)$$

$$v^2 = 2gs \qquad (4.9)$$

A set of general formulas for cases where initial or final velocities *are not zero* is easily obtained from Eqs. (4.3′), (4.5′), and (4.6′).

$$v_2 = v_1 + gt \qquad (4.7')$$

$$s = v_1 t + \tfrac{1}{2}gt^2 \qquad (4.8')$$

$$v_2^2 = v_1^2 + 2gs \qquad (4.9')$$

Fig. 4.9 The famous leaning tower of Pisa from which Galileo Galilei reportedly conducted his experiments with falling bodies.

It is emphasized that *these formulas are theoretical in that they neglect air resistance.*

Illustrative Problem 4.8 The hammer of a pile driver is released at the top of its boom and falls vertically a distance of 38 ft. What is its velocity as it hits the piling?

Solution The hammer starts from rest, so $v_1 = 0$; s is known, and $g = 32.2$ ft/sec². From Eq. (4.9)

$$v^2 = 2gs$$

Substituting,

$$v^2 = 2 \times 32.2 \text{ ft/sec}^2 \times 38 \text{ ft} = 2450 \text{ ft}^2/\text{sec}^2$$

$$v = \sqrt{2450 \text{ ft}^2/\text{sec}^2} = 49.5 \text{ ft/sec} \qquad \textit{ans.}$$

4.11 Vertical Motion Is Affected by Gravity

The vertical motion of objects thrown or projected upward is subject to *negative acceleration* due to gravity. The *vertical component* of all motions in the earth's gravitational field is likewise affected. Consider the following, for example.

Illustrative Problem 4.9 A baseball is thrown straight up, and 5.4 sec elapse before it hits the ground. (*a*) How high did it go? (*b*) With what velocity was it thrown? (See Fig. 4.10.)

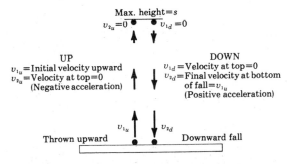

Max. height=s
$v_{2_u}=0$ $v_{1_d}=0$

UP
$v_{1_u}=$Initial velocity upward
$v_{2_u}=$Velocity at top=0
(Negative acceleration)

DOWN
$v_{1_d}=$Velocity at top=0
$v_{2_d}=$Final velocity at bottom of fall=v_{1_u}
(Positive acceleration)

v_{1_u} v_{2_d}
Thrown upward   Downward fall

Fig. 4.10 A ball thrown vertically upward with initial velocity v_{1_u} travels upward a distance s while it is being decelerated by the force of gravity. When its upward velocity is reduced to zero, it then becomes a freely falling body, starting from rest at its maximum height position. It falls vertically to the earth, being positively accelerated by the force of gravity. Its velocity as it arrives back at the level from which it was thrown is equal and opposite to the initial upward velocity.

Solution

(*a*) At the top of its flight its velocity, momentarily, will be $v_1 = 0$, and we can consider its return to earth as if it were a falling body dropped from that height. Let s stand for the maximum height the ball attains. The flight upward is *decelerated* by gravity over the upward distance, and the flight downward is accelerated by the same force over the same distance. The *time of rise* is equal to the *time of fall*, and the velocity as it returns to the ground (neglecting air resistance) is equal to that with which it was thrown upward. Let $T =$ total elapsed time, and $t =$ time of fall. Note that

$$t = \frac{T}{2} = \frac{5.4 \text{ sec}}{2} = 2.7 \text{ sec}$$

Select Eq. (4.8), since it involves t, g, and s:

$$s = \tfrac{1}{2}gt^2$$

Substituting,

$$s = \frac{9.81 \text{ m/sec}^2}{2} \times (2.7 \text{ sec})^2 = 35.8 \text{ m} \quad \textit{ans.}$$

(*b*) To find the velocity with which it was thrown, which, it will be recalled, is equal to the velocity with which it hits the ground, select Eq. (4.7),

$$v = gt$$

where again $t = T/2$.

$$v = 9.81 \text{ m/sec}^2 \times 2.7 \text{ sec} = 26.5 \text{ m/sec} \quad \textit{ans.}$$

Illustrative Problem 4.10 A baseball is thrown straight up from the top of a tall building with an initial velocity of 30.0 m/sec. As it comes back down it barely misses the railing and falls to the street 290 m below. (*a*) How far up does it go before starting to fall downward? (*b*) What is the total time of flight? (*c*) With what velocity does it hit the street? (Neglect air resistance.)

Solution

(*a*) The initial upward velocity is known. So is the value of g (9.81 m/sec²), and so is the final upward velocity, which is zero. The distance, s, is desired. Equation (4.9') involves all these factors, so use it and solve for s. Note that the acceleration is negative in this case since it is oppositely directed to the motion.

$$v_2^2 = 0 = v_1^2 + 2gs$$

or $\quad v_1^2 = -2gs \quad$ and $\quad s = \dfrac{v_1^2}{-2g}$

Substituting values we get,

$$s = \frac{(30.0 \text{ m/sec})^2}{(-2)(-9.81 \text{ m/sec}^2)}$$

$$= 45.9 \text{ m} \qquad\qquad ans.$$

(b) Total time of flight, T, equals the time in upward flight, t_u, plus the time of fall from the maximum height attained, t_f. But the time in upward flight, t_u, is equal to the time for the ball to fall back down to the level where it was thrown, t_d. We now know from part (a), so we can use Eq. (4.8) to find t_d.

$$s = \tfrac{1}{2}gt_d^2$$

Solving for t_d,

$$t_d = \sqrt{2s/g}$$

Substituting values,

$$t_d = \sqrt{\frac{2 \times 45.9 \text{ m}}{9.81 \text{ m/sec}^2}} = 3.06 \text{ sec}$$

But $t_d = t_u = 3.06$ sec, the time for upward travel.

Use the same equation to find t_f, recalling that the total distance of fall is now 290 m plus 45.8 m, or 335.8 m.

Substituting,

$$t_f = \sqrt{\frac{2 \times 335.9 \text{ m}}{9.81 \text{ m/sec}^2}}$$

$$= 8.28 \text{ sec}$$

Total time

$$T = 3.06 \text{ sec} + 8.28 \text{ sec} = 11.3 \text{ sec.} \qquad ans.$$

(c) Treat this part of the problem as a free fall from a height of 335.9 m, and use Eq. (4.9'):

$$v_2^2 = 2gs, \qquad \text{since} \qquad v_1 = 0$$

Substituting,

$$v_2^2 = 2 \times 9.81 \text{ m/sec}^2 \times 335.9 \text{ m}$$

$$= 6590 \text{ m}^2/\text{sec}^2$$

$$v_2 = \sqrt{6590 \text{ m}^2/\text{sec}^2} = 81.2 \text{ m/sec} \qquad ans.$$

4.12 The Flight of Projectiles

So far we have been dealing with objects going straight up or falling straight down under the influence of only one force, that of gravitational attraction. Now, the case of combined horizontal and vertical motion—motion in two dimensions but in a vertical plane—will be considered. Some extremely interesting and practical problems arise in motions of this sort.

An object which is propelled into space (thrown, catapulted, shot out of a gun, or launched from a moving airplane) and then left with no propelling or other forces acting on it except the force of gravity, is called a *projectile*. If air resistance is neglected, most of the balls with which modern games are played are projectiles. The third baseman, in "rifling" a throw to first to catch the runner, in effect is solving a problem in projectile motion in two dimensions. He must aim the ball upward at the correct angle to compensate for the acceleration of gravity so that with his throwing velocity and the distance to first base as factors, the ball will arrive there one foot off the ground in the absolute minimum of time. Examples from golf, basketball, and tennis could also be cited.

In the treatment to follow we shall assume that the distance traversed by a projectile is short enough for the earth's surface underneath to be considered a flat plane. Further, all such effects as projectile spin, air resistance, the earth's rotation, etc., will be neglected. The analysis of motions of idealized projectiles bears the technical name *ballistics*. *Ballistic missiles* are not projectiles until their propellant fuel has burned out, and a missile which retains a guidance system which controls it with rudders and ailerons acting against the atmosphere or with built-in miniature rockets is not a projectile in the technical sense of the term.

4.13 Horizontally Launched Projectiles

One familiar case of projectile motion is that which occurs when an object such as a ball, bullet, or bomb is projected or released with an initial horizontal velocity. It immediately begins acquiring vertical velocity downward under the pull of gravity, but its horizontal velocity is not affected thereby if air resistance is negligible.

Figure 4.11 shows such a situation recorded by high-speed flash photography. One golf ball is pushed out horizontally to the right at exactly the same instant that another is released for a vertical

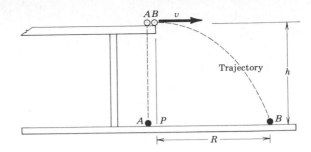

Fig. 4.12 Horizontal and vertical motions of a simple projectile (air resistance assumed zero).

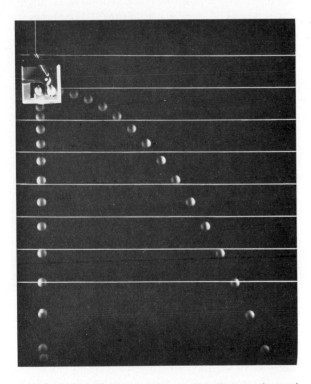

Fig. 4.11 Flash photo of two golf balls, one dropped vertically, and the other projected horizontally at the same instant. The horizontal wires shown in the photo are 6 in. apart, and the time interval between flashes is $\frac{1}{30}$ sec. Note (ball at right) that having a horizontal motion has no effect on vertical motion. The vertical velocity of the horizontally projected ball and the vertical distance it has traversed at any time are exactly the same as those same factors are for the ball that is merely dropped. (From *PSSC Physics*, D. C. Heath and Co., Boston, 1965.)

drop. Both balls, of course, immediately are accelerated downward by the force of gravity.

It is easy to set up and observe a simple experiment of this kind. Suppose two identical steel balls A and B are on the edge of a table (Fig. 4.12): B is projected horizontally, perhaps by a flip of the finger, while A is just rolled over the edge of the table (toward the reader) at the same instant. It will be observed that for any value of the horizontal velocity, v, fast or slow, the two balls hit the floor at the same instant. Listen for the impact on the floor. Both balls undergo *free fall* through the same vertical distance h, and the fact that ball B has a horizontal velocity and ball A does not, makes no difference in the vertical motion. Note that the flash photograph of Fig. 4.11 proves this. *The vertical and horizontal motions are independent of each other. Each takes place as if the other were zero.*

The actual flight path of ball B is called the *trajectory*, and if air resistance is neglected, this curve is a *parabola*.

We have already developed and used equations which describe the motion of ball A. Let us now do so for ball B, noting that they will then apply to any projectile with an initial horizontal velocity v which has no forces acting on it but the force of gravity.

First, the *time of flight* t for ball B is equal to that for ball A, and is simply the time required for the ball to fall through a vertical height h. Using Eq. (4.8) and writing h for s,

$$h = \tfrac{1}{2}gt^2$$

and the time of flight

$$t = \sqrt{2h/g} \qquad (4.10)$$

The ball B has a *horizontal range* R, as follows:

$$R = vt = v\sqrt{2h/g} \qquad (4.11)$$

Aircraft Bombing A military application of the above principles is in aircraft bombing, when bombs are released at high altitude with the aircraft in horizontal flight. Figure 4.13 shows the basic elements of the problem. At release, the bombs have a horizontal velocity v equal to that of the aircraft. The straight-line distance between the aircraft and the target is called the line of sight (LOS). The angle between the LOS and the horizontal flight path of the aircraft is called the *angle of depression* θ. In order to strike the target, the bomb must be released when the angle of depression θ is given by

$$\tan \theta = \frac{h}{R} = \frac{h}{v\sqrt{2h/g}}$$

or, simplifying, when

$$\tan \theta = \frac{1}{v}\sqrt{\frac{gh}{2}} \qquad (4.12)$$

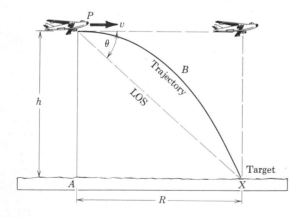

Fig. 4.13 The horizontal bombing problem. The falling bomb has both horizontal motion and vertical motion. They occur simultaneously but each is independent of the other.

It is again emphasized that wind and air resistance are not considered in the simplified treatment here given. Modern bombsights make speedy computations of such problems and take into account almost all conceivable sources of error. At night, or under conditions of poor visibility, they automatically compute the proper release point from radar information. So-called "smart bombs" have internal guidance systems and are not true projectiles.

4.14 Launcher and Target on the Same Horizontal Plane

In many games, such as baseball, golf, and tennis, the ball leaves a point near the ground at an angle and rises as it moves horizontally. It reaches a maximum altitude and then descends to a target on the same level—another player, the fairway, or the court. Figure 4.14 shows the basic elements of the problem. The ball at A has an initial velocity v at an angle θ with the horizontal. The horizontal component of its velocity is $v_x = v \cos \theta$, and the vertical component is $v_y = v \sin \theta$. The ball rises to a maximum height h at P. It comes back to the launch level at X. The horizontal distance traveled (range) is R. The curve APX is its trajectory and it would be a parabola if air resistance were absent. If the acceleration of gravity were zero, the ball would move out along the line AB indefinitely.

The horizontal component of the velocity remains constant, but the vertical component is always affected by the acceleration of gravity, and its value at any time t is expressed by

$$v_{y_t} = v \sin \theta - gt$$

where g is the acceleration of gravity and t is the number of seconds which have elapsed since the ball left point A. The height attained by the ball at any time t is, from Eq. (4.8′) (see Fig. 4.15),

$$y_t = vt \sin \theta - \tfrac{1}{2}gt^2 \qquad (4.13)$$

Now, when the ball arrives at the target X, $y = 0$ and $t =$ total time of flight, T. Putting $y_t = 0$ in Eq. (4.13) and solving for t, we obtain

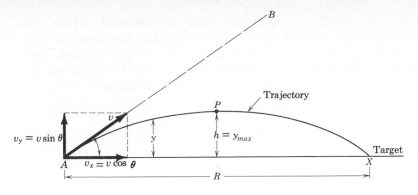

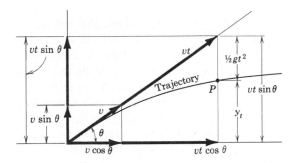

Fig. 4.15 Height y_t of projectile or ball at any time t after launch. With projectile (or ball) at P, after t sec, its height is $y_t = vt \sin \theta - \frac{1}{2}gt^2$.

$$t = T = \frac{2v \sin \theta}{g} \qquad (4.14)$$

the total time of flight. For the range R

$$R = v_x T = v \cos \theta \times \frac{2v \sin \theta}{g}$$

$$= \frac{2v^2 \sin \theta \cos \theta}{g}$$

or, since $2 \sin \theta \cos \theta = \sin 2\theta$,

$$R = \frac{v^2 \sin 2\theta}{g} \qquad (4.15)$$

which is the basic formula for *range* when *initial velocity* and *angle of elevation* are known. The maximum height to which the ball rises is given by

Fig. 4.14 Projectile flight when projectile launcher and target are on the same horizontal plane. This is often the case in games and sports, as well as in military gunnery.

$$h = \frac{v^2 \sin^2 \theta}{2g} \qquad (4.16)$$

In addition to the sporting applications of the above analysis, military applications are numerous also. A sample problem from both fields follows.

Illustrative Problem 4.11 A golfer selects a club which will give the ball, when hit properly, an angle of elevation of 35°. His target is the flag stick 160 yd away on the green. With what velocity (ft/sec) must the ball leave the club in order to land in the target area?

Solution Note that Eq. (4.15) contains the elements of the problem. First, solve it for v:

$$Rg = v^2 \sin 2\theta$$

from which

$$v^2 = \frac{Rg}{\sin 2\theta}, \qquad \text{and} \qquad v = \sqrt{\frac{Rg}{\sin 2\theta}}$$

Substituting values yields

$$v = \sqrt{\frac{160 \text{ yd} \times 3 \text{ ft/yd} \times 32.2 \text{ ft/sec}^2}{\sin 70°}}$$

$$= \sqrt{\frac{15460 \text{ ft}^2/\text{sec}^2}{0.94}} = \sqrt{16450 \frac{\text{ft}^2}{\text{sec}^2}}$$

$$= 128 \text{ ft/sec} \qquad \qquad \textit{ans.}$$

Illustrative Problem 4.12 It is desired to demolish an enemy command post whose horizontal range is 15,000 yd with 105-mm howitzer projectiles whose muzzle velocity is 2600 ft/sec. (*a*) Find the required theoretical angle of elevation of the gun. (*b*) What is the maximum range of this gun over level ground? Neglect air resistance.

Solution

(*a*) R, v, and g are known, and θ is desired. Select Eq. (4.15), and solve it for sin 2θ:

$$\sin 2\theta = \frac{Rg}{v^2}$$

Substituting,

$$\sin 2\theta = \frac{15{,}000 \text{ yd} \times 3 \text{ ft/yd} \times 32.2 \text{ ft/sec}^2}{(2600 \text{ ft/sec})^2}$$

$$= 0.214$$

From the table (Appendix III)

$$2\theta = 12°20'$$

and $\qquad\qquad \theta = 6°10' \qquad\qquad$ *ans.*

(*b*) To find maximum theoretical range, make use of Eq. (4.15):

$$R = \frac{v^2 \sin 2\theta}{g}$$

For any given value of v, R will be a maximum when sin 2θ is a maximum, since g is a constant. Sin 2θ is a maximum when $2\theta = 90°$, making sin $2\theta = 1$. Consequently, maximum range will

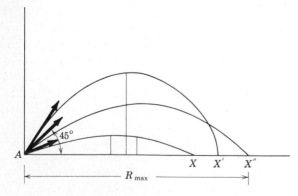

Fig. 4.16 The theoretical maximum range of a projectile is attained when it is fired at an angle of 45° (air resistance neglected).

occur when $\theta = 45°$. Solving,

$$R_{max} = \frac{(2600 \text{ ft/sec})^2 \times \sin 90°}{32.2 \text{ ft/sec}^2}$$

$$= 210{,}000 \text{ ft} = 70{,}000 \text{ yd} \qquad\qquad \textit{ans.}$$

This indicated maximum range is not realistic for such a gun, as air resistance would reduce the theoretical range appreciably. (See Figs. 4.16 and 4.17.)

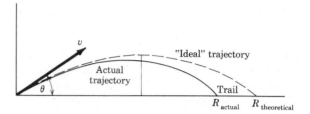

Fig. 4.17 Air resistance affects the flight of a projectile or a ballistic missile to a great extent. The ideal parabola (dashed line) is not even approximated by most projectiles. Heavy, slow-moving projectiles (like the shot used in track-and-field meets) are affected much less than lightweight, high-velocity projectiles.

THE RELATIONSHIP OF FORCE AND MOTION—NEWTON'S LAWS OF MOTION

4.15 Force Is the Cause of Change in Motion

We have been discussing some of the results of the force of gravity and you have learned that this force results in an important physical constant known as the acceleration of gravity g. By this time you know the metric and English values of g: 9.81 m/sec² and 32.2 ft/sec², respectively. You have also learned that *weights* are based on the earth's gravitational pull—that a one-pound weight is the force with which gravity pulls on a one-pound mass, and a one-kilo weight is the force with which gravity pulls on one kilogram of mass. Another force unit (the *newton*) and another mass unit (the *slug*) will be introduced shortly.

The force of gravity is not the only force which can act on bodies, and it is therefore necessary to discuss some *general laws of force and motion* which apply at any place and at any time—on earth or in space; today or yesterday; whether people are there to observe or not.

It is a matter of common experience that most ordinary objects are at rest with respect to the earth and that they tend to remain at rest unless acted upon by some force external to themselves. Aristotle and the ancient philosophers proclaimed that the natural state of bodies (on earth at least) was that of rest. This idea held sway through the Middle Ages until first Galileo, then Descartes, and finally Newton began to question it. Galileo pointed out that motion is just as "natural" as rest and that rest could be considered as that special case of motion in which the velocity is temporarily zero. Certainly the motion of the planets, the constant movement of the oceans, and the winds which continually blow over the earth's surface would support the contention that motion is fully as natural a state as rest is. Galileo first proposed the idea that forces cause or inhibit motion and that unless an external force acts on a body, it tends to remain at rest if it is at rest and tends to remain in motion if it is in motion. Newton (born in 1642, the year Galileo died) refined and sharpened these ideas about motion and rest, and formulated a basic law of motion which he called the *law of inertia*.

4.16   Force Is Necessary to Overcome Inertia

It is common knowledge that objects possessing mass resist being set in motion; or if already in motion, they resist efforts to stop them. This property of *resisting a change in motion* is known as *inertia*. In fact, *inertia* is the most outstanding property of *mass*. It is also a matter of common observation that an object which is moving in a straight line resists having the *direction* of its motion changed even if the speed is unchanged. Spinning wheels display *rotational inertia*, a fact which the gyroscope, with its many applications in compasses and gun-aiming devices, illustrates very well.

Newton assimilated the observable facts about *force*, *mass*, *motion*, and *inertia* and, in his writings in 1687, set down three general *laws of motion*.

4.17   Newton's First Law of Motion—Inertia

Newton's first law may be stated as follows:

A body which is at rest will remain at rest and a body which is in motion will remain in motion at the same speed and in the same direction unless acted upon by some unbalanced force.

This property of inertia is possessed by all matter and is independent of any forces (such as gravity) which may be acting. A body has inertia even out in interstellar space completely away from the gravitational forces of the earth or any other planet. *Mass* is the quantitative measure of *inertia*.

Some common examples of inertia will be useful in clarifying the concept. A stalled auto requires a great effort to *start* it moving, but once moving, only a modest push will keep it rolling. If an auto at high speed is braked to a stop suddenly, passengers, boxes, and any loose objects all tend to continue "in motion at the same speed and in the same direction." Seat belts can provide the force to nullify this inertia of motion of the passengers. Figures 4.18 and 4.19 (next page) illustrate the effects of inertia. One of the basic disadvantages of a reciprocating engine, compared to a turbine type (rotary) engine, is the *inertia of motion* possessed by the pistons during each and every stroke. The forces required of the engine parts to stop the pistons' motion in one direction and accelerate them in the other direction create strain, wear, vibration, noise, and inefficient operation of the reciprocating engine.

A further *distinction between mass and weight* is now possible, using the inertia concept. On earth the weight of a body is proportional to its mass, and at sea level, weight and mass are often

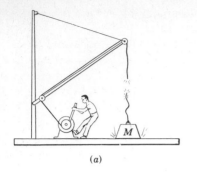

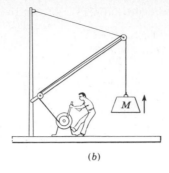

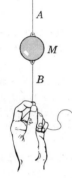

Fig. 4.19 Laboratory demonstration of inertia. M is a heavy ball suspended by string A. String B is of the same twine. If B is pulled slowly and steadily, A will eventually break; but if B is jerked suddenly, B will break. The ball M resists a change in motion, or as we say, "it has a good deal of inertia."

expressed *numerically* by the same units, in industry and commerce (the *pound* and the *kilo*, for example). But out in space, where a body has no measurable weight, it still has mass, and it still has inertia. Although a baseball would seem weightless in space, force would be required to throw it (overcoming its inertia of rest), and if caught, it would smack into the catcher's mitt and require force there to overcome its inertia of motion. A prankster might hand you one regular baseball and another made from solid lead around which the usual horsehide cover has been sewn. Holding them, you could not tell the difference since both would be *weightless* in space. Batting the regular one would give almost the same feel as batting a baseball on earth, since the same force would be required to overcome its inertia. Hitting the lead "baseball" would break the bat, perhaps sprain your wrists, and generally be quite a surprise. *Mass* and *inertia* are *universal* properties. Mass, wherever it is, resists a change in its motion, and this resistance to a change in motion Newton called *inertia*. Weight, on the other hand, is the *force of gravitational attraction.*

4.18 Newton's Second Law— Acceleration

When an unbalanced force *does* act on a body, a change in motion (acceleration) is produced. The greater the force acting, the greater the acceleration; but the greater the mass of the body, the greater its resistance to a change in motion (inertia). Newton's second law states that:

A body which is acted upon by an unbalanced force is given an acceleration in the direction of the force which is proportional to the force acting and inversely proportional to its mass.

Mathematically, this law may be stated as follows:

$$a \propto \frac{F}{m}$$

where
a = acceleration produced
F = force acting on the body
m = mass of the body

$\propto$ is read "is proportional to." By choosing units which are consistent with observed natural phenomena, we can write the equation for **Newton's Second Law:**

$$F = ma \tag{4.17}$$

Force = mass × acceleration

This formula is valid only when F, m, and a are expressed in consistent units. The following discussion will define the units required.

4.19 Units and Newton's Second Law

The metric unit of mass has already been defined as the *kilogram* (see Sec. 2.3). A kilogram of any matter has the same mass as the prototype platinum-iridium cylinder at the International Bureau of Weights and Measures at Sèvres, France. And, in like manner, one unit of mass in the English system has been defined as the *pound*, with a mass specified by the prototype at the Exchequer in London. (We will define another mass unit, the *slug*, later in this section.) Further, we have defined the unit of length as the *meter* in the metric system and as the *yard* in the English system, these units also being standardized by prototypes in France, Great Britain, and the United States. (It will be recalled that although the *yard* is the English-system *standard*, the *foot*, $\frac{1}{3}$ yd, is actually the unit in most common use.)

The concept of *acceleration* (change in velocity per unit time) is expressed in terms of length L and time T in metric units, as the meter per second per second (m/sec^2) and in English units as the foot per second per second (ft/sec^2). With *mass* and *acceleration* already possessing agreed-upon units in the metric system, all that remains to make Eq. (4.17) dimensionally correct is to define a *new unit of force*.

Metric-System Force Unit, the Newton, and the SI-Metric System of Units The fundamental unit of mass in the metric system is the *kilogram*; of length, the *meter*; and of time, the *second*. We now define a new unit of force, called the *newton*, which makes the equation $F = ma$ dimensionally correct.

> *One newton is that force which produces an acceleration of one meter per second per second when it acts alone on a mass of one kilogram.*

The newton (**N**) is the basic unit of force in the meter-kilogram-second (mks or SI-metric) system of units. It is an *absolute* or *universal* unit of force, in that it has no relationship to the earth's gravity or to the acceleration caused by gravity. Setting

$$F \text{ (newtons)} = m \text{ (kg)} \times a \text{ (m/sec}^2)$$

we see that dimensionally the newton is equivalent to 1 kg-m/sec^2.

The *weight* of 1 kg of mass at sea level is equal to about 9.81 *newtons of force*, since the acceleration of gravity at sea level is 9.81 m/sec^2. The older unit of force in the metric system, the *kilogram of force* (kg-f), is now used only as a measure of weight (the *kilo*) in metric countries. You will recall that when a *mass* of one kilogram is acted upon by a *force* of one kilogram, the acceleration produced is the acceleration of gravity, $g = 9.81$ m/sec^2. One kilogram of force (kg-f) equals 9.81 newtons (**N**). In time, it is probable that the *kilo* may become obsolete, and metric countries will weigh all commodities in newtons.

You may possibly come across references to the centimeter-gram-second (*cgs*) system of units. This system is no longer used in science and engineering, and we will not use it in this book.

English-Engineering System of Units: the Slug Engineers in English-system countries use another approach to the definition of consistent units for Newton's second law. Instead of choosing and defining a new *force unit* to fit $F = ma$, engineers decided instead to define a new *mass unit*. With the *pound-force* defined as the basic *force* unit and the foot per second per second as the basic unit of acceleration, a new mass unit, the *slug* (inertia is akin to being *sluggish*) is defined.

> *One slug is that unit of mass which when acted on by a force of one pound will be accelerated at the rate of one foot per second per second.*

Setting

$$F \text{ (lb)} = m \text{ (slugs)} \times a \text{ (ft/sec}^2)$$

we find the dimensions of the slug (sl) to be

$$1 \text{ slug} = 1 \text{ lb-sec}^2/\text{ft}$$

Since 1 lb force accelerates 1 slug mass at 1 ft/sec^2, and 1 lb force accelerates 1 lb mass at 32.2 ft/sec^2 (a body in free fall acted on only by the force of gravity), it follows that 1 slug mass is equal to about 32.2 lb mass.

In the English-engineering system of units forces are measured in pounds, masses in slugs, and accelerations in feet per second per second.

This book will make frequent use of both the mks (SI-metric) and the foot-slug-second (engineering) systems. These units are called *absolute units*.

It must be clearly understood that the equation $F = ma$ is dimensionally correct only when *absolute units* of force (newtons) are used, or when the slug of mass is used with the (English) engineering system.

Gravitational Units In contrast to the *absolute units* discussed above is the use of a system of units referred to the surface of the earth where the *earth's gravity* is an ever-present force factor.

In ordinary, everyday measures we say that a pound of mass *weighs* 1 lb, when what we really mean is that the pound of mass *is being attracted with a force of* 1 lb by the earth's gravity. The pound of force is therefore a *gravitational unit*. In like manner 1 kg of mass is acted on by a force of 1 kg at sea level, and we say that it weighs 1 kg. In order to understand clearly the distinction between *absolute units*, i.e., those units based on mass/inertia and acceleration considerations, and *gravitational units*, i.e., those based on the force of gravity at the earth's surface, we must recall that *weight* is the *force* exerted by gravity on a *mass*; and that *this force would give the mass an acceleration equal to g*, in the absence of any other forces.

In general, absolute units are used in scientific work and in engineering research and design, and gravitational units are used in commerce and industry.

The relationship of gravitational units to absolute units is governed by Newton's second law. It can be stated in words as follows:

Weight (force) equals mass times the acceleration of gravity.

As an equation, $w = mg$ (4.18)

and $m = \dfrac{w}{g}$ (4.18′)

Illustrative Problem 4.13 A crate of melons from Mexico is labeled 20 "kilos." (Although the kilogram is properly a mass unit, it is still used in commerce as a weight unit.) What would be the weight of this crate of melons in newtons?

Solution The crate of melons actually has a *mass* of 20 kg. Its weight in newtons would be given by Eq. (4.18).

$$w = 20 \text{ kg} \times 9.81 \text{ m/sec}^2$$

$$= 196 \text{ kg-m/sec}^2$$

$$= 196 \text{ N}$$

Note that since 1 N gives a mass of 1 kg an acceleration of 1 m/sec², *the newton has the dimensions of kg-m/sec²*. (Memorize this!).

At this point, for convenience and review, and *for ready reference throughout the study of applied physics*, Table 4.3 is provided, bringing together the important definitions, units, constants, and conversion factors that are essential to an understanding of mechanics.

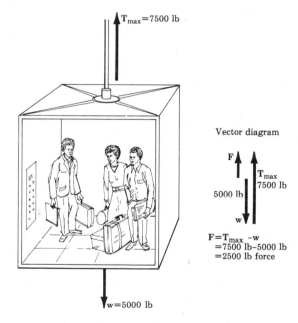

Fig. 4.20 Sketch of forces acting on an elevator, along with the associated vector diagram (Illus. Prob. 4.15).

Table 4.3 Definitions, units, constants, and conversion factors for the study of mechanics

Definitions of Basic Terms

Mass	The quantity of matter in a body. Mass resists a change in motion. This resistance is called *inertia*.
Force	A push or a pull which tends to cause a change in motion
Weight	The force of the earth's gravity on mass
Motion	Continuous change of position
Velocity	Rate of motion in a specified direction—a vector quantity
Speed	Rate of motion, no direction stated—a scalar quantity
Acceleration	Rate of change of velocity
Inertia	That property of mass which resists any change in motion

Mass Units and Their Definitions

Kilogram (kg)	The mass of the prototype cylinder at the International Bureau of Weights and Measures, Paris
Slug (sl)	A mass which, when acted upon by a force of 1 lb, is accelerated at the rate of 1 ft/sec². The English-engineering unit of mass. The slug is equivalent to 1 lb-sec²/ft. (Memorize this.)
Pound (lb)	Still used as a mass unit in some English-system commercial and industrial situations. We will use it as a mass unit in a few problems. Properly a force unit.

Force Units and Their Definitions

Absolute Units (For these units, $F = ma$)

SI-metric (mks)	newton (**N**)	1 **N** of force gives 1 kg of mass an acceleration of 1 m/sec²
Engineering	pound (lb)	1 lb of force gives 1 *slug* of mass an acceleration of 1 ft/sec²

Gravitational Units (*Weight*) (For these units, $F = \dfrac{w}{g} a$)

Metric	kilogram-force (kg-*f*)	1 kg-*f* gives 1 kg of mass an acceleration of 9.81 m/sec²
	gram-force (gm-*f*) (Now rarely used)	1 gm-*f* gives 1 gm of mass an acceleration of 981 cm/sec²
English	pound (lb)	1 lb-force gives 1 lb of mass an acceleration of 32.2 ft/sec²

Conversion Factors and Constants

1 kilogram-force (kg-*f*) = 9.81 **N**	= 2.2 lb	g_{earth} = 9.81 m/sec²
1 pound (lb)	= 0.454 kg-*f*	= 4.45 **N**	= 981 cm/sec²
1 slug (mass)	= 32.2 lb (mass)	= 14.59 kg	= 32.2 ft/sec²

Illustrative Problem 4.14 An auto has a mass of 1800 kg. What force in newtons would be required to give it an acceleration of 1.5 m/sec², assuming zero road friction and air resistance?

Solution Applying Newton's second law, the unbalanced force is given by

$$F = ma = 1800 \text{ kg} \times 1.5 \, \frac{\text{m}}{\text{sec}^2}$$

$$= 2700 \text{ N} \qquad\qquad ans.$$

Illustrative Problem 4.15 An elevator weighs 5000 lb fully loaded and is operated by a cable in which a tension of 7500 lb must not be exceeded. What is the greatest upward acceleration possible? (See Fig. 4.20.)

Solution Since the loaded elevator weighs 5000 lb (force), only 2500 lb force (7500 − 5000) of the allowable tension is available to accelerate the elevator upward. All units should first be changed to *absolute* units, in order that Newton's second law can be used. From Eq. (4.18′) the *mass* of the elevator is

$$m = \frac{w}{g} = \frac{5000 \text{ lb}}{32.2 \text{ ft/sec}^2} = 155.3 \text{ slugs}$$

When a force of 2500 lb is applied to this mass, the acceleration is

$$a = \frac{F}{m} = \frac{2500 \text{ lb}}{155.3 \text{ slugs}} = 16.1 \text{ ft/sec}^2 \quad ans.$$

4.20 Newton's Third Law— Action and Reaction

Although we have discussed the action of individual forces in the foregoing, it is a matter of fact that *forces never exist singly but always in pairs* (see Fig. 4.21*a*). As you walk, the force applied by your leg muscles through your shoes to the sidewalk is the cause of your motion. But the earth is pushing back on your feet with forces equal and opposite to those applied. These forces are due to the earth's *elasticity*, and the necessity for these *reactive* forces can easily be seen if you try to walk on a nonelastic surface such as deep mud or quicksand. If friction is absent, as on smooth ice, these reactive forces are not transmitted to your shoe soles, and floundering is the result. An automobile seems to push itself along, but actually the road under the wheels is exerting forces on the wheels, just as the wheels are exerting forces on the road (see Fig. 4.21*b*). It is this action-reaction pair of forces which moves the car.

Most of us have seen a squirrel in a revolving cage, running but not getting anywhere; and we know that when exploding powder expels a projectile from the muzzle of a gun, the gun always recoils. These, too, are examples of action and reaction.

The earth's gravitational pull (force) on the moon holds the moon in orbit, but the moon, in turn, has a reacting force on the earth, the most easily observed result being the tides in the oceans.

An occurrence all too frequently observed is the collision of two automobiles. Even if they are both traveling in the same direction and one overtakes the other and smashes the rear deck of the car in front, this *action* (a force) is accompanied by a *reaction* force back against the overtaking vehicle, damaging its grille, headlights, and engine hood.

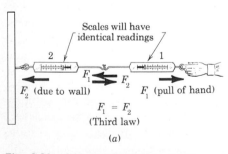

(b)

Fig. 4.21 Action and reaction. (*a*) Forces always exist in pairs, not singly. (*b*) The car's tires and wheels supply the action force, but without the reaction force provided by the road surface the auto would not change its state of motion. (Chrysler Corporation)

It should be noted that action-reaction pairs of forces never act on the *same* body.

In addition to the examples cited above, many steam and hot-gas turbines utilize the reaction principle in their design. Rotary lawn sprinklers are turned by the reactive force of the escaping water. Jet-propelled aircraft and the huge rockets used for putting earth satellites into orbit or for exploring the planets and outer space are pushed forward by reaction from the force of expanding gases accelerating to the rear (Fig. 4.22). There is still widespread misunderstanding of the principle of jet and rocket propulsion. One popular view is that the push of the escaping gases against atmospheric pressure creates the forward thrust. This is not the case at all; in fact, rockets and rocket-type aircraft *which carry their own oxygen for combustion* will operate at higher speeds and greater efficiency in interstellar space, where there is no atmosphere, than they will within the earth's atmosphere. The actual mechanics of jet engines and rocket engines is explained in Chapter 5.

4.21 Dimensional Analysis

In Sec. 2.1 we identified the fundamental units used in mechanics as length L, mass M, and time T. Some dimensions (units) of length were subsequently defined, as the meter, the foot, the centimeter, and the inch. A few of the units of mass were defined, as the kilogram, the slug, and (still to some extent) the pound; and some dimensions of time were given as the mean solar day, the hour, and the second. Mention was made of *derived units*, i.e., units derived from the fundamental units, and the student is by now familiar with such simple derived units as feet per second (velocity), and kilometers per second per second (acceleration), as well as a few more complex derived units such as the newton of force (kilogram-meters per second per second).

Throughout this book, emphasis will be placed on the importance of dimensions and units. Applied physics is a problem-solving science, and in the solution of any problem two

Fig. 4.22 Poseidon ballistic missile begins flight after being fired from a submerged nuclear-powered fleet submarine. The missile is propelled forward by the reaction force from the expanding gases being accelerated to the rear. This is an example of action and reaction—the third law of motion.

The fact that forces occur in equal and opposite pairs was recognized by Newton and formulated into another law of motion. **Newton's third law** of motion states:

When one body exerts a force on another body, the second body exerts an equal and opposite force on the first.

parts of the final answer are of equal importance: the numerical part, and the *dimensions*, or *units*. Needless to say, an answer is correct only if both the numerical value *and the dimensions* (units) are correct. The student will have noted that the illustrative problems distributed through the text have put as much emphasis on carrying out indicated operations with units as on the computing of the numerical answer.

Frequently it becomes necessary to change a given quantity from one unit of measurement to another. Two examples will be given here. Think them through carefully, since the principles demonstrated will apply to many of the problems you will solve in applied physics as well as those you will encounter later in your work.

Illustrative Problem 4.16   Change 45 mi/hr to its equivalent value in feet per second.

Solution

$$45 \text{ mi/hr} = \frac{45 \cancel{\text{mi}}}{1 \cancel{\text{hr}}} \times \frac{5280 \text{ ft}}{1 \cancel{\text{mi}}}$$

$$\times \frac{1 \cancel{\text{hr}}}{60 \cancel{\text{min}}} \times \frac{1 \cancel{\text{min}}}{60 \text{ sec}}$$

$$= 66 \text{ ft/sec} \qquad\qquad ans.$$

Note that we have really multiplied 45 mi/hr by $(1) \times (1) \times (1)$. Each of the units on the right side of the equation cancels out except the desired *ft* and *sec* units.

Illustrative   Problem   4.17   Convert   0.0753 tons/cubic foot to kilograms/cubic meter. Use the following   equivalencies: 1   ton = 2000 lb; 1 lb = 454 gm; 1 kg = 1000 gm; 1 ft = 30.5 cm; 1 m = 100 cm.

Solution

$$0.0753 \text{ ton/ft}^3 = \frac{0.0753 \cancel{\text{ton}}}{1 \cancel{\text{ft}^3}} \left(\frac{2000 \cancel{\text{lb}}}{1 \cancel{\text{ton}}}\right)\left(\frac{454 \cancel{\text{gm}}}{1 \cancel{\text{lb}}}\right)$$

$$\times \left(\frac{1 \text{ kg}}{1000 \cancel{\text{gm}}}\right)\left(\frac{1 \cancel{\text{ft}}}{30.5 \cancel{\text{cm}}} \times \frac{100 \cancel{\text{cm}}}{1 \text{ m}}\right)^3$$

$$= 2410 \text{ kg/m}^3 \quad ans.$$

Note how all units cancel out except those needed for a (correct) result.

A quick way of making a preliminary check on the solution of a problem is to see if the *dimensions* of the answer are consistent with the dimensions of the variables of the problem and with the known dimensions of the physical quantity which represents the answer desired. Treating dimensions as if they were algebraic quantities gives us a powerful tool for detecting errors in problem solutions. Some examples will serve to illustrate the point.

Illustrative Problem 4.18   A body whose weight is 4000 lb is acted on by a horizontal force of 250 lb. Neglecting all frictional forces, what is its acceleration?

Solution   Suppose it were forgotten that the acceleration formula, $F = ma$, requires that F and m be in *absolute* units. *Erroneously* (!) substituting values directly from the problem statement (as if weight were an absolute unit rather than a gravitational unit) gives

$$(\text{Erroneously}) \quad 250 \text{ lb} = 4000 \text{ lb} \times a$$

Solving for a,

$$a = \frac{250 \cancel{\text{lb}}}{4000 \cancel{\text{lb}}}$$

$$= 6.25 \times 10^{-2} \qquad \text{a dimensionless number!}$$

Knowing that the dimensions of the answer for a (acceleration) must be feet per second per second, we conclude that a mistake has been made and proceed to reanalyze the problem. When the proper relationships are recalled (in the engineering system, mass must be in *slugs* when force is in pounds), the problem is correctly set up as follows:

$$(\text{Correctly}) \quad 250 \text{ lb force} = \left(\frac{4000 \text{ lb}}{32.2 \text{ ft/sec}^2}\right)a$$

$$a = \frac{250 \text{ lb} \times 32.2 \text{ ft/sec}^2}{4000 \text{ lb}} = 2.01 \text{ ft/sec}^2 \quad ans.$$

Consider another example, with SI-metric units.

Illustrative Problem 4.19 An elevator has a mass of 1000 kg. What force (tension) in the hoisting cable must be applied to accelerate it upward at 2 m/sec²? Answer in newtons.

Solution Since the answer is required in newtons, we recall the dimensions of the newton: $1 \text{ N} = 1 \text{ kg-m/sec}^2$. Consequently, we know that unless the final answer has these dimensions, it cannot be correct. Suppose it were forgotten that a load (weight) is a *force*, and that the mathematical expression for weight is $w = mg$. In that case we might *erroneously* set up the problem as follows:

Tension in cable = tension to support load
+ force to accelerate

(Erroneously) $\quad T = 1000 \text{ kg} + 1000 \text{ kg}$
$\times 2 \text{ m/sec}^2$

Attempting to add the quantities on the right side of the equation is futile since they *do not have the same dimensions*. This very fact tells us that we have erred in setting up the problem. The "tension to support load" (weight) must be expressed in *absolute units*, since it is a force, not a mass, and it has to be added to the force to accelerate ($F = ma$), which expression requires absolute units.

(Correctly)

$T = mg + ma$

$= 1000 \text{ kg} \times 9.81 \text{ m/sec}^2 + 1000 \text{ kg}$
$\times 2 \text{ m/sec}^2$

$= 9810 \text{ kg-m/sec}^2 + 2000 \text{ kg-m/sec}^2$

$= 11{,}810 \text{ kg-m/sec}^2 = 11{,}810 \text{ N} \qquad ans.$

It should be emphasized that ending up with the correct units does not assure that the numerical value of the answer is correct; but, ending up with wrong or inconsistent units automatically indicates that the solution is wrong. Correct dimensions are a necessary condition for a correct answer, but they are not a sufficient condition.

We shall have occasion to use many more units and dimensions as the work in physics proceeds. Memorization of proper units and dimensions is often necessary; and in the solution of all but the very simplest problems, the student should carry all units and dimensions through the entire working of the problem. Units and dimensions obey the same algebraic laws with respect to compound fractions, cancellation, and exponents as numbers and literal algebraic terms do.

SUMMARY

A force is a push or a pull which tends to cause a change in motion.

The force of the earth's gravitational pull is the measure of weight.

The *absolute* units of *force* are:

SI-metric The *newton* (**N**) gives a mass of 1 kg an acceleration of 1 m/sec²

English The *pound* gives a mass of 1 slug an acceleration of 1 ft/sec²

The use of the pound as an absolute unit of force necessitates the definition of the *engineering unit of mass, the slug*: 1 lb (force) gives 1 slug (mass) an acceleration of 1 ft/sec². Dimensionally, a slug is 1 lb-sec²/ft.

The *gravitational* units of force (weight) are: for the metric system, the kilogram-force (kg-*f*), also called *kilo*; and for the English system, the pound and the ton.

Motion is defined as continuous change of position.

Velocity is the rate at which motion takes place; it is *uniform* when equal distances in the same direction are covered in each succeeding unit of time. It is a vector quantity.

Speed is the magnitude factor of velocity, without respect to direction; it is a scalar quantity.

Acceleration is defined as rate of change of velocity, and is a vector quantity.

The acceleration of gravity is caused by the force of gravity, and is a constant (approximately) over the surface of the earth.

$g = 9.81 \text{ m/sec}^2$ (metric) or 32.2 ft/sec^2 (English)

All bodies, heavy or light, fall with the same acceleration, g, under the influence of gravity (air resistance neglected).

The laws governing the flight of projectiles are merely extensions of two basic principles: (1) uniform motion horizontally and (2) vertical motion under the influence of the acceleration of gravity. These motions occur independently of one another.

Motion is caused by force, and motion may be increased, decreased, or stopped by force.

Mass resists any change of motion. This resistance to a change in motion is called *inertia*. Force is required to overcome inertia.

Sir Isaac Newton studied the relationship between force, motion, mass, inertia, and acceleration, and stated the laws of motion as follows:

Law I: Inertia. *A body which is at rest will remain at rest; and a body which is in motion will remain in motion at the same speed and in the same direction unless acted upon by an unbalanced force.*

Law II: Acceleration. *A body which is acted upon by an unbalanced force is given an acceleration in the line of the force which is proportional to the force acting and inversely proportional to its mass.*

Law III: Action and Reaction. *When one body exerts a force on another body, the second body exerts an equal and opposite force on the first.* Forces do not exist individually, but in pairs.

Newton's second law requires that a careful distinction be made between *weight* (a force) and *mass*. The relationship is expressed as follows:

$$w = mg, \quad \text{or} \quad m = w/g$$

Where Newton's second law or any of its derivatives are concerned, *absolute units* of *force* must be used.

For the SI-metric system	the *newton* (**N**)
For the English-engineering system	the *pound* (lb) (with mass in *slugs*)

Dimensional analysis is a powerful tool in the solution of problems in applied physics.

QUESTIONS AND EXERCISES

Note: In all the following questions, exercises, and problems the effects of friction and air resistance are to be neglected.

1. Exactly what is the difference between *velocity* and *acceleration*? Explain the units in which each is measured.

2. In the units of acceleration why does a unit of time to the second power appear in the denominator, as in $m/\sec^2$? Explain fully.

3. When you think about the units of mass, inertia, force, and acceleration and their definitions, it might seem that they are all defined in terms of one another. Which one(s) is (are) "pegged" to an external standard, and what is that standard?

4. From various fields of industry or engineering, list three examples of practical problems whose solution depends on the application of Newton's (*a*) first law, (*b*) second law, (*c*) third law.

5. Discuss completely, with sketches, why it is necessary to adjust the sights of a gun when shifting to a target at a different range.

6. Astronauts may experience "weightlessness" when sufficiently far from the gravitational field of the earth or the moon. How can they still have mass but no weight? List some advantages and disadvantages of being weightless. How would your life and actions be changed?

7. The story is told of the "stupid" hunter and the "smart" monkey. The hunter spied a monkey hanging from a tree limb a long way off and forgetting that his rifle was "sighted in" only for targets at point-blank (very close) range, he aimed directly at the monkey and fired. The monkey, accustomed to guns and hunters, watched for the flash at the gun barrel and immediately let go of the limb, becoming a "freely falling" object. What happened?

PROBLEMS

1. A supersonic aircraft travels at 2900 ft/sec. At this average speed how many hours would it take to cross the United States, a distance of 3100 miles?

2. The speedometer of an automobile registers 20 km/hr at a certain instant, and 7 sec later it is passing 70 km/hr. Find the acceleration in meters per second per second, assuming it to be uniform.

3. A catapult on an aircraft carrier deck starts a plane from rest and uniformly accelerates it to 105 mph at the instant of launching, 2.5 sec later. Find the acceleration in feet per second per second.

4. The barrel of a deer rifle is 24 in. long. The velocity of the bullet as it emerges from the end of the barrel is known to be 3000 ft/sec. Find (*a*) the time taken by the bullet to traverse the length of the gun barrel and (*b*) the acceleration of the bullet (assumed uniform) while it is in the barrel.

5. A motorist traveling at 90 km/hr sees trouble ahead and slams on the brakes. If the deceleration is 8 m/sec² and if 0.2 sec is allowed for his reaction time, how far ahead did he have to see the trouble in order to stop in time? (Answer in meters.)

6. A baseball pitcher can give the ball an acceleration of 600 ft/sec² (assumed uniform) throughout his delivery. The distance from the start of his delivery to the point where the ball leaves his fingers is 8.5 ft. With what velocity does the ball start toward the batter?

7. If an antiaircraft projectile is to reach a height of 4 km, what must be the vertical component of its initial velocity?

8. A rock is dropped off a sheer cliff into a lake. The splash is seen 5 sec after the rock is dropped. (*a*) How far is it down to the water? (*b*) What was the velocity of the rock as it hit the water? Answers in metric units.

9. A ball is thrown in to the catcher from left field, a distance of 250 ft, at an average *horizontal* velocity of 80 ft/sec. How high does it rise during its flight?

10. The quarterback on a football team gets off a perfect pass at an angle of 30° with the horizontal. His receiver gathers it in 40 yd downfield from the passer at the same height above ground as the passer's throwing hand. (*a*) What was the ball's velocity as it left the passer? (*b*) How high above the passer's outstretched arm did it rise?

11. If the system shown in Fig. 4.23 is released, what acceleration will result? (HINT: Remember that the entire mass has to be accelerated, but the pull of gravity on the mass hanging vertically is the only force available to do the accelerating.) Neglect friction. Be careful with units!

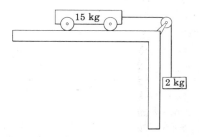

Fig. 4.23 Diagram for Problem 4.11.

12. A truck is towing a 3500-lb automobile on a level road and accelerating it at 3.4 ft/sec². What is the tension in the towline, in pounds?

13. It is desired to design an elevator so that it will be capable of an acceleration of 1.5 m/sec². Fully loaded, its mass is 6000 kg. What tension will exist in the cable as the elevator is being accelerated? (HINT: Remember the tension that already exists in the cable just to support the elevator.)

14. A thrust of 10,000 lb accelerates an airplane horizontally at 12 ft/sec². What is the mass of the airplane, in slugs?

15. A parachutist whose mass is 80 kg is in free fall at 70 m/sec. His parachute opens, and he slows to 6 m/sec in a vertical distance of 100 m. What force in newtons did his parachute harness exert on him?

16. A truck has a mass of 5000 kg. If frictional forces are negligible, what force would the engine have to produce to give the truck an acceleration of 1.8 m/sec² on a level road?

17. At what muzzle velocity (feet per second) must a projectile be shot if it is to strike a target 5000 yd away when the angle of elevation is 25°?

18. Logs from a greased chute are discharged horizontally at 40 ft/sec at the edge of a vertical cliff overhanging a millpond. The vertical drop to the water is 125 ft. How far from the base of the cliff do the logs hit water?

19. A 2000-kg auto is traveling at 80 km/hr. What is the least distance in which it can be stopped if the braking force is 8000 N?

20. A rotary oil-well drilling rig is handling 4200 ft of 6-in. pipe, which weighs 5 lb per foot of length. The pipe is being lowered into the hole at a uniform velocity of 10 ft/sec. The brake on the cable drum is applied, stopping the entire assembly in a distance of 12 ft. Calculate the maximum stress in pounds on the supporting hook.

21. A space vehicle whose total mass is 3.5 × 10⁶ kg is launched *vertically* from the earth's surface by a rocket engine which exerts a total thrust of 4.4 × 10⁷ N. Find the initial upward acceleration.

22. A cruiser's 8-in. guns are "on target" at 16,000 yd. If the muzzle velocity is 1500 ft/sec, (*a*) what should the angle of elevation of the guns be? (*b*) Find the time of flight of the projectiles (air resistance neglected).

23. A cable will withstand a tension of 2000 kg-*f*. If it is used to raise a load of 1500 kg from the ground to a platform 25 m high, what is the least permissible time required?

24. A baseball bat is in contact with the ball for 0.02 sec and changes the ball's velocity from 85 ft/sec northward to 150 ft/sec southward. The ball weighs 5.25 oz. What was the average force of the blow in pounds?

25. A horizontal bomber approaches on its bombing run at 48,000-ft elevation and 650 mph. (*a*) How far (horizontally) from the target is the correct release point (air resistance neglected)? (*b*) What is the angle of depression at release?

26. Figure 4.24 illustrates an apparatus called an *Atwood's machine*. If $m_1 = 2.5$ kg and $m_2 = 3.1$ kg, find (*a*) the acceleration of the mass-and-cord system when m_2 is released and (*b*) the tension in the cord while the system is running.

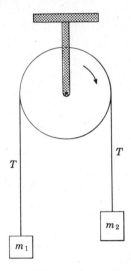

Fig. 4.24 Diagram of Atwood's machine (Problem 4.26).

5 Work, Energy, and Power

In industrialized nations most work is done by machines. For the majority of men and women in the work force the input to the job is more mental than manual. Even those jobs requiring manual skills utilize a great variety of machines and power tools. Many industrial and manufacturing processes are governed by computers or auto "feed-back" mechanisms which control both the quantity and quality of the output. We call this process "automation." The business world too, with its network of communications and computerized accounting, uses a great deal of mechanized and electronic "hardware." And agriculture today is as highly mechanized as heavy industry.

All of these machines, tools, and specialized instruments require vast amounts of energy, and until very recently there seemed to be an inexhaustible supply of it. Now, however, we realize that the near future may bring serious energy shortages, and along with the development of new energy resources, the conservation of existing energy supplies is a high priority.

Anyone who is contemplating a professional or technical career needs to know a great deal about energy, work, and power. This chapter will provide a basic foundation for understanding these concepts, and will illustrate a number of applications of energy principles. We repeat that energy is one of the twin themes of this book. You

Fig. 5.1 A computer, various sensors and gauges, and a cathode-ray tube are all used in this example of automation as applied to fuel and emissions testing in the auto industry. (Chrysler Corporation.)

can never learn too much about energy. It will concern you all of your life, both on and off the job.

WORK

5.1 Work Defined

Work performed by human muscle tends to be measured in terms of human fatigue, or in terms of the hours in a workday. The actual accomplishment of various workers may differ widely, and yet their pay for a day's "work" may be exactly the same. With machines, on the other hand, a more exact and technical definition of work is necessary, one which can be standardized to mean the same net accomplishment throughout industry. Technically, *work* is defined in terms of *force applied* and the *distance through which it acts*.

$$\text{Work} = \text{force} \times \text{distance}$$

or $$W = Fs \qquad (5.1)$$

In a scientific sense a force must act through a measurable distance before work is accomplished. You might exert the maximum muscular force of which you are capable in an effort to push a heavy truck, but if it does not move, no work is done, even though you soon become fatigued.

The analysis above and Eq. (5.1) both assume that the resulting motion has the same direction as the applied force. This is by no means always the case. The more general situation is illustrated in Fig. 5.2a, where the direction of the force pulling the sled is different from that of the sled's motion. A simple vector diagram (Fig. 5.2b) provides the component of the force (see Sec. 3.7) in the direction of the motion. *Useful work* is done *only by the component of the force in the direction of the displacement*. As a more general proposition then, work done is given by

$$W = F \cos \theta \cdot s = Fs \cos \theta \qquad (5.2)$$

where θ is the angle between the applied force vector and the resultant motion vector.

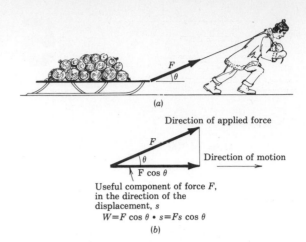

$$W = F \cos \theta \cdot s = Fs \cos \theta$$

(b)

Fig. 5.2 Force and work. (a) Useful work is done only by that component of the force which is in the direction of the motion. (b) Vector diagram of sled-hauling problem.

5.2 Units of Work. Metric Units

In the mks (SI-metric) system of units, forces are measured in *newtons* and distances in *meters*. The mks unit of work and of energy is therefore the *newton-meter*, abbreviated **N-m**. The newton-meter has been given the name *joule* (symbol **J**), in honor of James Prescott Joule, a British scientist who made many contributions to the branch of physics called *thermodynamics*.

$$1 \text{ joule} = 1 \text{ newton-m (N-m)}$$

One joule is the work accomplished (or energy expended) when a force of one newton acts through a distance of one meter.

Illustrative Problem 5.1 A hoist lifts a mass of 1500 kg through a distance of 18 m. How many joules of work are done?

Solution

$$W \text{ (J)} = F \text{ (N)} \times s \text{ (m)}$$

But $\quad F = mg$ (Recall that weight equals mass $\times g$; Eq. 4.18.)

$$= 1500 \text{ kg} \times 9.81 \text{ m/sec}^2$$

$$= 14{,}700 \text{ N}$$

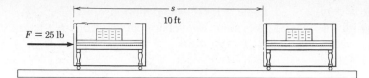

Fig. 5.3 Work equals (force applied) × (distance moved) in the line of the force. When a 25-lb force is applied and the piano moves 10 ft, the work done is $W = 10$ ft × 25 lb = 250 ft-lb.

Therefore

$$W = 14{,}700 \text{ N} \times 18 \text{ m}$$

$$= 2.65 \times 10^5 \text{ J (or N-m)} \qquad ans.$$

English System Units   Since forces are measured in pounds and distances in feet in the English system, the practical unit of work is the *foot-pound* (ft-lb). Suppose (Fig. 5.3) a force of 25 lb pushes an object 10 ft. The work accomplished is

$$W = Fs = 25 \text{ lb} \times 10 \text{ ft} = 250 \text{ ft-lb}$$

When objects are being lifted at constant velocity it is easy to compute the work done, since the weight of the object is actually the force of gravity pulling on it, and this must be overcome by the force applied to lift it. If a 60-lb casting is lifted from the floor to the bed of a milling machine 4 ft high, the work done in lifting (Fig. 5.4) is 4 ft × 60 lb = 240 ft-lb.

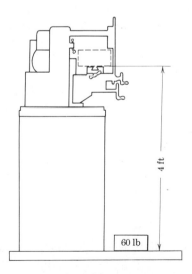

Fig. 5.4   Lifting requires work, since the force of gravity must be overcome. The net work done in lifting is equal to the *weight* of the object times the vertical distance through which it is lifted. W = 4 ft × 60 lb = 240 ft-lb.

If the mass of an object to be lifted is given in engineering units (slugs), the *force* to lift it is given by Eq. (4.18):

$$\text{Force (weight)} = mg = \text{slugs} \times 32.2 \text{ ft/sec}^2$$

It should be noted that, in computing the work done in lifting objects against the force of gravity, the assumption is that they are lifted at constant velocity. If they were jerked upward rapidly, they would be *accelerating* and, in addition to the force required to lift them ($F_w = mg$), an additional amount of force ($F_a = ma$) would be necessary to provide the acceleration.

ENERGY

Matter under certain conditions has the capacity or ability to do work even though none is being accomplished at the moment. A rushing stream or water behind a dam, for example, can be put to work at a variety of jobs. A heavy weight suspended at a great height could do work if it were allowed to fall; a compressed spring possesses a latent capacity for doing useful work and so does compressed air. Fuels like gasoline and natural gas do work as the result of combustion. *This capacity to do work is called energy.*

5.3   Classification of Energy

The running stream possesses energy of motion; the suspended weight and the impounded water, energy of position; the compressed spring, energy due to molecular arrangement; and fuels possess chemical energy. *Energy of motion* is called *kinetic energy* (KE), and *energy due to position or molecular arrangement* is called *potential energy* (PE). We will discuss *heat energy* in later chapters.

A body possessing kinetic energy does work when it is stopped or slowed down. A hammer possesses a considerable amount of kinetic energy as it strikes a nail, and this kinetic energy is converted into the useful work of driving the nail. Other examples of kinetic energy are a moving auto, a projectile in flight, a rotating flywheel, or a plunging fullback.

A body or system that possesses potential energy has had work done on it at some previous time, and its present potential energy is a measure of the energy once expended to give it its present position or molecular arrangement. The potential energy which it possesses can ordinarily be converted back into useful work (not without losses, however) by means of a suitable machine.

5.4 Measurement of Energy

Energy is expressed in the same units as work, namely, *joules* in the metric system, or *foot-pounds* in the English system. If an object possesses 10 ft-lb of energy, it is theoretically *capable* of doing 10 ft-lb of work. If work is expended on an object to lift it, potential energy is acquired as it is lifted. Then if it is allowed to fall, as it gains velocity the potential energy is converted into kinetic energy, which, in turn, can be converted into work as the falling body is stopped. (See Fig. 5.5.)

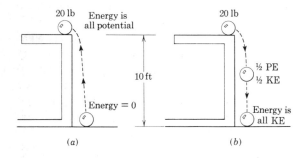

(a)   (b)

Fig. 5.5 Potential energy and kinetic energy compared. (*a*) Work done in lifting an object results in potential energy. If the 20-lb (wt) ball is lifted to a 10-ft shelf, potential energy is acquired. The amount of potential-energy gain is exactly equal to the work done. PE = 20 lb × 10 ft = 200 ft-lb. (*b*) As the ball falls, accelerated by the force of gravity, the potential energy is converted into kinetic energy. Midway in the fall the energy is half potential, half kinetic. What becomes of the kinetic energy as the ball hits the floor?

Table 5.1 lists the common units of energy and gives some conversion factors from metric to English-system units.

Table 5.1 Units of work and energy

English System	Metric System
Foot-pound (ft-lb) (engineering unit)	Joule (**J**) = newton-meter (**N-m**)

Conversion
1 ft-lb = 1.356 **J**
1 **J** = 0.738 ft-lb

5.5 Gravitational Potential Energy (GPE)

Potential energy often consists of the work done on a body to raise it to an elevated position with respect to some chosen level of zero energy (usually the earth's surface). It may therefore be readily reasoned that if an object has a mass m, and if it is slowly raised to a height h, the work done on it, and therefore the gravitational potential energy acquired, can be expressed as follows:

$$\text{GPE} = mgh \qquad (5.3)$$

Potential energy calculated from this formula will be expressed in the same units in which work is expressed, i.e., joules (mks system) and foot-pounds (English system).

One caution is necessary. The above equation is dimensionally correct for the English system only when mass is given in *slugs*. Since masses are often still given in *pounds* in nonscientific and industrial usage, you should be careful with units. Where that is the case, merely multiply the given mass in pounds (actually *weight*) by the height, to obtain the GPE in ft-lb, since (Eq. 4.18') $w = mg$. In other words, $\text{GPE}_{(\text{ft-lb})} = w_{(\text{lb})} \times h_{(\text{ft})}$. Also

$$\text{GPE}_{(\mathbf{J})} = w_{(\mathbf{N})} \times h_{(\text{m})}.$$

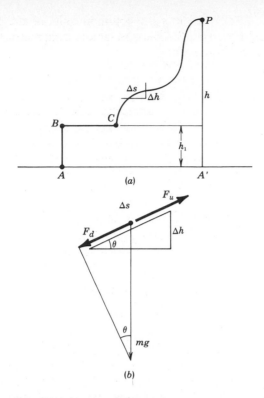

Fig. 5.6 Work done against gravity on a system, and GPE gained by the system do not depend on the path of the motion (friction neglected). (*a*) The work done in moving an object along path *ABCP* is equal to the work done in raising the same object vertically along path *A'P*. (*b*) Vector relations at any point along the curved path *CP*.

Path Makes No Difference If frictional and other forces are assumed to be zero, the path used to raise a body from one level to another makes no difference in the work done, or in the GPE gained. In terms of Fig. 5.6*a*, this means that if all forces except gravitational attraction are neglected, the work done on mass *A* by a force moving it upward along *any path* such as *ABCP* for example, is the same as that done if it were simply raised from *A'* to *P*. The net increase in potential energy from *A* to *P* is also the same, regardless of path. The following analysis should help you think this through.

From *A* to *B* there is a straight lift, and $W_{AB} = mgh_1$. From *B* to *C* no work at all is done

if friction is zero. On the curved path *CP*, consider any very small distance Δs*. The enlarged vector diagram of Fig. 5.6*b* shows the forces acting on the body as it is being moved through the distance Δs. The force F_u required to move the body up the slope without acceleration and in the absence of friction is exactly equal to F_d, the component of the force of gravity along the down slope. But $F_d = mg \sin \theta$. Therefore

$$F_u = mg \sin \theta.$$

The net work done for the segment Δs is $\Delta W = F_u \, \Delta s$. But

$$\Delta s = \frac{\Delta h}{\sin \theta},$$

and therefore

$$\Delta W = \frac{mg \, \sin\theta \, \Delta h}{\sin \theta} = mg \, \Delta h.$$

This is precisely the work which would have been done (and the gain in GPE which would result) if the object had been moved through the *vertical path* Δh instead of the curved path Δs.

5.6   Kinetic Energy

The *energy of motion* of a moving body is precisely the amount of work which would have to be done on it to accelerate it from rest to its present velocity or, conversely, the amount of work which the body could do against a resisting force which is decelerating it to zero velocity.

An expression for calculating kinetic energy can be obtained by recalling Newton's second law. The force F necessary to impart an acceleration a to a mass m is

$$F = ma \qquad (4.17)$$

Multiplying both sides of this equation by s,

$$Fs = mas$$

---

* The symbol Δ means "a very small amount of." Δs is read "delta s." It *does not* mean delta multiplied by s.

But $\qquad v^2 = 2as \qquad$ (4.6)

and therefore $\qquad as = v^2/2$

Substituting, $\qquad Fs = \frac{1}{2}mv^2$

But the product Fs represents the work done on a mass m to accelerate it from rest to velocity v and is therefore the energy of motion (kinetic energy) possessed by the mass m when it is moving at the velocity v. Or

$$KE = \frac{1}{2}mv^2 \qquad (5.4)$$

This equation can be used directly with SI-metric units. Just make sure that mass is in kilograms and velocity is in meters per second. The answer will be in newton-meters (joules), as this dimensional analysis shows:

KE units (metric):

$$\text{kg} \times \left(\frac{\text{m}}{\text{sec}}\right)^2 = \left(\frac{\text{kg-m}}{\text{sec}^2}\right) \times \text{m} = \text{N-m} = \text{J}$$

In the English (engineering) system however, you must be careful. The equation is derived, as noted above, from Newton's second law, and that law ($F = ma$) requires that all units be *absolute*, not gravitational units. It is necessary therefore, to use *slugs* for mass, not *pounds*. Dimensionally,

$$1 \text{ slug} = 1 \frac{\text{lb-sec}^2}{\text{ft}} \qquad \text{(see Sec. 4.19)}$$

Using slugs, kinetic energy will then be in foot-pounds, as this analysis shows:

KE units (English):

$$\frac{\text{lb-sec}^2}{\text{ft}} \times \left(\frac{\text{ft}}{\text{sec}}\right)^2 = \text{lb-ft} \quad \text{or} \quad \text{ft-lb}$$

In industrial practice, and sometimes even in engineering practice, design and operational problems start off with data in pounds, not in slugs. When this is the case, and when kinetic energy is being calculated, recall that $m = w/g$ (Eq. 4.18′) and use this alternative kinetic-energy formula (English system).

$$KE = \frac{1}{2} \cdot \frac{w}{g} v^2 \qquad (5.4')$$

If, in a metric country, weight in *newtons* is given (instead of mass in kilograms), the same form (Eq. 5.4′) should be used.

It should be apparent that in problems where both potential and kinetic energy are involved, both must be expressed in the same units, i.e. absolute or gravitational, before adding, subtracting, or otherwise combining them in any way.

Illustrative Problem 5.2 Suppose the hammer of a pile driver has a mass of 50 slugs and that it falls 20 ft before it hits the pile. What kinetic energy does the hammer have just before it hits?

Solution We will solve the problem by two different methods.

(*a*) *First method* (using energy principles, as discussed above): Remembering that the kinetic energy of the hammer at the bottom of its fall must be equal to the potential energy at the top, write Eq. (5.3):

$$PE = mgh$$

Remember that slugs have the dimension (lb-sec²)/ft, and write

$$PE = 50 \frac{\text{lb-sec}^2}{\text{ft}} \times 32.2 \frac{\text{ft}}{\text{sec}^2} \times 20 \text{ ft}$$

$$= 32{,}200 \text{ ft-lb} \qquad\qquad ans.$$

(*b*) *Second method* (using acceleration and velocity principles, as developed in Chap. 4). Before substitution of numerical values in the kinetic energy formula, the velocity at the moment of impact, v, must be found. Recall (Eq. 4.9) that the velocity of a freely falling body after falling a distance s is

$$v^2 = 2gs$$

Substitute given values,

$$v^2 = 2 \times 32.2 \frac{\text{ft}}{\text{sec}^2} \times 20 \text{ ft} = 1290 \left(\frac{\text{ft}}{\text{sec}}\right)^2$$

Now, substituting this value of v^2 in Eq. (5.4),

$$KE = \frac{50 \text{ slugs} \left(\text{that is, } \dfrac{\text{lb-sec}^2}{\text{ft}}\right) \times 1290 \dfrac{\text{ft}^2}{\text{sec}^2}}{2}$$

$$= 32{,}200 \text{ ft-lb} \quad \text{(as before)} \qquad \textit{ans.}$$

The comparative simplicity of the first method should be carefully noted. Solving problems in mechanics from energy considerations, when they are applicable, is usually preferable to applying acceleration principles to their solution. These considerations are often referred to as the *principle of work*, or the *work-energy principle*. This principle constitutes a powerful tool in the solution of problems in mechanics.

[handwritten: TEST ↗ when going against gravity use Energy principle, when can use acceleration prin.]

The Work-Energy Principle

The total mechanical energy of a body or system at any time t

> = *the total energy at some former time* t_0
> + *any energy added to the system*
> − *any work done against friction* (or other "useless" work) *in the time interval*

Illustrative Problem 5.3 A rocket with an attached satellite to be placed in orbit has a total mass of 100,000 kg. It is launched vertically with the rocket motors providing a constant thrust of 1.6 million **N**. What is the rocket's velocity as it passes through an altitude of 5000 m? Neglect effects of air resistance and loss of mass due to fuel used.

Solution
(*a*) *By the work-energy principle.* The total energy (KE + PE) at 5000 m equals the work done by the thrust force.

$$\tfrac{1}{2}mv^2 + mgh = Fs$$

Substituting,

$$\tfrac{1}{2}(100{,}000 \text{ kg}) \times v^2 + 100{,}000 \text{ kg}$$
$$\times 9.81 \frac{\text{m}}{\text{sec}^2} \times 5000 \text{ m}$$

$$= 1.6 \times 10^6 \frac{\text{kg-m}}{\text{sec}^2} \times 5000 \text{ m}$$

Simplifying,

$$v^2 = \frac{2 \times 5000 \text{ m}\left(1.6 \times 10^6 \dfrac{\text{kg-m}}{\text{sec}^2} - 981{,}000 \dfrac{\text{kg-m}}{\text{sec}^2}\right)}{100{,}000 \text{ kg}}$$

$$v = 249 \text{ m/sec} \qquad \textit{ans.}$$

(*b*) *By Newton's second law (acceleration principles).* The *net upward force* is the thrust of the rocket motor minus the *weight* (not mass) of the rocket and its payload. Substituting in $F = ma$,

$$1{,}600{,}000 \frac{\text{kg-m}}{\text{sec}^2} - 100{,}000 \text{ kg} \times 9.81 \frac{\text{m}}{\text{sec}^2}$$

$$= 100{,}000 \text{ kg} \times a$$

$$a = \frac{619{,}000(\text{kg-m})/\text{sec}^2}{100{,}000 \text{ kg}} = 6.19 \text{ m/sec}^2$$

Now, using

$$v^2 = 2as$$

and substituting,

$$v^2 = 2 \times 6.19 \frac{\text{m}}{\text{sec}^2} \times 5000 \text{ m}$$
$$v = 249 \text{ m/sec} \quad \text{(as before)} \qquad \textit{ans.}$$

5.7 Energy and Its Transformations

So far the discussion has involved two forms of energy, potential and kinetic, and we have shown how one can be transformed into the other, how potential energy is the result of work done, and how work can be done by kinetic energy. A swinging pendulum is a good example of potential energy being transformed into kinetic energy and back again to potential energy (see Fig. 5.7).

Illustrative Problem 5.4 A pendulum bob is pulled to the side, or displaced from its equilibrium position, until its center is 0.5 m higher than the "dead center," or equilibrium position. Find the velocity of the bob as it passes through dead center. (Neglect air resistance and friction; see Fig. 5.7.)

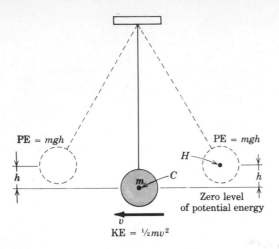

PE = mgh

PE = mgh

H

h

m   C

h

Zero level
of potential energy

v

KE = ½mv²

Fig. 5.7 A swinging pendulum illustrates the transformation from potential energy to kinetic energy and back again.

Solution The only energy possessed by the system as the bob is released is potential energy from the 0.5-m elevation of the bob above dead center. The kinetic energy as the bob passes the equilibrium point (no remaining potential energy) is equal to the potential energy as the bob was released.

$$\text{KE}_{(\text{point } C)} = \text{PE}_{(\text{point } H)}$$

In absolute units,

$$\frac{mv^2}{2} = mgh$$

(note that mass *m* cancels out—mass of bob makes no difference), and

$$v = \sqrt{2gh}$$

Substituting,

$$v = \sqrt{2 \times 9.81 \text{ m/sec}^2 \times 0.5 \text{ m}}$$

$$= 3.13 \text{ m/sec} \qquad\qquad ans.$$

Many forms of energy can readily be converted into other forms. The sun's heat energy is the result of nuclear and chemical-energy changes in its interior and on its surface. This radiant energy travels through space with the amazing speed of 2.998×10^8 m/sec (186,300 mi/sec).

When this radiation is stopped by the earth, it is reconverted into heat, which evaporates water from the oceans and causes winds and storms of tremendous energy. Rain and snow are deposited at higher elevations where the mass of water results in great potential energy. On its way to the sea this potential energy becomes kinetic energy. At hydroelectric plants it can be converted into electric energy, which can be reconverted into heat, light, sound, and mechanical energy at the flip of a switch.

From the viewpoint of industry today most of the energy we use comes from coal, oil, and natural gas—stored *chemical energy* which we convert into heat energy to fire boilers, run engines, make electricity, and power motorcars and trucks. Industrial use of atomic energy increased steadily until recently, when concerns about nuclear reactor safety and disposal of radioactive wastes slowed the construction of new plants. Later chapters will discuss atomic and nuclear energy at some length. New sources of energy—solar, wind, geothermal, and tidal—are currently under development in this country and abroad.

5.8   Law of Conservation of Energy

The energy transformations discussed above should stimulate your thinking about machines which utilize energy. Although energy can be changed from one form to almost any other form by the application of a suitable machine, no machine or device exists which can actually *create* energy. "Perpetual motion" machines (Fig. 5.8) have challenged the inventive genius of many generations, but none has ever operated successfully *to do useful work*. It seems to be an immutable law of nature that *energy can be changed in form but cannot be created or destroyed*. This important law was first clearly stated by the German physicist Hermann von Helmholtz (1821–1894) and is referred to as the *law of conservation of energy*.

Another way of stating the law of conservation of energy is:

In any isolated system the total energy is constant.

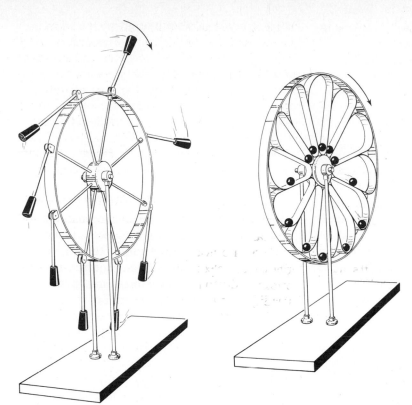

Fig. 5.8 Two models of a "perpetual motion" machine. The inventor's idea can be perceived by analyzing the construction. No such machine that will actually do any useful work has ever been devised. In fact, energy has to be supplied to such machines to overcome friction.

It is true that in energy transformations, some energy always *seems to be lost*. Careful analysis, however, reveals the fact that when energy of one form disappears, other forms of energy appear in amounts equal to the apparent discrepancy. For example, if the electric-energy input to a certain motor is carefully measured and compared with the mechanical-energy output, a loss of from 4 to 6 percent seems to have occurred. Actually, however, this "lost energy" has gone into heat, and the law of conservation of energy for this particular energy transformation could be stated as follows:

Energy input (electric)
= energy output (mechanical)
+ heat energy ("lost")

The heat energy is lost only in the sense that it is not useful for the purpose at hand.

Referring to Fig. 5.5, we can now see why the kinetic energy acquired by the falling ball must be exactly equal to the potential energy of the system as it rested on the edge of the shelf. No energy is created or destroyed; it is merely changed in form, from potential to kinetic.

MOMENTUM

A concept which is related to energy and work and also to Newton's laws of motion is the idea of *momentum*, which Newton called *quantity of motion*. The linear momentum of a body is defined as the product of its mass and its velocity. Momentum is designated by the letter p.

$$p = mv \qquad (5.5)$$

Momentum = mass × velocity

By comparison with Eq. (5.4) it is seen that both kinetic energy and momentum are proportional to the mass of the moving body. Further examination of the two equations shows that kinetic energy is proportional to the square of the velocity, while momentum increases as the first power of the velocity. Kinetic energy is a measure of the work a moving body will do as it is being stopped, while *momentum* is *a measure of the impulse or impact of a moving body when it hits another body*.

The quantity of motion possessed by a moving body is related to both its mass and its velocity. A heavy object even when moving slowly, e.g. a bulldozer, will have great momentum; and a small object moving with high speed, e.g. a bullet, also possesses a great deal of momentum. Objects possessing momentum exert forces on other objects with which they come in contact (the second law of motion), and in turn the objects they collide with will exert a reaction force (third law) back on them due to inertia (first law).

Common experience tells us that if a force is applied to an object for a short time, a small change in its motion will occur; and if the same force is applied for a longer time, a greater change in motion will take place. The all-important principle of follow-through, stressed by coaches of golf, tennis, and baseball, is recognition of this fact. The longer the time the club, racquet, or bat can be kept in contact with, that is applying force to, the ball, the greater will be the change in momentum and therefore the speed of the ball.

5.9 Impulse

The product of *force acting and the time which it acts* is called *impulse*. Comparing work with impulse, we have

$$\text{Impulse} = Ft \qquad (5.6)$$

and

$$\text{Work} = Fs \qquad (5.1)$$

To show that the momentum of a moving body is a measure of the impulse it surrenders when it hits another object, write Newton's second law,

$$F = ma \qquad (4.17)$$

Multiplying both sides by t gives

$$Ft = mat$$

But *change in velocity*

$$v_2 - v_1 = at \qquad (4.3)$$

and therefore

$$Ft = mv_2 - mv_1 \qquad (5.7)$$

or

$$\textit{Impulse} = \textit{change in momentum}$$

Equation (5.7) is called the *momentum equation*. You must remember that Newton's second law, $F = ma$, calls for absolute units of force, and since Eq. (5.7) is derived from it, the momentum equation also calls for absolute force units: newtons or pounds. For example, factoring Eq. (5.7),

$$F \times t = m(v_2 - v_1)$$

we have

$$\text{Newtons} \times \text{seconds} = \text{kilograms} \\ \times \text{meters per second}$$

$$\text{Pounds} \times \text{seconds} = \text{slugs} \times \text{feet per second}$$

An alternative form of the momentum equation for use with *gravitational* units is

$$Ft = \frac{w(v_2 - v_1)}{g} \qquad (5.8)$$

where F will be in pounds, w in pounds, and g is again the acceleration of gravity, 32.2 ft/sec². Dimensionally, Eq. (5.8) checks out (English system) as

$$F \text{ (lb)} \times t \text{ (sec)} = \frac{w \text{ (lb)} \times (v_2 - v_1)(\text{ft/sec})}{g \text{ (ft/sec}^2)}$$

or

$$\text{lb-sec} = \text{lb-sec}$$

Equation (5.8) should also be used if w and F are in newtons. In that case, dimensionally,

$$\text{N-sec} = \frac{\text{N(m/sec)}}{\text{m/sec}^2} \qquad \text{or} \qquad \text{N-sec} = \text{N-sec}$$

In the SI-metric system, recall that a newton is dimensionally equivalent to 1 kg-m/sec². In metric units, Eq. (5.7) may be analyzed dimensionally as follows:

$$F \left(kg \, \frac{m}{sec^2} \right) \times t \, (sec) = m \, (kg) \times v \left(\frac{m}{sec} \right)$$

or

$$\frac{kg\text{-}m}{sec} = \frac{kg\text{-}m}{sec}$$

Illustrative Problem 5.5   A 6-lb sledge hammer is moving at 40 ft/sec as it strikes a railroad spike. It is brought to rest in 0.025 sec. Find the average driving force on the spike.

Solution   Use the momentum equation, alternate form, Eq. (5.8):

$$Ft = \frac{w(v_2 - v_1)}{g}$$

Divide both sides by t and solve for F:

$$F = \frac{w(v_2 - v_1)}{gt}$$

Substitute:

$$F = \frac{6 \, lb \times (0 \, ft/sec - 40 \, ft/sec)}{32.2 \, ft/sec^2 \times 0.025 \, sec} = -298 \, lb$$

ans.

The minus sign merely means that the force reacting on the sledge hammer is in a direction opposite to its motion.

Illustrative Problem 5.6   A 15,000-lb rocket is to be fired vertically from rest. It is desired to give a velocity of 21,000 mi/hr during the burning time of the first-stage rocket motor, which is 35 sec. Calculate the average total thrust which the rocket motor must develop, in pounds. Assume that g is constant at 32.2 ft/sec^2 and ignore the loss of mass due to fuel consumed.

Solution   Using Eq. (5.8),

$$Ft = \frac{w(v_2 - v_1)}{g}$$

solve for F:

$$F = \frac{w(v_2 - v_1)}{gt}$$

Now,

21,000 mi/hr

$$= 30,800 \, ft/sec \left(\text{using the } \frac{44 \, ft/sec}{30 \, mi/hr} \text{ ratio} \right)$$

Substitute:

$$F = \frac{15,000 \, lb \times 30,800 \, ft/sec}{32.2 \, ft/sec^2 \times 35 \, sec}$$

$$= 410,000 \, lb \; \textit{thrust to accelerate}$$

Total thrust $= 410,000 \, lb + 15,000 \, lb$

$$= 425,000 \, lb \qquad \textit{ans.}$$

Illustrative Problem 5.7   A weather satellite in orbit has a mass of 200 kg and a velocity of 8500 m/sec. Find (*a*) its momentum and (*b*) its kinetic energy.

Solution
(*a*) The momentum, from Eq. (5.5), is

$$p = 200 \, kg \times 8500 \, m/sec$$

$$= 1.7 \times 10^6 \, kg\text{-}m/sec \qquad \textit{ans.}$$

(*b*) The kinetic energy, from Eq. (5.4) is

$$KE = \frac{200 \, kg \times (8500 \, m/sec)^2}{2}$$

$$= 7.23 \times 10^9 \, J \qquad \textit{ans.}$$

5.10   The Impact of Moving Fluids

The force of impact of a moving stream of water, or of a jet of steam, is of especial importance in the field of hydraulic- and steam-turbine design. Consider the following problem in this connection.

Illustrative Problem 5.8   If 250 cubic feet per second of water are flowing through the penstock of a turbine at a velocity of 25 ft/sec, calculate the average force (pounds) against the blades of the turbine rotor. Assume that the initial velocity of the water in the penstock (25 ft/sec) is reduced to zero by the turbine.

Solution Use Eq. (5.8):

$$Ft = \frac{w(v_2 - v_1)}{g}$$

and rewrite in the form

$$F = \frac{w(0 - v_1)}{tg}$$

You will note that w/t = weight of water per second striking the rotor blades. Now 1 ft³ water weighs 62.4 lb. Substituting given values, we obtain

$$F = \frac{(250 \text{ ft}^3/\text{sec} \times 62.4 \text{ lb/ft}^3) \times (-25 \text{ ft/sec})}{1 \text{ sec} \times 32.2 \text{ ft/sec}^2}$$

$$= -12{,}100 \text{ lb}$$

the force acting *back* on the water column. The equal and opposite force on the rotor blades is $+12{,}100$ lb. *ans.*

5.11 Momentum and Newton's Third Law

The third law of motion, the law of action and reaction, is directly involved with impulse and momentum. When a rifle is fired, the force of the rapidly expanding gases from the burning gunpowder accelerates the projectile and gives it *momentum* (see Fig. 5.9). During the same time that the bullet is acquiring momentum forward, the force of the expanding gases is *reacting* against the gun backward, and the momentum acquired by the bullet forward equals that acquired by the gun backward, since *equal and opposite forces act on each for the same length of time.* As an equation,

$$\underset{\text{(bullet)}}{Ft} = \underset{\text{(rifle)}}{-Ft}$$

therefore $\qquad m_B v_B = -m_G v_G \qquad$ (5.9)

or

$$\text{Bullet momentum} = -\text{rifle momentum}$$

These momentums will be in opposite directions, of course, as the *minus signs* indicate. Consider now a problem related to gunnery.

Illustrative Problem 5.9 An antiaircraft gun fires a 54-lb projectile with a muzzle velocity of 2650 ft/sec. The recoiling portion of the gun weighs 1450 lb. Find the theoretical velocity of recoil.

Solution *Velocity of recoil.* Making use of subscripts, let

$$w_p = \text{weight of projectile}$$
$$v_p = \text{velocity of projectile}$$
$$w_G = \text{weight of recoiling portion of gun}$$
$$-v_G = \text{velocity of recoil of gun}$$

Using these subscripts and Eq. (5.9),

$$-\frac{w_G}{g} v_G = +\frac{w_p}{g} v_p$$

$$v_G = -\frac{w_p}{w_G} v_p \qquad (g\text{'s cancel out})$$

Substituting,

$$v_G = -\frac{54 \text{ lb} \times 2650 \text{ ft/sec}}{1450 \text{ lb}} = -98.7 \text{ ft/sec} \quad \textit{ans.}$$

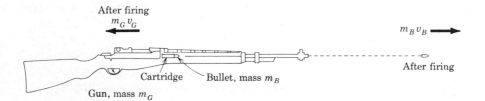

Fig. 5.9 Conservation of momentum. Before the gun is fired, the momentum of the gun-bullet system is zero. Immediately after firing, the bullet has acquired momentum $m_B v_B$ to the right, and the gun has acquired momentum $m_G v_G$ to the left. Newton's laws require that $m_B v_B = -m_G v_G$ after firing, or $m_B v_B + m_G v_G = 0$. Both before and after the firing, the momentum of the *gun-bullet system* is zero.

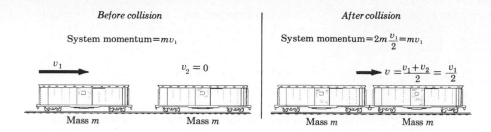

System momentum $= mv_1$

System momentum $= 2m\dfrac{v_1}{2} = mv_1$

v_1

$v_2 = 0$

$v = \dfrac{v_1 + v_2}{2} = \dfrac{v_1}{2}$

Mass m

Mass m

Mass m

Mass m

Fig. 5.10 Momentum is conserved at collision. Railway marshaling yards freely make use of this principle as long trains are made up from individual cars.

The negative sign indicates the direction of the recoil. Actually, the recoil would be controlled by strong springs or a hydraulic mechanism and would be much less than this theoretical value.

The example above (without any recoil control features) is one case of the more general statement known as *the law of conservation of momentum.* Each body, that is the projectile and the gun, underwent exactly the same change in momentum, but in *opposite directions.* Consequently there was no net gain or loss of momentum involved in firing the gun-projectile system.

In like manner, consider a moving freight car (Fig. 5.10) in a railway marshaling yard. Suppose it rolls slowly up to another car of *equal mass* which is at rest and collides with it. They lock together and both move forward, but with only half the speed of the original moving one. Since the same force acts on each car and they have the same mass, one will gain in velocity the amount which the other loses. Since their final common velocity is the same, it is therefore equal to one-half the velocity of the moving car before impact. Mathematically, we could state the momentum relations of the freight cars as follows, using symbols as indicated. Let

$v_1 =$ velocity of first car before impact
$v_2 = 0 =$ velocity of second car before impact
$m =$ mass of each car
$v_1/2 =$ velocity of cars (locked together) after impact

Then

$$\underset{\text{(first car)}}{mv_1} + \underset{\text{(second car)}}{mv_2} = \underset{\text{(total mass)}}{(m+m)} \times \underset{\substack{\text{velocity} \\ \text{after impact}}}{\left(\dfrac{v_1}{2}\right)}$$

Simplifying yields

$$mv_1 + 0 = \dfrac{2mv_1}{2}$$

$$mv_1 = mv_1$$

or

Total momentum before impact
= total momentum after impact

The law of conservation of momentum can be stated as follows:

The momentum of a system is conserved if the net external force on the system is zero.

5.12 Rockets and Jet Propulsion

The propulsion of jet aircraft, rockets, and missile-type weapons depends on the application of impulse and momentum principles and the law of action and reaction. True rockets and spacecraft carry both fuel and the oxidizing agent necessary to burn the fuel. They are therefore not limited in their operation to the earth's atmosphere. They can travel as far into space as their fuel and oxygen supply permits. Jet aircraft, on the other hand, depend on the earth's atmosphere for their oxygen supply, either scooping in sufficient quantities due to their high speed or

Fig. 5.11 Powerful rocket engine lifts Trident missile from launching pad in a Cape Canaveral test firing. The force which lifts and accelerates the rocket, giving it forward (and/or upward) momentum, is equal and opposite to the force, supplied by combustion, which accelerates the hot gases to the rear.

pumping it in with turbines driven by the jet engine (*turbojet* and *turbofan*). The forward thrust on the rocket or aircraft is obtained from the reaction force of the hot gases which are ejected backward at terrific speeds through the jet or *orifice* at the rear of the engine or rocket motor. It is to be emphasized that the forward momentum gained by the aircraft or spacecraft can be acquired only at the expense of backward momentum gained (which is forward momentum *lost*) by gases escaping to the rear. A familiar example may serve to illustrate this theory of propulsion. If you stand in a small rowboat and heave an anchor astern, what happens? The boat moves forward, of course, and its *momentum* forward equals that of the anchor backward. If you attempt to jump from a small boat to a dock you may find that the boat's backward velocity (the v part of $-mv$) is such that you land in the water and not on the dock.

An extended discussion of actual jet engines is given in a later chapter (see Sec. 16.21).

Elementary Principles of Rocket Propulsion Rockets of the type currently in use as vehicles for space exploration are powered by chemical energy. The fuel is usually a liquid like kerosene or hydrogen, and liquid oxygen or other oxidizing agent is carried by the rocket to oxidize (burn) the fuel. A high-temperature, high-velocity jet of gases roars out backward from the rocket motor, and the change in momentum of these gases rearward exerts an impulse forward on the rocket. These relationships can be analyzed as follows. Let

v_g = velocity of gases relative to motor
Δm = very small increment of mass of escaping gas
Δt = very small increment of time required for mass Δm to escape

Then

$$\frac{\Delta m}{\Delta t} = \text{rate at which mass is discharged (assumed constant)}$$

and,

$$v_g \frac{\Delta m}{\Delta t} = \text{time rate of change of momentum of escaping gases}$$

But (by Newton's second law) the time rate of change of momentum equals the force F given to the gases by the rocket motor, and (by Newton's third law) this force is equal and opposite to the force imparted to the rocket by the jet of escaping gas. Let F_R be the force imparted to the rocket. Then

$$F_R = -v_g \frac{\Delta m}{\Delta t} \qquad (5.10)$$

The minus sign indicates that the velocity of the hot gas is opposite in direction to the force on the rocket body. F_R is commonly called *thrust*.

Other forces, such as the gravitational weight of the rocket ($w = mg$) and air resistance, will be acting on the rocket during launch. The vector sum of these would be opposite in direction to F_R. Let F_G equal the aggregate of these forces. If M is the total mass of the rocket, Newton's second law gives

$$F_R - F_G = Ma \qquad (5.11)$$

M will vary during the flight as fuel is burned or as parts of the rocket are separated or jettisoned.

Illustrative Problem 5.10 A Saturn rocket is being used to put a space lab in orbit. The total gross weight is 6,200,000 lb. Its five engines produce a total thrust of 7,500,000 lb. The velocity of the hot gases shooting out of the jets is 6800 ft/sec. Find (*a*) the rate of ejection of hot gases in slugs per second, and (*b*) the acceleration of the rocket after 70 sec of flight.

Solution
(*a*) From Eq. (5.10)

$$\frac{\Delta m}{\Delta t} = \frac{F_R}{-v_g} = \frac{7,500,000 \text{ lb}}{-(-6800 \text{ ft/sec})}$$

$$= 1.1 \times 10^3 \frac{\text{lb-sec}}{\text{ft}}$$

$$= 1.1 \times 10^3 \text{ slugs/sec} \qquad ans.$$

(*b*) The loss of mass after 70 sec of firing is

$$70 \text{ sec} \times 1.1 \times 10^3 \text{ slugs/sec} = 77,000 \text{ slugs}$$

The remaining net mass of the vehicle at 70 sec is

$$M_{70} = \frac{6.2 \times 10^6 \text{ lb}}{32.2 \text{ ft/sec}^2} \text{(slugs)} - 77,000 \text{ slugs}$$

$$= 115,500 \text{ slugs}$$

To calculate the acceleration 70 sec after firing, neglect air resistance and assume the rocket is still close enough to the earth to equate $g = 32.2$ ft/sec². The weight at 70 sec is

$$w_{70} = mg = 115,500 \text{ slugs} \times 32.2 \text{ ft/sec}^2$$

$$= 3,720,000 \text{ lb}$$

The unbalanced force to cause acceleration is

$$F_{70} = 7,500,000 \text{ lb force} - 3,720,000 \text{ lb weight}$$

$$= 3,780,000 \text{ lb force}$$

From $F = ma$,

$$3,780,000 \text{ lb force} = 115,500 \text{ slugs} \times a$$

from which

$$a = 32.7 \text{ ft/sec}^2 \quad \text{or a little more than 1 "}g\text{"}$$
$$ans.$$

POWER

Industry has to be concerned with work accomplished and energy expended. But fully as important are the *time* required to do a job and the *rate* at which energy is transformed. *Power* is the technical term which takes the time element into consideration. Power is defined as *the rate at which work is done* or *the rate at which energy is transformed* or expended.

$$\text{Power} = \frac{\text{work}}{\text{time}} = \frac{\text{force} \times \text{distance}}{\text{time}}$$

$$P = \frac{W}{t} = \frac{Fs}{t} \qquad (5.12)$$

and since

$$\frac{s}{t} = \bar{v} \qquad (4.1)$$

$$P = F\bar{v} \qquad (5.13)$$

or $\qquad$ Power = force × velocity

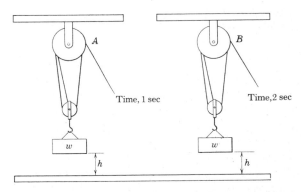

Fig. 5.12 Power equals work per unit time. Hoist *A* lifts weight *w* a distance *h* in 1 sec. It takes hoist *B* 2 sec to accomplish the same amount of work. Hoist *A* has twice the power of hoist *B*.

where both force and velocity have steady values and are in the same direction.

If one machine does a certain amount of work in half the time required by a second machine, the first has twice the power of the second (see Fig. 5.12). Time is often all important in production, and consequently engineers and technicians are continually interested, not only in what a machine or device can do, but also in the length of time it takes to do it.

5.13   Units of Power

In the SI-metric system, forces are measured in *newtons*, distances in *meters*, and time in *seconds*. The basic power unit is then

$$P_{(SI\ units)} = \frac{F\ (N) \times s\ (m)}{t\ (sec)}$$

$$= \frac{newton\text{-}meter}{sec}$$

$$= \frac{joule}{sec} \left(\frac{J}{sec} \right)$$

The unit *joule per second* has been given the name *watt* (symbol W), in honor of James Watt (1763–1819), who built the first commercially successful steam engine. The watt is a rather small unit of power, and the *kilowatt* (kW, 1000 watts) is the practical power unit in the metric system. The *megawatt* (mW, 1 million watts) and the *gigawatt* (gW, 1 billion watts) are also in common use. (See Table 5.2).

In the English (engineering) system, the basic power unit is defined as follows:

$$P_{(engineering\ units)} = \frac{F\ (lb) \times s\ (ft)}{t\ (sec)} = \frac{ft\text{-}lb}{sec}$$

This unit is read as *foot-pounds per second*. It is too small for practical purposes, however, and the *horsepower* (hp) is the unit commonly used to express the power rating of most types of machinery in the United States. Although in its original determination the horsepower was intended to indicate the rate at which the average horse could work, it is today defined in more exact terms. One horsepower equals 550 ft-lb/sec or 33,000 ft-lb/min by legal definition in the United States. In computing the horsepower ratings of machines either of these equivalents may be used.

$$1\ hp = 550\ ft\text{-}lb/sec = 33{,}000\ ft\text{-}lb/min \quad (5.14)$$

Modern automobile and truck engines are in the 25- to 400-hp range. Twenty-eight-cylinder reciprocating-type aircraft engines develop as much as 3500 hp. Fan-jet engines (see Fig. 16.1) for modern aircraft develop up to 20,000 hp, each horsepower delivering about 1 lb of thrust force. Huge rocket engines like the Saturn can develop several million horsepower at peak thrust. Steam turbines for marine and stationary use develop 100,000 hp or more. Electric motors can be made to develop 20,000 hp and more, but the trend in industrial plants is to use unit installations of small electric motors rather than a few large, powerful ones. Fractional-horsepower motors

Table 5.2   Units of power

English System	Metric System
ft-lb/sec	joule/sec = 1 watt
Horsepower (hp) = 550 ft-lb/sec	1 kilowatt (kW) = 1000 watts
= 33,000 ft-lb/min	1 megawatt (mW) = 1,000,000 watts
	1 gigawatt (gW) = 1 billion watts

Conversion
1 hp = 746 W = 0.746 kW
1 kW = 1.34 hp

and those up to 10 hp far outnumber the larger installations in modern industrial plants. Farm tractors range from 15 hp to 150 hp, rated at 2000 revolutions per minute (abbreviated rev/min or rpm), depending on function and auxiliary equipment to be powered by the tractor engine.

Illustrative Problem 5.11 On a construction job a hoist carries a 1000-kg bucket of concrete mix from ground level to the twelfth floor, a height of 50 m, in 1.5 min. What power P should the hoisting motor develop if there are no energy losses in the equipment?

Solution

$$P = \frac{W}{t} = \frac{F \times s}{t} = \frac{mg \times s}{t} \quad (5.12)$$

Substituting, we have

$$P = \frac{1000 \text{ kg} \times 9.81 \text{ m/sec}^2 \times 50 \text{ m}}{90 \text{ sec}}$$

$$= \frac{490,000 \text{ N-m}}{90 \text{ sec}}$$

$$= 5,440 \frac{\text{joules}}{\text{sec}} \text{ (watts)}$$

$$= 5.44 \text{ kW} \qquad ans.$$

Illustrative Problem 5.12 A 50-hp motor is installed to operate a water pump which lifts water from a depth of 85 ft (see Fig. 5.13). Assuming no power losses, how many gallons per minute (gal/min) will be pumped at full power?

Solution First find how many pounds per minute will be pumped, and then obtain gallons per minute from the relation, 1 gal water weighs 8.33 lb. Assuming no losses, write the equation which sets

Input power = output power

ft-lb/min input = ft-lb/min output

Let Q = pounds of water per minute that must be lifted 85 ft:

Fig. 5.13 Water pump driven by an electric motor. The pump itself is a turbine-type (see Chap. 11) of several stages, with the first stage set at the level of underground water. The motor shown is 50 hp, but 800- to 1500-hp motors are often used for deep-well irrigation pumps. (Pacific Gas and Electric Co.)

$$50 \text{ hp} \times \frac{33,000 \text{ ft-lb/min}}{1 \text{ hp}} = \frac{Q \text{ lb}}{\text{min}} \times 85 \text{ ft}$$

(Power input = power output)

Solve for Q:

$$Q = \frac{50 \text{ hp} \times \dfrac{33,000 \text{ ft-lb/min}}{1 \text{ hp}}}{85 \text{ ft}} = 19,400 \text{ lb/min}$$

Therefore the flow (gal/min) is

$$\frac{19,400 \text{ lb/min}}{8.33 \text{ lb/gal}} = 2330 \text{ gal/min} \qquad ans.$$

if 100% efficiency is assumed.

Illustrative Problem 5.13 An electric motor creates a tension of 3000 N in a hoisting cable and reels it in at the rate of 5 m/sec. What power is the motor supplying?

Solution From Eq. (5.12)

$$P = \frac{W}{t} = \frac{Fs}{t} = Fv$$

$$= 3000 \text{ N} \times 5 \text{ m/sec}$$

$$= 15{,}000 \text{ J/sec} = 15{,}000 \text{ W} = 15 \text{ kW} \quad ans.$$

5.14 Efficiency

All three of the foregoing examples assume that the entire power output of the motor is going into useful work accomplished. As a matter of fact, this is never the case, since in every actual machine, friction (see Chap. 6) causes some energy to be wasted in the form of heat. Any machine has certain bearings, gears, pulleys, and levers, whose frictional resistance causes a waste of energy. Consequently, *the useful work output of a machine is always less than the energy input to the machine.* Stated as an equation, the **Principle of Work** for actual machines is

Energy input − losses (due to friction
and other causes) = work output

The ratio of the work output of a machine to the energy input is called its *efficiency*. Since no real machine has a work output as great as the energy input to it, efficiencies are always less than 1 and are commonly expressed as percentages.

$$\% \text{ eff} = \frac{\text{work output of machine}}{\text{energy input to machine}} \times 100$$

$$= \frac{\text{power output}}{\text{power input}} \times 100 \qquad (5.15)$$

or, stated another way,

$$\text{Output} = \frac{\% \text{ eff} \times \text{input}}{100} \qquad (5.16)$$

Illustrative Problem 5.14 A farm pump motor uses electrical energy at the rate of 50 kW. The water level in the well is 200 ft below ground level. The actual water flow produced by the pump at ground level is 1000 gal/min. What is the efficiency of the motor-and-pump system?

Solution First, calculate the power output.

$$Q = \text{lb water/min} = 1000 \text{ gal/min} \times 8.33 \text{ lb/gal}$$

$$= 8330 \text{ lb/min}$$

$$\text{Power output} = \frac{8330 \text{ lb/min} \times 200 \text{ ft}}{\dfrac{33{,}000 \text{ ft-lb/min}}{1 \text{ hp}}}$$

$$= 50.5 \text{ hp}$$

$$\text{Power input} = 50 \text{ kW} \times 1.34 \text{ hp/kW}$$

$$= 67 \text{ hp} \quad \text{(see Table 5.2)}$$

From Eq. (5.15),

$$\text{Eff} = \frac{50.5 \text{ hp}}{67 \text{ hp}} \times 100 = 75 \text{ percent} \qquad ans.$$

5.15 Measuring Power

Manufacturers of all types of motors, engines, and turbines must subject their products to frequent test to be certain that they will actually perform up to the rated power on the nameplate. Power actually delivered to a drive shaft, flywheel, gearhead, pulley, or propeller is called *brake power* (bp) because it is ordinarily measured by a braking device such as that shown in Fig. 5.14. Such a test device is called a *Prony brake.* The engine or motor under test is coupled directly to the revolving drum (diameter *d*) of the Prony brake. The belt *B* is held on the drum *D* under tension by adjusting the spring balances at *A* and *C*. As the test engine turns at an observed speed in revolutions per second (rev/sec) the frictional force of the belt on the drum is measured by the difference between the readings of the balances. This difference between the readings is the force being applied by the test engine, and this force acts through the circumference of the drum during each revolution. The work done per second (power) is therefore

$$P = (F_1 - F_2) \times \pi d \times \text{rev/sec} \qquad (5.17)$$

If F_1 and F_2 are in newtons and d is in meters, power will be in joules/sec = watts.

Illustrative Problem 5.15 An electric motor is under test on a Prony brake. The spring balances read 840 N and 150 N, at 30 rev/sec. The drum diameter is 20 cm. Find the brake power (bp) of the motor.

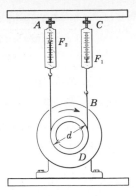

Fig. 5.14 The Prony brake in diagram form. An electric motor M is being tested for its horsepower rating.

Solution From Eq. (5.17)

$$P = (840 \text{ N} - 150 \text{ N}) \times \pi$$
$$\times 0.2 \text{ m} \times 30 \text{ rev/sec}$$
$$= 13{,}000 \text{ J/sec} \quad \text{(watts)}$$
$$= 13 \text{ kW} \qquad \qquad \qquad ans.$$

Fig. 5.15 A dynamometer in use in an auto test lab. The auto engine being tested, right, is driving an electric generator, center. The generator's magnetic field can be varied to absorb all the power the engine is capable of producing at any speed and under varying conditions. The test equipment can be programmed to simulate various operating conditions, such as altitude, temperature, load conditions and fuel quality. (Oldsmobile Division, General Motors Corp.)

When F_1 and F_2 are in pounds and d is in feet, P will be in foot-pounds per second. Therefore brake horsepower (bhp) will be given by the formula

$$\text{bhp} = \frac{(F_1 - F_2) \times \pi d \times \text{rev/sec}}{550} \qquad (5.18)$$

Illustrative Problem 5.16 A newly developed high-compression truck engine is on the test block. At 3500 rev/min by the tachometer, the spring balances of a Prony brake read 325 and 75 lb. The diameter of the drum is 2.0 ft. Find the brake horsepower of the engine.

Solution Using Eq. (5.18),

$$\text{bhp} = \frac{(F_1 - F_2) \times \pi d \times \text{rev/sec}}{550}$$

and substituting, changing rev/min to rev/sec,

bhp =

$$\frac{(325 \text{ lb} - 75 \text{ lb}) \times \pi \times 2 \text{ ft/rev} \times \dfrac{3500 \text{ rev/min}}{60 \text{ sec/min}}}{\dfrac{550 \text{ ft-lb/sec}}{1 \text{ hp}}}$$

$$= 167 \text{ hp} \qquad \qquad \qquad ans.$$

The Dynamometer Another common test device for measuring power output is the *dynamometer*. It is actually a heavy-duty electric generator whose magnetic field can be varied at the will of the operator, so that it can absorb all the power output of a test engine. The engine being tested is connected directly to the generator. The generator transforms mechanical energy into electric energy at a rate which can be measured by a suitable electric meter, in units of electric power (*watts*). Since 1 hp = 746 watts, and 1000 watts = 1 kW, it is easy to convert the measured output in kilowatts into horsepower. In actual practice, dynamometer dials are often calibrated to read horsepower directly. Figure 5.15 shows a dynamometer in use.

SUMMARY

Work is defined as *the product of a force times distance through which the force acts.* $W = Fs$ is the work equation.

Energy is the capacity to work. Energy may be classified as *kinetic* (KE), *the energy of motion,* or as *potential* (PE), *the energy of position or of molecular arrangement.* Potential energy from gravitational forces is termed *gravitational potential energy* (GPE).

The *principle of work* (or *work-energy principle*) states: *The total mechanical energy of a body or system at time t equals that at some former time t_0, plus any energy added, minus any "useless" work done in the time interval.*

Energy can neither be created nor destroyed, but may be changed in form. This is the general statement of the law of conservation of energy.

New sources of energy—solar, wind, tidal, geothermal, and nuclear—are being developed to supplement energy from coal, oil, and natural gas.

Momentum is defined as quantity of motion and equals mass times velocity. $p = mv$ is the momentum equation.

Impulse is defined as force times time. $Ft = mv_2 - mv_1$ is the impulse equation.

The law of conservation of momentum is a special case of Newton's third law of motion—the law of action and reaction. Momentum is conserved at impact if the net external force is zero.

Newton's laws of motion, especially the third law, action and reaction, are essential to the explanation of jet-propulsion and rocket motors.

The principle of work for actual machines is

Energy input − losses (frictional and other)
$$= \text{work output}$$

Power is the rate at which work is done. It is measured in watts, kilowatts, megawatts, or gigawatts in the metric system; and in foot-pounds per minute (or per second), and horsepower, in the English system.

Percent efficiency of a machine

$$= \frac{\text{power output}}{\text{power input}} \times 100\%$$

$$= \frac{\text{work output}}{\text{energy input}} \times 100\%$$

QUESTIONS AND EXERCISES

1. Define all of the following in technical terms, and after each definition give two examples chosen from actual practice in industrial, engineering, agricultural or medical fields: (*a*) work, (*b*) kinetic energy, (*c*) potential energy other than gravitational, (*d*) gravitational potential energy, (*e*) momentum, (*f*) impulse, (*g*) power, (*h*) efficiency.

2. Based on some library reading, write a short paper on "perpetual motion" machines. Describe and sketch one model of such a machine different from the one shown in Fig. 5.8. What happens when such a machine is applied to a load to do "useful" work? Could such a machine work on Jupiter where the force of gravity is many times greater than it is on earth?

3. Suppose your home is heated by electric radiant panels. Trace the energy transformations involved if the electricity is generated (*a*) at a steam plant with coal-fired boilers; (*b*) at a hydroelectric plant in the mountains. Start with the sun in each case.

4. Using the formula for kinetic energy, show that the SI-metric unit of energy is the newton-meter, or joule.

5. A right fielder makes a long, arching throw to home plate. The ball arrives a bit high, at the same level above the ground as his throwing hand. From energy principles, prove that its speed at home plate is the same as when it left his hand. (Neglect air friction.)

6. Team up with a friend and, with a good stop watch, measure each other's horsepower output, by running up a flight of stairs whose vertical rise can be measured. Weigh yourselves just before the trials.

7. What are some other forms of potential energy besides GPE? How can the amount of PE be determined? Give some ways in which machines can use these forms of PE.

8. In Fig. 5.10 the colliding railway cars absorbed the impact of collision without appreciable deformation of the cars. Momentum of the system was conserved. What about the kinetic energy of the two-car system? Was KE conserved at collision? If the cars were damaged at collision, would momentum be conserved? Explain your answers fully.

9. Explain in detail, with sketches, why the concept of "follow-through" is important in sports like golf, baseball, and tennis. What specific factor in the impulse equation is being maximized by a full "follow-through"?

PROBLEMS

Note: Assume that frictional and other losses are zero unless stipulated in the problem.

1. A hoist lifts an engine weighing 250 kilos to a mezzanine level 7 m from the floor. Calculate the work done in joules.

2. A switch engine exerts a force of 4 tons (drawbar pull) to overcome the rolling friction of a string of freight cars. How much work in foot-pounds is done in moving the cars 300 yd along a level track?

3. Calculate the potential energy per cubic meter of water just before it goes over the edge of a fall whose vertical drop is 200 m.

4. An industrial plant has a shop whose average power requirements are 500 kW. Assuming that it would be possible for human muscular energy to furnish this power, how many men would have to be employed if each man could work at a steady rate of 0.2 hp? (Contrast your answer with the 10 to 15 men which such a shop would probably employ to operate its modern machines.)

5. A 2-kg projectile is fired from an antiaircraft gun at a muzzle velocity of 500 m/sec. If the recoiling portion of the gun mechanism has a mass of 210 kg, what would be the velocity of uncontrolled recoil in meters per second?

6. A 65-ton freight car rolls along at 6 mph and collides with a stationary car weighing 50 tons, and the two lock and move off together. What is the velocity of the two-car train after the impact, in miles per hour?

7. A pendulum is 1.00 m long. If the bob is pulled aside until the cord makes an angle of 35° with the vertical and then released, with what speed does the bob pass through the rest position?

8. Assuming frictional forces are zero, determine the horsepower required to accelerate a 2500-lb auto from rest to 60 mi/hr in 10 sec, on a level road.

9. If a horse actually worked at the rate of 1 hp, how many pounds of ore could be lifted out of a mine shaft 300 ft deep in 1 min with suitable hoisting equipment?

10. If 1000 m³ of water is pumped from a reservoir into a tank 100 m higher than the reservoir in 2 min, find (*a*) the increase in potential energy and (*b*) the power delivered by the pump.

11. A construction elevator hauls a hopper of ready-mix concrete, total mass 1800 kg, to the pouring site 150 m above the ground in 30 sec. What is the power required?

12. An escalator in a department store is designed to carry 300 passengers per minute from the ground floor to the mezzanine, 10 m vertically higher. The design assumes an average weight per person of 70 kg. Allowing 30 percent loss for friction, calculate the power output which the driving motor must have.

13. An electric motor on a Prony brake-horsepower test turned at 1750 rpm. Spring balances attached to the belt read 54 and 16 lb. The diameter of the drum was 6 in. What horsepower did the motor develop?

14. A waterfall is 85 ft high, and 3000 ft³/sec flow over it driving a water wheel connected to an electric generator. If the overall efficiency is 22 percent, how many kilowatts does the generator develop?

15. An elevator is powered by a motor whose power input is 25 kW. The unbalanced mass of the elevator and its load is 1000 kg. If the overall efficiency of the hoisting system is 75 percent, how long does it take the elevator to rise to the fourteenth floor, a vertical distance of 90 m?

16. A force (thrust) of 4000 **N** is required to overcome road friction and air resistance in propelling an automobile at 80 km/hr. What power (kilowatts) must the engine develop?

17. An electric motor is found to have an output of 25 hp. An electric wattmeter is placed in the circuit, and reads 19.5 kW. What is the motor's efficiency?

18. A disabled truck is being towed at 25 mph. The tension (force) in the towing bar is known to be 350 lb. What horsepower is being expended to tow the truck?

19. The hammer of a pile driver weighs 750 lb and falls a distance of 20 ft before it hits the top of the pile. The blow drives the pile 8 in. deeper into the ground. What was the average force driving the pile, in pounds? (Solve from energy-work considerations.)

20. A ballistic missile has a total weight of 65,000 lb. Its solid-propellant engine gives a full-power thrust of 165,000 lb. The velocity of hot gases ejecting from the rocket motor is 6700 ft/sec. Find (*a*) the rate of gas ejection (in slugs per second) and (*b*) the acceleration 20 sec after firing. Assume air resistance negligible and *g* still = 32.2 ft/sec² at the 20-sec position. Rocket fired vertically.

21. A rocket engine ejects hot gases at a speed of 2000 m/sec. The rate of loss of mass, i.e., hot gases ejected, is 400 kg/sec. Find the thrust exerted on the rocket, in newtons.

22. A stream of water 2 in. in diameter, squirting from a fire hose, hits the side of a building nearby at a velocity of 25 ft/sec. What steady force does it apply to the wall?

23. A ball, mass 150 gm, moving at 30 m/sec, is struck by the bat and reverses direction, leaving the bat at 50 m/sec. What was the impulse applied to the ball?

24. A 50-gm bullet fired from a rifle at 750 m/sec embeds itself in a wooden block that is free to slide on the floor. The block's mass is 1 kg and the constant frictional force between block and floor is 20 N. How far will the block slide?

25. A space vehicle is on its way to Mars, at a point where it is still subject to some earth gravity, at a steady speed of 42,000 km/hr. Instruments aboard indicate that the thrust being delivered by the rocket motor is 5000 N. What power is the rocket motor delivering, in megawatts?

26. The roller coaster of Fig. 5.16 is assumed to roll without friction. If it passes point T at 5 m/sec, find its speed at point B and at point X.

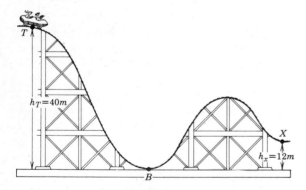

Fig. 5.16 Rollercoaster sketch for Problem 5.26.

27. A farm tractor pulls a gang plow at a steady speed of 3 mi/hr. If the engine is developing 80 hp, what is the "drawbar pull" on the plow, in pounds?

6 Basic Machines— Friction

We live in an age of computers, lasers, electronics, and nuclear energy. These somewhat glamorous industries are in the headlines nearly every day. But ordinary machines like pumps, levers, gears, cams, and pulleys actually have more effect on the lives of most of us than do the new "exotic" discoveries. And, for that matter, the exotic devices could not be manufactured nor would many of them operate without the basic machines that you will study in this chapter.

Technicians and engineers spend a great deal of their working time designing, adjusting, operating, and repairing machines. Your present or future job may necessitate a thorough familiarity with basic machines and machine components.

The word *machine* means different things to different people. A technician or an engineer may associate the word with an engine or a milling machine, or a turbine, while the businessman would think of an accounting machine, or an electric typewriter, or a stock-market ticker-tape machine. In physics, as usual, we have a very specific definition:

> *A machine is a device for the advantageous application of a force.*

BASIC MACHINES

The complex business and industrial machines mentioned above are merely clever combinations of what we shall call *basic machines*. Despite hundreds of different kinds of complex machines in use, there are actually only three basic machines, the *lever*, the *inclined plane*, and the *hydraulic press*. We shall consider the first two of these in this chapter (the hydraulic press will be explained in Chap. 11), and then see how they can be altered slightly to produce certain *simple machines* such as the pulley and the block and tackle; the wheel and axle, or windlass; the wedge and the cam; the screw jack, the gear, and gear combinations. We shall also investigate the effects of friction on machines and discuss a few ways in which simple machines can be combined to form *compound machines*.

All the basic and simple machines have certain common properties. It takes a force input to operate them, and there is an output force. There is an *energy input* to them, and they yield an *energy output*. Also, there are *energy losses* within the machine due to friction or other causes. In analyzing basic machines we will sometimes *assume* that these losses are zero, but actually they never are. In some machines they are so small, however, that for practical purposes they can be neglected without introducing serious error into the machine design.

Basic and simple machines obey the law of conservation of energy, namely:

Energy input = energy output

    + energy losses (including "useless work")

6.1 The Lever

A lever consists of any rigid bar or rod (it may be straight or bent) arranged in such a way that it can pivot about some definite point, so that a resistance w can be overcome by a force F. Figure 6.1a illustrates one such arrangement in which a lever in the form of a *crowbar* is being used to tip a heavy packing crate. The diagram of (b) shows the relationships involved, with the pivot point, which is called the *fulcrum*. If we designate the distance from the fulcrum to the resistance (load) as the *resistance arm r* and the distance from the fulcrum to the effort as the *effort arm e*, then the *law of the lever* can be stated thus:

Effort × effort distance

= resistance × resistance distance

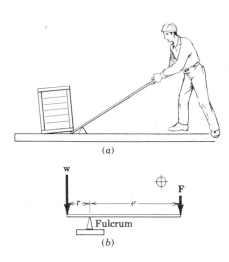

(a)

(b)

Fig. 6.1 The principle of the lever. (*a*) A lever used to move a heavy object. (*b*) The law of the lever is *Fe = wr*.

The mathematical formula for the law of the lever is

$$Fe = wr \qquad (6.1)$$

Another way of stating the law of the lever is evident from analyzing Eq. (6.1). *When a lever is in balance, the forces applied are inversely propor-*tional to their distances from the fulcrum, or mathematically $F/w = r/e$.

6.2 Mechanical Advantage

One of the important reasons for using a lever is to overcome a large resistance by exerting a relatively small effort. When used for this purpose, a lever *multiplies force*, and the ratio

$$\frac{\text{force overcome (resistance)}}{\text{force applied (effort)}}$$

is defined as the *mechanical advantage* of the lever. In fact, for *any machine* the *actual mechanical advantage* (AMA) is defined as *the number of times which the machine multiplies the force applied*; or, in other words,

$$\text{AMA} = \frac{\text{force output of the machine}}{\text{force input to the machine}} \qquad (6.2)$$

For the lever, Eq. (6.1) can be rewritten in the form

$$\frac{w}{F} = \frac{e}{r} \qquad (6.3)$$

The left side of this equation is seen to be the ratio of (force output)/(force input) and is therefore (from Eq. 6.2) apparently equal to the actual mechanical advantage. The right side is the ratio of the effort arm of the lever to the resistance arm. This ratio (e/r) is called the *theoretical mechanical advantage* (TMA). Note that Eq. (6.3) states, in effect, that for levers, theoretical mechanical advantage equals actual mechanical advantage. This is true only in situations where frictional and other losses are so small that they can be considered negligible. If there is friction, the effort force F, will have to be enough larger than its theoretical value (calculated from Eq. 6.3) to overcome the frictional resistance.

The idea of mechanical advantage has been introduced here in connection with the lever, but it is applicable to all simple machines. Remember that mechanical advantage is a ratio of *forces*, not a ratio of *energy output* to *energy input*. The latter ratio measures *efficiency*, and we will deal with that later in this chapter.

6.3 The Lever and the Principle of Work

Although *force* can be multiplied by levers such as the crowbar, and *speed* or *distance* can be multiplied by such lever arrangements as valve rocker arms, it is important to realize that in no case can *work* or *energy* be multiplied by a lever or any other known device! The principle of work was discussed in Chap. 5, and it holds true for levers and all other machines. Reference to Fig. 6.2 will show this to be so. The lever AB is 3 ft long, the effort arm e is 2 ft and the resistance arm r is 1 ft. An input force F_i, of 10 lb should therefore balance a resistance w of 20 lb, since from Eq. (6.1)

$$10 \text{ lb} \times 2 \text{ ft} = 20 \text{ lb} \times 1 \text{ ft} = 20 \text{ lb-ft}$$

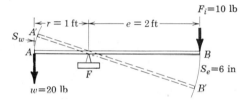

Fig. 6.2 The lever and the principle of work.

The TMA of this lever would be 2. Suppose we push down on the end B through a small arc ($S_e = 6$ in.) to the point B'. If the arc S_e is small compared to the length of the lever, BB' approximates a straight line. From simple geometry it is seen that the end A will have moved upward along its arc a distance $S_w = 3$ in. to the point A'. Since

$$\text{Work} = \text{force} \times \text{distance}$$

the work done by the input force F_i is

$$\text{Work input} = 10 \text{ lb} \times 6 \text{ in.} = 60 \text{ in.-lb}$$

The work accomplished at the left (A) end is

$$\text{Work output} = 20 \text{ lb} \times 3 \text{ in.} = 60 \text{ in.-lb}$$

Work output equals work input (assuming no losses), and we see that the principle of work holds true for the lever. *Force* was multiplied, but *distance* was sacrificed, so there was no net gain or loss of energy.

In contrast to the lever of Fig. 6.2, some levers are designed to multiply distance and speed at the sacrifice of force. See if you can name some levers of this type.

6.4 Examples and Types of Levers

The illustration of Fig. 6.1 shows the fulcrum located between the effort and the resistance. This arrangement is called a *Class I* lever. Other examples are oars, valve rocker arms, scissors, pliers, and "walking beams" of petroleum-pumping units (see Fig. 6.3a). It should be evident that there are two other possible combinations, i.e., resistance between effort and fulcrum (*Class II* levers), and effort between resistance and fulcrum (*Class III* levers). Figure 6.3b illustrates a Class II lever, and Fig. 6.3c a Class III lever. See how many more of each type you can name, particularly those that are found in industrial machinery.

Illustrative Problem 6.1 The valve spring of a certain engine holds the valve closed with a force of 70 lb. The rocker-arm arrangement is shown in the diagram of Fig. 6.4. What input force must the push rod exert to crack open the valve?

Solution Using Eq. (6.1) and noting that it is to be solved for the effort F_i, we obtain

$$F_i = \frac{wr}{e}$$

$$= \frac{70 \text{ lb} \times 4.5 \text{ in.}}{2.1 \text{ in.}}$$

$$= 150 \text{ lb} \qquad\qquad ans.$$

Note that this lever multiplies distance moved, and requires an input force larger than the output (resistance) force.

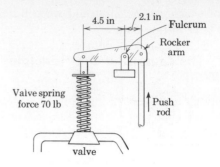

Fig. 6.4 Schematic diagram of a rocker-arm arrangement for an overhead-valve engine. A Class I lever.

6.5 Other Levers

The bars or rods of many levers are not perpendicular to the forces applied but, owing to some space requirement, are inclined or curved. They are, of course, still *rigid* members. Consider the wheelbarrow of Fig. 6.5. The load w is assumed concentrated at the center of gravity, (c.g.). The effort arm is the *perpendicular distance* from the fulcrum to the *line of action* of the input force, F_i. The shape of the wheelbarrow handles does not affect the "leverage" obtained, and $F_i e = wr$ as before.

Levers are in use in nearly all types of industrial machinery. Tractors, hoists, cranes, engines, pumps, presses, textile looms, bulldozers—the list of machines in which levers are involved is endless. Examine critically every machine you come in contact with to see how many levers you can identify.

6.6 The Inclined Plane

If you had to move a heavy safe from the ground to a truck bed h ft high without the aid of hoisting equipment, you might obtain some long planks and roll the safe up the incline, as shown

(a)

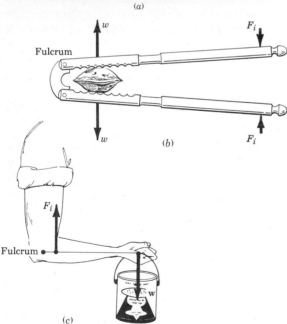

Fig. 6.3 Levers. (*a*) A Class I lever, the "walking beam" of a petroleum pumping unit. (Standard Oil Co. of California.) (*b*) Class II lever, the nutcracker. (*c*) Class III lever, the human forearm. The effort F_i is applied along the tendons of the upper forearm by the biceps muscle. The elbow is the fulcrum.

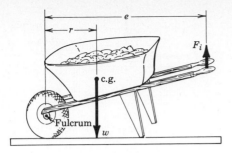

Fig. 6.5 The wheelbarrow, an example of a Class II lever.

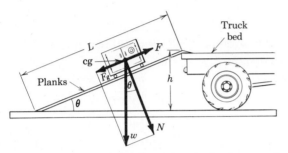

Fig. 6.6 The inclined plane used to load a heavy safe on a truck. The vector diagram is also shown.

in Fig. 6.6. If the planks available were L ft long, the inclined plane would be at an angle $\theta = \sin^{-1}(h/L)$, as shown in the diagram. The weight of the safe is indicated by the vector **w**, a force acting vertically downward as shown. This force has two measurable effects: (1) It tends to cause the safe to roll back down the planks, and (2) it causes a force to be exerted perpendicularly against the planks. These two *component* forces are shown in the diagram by the vectors $\mathbf{F}_g$ and $\mathbf{N}$, respectively. Note that, considered as a parallelogram of forces, **w** is the resultant of $\mathbf{N}$ and $\mathbf{F}_g$, and $\mathbf{N}$ and $\mathbf{F}_g$ are mutually perpendicular components of **w**. From trigonometry, N, the force pressing normally (perpendicularly) against the planks, is

$$N = w \cos \theta \qquad (6.4)$$

and the force with which the safe tends to roll back down the plane is

$$F_g = w \sin \theta \qquad (6.5)$$

If friction is neglected (we shall consider its effects later), the force F required to pull the safe up the plane *at uniform speed* is equal and opposite to the force with which it tends to roll down the plane. That is,

$$F = F_g = w \sin \theta$$

But $\sin \theta = h/L$, and therefore

$$F = w\frac{h}{L} \qquad (6.6)$$

The *theoretical* mechanical advantage (TMA) of the inclined plane is therefore

$$\underset{\text{(inclined plane)}}{\text{TMA}} = \frac{w}{F} = \frac{L}{h} \qquad (6.7)$$

The same result can also be obtained from the *principle of work*. Neglecting friction, the work done in pushing the safe up the plane is equal to the work which would be required if the safe were merely lifted through the vertical distance h. Stated mathematically,

$$FL = wh$$

Putting this equation in the form of a proportion, we obtain

$$\frac{w}{F} = \frac{L}{h}$$

which is the same result for TMA as obtained above from the vector analysis of the problem.

Illustrative Problem 6.2 In constructing a dam, heavy hoppers of concrete mix are slowly pulled up an incline (assumed frictionless), which rises 34 m for every 100 m along the incline. If the loaded hoppers have a mass of 1600 kg, what is the tension in the hoisting cable?

Solution Note that $h/L = \frac{34}{100}$. Substitution in Eq. (6.6) gives

$$F = w\frac{h}{L} = \frac{mgh}{L}$$
$$= \frac{1600 \text{ kg} \times 9.81 \text{ m/sec}^2 \times 34 \text{ m}}{100 \text{ m}} = 5340 \text{ N}$$

ans.

6.7 Roads Are Sometimes Inclined Planes

Highways in mountainous regions are laid out with curves and switchbacks. The reason is to prevent the steepness (*grade*) from exceeding a certain maximum specified by the highway engineer. Civil engineers measure steepness in terms of what they call *percent grade*. Distances are measured horizontally and vertically instead of along the incline itself. Consequently a 1 percent grade is an incline which rises 1 ft in 100 horizontal feet. Note from the diagram of Fig. 6.7 (not to scale) that the grade is $\frac{6}{100}$ or 6 percent. Also note that $\tan \theta = \frac{6}{100} = 0.06$. Thus θ is about 3.5°.

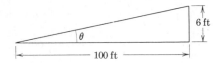

Fig. 6.7 The meaning of "percent grade" (not to scale).

FRICTION

6.8 Friction and Machines

Rarely, if ever, is friction absent when one object moves or slides over another. The surface of any material, no matter how smooth it may look or feel, is actually full of irregularities which oppose the sliding of some other surface. This force of opposition is called *friction. Friction is a force which always acts to oppose the relative motion of two sliding surfaces in contact.*

In the light of this definition of friction, and reasoning from *Newton's third law—action and reaction*—it follows that if there is no force tending to cause relative motion, then there is no force of friction. Consider two bodies, *A* and *B* (Fig. 6.8), pressed together by a force *N*. (The letter *N* is used to indicate a force *normal* (i.e. perpendicular) to the surfaces in contact). As a force *F* is applied to body *A*, tending to make it slide over body *B*, the force of *static friction* F_s, acting opposite to *F* and equal to *F*, increases as *F* increases

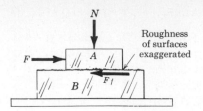

Fig. 6.8 The forces involved in sliding friction.

until motion actually impends. F_s has its maximum or limiting value just as *A* begins to slide on *B*, and F_s is zero when *F* is zero. *After* motion begins the force of friction decreases somewhat, and this reduced value is called the force of *sliding friction,* F_f.

6.9 The Nature of Friction

Friction has been investigated for more than 200 years, but there still remain many uncertainties and problems for current research. For *dry surfaces* the following laws of friction are applicable within wide limits:

1. The force of limiting or static friction (F_s) is not directly related to the area of contact but is directly related to the force pressing the two sliding surfaces together. Since this force acts perpendicularly on both surfaces, it is called the *normal force.*

2. When motion begins, the friction force (F_f) is usually somewhat less than the maximum value it has just before motion begins. In other words, *sliding friction is less than limiting or static friction.* The relative velocity of the sliding surfaces does not seem to affect the value of sliding friction for a considerable range of relatively low velocities. However, for high velocities, friction *decreases* as speed *increases.* It is a matter of common experience that the "stopping effect" of auto brakes is considerably less at high road speeds than at moderate speeds.

3. The force of friction is definitely related to the nature of the two surfaces sliding across each other. This constant of the sliding surfaces is called the *coefficient of sliding friction.* Values for almost any combination of surfaces can be found in engineering handbooks.

6.10 Analyzing Frictional Forces

We have defined three quantities involved in frictional problems, the force of static friction F_s, the force of sliding friction F_f, and the normal (perpendicular) force, which we designated by the letter N. A fourth factor, *coefficient of friction*, is designated by the Greek letter μ (mu) and is defined as

$$\mu = \frac{\text{force of sliding friction}}{\text{normal force}}$$

Mathematically the coefficient of *sliding* friction,

$$\mu = \frac{F_f}{N} \qquad (6.8)$$

and the coefficient of *static* or *limiting* friction,

$$\mu_s = \frac{F_s}{N} \qquad (6.8')$$

Illustrative Problem 6.3 A heavy machine is being skidded across a factory floor. If the machine weighs 1800 lb and the coefficient of sliding friction (steel on greased concrete) is 0.120, find the horizontal force required to slide it.

Solution Solve Eq. (6.8) for F_f and substitute given values.

$$F_f = \mu N = 0.120 \times 1800 \text{ lb}$$

$$F = 216 \text{ lb} \qquad \qquad \textit{ans.}$$

Illustrative Problem 6.4 A generator weighing 2600 lb is mounted on skids and is being pulled at

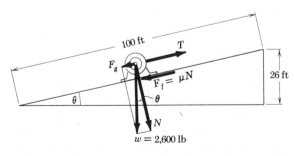

$$w = 2{,}600 \text{ lb}$$

Fig. 6.9 Vector analysis of an inclined plane with friction.

constant velocity up a ramp which rises 26 ft for every 100 ft along the incline. $\mu = 0.27$. What is the tension in the hoisting cable if the cable is parallel to the incline?

Solution
First method (using vectors, trigonometry, and the principles of equilibrium). Sketch the relations of the problem as shown in Fig. 6.9, showing the forces as vectors. Applying the principles of equilibrium, it is evident that the algebraic sum of the forces on the skid in any direction must be equal to zero. Consequently, in the line of the ramp,

$$\Sigma F = T - F_g - F_f = 0$$

From the vector diagram,

$$N = w \cos \theta$$

And from Eq. (6.8),

$$F_f = \mu N = \mu w \cos \theta$$

From the diagram,

$$F_g = w \sin \theta$$

Therefore

$$T = w \sin \theta + \mu w \cos \theta$$

$$\theta = \sin^{-1} \tfrac{26}{100} = 15.1° \quad \text{and} \quad \cos \theta = 0.966$$

Transposing and substituting numerical values,

$$T = 2600 \text{ lb} \times \tfrac{26}{100} + 0.27 \times 2600 \text{ lb} \times 0.966$$

$$= 676 \text{ lb} + 678 \text{ lb} = 1354 \text{ lb} \qquad \textit{ans.}$$

Second method (using the principle of work). The principle of work implies that the total work done by the hoisting cable tension equals the net work done against the force of gravity (increasing the GPE) plus the "wasted" work done against friction. Writing this relation in equation form, we have

$$100T = 2600 \text{ lb} \times 26 \text{ ft} + 100 \text{ ft} \times F_f$$

or $\qquad T = 26 \text{ lb} \times 26 \text{ ft} + F_f \times 1 \text{ ft}$

But $\quad F_f = \mu w \cos \theta$

$$= 0.27 \times 2600 \text{ lb} \times 0.966 = 678 \text{ lb}$$

Table 6.1 Approximate coefficients of sliding friction μ and static friction, μ_s (Average values for dry surfaces unless otherwise noted)

Materials	μ	μ_s	Materials	μ	μ_s
Hardwood on hardwood	0.25	0.5	Steel on concrete (smooth)	0.30	0.5
Rubber on dry concrete	0.75	1.0	Steel on babbit metal (dry)	0.14	0.20
Rubber on wet concrete	0.50	0.75	Steel on steel (oiled)	0.03	0.05
Metal on hardwood (varies)	0.40	0.6	Steel runners on ice	0.04	0.10 (depends on temperature)
Steel on steel (smooth)	0.10	0.15			

Substituting and evaluating, we have

$$T = 676 \text{ lb} + 678 \text{ lb} = 1354 \text{ lb} \qquad ans.$$

as before.

It is again pointed out that the principle of work constitutes a powerful tool in the analysis and solution of problems in mechanics.

Table 6.1 gives some average values of μ, the coefficient of sliding friction, and of μ_s, the coefficient of static friction.

6.11 Friction—Good or Bad

Friction is advantageous or disadvantageous depending entirely on the result desired. Certainly friction between rubber and concrete is desirable. And it would be very difficult even to walk if friction between the ground and shoe soles were absent. All automotive transportation is based on friction, and tires and roads are designed with a view to assure its presence. On the other hand, and again using the automobile as an example, the friction between pistons and cylinder walls is undesirable, and every effort is made to minimize the effects. The lubrication industry spends millions of dollars yearly in research to develop better lubricants to minimize friction in machines. The theory of lubrication is quite complex, but it is sufficient to say here that the presence of a film of oil or grease between two surfaces prevents them from sliding on themselves (if they are reasonably smooth to begin with) and substitutes fluid friction for the relatively greater friction which exists between solid surfaces.

6.12 Rolling Friction

The invention of the wheel was an important event in the history of man. Since the first wheels were probably cross sections cut from logs, possibly the idea for the wheel occurred when it was noticed how much easier it is to roll a log than to slide it. Be that as it may, man has known for centuries that *rolling friction* is much less than *sliding friction*. In addition to providing wheels on which to roll vehicles, modern industry provides ball and roller bearings for machine components in place of the older type *sleeve* bearings, whenever the factors of long life, efficiency, and cost justify the practice.

Analysis of the rolling of a wheel or ball on a hard surface (Fig. 6.10a) reveals that theoretically at least, the contact is a single point. Further, if the wheel or ball rolls without slipping, there is no relative motion between the point *A* (on the wheel) and the point *A* (on the surface) at the instant of contact. Consequently there is, theoretically, *no sliding friction*. Practically, however, either the wheel or the surface (or both) flattens a bit under the force of contact, and a situation which is shown in exaggerated proportions in Fig. 6.10b occurs. The surface indentation makes it necessary for the wheel (or ball) to roll slightly uphill all the time, and the flattening of the wheel itself implies that there is an appreciable *surface* of contact rather than a *point* of contact. The deformation is, of course, extremely small and only temporary. Consequently, there is in reality some relative motion between wheel and surface, and therefore some sliding friction involved.

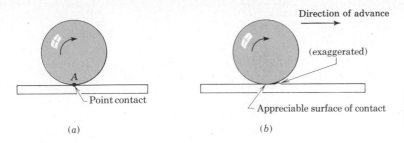

(a)

(exaggerated)

Point contact

Appreciable surface of contact

Fig. 6.10 Rolling friction. (*a*) Theoretical point contact. (*b*) Actual situation of contact with some temporary deformation.

The Coefficient of Rolling Friction Coefficients of rolling friction (μ_r) are subject to many variable conditions of load, wheel (or ball) diameter, rotation, speed, etc., but the student can form some idea of their relative magnitude from the values in Table 6.2. As before, $\mu_r = F_f / N$.

Table 6.2 Coefficient of rolling friction μ_r for dry surfaces

Situation	μ_r
Car wheels on rails	0.0045
Steel ball bearings on steel	0.0025
Steel roller bearings on steel	0.0003
Rubber tires on concrete	0.04

6.13 Sleeve Bearings vs. Ball (or Roller) Bearings

Bearings may be classified according to their use as (1) *guide bearings*, such as crankshaft main bearings or line-shaft bearings; (2) *journal bearings*, such as the bearings on a crankshaft "throw" or on a locomotive driving wheel; and (3) *thrust bearings*. A thrust bearing is one that is mounted in such a way that the bearing has to support the weight of a pump or rotor or counteract the thrust of a ship's screw or an airplane propeller. Depending on the factors previously referred to (load, speed, cost, etc.), all three of these classes of bearings are made in both *sleeve* form and in *ball* (or *roller*) form. The basic difference in performance can be seen from an analysis of Fig. 6.11, which shows a guide bearing in sleeve form (*a*) and in ball (or roller) form (*b*).

The ball (or roller) form will have less friction, is easier to lubricate, and keeps the shaft more nearly centered with less wear and greater efficiency. The initial cost of ball or roller bearings is several times the cost of sleeve bearings. Parts (*c*) and (*d*) of Fig. 6.11 show two types of commercially manufactured bearings.

6.14 Fluid Friction

Liquids flowing through pipes are subject to frictional forces between themselves and the pipe wall and to frictional forces (called *viscosity*) between the molecules of the liquid itself. Air or gases flowing in pipes or ducts also encounter frictional forces. Ships moving through or over water and aircraft moving through air experience frictional forces which we call *drag*. Both air resistance and liquid resistance are examples of *fluid friction*. Fluid friction tends to be much less pronounced than friction between solids. Pulling a raft through, or on, water requires less force than sliding the same raft over even a very smooth solid surface. And, as one might expect, frictional forces caused by air or gases are much less than those caused by liquids. Objects moving through air encounter very little friction at low speeds, but as speed increases markedly, air friction becomes a powerful retarding force. Friction with atmospheric air, for example, generates such heat that meteorites burn up before they hit the earth, causing the shooting stars often seen in the August sky. Spacecraft of the *Apollo* series were equipped with a special heat shield to prevent their being consumed by the heat of friction as they returned to the earth's atmosphere at speeds in excess of 30,000 mph.

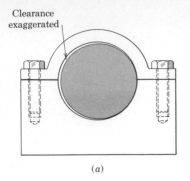

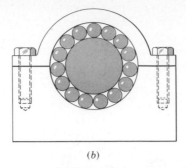

Clearance
exaggerated

(a)

(b)

Fig. 6.11 Guide bearings. (*a*) Sleeve bearing, showing shaft touching bearing over a considerable area. Sliding friction occurs. (*b*) Ball (or roller) bearing, showing shaft rolling on balls and balls rolling in the outer race. Only rolling friction is involved. Some commercial bearings, such as (*c*) ball-type guide bearing, and (*d*) roller-thrust bearing. (SKF Industries.)

(c)

(b)

Surface-Effect Machines Several modern developments make use of the comparatively low frictional forces associated with air when moderate speeds are concerned. One of these developments is in the field of transportation. *Surface-effect machines*, sometimes called *air-cushion vehicles*, or *Hovercraft*, have the ability to hover just above the surface of the ground or water and move with great rapidity on the cushion of air which is continually being forced out between the bottom of the vehicle and the ground or water surface. Figure 6.12 shows a surface-effect ship (SES) in operation.

A second example of the air-cushion effect comes from the physics laboratory itself. Apparatus for the laboratory verification of Newton's second law (Sec. 4.18) depended, until very recently, on observing the acceleration of a small car *rolling* on a steel track under the action of a known, unbalanced force. Results with such apparatus were usually in error from 3 to 5

Fig. 6.12 Experimental surface-effect ship in a speed trial. The SES-100B floats on a cushion or "bubble" of air, thus eliminating friction between water and hull. Speeds in excess of 100 miles per hour have been attained. Such vessels show great promise for operations in relatively mild seas, but their potential for heavy seas is yet to be demonstrated. (Official U.S. Navy photograph.)

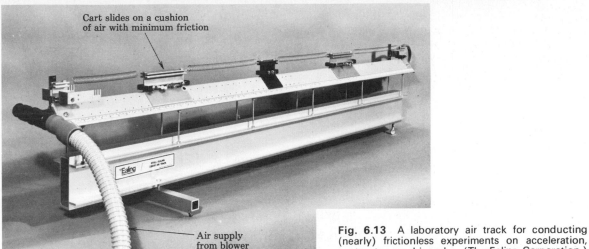

Cart slides on a cushion
of air with minimum friction

Air supply
from blower

Fig. 6.13 A laboratory air track for conducting (nearly) frictionless experiments on acceleration, momentum, and impulse. (The Ealing Corporation.)

percent because of friction. Now an apparatus which utilizes the air cushion principle, is available. The air cushion between the car and the track is provided by a blower much like an ordinary vacuum cleaner operated on the pressure side. The track has multiple pinholes through which the air emerges and forms the cushion between the car and the track. With such apparatus students can now obtain results in the first-year physics laboratory which heretofore were obtainable only in research laboratories. (See Fig. 6.13.)

6.15 The Effect of Friction on Machines

Force is required to overcome friction. When a force acts through a distance, work is done and energy must be expended to do this work. Therefore friction always results in energy expended, which is "wasted" in the form of heat within the machine. The student should recall the statement of the *principle of work* in Chap. 5:

Work output of a machine = energy input
$$\qquad\qquad\qquad - \text{losses (friction and other)}$$

and also the definition of efficiency,

$$\% \text{ eff} = \frac{\text{work output}}{\text{work input}} \times 100 \quad (5.15)$$

For the purposes of the present discussion it will be assumed that the only losses within machines are frictional losses, since we are not concerned with heat engines at this point.

6.16 Mechanical Advantage and Efficiency

We have defined the actual mechanical advantage (AMA) of a machine as the ratio (force overcome)/(force applied). Where frictional losses are small, as in most levers, the actual mechanical advantage is nearly equal to the theoretical mechanical advantage to be expected from the physical design of the machine. On the inclined plane, however, and in some other machines, friction is an item of considerable importance and cannot be neglected. In such cases, actual mechanical advantage is considerably less than theoretical mechanical advantage. These relationships are clarified in Fig. 6.14. *Theoretically* (by Eq. 6.6), we should be able to pull block *B* up the plane at constant velocity with a force $F = wh/L$. However, a greater force than *F* is actually required because of the friction between the block and the plane. This greater force actually required to pull the block up the plane is shown as *F'* in the diagram. The *actual* mechani-

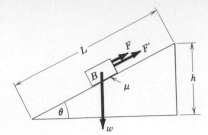

Fig. 6.14 Efficiency and the inclined plane.

cal advantage is then

$$\text{AMA} = \frac{w}{F'} \qquad (6.9)$$

The *net work output* in moving the block *B* through the vertical height *h* is, regardless of the path used, merely the product of the weight of the block *w* and the height *h*, or

$$\text{Work output} = w \times h$$

The *work input* in moving the block up the plane is equal to the force applied *F'* times the distance up the plane *L*, or

$$\text{Work input} = F' \times L$$

Therefore the efficiency of an inclined plane is given by

$$\text{Eff} = \frac{\text{work output}}{\text{work input}} = \frac{wh}{F'L} \qquad (6.10)$$

Now, noting that

$$\frac{\text{AMA}}{\text{TMA}} = \frac{w/F'}{L/h} = \frac{wh}{F'L} \qquad (6.11)$$

we can relate efficiency and mechanical advantage, from Eqs. (6.10) and (6.11), to obtain the basic relation

$$\text{Eff} = \frac{\text{AMA}}{\text{TMA}} \qquad (6.12)$$

Though derived from inclined-plane considerations, Eq. (6.12) gives the efficiency of any of the basic or simple machines.

Illustrative Problem 6.5 An inclined plane made of smooth planks is used to load heavy boxes onto a truck. The height of the truck bed is 4 ft, and the length of the planks is 10 ft. The coefficient of sliding friction μ between the boxes and plank surface is 0.30. Find (*a*) the theoretical mechanical advantage, (*b*) the actual mechanical advantage, and (*c*) the efficiency of this inclined plane.

Solution
(*a*) From Eq. (6.7),

$$\text{TMA} = \frac{L}{h} = \frac{10 \text{ ft}}{4 \text{ ft}} = 2.5 \qquad \textit{ans.}$$

(*b*) From Eq. (6.9),

$$\text{AMA} = \frac{w}{F'}$$

But, (see Fig. 6.9)

$$F' = F_g + F_f = w \sin\theta + \mu w \cos\theta$$
$$= w(\sin\theta + 0.3 \cos\theta)$$

From the given dimensions, $\sin\theta = 0.40$, from which $\theta = 23.6°$ and $\cos\theta = 0.917$. Therefore,

$$F' = w(0.40 + 0.3 \times 0.917) = 0.642w$$

$$\text{AMA} = \frac{w}{F'} = \frac{w}{0.642w} = 1.56 \qquad \textit{ans.}$$

(*c*) From Eq. (6.12),

$$\text{Eff} = \frac{\text{AMA}}{\text{TMA}} = \frac{1.56}{2.5} = 0.62 = 62\% \qquad \textit{ans.}$$

MODIFICATIONS OF THE INCLINED PLANE

6.17 The Wedge

The *wedge* (Fig. 6.15) is actually *a double inclined plane*. Its mechanical advantage is, in terms of the dimensions shown (and neglecting friction),

$$\text{TMA}_{\text{(wedge)}} = \frac{L}{t} \qquad (6.13)$$

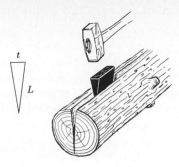

Fig. 6.15 The wedge.

Friction is considerable and desirably so, else the wedge might be forced back out by the pressures acting on it. The actual mechanical advantage of a wedge is much less than the theoretical mechanical advantage and depends on the friction between the contacting surfaces. Wedges are used in splitting wood and in leveling heavy machinery before cementing the base ("grouting in") to the foundation. The wedge principle is the basis for all cutting tools—knives, axes, chisels, lathe tools, and the cutting tools of planers and milling machines. The sharper the tool, the greater the ratio L/t (TMA). The rotating cams of internal combustion engines (see Fig. 6.16) are based on the wedge principle.

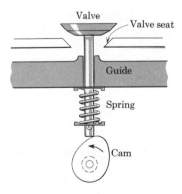

Fig. 6.16 The cam is basically a rotating wedge. The diagram shows one arrangement for a cam-operated engine valve.

6.18 Screws

Another variation of the inclined plane is the *screw*. The principle of the screw has been known at least since the time of Archimedes (about 250 B.C.). It is possible to obtain extremely high mechanical advantages with screws. Actually, a screw is merely an inclined plane wrapped around a cylinder so that a helical form results. Figure 6.17a shows the basic nomenclature applied to screws. The alternate crests and valleys are called *threads*, and the distance from one crest straight across to the next crest is called the *pitch* (p in the diagram) of the screw. This is the distance the screw advances as it is turned through one full revolution. Most screws are *right-hand threaded*; i.e., upon being turned to the right (clockwise) they advance away from the operator. Some bolts and screws are left-hand threaded. This arrangement is used if the normal rotation of the machine would tend to loosen a right-hand-threaded bolt. Bolts and screws have large TMA's, but are purposely designed to have a great deal of friction. Since they are fastening devices, the reaction forces of the materials they are compressing must not be sufficient to "back them off."

6.19 The Screw Jack

The screw principle is used in the design of the common screw jack used for jacking up houses, autos, trucks, heavy machinery, etc. The common form is illustrated in Fig. 6.17b. The weight w being lifted acts down on the cap C, which does not rotate, although the head H does. Lubrication between C and H reduces friction. The effort F is applied at the end of the handle, whose length, to the center of the screw, is L. The screw itself is usually of the square-thread type, lubricated to reduce friction somewhat, but designed so that friction will always be sufficient to keep the jack from "running down" under the load. In practice this means that screw jacks are less than 50 percent efficient.

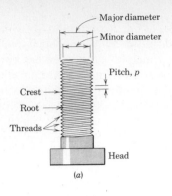

Major diameter

Minor diameter

Pitch, p

Crest

Root

Threads

Head

(a)

w

C

F

H

L

p

B

(b)

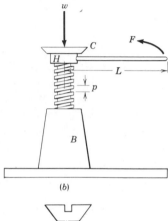

p

(c)

Fig. 6.17 The screw and its nomenclature. (*a*) A right-hand threaded bolt. (*b*) The screw jack. (*c*) A right-hand threaded wood screw.

6.20 Mechanical Advantage and Efficiency of the Screw Jack

Just as for any machine, the *actual* mechanical advantage of a screw jack is the ratio resistance/effort, or, in terms of the diagram of Fig. 6.17*b*,

$$\underset{\text{(screw jack)}}{\text{AMA}} = \frac{w}{F} \qquad (6.14)$$

The *theoretical* mechanical advantage is derived in terms of distances moved by the effort and the resistance. If the effort moves through one revolution, the resistance is raised a distance p, equal to the pitch of the screw. Defining the theoretical mechanical advantage as

$$\text{TMA} = \frac{\text{distance effort moves}}{\text{distance resistance moves}}$$

we have

$$\underset{\text{(screw jack)}}{\text{TMA}} = \frac{2\pi L}{p} \qquad (6.15)$$

Owing to the necessarily large friction, screw jacks rarely have efficiencies higher than 30 to 40 percent. They might well have theoretical mechanical advantages of 500 or more, however. The efficiency of a screw jack can be calculated from the formula

$$\underset{\text{(screw jack)}}{\text{Eff}} = \frac{\text{AMA}}{\text{TMA}} = \frac{w/F}{2\pi L/p}$$

$$= \frac{wp}{2\pi LF} \qquad (6.16)$$

Illustrative Problem 6.6 A screw jack has a screw whose pitch is 0.25 in. Its handle length is 18 in., and a load of 1.5 tons is being lifted. (*a*) If friction is neglected, what effort is required at the end of the handle? (*b*) What is the theoretical mechanical advantage of this jack? (*c*) If the force actually required at the end of the jack handle is 65 lb, what are the actual mechanical advantage and the efficiency of the jack?

Solution
(*a*) If friction is neglected,

$$\text{TMA} = \text{AMA}$$

or

$$\frac{2\pi L}{p} = \frac{w}{F}$$

from which

$$F = \frac{pw}{2\pi L}$$

$$= \frac{0.25 \text{ in.} \times 1.5 \text{ tons} \times 2000 \text{ lb/ton}}{2\pi \times 18 \text{ in.}}$$

$$= 6.63 \text{ lb} \qquad\qquad ans.$$

(**b**) $\quad$ TMA $= \dfrac{2\pi L}{p} = \dfrac{2\pi \times 18 \text{ in.}}{0.25 \text{ in.}} = 450 \quad ans.$

(**c**) $\quad$ AMA $= \dfrac{w}{F} = \dfrac{1.5 \text{ tons} \times 2000 \text{ lb/ton}}{65 \text{ lb}}$

$$= 46 \qquad\qquad ans.$$

$$\text{Eff} = \frac{\text{AMA}}{\text{TMA}} = \frac{46}{450} = 0.102 = 10.2\% \quad ans.$$

6.21 Other Applications of the Screw Principle

The most familiar screws are the ordinary *carpenter's wood screw* and the *bolts* used in the assembly of machinery. *Vises* and screw-type *presses* also utilize the large mechanical advantages possible with the screw principle. The cutting edges of drills and carpenters' bits are wedges, but the helical form which the metal chips or wood cuttings take is in the form of a screw. Turbine-type pumps also utilize the principle of the screw.

Ships' propellers are technically called *screws*. In their rotation they "bite" into and throw water to the rear in the same way that a screw or bit advances into wood as cuttings move up the spiral. In the case of the ship the force which accelerates the water to the rear has a reaction force which pushes the ship forward. The same principle applies to propeller-driven aircraft.

SIMPLE MACHINES BASED ON THE LEVER PRINCIPLE

For the past several pages we have been considering machine applications based on the inclined plane. We now discuss some machines which owe their design and operation to the basic lever principle.

6.22 The Wheel and Axle

The simple lever, as discussed previously, has the limitation of a restricted arc of motion. This limitation can be removed by designing a lever system capable of continuous rotation. Such a device is called a *wheel and axle*. As shown in Fig. 6.18, a heavy load w is the resistance whose lever arm is the relatively small radius r of the *axle*.

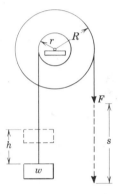

Fig. 6.18 The wheel-and-axle system is a rotating lever.

The force F is applied to the circumference of the attached *wheel*, whose radius is the larger distance R. The *actual* mechanical advantage of such a device is AMA $= w/F$. To obtain the *theoretical* mechanical advantage apply the principle of work. If the force F moves down through the distance s, the load w will be lifted through a smaller distance h. The principle of work requires that $Fs = wh$, or work input equals work output. In one revolution of the wheel and axle the work input is $Fs = 2\pi RF$ and the work output is $wh = 2\pi rw$. Equating these,

$$2\pi RF = 2\pi rw$$

and, rearranging,

$$\frac{w}{F} = \frac{R}{r} \tag{6.17}$$

or
$$\text{TMA} \atop \text{(wheel and axle)} = \frac{R}{r} = \frac{D}{d} \qquad (6.18)$$

or, in words, the *theoretical mechanical advantage of a wheel and axle is the ratio of the radius (or diameter) of the wheel to the radius (or diameter) of the axle.*

Illustrative Problem 6.7 A wheel and axle system is used to lift a 2200-kg load. The wheel diameter is 150 cm and the axle diameter is 15 cm. The operator has to exert a force of 2250 N. Find the following: (*a*) TMA, (*b*) AMA, (*c*) efficiency.

Solution

(*a*) $\quad \text{TMA} = \dfrac{R}{r} = \dfrac{D}{d} = \dfrac{150 \text{ cm}}{15 \text{ cm}} = 10 \qquad ans.$

(*b*) $\quad \text{AMA} = \dfrac{w}{F} = \dfrac{2200 \text{ kg} \times 9.81 \text{ m/sec}^2}{2250 \text{ (kg-m)/sec}^2} = 9.58$

$$ans.$$

(*c*) $\quad \text{Eff} = \dfrac{\text{AMA}}{\text{TMA}} = \dfrac{9.58}{10} = 0.958 = 95.8\%$

$$ans.$$

6.23 Pulleys and Pulley Systems

A pulley is merely a continuous lever which has equal lever arms. As Fig. 6.19 shows, the only advantage of a *single fixed pulley* is whatever benefit might result from a change of direction. Since the effort arm e equals the resistance arm r, the theoretical mechanical advantage is

$$\text{TMA} \atop \text{(single fixed pulley)} = \frac{e}{r} = 1 \qquad (6.19)$$

Even though they multiply neither force nor speed, single *fixed* pulleys are frequently used for the convenience of direction change as pointed out above.

A *single movable pulley*, diagramed in Fig. 6.20, is frequently inconvenient as concerns the direction of application of the effort, but the following analysis shows it to have a TMA of 2. To move the load w up a distance h, both the supporting cords must be shortened by the amount

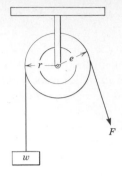

Fig. 6.19 The single fixed pulley affords a change in direction, but does not change the mechanical advantage.

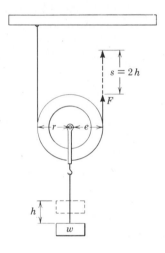

Fig. 6.20 The single movable pulley gives a TMA of 2.

h. This result is obtained by moving the applied force F a distance equal to $2h$. Applying the principle of work, we write

$$Fs = wh$$

$$F \times 2h = wh$$

from which $\qquad \dfrac{w}{F} = \dfrac{2h}{h} = 2$

Consequently $\qquad \text{TMA} \atop \text{(single movable pulley)} = 2 \qquad (6.20)$

6.24 Compound Pulley Arrangements

In order to secure both the advantage of multiplying force or speed and the convenience of a selected direction of application of the force, pulleys are usually arranged in combinations called a *block and tackle*. The pulleys, which are commonly called *sheaves* (pronounced "shivs"), are arranged two or more to a block. The blocks may be made of steel or wood and the sheaves turn on a common lubricated shaft. Standard types are fitted with ordinary sleeve bearings, but those subjected to heavy-duty or high-speed operation have ball or roller bearings. Figure 6.21*a* is a diagram of a block and tackle with four supporting ropes to the movable block.

In order to illustrate the action of a block and tackle, Fig. 6.21*b* shows the blocks separated. Here there are three sheaves on each block and you can trace the path of the rope or cable over each one. Six strands support the movable block. Assuming the pulleys to be frictionless, the tension throughout the rope or cable is constant, and each pulley permits a reversal of direction of the rope. With the arrangement shown (Fig. 6.21*b*) it can be seen that a force F on the "hauling part" contributes an equal force F in *each of the six strands which support the movable block*. Consequently the ratio $w/F = 6$, and in general,

$$\underset{\text{(block and tackle)}}{\text{TMA}} = \frac{\text{No. of strands supporting or}}{\text{pulling on the movable block}} \quad (6.21)$$

The same result can be obtained using the principle of work. This is left as an exercise in analysis for the student.

It should be noted that it would be possible to obtain a TMA of 7 from the block and tackle

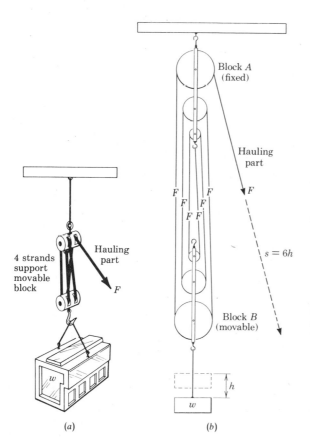

(a) *(b)*

Block A (fixed)

Hauling part

F

$s = 6h$

Block B (movable)

h

w

4 strands support movable block

Hauling part

F

w

Fig. 6.21 Block and tackle arrangements. (*a*) Sketch of a block and tackle with four strands supporting the movable block. (*b*) Schematic diagram of a block and tackle with sheaves separated to show details. (*c*) Photo of a block and tackle in use. Note eight strands to movable block. (Official U.S. Navy photo.)

(c)

of Fig. 6.21*b* if the rope were attached initially to block *B* instead of to block *A*. What might be the disadvantage of this arrangement?

6.25 The Chain Fall, or Differential Pulley

Wherever heavy machinery has to be handled, there is likely to be found suspended from an overhead beam or monorail a device called a *chain fall* or *differential pulley*. Reference to the diagram of Fig. 6.22 reveals that a chain fall is actually a combination of a wheel and axle and a single movable pulley. An endless chain runs over the wheel and axle (R, r) and the movable pulley *P*, which is attached to the load at the bottom. The wheels of radii *R* and *r* turn together on the same axis and are slotted to fit the chain links so that no slippage will occur. Such an arrangement will cause friction, of course, so that the AMA of such a device is considerably less than its TMA. However, even with frictional losses, the AMA of chain falls can be made quite large, as will be shown.

Let the force *F* pull the chain down far enough to turn the fixed wheel and axle through one revolution. This will be a distance of $2\pi R$, and the work done in one revolution will be $2\pi RF$. Now the length of the chain unwound off the small wheel of radius *r* in one revolution will

be $2\pi r$. The load *w* will therefore be raised a distance equal to $\frac{1}{2}(2\pi R - 2\pi r)$ or $\pi(R - r)$. The work done on the load will be $w\pi(R - r)$. Neglecting friction and applying the principle of work, we write

$$2\pi RF = w\pi(R - r)$$

work input = work output

Rearranging gives

$$\frac{w}{F} = \frac{2\pi R}{\pi(R - r)} = \frac{2R}{R - r}$$

or
$$\mathrm{TMA}_{\text{(chain fall)}} = \frac{2R}{R - r} = \frac{2D}{D - d} \qquad (6.22)$$

The smaller the difference between the radii *R* and *r*, the greater the TMA becomes. In some models, friction is purposely planned to be great enough to prevent the load from running down when the force *F* is released. In other forms of the device a ratchet or catch holds the chain from running backward.

6.26 Pulleys, Belts, and Chains Used to Transmit Power

In many machines the driving motor is connected to the machine to be driven by means of *pulleys* and *V-belts*, or by a chain-and-sprocket drive. The designation *V-belt* comes from the shape of the cross section of the belt (see Fig. 6.23*a*). The relative sizes of belt and pulley groove are so selected that the belt does not touch the bottom of the groove. The greater area of contact afforded by the sides of the V is utilized for the friction required to prevent slipping.

A V-belt drive arrangement such as might be used to drive an air compressor is diagramed in Fig. 6.23*b*. Note that both pulleys will turn *in the same direction*. The motor pulley *A* turns at motor speed, usually either 1150 or 1750 rev/min, depending on the type of motor and the nature of the electric power being used. Pulley *B*, on the air compressor, has a diameter (or radius) 2.5 times that of pulley *A*, and therefore the TMA of the system is 2.5. The *speed reduction* is therefore

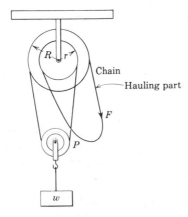

Fig. 6.22 The chain fall, or differential pulley.

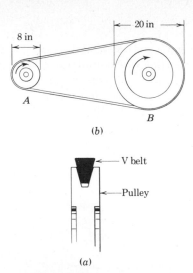

Fig. 6.24 V-belt drive for heavy machinery. Diesel engine (left) driving air compressor (right). (General Motors Corp. and Gardner-Denver Co.)

Fig. 6.23 (*a*) Cross-sectional view of a V-belt and pulley. (*b*) V-belt drive diagram.

2.5 : 1. This arrangement would be used when it is desired to multiply turning moment (technically called *torque* and discussed in detail in Chap. 8) at the sacrifice of speed. Reversing the procedure and making *B* the driving pulley would result in a *speed multiplication* of 2.5 : 1 at the sacrifice of torque.

In general, for pulley and belt drives, in terms of pulley diameters,

$$\text{Speed ratio}_{\text{(belt-pulley drives)}} = \frac{D}{d} \qquad (6.23)$$

For heavy machinery which requires a great deal of power for its operation, pulleys may have multiple grooves, each groove on the driving pulley being connected to a corresponding groove on the driven pulley by a V-belt (see Fig. 6.24). On lathes, drill presses, and similar machines where it is frequently desired to change speed ratios, a set of *step pulleys* is provided. These commonly have three or four matched pairs of pulleys, permitting changing the belt from one pair to another to get the desired speed ratio (see Fig. 6.25).

Chain drives If, in Fig. 6.23*b*, *sprockets* were used instead of pulleys, and a *link chain* instead of

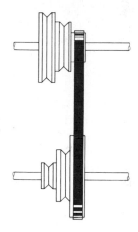

Fig. 6.25 A V-belt step-pulley drive.

a V-belt, a chain drive would result. In terms of the number of teeth on the sprockets

$$\text{Speed ratio}_{\text{(chain drive)}} = \frac{N}{n} \qquad (6.23')$$

where N and n refer to the large and the small sprockets, respectively. Chain drives have the advantage of being able to transmit greater torque without slippage.

Link-belt drives These combine the flexibility and low noise level of V-belts with the nonslip characteristics of chain drives.

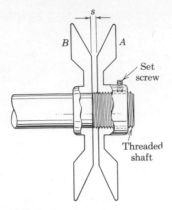

Fig. 6.26   A variable-pitch pulley.

6.27   Variable-pitch Pulleys

To effect relatively small speed-ratio changes, *variable-pitch* driver pulleys are frequently used. These pulleys are made in two pieces, split down the middle as shown in cross section in Fig. 6.26. The *B* part is keyed to the shaft and remains fixed. The *A* part is fixed to the shaft by setscrews only. These can be loosened, permitting the *A*

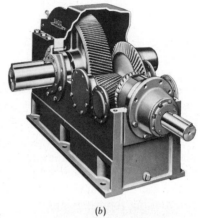

(b)

(c)

(a)

Fig. 6.27   Gears and gear boxes. (*a*) Huge spur-and-pinion gear combination for a very large press. (*b*) Bevel gear and helical gears in a speed reducer. (*c*) Worm gear combination in a speed reducer. Note roller thrust bearings. (The Falk Corporation.)

half to be screwed left or right on the threaded shaft. The space s can thus be made very narrow, forcing the V-belt to ride high in the groove, or it can be made wide, allowing the belt to ride low in the groove. Changing the width of s changes, in effect, the diameter of the pulley, which in turn changes the speed ratio, since the diameter of the driven pulley remains constant. Provision is also made for adjusting belt tension.

6.28   Gears and Gear Combinations

In cases where absolutely no slippage can be tolerated and in situations where large torque must be transmitted at low speeds, pulleys and belts are replaced by gears and gear combinations. There is a tremendous variety of gears of all sizes and shapes, but they are ordinarily classified under four general headings: *spur gears*, *bevel gears*, *worm gears*, and *helical gears* (see Fig. 6.27).

Spur Gears   Spur gears are used to transmit rotary motion between parallel shafts. The teeth are cut parallel to the axis. Timing gears of engines and drive gears for lathes and presses are examples. When two spur gears are in mesh, a situation very similar to that of two pulleys with V-belt drive results. The significant difference is that the gears rotate in opposite directions, while belt-connected pulleys rotate in the same direction. The speed-reduction ratio (called *gear ratio* in this case) can be expressed either as the ratio of the diameters of the gear wheels or more commonly as the ratio of the number of gear teeth on one gear to the number on the other. Let N be the number of teeth on the large (driven) gear (see Fig. 6.28) and n the number on the small (driving) gear. The speed change obtained is as follows:

$$\frac{\text{rev/min driving gear}}{\text{rev/min driven gear}} = \frac{N}{n} \qquad (6.24)$$

Or, the rotational speeds are *inversely proportional* to the number of teeth on the gears. It is emphasized that, just as with pulleys, *if input power is constant*, a speed gain means a corre-

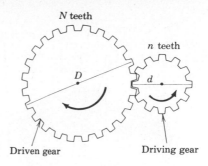

Fig. 6.28   Gear ratio of spur gears.

sponding torque loss; and conversely, a speed reduction results in a torque increase. Both "speed gears" and "reduction gears" are common in industrial practice.

Gear Trains   If it is desired to effect a speed reduction (or torque multiplication) greater than can be conveniently obtained by one pair of gears, a *gear train* is used. Figure 6.29 shows one simple arrangement. The overall gear ratio of such a train can be obtained by analysis as follows. Consider gears A and B as constituting one system. The gear ratio of this system is N_1/n_1. Gears C and D constitute a second system whose gear ratio is N_2/n_2. Since gear C turns at the same speed as gear B (both are keyed to the same shaft), the overall speed reduction formula is

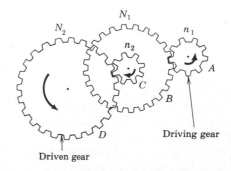

Fig. 6.29   Gear ratio of a gear train for speed reduction (torque multiplication). If speed multiplication is desired, D could be made the *driving* gear and A the *driven* gear.

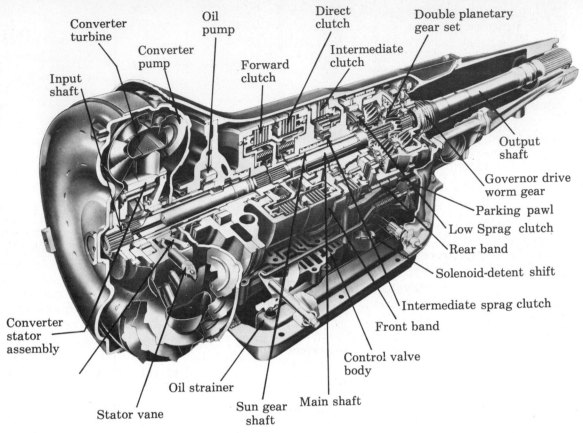

Fig. 6.30 Automatic transmission of a modern automobile. It is a veritable maze of gears, valves, pumps, bearings, and friction devices.

Torque multiplication $\left(\underset{\text{(gear train)}}{\text{speed reduction}}\right)$

$$= \frac{N_1}{n_1} \times \frac{N_2}{n_2} \times \cdots \times \frac{N_n}{n_n} \quad (6.25)$$

The analysis and the formula would be reversed if speed multiplication were desired.

Planetary Gearing   Some makes of automobiles with automatic transmissions have very complex gearing systems which are frequently hydraulically operated. One such arrangement is *planetary gearing*, so called because small spur gears are mounted on a ring, and one of the possible modes of motion is such that they revolve around the center ("sun") gear as the planets of our solar system do around the sun (see Fig. 6.30).

6.29   Compound Machines

We have been studying the basic machines and the several simple machines derived from them. We have analyzed the effects of friction on the efficiency of their operation, and have seen how the principle of work applies in all cases. The basic law that "you can't get something for nothing" certainly holds true for machines. If speed is to be multiplied, torque must be sacrificed, and conversely. The power output of a machine can never be greater than the power input; in fact, because of friction, the output is always less than the input. Machines may also have other losses, such as those from heat, magnetism, and similar effects.

The *principle of work* for machines can be

Fig. 6.31 The huge power shovels used in mining operations illustrate the combination of basic and simple machines to form the complex machines of industry. Note the relative size of the man in front of the truck. (Harnischfeger Corporation.)

stated:

Power (or energy) output

> = power (energy) input
> − power (energy) losses

On actual industrial jobs the basic and simple machines analyzed in this chapter are rather infrequently used alone. Rather, they are combined in many ingenious and complex ways to produce the *compound machines* of industry. It is beyond the scope of a physics book to treat compound machines in detail, but in closing this discussion let us consider the power shovel of Fig. 6.31 as a typical example of a compound machine. Notice the horns on the shovel itself. They are *wedges*—a form of inclined plane. The boom is a huge *lever*, operating with tremendous mechanical advantage. The forces which operate the boom are carried by the cables which, with the sheaves over which they pass, constitute a huge *block-and-tackle* system. Inside the cab the drum on which the cable is wound constitutes a *wheel and axle*. The wheel is a large *gear*, driven by a *gear train* from the source of power. The operator controls the machine by a set of *levers* at his seat. The cab and boom are rotated by a ring-and-pinion gear arrangement.

You can see from this example how the design, construction, and operation of complex machines depend on a thorough understanding of the principles of basic and simple machines.

SUMMARY

A machine is *a device for the advantageous application of a force.*

There are actually only three basic machines: the *lever*, the *inclined plane*, and the *hydraulic press*. (The hydraulic press is discussed in Chap. 11.)

Examples of simple machines are (1) the wedge, the cam, and the screw, all based on the inclined plane; and (2) the pulley, the wheel and axle, and gears, all based on the lever principle.

The *actual* mechanical advantage (AMA) of *any* machine is the ratio of the force output to the force input:

$$\text{AMA} = \frac{w}{F}$$

The *theoretical* mechanical advantage (TMA) of a machine is calculated from various dimensions, and neglects losses due to friction.

The efficiency of a machine is the ratio

$$\frac{\text{work output}}{\text{work input}} \quad \text{or} \quad \frac{\text{power output}}{\text{power input}}$$

Efficiencies of machines are always less than 100 percent. In terms of mechanical advantage,

$$\text{Eff} = \frac{\text{AMA}}{\text{TMA}}$$

The principle of work is merely the law of conservation of energy applied to machines. For machines,

Energy input − losses (frictional and other)
= net work output

Friction is *a force which always acts to oppose the relative motion of two sliding surfaces in contact.*

The coefficient of sliding friction is defined as the ratio of the force of friction to the normal force pressing the two sliding surfaces together:

$$\mu = \frac{F_f}{N}$$

The effect of friction on machines is to cause energy to be expended ("wasted") within the machine. This results in lowered efficiency. However, in some machines, friction is a necessity, and is built into the design.

Pulleys, belts, chains, and gears are machine components frequently used in transmitting torque and power.

The compound machines of modern industry are merely complex and ingenious combinations of the basic and simple machines discussed in this chapter.

QUESTIONS AND EXERCISES

1. Diagram the three classes of levers and list five practical applications (some from the human body) of each.

2. If a simple machine such as a lever is operated "backward," what change occurs in the AMA? in the efficiency? (Friction negligible.)

3. A screw jack is rusty and corroded. Before starting to use it you decide to lubricate it thoroughly. How does this affect its efficiency? its theoretical mechanical advantage (TMA)? its actual mechanical advantage (AMA)?

4. Based on some library research write a brief paper on the theory of friction in machine parts. Touch on all the following: (*a*) the effect of smoothness of surface, (*b*) the choice of alloys for sleeve bearings, (*c*) the effect of relative velocity of the rubbing parts, (*d*) the general theory of lubrication in reducing friction, (*e*) the "oil-wedge" theory of lubrication of sleeve-type bearings, and (*f*) the advantages of ball and roller bearings over sleeve-type bearings.

5. Some models of automatic transmissions for autos have small radiators through which the transmission oil is circulated to cool it. Why does the oil get so hot?

6. *A* and *B* are two bicycle riders. *A* has a 10-speed bike which he shifts into low gear as they approach a steep hill. *B* keeps his in "regular" coaster-brake gear. They start even and race to the top of the hill, finishing in a dead heat. (*a*) Who pedaled faster? (*b*) Who applied more force on each down stroke of the pedal? (*c*) Who generated the greater power output? Who did the most work? (Assume same friction for both.)

7. A block-and-tackle system is to be used to pull a stalled pickup truck out of the mud. There are two blocks of three sheaves each and plenty of strong wire rope. Diagram the arrangement which (*a*) would give the greatest TMA; (*b*) would probably be used in practice because of convenience. (There is a convenient tree to "anchor" one end of the equipment.)

8. Refer to the diagram of the differential pulley on p. 124. As force *F* pulls on the hauling part for one full revolution of the differential pulley, a length of chain $2\pi R$ passes over the large pulley. The small pulley is fixed to the large pulley so that the two turn as one, and in one full revolution a much shorter length of chain, $2\pi r$, goes up and over this pulley. Explain the "mystery" of the "ever-lengthening" chain.

9. The sketch of Fig. 6.32 shows a device called a turnbuckle. One bolt is right-hand threaded and the other is left-hand threaded. The yoke can be turned by hand, pulling the two bolts into the center at the same rate and thus putting tension on the cable, *C*. If the pitch of the screw(s) is *p*, and the diameter of the yoke (to which one's hand applies the forces *F* − *F*) is *D*, what is the TMA of the turnbuckle?

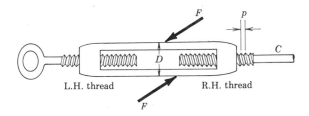

Fig. 6.32 Diagram of a turnbuckle (Exercise 6.9).

PROBLEMS

Note: Drawing a careful diagram for each problem, and showing the various forces as vectors, will help in understanding and solving the problems.

1. A 185-lb man sits on a seesaw 4 ft from the fulcrum. Where must an 85-lb boy sit to balance the seesaw?

2. A resistance of 800 N must be overcome through a lever whose resistance arm is 18 cm long and whose effort arm is 5 cm long. (*a*) What is the magnitude of the effort force? (*b*) For what purpose would such a lever be used?

3. A shovel is used to lift 20 kg of concrete mix. The workman's right hand is at the end of the shovel handle 1.3 m from the center of the load. The left hand acts as a fulcrum 45 cm from the center of the load. Neglect the weight of the shovel and find the force exerted by each hand.

4. The wheelbarrow of Fig. 6.5 contains 250 lb of gravel. If *r* is 18 in. and *e* is 4 ft, what total force *F* must be applied at the handles?

5. A pair of wire-cutting pliers is gripped in the hand so that the force applied is 20 cm from the pivot point of the jaws. A wire to be cut is 1.6 cm from the pivot axis. If a force of 300 N is exerted by the gripping hand as the wire is cut, what is the resistance offered by the wire?

6. Four seamen are 6 ft from the center of a capstan 18 in. in diameter, pushing with a force of 90 lb each. If the efficiency of the capstan is 60 percent and the men just weigh the anchor, how heavy is it? (Neglect buoyancy of the water.)

7. A windlass has a drum 15 cm in diameter and a crank 60 cm long. A man pulls up a bucket of concrete whose mass is 100 kg. What force must he exert on the crank if the device is 90 percent efficient?

8. A loaded dolly weighing 450 lb is rolled up a ramp 10 ft long, which rises 2.5 ft vertically. Neglecting friction, (*a*) what force parallel to the ramp is required? (*b*) Find the *normal* force pressing against the ramp.

9. The TMA of an inclined plane is 5.6. What percent grade is this incline?

10. The cable of a ski lift pulls up a 25° slope at uniform speed. If the total mass of chairs, skiers, and equipment is 4000 kg, what is the tension in the cable? (Neglect friction.)

11. An auto engine is propelling a 2000 kg car up a 20 percent grade at 20 km/hr. Assuming an overall efficiency of 65 percent from crankshaft to rear wheels and pavement, (*a*) what power (kW) is the engine developing? (*b*) What is the thrust at each rear wheel in newtons?

12. What force is required to pull a loaded sled, mass 2500 kg, across hard-packed snow (level) if the coefficient of sliding friction between runners and snow is 0.035?

13. A steel loading chute slants downward at 45° from a warehouse to a freight siding. It is 60 ft long. The coefficient of sliding friction between cardboard and steel is 0.45. What is the speed of the cartons as they arrive at the bottom of the chute? Work the problem two ways: (*a*) from acceleration considerations, and (*b*) using energy principles.

14. The system shown in Fig. 6.33 consists of a 10-kg mass acted on by the force of gravity, a frictionless pulley and cord, and a 4-kg hardwood block on a hardwood ramp inclined upward at an angle of 35°. What is the acceleration of the system? (See Table 6.1 for μ.)

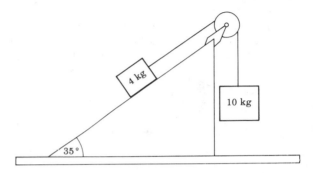

Fig. 6.33 Friction on an inclined plane (Problem 6.14).

15. If the human forearm acts as a Class III lever, as in Fig. 6.3*c*, calculate the force *F* which must be applied by the muscles in order to lift a bucket of paint weighing 18 lb. Assume the arm muscles apply the force at a point 1.25 in. from the elbow, and that the distance from elbow to the bucket bail in the hand is 14 in.

16. A 150-kilo cake of ice is being pulled steadily up a steel ramp to the loading dock. The slope is inclined 25° to the horizontal. The force required is noted to be 820 N. Find μ between ice and steel.

17. A truck engine is delivering 195 hp. Assuming a 25 percent loss between engine crankshaft and rear wheels, find the coefficient of rolling friction for rubber tires on concrete if the loaded truck (total weight 15 tons) is making 16 mph up a 6 percent grade. (Neglect air resistance.)

18. A screw jack is being used on a house-moving job. The pitch of the screw is 0.25 in., and the handle is 3 ft long. (*a*) What is the theoretical mechanical advantage? (*b*) If the jack is only 46 percent efficient, what load could be lifted by a force of 125 lb applied to the end of the handle?

19. The hoisting gear of an oil-well drilling rig (called the *draw works*) has an outer chain-driven sprocket of 3-ft diameter. The hoisting cable is wound around the drum, which is 10 in. in diameter. If the drive chain applies a force of 15,000 lb to the rim of the sprocket, what tension will be created in the hoisting cable?

20. The "crown block" (at the top of the derrick) and "traveling block" arrangement used in pulling oil-well casing has six sheaves per block. If it is desired to pull casing out of the hole at a linear speed of 10 ft/sec, at what linear velocity must the cable be pulled in by the draw works? (NOTE: The "hauling part" comes down from the crown block to the draw works.)

21. A chain fall is being used to lift a 1750-lb aircraft engine from the overhaul stand to the nacelle for installation. The radii of the large and small wheels are, respectively, 18 and 17 in. If the efficiency is 40 percent, what force F must be applied to the hauling part of the chain?

22. A refrigeration compressor comes equipped with a 30-cm diameter pulley and is designed to operate at 500 rpm. What should be the diameter of the motor pulley if the motor speed is 1750 rpm?

23. An auto differential (at the rear axle) has a gear ratio of 3.6 : 1. The transmission is in low gear, where the ratio is 7.6 : 1. How many turns does the engine make for each revolution of the rear wheel?

24. In the gear train of Fig. 6.29, gear A has 16 teeth, gear B 48 teeth, gear C 10 teeth, and gear D 64 teeth. If the shaft of A is turning 1750 rev/min, at what speed is the shaft of D turning?

25. A bicycle has a pedal sprocket 9 in. in effective diameter, and the driven (high-speed) sprocket on the rear wheel is 4 in. in diameter. The bike wheels are 26 in. in diameter. If a person wants to make 15 mi/hr, how many rev/min must the pedals be turned?

7 Circular Motion and Satellite Mechanics

Up to this point we have studied velocity, acceleration, and the laws of motion mostly as they are involved in straight-line motion. Also, we have assumed that *g*, the acceleration of gravity, is a constant in both magnitude and direction. However, many objects move in circular or elliptical paths and their direction is continually changing. Since acceleration is a *vector* quantity, this change of direction means that their accelerations are not constant. Rotating machinery, the planets in their orbits, and man-made satellites in orbit around the earth are all examples of this kind of motion.

Newton's laws of motion, which we studied in Chapter 4, were the outgrowth of his inquiries into the motions of the planets around the sun. He had at his disposal the observations made by such scientists as Nicolaus Copernicus, Tycho Brahe, Johannes Kepler, and Galileo Galilei, over a long period of time. All of these men lived and worked in the century prior to Newton's birth in 1642.

In this chapter we study circular motion and satellite mechanics. We start with *uniform circular motion* and Newton's *law of universal gravitation.*

CIRCULAR MOTION

7.1 Rotation Contrasted With Translation

For a body to have rotary motion, two requirements must be met:

1. Every particle of the body must move in a circular path.

2. All these circles must have their centers on the same straight line. The straight line is said to be the *axis of rotation.*

In contrast, motion in which no imaginary straight line in a body changes direction as the body moves from one place to another, is called *translation* (see Fig. 7.1).

Translation may be either *straight-line* or *curvilinear*, depending on the path of the body. The straight-line distance between the starting point and the end point of the translation is called the *displacement.*

Objects often have complex combinations of rotary motion and translation. This is true of the rotating parts of an auto engine when the car is in motion along the highway. We shall begin with some elementary considerations of rotary motion and gradually lead to some of the more difficult problems with important scientific and industrial applications.

7.2 Angular Measurement

Since rotary motion is motion in a circular path, you should recall the geometry of the circle. The linear distance around a circle is the *circumference*, and the distance from the center to the circle is the *radius*. The angular displacement through which a radius turns if pivoted at the center is measured in *degrees* (°), and there are 360° in the

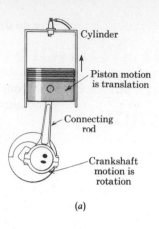

Cylinder

Piston motion
is translation

Connecting
rod

Crankshaft
motion is
rotation

(a)

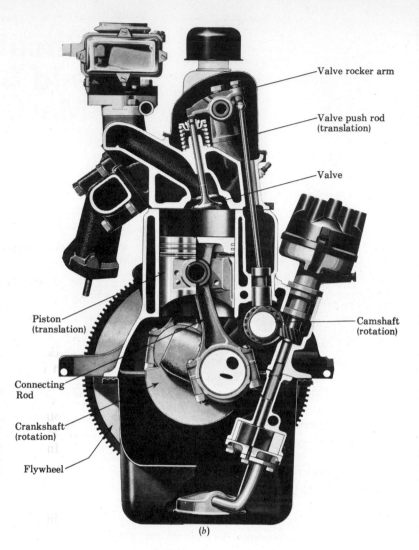

Valve rocker arm

Valve push rod
(translation)

Valve

Camshaft
(rotation)

Piston
(translation)

Connecting
Rod

Crankshaft
(rotation)

Flywheel

(b)

Fig. 7.1 Translation and rotation compared. (*a*) The diagram shows how the connecting rod and crankshaft of an engine convert back-and-forth (reciprocating) translation of the piston into rotary motion of the crankshaft. (*b*) Cutaway view of an auto engine showing component parts, some of which are in rotation when the engine is running, and some in translatory motion. (Ford Motor Co.)

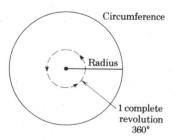

Circumference

Radius

1 complete
revolution
360°

Fig. 7.2 Circle relationships. Circumference = $2\pi \times$ radius.

total angular distance around a point (see Fig. 7.2).

Another measure of angular displacement is the *revolution.* One revolution is one complete circular trip around the center or axis of rotation. One revolution is thus equal to 360°. In science and technology there is still another unit of angular measure. It is larger than the degree but smaller than the revolution, and is called the *radian* (abbreviated rad). *The radian is the measure of an angle whose arc along the circumference is just*

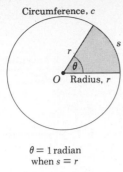

Circumference, c

Radius, r

$\theta = 1$ radian
when $s = r$

Fig. 7.3 Diagram showing radian measure. $\theta = 1$ radian when $s = r$. 2π radians = 360 degrees.

equal to the radius, as shown in Fig. 7.3. *The angle θ is defined as one radian when s, the arc length, is equal to r, the radius.* Now, since $c = 2\pi r$, we have 2π rad = 360°, and 1 rad = 360°/2π. This makes one radian approximately equal to 57.3°. Since $\theta = 1$ rad when $s = r$ (by definition), $\theta = 2$ rad when $s = 2r$. In general,

$$\theta = \frac{s}{r} \quad \text{or} \quad s = r\theta \quad (7.1)$$

or, in words

Arc length = radius

$\times$ angular displacement in radians

This is the basic equation for radian measure. It is one of the fundamental equations for motion in a circular path. The radian, as a unit of angular measurement, is a dimensionless quantity, since it is the ratio of two lengths.

7.3 Angular Velocity

If rotary motion is *uniform*, i.e., at constant angular speed, it is often described in terms of revolutions per minute (rev/min or rpm) or revolutions per second (rev/sec or rps). For many industrial purposes, these measures of angular velocity are perfectly satisfactory. However, we often need to know the *linear speeds* of points that are in rotary motion, e.g., tip speeds of airplane propellers or rim speeds of high-speed turbine rotors. Angular

velocities in revolutions per minute give no direct information on tip speed. Engineers and technicians therefore prefer to express angular velocities in *radians per second*, denoted by the symbol ω (the Greek letter omega). Since $\theta = s/r$ from Eq. (7.1), if we divide both sides by t, we obtain

$$\frac{\theta}{t} = \frac{s}{rt} \quad (7.1')$$

Now θ/t is angular displacement per unit time, or the *angular velocity*, ω, measured in rad/sec. And s/t is the linear speed, v, of a point on the circle. Consequently, from Eq. (7.1'),

$$\text{Angular velocity} = \frac{\text{linear speed}}{\text{radius}}$$

As an equation,

$$\omega = v/r$$

or $$v = \omega r \quad (7.2)$$

Equation (7.2) is the basic relationship between the linear speed of a point on a rotating object and the angular velocity of the object. Either angular velocity (ω) or linear speed (v) can be obtained if the other is known and if the radius can be measured.

In SI-metric units,

$$\omega = \frac{\text{rad}}{\text{sec}} = \frac{\text{m/sec}}{\text{radius (m)}}$$

In English units,

$$\omega = \frac{\text{rad}}{\text{sec}} = \frac{\text{ft/sec}}{\text{radius (ft)}}$$

You should note that angular velocity, ω, has the dimension of time^{-1} (that is, sec^{-1}), since the length units on the right side of these equations cancel out.

Angular velocity in radians per second is related to revolutions per minute as follows:

$$\omega \ (\text{rad/sec}) = 2\pi \times \frac{\text{rev/min}}{60} \quad (7.3)$$

As an exercise you should try to prove this.

7.4 Angular Displacement Related to Angular Velocity

If a rotating object turns through an angle of θ rad in a time t sec, its average angular velocity is given by the relation $\omega = \theta/t$. This can be written in the form

$$\theta = \omega t \qquad (7.4)$$

Equation (7.4) expresses angular displacement in terms of angular velocity and time. It is very much like Eq. (4.1″), which gives linear displacement in terms of linear velocity and time ($s = vt$), for translation.

Illustrative Problem 7.1 A bicycle wheel is 26 in. in diameter. The rider is pedaling at 30 ft/sec. What is the wheel's angular velocity in (a) radians per second, and (b) revolutions per minute?

Solution (*a*) Assume no slippage of tire on pavement. Then the linear rim speed of the wheel must be 30 ft/sec. Select Eq. (7.2) which relates linear velocity, angular velocity, and radius, and write

$$v = \omega r$$

Substituting,

$$30 \text{ ft/sec} = \omega \times \tfrac{13}{12} \text{ ft}$$

Solving,

$$\omega = \frac{30 \text{ ft/sec}}{\tfrac{13}{12} \text{ ft}} = 27.7 \text{ rad/sec} \qquad ans.$$

(*b*) To solve for revolutions per minute, write Eq. (7.3) in this form:

$$\text{rev/min} = \frac{60\omega}{2\pi}$$

Substitute the value of ω from (*a*) above, and solve:

$$\frac{\text{rev}}{\text{min}} = \frac{60 \text{ sec/min} \times 27.7 \text{ rad/sec}}{6.28 \text{ rad/rev}} = 265 \frac{\text{rev}}{\text{min}}$$

$$ans.$$

Illustrative Problem 7.2 A cargo plane has propellers whose arcs are 5 m in diameter. What is the propeller-tip speed in meters per second when the shaft is turning 1200 rev/min?

Solution Use Eq. (7.3) and solve it for ω, the angular velocity.

$$\omega = \frac{2\pi \times \text{rev/min}}{60}$$

Substitution gives

$$\omega = \frac{2\pi \text{ rad/rev} \times 1200 \text{ rev/min}}{60 \text{ sec/min}} = 126 \text{ rad/sec}$$

Substitute this value of ω in Eq. (7.2), recalling that the radius $r = \tfrac{5}{2} = 2.5$ m. This gives

$$v = \omega r = 126 \text{ rad/sec} \times 2.5 \text{ m} = 315 \text{ m/sec} \quad ans.$$

NOTE: Propeller-tip speeds are quite important. As the speed of sound (approximately 330 m/sec or 1130 ft/sec at standard sea-level conditions) is approached, shock waves are produced which materially decrease propeller efficiency.

7.5 Uniform Circular Motion

When a body moves in a circular path at constant speed, it is said to be in *uniform circular motion*. An example is a tooth on a gear which is revolving about a fixed axis at uniform angular velocity. In the diagram of Fig. 7.4*a*, consider a particle at position P on a rotating wheel, at which instant its linear velocity is represented by the vector $\mathbf{v}_1 = \omega r$. A short time (t sec) later the particle is at P', and the *magnitude* of its velocity vector is still ωr. However, as the diagram shows, the *direction* of the velocity has changed. Therefore, since velocity is a quantity which has both magnitude and direction, the *linear velocity* of the particle is continually changing, even though its *linear speed* remains constant and equal to ωr. A changing velocity means *acceleration*. In this section we determine the magnitude and direction of the acceleration in uniform circular motion.

In Fig. 7.4*b* the two velocity vectors of Fig. 7.4*a* have been laid out from a common point A, parallel to their original directions. The vector $\mathbf{pp'}$, which closes the vector triangle, represents the small change in velocity in the short time t sec, and can be designated as $\Delta\mathbf{v}$.

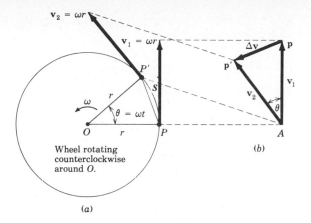

Fig. 7.4 Relationships involved in uniform circular motion. (a) Space diagram showing particle P on the rim of a wheel, rotating around the center O at a distance r, with constant angular velocity ω. P moves arc distance s and angular distance θ in time t sec. (b) Vector diagram, showing vector $\mathbf{pp}' = \Delta\mathbf{v}$ as the vector difference between $\mathbf{v}_2$ and $\mathbf{v}_1$. As a vector equation, $\mathbf{v}_1 + \Delta\mathbf{v} = \mathbf{v}_2$, or $\Delta\mathbf{v} = \mathbf{v}_2 - \mathbf{v}_1$.

NOTE: Remember that the symbol Δ means "a very small amount of," and Δv is read "delta v." It does not mean delta multiplied by v.

Vector diagrams like Fig. 7.4b can be drawn for smaller and smaller values of t. As t approaches zero as a limit, it is found that the direction of the vector $\Delta\mathbf{v}$ is always toward the center of the circle. As t approaches zero as a limit, the arc distance s becomes essentially equal to the chord PP', and triangle OPP' is similar to triangle App'. Consequently, since corresponding sides of similar triangles are proportional,

$$\frac{s}{r} = \frac{\Delta v}{v}$$

Now Δv represents a small change in velocity, and is therefore the result of an acceleration over a small time interval Δt. It can therefore be represented by the equation $\Delta v = a\,\Delta t$. Further, the arc distance moved by the body in time Δt is $s = v\,\Delta t$. Making these substitutions,

$$\frac{v\,\Delta t}{r} = \frac{a\,\Delta t}{v}$$

Solving for a gives

$$a = \frac{v^2}{r}$$

Since the acceleration in uniform circular motion *is always toward the center*, we call it *centripetal* ("center-seeking") *acceleration*. It is defined by the equation

$$a_c = \frac{v^2}{r} \qquad (7.5)$$

Or, since $v = \omega r$,

$$a_c = r\omega^2 \qquad (7.5')$$

FORCES INVOLVED IN CIRCULAR MOTION

7.6 Centripetal Force

It was noted in Chapter 4 that when an unbalanced force acts on a body, acceleration results (Newton's *second law*). Since objects or particles in circular motion undergo a continual change in direction of their motion, they are being constantly accelerated. This acceleration is always toward the center of rotation.

All of this leads to an important question: What causes the acceleration toward the center? According to Newton's *first law of motion*, the particle of Fig. 7.4a must be acted upon by an unbalanced force before it will change either its speed or its direction. The force causing centripetal acceleration is supplied by the molecular forces of cohesion within the material of which the wheel is made. This force which causes centripetal acceleration is called *centripetal force*. If the molecular forces of cohesion were destroyed, the centripetal force would become zero. The particle would immediately *move off at a tangent*, in the line of its motion at the instant the centripetal force was destroyed.

Using a simple illustration, let us look at the factors that determine centripetal force. Consider a ball of mass M being whirled in a circle about a center O, attached to a string whose length is r, as diagramed in Fig. 7.5. Everyone has, at one time or another, performed this simple trick. The tension in the string supplies the force necessary to

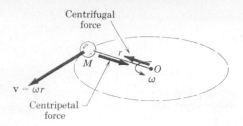

Fig. 7.5 Centripetal vs. centrifugal force. Centripetal force holds the ball in orbit. Centrifugal force exerts an outward "pull" on the hand of the person whirling the ball. The string is in tension, supplying the centripetal force.

hold the ball in its circular orbit, i.e., to keep it accelerating toward the center.

It is Newton's *second law of motion* ($F = ma$) which relates force to mass and acceleration. Remember that a *force of* 1 **N** is necessary to give *a mass of* 1 *kg an acceleration of* 1 *m/sec²*. Applying the *second law* to uniform circular motion gives

$$F_c = ma_c \qquad (7.6)$$

or

Centripetal force

$$= \text{mass} \times \text{centripetal acceleration}$$

But

$$a_c = \frac{v^2}{r} \qquad (7.5)$$

and therefore

$$F_c = \frac{mv^2}{r} = \frac{wv^2}{gr} \qquad (7.7)$$

This equation expresses centripetal force in newtons when m is in kilograms, linear velocity v is in meters per second, and r is in meters. Or, in the English system, F_c is in pounds when m is in slugs, linear velocity v is in feet per second, and r is in feet. Equation (7.7) is often considered the basic formula for centripetal force, but a second form, equally useful, may be written where we recall that $v = \omega r$. Substituting and simplifying, we obtain

$$F_c = mr\omega^2 = \frac{w}{g} r\omega^2 \qquad (7.7')$$

which is a more convenient form to use when the *angular velocity* is known.

7.7 Centrifugal Force

You will recall from our study of Newton's laws that forces never exist singly, but always in pairs. For every *action force* there is an equal and opposite *reaction force*. In this case, the centripetal (action) force is exerted by the string on the ball, and provides the "center-seeking" acceleration which holds it in a circular orbit. The reaction force is the outward force which the ball exerts on the string. The string transmits this force to point *O*, or to a person's hand. The inertia of the ball, its resistance to a change in direction, is the cause of this reaction force. Since this force is directed away from the center it is called *centrifugal* ("center-fleeing") *force*. It is important to note that centripetal force and centrifugal force, like any action-reaction pair of forces, act on *different* bodies, not on the same body.

What happens if the ball is whirled too fast? The string's tensile strength will eventually be exceeded; it breaks and the ball flies off in straight-line motion tangent to the arc at the instant of the break, and at the linear velocity v at that instant. You should be able to show, from Newton's laws, why the ball moves off tangentially.

7.8 Applications of Centripetal Force

Any rotating machine part, such as a turbine rotor, motor armature, engine flywheel, or grinding wheel (Fig. 7.6) is subject to internal stresses caused by the rotation. These stresses are, as shown by Eq. (7.7′), directly proportional to the mass of the rotating object, to the square of the angular velocity, and to the radius of the motion. All materials which go into the manufacture of rotating parts must be carefully tested to make sure that they are strong enough to supply the stresses required for centripetal force. Rotational speeds of 10,000 to 50,000 rev/min are common for centrifuges and for spin-stabilized instruments such as gyrocompasses and inertial navigation systems. Slower rotational speeds, for example,

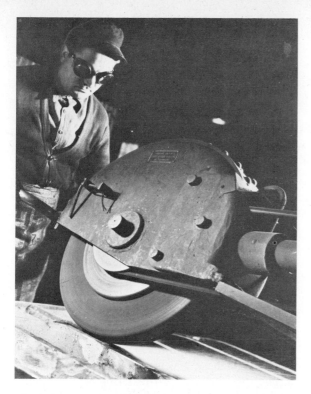

Fig. 7.6 High-speed grinding wheels have been known to fly apart at high rotational speeds. Note protective cover over the wheel. (*Steelways Magazine.*)

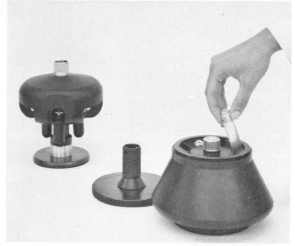

Fig. 7.7 Rotors from a medical research laboratory centrifuge. These rotors turn at speeds of up to 65,000 rev/min and set up centripetal accelerations equal to 300,000 times the acceleration of gravity.

1500 to 3600 rev/min, are required of steam-turbine rotors and aircraft propellers, but their large diameters (8 to 16 ft) and huge masses set up tremendous internal stresses within the material.

7.9 The Centrifuge

The centrifuge is a high-speed rotating device (Fig. 7.7). Liquids containing microscopic particles are placed in a small vessel and whirled at the incredibly high speed of 50,000 rev/min or more. The tremendous centripetal forces have the result that the particles collect or "sediment" to the outside of the vessel. In medical research, centrifuges are used in determining the sedimentation rates of blood samples. They are also used for concentrating solutions thought to contain viruses, hormones, etc., and for separating impurities from pharmaceuticals. The dairy industry

uses the centrifuge principle both in making laboratory butterfat tests and in the separation of cream from milk.

7.10 Applications of Centripetal Force to Aerobatics

Even in this era of rockets and jet propulsion, some attention is given to the maneuverability of military aircraft, since the ability to climb rapidly and make "tight" turns is of great importance in evasive tactics. Frequently, in such *aerobatics* it is necessary to pull out of a dive as illustrated in Fig. 7.8. If a sharp loop is attempted, a tremendous centripetal force will be required to hold the plane in the chosen arc. This force has to be supplied by the action of the air on the airfoil surfaces (wing and fuselage) of the plane. If these forces become too great, the wings may actually shear off in flight. Further, the pilot himself is in danger of "blacking out," since his body, i.e., the flesh and bone, is being accelerated toward the center of the loop while his blood, being fluid, tends to flow along the tangent to the loop at any given instant. The result is a draining of the blood from the head along with loss of vision. Uncon-

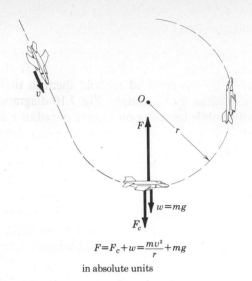

$$F = F_c + w = \frac{mv^2}{r} + mg$$

in absolute units

Fig. 7.8 The vertical loop problem in aerobatics.

sciousness or even death may follow if the centripetal acceleration is too great and lasts too long.

In the situation of Fig. 7.8, when the plane is at the pullout point, the force F must be sufficient to furnish the required centripetal force and also balance the force of gravity on the plane. Since the expression $F = mv^2/r$ for centripetal force gives F in absolute units, we must express the force of gravity in absolute units also. The total reaction force of the air against the plane at the bottom of a vertical loop is then

$$F = \frac{mv^2}{r} + mg \quad (m \text{ in kg or slugs}) \quad (7.8)$$

or

$$F = \frac{wv^2}{gr} + w \quad (w \text{ in N or lb}) \quad (7.8')$$

where g is the acceleration of gravity (9.81 m/sec² or 32.2 ft/sec²).

Illustrative Problem 7.3 A 180-lb pilot is executing a vertical loop of radius 2000 ft at 350 mi/hr. With what force does the seat press upward against him?

Solution Use Eq. (7.8′) since absolute units must be used. Change mi/hr to ft/sec:

$$350 \; \frac{\text{mi}}{\text{hr}} \times \frac{44 \text{ ft/sec}}{30 \text{ mi/hr}} = 513 \text{ ft/sec}$$

Substitute values:

$$F = \frac{180 \text{ lb} \times (513 \text{ ft/sec})^2}{32.2 \text{ ft/sec}^2 \times 2000 \text{ ft}} + 180 \text{ lb} = 915 \text{ lb}$$

ans.

This tremendous force is more than five times the actual weight of the pilot. It is no wonder that mental and physical efficiency is impaired during such pullouts. One's hands and arms, for instance, become so "heavy" that it is impossible to lift them. The jaw sags and the cheeks and jowls are pulled down in such a way as to render the face unrecognizable.

Physiological Limits The actual force acting on an airman depends on his mass. It is thus simpler to measure these effects in terms of centripetal *acceleration* rather than in terms of centripetal *force*. Further, since any person at the earth's surface, no matter what his or her mass, is subject to an acceleration of 1 g due to the earth's gravity, it has become the custom to evaluate centripetal acceleration in terms of g's.

Illustrative Problem 7.4 How many g's must the pilot of the preceding problem withstand at the bottom (pullout) of the loop?

Solution From Eq. (7.5),

$$a_c = \frac{v^2}{r}$$

Substituting,

$$a_c = \frac{(513 \text{ ft/sec})^2}{2000 \text{ ft}} = 132 \text{ ft/sec}^2$$

Dividing this result by g (32.2 ft/sec²), we obtain

$$a_c = \frac{132 \text{ ft/sec}^2}{32.2 \text{ ft/sec}^2} = 4.1 \; g$$
$$\overline{ 1 \; g }$$

Fig. 7.9 Astronaut riding a prototype moon bike under one-sixth "g" conditions aboard a KC-135 aircraft. He is wearing a pressurized space suit which protects him throughout a range from zero-gravity up to 10 "g's". (National Aeronautics and Space Administration.)

7.11 Highway and Rail Travel Depend on Centripetal Force

When high-speed trains, autos, and trucks travel around curves, tremendous forces must be supplied by the roadbed to hold them in their curving paths. For example, Fig. 7.10 diagrams an automobile traveling on a curve of radius r at speed v. If the roadbed is level, the only force which can supply the required centripetal force is the force of friction between tires and pavement. Recall the analysis of frictional forces in Chapter 6. We write an equation for the instant when skidding is impending, i.e., at the instant when the force of static friction has reached its maximum. Setting friction force equal to centripetal force, we have

$$F_f = F_c$$

or, in absolute units,

$$\mu_s mg = mv^2/r$$

or $$\mu_s = v^2/rg \qquad (7.9)$$

Note that the mass of the auto makes no difference, since m's cancel out in obtaining Eq. (7.9).

Illustrative Problem 7.5 What is the maximum velocity permissible for an automobile rounding a level curve of 200-ft radius if μ_s between tires and roadbed is 1.0. Answer in miles per hour.

Adding 1 g for the effect of gravity, we find the total acceleration on the pilot is

$$a = 5.1\ g \qquad \qquad ans.$$

Note that this number of g's is also the ratio of the centripetal force to the pilot's weight.

Research has shown that young people in good physical condition can tolerate accelerations up to 7 or 8 g's without blacking out. Efficiency is impaired, of course, anywhere beyond 3 or 4 g's. To minimize this impairment and to permit exceeding the 8-g limit in emergencies, special flying suits called *g suits* or *space suits* are used. These suits apply pressure at critical body areas and help prevent blood drainage and organic damage during aerobatics and space flights (see Fig. 7.9).

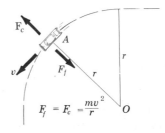

Fig. 7.10 Road friction can supply some centripetal force to reduce the tendency to skid on highway turns. Skidding impends when the required centripetal force is just equal to the maximum value of the force of static friction between tires and the road surface.

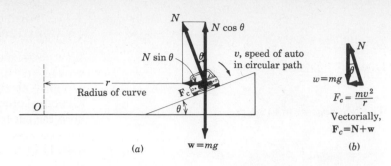

$$F_c = \frac{mv^2}{r}$$

Vectorially,
$$\mathbf{F}_c = \mathbf{N} + \mathbf{w}$$

(a)   (b)

Fig. 7.11 Banking of highway curves. The angle θ is called the *banking angle*. (*a*) Geometric and (*b*) vector relationships, assuming frictional forces are zero. (*c*) Banking of speedway on auto testing grounds. (Ford Motor Co.)

(c)

Solution Use Eq. (7.9) and solve it for v:

$$v = \sqrt{\mu_s g r}$$
$$= \sqrt{1.0 \times 32.2 \text{ ft/sec}^2 \times 200 \text{ ft}} = 80 \text{ ft/sec}$$
$$= 80 \text{ ft/sec} \times \frac{30 \text{ mi/hr}}{44 \text{ ft/sec}} = 54.5 \text{ mi/hr} \qquad ans.$$

This seems a reasonable speed for a curve of 200-ft radius, but dry pavement and good tire treads are not always present. On a slick or icy pavement μ_s might well be less than 0.2 and a safe value of v would then be only 24 mi/hr!

7.12 Banking Roadbeds

The value of μ_s is a variable, dependent on such factors as weather and tire condition. Highway engineers therefore attempt to provide another and more constant factor to supply the required centripetal force on curves of short radius. This is

done by inclining, or "banking," the roadbed. Suppose the auto in Fig. 7.11a is coming toward you at speed v around a curve of radius r. The center of the curve O is toward the left. Neglecting frictional forces entirely, we analyze the two forces, $w = mg$ (the weight of the car acting vertically downward) and N (the normal reaction force of the roadbed on the car). The value of the force N is not dependent on friction. Since the road is *banked* at an angle θ with the horizontal, N is inclined to the vertical by this same angle. Its horizontal and vertical components are, respectively, $N \sin \theta$ and $N \cos \theta$, as shown in Fig. 7.11a. The vertical component, $N \cos \theta$, is equal and opposite to the weight w. If it were not, the car would have acceleration in a vertical plane. The horizontal component, $N \sin \theta$, must supply the centripetal force F_c to hold the car in its circular path, if there is to be no reliance on friction to do so. Consequently (see vector diagram, Fig. 7.11b),

$$w = mg = N \cos \theta \qquad (7.10)$$

and $\qquad F_c = mv^2/r = N \sin \theta \qquad (7.11)$

Dividing Eq. (7.11) by Eq. (7.10) results in

$$\tan \theta \doteq v^2/rg$$

The *banking angle* is, then,

$$\theta = \tan^{-1} \frac{v^2}{rg} \qquad (7.12)$$

where $v =$ linear speed of vehicle, m/sec or ft/sec
$r =$ radius of curve, m or ft
$g =$ acceleration of gravity = 9.81 m/sec^2
or 32.2 ft/sec^2.

Note that w, the weight of the car, does not affect the value of the banking angle. However θ does depend on the value of the velocity v. Consequently, a turn can be banked correctly only for a speed chosen in advance. Engineers try to make the best estimate of the speed at which motorists may drive the curve, or they may bank it for the legal speed limit. Friction will also be present, even though it is assumed zero when determining the banking angle. This adds a factor of safety for the fast driver except when the road is icy. Figure 7.11c shows the banking of a speedway curve at an automotive proving ground.

Illustrative Problem 7.6 A roadway whose radius of curvature is 200 m is to be "banked." Traffic is expected to move around the curve at 90 km/hr. What should be the value of the banking angle if no dependence is to be placed on friction?

Solution Using Eq. (7.12),

$$\theta = \tan^{-1}(v^2/rg)$$

Substitute the given data, changing km/hr into m/sec:

$$\theta = \tan^{-1} \frac{\left(\dfrac{90 \text{ km/hr} \times 1000 \text{ m/km}}{3600 \text{ sec/hr}} \right)^2}{200 \text{ m} \times 9.81 \text{ m/sec}^2}$$

$$= \tan^{-1} 0.319$$

From the tables, or by calculator

$$\theta = 17.7° \qquad \qquad ans.$$

You should note again that since r and g are constants of the problem, θ, the *banking angle*, is correct only for the particular speed v for which it is determined. In the illustrative problem just solved, if the speed were in excess of 90 km/hr, frictional force would have to be relied upon to keep the car from skidding up the bank, whereas if an overly cautious driver negotiated the curve at a slow speed, frictional force would have to be relied upon to prevent the car from sliding down the banked roadway. The latter condition is sometimes a problem on banked curves after a winter ice storm.

GRAVITATION AND SPACE EXPLORATION

7.13 Newton's Law of Universal Gravitation

In the century before Newton many scientists and philosophers contributed to knowledge of natural phenomena. Galileo studied freely falling bodies and formulated some of the general laws of accelerated motion. He also devised the first telescope for the magnification of celestial bodies. Tycho Brahe made determinations of the daily motions of the planets and some of the stars over a period of 30 years. Copernicus formulated the heliocentric theory of the solar system (the ancients had believed that the sun and the planets all revolved about the earth). Kepler, after years of studying Brahe's data and Copernicus' papers, formulated his ideas of planetary motion into what have since been known as *Kepler's laws*. They are three in number. Two of them are illustrated by Figs. 7.12 and 7.13.

> *The orbits of the planets are elipses (not circles), with the sun at one focus.*
>
> *The speed of the planets in their orbits around the sun varies in such a manner that a line joining the planet and the sun sweeps out equal areas in equal periods of time.*
>
> *The squares of the orbital periods of the planets around the sun are proportional to the cubes of their mean distances from the sun.*

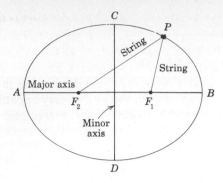

Fig. 7.12 An ellipse, showing major and minor axes and the two foci, F_1 and F_2. An ellipse can be constructed by setting two pins, F_1 and F_2, firmly in place and fastening the ends of a piece of string to the pins. A pencil P is then used to draw the string taut, forming triangle F_1PF_2. The pencil is then swept around, keeping the string taut, as the pencil slides along the string.

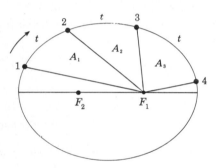

Fig. 7.13 Kepler's second law. Points 1, 2, 3, and 4 represent four different positions of a planet in its orbit around the sun, which is at F_1. The speed of the planet in its orbit varies in such a way that a line joining the planet to the sun sweeps out equal areas in equal time intervals, t. Area A_1 = area A_2 = area A_3, etc. The eccentricity of the planetary orbit is exaggerated in the diagram.

The paths of the planets around the sun and the orbit of the moon around the earth are actually ellipses rather than circles. But the deviation is so small that for our purposes we shall treat planetary and lunar orbits as circular. In actual space navigation however, such a simplification would not be possible. You can see that in drawing ellipses by the method of Fig. 7.12, when the pins F_1 and F_2 are brought

closer and closer together, a circle results in the limiting case when F_1 and F_2 merge. From years of analytical study and experimentation, Newton was able to propose a unifying theory of gravitation. Using the solar system as a starting point, Newton generalized a theory of gravitational attraction which he felt would apply anywhere in the universe. It can be stated as follows:

Newton's Law of Universal Gravitation

Any two bodies or particles in the universe attract each other with a force which is directly proportional to the product of their masses and inversely proportional to the square of the distance between them.

As a mathematical expression, it can be written

$$F \propto \frac{m_1 m_2}{d^2}$$

This expression describes the nature of gravitational attraction, but it cannot evaluate the force F between two bodies. In order to do that the proportionality sign must be replaced with an equals sign and a constant which will validate the equals sign, as follows:

$$F = \mathbf{G} \frac{m_1 m_2}{d^2} \qquad (7.13)$$

The constant **G** is called the *universal gravitation constant*. It must not be confused with g, the acceleration of gravity. Scientists have been measuring the value of **G** for over 200 years, and the following values are now agreed upon.

SI-Metric Units   If F is in newtons, m_1 and m_2 in kilograms, and d in meters, **G** has the value

$$\mathbf{G} = 6.67 \times 10^{-11} \text{ N-m}^2/\text{kg}^2 \qquad (7.14a)$$

or, since the dimensions of the newton are kg-m/sec²,

$$\mathbf{G} = 6.67 \times 10^{-11} \text{ m}^3/(\text{kg})(\text{sec}^2) \quad (7.14b)$$

English-Engineering Units   If F is in pounds, m_1 and m_2 in slugs, and d in feet, **G** has the value

$$\mathbf{G} = 3.44 \times 10^{-8} \text{ lb-ft}^2/\text{slug}^2 \qquad (7.14c)$$

The distance d in Eq. (7.13) is the *straight-line distance between the centers of mass* of the two masses m_1 and m_2.

Among the planets, gravitational forces are, of course, tremendous. But, for objects of our ordinary experience, the forces are small indeed, as the following problem shows.

Illustrative Problem 7.7 Two 16-lb shot spheres (as used in track meets) are held 2 ft apart. What is the force of attraction between them?

Solution In engineering units,

$$16 \text{ lb} = 0.497 \text{ slug}$$

Substituting in Eq. (7.13) using the value of G from Eq. (7.14*c*),

$$F = 3.44 \times 10^{-8} \frac{\text{lb-ft}^2}{\text{slug}^2}\left(\frac{0.497 \text{ slug} \times 0.497 \text{ slug}}{(2 \text{ ft})^2}\right)$$

$$= 2.12 \times 10^{-9} \text{ lb force} \qquad\qquad ans.$$

This very small force would be detectable only by precision scientific apparatus in a research laboratory.

It is again emphasized that the *universal gravitation constant* G is a completely different concept and quantity from the *acceleration of gravity* g. However, the two constants can be related mathematically as follows: Let

M = mass of earth
w = force of gravitation on a small mass m at or near the *surface* of earth
r = earth's radius, the distance to the center from the surface.

The substitution of these symbols in Eq. (7.13) gives

$$w = G\frac{Mm}{r^2} \qquad\qquad (7.15)$$

But from Newton's second law, $g = w/m$, and consequently

$$g = \frac{GM}{r^2} \qquad\qquad (7.16)$$

Equation (7.16) shows that g decreases as r increases. In other words, as we get farther from the earth's *center*, g diminishes rapidly. However, since the earth's radius is 4000 miles, one has to attain a considerable distance from the earth's *surface* before an appreciable diminution in the value of g can be observed by ordinary apparatus. Since the earth's radius is not everywhere the same, g varies slightly over the earth, being less at the equator than it is at the poles.

7.14 Mass of the Earth

The diagram of Fig. 7.14 (not to scale) shows the relationships involved in the mutual attraction between the earth and a 1-kg mass on its surface. We will use it to calculate the mass of the earth. We know that the earth imparts an acceleration $g = 9.81$ m/sec^2 to any small mass at sea level. The radius of the earth is 6.37×10^6 m. G is given by Eq. (7.14*b*) as 6.67×10^{-11} m^3/(kg)(sec^2). Solving Eq. (7.16) for M, the mass of the earth, gives

$$M = \frac{gr^2}{G}$$

$$= \frac{9.81 \text{ m/sec}^2 \times (6.37 \times 10^6 \text{ m})^2}{6.67 \times 10^{-11} \text{ m}^3/(\text{kg})(\text{sec}^2)}$$

Mass of earth $= M = 5.96 \times 10^{24}$ kg

which is equivalent to a weight of about 6.6×10^{21} tons (at 1 g).

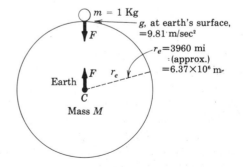

Fig. 7.14 Mutual attraction between the earth and a 1-kg mass on the earth's surface (not to scale).

Illustrative Problem 7.8 The moon's mass is 7.37×10^{22} kg (see Table 7.1) and its radius is 1740 km. Find (a) the acceleration of gravity on the surface of the moon and (b) the apparent "weight" on the moon of a man who weighs 80 kilos (kg-f) on earth.

Solution The equations above apply to *universal* gravitation, and they can therefore be used anywhere in the universe.

(a) Using Eq. (7.16),

$$g_{moon} = \frac{GM_{moon}}{r^2}$$

and substituting values for the moon,

$g_{moon} =$

$$\frac{6.67 \times 10^{-11} \text{ m}^3/(\text{kg})(\text{sec}^2) \times 7.37 \times 10^{22} \text{ kg}}{(1.74 \times 10^6 \text{ m})^2}$$

$$= 1.62 \text{ m/sec}^2 \qquad \qquad ans.$$

or roughly one-sixth that at the earth's surface.

(b) The apparent weight of a man on the moon equals his weight on earth times the ratio g_{moon}/g_{earth}.

$$w_{moon} = 80 \text{ kg-}f \times \frac{1.62 \text{ m/sec}^2}{9.81 \text{ m/sec}^2}$$

$$= 13.2 \text{ kg-}f$$

$$= 130 \text{ N} \qquad \qquad ans.$$

Table 7.1 presents selected data for some of the bodies in the solar system.

7.15 Orbiting Satellites

Launching and putting into orbit an earth satellite is a very complex operation. For successful orbiting, the final-stage rocket motor must impart exactly the right velocity for the intended radius of the orbit. It was pointed out earlier that orbital paths of planets around the sun are actually ellipses. The orbits of man-made satellites can be adjusted by firing small propulsion rockets. These adjustments enable the satellite to be maneuvered into a nearly circular orbit at a selected distance from the surface of the earth, moon, or planet. The discussions to follow will assume circular orbits.

When orbits are assumed circular in satellite mechanics, the centripetal force needed for uniform circular motion is provided by gravitational attraction. We neglect in the following treatment the *drag* forces due to "air" resistance, which, although satellites operate in outer space, are sometimes of sufficient magnitude to cause the orbit to "decay" over a period of weeks, months, or years, as happened with *Skylab* in 1979.

In Fig. 7.15 a man-made satellite is shown circling a planet of mass M in an orbit of radius r_s. Let

r_p = radius of planet
F = force of gravitational attraction between planet and satellite
g_s = value of the acceleration of gravity at distance of orbit
v_s = speed of satellite in orbit

Table 7.1 Selected data for the solar system

Celestial Body	Mass, kg	Mean Radius, km	g at Surface, m/sec²
Earth	5.97×10^{24}	6,370	9.81
Moon	7.37×10^{22}	1,740	1.62
Sun	1.97×10^{30}	695,000	274.0
Mars	6.37×10^{23}	3,400	3.90
Jupiter	1.88×10^{27}	71,800	26.45

NOTE: $G = 6.67 \times 10^{-11}$ N-m²/kg² or 6.67×10^{-11} m³/kg-sec²

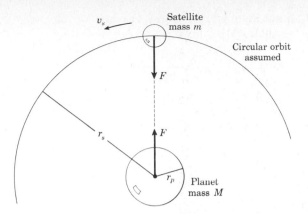

Fig. 7.15 Diagram (not to scale) of a man-made satellite of mass m revolving around a planet of mass M in a circular orbit of radius r_s.

Fig. 7.16 The Landsat-3 spacecraft, or satellite, for use in observing and recording earth data. The photo shows the satellite on a test stand completely assembled just as it will be later when it is in orbit. The solar arrays are extended and the technicians are making final adjustments on the battery of solar cells that will convert the sun's rays into electric power to operate the satellite's observation and communications systems. (National Aeronautics and Space Administration.)

Then $F = mg_s$. But from centripetal-force considerations (Eq. 7.7), the gravitational force of attraction must supply the centripetal force to hold the satellite in orbit, or

$$F = mg_s = \frac{mv_s^2}{r_s}$$

Therefore if the satellite is to remain in its orbit, its velocity must be

$$v_s = \sqrt{g_s r_s} \qquad (7.17)$$

Note that the mass of the orbiting object does not affect the value of the required velocity.

For a satellite orbiting about any given planet, e.g., the earth, $g = GM/r^2$ (Eq. 7.16). Note that G and M are both constants (G is a universal constant, and M is a constant for any one celestial body). Consequently, the relationship between the value of g at the satellite altitude and the value of g at the planet's surface is given by

$$\frac{g_s}{g_p} = \frac{r_p^2}{r_s^2} \qquad (7.18)$$

where g_s = value of g at satellite orbit distance
g_p = value of g on surface of planet
r_s = radius of satellite orbit
r_p = radius of the planet

In other words, for a satellite orbiting any one celestial body, g is inversely proportional to r^2, where r is the distance from the center of the celestial body to the satellite's orbital path.

Illustrative Problem 7.9 A "Surveyor" satellite is to be put into a circular orbit 400 miles above the earth's surface (see Fig. 7.16). If the mean radius of the earth is 3960 mi and g at the earth's surface is 32.2 ft/sec², find (*a*) the required speed for the satellite to stay in orbit and (*b*) the time required for a complete "pass" around the earth.

Solution

(*a*) $r_s = 3960 + 400 = 4360$ miles. From Eq. (7.18),

$$g_s = \frac{g_p r_p^2}{r_s^2} = 32.2 \text{ ft/sec}^2 \times \frac{(3960 \text{ miles})^2}{(4360 \text{ miles})^2}$$

$$= 26.6 \text{ ft/sec}^2$$

Equation (7.17) then gives the necessary orbital velocity:

$$v_s = \sqrt{g_s r_s}$$

$$= \sqrt{\frac{(26.6 \text{ ft/sec}^2) \times 4360 \text{ mi}}{5280 \text{ ft/mi}}}$$

$$= 4.69 \text{ mi/sec} = 16,900 \text{ mi/hr} \qquad ans.$$

(*b*) From Eq. (7.2), we find the satellite's angular velocity:

$$\omega = v/r_s$$

Substituting from (*a*), we get

$$\omega = \frac{16,900 \text{ mi/hr}}{4360 \text{ miles}} = 3.87 \text{ rad/hr}$$

Since there are 2π rad/rev, the angular velocity in revolutions per hour is

$$\frac{3.87 \text{ rad/hr}}{6.28 \text{ rad/rev}} = 0.616 \text{ rev/hr}$$

The period, or time for one revolution, is

$$\frac{1}{0.616 \text{ rev/hr}} = 1.623 \text{ hr/rev} = 97.4 \text{ min.} \quad ans.$$

For *any satellite* orbiting around *any celestial body* in a circular orbit, we can derive a useful formula for the velocity. From Eqs. (7.7) and (7.13) we can write

$$F = \frac{mv_s^2}{r_s} = \mathbf{G}\frac{Mm}{r_s^2}$$

Dividing both sides by m and multiplying both sides by r_s gives

$$v_s^2 = \frac{\mathbf{G}M}{r_s}$$

and

$$v_s = \sqrt{\frac{\mathbf{G}M}{r_s}} \qquad (7.19)$$

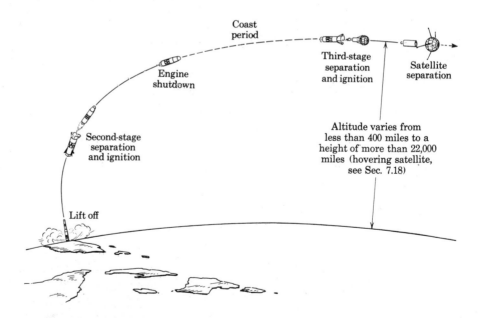

Fig. 7.17 Artist's conception of the launch sequence of a three-stage rocket with satellite payload, showing the various stages of lift-off and satellite separation.

Coast period

Engine shutdown

Third-stage separation and ignition

Satellite separation

Second-stage separation and ignition

Altitude varies from less than 400 miles to a height of more than 22,000 miles (hovering satellite, see Sec. 7.18)

Lift off

This is the required orbiting velocity for *any satellite* to remain in a circular orbit of radius r_s around *any celestial body*.

Illustrative Problem 7.10 A satellite is to circle Mars at a distance of 200 km from the surface. What is the required orbital speed?

Solution From Table 7.1,

$$M_{\text{Mars}} = 6.37 \times 10^{23} \text{ kg}$$

$$\text{Radius} = 3400 \text{ km}$$

The radius of the satellite orbit is

$$r_s = 3400 + 200 = 3600 \text{ km}$$

From Eq. (7.19), the velocity required is

$$v_s = \sqrt{\frac{GM}{r_s}} =$$

$$\sqrt{\frac{6.67 \times 10^{-11} \text{ m}^3/(\text{kg})(\text{sec})^2 \times 6.37 \times 10^{23} \text{ kg}}{3.6 \times 10^6 \text{ m}}}$$

$$= 3.44 \times 10^3 \text{ m/sec} \qquad\qquad ans.$$

7.16 Weightlessness

Astronauts during space flights may encounter *zero gravity* or "weightlessness." A rapidly

Fig. 7.18 Zero gravity or "weightlessness" in an elevator. In this case, the elevator's downward acceleration is momentarily greater than *g*.

descending elevator, if its acceleration is near 9.81 m/sec², creates a similar condition (see Fig. 7.18). During astronaut training on earth, zero-gravity conditions must be created artificially. One convenient method is to use the principles of centripetal force.

Astronaut trainees and their equipment are taken aloft in an airplane with a large cabin space which has been properly prepared in advance. At a suitable altitude the aircraft executes an outside loop (see Fig. 7.19) with a proper combination of curvature of path and speed which allows everybody and everything in the aircraft to be a "freely falling" body, that is, unsupported by the airframe.

Let *A* (Fig. 7.19) be a man of mass *m* sitting on the floor of the aircraft. Two forces will act on him: (1) the force of gravity, $F_g = mg$ (his weight), acting downward, and (2) the force *N*, with which the floor supports him. The condition for "weightlessness" is that *N* = zero.

Now, the resultant of *N* and *mg* is the centripetal force F_c that "holds" the aircraft (and the man) in the curved path. In equation form,

$$F_c = mg - N$$

With *N* = 0,

$$F_c = mg$$

But

$$F_c = mv^2/r$$

and we write

$$mv^2/r = mg$$

from which

$$v = \sqrt{gr} \qquad\qquad (7.20)$$

which is the condition for zero gravity. Note that the *m*'s cancel out (all persons in the plane will be "weightless" at the same time regardless of their differences in size).

It should be noted that the crew and objects are not *really* weightless. They still have mass and the force of earth's gravity still acts on them $(w = mg)$. They are actually in free fall and are

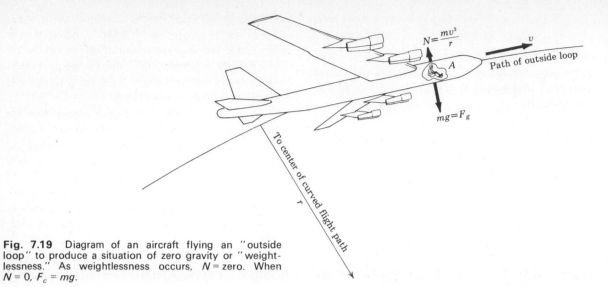

$$N = \frac{mv^2}{r}$$

v

Path of outside loop

A

$mg = F_g$

To center of curved flight path

r

Fig. 7.19 Diagram of an aircraft flying an "outside loop" to produce a situation of zero gravity or "weightlessness." As weightlessness occurs, N = zero. When $N = 0$, $F_c = mg$.

unsupported, just like the person in the elevator of Fig. 7.18.

Illustrative Problem 7.11 Some astronauts are undergoing zero-gravity training in a jet transport plane, as illustrated in Fig. 7.20. The pilot is instructed to fly at 450 mi/hr. What is the radius of curvature of the outside loop when "weightlessness" occurs?

Solution Square both sides of Eq. (7.20), obtaining

$$v^2 = gr \qquad \text{or} \qquad r = \frac{v^2}{g}$$

Since 450 mi/hr = 660 ft/sec,

$$r = \frac{(660 \text{ ft/sec})^2}{32.2 \text{ ft/sec}^2}$$

Fig. 7.20 Astronauts engaged in zero-gravity training. The aircraft is flying an outside loop along a parabolic path.

$$= 1.35 \times 10^4 \text{ ft} \qquad ans.$$

If the curvature were more gradual, the men would remain sitting on the floor; if a tighter loop were negotiated, they would appear to "float upward" and would bump against the ceiling of the aircraft.

7.17 Landing on Another Planet

The problem of landing on the moon or on another planet is far more complex than that of orbiting the earth with a satellite. First, the vehicle must "escape" from the earth along a flight path which will result in its intercepting in space, perhaps many days later, the planet intended. This interception is in itself a complex problem of relative motion. If the "target" is one of the planets of our solar system, the sun may be used as the frame of reference for solving the relative-velocity problem of earth, rocket, and target planet.

Escape Velocity As a *projectile* climbs vertically upward against the force of gravity, its kinetic energy diminishes as the potential energy increases. In order to "escape" the earth's gravity, the initial velocity must provide an initial kinetic energy equal to or greater than the potential energy the projectile-earth system will have when it reaches a height where the earth's gravity is zero. This is called *escape velocity*. Since the derivation of a formula for escape velocity requires the use of the calculus, the relationship is given here without derivation. If r_e is the radius of the earth and g the acceleration of gravity at the earth's surface, the velocity which must be imparted to a *projectile* at the earth's surface in order for it to escape is given by

$$v_{\text{escape}} = \sqrt{2gr_e} \qquad (7.21)$$

It is interesting to note that Eq. (7.21) gives the escape velocity for *any direction* of launching. (Air resistance is neglected throughout this discussion.)

It should be noted that we have been describing the escape velocity for a *projectile*. The same general principles apply to rocket-propelled vehicles. A rocket *starts* from zero velocity but it has a continued push, and it is common practice to accelerate a rocket-propelled vehicle to reach escape velocity soon after it leaves the earth's atmosphere, or, if it has orbited the earth preparatory to embarking on its space flight, soon after leaving the "parking orbit." The continued thrust of rocket propulsion also prevents air resistance from slowing the vehicle.

Escape velocity from earth is of the order of 25,000 mph, as compared to the velocity required to stay in a low-altitude orbit around the earth, which is of the order of 17,500 mph. The required *orbital velocity* for a low-altitude satellite is (Eq. 7.17) $v_s = \sqrt{g_s r_s}$. The *escape velocity*, v_{escape} (Eq. 7.21), is

$$v_{\text{escape}} = \sqrt{2gr_e} \quad \text{(from earth)}$$

Therefore escape velocity equals $\sqrt{2}$ times the orbital velocity for a low-altitude orbit.

As the probe or spaceship nears its intended target, one of three outcomes is possible:

1. Its speed and direction with respect to the target may be such that the centripetal force due to the target's gravitational attraction never becomes great enough to result in "capture." In this case the probe or spaceship will make one pass in a curved path near the planet, and then fly off into space.

2. The speed and approach path may be such that the centripetal acceleration (g_s) just satisfies Eq. (7.17), and the spaceship will be "captured" by the planet and remain in orbit around the planet at a distance r_s.

3. The speed and path may have been such as to bring the vehicle so close to the target planet (or moon) that the force of gravitational attraction is greater than that required as centripetal force for an orbit, and the spaceship will be "captured" in the very real sense that it crashes on the planet.

These three possibilities can be varied by firing additional guidance and control rockets on the spaceship, thus making course corrections in

Fig. 7.21 Artist's conception of Space Shuttle vehicle returning for an earth landing after a space mission. These vehicles have been successfully test flown and have landed just as more or less conventional aircraft. They are expected to serve as commuter craft to and from future space satellites. (National Aeronautics and Space Administration.)

flight. Such mid-course corrections have been common practice on manned space missions.

Landing on another celestial body is usually preceded by going into orbit around it for several "passes" in order to accomplish such objectives as visual and photographic surveillance; measurement of magnetic fields, gravitational forces, radiation, and meteorite intensity; and making sure that all is in readiness for the descent to the surface. Landing on the planet requires the firing of retrorockets and course-correction rockets. Finally, the spaceship (or a part of it) must return to earth. The return trip is no simpler than the outbound trip. It, too, requires escape, interception, orbit, and a safe landing.

Guidance rockets for mid-course corrections put the returning vehicle into the proper "window" for a return through the earth's atmosphere to the preselected landing area. The atmosphere supplies the braking force to slow the vehicle from its 30,000 mph (plus) speed to a manageable speed for a landing. The Apollo Program vehicles were landed by parachute, but the Space Shuttle (see Fig. 7.21) is an aircraft, and will be piloted to its landing. The heat generated as the vehicle streaks through the atmosphere is tremendous, and a special heat shield is necessary

to dissipate the heat and prevent a buildup of heat which might consume the vehicle like a fiery meteor.

7.18 Practical Uses of Satellites

Space probes and earth satellites have many valuable uses. *Weather satellites* bring us worldwide meteorological information; *scientific satellites* collect information on magnetic fields, radiation from outer space, meteorites, and the sphericity of the earth. *Surveillance satellites* have been used to gather information on military installations, and *communications satellites* have made a significant contribution to worldwide communications. Such early satellites as Telstar and Early Bird revolutionized intercontinental television in 1962, when Telstar I was launched. Early Bird was placed in commercial service in 1965 and was reported to have a capability equivalent to that of 240 transatlantic telephone channels (see also Chap. 32).

Mammoth space laboratories are already in a preliminary design stage. They will orbit the earth at altitudes of several hundred miles, and may house research laboratories or specialized manufacturing operations that must take place in vacuum or under "zero-gravity" conditions. Space shuttles, like that in Fig. 7.21, will be used to commute between these satellites and the earth. There is even talk of "space colonization," but the line between practical reality and science fiction is difficult to draw in these matters. Energy shortages may force cancellation of some space exploration projects.

Hovering Satellites Most satellites are in orbits which require relative motion between them and the earth's surface. However, satellites can be placed in orbits where the required orbital speed is exactly equal to the surface rotational speed of the earth, and thus the satellite "hovers" over one location on the earth's equator. Such satellites are said to be in a synchronous orbit, since their motion is synchronized with the earth's daily rotation. These satellites must be at altitudes of about 22,250 miles to satisfy Eq. (7.17), and they are used for international communications, space-program communications, and for television programs and other message traffic between the United States and other nations of the world.

SUMMARY

Translation is *motion in which no straight line in a body changes direction as the body moves from one place in space to another.*

Rotation is *motion in which all particles of a body move in circles whose centers lie on the same straight line.* This straight line is called the *axis of rotation.*

The radian is the engineering unit of angular displacement and the radian per second the engineering unit of angular velocity. Another unit of angular velocity is the revolution per minute:

$$\omega = \frac{rad}{sec} = 2\pi \times \frac{rev/min}{60}$$

The relationship between linear speed in a circular path and angular velocity is given by the equation $v = \omega r$.

A particle is in *uniform circular motion* if a radius from the point to the center of the circle is rotating at a constant angular velocity.

A particle in uniform circular motion has a constant linear speed, but the *velocity* is not constant, since the *direction* is continually changing. The particle therefore has *acceleration,* which is directed toward the center of rotation and is called *centripetal acceleration*

$$a_c = \frac{v^2}{r} \qquad \text{or} \qquad a_c = r\omega^2$$

Centripetal force, the force necessary to hold an object in uniform circular motion, is

$$F_c = \frac{mv^2}{r} \qquad \text{or} \qquad F_c = mr\omega^2$$

Centripetal force is a *center-seeking* force. Centrifugal force is a *center-fleeing* force. It is the result of the inertia of a mass which is forced to move in a circular path by centripetal force.

Centripetal and centrifugal forces have frequent practical applications in many fields of engineering and industrial design. A few examples are stresses in machine parts which have high rotational speeds, and centrifuges and centrifugal separators. Other examples are the material and physiological aspects of aerobatics, and the banking of roadways for auto and rail transportation.

Centripetal force is closely related to problems of space technology.

Newton's law of universal gravitation states that *any two bodies in the universe attract each other with a force proportional to the product of their masses and inversely proportional to the square of the distance between them.* As a formula,

$$F = G\frac{m_1 m_2}{d^2} \qquad (7.13)$$

where **G** is the *universal gravitation constant.*

The problems encountered in satellite mechanics can, with certain assumptions and simplifications, be solved by utilizing the principles of *uniform circular motion.*

The orbiting velocity for *any satellite* around *any celestial body* (assuming a circular orbit) is given by

$$v_s = \sqrt{\frac{GM}{r_s}} \qquad (7.19)$$

"Weightlessness," or zero gravity, is a characteristic of outer space, but it can be artificially created on earth by a carefully controlled use of centripetal-force and free-fall principles. The condition for "zero gravity" at or near the earth's surface is to relate linear velocity to the radius of an aerobatic "loop" in accordance with:

$$v = \sqrt{gr} \qquad (7.20)$$

where v = velocity in curved path
r = radius of curvature of path being flown while executing an outside loop
g = acceleration of earth's gravity

Escape velocity from the earth is of the order of 25,000 mph and is given by the formula

$$v_{escape} = \sqrt{2gr_e} \qquad (7.21)$$

Earth satellites have many practical uses. Hovering satellites are already an important element in world-wide communcations.

QUESTIONS AND EXERCISES

1. A ship's propeller rotates on the propeller shaft as its axis. But the ship also moves forward. Describe what kind of motion the propeller has with respect to the earth. What technical term describes the path of the motion of the propeller?

2. Analyze three common machines (for example, an auto, a lawn mower, a bulldozer—but you pick any three you like) and classify the movement of their main parts as to whether it is translation, rotation, or a combination of both. Simplify by neglecting any forward motion of the machine itself.

3. What is the essential difference between a projectile and a rocket-powered vehicle? Does the rocket ever become a projectile?

4. Explain carefully the distinction between G and g. Is either ever zero? Does G vary over the surface of the earth? Does g? Explain.

5. A manned landing on Mars is being planned. Explain each phase of the operation, step by step, including the process of bringing the space vehicle (or a part of it) with the astronauts back to earth safely.

6. Orbiting satellites revolve around the earth (like a stone being whirled around on a string) with the earth's gravitational pull supplying the centripetal force. If they "slow down," their orbits "decay" and they eventually crash. How, then, is it possible to have "hovering satellites"—satellites that stay "parked" over one spot on the earth's equator?

7. Everyone who drives an auto is aware of the danger of speeding around a banked curve if the roadway is wet or icy. Under what conditions might it be dangerous to drive too slowly?

8. Equation (7.16) relates g and G as follows: $g = GM/r^2$. G and M are constants (G is a universal constant, and M is a constant for earth problems). Is it correct to deduce that g diminishes with the square of the altitude above the earth's surface?

9. Look at Eq. (7.7), $F_c = mv^2/r$, and at Eq. (7.7'), $F_c = mr\omega^2$. They are equivalent expressions for centripetal force, one in terms of linear (orbital) speed, and the other in terms of angular velocity. How is it possible that in Eq. (7.7), centripetal force is *inversely* proportional to the radius of revolution, while in Eq. (7.7'), it is *directly* proportional to the radius of revolution?

10. The term "weightlessness" is sometimes used to describe the apparent "zero gravity" caused by orbital motion around the earth. Explain why both of these terms are incorrect. Use the basic relation, $w = mg$ in your explanation. When an object is apparently "weightless" at a point above the earth where g is definitely not zero, what is really the state of motion of the body?

PROBLEMS

1. The earth rotates on its axis once every 24 hr. Its radius is 6370 km (Table 7.1). (*a*) What is the earth's angular velocity in rad/sec? (*b*) If an airplane flies due west at the equator, at what speed (km/hr) must it fly to just "keep up with the sun"?

2. The average distance of the earth from the sun is 93 million mi. Taking the period of revolution around the sun (1 yr) as an even 365 days, what is the earth's linear velocity in miles per hour? (Assume a circular orbit.)

3. Standard industrial electric motor speed is 1750 rev/min. How many radians per second is this?

4. A steam turbine rotor is 12 ft in diameter and turns at 1200 rev/min. How many miles does a point on its rim travel in a 24-hr day?

5. To start an outboard motor, a rope which is wound on a 10-cm drum is given a strong pull. At what linear speed is the rope pulled if the engine is rotated at 600 rev/min? (Answer in meters per second.)

6. A shaft 6 in. in diameter is being turned down on a lathe at 200 rev/min. The cutting tool takes off a continuous ribbon of metal. How many linear feet of the metal ribbon will be cut off in 5 min?

7. A small drive gear with 17 teeth, turning at 1700 rev/min, drives a larger gear with 96 teeth. On the same shaft with the large gear is a V-pulley 6 in. in diameter. Find the linear speed (in feet per second) of the V-belt which this pulley drives. Assume no belt slippage.

8. How much total force (newtons) must a set of bolts be able to withstand if they are to hold a 30-kg counterweight in place on the rim of a flywheel 3 m in diameter which is turning at 250 rev/min? (See Fig. 7.22.)

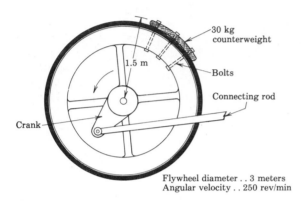

Flywheel diameter . . 3 meters
Angular velocity . . 250 rev/min

Fig. 7.22 Stresses in rotating machinery provide centripetal force (Problem 7.8).

9. A motorcycle and rider perform the stunt of circling in a vertically walled cylindrical track 30 ft in diameter. If the machine makes one revolution every 3 sec and the combined weight of the machine and rider is 580 lb, with what force (pounds) does the wall push in on the cycle's tires? (Neglect the earth's gravitational force.)

10. In a centrifuge, liquid suspected of containing a virus is being rotated at high speed in a circular path of 4-in. radius. If a centripetal acceleration of 100,000 g's is required for separation of the virus, at what speed (rev/min) must the centrifuge be driven?

11. An auto rounds an unbanked curve of 300-ft radius at 30 mi/hr. What is the minimum value of μ_s required between tires and pavement to prevent skidding?

12. Standard-gauge rails are 4 ft 8½ in. apart in the United States. On a curve of 1700-ft radius, how far (in inches) above the inner rail should the outer rail be elevated to allow trains to negotiate the curve at

50 mi/hr without relying on friction or wheel flanges to hold the train on the tracks?

13. How many g's does a pilot experience in pulling out at the bottom of a vertical dive at 400 mi/hr if the loop radius is 3000 ft?

14. What is the minimum speed (km/hr) at which a pilot can execute an inside vertical loop of a 700-m radius and still be pressing against his seat at the top of the loop when his head is hanging down toward the earth?

15. An oval auto race track has semicircular ends designed on a 400-ft radius. If these turns are to be driven at 150 mi/hr, what must be the banking angle if no reliance is to be placed on friction?

16. A small earth satellite is to orbit at a height of 300 km. If the orbit is assumed circular, how many minutes are required for a complete trip around the earth? Would a very large satellite require more or less time?

17. It is desired to place in circular orbit around the earth a 20,000-lb satellite whose path (assumed circular) will be 300 mi from the earth's surface. What must be the terminal velocity of the satellite vehicle in miles per hour as the last-stage rocket motor disengages itself?

18. At what altitude (km) above the earth's surface would the numerical value of g be half that at the earth's surface?

19. In an amusement park a roller coaster runs over the crest (top) of a curve whose radius of curvature is 10 m at a speed of 7.0 m/sec. If a passenger "weighs" 55 kilos under normal conditions, what is her *apparent* weight at the top of the ride?

20. If the command module of a lunar expedition is in a circular orbit around the moon, 20 km from the moon's surface, find (*a*) the module's linear speed in orbit and (*b*) the length of time required for one complete orbit.

21. It is desired to orbit a manned vehicle around Mars at a distance of 80 km from the surface. Find (*a*) the required orbital speed and (*b*) the period of the orbit, in hours.

22. Calculate the force of mutual attraction, in newtons, between the earth and the moon, given the average distance between their centers as 384,500 km. (See Table 7.1.)

23. Calculate the centripetal acceleration of the moon as it revolves around the earth, making a complete revolution in 27.3 days. Its mean distance from the earth is 384,500 km (answer in m/sec²).

24. For a "zero-gravity" exercise to be conducted in a military jet transport aircraft, it is decided to fly an outside loop whose radius at the point of "zero gravity" is 5000 m. Find the required speed of the aircraft, in kilometers per hour.

25. If an astronaut weighs 120 lb on earth, what would be her apparent weight on Mars? (See Table 7.1.)

26. In order to remain in a circular orbit 500 km above the surface of the planet Jupiter, what orbital speed would be required? (See Table 7.1.)

27. Calculate the escape velocity in miles per hour from the earth's surface for a space vehicle whose total weight is 2.4 million lb.

28. Find the escape velocity in miles per hour from the surface of the moon.

29. Prove mathematically that a satellite can hover over the same spot on the equator if it is in orbit at an altitude of 22,250 miles above the earth's surface.

30. A stunt bicyclist "loops the loop" as shown in the diagram of Fig. 7.23. From what minimum height h must he start in order to roll down the incline and around the loop successfully? (Neglect friction. Weight of bike and rider not required. Why?)

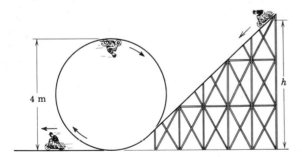

Fig. 7.23 "Looping the loop" (Problem 7.30).

8 Torque, Power Transmission, and Rotational Dynamics

One of the most important applications of rotary motion is in the transmission of energy and power. Electric motors, steam turbines, hydroelectric generators, and turbojet engines are all examples of machines that convert energy into usable power by means of rotary motion.

The turning effect in rotary motion is technically called *torque*. The force which causes torque also tends to cause *rotational acceleration*, just as a force applied in straight-line motion tends to cause *translational acceleration*, which we studied in Chapter 4.

The present chapter first takes up torque and power transmission. Following that, rotational acceleration will be explained. The closing sections of the chapter deal with momentum and kinetic energy—the *dynamics* of rotating bodies. In learning about rotational dynamics you will note that Newton's laws of motion apply to all aspects of rotary motion, just as they do to translational motion.

TORQUE

It was pointed out in previous discussions that force is the cause of motion. This is true not only of translation, but of rotation as well. It is a matter of common experience that it takes force to start a body rotating and also to slow its rotation if it is already spinning.

8.1 Torque Is the Cause of Rotary Motion

If the flywheel of Fig. 8.1 is at rest, a force F applied at the rim will start it rotating clockwise around its axis. Or, if it is already rotating clockwise, a force F' applied in the opposite direction will gradually bring it to rest. However, force alone is not the only factor affecting the rotation of the wheel. The manner of applying the force is just as important as the magnitude of the force itself. A force is most effective in producing rotation of a body when (1) *its line of application is perpendicular to a radius at the point of application* and (2) *the force is applied as far as possible from the center of rotation.* Stated another way, the *turning effect of a given force is proportional to the*

Fig. 8.1 Torque causes a change in rotational motion (rotational acceleration). The flywheel possesses rotational inertia, and torque is necessary to overcome that inertia.

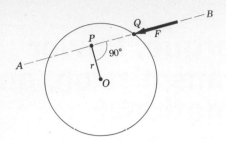

Fig. 8.2 The effect of moment arm on torque. *OP* is the *moment arm* of the force *F*. The product of force and its moment arm is *torque* : τ = Fr.

Fig. 8.3 Mechanic's torque wrench being used in the assembly of electrical apparatus. (P.A. Sturtevant Co.)

perpendicular distance from the line of application of the force to the center of rotation. Thus in Fig. 8.2, *AB* is the line of application of force *F*, *Q* is its point of application, and *OP* is the perpendicular distance from the center of rotation to the line of application of the force. The distance *OP* is called the *moment arm* of the force. The *turning effect* produced by the force is defined as *the product of the force itself and its moment arm.* The turning effect is called *torque.*

If we let the Greek letter τ (tau) stand for torque and designate the moment arm by *r*, we have

$$\tau = Fr \qquad (8.1)$$

Torque = force × moment arm

Units of Torque   *F* is most commonly measured in newtons (SI-metric), or in pounds (English-engineering). Moment arm *r* is expressed in meters or feet. The unit of torque in the metric system is the newton-meter, and the common engineering-industrial unit of torque in the U.S. is the pound-foot (lb-ft). The pound-inch (lb-in.) is also sometimes used. Note that although these English-system units involve the same quantities as the units of *work*, previously defined, the order of the quantities is reversed to avoid confusing one unit with the other. Torque and work are not the same concept by any means.

The mechanic's torque wrench (Fig. 8.3) is a common industrial example of the use of torque in assembly operations. Nuts and bolts must be made up tight enough to hold the assembled parts against design stresses, but "strong-arm" methods often strip threads, shear bolts, or set up unequal stresses in the machine parts which cause trouble later. Some torque wrenches have dials and pointers which read directly in pound-inches. Others (especially motorized or compressed-air wrenches) have a ratchet which begins slipping when the torque reaches a value preset for the particular assembly operation.

POWER TRANSMISSION

In the early days of steam power, the first steam engines were capable only of back-and-forth or *reciprocating motion.* It soon became evident, however, that this new source of power from steam would be of greater value if the reciprocating motion of the piston could somehow be converted to *rotary motion.* The invention of the *connecting rod* (see Fig. 7.1*a*) made this conversion to rotary motion possible. Today almost all industrial power is converted into rotary motion, whether its source is steam, electricity, water, wind, or nuclear energy.

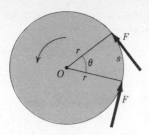

Fig. 8.4 Work done in rotary motion. The work done by the torque, $\tau = Fr$, in turning the shaft through an angular distance θ, is $W = Fr\theta = \tau\theta$.

8.2 Work and Power in Rotary Motion

The transmission of power is one of the most important applications of rotary motion. Consider the force F acting on the shaft whose cross section is shown in Fig. 8.4. Let F turn the shaft counterclockwise through an arc distance s. The work done is

$$W = Fs$$

But (see Sec. 7.2)

$$s = r\theta \qquad (7.1)$$

and therefore $\qquad W = Fr\theta$

Also, by definition, $\quad Fr = \tau \qquad (8.1)$

Substituting, we obtain

$$W = \tau\theta \qquad (8.2)$$

or, in words,

Work = torque × angular displacement

Now, dividing both sides of Eq. (8.2) by t, we obtain

$$\frac{W}{t} = \tau\frac{\theta}{t}$$

But W/t is the rate of doing work, or *power*, and θ/t is *angular velocity*. There results the very useful formula

$$P = \tau\omega$$

Power = torque × angular velocity $\quad (8.3)$

This is the fundamental formula for power transmission by shafts, pulleys, and gears. An analysis of the formula reveals that for a given value of the power P, the torque τ is inversely proportional to the angular velocity. In other words, to transmit a given amount of power at a high rotational speed requires relatively little torque, while at a low speed high torque is necessary.

The units for Eq. (8.3) can be clarified by recalling that the unit of power is the watt, which is a joule/sec, or a newton-meter/sec. On the right-hand side of the equation, the unit of torque is the newton-meter, and the unit of angular velocity is $\sec^{-1}$ (see Sec. 7.3). A similar analysis can readily be done for English units.

Illustrative Problem 8.1 An industrial engine is delivering 25 hp driving a small generator. The V-belt pulley has a mean diameter of 14 in., and the engine is turning at 1800 rev/min. Assuming no belt slippage, find the difference between the tension of the belt on the drive side of the pulley and that on the slack side (answer in pounds).

Solution The difference in belt tension is actually the net force applied to the rim of the pulley. It is applied with a moment arm of the given radius of 7 in. Using Eq. (8.3),

$$P = \tau\omega = Fr\omega$$

Solve for F:

$$F = \frac{P}{r\omega}$$

Since the answer is to be in pounds, substitute as follows, and remember that the radian is a dimensionless number.

$$F = \frac{25 \text{ hp} \times \dfrac{550 \text{ ft-lb/sec}}{1 \text{ hp}}}{\dfrac{7}{12} \text{ ft} \times \dfrac{1800 \text{ rev/min} \times 2\pi \text{ rad/rev}}{60 \text{ sec/min}}}$$

$$= 125 \text{ lb} \qquad\qquad ans.$$

Illustrative Problem 8.2 An electric motor is delivering 100 kW of power through a set of gears to the drum of a winch. A steel cable is wrapped around the winch, whose diameter is 50 cm. If the winch winds in the cable at the rate of 10 m/sec, what is the tension in the cable? Neglect all frictional losses and assume no slippage.

Solution   First, find the angular velocity ω of the drum. Since there is no slippage between cable and drum, the drum surface velocity is the same as that of the cable, 10 m/sec. From Eq. (7.2),

$$\omega = \frac{v}{r} = \frac{10 \text{ m/sec}}{0.25 \text{ m}} = 40 \text{ rad/sec}$$

Next, use Eq. (8.3), recalling that $\tau = Fr$ and therefore $P = Fr\omega$. Solving for F and substituting values gives

$$F = \frac{P}{r\omega} = \frac{100,000 \text{ N-m/sec}}{0.25 \text{ m} \times 40 \text{ rad/sec}}$$

$$= 10,000 \text{ N} \qquad\qquad ans.$$

Engineers use a handy approximate formula to relate torque, horsepower, and angular velocity in revolutions per minute (rev/min). If you pay careful attention to units, you can derive this formula:

$$\text{Torque (lb-ft)} = \frac{5250 \times \text{hp}}{\text{rev/min}} \qquad (8.4)$$

Such a formula enables design engineers to estimate the approximate horsepower requirements of rotating machinery if shaft revolutions per minute and the required torque are known.

Illustrative Problem 8.3   A hay mower is to operate from the power takeoff shaft of a farm tractor. If the required torque is 80 lb-ft when the drive shaft to the mower is turning at 500 rev/min, what horsepower must the tractor engine furnish to operate the mower?

Solution: Solve Eq. (8.4) for horsepower:

$$\text{hp} = \frac{\text{Torque} \times (\text{rev/min})}{5250}$$

Substitute values:

$$\text{hp} = \frac{80 \times 500}{5250} = 7.62 \text{ hp} \qquad ans.$$

8.3   Power Transmission by Drive Shafts

There is no such thing as an absolutely rigid body. Cables stretch under load; beams bend or are compressed, depending on the nature of the stresses they are subjected to; and shafts twist as torques are applied. In every case, the amount of stretch, bending, or twist is proportional to the stress applied. In other words, *stress is proportional to strain* within the limits of elasticity defined by Hooke's law (see Chap. 10).

As discussed in Chap. 6, power transmission by V-belts and pulleys is common practice in shops and industrial plants. The transmission of power by shafts is of tremendous importance because it is the method of transmitting power in automotive and truck transportation, in water- and oil-well drilling, in marine transportation, and in many other applications.

The solution of practical power-transmission problems involving drive shafts is related to the concept of *shearing stress* which will be discussed in Chapter 10.

ROTATIONAL KINEMATICS

The term *kinematics of rotation* is used to describe angular velocity and angular acceleration without reference to force, mass, and inertia considerations. When force, mass, and inertia *are* considered, the term *dynamics of rotation* is used. A later section in this chapter treats rotational dynamics. We introduced the concepts of angular *displacement* and angular *velocity* in Chap. 7, and will now discuss some aspects of angular *acceleration*.

8.4   Angular Acceleration

A rotating body may have a changing angular velocity ω, just as a body in linear motion may have a varying velocity v. If the angular velocity of a body changes from ω_1 to a new value ω_2 in a time interval t, the angular acceleration α is given by the expression

$$\alpha = \frac{\omega_2 - \omega_1}{t} \qquad (8.5)$$

This equation defines angular acceleration as the change of angular velocity per unit time. It is just as often seen in an alternative form,

$$\omega_2 = \omega_1 + \alpha t \qquad (8.6)$$

In order to derive an expression for the *total acceleration* of a body in accelerated angular motion, it will be recalled (see Fig. 7.4) that for a particle rotating at distance r from the center of rotation

$$\omega_1 = \frac{v_1}{r} \quad \text{and} \quad \omega_2 = \frac{v_2}{r} \quad (7.2)$$

Subtracting the first of these expressions from the second, gives

$$\omega_2 - \omega_1 = \frac{v_2 - v_1}{r}$$

Dividing both sides of this equation by t gives

$$\frac{\omega_2 - \omega_1}{t} = \frac{v_2 - v_1}{rt}$$

But (see Fig. 8.5) $(\omega_2 - \omega_1)/t$ is *angular acceleration* α; and $(v_2 - v_1)/t$ is the linear acceleration a_t of the particle *tangent to its momentary path* at P. Combining these relationships results in the following expression:

$$\alpha = \frac{a_t}{r}$$

or

$$a_t = \alpha r \qquad (8.7)$$

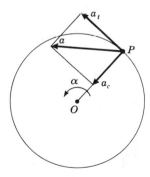

Fig. 8.5 Tangential and centripetal acceleration of a particle in a circular path undergoing angular acceleration.

Equation (8.7) shows that tangential acceleration is equal to the product of angular acceleration and the radius of the rotary motion.

The point whose tangential acceleration at a given instant is a_t also has an instantaneous centripetal acceleration $a_c = r\omega^2$ (Eq. 7.5′). From Fig. 8.5 it should be noted that the tangential acceleration a_t and the centripetal acceleration a_c of a particle P undergoing rotational acceleration are perpendicular to each other and that they have a resultant a, called *total linear acceleration*. From the geometry of Fig. 8.5 the total instantaneous acceleration is

$$a = \sqrt{a_t^2 + a_c^2} \qquad (8.8)$$

Illustrative Problem 8.4 The angular velocity of a turbine rotor is 20 rad/sec at time $t_1 = 0$. At time $t_2 = 50$ sec the angular velocity is 100 rad/sec. Find the angular acceleration, assuming it to be uniform.

Solution From Eq. (8.5), angular acceleration is

$$\alpha = \frac{\omega_2 - \omega_1}{t}$$

substituting values gives

$$\alpha = \frac{(100 - 20) \text{ rad/sec}}{50 \text{ sec}}$$

$$= 1.6 \text{ rad/sec}^2 \qquad \textit{ans.}$$

Illustrative Problem 8.5 A particle on the rim of a high-speed grinding wheel (see Fig. 7.6) is 40 cm from the center of rotation. If the instantaneous angular velocity is 5π rad/sec and the angular acceleration of the wheel is 50 rad/sec², find the tangential and total linear accelerations of the particle.

Solution From Eq. (8.7) the tangential acceleration is

$$a_t = \alpha r$$

$$= 50 \text{ rad/sec}^2 \times 0.4 \text{ m}$$

$$= 20 \text{ m/sec}^2 \qquad \textit{ans.}$$

Centripetal acceleration is

$$a_c = r\omega^2 \quad \text{(Eq. 7.5')}$$

$$= 0.4 \text{ m} \times (5\pi \text{ rad/sec})^2$$

$$= 98.7 \text{ m/sec}^2$$

Total linear acceleration is

$$a = \sqrt{a_t^2 + a_c^2}$$

$$= \sqrt{(20 \text{ m/sec}^2)^2 + (98.7 \text{ m/sec}^2)^2}$$

$$= 101 \text{ m/sec}^2 \qquad \qquad ans.$$

Generalized Equations of Angular Acceleration   The discussion of Sec. 7.4 related angular distance to angular velocity by the equation

$$\theta = \omega t \qquad \qquad (7.4)$$

The analogous equation relating linear distance to linear velocity is, from Sec. 4.4,

$$s = vt \qquad \qquad (4.1'')$$

An entire set of equations describing angular acceleration have the same general form as those developed in Sec. 4.8 to describe linear acceleration. Compare the following:

Linear acceleration equations (all from Sec. 4.8):

$$v_2 = v_1 + at$$

$$s = v_1 t + \tfrac{1}{2}at^2$$

$$v_2^2 = v_1^2 + 2as$$

Angular acceleration equations of the same general form are:

$$\omega_2 = \omega_1 + \alpha t \qquad \qquad (8.6)$$

$$\theta = \omega_1 t + \tfrac{1}{2}\alpha t^2 \qquad \qquad (8.9)$$

$$\omega_2^2 = \omega_1^2 + 2\alpha\theta \qquad \qquad (8.10)$$

Angular acceleration, like linear acceleration, may be either positive or negative—positive if ω is increasing, negative if ω is decreasing.

Some examples follow, illustrating the use of these relationships for solving problems in situations involving angular acceleration.

Illustrative Problem 8.6   A ship's propeller shaft is turning at 100 rev/min. The order is given to the engine room to increase the angular speed at a uniform rate to 400 rev/min and to reach that angular speed in exactly 1 min. Find the angular acceleration required.

Solution

$$\omega_1 = 100 \text{ rev/min} = 10.47 \text{ rad/sec}$$

(Recall that 1 rev $= 2\pi$ rad.)

$$\omega_2 = 400 \text{ rev/min} = 41.88 \text{ rad/sec}$$

From Eq. (8.5),

$$\alpha = \frac{\omega_2 - \omega_1}{t} = \frac{41.88 \text{ rad/sec} - 10.47 \text{ rad/sec}}{60 \text{ sec}}$$

$$= 0.52 \text{ rad/sec}^2 \qquad \qquad ans.$$

Illustrative Problem 8.7   The rear wheels of an auto are turning at 60 rad/sec. The brakes are applied for 5 sec, giving a uniform (negative) angular acceleration of 8 rad/sec². Find (*a*) the final angular velocity and (*b*) the number of revolutions turned by the rear wheels during the braking period.

Solution
(*a*) From Eq. (8.6),

$$\omega_2 = \omega_1 + \alpha t$$

$$= 60 \text{ rad/sec} + (-8 \text{ rad/sec}^2) \times 5 \text{ sec}$$

$$= 20 \text{ rad/sec} \qquad \qquad ans.$$

(*b*) Angular displacement,

$$\theta = \omega_1 t + \tfrac{1}{2}\alpha t^2 \qquad \qquad (8.9)$$

$$= 60 \text{ rad/sec} \times 5 \text{ sec}$$

$$+ \tfrac{1}{2} \times (-8 \text{ rad/sec}^2) \times (5 \text{ sec})^2$$

$$= 200 \text{ rad}$$

$$= \frac{200 \text{ rad}}{6.28 \text{ rad/rev}} = 31.8 \text{ rev} \qquad ans.$$

ROTATIONAL DYNAMICS

We have considered angular acceleration and have developed some principles and equations which relate angular acceleration to linear acceleration. But we have not yet looked into the *cause* of angular acceleration. Just as an *unbalanced force* is the cause of linear acceleration, so an *unbalanced torque* is the cause of angular acceleration. And just as the magnitude of the linear acceleration produced by a force depends on the mass of the object being accelerated (Newton's laws), the angular acceleration due to a given torque depends on mass considerations. When angular acceleration, torque, and mass are all involved, the term *rotational dynamics* is used.

8.5 Moment of Inertia

Figure 8.6 shows a small mass m at the end of a slim rod, rotating with angular acceleration α about point O at a distance r. It is being accelerated by a force F acting perpendicularly to r. Let the rod of length r be of such small mass that it is negligible compared to m. We can write Newton's second law for the linear (tangential) acceleration, as

$$F = ma_t$$

Multiplying both sides by r gives

$$Fr = ma_t r$$

But $Fr = \tau$, the torque, from Eq. (8.1) and

$$a_t = \alpha r$$

the tangential acceleration. Therefore,

$$\tau = mr^2\alpha$$

But m and r^2 are constants for any given rotating body, and their product can be replaced by a

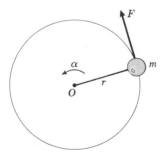

Fig. 8.6 Moment of inertia I of a mass m revolving around a point O is given by $I = mr^2$, if the entire mass can be considered as effectively located at the distance r from the center O.

single constant. The quantity mr^2 is called the *moment of inertia* (symbol I) of a rotating body in which all the mass can be considered as being concentrated at a distance r from the center of rotation. It is also assumed here, and in the following discussions, that all parts of the body have the *same* angular acceleration. As a general law of rotational dynamics, then,

$$\tau = I\alpha \tag{8.11}$$

Torque = moment of inertia

× angular acceleration

Note the parallelism here with translational dynamics, from Chap. 4:

$$F = ma \tag{4.17}$$

Force = mass (a measure of inertia)

× linear acceleration

Equation (8.11) can be regarded as a mathematical statement of Newton's second law for rotation. Just as *mass* is a measure of resistance to a change in translational motion (inertia), *moment of inertia* is a measure of resistance to change in rotary motion.

Moment of inertia is a rather complex quantity in mechanics. The case referred to above, in which all the mass of a rotating body can be considered to be located at the same distance from the center of rotation, is rarely, if ever, encountered in practical industrial and engineering situations. Moment of inertia is actually a concept which involves not only the total mass rotating, but the manner in which that mass *is distributed* with respect to the center of rotation (see Fig. 8.7). In general, as mass is distributed farther from the axis of rotation, rotational inertia (i.e. moment of inertia) increases.

In order to "visualize" the moment of inertia of a rigid body rotating about an axis, we can imagine the body being composed of an infinite number of small mass elements, each at its own distance from the center of rotation. Each, then, has its own small but finite moment of inertia mr^2. We can obtain the total moment of inertia of the entire body by summing up these small and

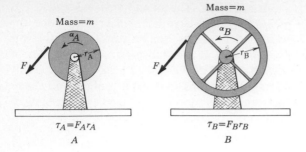

Mass=m

α_A

F

r_A

$\tau_A = F_A r_A$

A

Mass=m

α_B

F

r_B

$\tau_B = F_B r_B$

B

Fig. 8.7 Moment of inertia involves not just total mass, but the manner in which the mass is distributed with respect to the axis of rotation. Two flywheels of equal mass are shown. A has its mass concentrated near the axis of rotation. B has spokes so that most of its mass can be located in the rim. A much greater torque $\tau_B = F_B r_B$ is required to set B in motion (or to stop it if it is already in motion) than is required for A ($\tau_A = F_A r_A$). Since $\tau = I\alpha$ in each case, I_B is greater than I_A, even though the masses of the flywheels are equal.

discrete moments of inertia, as follows:

$$I = \Sigma\, mr^2 = m_1 r_1^2 + m_2 r_2^2 + m_3 r_3^2 + \cdots + m_n r_n^2 \tag{8.12}$$

where the Greek Σ (sigma) means "the sum of."

For rotors, shafts, flywheels, propellers, and other complex shapes, the use of the calculus is required to evaluate moments of inertia. However, for a simple thin ring whose total mass is M and whose entire mass can effectively be considered to be at a distance r from the center, we can derive an expression for the moment of inertia by the summing-up method of Eq. (8.12). In Fig. 8.8, since each of the small masses is essentially at the same distance from O, we can sum up as follows:

$$\begin{aligned}
I &= m_1 r_1^2 + m_2 r_2^2 + m_3 r_3^2 + \cdots \\
&= m_1 r^2 + m_2 r^2 + m_3 r^2 + \cdots \\
&= (m_1 + m_2 + m_3 + \cdots) r^2
\end{aligned}$$

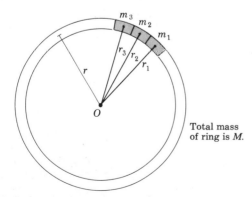

m_3 m_2 m_1

r_3 r_2 r_1

r

O

Total mass of ring is M.

Fig. 8.8 Division of a thin-walled ring into small masses for determination of moment of inertia. $I = Mr^2$ for a thin-walled ring.

But,

$$\Sigma\,[m_1 + m_2 + m_3 + \cdots] = M$$

the total mass of the ring. Therefore the moment of inertia of a thin ring

$$I = Mr^2 \tag{8.13}$$

The *dimensions* of moment of inertia I are kg-m^2 or slug-ft^2. Occasionally, in industrial work where the pound (lb) is still used as a unit of mass, I may be expressed in lb-ft-sec^2. The usual care must be taken when using the equation $\tau = I\alpha$, since Newton's second law requires *absolute units* of force. The *newton* of force (SI-metric) and the *slug* of mass (English-engineering system) are the units to use when computing $\tau = Fr = I\alpha$.

Moments of inertia of some common shapes are given in Fig. 8.9 without derivation, since calculus methods are necessary for developing these formulas.

Illustrative Problem 8.8 A flywheel approximates the shape of a solid disk. Its mass is 6 slugs and its diameter is 4 ft. Find its moment of inertia.

Solution From Fig. 8.9, for a solid disk,

$$\begin{aligned}
I &= \tfrac{1}{2}Mr^2 \\
&= \tfrac{6}{2} \text{ slugs} \times (2 \text{ ft})^2 \\
&= 12 \text{ slug-ft}^2 \qquad\qquad ans.
\end{aligned}$$

Illustrative Problem 8.9 A bowling ball weighs 4 kilos and has a diameter of 25 cm. Find its moment of inertia around an axis through its center.

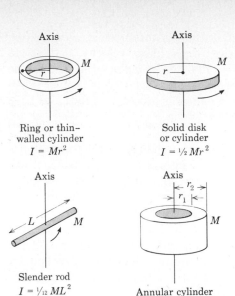

Ring or thin-
walled cylinder
$I = Mr^2$

Solid disk
or cylinder
$I = \frac{1}{2} Mr^2$

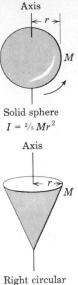

Solid sphere
$I = \frac{2}{5} Mr^2$

Slender rod
$I = \frac{1}{12} ML^2$

Annular cylinder
$I = \frac{1}{2} M(r_2^2 + r_1^2)$

Right circular
cone
$I = \frac{3}{10} Mr^2$

Fig. 8.9 Moments of inertia of some common bodies around the axes indicated.

Solution From Fig. 8.9 (solid sphere),

$$I = \frac{2}{5} Mr^2$$
$$= \frac{2}{5} \times 4 \text{ kg} \times (0.125 \text{ m})^2$$
$$= 0.025 \text{ kg-m}^2 \qquad \textit{ans.}$$

Illustrative Problem 8.10 A solid cylinder 20 cm
in diameter is mounted in a frictionless bearing at
O as shown in Fig. 8.10. A 100-gm mass is attached as shown. When the system is released the
100-gm mass has a downward acceleration of
0.05 m/sec². Find the mass of the solid cylinder.

Solution If I were known, the mass of the cylinder could be calculated from $I = \frac{1}{2} Mr^2$. The problem is to determine I. Two steps are
necessary—one to analyze the motion of the
100-gm mass as it is accelerated downward by
gravity, and the other to analyze the angular
motion of the cylinder produced by the torque
supplied by the tension T in the cord.

First, isolate the 100-gm mass, and write, for
the accelerating force,

$$F_a = mg - T = ma$$

Substituting gives

$$0.10 \text{ kg} \times 9.81 \text{ m/sec}^2 - T$$
$$= 0.10 \text{ kg} \times 0.05 \text{ m/sec}^2$$

from which

$$T = 0.981 \frac{\text{kg-m}}{\text{sec}^2} - 0.005 \frac{\text{kg-m}}{\text{sec}^2}$$

$$= 0.976 \frac{\text{kg-m}}{\text{sec}^2} \text{ (or newtons)}$$

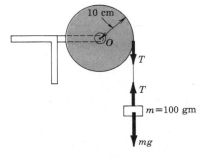

Fig. 8.10 Rotational acceleration of a solid cylinder
(Illustrative Problem 8.10).

Next, isolate the cylinder and write

$$\tau = I\alpha \quad \text{or} \quad I = \frac{\tau}{\alpha} \qquad (1)$$

Now

$$\tau = Fr \quad \text{(or } Tr \text{ in this case)}$$

Substituting

$$\tau = 0.976\,\frac{\text{kg-m}}{\text{sec}^2} \times 0.10 \text{ m}$$

$$= 0.0976\,\frac{\text{kg-m}^2}{\text{sec}^2} \quad (\textbf{N-m})$$

Now, recall Eq. (8.7), which relates tangential acceleration a_t to angular acceleration α.

$$a_t = \alpha r \qquad (8.7)$$

from which

$$\alpha = \frac{a_t}{r}$$

$$= \frac{0.05 \text{ m/sec}^2}{0.10 \text{ m}}$$

$$= 0.5 \text{ rad/sec}^2$$

Substitute these values of τ and α in (1) above:

$$I = \frac{0.0976\,\dfrac{\text{kg-m}^2}{\text{sec}^2}}{0.5\,\dfrac{\text{rad}}{\text{sec}^2}} = 0.195 \text{ kg-m}^2$$

Finally, since $I = \frac{1}{2}Mr^2$,

$$M = \frac{2I}{r^2} = \frac{2 \times 0.195 \text{ kg-m}^2}{(0.10 \text{ m})^2}$$

$$= 39 \text{ kg} \qquad\qquad ans.$$

8.6 Angular Momentum

The rotational counterpart of linear momentum is *angular momentum*. It will be recalled that linear momentum was defined by Newton as "the quantity of motion" possessed by a body. It is given by the expression

$$p = mv \qquad (5.5)$$

Linear momentum is the product of mass (the inertia factor) times linear velocity. It therefore follows logically that *angular momentum* is the product of *moment of inertia* times *angular velocity*. Denoting angular momentum by L, we have

$$L = I\omega \qquad (8.14)$$

Linear momentum of a body in translation is a measure of the impact or impulse which it will impart if it acts on some other object. And angular momentum is a measure of the angular impulse which a rotating body can deliver to a shaft or a brake. Practical problems of angular momentum and angular impulse are encountered in automotive transmissions, spin-dry laundry equipment, impulse wrenches, and torque converters for heavy industrial and marine machinery, to list only a few examples.

Illustrative Problem 8.11 The flywheel of a stationary diesel engine is in the form of a flat circular disk whose radius is 2 ft and whose weight is 200 lb.* It is rotating at 450 rev/min. Find its angular momentum.

Solution Angular velocity is

$$\omega = \frac{2\pi \text{ rad}}{\text{rev}} \times 450\,\frac{\text{rev}}{\text{min}} \times \frac{1 \text{ min}}{60 \text{ sec}} = 47.1\,\frac{\text{rad}}{\text{sec}}$$

$$I_{\text{disk}} = \frac{1}{2} \times \frac{200 \text{ lb}}{32.2 \text{ ft/sec}^2} \times (2 \text{ ft})^2$$

$$= 12.42 \text{ lb-ft-sec}^2$$

$$L = I\omega = 12.42 \text{ lb-ft-sec}^2 \times 47.1 \text{ rad/sec}$$

$$= 585 \text{ lb-ft-sec} \qquad\qquad ans.$$

Conservation of Angular Momentum It will be recalled that in interactions between bodies which make up a system in translatory motion, the total momentum of the system of interacting bodies is conserved (see Sec. 5.11). It is only in the case where an *external force* acts on the system that momentum is changed. When Newton's second law, $Ft = mv$, is expressed in the form

---

* Note that engineers may prefer to express mass in slugs, giving I in slug-ft^2, and L in slug-ft^2/sec.

Large I
Small ω

Small I
Large ω

(a)

$$L = I_1\omega_1 = I_2\omega_2 = \text{constant}$$

Large I

Small ω

ω_1

Small I

Large ω

ω_2

(b)

$$I_1\omega_1 = I_2\omega_2$$

Fig. 8.11 Examples of conservation of angular momentum. (a) Ballet dancers make use of conservation of angular momentum. (b) A laboratory demonstration of conservation of angular momentum. Changing the distribution of mass of a rotating body results in a change of angular *velocity*, but the angular *momentum* is constant. $I_1\omega_1 = I_2\omega_2$.

$F = mv/t$, it states, in effect, that a force acting on a body is measured by the time rate of change of momentum of the body. In like manner, *torque* is measured by the *time rate of change of angular momentum* of a rotating body. As an equation

$$\tau = \frac{I(\omega_2 - \omega_1)}{t}$$

or

$$\tau t = I(\omega_2 - \omega_1) \qquad (8.15)$$

Eq. (8.15) is the rotational counterpart of Eq. (5.7).

The total angular momentum of a body or system of bodies is a constant unless the body or system of bodies is acted upon by an *external* torque. Rotating bodies which are nonrigid may therefore undergo changes in mass distribution, i.e., in moment of inertia, if at the same time the angular velocity changes in such a way that the product $I\omega$ remains a constant. There are a number of interesting applications of conservation of angular momentum in the fields of sports and entertainment. Figure skaters, ballet dancers, and high-divers start into a spin at slow speed with arms and legs outstretched (I at a maximum); then, pulling in arms and legs (making I smaller), they spin at a much greater angular velocity, with *angular momentum unchanged* (Fig. 8.11a). The phenomenon can be illustrated in the laboratory or lecture room with the arrangement shown in Fig. 8.11b.

The discus thrower winds up in a tight, high-speed spin (small I, large ω), then allows the discus and his arm to spiral out, resulting in a larger value of I (and increased *linear* speed of the discus) at the sacrifice of *angular velocity* ω. In the final stages of his throw the inertial reaction force of the accelerating discus, acting backward through his hand and arm exerts a torque which stops the thrower's angular motion entirely. The quantity τt in Eq. (8.15) is called *angular impulse*; it equals the change in *angular momentum* of the body about the same axis.

Other examples of conservation of angular momentum can be found in many of the rotary-mechanical rides at amusement parks.

Angular momentum is a vector quantity Earlier, we found that linear velocity, acceleration, and momentum are vector quantities. So are angular velocity, angular acceleration, and angular momentum. *Angular velocity* can be represented by a vector whose length to scale is equal to the rotational speed in rad/sec, and whose direction is along the axis of rotation in the direction of advance of a right-hand-threaded screw (see Fig. 8.12a).

Angular momentum is represented by a vector whose magnitude is equal to $I\omega$, and whose direction is also given by the advance of a right-hand-

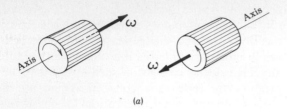

(a)

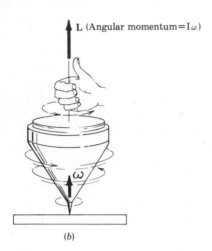

L (Angular momentum=Iω)

ω

(b)

Fig. 8.12 Vector considerations in rotational motion. (*a*) Vector representation of angular velocity. The vector direction is that in which a right-hand threaded screw would advance. (*b*) The angular momentum vector of a spinning top. If one imagines the fingers of the right hand wrapped around the axis of rotation in the direction of the rotation, the extended thumb shows the direction of the angular momentum vector, **L**.

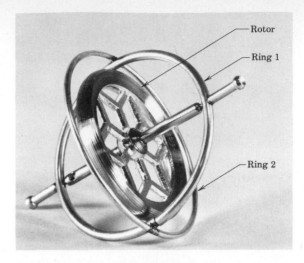

Fig. 8.13 A simple gyroscope for laboratory demonstration. The rotor has great angular momentum when spinning at high speed. Note that its shaft has bearings in Ring 1. Ring 1 can also be fitted with bearings and so can Ring 2. When so arranged the gyro is said to have three degrees of freedom. The rotor will tend to continue to spin in its *original* plane of rotation regardless of changes in direction or erratic motions of the vehicle in which it is installed. The original plane of rotation is also unaffected by earth motions or planetary influences. Consequently, gyros are much used in navigational and guidance systems. Bearing friction is a problem of course, and a means must be provided to keep the rotor spinning at the design angular velocity.

threaded screw. The fingers and thumb of the right hand can also be used, as in Fig. 8.12*b*, for a "right-hand-thumb rule" to determine the direction of the vector.

Since angular momentum is a *vector quantity*, an externally applied torque is required to change the *direction* of the axis of rotation of a spinning wheel, whether or not the speed of rotation is changed. Another way of stating this gives a rotational counterpart of Newton's first law of motion:

A rigid body rotating about a fixed axis will continue rotating about that axis with constant angular speed, unless acted upon by an unbalanced torque.*

The spinning top is a simple example of angular momentum. Its "spin stabilization" keeps its axis from wobbling until, due to friction, the angular velocity and the angular momentum fall to near zero. Spin is purposely imparted to gun projectiles and to footballs to keep them from wobbling in flight. The gyroscope (Fig. 8.13) is a device especially designed to utilize the vector properties of angular momentum. Gyroscopes are used in spin-stabilized navigation equipment

---

*   Note that this is not true for nonrigid bodies, i.e. bodies where mass distribution can be changed while the body is rotating.

for aircraft, submarines, and aerospace vehicles, and also in some ballistic missile guidance systems. Once set spinning in a given plane of rotation, the gyro rotor resists any effort to change its rotation plane.

8.7 Rotational Kinetic Energy

For a body in translation, the energy of motion, or *kinetic energy*, is expressed as $KE = \frac{1}{2}mv^2$. By analogy, the kinetic energy of a rotating body is

$$KE_{rot} = \frac{1}{2}I\omega^2 \qquad (8.16)$$

As a ball, cylinder, or hoop rolls along a flat surface, it has both kinetic energy of translation and kinetic energy of rotation. The cylinder of Fig. 8.14, for example, rotates around C and has a rotational kinetic energy given by

$$KE_{rot} = \frac{1}{2}I\omega^2$$
$$= \frac{1}{2}(\frac{1}{2}mr^2)\omega^2$$
$$= \frac{mr^2\omega^2}{4} \quad \text{(for a solid cylinder)}$$

In addition, its center of mass C moves to the right with *linear velocity v*, and the cylinder has $KE_{trans} = \frac{1}{2}mv^2$. The *total kinetic energy* of a body which has both angular velocity and linear velocity (any rolling object) is then

$$KE_{tot} = \frac{1}{2}I\omega^2 + \frac{1}{2}mv^2 \qquad (8.17)$$

In the absence of dissipative forces kinetic energy will be conserved (law of conservation of energy). This leads to the explanation of tactics frequently used by athletes, acrobats, and stunt men, to convert kinetic energy of translation into kinetic energy of rotation and thus *decrease linear velocity* as *angular velocity is increased*. A football player rolls across the turf after a missed tackle, a barrel jumper goes into a spin as he clears the last barrel and hurtles toward the ice, and movie stunt men literally roll out on the ground like hoops after intentional falls from horses or speeding automobiles. The energy transformation from linear to rotational motion lessens the chances of serious injury, which would be quite likely to occur if the translational kinetic energy were absorbed entirely through collision with the ground, a fence, or other unyielding matter.

Flywheels for energy storage The use of heavy flywheels to "smooth out" the flow of mechanical energy from reciprocating engines has been common for more than a century. The flywheel gains kinetic energy as the engine supplies torque, "stores" the energy temporarily as rotational KE, and then provides it to the load during the interim when the engine is getting ready for its next power stroke.

There is currently considerable interest in the potential-energy storage capabilities of really mammoth flywheels. As solar power, wind power, and tidal power—all of them intermittent in nature—are being developed, the problem of energy storage will become critical. Huge flywheels may provide one means of "stockpiling" energy as it is produced, and storing it in the form of rotational KE for delivery at another time of day or even in a different season of the year.

8.8 Linear and Angular Motion Compared

In concluding the chapter, Table 8.1 is provided, to bring together, for ready reference, the principal equations and symbols of linear and angular motion.

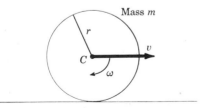

Fig. 8.14 A rolling cylinder has both kinetic energy of rotation and kinetic energy of translation. $KE_{tot} = KE_{rot} + KE_{trans} = \frac{1}{2}I\omega^2 + \frac{1}{2}mv^2$.

Table 8.1 Comparison of symbols and equations for translational and rotational motion

Concept	Linear (Translation)	Angular (Rotation)
Distance (displacement)	s	θ
Velocity	$v = \dfrac{s}{t}$	$\omega = \dfrac{\theta}{t}$
Acceleration	$a = \dfrac{v_2 - v_1}{t}$	$\alpha = \dfrac{\omega_2 - \omega_1}{t}$
Uniformly accelerated motion	$v_2 = v_1 + at$ $s = v_1 t + \tfrac{1}{2}at^2$ $v_2^2 = v_1^2 + 2as$	$\omega_2 = \omega_1 + \alpha t$ $\theta = \omega_1 t + \tfrac{1}{2}\alpha t^2$ $\omega_2^2 = \omega_1^2 + 2\alpha\theta$
Work	$W = Fs$	$W = \tau\theta$
Power	$P = Fv$	$P = \tau\omega$
Momentum	$p = mv$	$L = I\omega$
Impulse	$Ft = m(v_2 - v_1)$	$\tau t = I(\omega_2 - \omega_1)$
Newton's Second Law	$F = ma$	$\tau = I\alpha$
Kinetic energy	$KE_{trans} = \tfrac{1}{2}mv^2$	$KE_{rot} = \tfrac{1}{2}I\omega^2$

SUMMARY

Torque is the cause of rotary motion. Torque is defined as the product of a force applied to a body times the perpendicular distance from the point of application of the force to the center of rotation of the body. As a formula

$$\tau = Fr \qquad (8.1)$$

Power is transmitted by rotating shafts. They deliver torque supplied by a motor or engine or turbine to the load at the other end of the shaft.

Power = torque × angular velocity

$$P = \tau\omega \qquad (8.3)$$

Just as mass is the inertial property of matter in translational motion, *moment of inertia* is a property which resists a change in rotational motion. In translational motion, $F = ma$, but in rotational motion,

$$\tau = I\alpha \qquad (8.11)$$

Torque = moment of inertia × angular acceleration

Moment of inertia involves not just total mass, but the way in which the mass is distributed with respect to the axis of rotation.

Rotating bodies may undergo changes in *angular velocity*, just as bodies in translation undergo changes in linear velocity.

The basic expression for *angular acceleration* is

$$\alpha = \dfrac{\omega_2 - \omega_1}{t} \qquad (8.5)$$

A particle subject to angularly accelerated motion has two accelerations: (1) its *centripetal acceleration* $a_c = r\omega^2$ and (2) the *acceleration tangent to its momentary path*, $a_t = \alpha r$, called *tangential acceleration*.

The total instantaneous linear acceleration of a particle undergoing angular acceleration is the vector sum of its *centripetal acceleration* and its *tangential acceleration* and is given by the formula

$$a = \sqrt{a_t^2 + a_c^2} \qquad (8.8)$$

The angular momentum of a rotating body is given by the relation

$$L = I\omega \qquad (8.14)$$

Angular momentum = moment of inertia

× angular velocity

Bodies or systems of bodies in rotational motion, unless acted on by an external force, exhibit *conservation of angular momentum*; i.e., the product $I\omega$ is a constant. It is this factor which explains the ability of skaters, dancers, discus throwers, and divers to change their speed of rotation, without supplying additional torque.

Angular momentum is a vector quantity—that is, it is directional. An external force or torque is required to change the direction of the axis of rotation of a spinning body. Gyroscopes illustrate the vector nature of angular momentum.

Torque is measured by time rate of change of angular momentum

$$\tau = \frac{I(\omega_2 - \omega_1)}{t} \tag{8.15}$$

The *law of conservation of energy* holds for rotational motion just as it does for linear motion.

$$KE_{tot} = KE_{rot} + KE_{trans}$$

or $\hspace{8cm}$ (8.17)

$$KE_{tot} = \tfrac{1}{2}I\omega^2 + \tfrac{1}{2}mv^2$$

Heavy flywheels possess large amounts of rotational kinetic energy when spinning rapidly. The potential for energy storage of mammoth flywheels is currently being investigated as a means of "stockpiling" energy from intermittent energy sources such as the sun, wind, and tides.

QUESTIONS AND EXERCISES

1. $F = ma$ is the mathematical statement of Newton's Second Law for translation. What is the equivalent equation for rotation? Define each term in the equation.

2. A particle is in uniform circular motion; i.e. its angular velocity is constant. Does it have any acceleration? Explain your answer by a simple vector diagram.

3. If the power being transmitted by a shaft has to be calculated, what two factors must be known or measured?

4. How is moment of inertia different from mass? In what way(s) is it similar to mass?

5. If you were building a coaster car for the "roller derby" (the cars start from rest and coast down a hill to the finish line) would you use relatively heavy, large-diameter wheels and tires or lighter, smaller ones? (Total mass of car same in either case. Assume friction equal also.)

6. In designing flywheels, as much as possible of the total mass is placed in the outer rim. Why?

7. Explain why the rifling of a gun barrel tends to increase the accuracy of the bullet's flight. Use the vector property of angular momentum in your discussion.

8. A hoop (a thin ring with all the mass distributed around the rim) of mass m rolls down an inclined plane. Prove that when it arrives at the bottom, half its KE is rotational and half translational.

9. A solid sphere, a solid cylinder, and a hollow cylinder all start simultaneously from rest and roll down a smooth incline. Assuming no friction and no slippage, in what order will they reach the bottom? Explain.

10. If a mass m is hung from a string secured to the ceiling, the tension in the string is mg. If the string is wound around the rim of a wheel of mass m which is free to turn and which is then released, will the tension still be mg? Why? When the mass hits the floor, what has happened to the potential energy it had just before it was released to unwind?

11. A solid cylinder and a solid block, both made of steel and both of the same mass, are released simultaneously at the top of a smooth incline. Assume no friction for the sliding block and no slippage of the rolling cylinder. Which arrives at the bottom first? What if the cylinder had only half the mass of the block? What if a bicycle wheel of the same mass were substituted for the cylinder?

PROBLEMS

1. A force of 50 N is applied at the rim of a wheel 1 m in diameter. The line of action of the force is perpendicular to the radius of the wheel at that point. What torque is exerted on the wheel?

2. Wrenches used by final-assembly personnel in industry have built-in torquemeters. If the wrench handle is 18 in. long and the torquemeter reads 500 lb-in., what force is the mechanic applying (a) when he pulls perpendicular to the wrench handle? (b) when the position is such that he has to pull at an angle of 50° with the wrench handle?

3. Find the horsepower being delivered by an engine which exerts a torque of 4000 lb-ft on a rotating kiln in a cement factory and turns it at 30 rev/min.

4. What is the power transmitted by a shaft turning at 3000 rev/min if the torque is found to be 1600 N-meters?

5. A steam turbine in a power-generating plant starts from rest and accelerates at a uniform rate of 3.4

rad/sec². Find its angular velocity in rev/min after 70 sec.

6. If the allowed angular acceleration of a turbojet engine is 60 rad/sec², and if idling speed is 600 rev/min, how long does it take to attain a speed of 5000 rev/min? (Assume uniform acceleration.)

7. An engine governor is rotating at 2500 rev/min and is slowed to 500 rev/min in 40 sec. Assuming uniform deceleration, calculate (a) the deceleration in radians per second per second and (b) the number of revolutions turned by the governor during the 40-sec period.

8. What is the magnitude of the total linear acceleration of a particle on the rim of a large air-compressor flywheel 2 m in diameter when the instantaneous angular velocity is 3π rad/sec and the angular acceleration is 5 rad/sec²? What angle does the direction of the total linear acceleration make with the radius whose end point is the particle?

9. A merry-go-round (carousel) has a top angular speed of 10 rev/min. (a) In order to brake it to a stop in 20 sec, what (uniform) negative angular acceleration is necessary? (b) What total angle in radians is turned during the braking process?

10. An elevator is designed to lift a total unbalanced load of 2500 kg at a steady rate of 300 m/min. The hoisting drum around which the elevator cable is wound is 1 m in diameter. (a) What is the angular velocity of the drum during lift? (b) What power (kW) is required?

11. Find the moment of inertia of a large cylindrical roller in a steel-rolling mill from the following data; diameter, 4 ft; length, 6 ft; weight, 2800 lb.

12. What is the moment of inertia of a squirrel-cage rotor of a large electric motor if it is in the form of an annular cylinder whose outer radius is 80 cm and whose inner radius is 50 cm. Its mass is 50 kg. Neglect weight of spokes.

13. Consult Table 7.1 for data and calculate (a) the moment of inertia of the earth and (b) the force in kilograms required to stop the earth from rotating in a 24-hr period if it were applied along the equator and effectively used as a brake. Consider the earth as a sphere of uniform density.

14. A flywheel in the form of a solid disk of diameter 0.8 m has a mass of 100 kg. In order to accelerate it from an initial speed of 50 to 2000 rev/min in 10 sec, what torque must be applied?

15. An engine has a flywheel whose moment of inertia is 50 slug-ft². What unbalanced torque will accelerate the flywheel from rest to 700 rev/min in 20 sec?

16. A heavy wheel, weight 400 lb, is in the form of an annular cylinder, $r_2 = 1.8$ ft, $r_1 = 1.2$ ft. (Neglect spokes.) If it is spinning at 500 rev/min and a load is applied which brings it uniformly to a stop in 30 sec, what torque is exerted?

17. A figure skater has just gone into a pirouette on one skate with both arms and the other leg and foot extended. Her angular velocity in this position is 1 rev/sec, and her moment of inertia is 4.5 kg-m². Neglecting friction, what will her angular velocity be when she pulls in to the fast-spin position, where her moment of inertia is only 1.2 kg-m²?

18. The earth's linear speed in its orbit around the sun is 9.85×10^4 ft/sec. In addition to this translational motion, the earth spins on its polar axis at a rate of one revolution every 24 hr. Find (a) the earth's rotational kinetic energy and (b) its kinetic energy of translation. Compare the two. (See Table 7.1 for data.)

19. A bowling ball has a mass of 7.5 kg and a diameter of 20 cm. It is rolling without slipping along the floor with a translational velocity of 6.5 m/sec. Find its translational, rotational, and total KE.

20. A small airplane propeller approximates a slender rod and rotates around an axis as shown in Fig. 8.9. It is 7 ft long and has a mass of 2.5 slugs. What uniform torque must be exerted to bring its angular velocity from zero to 1800 rev/min in an elapsed time of 10 sec?

21. An "energy-storage" flywheel is being designed. It will be in the form of a huge ring, with all the mass assumed to be located in the ring at a mean distance of 5 m from the axis of rotation. The design maximum angular velocity is 380 rev/min. At 20 rev/min, drawing additional power from the device is impractical. If the mass of the wheel is to be 50,000 kg, (a) how much energy (joules) will be available between the above angular velocity limits? (b) If this energy were drawn from the wheel at a steady rate over a 4-hr period, what power (kW) would be available? Neglect all friction and other losses.

PART **3** Mechanical Properties of Matter

Hydraulic press forge squeezes a mass of heated steel, formerly an ingot, between dies under enormous pressure to form a giant shaft. (Bethlehem Steel Photo)

9 The Structure of Matter—Atoms, Molecules, Elements

Thus far we have been studying the motions and energies of bodies large enough to be observed with the naked eye. All physical objects are composed of a substance called *matter*. Matter possesses the property of inertia, and this property has been described by the term *mass*. Matter in motion possesses *kinetic energy* and *momentum*. Matter also possesses *potential energy* as determined by its position or internal configuration. Material objects attract each other with forces called *gravitational forces*.

The ultimate structure of matter is not known, but the subject has been and still is an area of active research. However, it is generally agreed that matter is composed of extremely large numbers of tiny particles called molecules. The size of most molecules is so small that they cannot be seen by even the most powerful optical microscope. Some larger molecules can be seen by an electron microscope. Their masses are too minute to be detected by the most sensitive analytical balances. However, there is plenty of indirect physical and chemical evidence to establish the molecular nature of matter.

In this and succeeding chapters we will discuss how the molecular theory satisfactorily explains the many physical properties of solids, liquids, and gases. The molecular nature of matter will also be used to explain heat, sound, and electrical energies in later chapters.

9.1 The Kinetic-Molecular Theory

The kinetic-molecular theory helps to explain the properties of matter. It theorizes that all matter, regardless of the state in which it is found, is comprised of extremely small particles called *molecules*. All molecules of one substance are alike, while those of different substances are never alike. Thus, a molecule is the smallest particle of a substance that has all the chemical and physical properties of the substance.

Molecules are so small that on the average about a thousand of them side by side could barely be visible with the best optical microscope. In a 1-liter flask of any gas under ordinary conditions of temperature and pressure there are approximately 26×10^{21} molecules. This is an extremely large number; so large that if a million of these molecules could flow out of the liter container each second, it would take almost 850 million years to empty the flask!

In the kinetic-molecular theory, the molecules are constantly in motion with varying speeds. One of the most direct evidences of this motion can be observed by the student in the laboratory. A device illustrated in Fig. 9.1 draws smoke from a burning match into a small box by the compression and release of a rubber bulb. A strong ray of light entering through a hole from one side of the smoke box will illuminate the

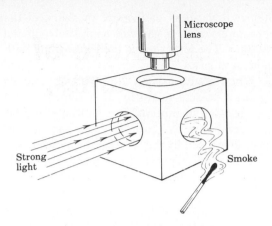

Fig. 9.1 Brownian movement apparatus.

smoke particles in the box. The motion of the particles can be observed by a high-powered microscope. The tiny particles will appear to dart first one way and then another. A typical (random) movement pattern of one of the particles is illustrated in Fig. 9.2. Such movements are called *Brownian movements*, after an English botanist, Robert Brown, the man who first observed them in 1827.

The particles of smoke are very much larger than the molecules of air in which they are suspended. The motion of the particles is caused by millions of collisions per second with the air molecules. When more of these collisions occur on one side of the particle than on the other sides, the smoke particle will dart away from the excess collisions.

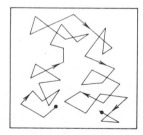

Fig. 9.2 Typical random movement of a gas molecule.

Molecules are assumed to be perfectly elastic. When two molecules interact, they rebound without any net loss of energy or momentum, for the two-molecule system. This means that, for a given equilibrium state, the total amount of kinetic energy and the total momentum of the molecules in a mass of matter does not change. We will discuss in Chap. 13 how increasing the temperature of a substance will result in increased molecular speeds and, therefore, increased kinetic energy of the molecules.

Molecules attract each other very little when they are far apart. When they are close together they attract each other very strongly. But when the molecules are too close together, they repel each other, often with tremendous forces. These forces of attraction and repulsion determine many properties of matter. One of these is the *state* of matter.

9.2 The Three States of Matter

In dealing with the physical properties of matter it is convenient to divide matter into three general classes, or states: *solids*, *liquids*, and *gases*. (A "fourth state"—*plasmas*—is suggested by some modern physicists. Plasmas will be discussed in Chap. 33.) The same kind of matter may exist in each of the three states. As an example, liquid water may be frozen to a solid (ice) or vaporized to a gaseous state. In each of these forms it is composed of identical molecules. The different states of matter are explained by the relative positions of these molecules and the kinds of freedom of their motions.

Matter is said to be in a solid state if it retains a definite shape and a definite volume. This volume and shape, however, may be subjected to change under varying external conditions such as temperature or the application of tensional or compressional forces.

In the solid state it is probable that the molecules oscillate or vibrate about certain fixed points. The molecules are held together by relatively strong molecular forces, called *cohesive*

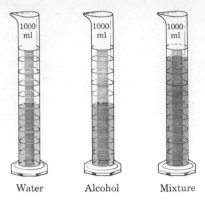

Water   Alcohol   Mixture

Fig. 9.3 Adding ½ liter of alcohol to ½ liter of water produces less than 1 liter of mixture.

forces, in such a way that they cannot move far from their positions with reference to other molecules in the body. It is the strong molecular forces of cohesion that cause solids to retain a definite shape and volume.

Liquids have a definite volume but an indefinite shape. For example, a gallon of water maintains a constant volume under constant temperature conditions but varies its shape to conform to that of the containing vessel. Liquids are nearly incompressible.

In liquids the molecules are not held together in rigid patterns. Liquid molecules are, in general, almost as close together as the molecules of solids, but they slip over each other with ease and hence have no fixed positions. Molecular motion in liquids can be evidenced by the fact that a drop of ink released in a glass of water can be observed to diffuse quite rapidly until a uniform mixture results. The existence of empty spaces between molecules of a liquid can be easily demonstrated by mixing ½ liter of alcohol with ½ liter of water (see Fig. 9.3). The mixture occupies a volume considerably less than 1 liter. Apparently the molecules of one liquid fill in some of the spaces between the molecules of the other.

Gases differ from solids and liquids in that they have neither a definite shape nor a definite volume. Regardless of the size or shape of the containing vessel, a gas will completely fill the container. The term *fluid* is used to include both liquids and gases.

When matter exists in the gaseous state, its molecules possess relatively high velocities and great freedom of motion. The molecules are far apart compared to those in solids and liquids. This is evidenced by the fact that the molecules of 1 liter of water separate to occupy about 1600 liters when the liquid is changed to saturated steam at 100°C.

In a change of state, the size of the molecules remains the same, but the space between the molecules changes. While the molecules of a solid or liquid are closely spaced, the molecules of a gas are relatively far apart, separated by a void. Molecules of a gas exert almost no cohesive force upon one another. This lack of mutual attraction of gaseous molecules, coupled with the fact that they may have a velocity exceeding in many cases that of the fastest rifle bullet, helps explain the relatively rapid *diffusion* of gases.

One of the most striking properties of a gas is that of unlimited expansibility. No matter how small an amount of gas is placed in a vessel, the gas will expand until it completely fills the vessel. If only half as much gas had been placed in the vessel, the vessel would still be *full*, but at a lower pressure. For this reason it is impossible to get a complete vacuum in a container. No matter how much air is "pumped" from a container, the remaining air will almost instantly redistribute itself throughout the entire vessel. The molecules tend to expand indefinitely if unconfined.

Gases are assumed to be perfectly elastic and can readily be compressed. This combination of perfect *elasticity* and *compressibility* gives gases the property called *resiliency*, i.e., the ability to yield to a force and to return promptly to original condition when that force is removed. It is this characteristic property of air that is utilized in the pneumatic tire, tennis balls, basketballs, and the like. Elasticity and compressibility will be covered in detail in Chap. 10.

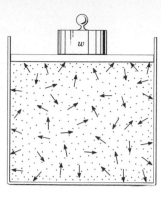

Fig. 9.4 Molecular bombardment creates the force which holds up the weight *w*.

9.3 Molecular Motion and Gas Pressure

Daniel Bernoulli (1700–1782), a Swiss scientist, first proposed the hypothesis that has now been generally accepted as the *kinetic theory of gases*. In his attempt to account for the compressibility peculiar to gases, he conceived of gas pressure as being caused by countless molecular impacts against the walls of the container.

In order to discuss this theory more fully, let us assume an imaginary cylindrical vessel filled with a gas (Fig. 9.4), which we shall show by a great number of small dots, each of which represents a molecule moving with extremely high velocity. Let us imagine further that the vessel is fitted with a movable frictionless piston *P* of negligible weight, upon which is placed a weight *w*. It is the continual bombardment against the piston by the moving molecules that sustains the weight *w*. In order to compress the gas, a greater weight would have to be added to compensate for the resultant greater number of impulses imparted to the piston by more frequent impacts of the compressed gas. In like manner, by diminishing the weight *w*, the bombardment on the piston would force it to rise, letting the gas expand until a state of equilibrium is reached in which the decreased pressure due to fewer impacts of

the less dense gas on the piston is balanced by the lighter weight.

Pressures exerted by gases on the walls of the vessels which contain them are due to the continuous bombardment of the walls by the gas molecules. The greater the number of molecules, the greater the pressure. Thus, by pumping more (molecules of) air into a container, we increase the pressure in the container.

9.4 Molecular Structure

The smallest particle into which a substance can theoretically be divided and maintain its chemical properties is a molecule. But one may ask: What are the molecules made of? What would happen if we could divide the molecule into smaller parts?

Scientists are now generally agreed that molecules of all types are built up of more elemental particles, called *atoms*. There are only 92 different kinds of atoms *found in nature*, although atomic physicists have produced a number of additional elements in the laboratory by nuclear bombardment. (A more detailed discussion of atomic structure will be found in Chaps. 33 and 34.) These basic substances are called *elements*. Some common elements are carbon, oxygen, sulfur, iron, hydrogen, phosphorus, and calcium. Various combinations of these atoms result in the multitude of kinds of molecules which make up our universe. A molecule of water, for example, consists of two atoms of hydrogen chemically combined with one atom of oxygen. A molecule of ordinary sugar is made of 12 atoms of carbon, 22 atoms of hydrogen, and 11 atoms of oxygen. Hundreds of thousands of substances result from different chemical combinations of the atoms of just the five common elements carbon, hydrogen, oxygen, nitrogen, and sulfur.

Most molecules have two or more atoms. However, in some instances molecules consist of single atoms. For example, molecules of metallic elements consist of single atoms. Also, there are a few gases, called inert gases, that have one atom in each molecule. Examples of inert gases are argon, neon, and helium.

9.5 Mixtures and Compounds

Elements contain atoms of only one kind. The atom of a given element differs from that of any other element, but the number of different kinds of molecules is legion. If two or more different molecules are grouped together and each molecule maintains its chemical identity, the result is called a *mixture*. Examples of mixtures are soil, air, milk, and such combinations as iron and sulfur, and sand and water. In mixtures, the amounts of the constituent parts can be varied in any proportions. However, if two or more atoms or molecules combine, with a resulting product which is *chemically different* from the original ingredients, a *compound* is formed. Water is a compound of hydrogen and oxygen. Alcohol is a compound of carbon, hydrogen, and oxygen—the same elements found in the compound sugar, but in different proportions and united in a different way.

The smallest particle into which a compound can be broken and yet maintain its identity is the *molecule*. In the molecule a certain number of atoms of one or more of the elements combine in a fixed proportion. Always this proportion is the same for each molecule of any one compound. Each molecule of table salt, for example, will be found to contain one atom of sodium combined chemically with one atom of chlorine.

9.6 Molecular Forces of Attraction

Forces of attraction between *like* molecules are called *cohesive forces*. It is this molecular attraction which gives solids rigidity and strength. For example, it requires approximately 50 tons of force to pull apart a round bar of hard steel 1 in. in diameter. Since various substances differ in the magnitude of the force required to pull them apart, it is evident that the forces of mutual attraction between some kinds of molecules are greater than those between other kinds of molecules.

Cohesion is greatest in solid substances. These forces not only hold the object together, but they maintain the definite shapes of solid objects. The cohesive forces in liquids are weaker than those in solids, and while appreciable, they are not sufficient to give the liquid a shape of its own. The cohesive forces among the molecules of a gas are negligible.

Forces of attraction between *unlike* molecules are called *adhesive forces*. Such forces are utilized in binding surfaces together with glue. In many cases, adhesive forces are greater than cohesive forces. These relative strengths are important in industry. In gluing together pieces of furniture, the adhesive forces between the glue and the wood and the cohesive attraction of glue for glue is a vital problem. Since, in this case, the adhesive forces exceed the cohesive forces, the layer of glue should be made as thin as possible. Adhesive forces come into play in the use of such common objects as pencils, ink, chalk, cement, paint, shellac, and tar.

9.7 Molecular Forces of Repulsion

To compress a solid such as a steel beam or a liquid such as oil or water is not easy. It can be done, but even large forces applied to a solid or a liquid produce relatively small compressions. The molecules of solids and liquids apparently resist being pressed closer together by repelling each other.

The force exerted by one molecule upon another depends upon the distance between them. Experiments have shown that molecules have maximum attraction for each other when they are about 10^{-10} m apart. They repel each other when they are closer than that, and their mutual attraction decreases rapidly when they are more than 10^{-10} m apart. A graph illustrating how the force between molecules varies with the distance between them is shown in Fig. 9.5. As the molecules come closer together, the attractive forces increase until they reach a maximum, shown at *B*. As the molecules move still closer, repulsive forces appear which oppose the forces

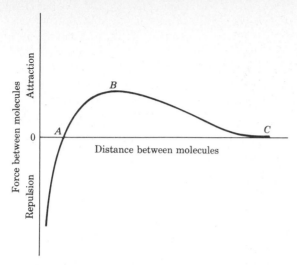

Fig. 9.5 Forces between molecules. When the molecules are far apart at *c* (for gases), there is almost no net forces between them.

of attraction. At point *A* no net force acts between the molecules. As molecules move still closer, the net repulsion forces between the molecules rise sharply.

At first one might be led to believe that forces of attraction between molecules are gravitational, but physicists have concluded that attractive and repulsive molecular forces are primarily a result of electric fields associated with the atoms which make up the molecules.

SUMMARY

All matter may be classified into three states: *solids*, *liquids*, and *gases*. (A "fourth state"—*plasmas*—is suggested by some modern physicists).

Solids have a definite volume and a definite shape.

Liquids have a definite volume but their shape depends on the container.

Liquids are nearly incompressible.

Gases possess the property of unlimited expansibility.

The *kinetic-molecular theory of matter* assumes the following:

1. All matter is composed of tiny particles called *molecules*.

2. Molecules are in continuous motion at various speeds.

3. Momentum and kinetic energy are conserved when molecules interact. That is, collisions between molecules are perfectly *elastic*.

4. Molecules gain kinetic energy as their temperatures rise.

5. Molecules have very little mutual attraction when they are far apart. When they are close together, there is a strong mutual attraction between them. But when they come too close to each other, they repel each other.

Molecules are made up of *atoms*. The more than 100 kinds of atoms combine chemically in various ways to form hundreds of thousands of different kinds of molecules.

Forces of attraction between like molecules are called *cohesive* forces.

Forces of attraction between unlike molecules are called *adhesive* forces.

Fluids is the term used to include both liquids and gases.

QUESTIONS

1. Are atoms ever molecules? Explain.

2. What is a fluid?

3. What are the basic assumptions of the kinetic-molecular theory?

4. List three evidences that support the theory that molecules are in continual motion.

5. Explain the fact that however rapidly the piston of an air pump is drawn up, air always appears to follow it almost immediately.

6. If gases possess the property of unlimited expansibility, why does the atmosphere not leave the earth?

7. Sugar is heavier than water. Why does not all the sugar in a cup of coffee settle to the bottom?

8. Left uncovered, fish stored in the refrigerator will soon impart its flavor to other food in the refrigerator. Explain in the light of molecular motion theory.

9. Why does a damp piece of cloth absorb water more quickly than a dry one?

10. The particles displaying Brownian movements in Fig. 9.1 are not molecules, but much larger particles. How do you explain their motion?

11. In terms of adhesion and cohesion, explain how glue works.

12. A smooth block of gold was placed on a smooth block of lead. Several months later traces of gold were detected in the lead, and traces of lead were found in the gold. Give an explanation for what happened.

13. A silk umbrella is more effective in shedding rain than one made of cotton. Give a possible explanation in terms of forces of adhesion and cohesion.

14. Explain why oil poured on water spreads into a very thin film on the water surface.

15. When a patch is removed from a bicycle inner tube the rubber is often torn before the patch comes off. Explain.

10 Properties of Solids

In our study this far we have dealt with the transmission of forces from one body to another by means of such media as beams, metal rods, wedges, steel balls, bearings, and metal screws. In each of these cases the objects were treated as rigid bodies. We have assumed that under the action of applied forces the body moves as a whole with translation, rotation, or combinations of both types of motion.

Actually there is no such thing as a completely "rigid" body. Every real substance yields to some extent under the influence of applied forces. Whenever a body is acted upon by an external force, some change in shape, or *distortion* (however small), of the body results.

When an external force acts on a body the forces are not confined to the outer surface of the body. The force is transmitted from molecule to molecule throughout the body. Considering any element of area in the interior of the body, there are, in general, distributed forces acting over the area of the element.

Thus the ability of a steel rod to support a heavy load depends upon the attraction of millions and millions of steel molecules for millions and millions of adjacent steel molecules. This mutual attraction results in their resistance to separation.

Industry and government today spend billions of dollars annually on continuing research into the structure and properties of matter. Their efforts have transformed our lives beyond the wildest dreams of our forefathers. Some of the properties of solids which we shall study in this chapter are *density*, *elasticity*, *tensile strength*, *hardness*, and the ability to withstand extreme temperatures. Engineering handbooks tabulate the results of many studies that have been made of these properties.

10.1 Two Kinds of Density

Gold is often described as a *heavy* metal, while aluminum is referred to as one of the *lighter* metals. What is really meant is that the gold has a greater density than aluminum. The *density* of matter is defined in metric units as the *mass per unit volume*. Algebraically, this can be expressed as

$$\text{Mass density} = \frac{\text{mass}}{\text{volume}}$$

$$\rho = \frac{m}{V} \tag{10.1}$$

Mass densities, which will be represented in this text by the Greek letter ρ (rho), are usually expressed in kilograms per cubic meter or in grams per cubic centimeter.

Mass density in the English system is expressed in slugs per cubic foot. Engineers and technicians also often employ units of density based upon weight units. *Weight density* is defined as the *weight per unit volume*. We shall use the English letter D to represent weight densities. The unit of weight density in the English system is the *pound per cubic foot*. In the metric system it is the *newton per cubic meter*. Thus,

Table 10.1 Densities of common substances

Material	Mass Density ρ		Weight Density D, lb/ft³ at sea level
	kg/m³	gm/cm³ (and specific gravity)	
Solids:			
Aluminum	2,700	2.7	168.5
Brass	8,600	8.6	540
Copper	8,890	8.89	555
Gold	19,300	19.3	1,205
Ice	910	0.91	57.2
Iron	7,900	7.9	493
Lead	11,400	11.4	712
Platinum	21,400	21.4	1,330
Steel	7,830	7.83	489
Liquids:			
Alcohol, ethyl (20°C)	790	0.79	49.4
Benzene	900	0.90	56.1
Gasoline	690	0.69	42
Kerosene	820	0.82	51
Linseed oil	940	0.94	59
Mercury	13,600	13.6	849
Turpentine	870	0.87	54
Water at 0°C	1,000	1.00	62.4
Ocean	1,025	1.02	64
Gases (32°F, 1 atm):			
Air	1.29	0.00129	0.0807
Ammonia	0.76	0.00076	0.0481
Carbon dioxide	1.96	0.00196	0.1234
Carbon monoxide	1.25	0.00125	0.0781
Helium	0.18	0.00018	0.0111
Hydrogen	0.090	0.00009	0.0056
Nitrogen	1.25	0.00125	0.0781
Oxygen	1.43	0.00143	0.0892
Propane	2.02	0.00202	0.1254

$$\text{Weight density} = \frac{\text{weight}}{\text{volume}}$$

$$D = \frac{w}{V} \qquad (10.2)$$

Since weight is equal to the product of mass and acceleration of gravity ($w = mg$), weight density is related to mass density by the formula

$$D = \frac{w}{V} = \frac{mg}{V} = \frac{m}{V}g$$

$$D = \rho g \qquad (10.3)$$

or

weight density = mass density
$\times$ acceleration of gravity.

It should be noted that since weight may vary from place to place, the weight density also varies but the mass density does not. The weight density of iron on the surface of the earth is about 77,000 N/m³ (490 lb/ft³), while it would be about 13,000 N/m³ (82 lb/ft³) on the surface of the moon.

The definition for density of matter is the same whether its state is solid, liquid, or gas. Weight density is commonly used when we are interested in the effects of forces on material, while mass density is used when mass is to be considered.

Table 10.1 gives the densities of some common substances.

Illustrative Problem 10.1 What is the mass density of a block of granite whose dimensions are 2 m by 75 cm by 25 cm if the block has a mass of 1020 kg?

Solution Converting the dimensions of the block to meters and using the formula $V = $ length $\times$ width $\times$ height, we have

$$V = 2 \text{ m} \times 75 \text{ cm} \times \frac{1 \text{ m}}{100 \text{ cm}} \times 25 \text{ cm} \times \frac{1 \text{ m}}{100 \text{ cm}}$$

$$= 0.375 \text{ m}^3$$

Substituting values in Eq. (10.1) yields

$$\rho = \frac{m}{V} = \frac{1020 \text{ kg}}{0.375 \text{ m}^3} = 2720 \text{ kg/m}^3 \qquad ans.$$

Illustrative Problem 10.2 An irregularly shaped casting weighing 234 lb is lowered into a barrel which is completely full of water. The overflow water is carefully collected and weighed. If the overflow water weighs 31.8 lb, what is the weight density of the casting?

Solution It is apparent that the volume of the casting will equal the volume of the water it displaces. The volume of the overflow water must first be determined. From Table 10.1 we observe that the weight density of water is 62.4 lb/ft³. Solving Eq. (10.2) for V gives

$$V = \frac{w}{D}$$

Now, for the overflow water (and for the casting),

$$V = \frac{31.8 \text{ lb}}{62.4 \text{ lb/ft}^3} = 0.51 \text{ ft}^3$$

Substitute in Eq. (10.2), and solve for the density of the casting:

$$D = \frac{w}{V} = \frac{234 \text{ lb}}{0.51 \text{ ft}^3} = 459 \text{ lb/ft}^3 \qquad ans.$$

10.2 Density and Specific Gravity

The statement "ice is lighter than water" is meaningless unless it implies that weights of equal volumes of ice and water are compared. A cubic centimeter is a convenient volume for comparison. Thus 1 cm³ of water has a mass of 1 gm, and 1 cm³ of ice has a mass of 0.91 gm. Since the weights are proportional to the masses, we can say that for equal volumes ice is 0.91 times as heavy as water.

It is customary to compare weights or masses of substances with the weight or mass of an equal volume of water, since water is the most abundant common liquid. The concept of *specific gravity* comes from such comparisons.

The *specific gravity* (sp gr) of a substance is the ratio of its density to that of water at 4°C (39.2°F). Water has its maximum density at this temperature (see Sec. 13.19).

If ρ_s (or D_s) is the density of a substance and ρ_w (or D_w) is the density of water, the specific gravity of the substance is found by the formula

$$\text{sp gr} = \frac{\rho_s}{\rho_w} = \frac{D_s}{D_w} \qquad (10.4)$$

A sometimes more convenient and equivalent way of defining specific gravity is

sp gr =

$$\frac{\text{mass (or weight) of substance}}{\text{mass (or weight) of an equal volume of water}}$$

$$(10.5)$$

The distinction between density and specific gravity must be clearly understood. *Density* refers to *mass or weight per unit volume* of a given substance. The numerical value of the density of a substance will vary with the units of weight and volume used. Aluminum, for example, has densities of 2700 kg/m³, 2.7 gm/cm³, 168.5 lb/ft³, and 0.097 lb/in.³.

The *specific gravity* of a substance is a pure number that has no units. Since, in the metric system, the mass density of water is 1 gm/cm³, *the specific gravity of any substance is equal numerically to the mass density of that substance expressed in grams per cubic centimeter.* Hence in the previous example we would know that aluminum has a specific gravity of 2.7, or that it is 2.7 times as heavy as an equal volume of water. Other specific gravities can be found by referring to Table 10.1. The column labeled " gm/cm³ " can also be read "specific gravity."

Solving Eq. (10.4) for densities, we get

$$\rho_s = (\text{sp gr}) \times \rho_w \qquad (10.4')$$

and

$$D_s = (\text{sp gr}) \times D_w \qquad (10.4'')$$

For mass density, therefore,

$$\text{Density (kg/m}^3) = \text{sp gr} \times 1000 \text{ kg/m}^3$$

or Density (gm/cm³) = sp gr × 1 gm/cm³

$$(10.4''')$$

For weight density, since the density of water is 62.4 lb/ft³,

$$\text{Density (lb/ft}^3) = \text{sp gr} \times 62.4 \text{ lb/ft}^3 \quad (10.4^{IV})$$

Illustrative Problem 10.3 A metal casting whose mass is 35 kg displaces 13 liters of water. What is (*a*) the density, and (*b*) the specific gravity of the casting?

Solution The volume V of the casting is 13 liters, since it displaces 13 liters of water. Expressing this value in cubic meters,

$$V = 13 \text{ liters} \left(\frac{1000 \text{ cm}^3}{1 \text{ liter}} \right) \left(\frac{1 \text{ m}}{100 \text{ cm}} \right)^3 = 0.013 \text{ m}^3$$

(*a*) Using Eq. (10.1),

$$\rho = \frac{m}{V} = \frac{35 \text{ kg}}{0.013 \text{ m}^3} = 2700 \text{ kg/m}^3$$

(*b*) Then

$$\text{sp gr} = \frac{\rho_s}{\rho_w} = \frac{2700 \text{ kg/m}^3}{1000 \text{ kg/m}^3} = 2.7 \qquad ans.$$

10.3 Elasticity

Solid bodies tend to maintain their shape and volume. However, whenever a body is acted upon by an external force, a change in shape, or *distortion*, of the body results. That property of matter which, upon the removal of these distorting forces, will cause the body to resume its original shape or volume is called *elasticity*. Every substance has some degree of elasticity, but the term *elastic* is usually applied to those bodies that return *readily* to their original shape or volume when the distorting forces are removed. Contrary to common belief, rubber (which the word elastic commonly suggests) is not a highly elastic material. Highly tempered steel, drawn tungsten, and ivory are much more elastic than rubber.

Whenever the shape of a solid is distorted, the molecules are rearranged: some are separated, and others are pushed closer together. When the forces of distortion are removed, the forces of attraction and repulsion between the molecules tend to cause the original shape to be restored. It can be shown that the forces between the molecules are primarily electric. Materials such as putty, sealing wax, and lead are called *inelastic*.

Industry is continually attempting to increase personal comfort and safety by utilizing the property of elasticity. The innerspring mattress, the various springs in autos and trains, the balls used in many sports, and the tires on automobiles are just a few examples where this property of elasticity is of major importance.

10.4 Stresses in Solids

Whenever a solid body is deformed by the action of external forces, there are set up within that body internal molecular forces which tend to resist changes in shape and/or volume. These internal forces are called forces of *stress*. In specifying a girder, a column, or a machine part, engineers must know the forces it is expected to resist. There are four kinds of stresses: *tension, compression, volume* and *shear* (see Fig. 10.1). The stress can be expressed as the ratio of the applied

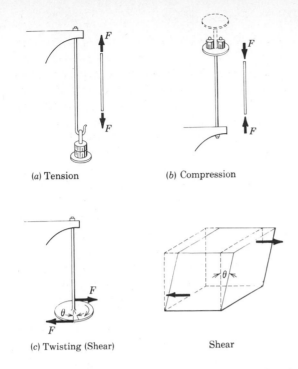

(a) Tension (b) Compression

(c) Twisting (Shear) Shear

Fig. 10.1 Different kinds of stress. (*a*) *Tension:* Equal and opposite forces act away from each other along the same line of action. (*b*) *Compression:* Equal and opposite forces act toward each other along the same line of action. (*c*) *Shear:* Equal and opposite forces are exerted along different lines of action. Volume stresses will be discussed later in connection with liquids and gases, in this and subsequent chapters.

external force creating a distortion to the area over which the force acts.

$$\text{Stress} = \frac{\text{external force acting}}{\text{area over which force acts}}$$

$$S = \frac{F}{A} \qquad (10.6)$$

Common units of stress are newtons per square meter, and pounds per square inch.

When the actions of an external force tend to stretch or pull the parts of a body apart, the body is said to be under *tension*. Tension, for example, occurs when a rope or cable is tied to a rigid support and sustains a weight. The belt that is transmitting power from one pulley to another is in a state of tension.

The column of a building, the pier of a bridge, and the foundation of a house serve a different function. They resist external forces which tend to push their parts together, thereby shortening them. These members are said to be under *compression* (see Fig. 10.1*b*).

10.5 Stress Causes Strain

No matter how small a stress is applied to a body, the body will yield a little. A steel rail, for example, will bend ever so slightly even when a bird alights on it. The yielding of bodies under stress may be in the form of an elongation, shortening, twisting (torsion), or change in volume and is called *strain*. Strain is a necessary consequence of stress.

Suspend a helical (spiral) spring from a rigid support (Fig. 10.2). Attach a weight holder to the bottom end of the vertical spring. Place a weight on the holder so that its weight combined with that of the holder is w_0. Adjust a vertical scale next to a pointer P on the holder so that the pointer reads zero on the scale. Next add weights one by one to the holder and record corresponding elongations of the spring. Then take off the weights in the reverse order and record the elongations.

Results of this experiment indicate that upon removal of the weights the pointer P will return to its original position. If we then compute the corresponding elongation e for each weight w applied, we shall find that all the values of w/e are approximately equal. In other words w/e is a constant, or $w = Ke$. That is, the elongation is proportional to the force (weight) producing it, and K is the constant of proportionality.

A similar experiment can be set up in which the shortening or compression of a beam or spring can be measured for various loads or forces. It can be shown that within certain limits the compression (strain) is proportional to the force (stress) applied. Likewise, the amount of twisting of a rod or bar is proportional to the stress causing it. This fact was first expressed as a scientific truth in 1660 by Robert Hooke (1635–

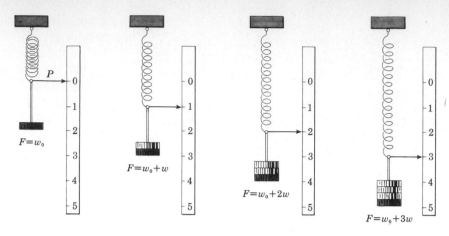

Fig. 10.2 An experiment which illustrates Hooke's law.

1703), an English physicist, and the law is now named after him.

Hooke's Law

If the applied forces on a body are not too large, the deformations resulting are directly proportional to the forces producing them.

Or, more simply, *strain is proportional to stress.* Hooke's law applies to all kinds of strains if the stresses are not too great. (What happens when the forces *are* too great is discussed in Sec. 10.7.)

10.6 Modulus of Elasticity

Hooke's law states that, within certain limits, the strain produced in a body is proportional to the stress, or stress/strain = a constant K. This constant for a particular substance is called its *modulus of elasticity.* In order to arrive at a value of K, we must first express stress and strain in terms of some common measurable units. The *stress*, which is the internal resistance of a body to a change of shape, is expressed by Eq. (10.6):

$$\text{Stress} = \frac{\text{force applied}}{\text{area of cross section}}$$

$$\text{Stress} = \frac{F}{A} \qquad (10.6)$$

The *strain* is the amount of change in size of the material per unit original size. The size may be expressed in terms of length, area, or volume units.

Tension and Compression Tensile stresses produce tensile strains, expressed as

$$\text{Tensile strain} = \frac{\text{elongation}}{\text{original length}}$$

$$\text{Tensile strain} = \frac{e}{L} = \frac{\Delta L}{L} \qquad (10.7)$$

where ΔL is called the *increment* of L and is the change in L as it passes from one value to a second. ΔL is found by subtracting the first value from the second. Remember that the increment of L is read "delta L." Thus, ΔL signifies a small change in L. Note also that ΔL may be either positive or negative.

The ratio of stress to strain for cases of tension or compression in a given material is called *Young's modulus* of elasticity for that material. It is usually denoted by the letter Y.

$$Y = \frac{\Delta F/A}{\Delta L/L}$$

from which Young's modulus becomes

$$Y = \frac{\Delta F \, L}{A \, \Delta L} \qquad (10.8)$$

In Eq. (10.8) ΔL is the increment of L which corresponds to an increment ΔF of F. Young's modulus is expressed in newtons per square meter (N/m^2) or in pounds per square inch (lb/in^2). Table 10.2 gives values of Y for some common materials.

As can be seen from Table 10.2, the numerical value for Young's modulus varies for each material and may vary for the same kind of substance under differing conditions of physical structure. The heat treatment given a substance affects this modulus. This constant is also affected by the temperature of the material. For example, the modulus for a rubber band will increase with a rise in temperature, while that for steel will decrease as the temperature rises.

Table 10.2 Approximate values of Young's modulus

Material	Y	
	N/m^2	$lb/in.^2$
Aluminum	7.0×10^{10}	10×10^6
Brass, cast	9×10^{10}	13.1×10^6
drawn	13×10^{10}	18.9×10^6
Copper	12.5×10^{10}	18×10^6
Iron, cast	9.1×10^{10}	13×10^6
wrought	18.3×10^{10}	26×10^6
Lead, rolled	1.6×10^{10}	2.3×10^6
Steel, drawn	20.0×10^{10}	29×10^6
mild	17.2×10^{10}	25×10^6
Rubber, vulcanized	14.0×10^4	20
Tungsten, drawn	35.0×10^{10}	51×10^6

Illustrative Problem 10.4 How much will a steel wire 10 ft long and 0.06 in. in diameter stretch when it is subjected to a force of 80 lb?

Solution Solving Eq. (10.8) for ΔL,

$$\Delta L = \frac{\Delta F L}{A Y}$$

Substituting numerical values, using $Y = 29 \times 10^6$ lb/in.^{2} from Table 10.2,

$$\Delta L = \frac{80 \text{ lb} \times 10 \text{ ft} \times 12 \text{ in./ft}}{3.14 \times (0.03 \text{ in.})^2 \times 29 \times 10^6 \text{ lb/in.}^2}$$

$$= 0.12 \text{ in.} \qquad\qquad ans.$$

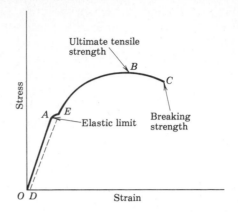

Fig. 10.3 Stress-strain diagram for mild steel.

The modulus of elasticity in cases of compression can be computed in a similar way as long as no bending is involved. In this case, ΔL is the amount the member is shortened by the increment ΔF of the compressing force.

Illustrative Problem 10.5 How much will a wrought-iron bar 5 by 12 cm in cross section and 2 m long shorten under a compression load of 2500 N? (Assume $Y = 18.3 \times 10^{10}$ N/m^2.)

Solution From Eq. (10.8),

$$\Delta L = \frac{\Delta F L}{A Y}$$

Substitution yields

$$\Delta L = \frac{2500 \text{ N} \times 2 \text{ m}}{0.05 \text{ m} \times 0.12 \text{ m} \times 18.3 \times 10^{10} \text{ N/m}^2}$$

$$= 4.6 \times 10^{-6} \text{ m} = 4.6 \times 10^{-4} \text{ cm}$$

$$= 4.6 \ \mu\text{m} \qquad\qquad ans.$$

10.7 Elastic Limit, Ultimate Strength

In the discussion to this point we have assumed that upon removal of the forces which produced the various strains the bodies would resume their original shapes or volumes. However, if in the experiment on tensile stress (Fig. 10.2) a sufficiently heavy load is hung from the spring, the latter will not return to the original length when the load is removed. The same situation can be made to develop in each of the other three types of elasticity mentioned if the stresses are made large enough. *Hooke's law*, then, *holds only for values of stresses less than the elastic limit.*

The relation between stress and strain for a body under tension can be represented by a *stress-strain diagram*. Figure 10.3 represents the stress-strain diagram for mild steel. A body that has been deformed under stress will regain its original size and shape on removal of the stress unless the stress has exceeded a certain limit. Thus, starting at *O*, the strain increases in direct proportion to the stress. That is, the graph is a straight line. The maximum stress from which the material will completely recover is called the *elastic limit* (point *A* on the graph). If the stress is not increased beyond the elastic limit, the curve will be retraced as the strain becomes zero when the stress is removed.

If the stress exceeds the elastic limit, the strain will increase more rapidly (the curve flattens out). The region represented by *AE* is called the region of *plastic flow*. The elastic properties of the metal will have been changed. If the stress is then decreased, the strain will decrease as shown by the dashed line *ED*. The strain will not return to zero when the stress is reduced to zero. The material will have acquired a *permanent set*, represented by *OD* in the graph.

Fig. 10.4 Electromechanical universal testing machine determines a wide range of mechanical properties under tensile (shown) and compressive loads. Stress-strain curves are plotted by recorder; loads are indicated digitally. (Tinius Olsen Testing Machine Co., Inc.)

Table 10.3 Elastic limit and ultimate tensile strength

Material	Elastic Limit		Ultimate Tensile Strength	
	$N/m^2 \times 10^8$	$lb/in.^2 \times 10^3$	$N/m^2 \times 10^8$	$lb/in.^2 \times 10^3$
Aluminum, cold rolled	1.3	19	1.4	21
Brass, cast	3.8	55	4.7	67
Copper-aluminum-nickel, rolled	7.9	115	8.2	120
Iron, annealed	1.4	20.3	2.9	42
drawn	7.7	111	7.8	114
wrought	1.7	25	2.4	50
Steel, hard	2.7	40	5.5	80
mild	2.1	30	4.1	60
spring-tempered	11.7	170	13.8	200
Tungsten, drawn			41.2	600

If the stress had been continually increased beyond the elastic limit, eventually a point *B* would have been reached where the strain would increase with no increase in stress. Indeed, beyond this point loads may even be lightened without halting the continuing deformation. The greatest stress that can be applied is called the *ultimate strength* of the material. Ultimately a point will be reached where rupture occurs. This point is called the *breaking strength*.

Each substance will have a different stress-strain diagram. But, in general, each graph will have the same characteristics. A substance for which *EC* is relatively long is called *ductile* (see Sec. 10.15). Such a substance can be subjected to a large increase in length before breaking. Substances for which *EC* is quite short are called *brittle* (see Sec. 10.14).

Table 10.3 lists values of elastic limit and ultimate tensile strength for a number of common metals.

10.8 Factor of Safety

Engineers in designing a bridge or machine always plan to make each member strong enough to withstand several times as much load as will ever be imposed upon it. This is to allow for any unexpected overloading of the structure or for possible overlooked or imperceptible flaws in the material. The *ratio of the ultimate strength of any material to its maximum expected stress* is called the *factor of safety* of that material. Stated in another way, the factor of safety of a structure is the ratio of the maximum load the structure will maintain to the maximum load that it is ever *expected* to maintain. For example, if an elevator is built to withstand a load of 5 tons but at no time is expected to carry more than a load of 1 ton, its factor of safety is 5.

The factor of safety varies with the material. For example, a factor of safety of 10 is common practice in brick structures, while steel structures use a factor of about 4. The nature of the load also affects the factor of safety to be recommended, as do wind and flood conditions and the possibility of earthquakes.

Illustrative Problem 10.6 Determine the factor of safety involved in using a cable 0.64 cm in diameter to support a load of 540 kg. The ultimate tensile strength of the cable is 5.5×10^8 N/m².

Solution Use Eq. (10.6),

$$F = 540 \text{ kg} = 540 \text{ kg}\left(\frac{9.81 \text{ N}}{1 \text{ kg}}\right) = 5300 \text{ N}$$

and

$$A = \frac{\pi d^2}{4} = 3.14 \frac{[0.64 \text{ cm} \times (1 \text{ m}/100 \text{ cm})]^2}{4}$$

$$= 3.2 \times 10^{-5} \text{ m}^2$$

Applied stress,

$$S = F/A = \frac{5300 \text{ N}}{3.2 \times 10^{-5} \text{ m}^2}$$

$$= 1.65 \times 10^8 \text{ N/m}^2$$

The ultimate (breaking) stress is 5.5×10^8 N/m². Hence the factor of safety is

$$\frac{5.5 \times 10^8 \text{ N/m}^2}{1.65 \times 10^8 \text{ N/m}^2} = 3.3 \qquad ans.$$

10.9 Metal Fatigue

When metallic material is subjected to repeated stresses over long periods of time, the internal structure of the material will change. As a result, certain regions are weakened and the metal may rupture under the repeated applications of stresses. Failure usually occurs around areas where there is a flaw in the material. This loss of strength due to repeated applications of stress is known as *fatigue*.

Fatigue is a common cause of failure in machinery that is subjected to cyclic stresses. Thus it is important to discover flaws in a machine part before it is installed. In many manufacturing plants x-rays are used to detect hidden flaws in parts of complicated machinery.

Any discontinuity in a section of material, e.g., a bend, groove, or hole will increase the chance of failure. Stresses at these sections of discontinuity will reach values three or more times the value of the average stress in the member. Such weakening is used sometimes by the person who wants to cut a piece of wire but has no wire cutters. Whereas he may be unable to rupture the wire by pulling on it, he can produce a break in the wire by repeated bendings at the desired point of rupture.

10.10 Bulk Modulus and Compressibility

All substances can be diminished in size under sufficient pressure. The decrease in volume of a solid or liquid is proportional to the pressure exerted on its outer surface. Uniform pressure can be placed on a solid if it is submerged in a liquid, for the liquid pressure will be transmitted undiminished and will act perpendicularly to each surface. (Such hydrostatic pressures will be discussed in Chap. 11.) The pressure applied will be the force acting on a unit area.

Consider a cube and a sphere of some material on which forces are applied normal to each surface and uniformly distributed over the surface (Fig. 10.5). The solids will be in equilibrium under the action of these external forces, but the volumes will be reduced. If the original volume of a solid V decreases by an increment ΔV, the *volume strain* will be $\Delta V/V$. The *volume stress* is the normal force per unit area; i.e., volume stress is $\Delta F/A = \Delta P$, where ΔP is the increase in pressure on the surface.

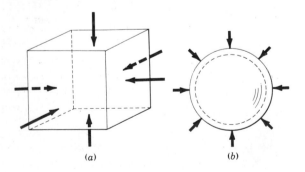

(a)　　　　　　　　　(b)

Fig. 10.5 Volume stresses. Forces are normal to the surfaces of (a) a cube and (b) a sphere.

Table 10.4　Bulk modulus for solids and liquids

Material	Bulk Modulus B	
	N/m² × 10¹⁰	lb/in.² × 10⁶
Solids:		
Aluminum	7.7	11
Brass	6.1	8.5
Copper	14	20
Glass	3.7	5.3
Iron, cast	9.6	14
Lead	0.77	1.1
Steel	16	23
Liquids:		
Ethyl alcohol	0.110	0.16
Kerosene	0.13	0.19
Lubricating oil	0.17	0.25
Mercury	2.8	4.0
Water	0.21	0.31

The ratio of stress to strain in these cases is called the *bulk modulus* (*B*) or the *coefficient of volume elasticity*. Thus

$$B = \frac{\text{stress}}{\text{strain}} = -\frac{\Delta F/A}{\Delta V/V} - \frac{\Delta P}{\Delta V/V} = -\frac{\Delta P\,V}{\Delta V} \tag{10.9}$$

The negative sign is used here because an increase in pressure causes a corresponding decrease in volume.

The type of deformation which involves only volume changes applies particularly to fluids, since they offer resistance only to change of volume. They can have only bulk moduli of elasticity. The bulk moduli of solids and liquids (see Table 10.4) are relatively large numbers, indicating that large forces are needed to produce even minute changes in volume. Gases are more easily compressed and have correspondingly smaller bulk moduli.

The *compressibility c* of a material is the reciprocal of its bulk modulus.

$$\text{Compressibility } c = \frac{1}{B} = \frac{\Delta V}{\Delta P\,V} \tag{10.10}$$

Illustrative Problem 10.7 Find the decrease in volume of 7.5 liters of water under a pressure of

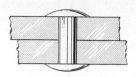

Single rivet lap joint

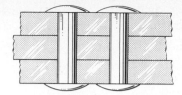

Double riveted butt joint

Failure of rivet due
to single shearing

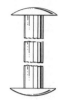

Failure of rivet due
to double shearing

Fig. 10.6 Single and double shearing of a rivet.

$2.3 \times 10^4 \text{ N/m}^2$ (use Table 10.4). Answer in cubic centimeters.

Solution Solve Eq. (10.9) for ΔV:

$$\Delta V = -\frac{\Delta P \; V}{B}$$

From Table 10.4, $B = 0.21 \times 10^{10}$ N/m² for water. Substituting proper converted units in the above equation gives

$$\Delta V = -\frac{2.3 \times 10^4 \text{ N/m}^2 \times 7.5 \text{ l} \times (10^3 \text{ cm}^3/1 \text{ l})}{0.21 \times 10^{10} \text{ N/m}^2}$$

$$= -\frac{2.3 \times 7.5 \times 10^7}{0.21 \times 10^{10}} \text{ cm}^3$$

$$= -0.082 \text{ cm}^3 \qquad \qquad ans.$$

The negative sign merely means that volume has decreased.

Illustrative Problem 10.8 Determine the compressibility of steel (use Table 10.4). What does the answer actually mean?

Solution The bulk modulus of steel is 23×10^6 lb/in.². Substituting this value for B in Eq. (10.10) gives for compressibility

$$c = \frac{1}{B}$$

$$= \frac{1}{23 \times 10^6 \text{ lb/in.}^2} = 4.35 \times 10^{-8} \text{ in.}^2/\text{lb}$$

$$ans.$$

This means that a hydrostatic pressure of 1 lb/in.² will cause an original volume V_0 to decrease by $0.0000000435 \; V_0$; that is ΔV_0 (the decrease in volume) $= 0.0000000435 \; V_0$ for each lb/in.² increase in the pressure.

10.11   Shear

Another form of stress is that of *shear*, in which certain particles of a body tend to slide over the particles in that body adjacent to them. An example of this type of stress is that set up in rivets of a steel structure. A rivet has to keep one plate from sliding over another, as well as hold the plates together (see Fig. 10.6). This same type of stress is set up when a piece of sheet metal is cut with tin shears, or when steel plates are cut with heavy power shears in steel-fabricating plants.

Illustrative Problem 10.9 What force must be applied in punching out $\frac{1}{2}$-in.-diameter holes in a steel plate $\frac{3}{8}$ in. thick if the shearing strength (stress) of the material is 38,000 lb/in.²?

Solution The force must act over an area equal to the lateral surface area of a cylindrical hole punched out. Use the formula for the lateral surface area of a cylinder:

$$A = 2\pi r h$$

$$= 2(3.14) \times \tfrac{1}{4} \text{ in.} \times \tfrac{3}{8} \text{ in.} = 0.589 \text{ in.}^2$$

the lateral area over which the shearing force acts. Now, selecting Eq. (10.6) which relates force, stress, and area,

$$\text{Stress} = \frac{F}{A}$$

Solving for F,

$$F = \text{stress} \times A$$
$$= 38{,}000 \text{ lb/in.}^2 \times 0.589 \text{ in.}^2 = 22{,}400 \text{ lb}$$
ans.

Illustrative Problem 10.10 Suppose two of the plates of Prob. 10.9 are riveted together; how much force per rivet, applied at right angles, would be necessary to shear off the rivets? Assume the shearing strength of the rivet material to be 30,000 lb/in.². Neglect the friction between the plates.

Solution Here the area to be sheared is the cross-sectional area of the rivet, which is assumed to have the same diameter as the holes ($\tfrac{1}{2}$ in.). Selecting Eq. (10.6) again,

$$F = \text{stress} \times A$$
$$= S \times \pi r^2$$
$$= 30{,}000 \text{ lb/in.}^2 \times 3.14 \times (\tfrac{1}{4} \text{ in.})^2 = 5900 \text{ lb}$$
ans.

10.12 Shear Modulus or Torsion Modulus

When an object is subjected to a pair of equal forces which act in opposite directions but not along the same line, the shape of the body will be changed. Consider a pair of forces F, F acting on the upper and lower faces of a rectangular body, as in Fig. 10.7. Assume the pair of forces shears the block so that face *BCDE* becomes the paral-

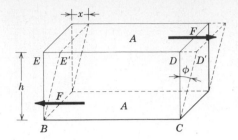

Fig. 10.7 Effect of shearing forces on rectangular body.

lelogram *BCD'E'*, with edges *BE* and *CD* turning through the small angle ϕ.

Upon removal of the shearing forces, the body will return to its original shape if the elastic limit of the material has not been exceeded. This type of elasticity is called *elasticity of shear*. It is of great significance in structural design and in power transmission.

The block may be considered as made up of many layers of horizontal sheets of area A, each of which under shearing forces slides slightly with respect to the adjacent layers. The *shearing stress* is defined as the tangential force per unit area producing the sidewise motion. Hence, the shearing stress is $\Delta F/A$, where A is the area of the upper or lower face of the block. The *strain* is defined as the amount of the lateral displacement per unit height. Thus, in Fig. 10.7, if h is the height of the rectangular block and x is the lateral displacement of the top with respect to the lower face, the strain is x/h. The shear modulus n is the ratio of shearing stress to shearing strain.

$$n = \frac{\Delta F/A}{x/h} = \frac{h\,\Delta F}{xA} \qquad (10.11)$$

For very small angles $\tan \phi = x/h \approx \phi$, where ϕ is in radians. Hence,

$$n = \frac{\Delta F/A}{\phi} = \frac{\Delta F}{A\phi} \qquad (10.11')$$

Shear is also involved in the twisting of a straight rod with circular cross section. The drive shaft of a car or any shaft which transmits rotary motion throughout its length is subjected to

Fig. 10.8 Oil-field rotary drilling table in operation. Strings of drill pipe, often more than a mile long, are subjected to tremendous twisting stresses throughout their entire length. The heavy hose at left carries a special drilling "mud" into the well. (Standard Oil Co. of California.)

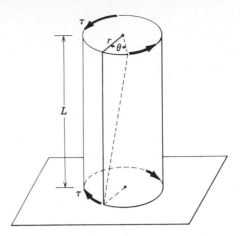

Fig. 10.9 Shearing-stress effect on a cylindrical body.

twisting forces when under load (see Fig. 10.8). Within the shaft, *twisting stresses* are set up to counteract the external forces applied.

If one end of a cylindrical solid rod or shaft is clamped in a fixed position and the other is turned by an applied torque τ (see Fig. 10.9), successive sections of the rod move over each other. The resultant twist θ produced by the applied torque τ depends on the length L and the radius r of the rod. It can be shown that for a solid rod,

$$\theta = \frac{2\tau L}{\pi n r^4} \tag{10.12}$$

where θ is in radians, τ is in newton-meters, L in meters, r in meters, and n in newtons per square

meter. In English units, τ is in pound-feet, L in feet, r in feet, and n in pounds per square foot.

The angle of twist θ for a hollow cylinder of radius r and wall thickness t is given by

$$\theta = \frac{\tau L}{2\pi n r^3 t} \quad \text{Pipe} \tag{10.13}$$

Values of the shear modulus n for some common materials are given in Table 10.5.

Illustrative Problem 10.11 Through what angle will the free end of a copper rod be twisted if it is

Table 10.5 Shear modulus for selected solids

Material	Shear Modulus n	
	$N/m^2 \times 10^{10}$	$lb/in.^2 \times 10^6$
Aluminum, rolled	2.37	3.44
Brass, cold rolled	3.53	5.12
Copper	4.24	6.14
Lead, rolled	0.54	0.78
Nickel	7.30	10.6
Platinum, pure drawn	6.42	9.32
Steel	8.04	11.7
Tungsten, drawn	14.8	21.5

110 cm long and has a diameter of 0.68 cm, if a torque of 320 N-cm is applied to it? (Use Table 10.5.)

Solution Convert the given data to units consistent with Eq. (10.12) and substitute:

$$\theta = \frac{2(320 \text{ N-cm}) \times (110 \text{ cm})}{3.14(4.24 \times 10^{10} \text{ N/m}^2)(1 \text{ m}/100 \text{ cm})^2(0.34 \text{ cm})^4}$$

$$= 0.40 \text{ rad} \cong 23° \qquad\qquad ans.$$

Illustrative Problem 10.12 Each of the two alloy-steel drive shafts of a Navy ship is 60 ft long and 8 in. in diameter. If the modulus of rigidity $n = 11.7 \times 10^6$ lb/in.^{2}, and the shaft is delivering 25,000 hp at 400 rev/min, what is the total twist produced in the shaft, in radians?

Solution Use Eq. (10.12),

$$\theta = \frac{2\tau L}{\pi n r^4}$$

and note that torque τ must first be evaluated. From Eq. (8.3), solving for τ and substituting numerical values, we have

$$\tau = \frac{25,000 \text{ hp} \times (550 \text{ ft-lb/sec}/1 \text{ hp})}{400 \text{ rev/min} \times (1 \text{ min}/60 \text{ sec}) \times 2\pi \text{ rad/rev}}$$

$$= 3.28 \times 10^5 \text{ lb-ft}$$

Substituting this value in Eq. (10.12), with careful attention to units, gives

$$\theta = \frac{2 \times 3.28 \times 10^5 \text{ lb-ft} \times 60 \text{ ft}}{\pi \times 11.7 \times 10^6 \text{ lb/in.}^2 \times 144 \text{ in.}^2/\text{ft}^2 \times (\frac{4}{12} \text{ ft})^4}$$

$$= 0.60 \text{ rad}$$
$$\approx 34° \qquad\qquad ans.$$

10.13 Hardness, Malleability, and Ductility

Engineers, in selecting suitable materials, must consider many other properties in addition to those discussed up to this point. Among them are hardness, *malleability*, and *ductility*. One of the methods employed in determining the relative hardness of various substances is the *scratch test*. At slow speeds, the harder of two substances can always scratch the less hard material. An arbitrary scale showing the positions of 10 substances in the scratch test (see Table 10.6) has been made, ranging from soft talc to the hardest of all known substances, diamond.

Other materials are classified for hardness by comparing, by scratching, with these known substances. Mohs' scale of hardness of some common structural materials is as follows: aluminum, 2 to 2.9; brass, 3 to 4; emery, 7 to 9; iron, 4 to 5; lead, 1.5; marble, 3 to 4; steel, 5 to 8.5; tin, 1.5 to 8; wax (0°C), 0.2.

Table 10.6 Mohs' scale of hardness

1. Talc
2. Rock salt or gypsum
3. Calcite
4. Fluorite
5. Apatite
6. Feldspar
7. Quartz
8. Topaz
9. Corundum
10. Diamond

10.14 Hardness by the Brinell Scale

One industrial method of measuring the degree of hardness of metals involves pressing a 10-mm hardened chrome-steel ball into the metal to be tested, with a force exerted by a mass of 3000 kg. The diameter of the resulting indentation is used as the measure of hardness. This is called the *Brinell method* (Fig. 10.10). Each solid is given a *Brinell number* (see Table 10.7 for examples). It is found by dividing the load (in kilograms-force) by the surface area of the impression (in square millimeters). The larger the Brinell number, the harder the material.

The Rockwell hardness machine is designed to test materials of widely varying hardness. This

Table 10.7   Brinell hardness of materials

Material	Brinell Number
Aluminum, annealed	16
Chromium	91
Copper-aluminum (11.73%), hard-rolled	269
Iron, annealed	77
Lead, cast	4.2
Platinum, drawn	64
Carbon steel	460
Alloy steel-nickel-vanadium-carbon-manganese-silicon	627

Fig. 10.10   Metal hardness tester with a system of digital readout of Brinell values. (Tinius Olsen Testing Machine Co.)

Fig. 10.11   Coils of aluminum foil feedstock are shown being checked before entry into an annealing oven. The annealing process eliminates work hardness and softens the metal to allow further thickness reduction to foil gauges. These large 30,000-lb coils, when rolled on a mill to foil gauge, would stretch about 265 mi. (Reynolds Metals Co.)

is done by changing indenters on the machine. For very hard materials a diamond cone is used. Hard steel balls are used for the softer materials.

Steel can be hardened by heating it to a high temperature and then cooling suddenly by plunging it into water or oil. However, such steel becomes too brittle for most uses. It can then be tempered (toughened) by reheating and cooling slowly. As the metal loses hardness, it gains toughness. Thus, if the metal were allowed to cool slowly and completely, it would be left soft and tough but not brittle. This process is called *annealing*.

Hardness must not be confused with brittleness. For example, steel is hard and tough, while glass is hard but brittle.

10.15   Ductility

The *ductility* of a material is that property which permits it to be drawn into a wire. The smaller the diameter of the wire to which it can be drawn, in general, the greater the ductility. Many common metals are tenacious enough to allow rods or large wires to be drawn through a hole in a

Fig. 10.12 Heated steel billets pass through a succession of reducing stands on the 11-in. reinforcing bar mill at speeds of 2,500 ft/min. (Bethlehem Steel Corporation.)

hard steel plate (*die*) that has a smaller diameter than that of the original rod or wire. Gold, silver, and copper are common metals with a high degree of ductility. Platinum can be drawn into wires 0.00003 in. in diameter. Glass, when heated to softness, can be drawn into such fine threads that garments and curtains can be woven from them. Iron and steel are also ductile, but not to the extent that characterizes the above materials. In *wiredrawing* steel, it is common practice to resort to heat treatment (annealing) between successive passes through the die.

Ordinary wire for fencing and utility use is *hot-drawn*. Piano wire and other special wires are *cold-drawn* so that their hardness will be retained.

10.16   Malleability

The property of a material that makes it capable of being rolled or hammered into thin sheets of various shapes is called *malleability*. Gold is ex-

tremely malleable and has been rolled into sheets that are 1/300,000 in. thick—so thin that they will actually transmit diffused light. Copper, aluminum, and tin are quite malleable—one reason why we have so many pots, pans, and appliances made from them. Lead is extremely malleable and was the first metal used for pipes. In fact, our word "plumbing" comes from the Latin *plumbum*, meaning lead. Aluminum foil is sheet aluminum that has been rolled very thin. It is an excellent wrap for food, candy, gum, cigarettes, and other packaged products.

The capability of industry to roll steel into girders of different shapes or cross sections is important to the engineer. The malleability of various steels is a matter of extreme importance in the automobile industry because of the *streamlined* shapes of auto tops, bodies, and fenders demanded by modern design.

SUMMARY

Solids have definite shapes and volumes because their molecules are confined to a small space between neighboring molecules. The molecules are fixed in relation to each other.

Mass density is the mass per unit volume of a substance:

$$\rho = \frac{m}{V} \tag{10.1}$$

Weight density is the weight per unit volume of a substance:

$$D = \frac{w}{V} \tag{10.2}$$

Mass density and weight density are related by the formula

$$D = \rho g \tag{10.3}$$

The specific gravity of a solid or liquid is the ratio of the density of that substance to the density of water at 4°C.

$$\text{sp gr} = \frac{D_s}{D_w} = \frac{\rho_s}{\rho_w} \tag{10.4}$$

or

$$\text{sp gr} = \frac{\text{weight (or mass) of substance}}{\text{weight (or mass) of equal volume of water}} \tag{10.5}$$

Elasticity is that property of matter which enables it to resist forces that tend to change its size and shape and by virtue of which it will recover from a deformation upon removal of the distorting forces. All solids have some degree of elasticity, varying from steel and ivory, which are highly elastic, to lead, which shows very little elasticity.

Stress is defined as the ratio of the force producing a deformation in a body to the area over which the force is applied.

$$S = F/A \tag{10.6}$$

Strain is defined as the ratio of the change in a dimension of a body to the total value of the dimension in which the change occurs. For tension

$$\text{Tensile strain} = \frac{e}{L} = \frac{\Delta L}{L} \tag{10.7}$$

Hooke's law expresses the fact that *within the elastic limit of any material the ratio of the stress to the strain produced is a constant*:

$$\frac{\text{Stress}}{\text{Strain}} = K$$

The modulus of elasticity for a given substance is found by dividing the stress applied to the body by the strain produced on it. Moduli of elasticity include those for tensile, compressional, bulk, and shearing elasticity. Young's modulus for tensile and compressional stresses is found by

$$Y = \frac{\Delta F}{A} \frac{L}{\Delta L} \tag{10.8}$$

The maximum unit stress a given material is capable of sustaining before breaking is called the *ultimate strength* of that material.

The ratio of the breaking load of a structure to the maximum load which is expected to be applied to it is called the *factor of safety*.

The bulk modulus B of elasticity of a substance is expressed by the formula

$$B = -\frac{\Delta F/A}{\Delta V/V} = -\frac{\Delta P \, V}{\Delta V} \tag{10.9}$$

The compressibility c of a material is the reciprocal of its bulk modulus:

$$c = \frac{1}{B} \tag{10.10}$$

The shear modulus n is the ratio of shearing stress to shearing strain:

$$n = \frac{\Delta F/A}{x/h} = \frac{h \, \Delta F}{xA} \tag{10.11}$$

The angle of twist resulting when a torque τ is applied to a solid cylindrical rod of length L and radius r is found by the formula

$$\theta = \frac{2\tau L}{\pi n r^4} \tag{10.12}$$

The angle of twist of a hollow cylinder with wall thickness t is given by

$$\theta = \frac{\tau L}{2\pi n r^3 t} \tag{10.13}$$

The Brinell number indicates the relative hardness of materials.

A ductile substance is one that is capable of being drawn into wires.

A malleable substance is one that is capable of being extended or shaped by beating with a hammer or by the pressure of rollers or dies.

QUESTIONS AND EXERCISES

1. When is an atom a molecule?

2. How does a change in elevation, above the earth's surface affect (a) the mass density and (b) the weight density of a fixed volume of a given solid?

3. Name four properties possessed only by solids. Show why, on the basis of the molecular theory, liquids and gases would not be expected to exhibit these properties.

4. The manufacture of springs and their applications to comfort and safety are big business. Discuss some of the factors involved in selecting materials from which to make springs. (Refer to Tables 10.2 and 10.3.)

5. What is the name given to the force of attraction between (a) like molecules, (b) unlike molecules.

6. Which is the more elastic, (a) rubber or cast iron, (b) wrought iron or cast iron?

7. Discuss the difference in behavior of the molecules of a solid when in tension and compression.

8. Discuss why molecular collisions are assumed to be elastic.

9. What is meant by tensile strength? In what units is it expressed?

10. What is meant by the modulus of elasticity? State the formula for the modulus and identify each term in the formula.

PROBLEMS

1. The pointer on a spring balance is displaced $1\frac{1}{4}$ in. by a load of 10 lb. How much will 24 lb displace the spring? (Assume Hooke's law is obeyed.)

2. In Fig. 10.3 a load of 1 kg added to w_0 will cause 0.7 cm elongation as read on the scale. Total loads of 2, 4, 5, and 8 kg are placed successively on the pan. (a) Make a table showing the corresponding elongations for each of these loads. (b) Plot a graph showing the relation between these loads and the corresponding elongations. (c) What load would be represented by an elongation of 5.0 cm on the scale? (Assume Hooke's law obeyed throughout.)

3. How much will a 6-ft long cast-iron pipe weigh if it has an outside diameter of 4.0 in. and an inside diameter of 3.6 in.? (Assume the density of cast iron to be 445 lb/ft^3.)

4. A piece of nylon rope 1 cm in diameter breaks under a load of 13,000 N. What is its breaking stress?

5. Find the stress developed in a steel rod which has a diameter of 0.75 in. and which is hung vertically and supports a load of 8000 lb.

6. A steel plate $\frac{1}{2}$ in. thick has an ultimate shearing strength of 40,000 lb/in.^{2}. What force must be applied to punch a $\frac{3}{8}$-in. hole through the plate?

7. Suppose four $\frac{1}{4}$-in. wrought-iron rivets were used to rivet together two of the plates of Prob. 6. What force would have to be applied before the four rivets would shear off at right angles? Let the shearing strength of the rivets be 35,000 lb/in.^{2}. (Neglect friction between the plates.)

8. A wrought-iron member of a small highway-bridge truss is built to support expected tensile loads of 250,000 N. The factor of safety used in the truss is 5. What should be the cross-sectional area of the member? The ultimate tensile strength of wrought iron is 3.2×10^8 N/m^2.

9. What is the factor of safety in effect if the maximum tensile load to be supported by a steel rod $\frac{3}{4}$ in. in diameter is 8000 lb? Ultimate tensile strength of the steel $= 160 \times 10^3$ lb/in.^{2}.

10. A wire is 0.14 in. in diameter and 6 ft long and stretches 0.013 in. when a force of 50 lb is applied. Find Young's modulus for the material.

11. A steel rod 2 m long whose cross-sectional area is 9.8 cm^2 is subjected to a tensile load of 6500 N. Find the resultant elongation if Young's modulus for steel is 20.0×10^{10} N/m^2.

12. Rubber used in a certain shock absorber has a cross-sectional area of 13 cm^2 and is 6.4 cm long. The rubber has a Young modulus of 2.4×10^8 N/m^2. How much will it support without being compressed more than 0.48 cm?

13. How much will a copper wire 5 m long stretch when a mass of 3 kg is suspended from one end of the wire if the cross-sectional area is 0.04 cm^2?

14. A wire that is 4 m long and has a diameter of 1.0 mm is elongated 1.2 mm when a 5-kg mass is hung from its end. What is the value of Young's modulus for the material of the wire?

15. A rod 3.2 m long with cross section 0.75 cm^2 hangs vertically. When a load of 200 kg is applied to the wire, it stretches 0.049 cm. Find Young's modulus of elasticity for the rod.

16. Four 2-in. diameter cast-iron rods support under compression a load of 3 tons. Each rod has a pre-loaded length of 6 ft. Find the decrease in length of each rod produced by this load.

17. A drawn-brass wire with diameter of 0.162 in. is 6 ft long when supporting a load of 40 lb. How much longer will the wire be when the load is increased to 120 lb?

18. A wire made of mild steel is 1.5 m long and has a cross-sectional area of 2.4 mm^2. (a) What is the greatest load the wire can support without exceeding its elastic limit? (b) If the wire is fastened at its upper end, how far can it be stretched without exceeding its elastic limit? (c) What is the ultimate tensile strength of the wire?

19. What decrease in length of a 6.2-m steel girder whose cross-sectional area is 62 cm^2 will result when it is subjected to a compressional load of 48,000 kg? (Assume $Y = 21 \times 10^{10}$ N/m^2.)

20. A drawn-steel wire 300 m long whose cross-sectional area is 0.65 cm^2 hangs vertically in a deep well. How much does it stretch under its own weight? (Take the density of steel as 7800 kg/m^3. Find the average force acting throughout the length of the wire.)

21. How many steel cables of 0.75-in.^{2} cross section should there be in the hoisting mechanism of a 6-ton elevator (no counterweights) which will have a maximum upward acceleration of 12 ft/sec^2? Assume that the factor of safety is 5 and that the tensile strength of the cable steel is 40,000 lb/in.^{2}

22. A lead ball having a volume of 1600 cm^3 is subjected to a uniform pressure of 7.5×10^6 N/m^2. Find the change in volume that takes place. (Answer in cubic centimeters.)

23. A hydraulic press contains 0.30 m^3 of oil when under no load. What is the decrease in volume of the oil when it is subjected to a pressure of 3.5×10^7 N/m^2 if the compressibility of the oil is 2.0×10^{-10} m^2/N?

24. A marine engine delivers 250 hp at 1000 rev/min to a solid steel drive shaft 3 m long and 3 cm in diameter. Through what angle is the shaft twisted?

25. The hollow drill pipe on an oil rig (see Fig. 10.8) has an outside diameter of 4 in. and a thickness of $\frac{1}{4}$ in. A 500-hp power source turns the pipe at 180 rev/min. Find the angle through which the steel pipe twists if it is 14,000 ft long. (Use the modulus of rigidity, $n = 11.7 \times 10^6$ lb/in.^{2}.)

11 Properties of Liquids

For centuries man has been blessed with more than enough clean water for his needs. At times he has had to develop means of transferring the water from regions of surplus water to farming lands and population centers where shortages existed.

Waterways of the high seas, rivers, and canals have been used to transport goods throughout the world.

With the development of improved methods of producing electric power through electric generators and distribution systems, commensurate efforts have been made to harness the tremendous hydraulic power available, as water in streams and rivers flows to lower levels.

The industrial growth of nations, however, has produced new and serious problems. No longer can man accept an abundance of clean water as an ever-present fact of life. Factories, which once meant prosperity and economic development for a nation, are now seen by some people as a possible peril to the health of its workers and the inhabitants of the communities in which they are located.

Cleaning our streams and waterways is one of the greatest challenges for the technology of today. We are faced with the problem of cleaning up the environment without imperiling the nation's economy.

11.1 Liquids

Liquids in nature are both abundant and rare. Water (a liquid) covers three-fourths of the earth's surface and makes up 60 percent of the weight of our bodies. Yet water and petroleum from the earth's crust are the only two liquids which occur in large quantities in nature. Except for such liquids as molten lava, blood (mostly water), and the sap of trees and plants (also mostly water), most of the liquids in the world today have been created by man, e.g., vegetable oils, some alcohols, gasoline, glycerin, synthetic detergents, and the like.

The study of liquids is divided, for convenience, into two main parts: *hydrostatics*, or liquids at rest, and *hydraulics*, or liquids in motion. The properties of liquids at rest can very often be expressed by means of simple relationships. However, liquids in motion present problems which are usually more difficult to solve than those experienced in hydrostatics. This is due to the presence of frictional and other disturbances whose actions often cannot be expressed in simple mathematical terms.

Liquids are almost incompressible. For example, a pressure of 16,400 lb/in.^{2} (1.13×10^8 N/m^2) (Fig. 11.1) will cause a given volume of water to decrease by only 5 percent from its volume at atmospheric pressure. Practically the

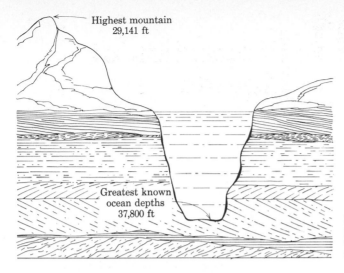

Highest mountain
29,141 ft

Greatest known
ocean depths
37,800 ft

Fig. 11.1 At the greatest known ocean depths, pressures equal 16,400 lb/in² (1120 atm), but the volume of a given amount of water decreases by only 5%.

only type of elasticity possessed by a liquid is that of volume elasticity. The volume elasticity of water is almost perfect. If a tremendous pressure is applied to a volume of water, the water will be compressed slightly, but on removal of the pressure the volume will almost instantly return to its original value.

11.2 Surface Tension

The kinetic-molecular theory (Sec. 9.6) states that although molecules of a substance are in continual motion, they possess mutual attractions for other neighboring molecules. Forces between like molecules are termed *cohesive* forces. Forces between unlike molecules are called *adhesive* forces.

Cohesive and adhesive forces at the interface between two different materials give rise to interesting phenomena. One of these is that of surface tension for liquids. The forces acting on the molecules constituting the surface layer of a liquid are different from those acting on the remaining molecules. No net average cohesive force can act on a molecule in the interior of a body of liquid (Fig. 11.2) because other molecules are symmetrically arranged around it. Consequently, the cohesive forces have the same magnitude in all directions.

Fig. 11.2 The cohesive forces are greater than the adhesive forces at the surface.

However, at the boundary with air, the surface molecules experience a net cohesive force from the other liquid molecules below them which is greater than the net adhesive forces between water molecules and air molecules. The net effect is a greater compacting of molecules at the surface, which causes the surface to act like a stretched membrane. This effect is called *surface tension*. The inward pull on the surface layer tends to make the surface as small as possible. It is this tension that causes dewdrops, raindrops, soap bubbles, etc., to assume the shape of spheres, since for a given volume the shape having minimum surface area is the sphere. Lead shot is man-

Fig. 11.3 Surface tension acting on a paintbrush.

Fig. 11.4 A razor blade and a needle can be made to float on the water surface.

ufactured by passing molten lead through a sieve from a high tower and allowing it to fall into water. The descending molten-lead particles assume a spherical form and solidify in this shape before hitting the water.

When a paintbrush is placed in paint, the bristles spread out, but when it is lifted out, the film of liquid on the bristles contracts because of surface tension, and that pulls the bristles together, making a fine point (Fig. 11.3).

Industry has manufactured many products designed to overcome surface tension. The effectiveness of soap and *detergents* depends upon the ability of these compounds to break down the cohesion and surface tension of the liquid and thus permit the liquid to penetrate under dirt and grease to dissolve it. These substances are now commercially used in washing powders, shampoos, dentifrices, in many germicides, and in agricultural sprays.

11.3 Measurement of Surface Tension

A common sewing needle or razor blade lowered gently to the surface of water in an open vessel can be made to float, as shown in Fig. 11.4. The metal will float on the surface even though it may be 7 times as dense as water.

If next the needle is dipped into the water, cohesive and adhesive forces will cause a thin film of water to adhere to the needle. To pull the needle upward through the surface requires a force in excess of the weight of the needle. This phenomenon is due to surface tension. All liquids have the property of exerting a tension upon adjacent portions of the liquid.

Surface tension is defined as the force per unit length which must be applied to overcome the molecular forces of attraction of the liquid. It can be measured by observing the force needed to pull an inverted U-shaped wire (Fig. 11.5) upward through the surface of the liquid.

Suppose a force F (not including weight of the wire) is required to pull a wire of length L through the surface layer. As the wire leaves the surface, two films of liquid are attached to it, making the effective length $2L$. The two films exert a downward force on the wire equal to $2TL$, where T represents the surface tension of the liquid in newtons/meter of contact between the liquid and the wire. Thus

$$F = TL + TL = 2TL \qquad (11.1)$$

or

$$T = \frac{F}{2L} \qquad (11.2)$$

In general, the surface tension of liquids decreases with rise in temperature. Values of the surface tension of a few common liquids measured in air are given in Table 11.1.

Illustrative Problem 11.1 A fine wire in the shape of a ring of 3.0 cm in diameter is placed horizontally in a vessel containing olive oil. What force in addition to the weight of the wire will be required to pull the wire from the oil surface?

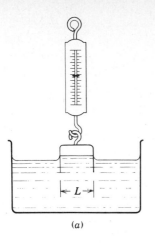

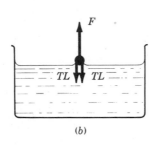

(a)

(b)

Fig. 11.5 Measurement of surface tension:
(a) plan view; (b) force vectors involved.

Table 11.1 Surface tension of some common liquids in contact with air

Liquid	T, newtons/m
Alcohol at 0°	0.024
at 20°	0.022
Benzene at 20°	0.029
Mercury at 20°	0.480
Olive oil	0.035
Soap solution	0.026
Water at 20°	0.073
at 100°	0.059

Solution Use Eq. (11.1) for F:

$$F = 2TL$$

where L is the circumference of the ring.

$$L = \pi d = 3.14 \times 3.0 \text{ cm} = 9.42 \text{ cm}$$

Then $F = 2 \times 0.035 \text{ N/m} \times 9.42 \text{ cm} \times \dfrac{1 \text{ m}}{100 \text{ cm}}$

$$= 6.6 \times 10^{-3} \text{ N}$$

ans.

11.4 Capillary Action

Adhesive, cohesive, and surface-tension forces combine to give a liquid the characteristics which, depending upon the liquid used, will cause it to be raised or depressed in a tube of small bore. These tubes are called *capillary tubes*.

Consider a clean glass tube placed in a vessel of water, as shown in Fig. 11.6. Since the water wets the walls of the tube, the forces of adhesion between the molecules of water and glass must be greater than the forces of cohesion between the water molecules. Consequently a thin film of water will quickly be drawn up the side of the tube (Fig. 11.6a). Surface tension causes this film to contract (Fig. 11.6b), with the net result that the water column rises in the tube. As the water rises in the tube, adhesive forces continue to wet more surface inside the tube. The water will continue to rise in the tube until the upward pull of the adhesive and surface-tension forces just balances the downward pull of gravity on the column of water lifted (Fig. 11.6c). Since, when

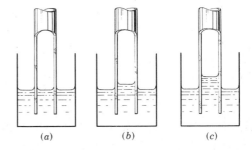

(a)   (b)   (c)

Fig. 11.6 Capillary action.

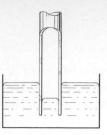

Fig. 11.7 Capillary tube placed in a container of mercury.

the diameter of the tube is small, the weight of the water lifted will be correspondingly light, it can be shown that the smaller the tube, the greater the rise of liquid due to this capillary action. In a glass tube with a bore the size of a human hair, water will rise to a height of about 12 in.

When a liquid does not wet the surface of the tube (cohesion greater than adhesion), such as mercury in glass, the cohesion of the particles of mercury for one another makes it appear as if there were repulsion between the glass and mercury. The surface tension pulls downward on the surface, and the liquid inside the tube is depressed until it stands at a lower level (Fig. 11.7).

Many common phenomena are due in part to capillary action. The rise of oil in lampwicks, the effectiveness of a blotter or towel, the rise of water in the soil, the flow of sap in a plant, and the circulation of the blood in the capillaries of the human body are examples of this action.

11.5   Volatility

Another important property of liquids is that of *volatility*. A liquid is said to have a high degree of volatility if it can be readily changed to a vapor state. Perfumes and many food flavors depend upon this property. The effectiveness of gasoline as a fuel in an internal-combustion engine depends to a great extent upon its volatility, since the liquid gasoline must be quickly converted to a vapor state before proper combustion can occur. On the other hand, too high a volatility can be a disadvantage in an internal-combustion engine. If the volatility is too high, the combustion is not smooth throughout the stroke of the piston and the engine *pings* or *knocks*. Refinery chemists are

constantly striving to produce a properly balanced fuel, one that has enough volatile *lighter fractions* to assure proper and quick starting but enough *heavier components* to result in slow burning and smooth engine performance.

11.6   Viscosity

The internal friction, or resistance to flow, set up within a liquid is called *viscosity*. The water on the surface of a river moves more rapidly than the water near the bottom or sides, and the water in the center of a pipe also moves with greater velocity than the water next to the pipe surface. This difference in velocity is due in part to the friction between the water and the river bed or the pipe, which causes the adjacent layers of water to move slowly. These slowly moving layers of water in turn, because of internal friction (viscosity), tend to retard the motion of adjacent layers. Hence viscosity, or internal friction, is a very important factor in fluid flow.

This property of viscosity is especially significant in the study of oils. In industry, oils with the higher viscosity are the " heavier " oils. It should be pointed out that the term heavier as here used does not refer to density. An increase in temperature of the oil is accompanied by a decrease in its viscosity, and vice versa. Viscosity is responsible for most of the dissipation of energy, and hence for most of the expense, in transporting liquids through pipelines. Thus it is found that a great deal of energy is expended in transporting crude oil in pipelines—the colder the weather, the greater the energy expended. Sometimes the oils must be heated before they can be pumped.

The effectiveness of lubricating oils depends, among other things, upon their viscosity. Lubricating oils should be viscous enough so that they will not be squeezed out of the bearing, yet not so viscous as to offer needless resistance to the motion of the parts being lubricated.

The viscosity of a fluid can be determined by measuring the time required for a given quantity of the fluid to escape through a long tube of small diameter. For determining the viscosity of paints,

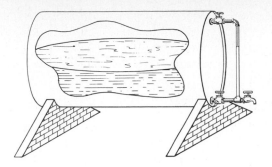

Fig. 11.8 The water in the gauge stands at the same level as that in the tank.

varnishes, lacquers, and similar liquids, the test often consists in timing the flow of the liquid through a standard orifice. The orifice is screwed into the conical bottom of a heavy polished-metal cup. The instruments used in these tests are called *viscosimeters*.

11.7 A Liquid Seeks Its Own Level

When a liquid is poured into a vessel, it flows until it takes the shape of the vessel. The height of the liquid in the vessel will depend on the volume of the liquid and on the shape of the vessel. A liquid, like a gas, offers no resistance to a shearing stress.

A liquid will tend to keep its surface level. When the surface is not level, the liquid will flow in whatever direction will make the surface level. "Water runs downhill" is a common expression for this property. "Water seeks its own level" is another (see Fig. 11.8).

HYDROSTATICS

11.8 Pressure Is Force per Unit Area

A fluid at rest will exert forces on the walls of the container which are perpendicular to the containing surface. The normal force per unit area is called *pressure*:

$$\text{Pressure} = \frac{\text{total force}}{\text{area}}$$

$$P = \frac{F}{A} \tag{11.3}$$

Units of pressure are obtained as ratios of force units and area units. Common units for pressure are the newton per square meter, called a *pascal* (Pa), in the SI-metric system; and the pound per square inch in the English (Engineering) system.

Units of pressure

SI-metric unit   pascal (Pa) = 1 N/m^2

English (Engineering) unit   lb/in^2 (psi)

$$1 \text{ psi} = 6895 \text{ Pa}$$

The *atmosphere*, another pressure unit, represents the average pressure exerted by the earth's atmosphere at sea level. One atmosphere is equal to 1.013×10^5 Pa or 14.7 lb/in.^{2} The *millibar* (mb) is used in meteorology and is equal to 100 N/m^2 or 100 Pa.

Pressures are also measured in millimeters of mercury (mm Hg) and in in Hg. The unit mm Hg is called the *torr* (see Sec. 12.10).

It is essential that a careful distinction be made between the terms *force* and *pressure*. Let

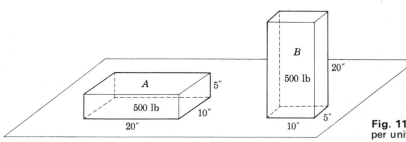

Fig. 11.9 Explanation of pressure as force per unit area.

us illustrate the distinction by considering two rectangular blocks 5 by 10 by 20 in. weighing 500 lb (see Fig. 11.9) resting on a table. Each block exerts a downward force of 500 lb on the surface of the table. Block A will exert a pressure given by

$$P_A = \frac{500 \text{ lb}}{20 \text{ in.} \times 10 \text{ in.}} = 2.5 \text{ lb/in.}^2$$

while block B will exert a pressure given by

$$P_B = \frac{500 \text{ lb}}{10 \text{ in.} \times 5 \text{ in.}} = 10 \text{ lb/in.}^2$$

Thus the pressure exerted by block B is four times that exerted by block A. Equation (11.3) can be solved for F to give

$$F = P \times A \qquad (11.3')$$

Total force equals the product of pressure and area.

11.9 Hydrostatic Pressure Increases with Depth of Liquid

Anyone who dives under the surface of water notices that the pressure on the eardrums at a depth of even a few feet is quite noticeable. Deep-sea divers wear special gear to protect them against extreme pressures.

In January 1960 Jacques Piccard and Lt. Donald Walsh, USN, descended in a specially constructed underwater vessel to a depth of 37,800 ft off the coast of Guam. Pressures on the vessel reached 16,400 lb/in.², requiring steel walls 5 in. thick for the underwater craft.

Careful measurements show that the *pressure of a liquid* is directly proportional to the depth and for a given depth the liquid exerts the same pressure in all directions.

11.10 The Pressure-Force Formulas for Liquids

Consider a cylindrical container with vertical sides filled with a liquid. Let A be any horizontal cross-sectional area of a column of the liquid and h the height (depth) of the column. A can be the cross-sectional area of the bottom of the container (Fig. 11.10a) or the cross-sectional area of

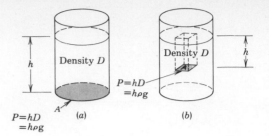

Fig. 11.10 The same formula, $P = hD = h\rho g$, holds at (a) the bottom of the tank as at (b) any depth h below the liquid surface.

any (imaginary) column within the liquid (Fig. 11.10b). The entire weight of the column of liquid above A will be pressing down on the area A. The volume V of the column is Ah. The weight of the column of liquid can be found if we know the density and volume. The density is found from Eq. (10.2), which states that

$$D = w/V$$

From Eq. (10.2) and the formula $V = Ah$, we get

$$w = DV = DAh$$

Equation (10.3) relates weight density and mass density: $D = \rho g$. Substituting ρg for D in the equation for w, gives

$$w = DAh = \rho g Ah$$

Since the force F acting on area A is equal to the weight w of the column of liquid,

$$F = DAh = \rho g Ah \qquad (11.4)$$

which gives the force of a liquid on a horizontal area below a free surface.

If we divide each member of Eq. (11.4) by A, we get

$$\frac{F}{A} = \frac{DAh}{A} = \frac{\rho g Ah}{A}$$

We can now substitute pressure P for F/A to get the formula for *liquid pressure* at a given depth below the surface of the liquid.

$$P = hD = h\rho g$$

Pressure = depth × weight density   (11.5)

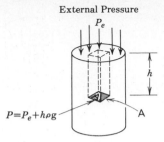

External Pressure
P_e

$P = P_e + h\rho g$ A

Fig. 11.11 The total pressure at A is the sum of the fluid pressure due to the liquid depth h, and the external pressure, P_e.

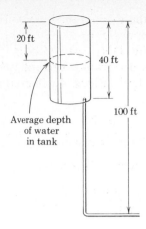

20 ft

40 ft

100 ft

Average depth
of water
in tank

Fig. 11.12 Illustrative Problem 11.2.

It is essential that depths and densities be expressed in consistent units when using Eq. (11.5). When D is in pounds per cubic foot and h is in feet, P will be in pounds per square foot. When ρ is in kilograms per cubic meter, h in meters, and g is 9.81 m/sec², P will be in newtons per square meter or pascals.

Since $D = 62.4$ lb/ft³ (from Table 10.1) for water, it follows that if height h is one foot,

$P = hD$

$= 1$ ft $(62.4 \text{ lb/ft}^3) = 62.4 \text{ lb/ft}^2$

$= 62.4 \text{ lb/ft}^2 (1 \text{ ft/12 in.})^2 = 0.433 \text{ lb/in.}^2$

This is equivalent to saying that the weight of a column of water 1 ft high and 1 in.² in cross-sectional area is 0.433 lb.

Using the same approach for metric units we find that, since ρ for water is 1000 kg/m³, if height h is one meter, then

$P = h\rho g$

$= 1$ m $(1000 \text{ kg/m}^3)(9.81 \text{ m/sec}^2)$

$= 9.81 \times 10^3 \dfrac{\text{kg}}{\text{m-sec}^2}\left(\dfrac{1 \text{ N}}{1 \dfrac{\text{kg-m}}{\text{sec}^2}}\right)$

$= 9.81 \times 10^3 \text{ N/m}^2 = 9.81 \times 10^3 \text{ Pa}$

This, in turn, is equivalent to saying that the weight of a column of water 1 m high and 1 m² in cross-sectional area is 9.81×10^3 N. It should be noted that Eq. (11.5) is used to determine the pressure due to the liquid alone. The *total* pres-

sure within a fluid also depends upon the external pressure P_e at the surface (see Fig. 11.11). To obtain the total pressure P, the external pressure P_e must be added to the fluid pressure.

$$P = P_e + hD = P_e + h\rho g \qquad (11.6)$$

The external pressure on the surface of a liquid may be the pressure of the atmosphere or pressure caused by a piston. In general the atmospheric pressure is equal to 14.7 psi or 1.013×10^5 N/m² (pascal).

Illustrative Problem 11.2 A cylindrical water tank 40 ft high and 20 ft in diameter is filled with water (Fig. 11.12). (*a*) What is the water pressure on the bottom of the tank? (*b*) What is the total force on the bottom? (*c*) On the vertical wall? (*d*) What is the pressure (pounds per square inch) in a water pipe at street level which is 100 ft below the water surface?

Solution

(*a*) $h = 40$ ft

$D = 62.4 \text{ lb/ft}^3$

$P = h \times D$

$= 40 \times 62.4$

$= 2500 \text{ lb/ft}^2$ *ans.*

(b)
$$F = P \times A$$

$$= 2500 \text{ lb/ft}^2 \times 3.14 \times (10 \text{ ft})^2$$

$$= 780{,}000 \text{ lb (on bottom)} \qquad ans.$$

(c) In calculating the total outward force against the tank wall, it must be realized that the pressure along the sides varies with the depth. Since the pressure varies directly with the depth, we can use the pressure at the midpoint of the tank as an average value with which to compute force against the vertical wall. The lateral surface area of a cylinder is given by substituting

$$S = 2\pi rh$$

$$= 2 \times 3.14 \times 10 \text{ ft} \times 40 \text{ ft}$$

$$= 2510 \text{ ft}^2$$

Then, substituting in Eq. (11.5), and taking h as *average depth*, or 20 ft:

$$F = 2510 \text{ ft}^2 \times 20 \text{ ft} \times 62.4 \text{ lb/ft}^3$$

$$= 3{,}130{,}000 \text{ lb} \qquad ans.$$

(d) $\qquad h = 100 \text{ ft}$

$$P = \frac{100 \text{ ft} \times 62.4 \text{ lb/ft}^3}{144 \text{ in.}^2/\text{ft}^2}$$

$$= 43.3 \text{ lb/in.}^2 \qquad ans.$$

Illustrative Problem 11.3 Determine the pressure, in pascals, due to a column of mercury 75.8 cm high.

Solution From Eq. (11.5), $P = h\rho g$. Also, $1 \text{ Pa} = 1 \text{ N/m}^2$. The density of mercury is 13,600 kg/m³ (Table 10.1). Thus

$$P = h\rho g$$

$$= 75.8 \text{ cm}(1 \text{ m}/100 \text{ cm})(13{,}600 \text{ kg/m}^3)$$

$$\times (9.81 \text{ m/sec}^2)$$

$$= 10.1 \times 10^4 \text{ N/m}^2$$

$$= 10.1 \times 10^4 \text{ Pa} \qquad ans.$$

Illustrative Problem 11.4 What horsepower motor will be needed to lift water 100 ft and de-

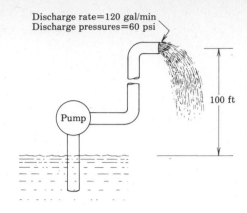

Discharge rate=120 gal/min
Discharge pressures=60 psi

Pump

100 ft

Fig. 11.13 Illustrative Problem 11.4.

liver it at a rate of 120 gal/min at a discharge pressure of 60 psi (see Fig. 11.13)? (A review of Secs. 5.13 and 5.14 may be helpful in solving this problem.)

Solution Let P_1 be the pressure equal to a 100-ft column of water and P_2 equal the discharge pressure. Then

$$P_1 = hD = 100 \text{ ft}(62.4 \text{ lb/ft}^3) = 6240 \text{ lb/ft}^2$$

$$P_2 = 60 \frac{\text{lb}}{\text{in.}^2} \left[\frac{12 \text{ in.}}{1 \text{ ft}}\right]^2 + 8640 \text{ lb/ft}^2$$

The total water pressure P at the pump will be

$$P = P_1 + P_2 = 6240 \text{ lb/ft}^2 + 8640 \text{ lb/ft}^2$$

$$= 14{,}880 \text{ lb/ft}^2$$

Water is lifted at the rate of

$$120 \frac{\text{gal}}{\text{min}} = 120 \frac{\text{gal}}{\text{min}} \left[\frac{1 \text{ ft}^3}{7.48 \text{ gal}}\right]$$

$$= 16.0 \text{ ft}^3/\text{min}$$

Eq. (5.12) states that

$$\text{Power} = \frac{\text{force} \times \text{distance}}{\text{time}} = \frac{Fs}{t}$$

Eq. (11.3′) states that $F = PA$. By proper substitution, we can get

$$\text{Power} = \frac{PAs}{t}.$$

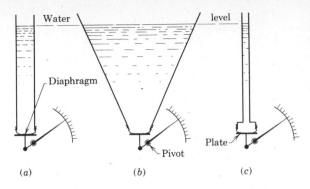

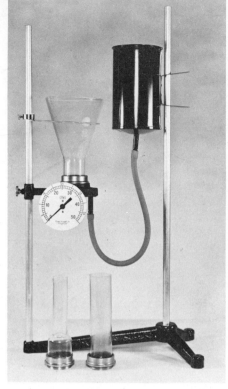

(a)   (b)   (c)

Water   level

Diaphragm

Pivot

Plate

(d)

Fig. 11.14 Pascal's vases. (*a*) to (*c*) Despite varying quantities of water in the vessels, the pressure at the bottom is the same. (*d*) A typical arrangement of Pascal's vases used in the physics laboratory.

But Λs = volume V. Hence

$$\text{Power} = \frac{PV}{t}$$

and the horsepower needed is

$$HP = 14,880\ \frac{\text{lb}}{\text{ft}^2}\left(16.0\ \frac{\text{ft}^3}{\text{min}}\right)\left[\frac{1\ \text{hp}}{33,000\ \text{ft-lb/min}}\right]$$

$$= 7.21\ \text{hp} \qquad\qquad ans.$$

11.11 Shape and Size of Container Related to Liquid Pressure

Equation (11.5) indicates that the pressure due to the weight of a liquid is independent of the size and shape of the reservoir or container; the pressure depends only upon the depth below the free surface and on the density of the liquid.

Blaise Pascal (1623–1662) was the first to prove experimentally that shape and volume of a container do not affect pressure. Figure 11.14 shows the essential parts of an apparatus to illustrate this principle. Three vases with different shapes but with equal cross-sectional openings at the bottom have a flexible diaphragm stretched across the bottom openings. These containers are then placed on a solid plate which by its vertical motion imparts rotary motion to a pointer pivoted about a fixed point. When each of the vases is filled with water to the same height, the water forces acting on the diaphragm are found to be equal by identical readings on the scale—this despite the fact that the weight of water in vase *b* may exceed by many times that in either *a* or *c*. Since the forces at the bottom in each of the vases were shown to be the same and since the cross-sectional areas are the same, the quotient

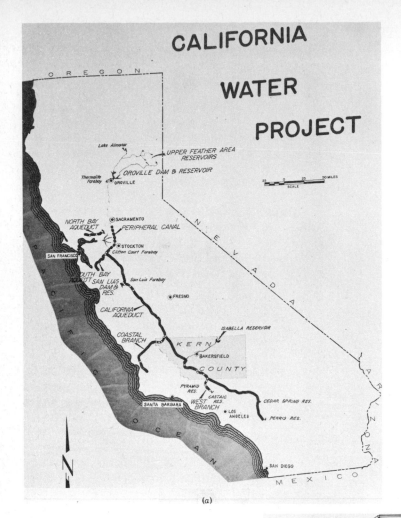

(a)

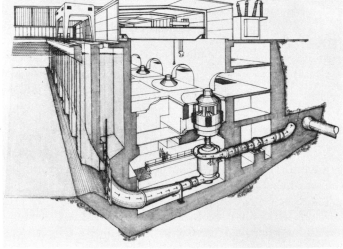

(b)

Fig. 11.15 (a) The map shows how surplus water from northern California is conveyed through a series of dams, reservoirs, aqueducts, and canals to southern California, which has an arid climate and large population needs for water. (b) The drawing shows one of the 14 four-stage 80,000-hp centrifugal pumps used at the Edmonston pumping plant of the California State Water Project. The combined pumps produce a water flow of 4100 ft³/sec. Each assembly consists of four pump impellers mounted on a common vertical shaft inside a housing with driving motor on top. The pump impellers turn at 600 rev/min. Each assembly is about 65 ft high and weighs nearly 300 tons. The pumps boost the pressures of the water sufficiently to push it 750 ft higher than the Empire State Building.

F/A (the pressure) must in consequence be the same. Figure 11.14*d* shows a typical arrangement of Pascal's vases found in the physics laboratory.

11.12   Water Dams

Water is impounded behind dams for four main reasons: for flood control and conservation purposes; for the development of electric power; for the distribution of water to cities, towns, and farms; and for recreation.

The California State Water Project illustrates how modern engineering makes it possible to convey surplus water from one part of the country to areas of need and at the same time provide flood control (see Fig. 11.15*a*). While 85 percent of the people of California live between Sacramento and the Mexican border, 70 percent of the state's water supply originates north of the latitude of San Francisco Bay. Moreover, throughout the state, the bulk of the rainfall occurs in a few winter months, while the summers, when water needs are greatest, are long and dry.

The State Water Project is designed to provide 4.23 million acre-ft of water annually to the water-deficient areas of central and southern California.

The project includes eight power plants which can produce a maximum annual output of 6233 million kW-hr; 21 pumping plants which will have an annual energy requirement of 13,646 million kW-hr; 21 major dams and reservoirs, and 700 miles of canals, siphons, and tunnels.

The Oroville Dam, which is the key water-conservation facility, towers 770 ft above its foundation and is the highest dam in the United States. The Hyatt power plant deep within the abutment of Oroville Dam produces enough power for a city of 1 million people.

The Edmonston pumping plant (Fig. 11.15*b*) lifts more water higher than any other such plant in existence: 110 million gallons of water per hour in a single lift more than 1900 ft up the northern face of the Tehachapi Mountains.

11.13   Transmission of Liquid Pressure —Pascal's Law

The pressure of a liquid in most of the cases cited to this point has been due to the weight of the liquid. Liquid pressures may also result from the application of pressures on the liquid from without. Consider the following experiment. Figure 11.16 represents a container completely filled with liquid. *A, B, C, D,* and *E* represent pistons of equal cross-sectional areas fitted into the walls of the vessel. There will be forces acting on the pistons *C, D,* and *E* because of the different depths of liquid. Assume that the forces on the pistons due to the weight of the liquid are as follows: *A*, 0 N; *B*, 0 N; *C*, 10 N; *D*, 30 N; and *E*, 25 N. Now let an external force of 50 N be applied on piston *A*. If only piston *B* is allowed to move, it will be pushed up with a force of 50 N. If only piston *C* is permitted to move, it will be pushed to the right with a force of 50 + 10 or 60 N. In like manner piston *D* would be pushed down with a force of 80 N and piston *E* to the left with a force of 75 N. That is, when a downward force *F* is applied at piston *A*, this force *F* will be transmitted undiminished to all the other pistons. Since the transmitted forces on, and the areas of, the pistons are the same, the pressures transmitted must be the same. This fact was first deduced and stated by Pascal in 1653. The principle is commonly known as *Pascal's law* and applies to all fluids which are confined and at rest.

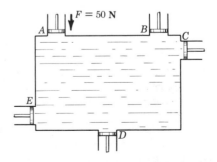

Fig. 11.16   Pressure applied at any piston is transmitted undiminished to each of the other pistons.

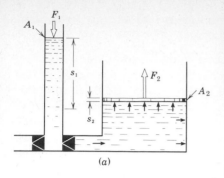

(a)

520 HP hydraulic system operating at operating pressure of 2825 psi

Height above floor: 36 ft

60 in diameter single-acting main ram

Stroke of moving platen: 60 in

(b)

Fig. 11.17 (a) Principle of the hydraulic press. (b) The compression molding press is rated at 4000 tons and is used to produce hood units for heavy-duty trucks. (Erie Press Systems.)

Pascal's Law

Pressure applied to a confined fluid is transmitted undiminished throughout the confining vessel or system.

The pressure exerted by a fluid on the walls of a containing vessel is always at right angles to the surface.

11.14 The Hydraulic Press

An important industrial application of Pascal's law is found in the *hydraulic press*. This machine makes it possible to obtain an enormous force by exerting a relatively small effort. In Fig. 11.17a if a small force F_1 is applied to the smaller piston of area A_1, the pressure F_1/A_1 is transmitted undiminished throughout the confined liquid. This pressure acting on the larger-area piston A_2 will exert a total force on it equal to the product of the area A_2 times the pressure. Hence,

$$F_2 = PA_2 = \frac{F_1}{A_1} A_2$$

or, rearranging,

$$F_2 = \frac{A_2}{A_1} F_1 \qquad (11.7)$$

The force that can be exerted by the large piston equals the force exerted on the small piston times the ratio of the areas of the pistons.

Thus by simply changing the ratio of A_2 to A_1, the force exerted on the larger piston for a given effort on the smaller piston can be made as large as desired. It should be emphasized that where force is gained, distance is lost. For example, if the force obtained at the larger piston is 1000 times the effort applied on the smaller one, the distance through which the larger piston will move is only 1/1000 the distance the smaller one travels. That is, the product of the force and the distance is the same for both pistons (neglecting friction):

$$F_1 s_1 = F_2 s_2 \qquad (11.8)$$

Equation (11.8) illustrates the principle of work which was discussed at length in Chap. 6.

Representing the forces by F_1 for the small piston and F_2 for the large piston, the areas by A_1 and A_2, respectively, and the diameters by d_1 and d_2, we may write the following relation for the hydraulic press.

$$\frac{F_1}{F_2} = \frac{A_1}{A_2} = \frac{d_1^2}{d_2^2} \qquad (11.9)$$

It is left to the student to prove that A_1 and A_2 are proportional to d_1^2 and d_2^2.

Hydraulic presses have many industrial uses. They are used for baling paper and cotton; stamping out crankcases, tops, and other sheet metal parts for automobiles; for forcing lead through dies; and punching holes through iron plates. Other common hydraulic applications occur in barbers' or dentists' chairs, in the brake systems of autos and airplanes, and in presses for extracting oil from seeds.

Illustrative Problem 11.5 The smaller and larger pistons of a hydraulic press have diameters of 7.50 cm and 100 cm, respectively. (*a*) What force must be applied by the smaller piston in order to develop a compressive force of 1.11×10^6 N at the larger piston? (*b*) How far must the small piston travel to move the large piston 2.54 cm? Ignore friction.

Solution
(*a*) $d_1 = 7.50$ cm, $d_2 = 100$ cm, $F_1 =$ force on smaller piston, $F_2 = 1.11 \times 10^6$ N.   Using

Eq. (11.9), we have

$$\frac{F_1}{F_2} = \frac{d_1^2}{d_2^2}$$

$$\frac{F_1}{1.11 \times 10^6 \text{ N}} = \left[\frac{7.50 \text{ cm}}{100 \text{ cm}}\right]^2$$

Then

$$F_1 = \left[\frac{7.50}{100}\right]^2 (1.11 \times 10^6 \text{ N})$$

$$= 6.24 \times 10^3 \text{ N} \qquad \qquad ans.$$

(*b*) Letting $s_1 =$ distance small piston travels and $s_2 = 2.54$ cm, we can use Eq. (11.8):

$$F_1 s_1 = F_2 s_2$$

$$6.24 \times 10^3 \text{ N} \times s_1 \text{ cm}$$

$$= 1.11 \times 10^6 \text{ N} \times 2.54 \text{ cm}$$

$$s_1 = \frac{1.11 \times 10^6 \text{ N} \times 2.54 \text{ cm}}{6.24 \times 10^3 \text{ N}}$$

$$= 452 \text{ cm} = 4.52 \text{ m} \qquad ans.$$

The smaller piston moves a total distance of 4.52 m in a series of short strokes.

11.15 Archimedes' Principle

Buoyancy We all have had numerous opportunities for observing the buoyant effect of liquids. When we go swimming, our bodies are held up almost entirely by the water. Wood, ice, and cork float on water. When we lift a rock from a stream bed, it suddenly seems heavier on emerging from the water. Boats rely on this buoyant force to stay afloat.

The amount of this buoyant effect was first computed and stated by the Greek philosopher Archimedes (287–212 B.C.).

Archimedes' Principle

When a body is placed in a fluid, it is buoyed up by a force equal to the weight of the displaced fluid.

If the body weighs more than the liquid it displaces, it will obviously sink but will *appear* to

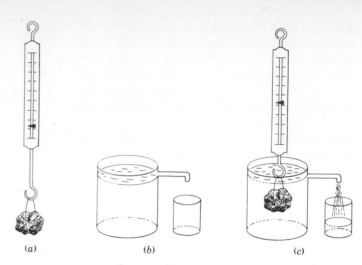

Fig. 11.18 Experimental verification of Archimedes' principle.

(a) (b) (c)

lose an amount of weight equal to that of the displaced liquid. If the body weighs less than that of the displaced liquid, the body will rise to the surface, eventually floating at such a depth that it will displace a volume of liquid whose weight will just equal its own weight. *A floating body displaces its own weight of the fluid in which it floats.* This is an alternative statement of Archimedes' principle, sometimes called the *law of flotation.*

Archimedes' principle can easily be verified experimentally in the laboratory (see Fig. 11.18). (1) Weigh a solid object in air. (2) Fill an overflow vessel to the brim with water and weigh a small can to be used to catch overflow water. (3) Slowly submerge the solid object in the water and catch the overflow water in the small can. Read from the scale the apparent weight of the solid when it is submerged in the water. A definite decrease in weight of the solid will be observed. Now, weigh the small can with the overflow water in it. It will be noted that the increase in weight of the small can (the weight of the displaced water) will be equal to the apparent weight "loss" of the solid.

11.16 Theoretical Proof of Archimedes' Principle

Archimedes' principle can be derived from the laws of liquid pressure as follows. Consider a rec-

tangular block of height h and cross-sectional area A completely immersed in a liquid of density D (or ρg) so that its top is h_1 units below the surface (see Fig. 11.19).

The block has six rectangular surfaces. On the vertical faces the liquid exerts horizontal forces which are in equilibrium (why?). On the top face the liquid exerts a downward force $Ah_1 D$ and on the bottom face an upward force $Ah_2 D$. Since h_2 is greater than h_1, the liquid will exert a net force upward equal to

$$F = Ah_2 D - Ah_1 D$$
$$= AD(h_2 - h_1)$$
$$= AhD \quad \text{(or } Ah\rho g\text{)}$$

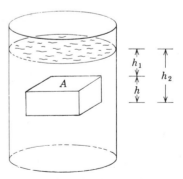

Fig. 11.19 Diagram for theoretical proof of Archimedes' principle.

But Ah is the volume of the block and hence also equal to the volume of the displaced liquid. Substituting V for Ah, we get, for the net upward force,

$$F = VD = V\rho g = \text{weight of the liquid displaced}$$

This indicates that the resultant upward (buoyant) force of the liquid on the solid is truly equal to the weight of the displaced liquid. Although we have proved Archimedes' principle only for a regularly shaped block, the proof is equally valid for an irregular one, since we can think of an irregular solid as being made up of a large number of very small rectangular blocks.

Archimedes' principle has many applications. Although made of steel, which is much heavier than water, ocean liners float because the steel ship encloses a large volume and the average density of the entire volume of the ship fully loaded is less than that of water.

The size of a ship is generally described in terms of the weight of the seawater it will displace when afloat. The world's largest merchant ship is an oil tanker constructed in France in 1977 with a seawater displacement of 555,031 tons. It is 1359 ft long and 206′ 9″ wide.

Submarines and deep-sea vessels apply Archimedes' principle in their design. The submarine submerges by admitting seawater into tanks, thus counteracting the net upward force of buoyancy on the ship. To raise the submarine, compressed air or pumps are used to force the water out of the tanks.

Archimedes' principle holds for all fluids. A gas balloon will ascend if the combined weight of bag, gas contained in the bag, and the balloon's load is less than the weight of the air it displaces. As the balloon rises into the atmosphere, the external air pressure decreases, causing the balloon to expand. However, at the same time the density of the air displaced diminishes, so that the buoyancy per unit volume is decreased. The balloon will continue to rise until the weight of the air it displaces is no longer any greater than the total weight of the balloon and its contents.

11.17 Specific Gravity of Solids

There are several methods used in finding the specific gravity of solids. Each method employs the same general rule: (1) Weigh the object. (2) Find the weight of an equal volume of water. (3) Divide the weight of the object by the weight of an equal volume of water. Archimedes' principle must often be utilized in obtaining the weight of an equal volume of water.

Illustrative Problem 11.6 A piece of cast iron weighs 23.7 lb in air and 20.6 lb when submerged in water. (*a*) What is the specific gravity of the iron? (*b*) What is the weight density of the iron in pounds per cubic foot?

Solution
(*a*) By Archimedes' principle, we know that the weight of water equal in volume to that of the cast iron is $23.7 - 20.6$ or 3.1 lb, since this is the apparent loss in weight.

Hence from Eq. (10.4), the specific gravity of the iron is,

sp. gr.

$$= \frac{\text{wt of iron}}{\text{wt of equal volume of water}} = \frac{23.7 \text{ lb}}{3.1 \text{ lb}} = 7.65$$

(*b*) Using Eq. (10.4″), the density of the iron is

$$D = 62.4 \text{ lb/ft}^3 \times (\text{sp. gr. of iron})$$
$$= 62.4 \text{ lb/ft}^3 \times 7.65$$
$$= 480 \text{ lb/ft}^3 \qquad\qquad ans.$$

11.18 Specific Gravity of Liquids

There are several methods of determining the specific gravity of liquids. Three of these will be discussed here.

The comparison of the weight of a liquid with that of an equal volume of water is made with moderate accuracy and facility by the use of special specific-gravity bottles, called *pycnometers*. The bottle is carefully weighed, first empty, then full of the liquid, and then full of water. By subtracting the weight of the empty

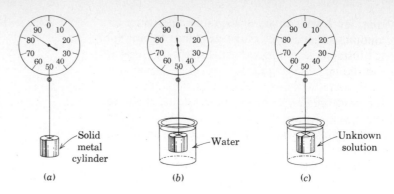

Solid metal cylinder

Water

Unknown solution

(a) (b) (c)

Fig. 11.20 Determining specific gravities of solids and liquids.

bottle in each case, the weights of equal volumes of the liquid and water are obtained. Specific gravity is then obtained from Eq. (10.5):

$$\text{sp. gr.} = \frac{\text{weight of liquid}}{\text{weight of equal volume of water}}$$

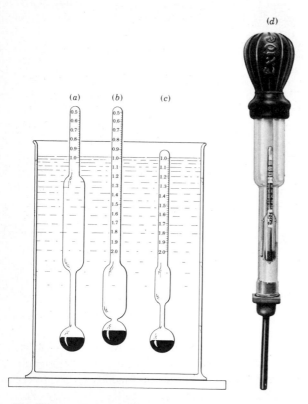

(d)

(a) (b) (c)

Fig. 11.21 Hydrometers for (a) light liquids, (b) light or dense liquids, (c) dense liquids. All three are shown in water. (d) Battery-tester hydrometer. The rubber bulb and outer glass are for drawing a sample of electrolyte out of the battery being tested. (Exide Sales, Electric Storage Battery Co., Automotive Division.)

The second method can be explained by describing an imaginary laboratory problem (see Fig. 11.20). Weigh a solid (1) in air, (2) in water, and (3) in the solution whose specific gravity is desired. Assume the solid weighs 0.85 N in air, 0.49 N in water and 0.61 N in the solution of unknown specific gravity.

The difference, 0.85 N − 0.49 N = 0.36 N, is the buoyant force exerted by the water on the submerged cylinder and therefore (by Archimedes' principle) the weight of the displaced water. Thus, the specific gravity of the solid is

$$\text{sp. gr. of solid} = \frac{0.85\ N}{0.36\ N} = 2.4$$

In like manner, the difference 0.85 N − 0.61 N = 0.24 N, is the weight of the unknown solution which is equal in volume to that of the cylinder and hence also to that of the displaced water.

The specific gravity of the unknown liquid is found to be

$$\text{sp. gr. of unknown liquid} = \frac{0.24\ N}{0.36\ N} = 0.67$$

The quickest and most common method employed in determining specific gravities of liquids is by using a *hydrometer*. The hydrometer is usually a sealed glass tube with an enclosed scale in a cylindrical stem and a bulb weighted with mercury or shot at the bottom (Fig. 11.21). The depth to which a given hydrometer will float in a liquid is determined by the specific gravity of the liquid. In every case the hydrometer will sink to that level which will cause it to displace an amount of the liquid whose weight equals that of the instrument. The greater the specific gravity of the liquid, the higher out of the liquid the hydrometer will float. Because the depth at which the instrument floats is inversely related to the specific gravity of the liquid, a scale which reads specific gravity directly can be mounted vertically along the instrument.

Hydrometers have numerous commercial uses. They are used to measure the specific gravities of alcohols, gasoline, other petroleum products, acids, antifreeze mixtures, sugar solutions, and the like. A common form of hydrometer is the battery-acid tester (Fig. 11.21d).

HYDRAULICS

11.19   Fluids in Motion

Fluids that are in motion exhibit several characteristic differences from fluids at rest. Frictional resistances within the fluid itself (called *viscosity*) and inertia contribute to these differences. Inertia, it will be recalled, means the resistance which a mass offers to a change in its motion. When a fluid is at rest, the pressure is the same at all points of the same elevation. However, this is not

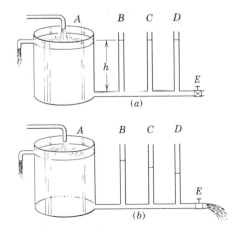

Fig. 11.22   Fluid friction reduces pressure head. (a) Valve E is closed, liquid is not flowing, and gauges B, C, and D all indicate the same pressure. (b) Steady flow is assumed, and friction causes a drop in pressure along the pipe.

true when the liquid is in motion. Figure 11.22a represents the condition of a liquid at rest. When valve E is closed, the levels of liquid in A, B, C, and D are the same. When the valve is opened and liquid flows (Fig. 11.22b), the liquid level in each tube will fall in such a way as to be progressively lower than that in the preceding one. Should the valve be opened wider to permit more liquid to flow, the difference in levels will be further increased. This drop in pressure is therefore seen to be caused in some manner by the *motion* of the liquid.

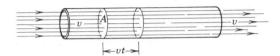

Fig. 11.23   The rate of flow Q of liquid through a pipe is Av. Q = Av is the flow equation.

The Flow Equation   When an incompressible liquid flows in a pipe or channel under steady-state flow conditions, the quantity of flow past all points in the pipe or channel is a constant. If v is the average speed of the liquid in a section of the pipe in Fig. 11.23, the distance through which the liquid will move in time t is vt. If A is the cross-sectional area of the pipe, then the volume of flow V in time t is equal to the volume of the cylinder with base area A and height equal to vt. The volume equals $A(vt)$. The rate of flow Q of the liquid is given by

$$Q = \frac{V}{t} = \frac{Avt}{t} = Av \qquad (11.10)$$

This is the basic *flow equation* for fluids. Q is commonly expressed in m^3/min, ft^3/min or ft^3/sec units.

The steady-flow equation for an incompressible liquid flowing in a pipe with varying cross-sectional areas (see Fig. 11.24) is

$$Q = A_1 v_1 = A_2 v_2 \qquad (11.11)$$

or since $A = \pi d^2/4$,

$$\frac{v_1}{v_2} = \frac{d_2^2}{d_1^2} \qquad (11.11')$$

It will be noted that the velocity varies *inversely* as the cross-sectional area of the pipe or inversely as the square of the diameter if it is a round pipe.

Illustrative Problem 11.7 Water in a pipe 12 cm in diameter flows with a velocity of 8 m/sec. This pipe is joined to a second pipe that is constricted down to 3 cm in diameter. What is the flow velocity in the 3-cm pipe?

Solution Substitute in Eq. (11.11′) to get

$$v_1 = \left(\frac{d_2^2}{d_1^2}\right)v_2$$

$$= (12 \text{ cm})^2/(3 \text{ cm})^2 \times 8 \text{ m/sec}$$

$$= 130 \text{ m/sec} \qquad\qquad ans.$$

11.20 Liquid Heads in Fluid Flow

In the study of hydraulics the vertical depth h_p (in feet) of any point below the free surface of the liquid is called the *pressure head* or *hydrostatic head*. In engineering practice, the total head p is commonly expressed in *feet of water* or in pounds per square inch. *If the liquid is water*, using Eq. (11.4), we find that if h_p is measured in feet, the pressure

$$p = h_p D$$

$$= \left(h_p \text{ ft} \times \frac{12 \text{ in.}}{1 \text{ ft}}\right)$$

$$\times \left(62.4 \text{ lb/ft}^3 \times \frac{1 \text{ ft}^3}{1728 \text{ in.}^3}\right)$$

$$= 0.433 h_p \text{ lb/in.}^2$$

$$P \text{ (lb/in.}^2) = 0.433(h_p \text{ ft}) \qquad (11.12)$$

Drop in pressure due to friction is called the *friction head* (see Fig. 11.22). This resistance to flow due to friction is (approximately) directly proportional to the total area of the rubbing surface, i.e., the product of the length of the channel and its wetted circumference, for round pipes. The resistance also increases with increased roughness of the internal channel surface and with increased velocities of the liquid. Abrupt changes in the cross-sectional area of the channel and changes of direction of flow increase the resistance to flow. With extreme velocities, the frictional resistance set up is increased many times because of the setting up of surging and eddying currents of the fluid, called *turbulent flow*. Friction head results in a drop in pressure head.

The pressure in a pipe or channel due to the velocity of a liquid is called the *velocity head*. It can be shown that if v is the flow velocity and g the acceleration of gravity, the velocity head is determined by the formula

$$h_v = \frac{v^2}{2g} \qquad (11.13)$$

The value of h_v in Eq. (11.13) indicates a pressure due to the velocity of the liquid particles which is equivalent to a hydrostatic head of h_v ft of the liquid. To convert this to pounds per square inch use Eq. (11.12).

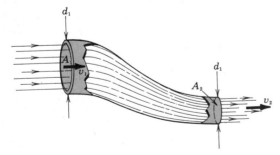

Fig. 11.24 Fluid flow in a tube or pipe of variable diameter.

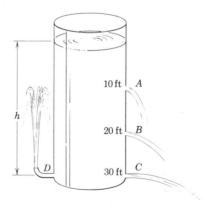

Fig. 11.25 Velocity of water flow varies with pressure head.

Let us consider a tank (Fig. 11.25) filled with water. Openings are placed at depths below the water surface of 10, 20, and 30 ft as shown. The *pressure head* at A is 10 ft or, using Eq. (11.12),

$$P_{10} = 0.433 \times 10 \text{ ft} = 4.33 \text{ lb/in.}^2$$

In like manner, the pressure head at *B* is 8.66 lb/in.^{2} and at *C*, 12.99 lb/in.^{2}.

If Eq. (11.13) is solved for *v*, the formula for the velocity of flow at the various openings is obtained:

$$\text{Velocity of flow} = \sqrt{2g \times \text{pressure head}}$$
$$v = \sqrt{2gh} \qquad (11.14)$$

Substituting values of 10, 20, and 30 ft for *h* in Eq. (11.14), we get values for velocities of flow at *A*, *B*, and *C* of 25.3, 35.9, and 44.0 ft/sec, respectively. The direction of flow does not affect the velocity. Hence if *C* and *D* are at the same depth, the velocity of flow will be the same. It must be made clear, however, that Eqs. (11.13) and (11.14) are theoretical and would be strictly true only if frictional forces within the liquid and at the walls and openings could be neglected. If there were no friction, the velocity head at *D* would be just the exact amount required to force a jet of water back up to the level of the water surface in the tank. The diagram shows it falling short, since friction is always present.

11.21 Bernoulli's Principle

The foregoing paragraphs have shown that several factors affect the flow of liquids in pipes. The diagram of Fig. 11.26 will serve to relate these factors and to introduce a basic law of liquid flow.

Let water be maintained in a reservoir at a constant level *a* when valve *V* is open with steady-state flow conditions in effect. The *total head* is indicated by *h*. Water leaves the reservoir at *A* and acquires velocity *v* as it reaches *B*. The drop in level from *a* to *b* is due to this acquisition of velocity. The drop in head h_v is equal to $v^2/2g$ in accordance with Eq. (11.13). Note that *where friction is neglected*, total head *h* is equal to *pressure head* h_p plus *velocity head* h_v.

There is no change in velocity from *B* to *C*, but a further drop in pressure occurs from level *b* to level *c*. This drop is caused by friction in the pipe length *BC* and is indicated by h_f in the diagram. It should be noted that *where friction in pipes is taken into account*, total head equals pressure head plus velocity head plus friction head.

$$h = h_p + h_v + h_f \qquad (11.15)$$

If a smaller pipe or constriction (section *DE*) is placed in the pipe, flow velocity must of necessity increase, since the flow *rate* is a constant. This increased velocity v_1 results in a further

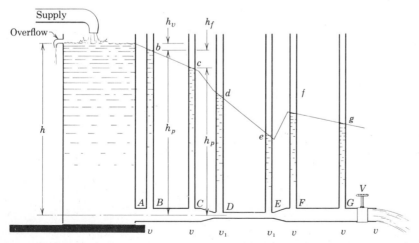

Fig. 11.26 Total head related to velocity head, friction head, and pressure head. Note that where velocity increases, pressure decreases, and where velocity decreases, pressure increases.

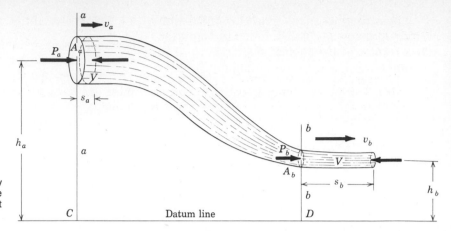

Fig. 11.27 Liquid flow through a tube of variable cross section at different heights.

pressure drop, down to level d, and friction in the narrow pipe results in another drop to e. As the water enters the section FG, whose diameter is the same as that of the original pipe, the flow velocity reduces (since flow *rate* is constant) and pressure rises to level f. The drop from f to g is caused by friction in the section FG.

A study of these flow considerations by Daniel Bernoulli resulted in a fundamental theorem of fluid flow known as *Bernoulli's principle*. One way to state Bernoulli's principle is:

> *When fluids flow in pipes or channels under steady-state conditions, where the velocity is high the pressure is low, and where the velocity is low the pressure is high.*

Actually, Eq. (11.15) is also a statement of Bernoulli's principle. (Another statement will be given shortly.)

The path traced by a particular molecule of a fluid flowing with steady motion (without turbulence) is called the *streamline* of flow. Consider a volume V of fluid with density ρ flowing between planes aa and bb (Fig. 11.27). Assume that (1) the liquid is incompressible, (2) the flow is streamlined, and (3) there is no fluid friction (zero viscosity). As the liquid passes plane a it has potential energy $h_a \rho V g$ and kinetic energy $\frac{1}{2}mv_a^2 = \frac{1}{2}\rho V v_a^2$. The relationship between V, A_a, and s_a is $V = A_a s_a$, or $s_a = V/A_a$. In like manner, $s_b = V/A_b$. The force pushing the volume V of the

fluid past plane a is $P_a A_a$, and that pushing the volume V past plane b is $P_b A_b$.

Thus the work done to push volume V a distance s_a past a is $P_a A_a(V/A_a) = P_a V$. Similarly, at plane b the volume V has potential energy $h_b \rho V g$ and kinetic energy $\frac{1}{2}\rho V v_b^2$, and $P_b V$ units of work are required to push it a distance s_b units past plane b.

If we can neglect viscous losses of energy (we assume streamline flow), the sum of the *three energies* for a unit volume V is the same at a as at b:

$$P_a V + \tfrac{1}{2}\rho V v_a^2 + h_a \rho V g = P_b V + \tfrac{1}{2}\rho V v_b^2 + h_b \rho V g$$
$$(11.16)$$

Dividing each term of Eq. (11.16) by V, we get

$$P_a + \tfrac{1}{2}\rho v_a^2 + h_a \rho g = P_b + \tfrac{1}{2}\rho v_b^2 + h_b \rho g \quad (11.17)$$

You will note that the potential energies were determined by measuring h_a and h_b from an arbitrarily chosen datum line CD. Another form of Bernoulli's equation can be developed by dividing each term of Eq. (11.17) by ρg, yielding an equation with dimensions of a length:

$$\frac{P_a}{\rho g} + \frac{v_a^2}{2g} + h_a = \frac{P_b}{\rho g} + \frac{v_b^2}{2g} + h_b \quad (11.18)$$

Each term of Eq. (11.18) is called a *head*: $P/\rho g$ is the *pressure head*, $v^2/2g$ the *velocity head*, and h the *elevation head*.

In terms of the three heads discussed above (neglecting friction head), Bernoulli's principle may then be stated another way:

At any section of a pipe in which a fluid is flowing under steady-state conditions without friction, the total head is constant; whatever pressure head is lost appears as a gain in velocity head.

Illustrative Problem 11.8 Water flows through the tube of Fig. 11.27 at the rate of 200 ft³/min. The diameter of the pipe at a is 10 in., and P_a is 18 lb/in.². What is the pressure at b, where the diameter is 6.0 in., if the center of the pipe at a is 2 ft higher than the center of the pipe at b? Assume the density of water is 62.4 lb/ft³.

Solution

$$A_a v_a = A_b v_b$$

$$= (200 \text{ ft}^3/\text{min})\left(\frac{1 \text{ min}}{60 \text{ sec}}\right) = \tfrac{10}{3} \text{ ft}^3/\text{sec}$$

$$A_a = \pi\left(5.0 \text{ in.} \times \frac{1 \text{ ft}}{12 \text{ in.}}\right)^2 = \tfrac{25}{144}\pi \text{ ft}^2$$

$$A_b = \pi\left(3.0 \text{ in.} \times \frac{1 \text{ ft}}{12 \text{ in.}}\right)^2 = \tfrac{1}{16}\pi \text{ ft}^2$$

Then

$$\left(\tfrac{25}{144}\pi \text{ ft}^2\right)v_a = \tfrac{10}{3} \text{ ft}^3/\text{sec}$$

$$v_a = \left(\tfrac{10}{3} \text{ ft}^3/\text{sec}\right) \times \frac{144}{25\pi \text{ ft}^2}$$

$$= \frac{96}{5\pi} \text{ ft/sec}$$

and

$$\left(\frac{\pi}{16} \text{ ft}^2\right)v_b = \tfrac{10}{3} \text{ ft}^3/\text{sec}$$

$$v_b = \left(\tfrac{10}{3} \text{ ft}^3/\text{sec}\right) \times \frac{16}{\pi \text{ ft}^3}$$

$$= \frac{160}{3\pi} \text{ ft/sec}$$

$$P_a = \left(18 \ \frac{\text{lb}}{\text{in.}^2}\right) \times \frac{144 \text{ in.}^2}{1 \text{ ft}^2}$$

$$= 2590 \text{ lb/ft}^2$$

$$\rho = \left(62.4 \ \frac{\text{lb}}{\text{ft}^3}\right) \times \frac{1 \text{ slug}}{32 \text{ lb}}$$

$$= 1.95 \text{ slugs/ft}^3$$

From Eq. (11.17),

$$P_a + \tfrac{1}{2}\rho v_a^2 + h_a \rho g = P_b + \tfrac{1}{2}\rho v_b^2 + h_b \rho g$$

$$P_b = P_a + \tfrac{1}{2}\rho(v_a^2 - v_b^2) + \rho g(h_a - h_b)$$

$$= 2590 \text{ lb/ft}^2 + \tfrac{1}{2}(1.95 \text{ slugs/ft}^3)$$

$$\times \left[\left(\frac{96}{5\pi}\right)^2 - \left(\frac{160}{3\pi}\right)^2 \text{ ft}^2/\text{sec}^2\right]$$

$$+ 62.4 \text{ lb/ft}^3 \times 2 \text{ ft}$$

$$= (2470 \text{ lb/ft}^2) \times \frac{1 \text{ ft}^2}{144 \text{ in.}^2}$$

$$= 17 \text{ lb/in.}^2 \qquad\qquad\qquad ans.$$

Illustrative Problem 11.9 Water flows in the pipe illustrated in Fig. 11.27 at the rate of 8.50 m³/min. The diameter at a is 30.4 cm. The diameter at b is 15.2 cm. The pressure at a is $1.03 \times 10^5 \text{ N/m}^2$. What is the pressure at b if the center at b is 60.4 cm lower than the center at a?

Solution

$$A_a v_a = A_b v_b$$

$$= 8.50 \text{ m}^3/\text{min } (1 \text{ min}/60 \text{ sec})$$

$$= 0.142 \text{ m}^3/\text{sec}$$

$$A_a = \pi(0.304 \text{ m}/2)^2 = 7.26 \times 10^{-2} \text{ m}^2$$

$$A_b = \pi(0.152 \text{ m}/2)^2 = 1.81 \times 10^{-2} \text{ m}^2$$

Thus

$$v_a = \frac{0.142 \text{ m}^3/\text{sec}}{A_a} = \frac{0.142 \text{ m}^3/\text{sec}}{7.26 \times 10^{-2} \text{ m}^2}$$

$$= 1.96 \text{ m/sec}$$

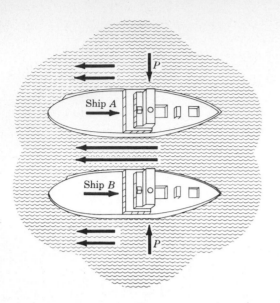

Fig. 11.28 Bernoulli's principle in a practical situation. Pressure is greater on the outboard side of the two ships.

Bernoulli's theorem explains many phenomena which at first may seem strange to the layman. Suppose two ships steaming on a parallel course in still water decide to travel in close formation (see Fig. 11.28). The relative motion of the ships with respect to the water would be such as to cause the water between the ships to "speed up" relative to the ships because of the narrowed space between them. The pressure in the water between the ships will therefore be diminished and will become less than the water pressure on the far sides of the ships. This difference will cause the ships to come closer and closer together, with danger of collision if care in steering is not taken.

The Bernoulli effect is even more pronounced with gases than it is with liquids (see Sec. 12.21).

11.22 Measurement of Liquid Flow

Bernoulli's theorem provides a ready means for measuring the flow of a liquid through a pipe. The *venturi meter*, illustrated in Fig. 11.29, consists of a horizontal section of pipe containing a constriction, or throat, having properly designed tapers to avoid turbulence and assure streamline flow. Bernoulli's equation applied to the wide (a) and constricted (b) portions of the pipe becomes

$$P_a + \tfrac{1}{2}\rho v_a^2 = P_b + \tfrac{1}{2}\rho v_b^2$$

The h_a and h_b terms of Eq. (11.17) drop out since the meter is level.

and

$$v_b = \frac{0.142 \text{ m}^3/\text{sec}}{A_b} = \frac{0.142 \text{ m}^3/\text{sec}}{1.81 \times 10^{-2} \text{ m}^2}$$

$$= 7.85 \text{ m/sec}$$

For water, $\rho = 1000 \text{ kg/m}^3$. From Eq. (11.17),

$$P_b = P_a + \tfrac{1}{2}\rho(v_a^2 - v_b^2) + \rho g(h_a - h_b)$$

Then

$$P_b = 1.03 \times 10^5 \text{ N/m}^2 + \tfrac{1}{2}(1000 \text{ kg/m}^3)$$

$$\times [(1.96 \text{ m/sec})^2 - (7.85 \text{ m/sec})^2]$$

$$+ 1000 \text{ kg/m}^3 (9.81 \text{ m/sec})(0.604 \text{ m})$$

$$= 1.03 \times 10^5 \text{ N/m}^2 - 0.289 \times 10^5 \frac{\text{kg-m}}{\text{sec}^2}\Big/\text{m}^2$$

$$+ 0.0592 \times 10^5 \frac{\text{kg-m}}{\text{sec}^2}\Big/\text{m}^2$$

$$= 8.00 \times 10^4 \text{ N/m}^2 \qquad\qquad ans.$$

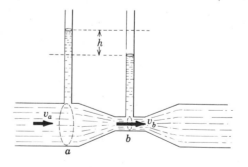

Fig. 11.29 Venturi meter.

Since velocity v_b is greater than v_a, the pressure P_b is less than pressure P_a. Consequently the liquid in the throat manometer will not rise as high as in the pipe manometer. The difference in manometer heights is a measure of the difference in pressures. Thus, if h is the difference in manometer heights,

$$P_a - P_b = \rho g h$$

Assuming the liquid to be incompressible, we know

$$A_a v_a = A_b v_b$$

or

$$v_b = \frac{A_a}{A_b} v_a$$

Bernoulli's equation then becomes

$$P_a - P_b = \rho g h = \tfrac{1}{2}\rho(v_b^2 - v_a^2)$$

$$= \tfrac{1}{2}\rho\left(\frac{A_a^2}{A_b^2} v_a^2 - v_a^2\right)$$

Solving for v_a^2,

$$v_a^2 = \frac{\rho g h}{\tfrac{1}{2}\rho(A_a^2/A_b^2 - 1)}$$

and then for v_a,

$$v_a = \sqrt{\frac{2gh}{A_a^2/A_b^2 - 1}}$$

Since the discharge rate $Q = A_a v_a$, we get

$$Q = A_a\sqrt{\frac{2gh}{A_a^2/A_b^2 - 1}} \qquad (11.19)$$

Hence, if the diameters of the pipe and the throat are known, one need only read the difference in the heights of the liquid in the two tubes to have sufficient data to determine the volume flow rate.

The reduction of fluid pressure at a constriction finds application in such devices as the carburetor of an auto, the aspirator or spray nozzle (see Sec. 12.21), and the filter pump.

Illustrative Problem 11.10 Compute the discharge rate of water through a venturi meter having a pipe diameter of 10 cm and a throat diameter of 5.0 cm if the difference of the liquid heights in the manometer tubes is 6 cm.

Solution

$$A_a = 25\pi \text{ cm}^2 \qquad A_b = 6.25\pi \text{ cm}^2 \qquad h = 10 \text{ cm}$$

$$\frac{A_a^2}{A_b^2} = \left(\frac{25\pi \text{ cm}^2}{6.25\pi \text{ cm}^2}\right)^2 = 16$$

Substitute known values in Eq. (11.19):

$$Q = 25\pi \text{ cm}^2 \sqrt{\frac{2 \times 981 \text{ cm/sec}^2 \times 6 \text{ cm}}{16 - 1}}$$

$$= 2200 \text{ cm}^3/\text{sec} \qquad\qquad ans.$$

11.23 Water Pumps

Many early crude devices were invented to raise water from one level to another, usually from rivers or streams to the higher, fertile agricultural lands nearby. These were followed by the lift pump and the force pump, which were in common use up to several generations ago. Lift pumps and some force pumps depend on atmospheric pressure for their operation, and are therefore quite limited in their effectiveness.

11.24 Modern Pumps

Today there are many types of efficient water pumps, each designed to meet a specific need. These can be generally classified into either the *reciprocating type* or the *rotary type*. Water pumps where high pressures are needed, as in fire engines, are usually of the *double-action* reciprocating type illustrated in Fig. 11.30. As the piston P moves to the right, water enters at the intake and passes through the open valves at D. On the return stroke of the piston, the valves at D are forced closed and those at A open. At the same time, the valves close at B and open at C, permitting water from the intake to flow into the chamber to the right of the piston. As water is alternately forced through valves A and B, it passes into an air chamber E, which acts as a

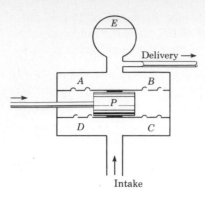

Fig. 11.30   Reciprocating pump.

cushion to maintain a steady stream through the delivery end of the pump.

11.25   Centrifugal Pumps

When a large volume of liquid is to be lifted a short distance or pumped against relatively low pressure, the centrifugal pump is generally used. The essential parts of this pump are shown in Fig. 11.31. The liquid is admitted at the center part of a rotating wheel carrying curved blades, called *impellers*, and is caught between them and thrown outward in such a direction as to force the liquid out the discharge tube against moderate pressures. Centrifugal pumps are made in various sizes, some as large as 16 ft in diameter (see Fig. 11.15*b*).

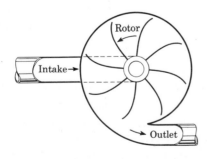

Fig. 11.31   Centrifugal pump.

11.26   Turbine Pumps

Pumps which effectively lifted water from the first or *surface* stratum had to be improved to a point where water could be lifted from much greater depths. The original *dug wells* are now largely replaced by wells which are *drilled* by machines. These drilled wells are of much smaller diameter and are encased with cylindrical steel well casing, which follows the drilling tools through the various strata of geological formations until a sufficient number of water-bearing strata have been penetrated to provide an ample supply of water.

The well casing is then perforated opposite the water-bearing stratum. A *turbine pump* is placed inside this casing, as shown in Fig. 11.32. The turbine pump consists of three major parts. The first part is the pump head and is usually the only part that is visible once the pump is installed. The second major part of the deep-well pump is the discharge column. It consists of a pipe (called the *eduction pipe*) which serves as a means for conveying the water to the surface and also as a supporting member of the third part, the pumping element or bowl assembly. The pumping element (*pump bowls*) consists of a rotating shaft to which the impellers are rigidly attached. These are provided with vanes which react the same way as the blades of an electric fan. The vanes of the water-pump impellers by their rotary motion force the water up the discharge column. The pumping element frequently consists of a system of multistage impellers rotating within the bowls or diffusion cases which in themselves are stationary but which serve to redirect the water from the discharge end of one impeller to the intake portion of the next impeller. Multistage turbine pumps are especially designed for deep-well operations from 200 to 2000 ft.

Submersible pumps are usually used for home, farm, and small industrial water systems, where the water is pumped from more moderate depths and in lesser quantities. For the submersible pump, both the motor and the pump are at the bottom of the well in the water.

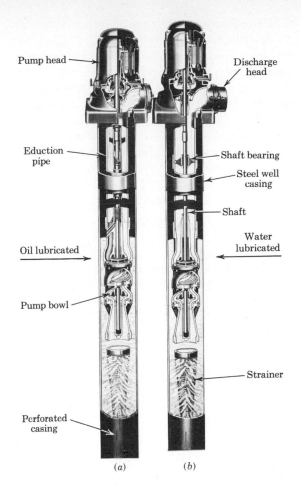

Pump head

Discharge
head

Eduction
pipe

Shaft bearing

Steel well
casing

Shaft

Oil lubricated

Water
lubricated

Pump bowl

Strainer

Perforated
casing

(a)   (b)

Fig. 11.32 Deep-well turbine pump. (*a*) The oil-lubricated type embodies a sealed line shaft in which the bearings are lubricated by an oil-transmission drip method. (*b*) The water-lubricated type eliminates use of oil in the well, ensuring freedom from contamination. (Peerless Pump Co.)

SUMMARY

Liquids have a definite volume but their shape depends on the container.

Liquids are nearly incompressible.

Cohesion is the attraction between like molecules; adhesion is the attraction between unlike molecules.

The surface of a liquid tends to contract and become as small as possible. This contracting force is called *surface tension*. Soaps and detergents are used for the express purpose of minimizing surface tension.

Surface tension of liquids in contact with air is found by the formula:

$$T = F/2L \qquad (11.2)$$

Liquids rise in capillary tubes which they are capable of wetting and are depressed in tubes which they do not wet.

Internal friction (resistance to flow) set up within a liquid is called *viscosity*.

Hydrostatics deals with liquids at rest; hydraulics, with liquids in motion.

Pressure is force per unit area:

$$P = F/A \qquad (11.3)$$

Pressure under a free surface liquid equals the product of the depth times the density:

$$P = hD = h\rho g \qquad (11.4)$$

The hydrostatic force F of a liquid on an area A is equal to

$$F = AhD = Ah\rho g \qquad (11.5)$$

For liquids flowing in pipes (steady-state, incompressible, streamline flow), the flow equation is

$$Q = Av \qquad (11.10)$$

Pressure applied on any part of a confined liquid is transmitted undiminished in all directions throughout the liquid (Pascal's law).

When a body is placed in a liquid, the buoyant force of the liquid on that body equals the weight of the liquid displaced (Archimedes' principle).

Fluids in motion may possess three different kinds of head: *pressure head*, *friction head*, and *velocity head*.

The velocity head is the pressure in a pipe due to the velocity of a liquid and is found by the formula

$$h_v = v^2/2g \qquad (11.13)$$

Whenever the velocity of a fluid in a closed pipe is increased, the pressure is decreased, and vice versa (Bernoulli's principle).

At every section of a steady stream of fluid (friction neglected) the total head is constant; whatever head is lost as pressure is gained as velocity (an alternative statement of Bernoulli's principle).

The Bernoulli equation (for any given pipe flowing full, neglecting friction) is

$$\frac{P_a}{\rho g} + \frac{v_a^2}{2g} + h_a = \frac{P_b}{\rho g} + \frac{v_b^2}{2g} + h_b = \text{constant} \quad (11.18)$$

where
$$P/\rho g = \text{pressure head}$$
$$v^2/2g = \text{velocity head}$$
$$h = \text{elevation head}$$

The venturi meter is used to measure the quantity Q of liquid flow in a pipe. It's equation is:

$$Q = A_a \sqrt{\frac{2gh}{A_a^2/A_b^2 - 1}} \quad (11.19)$$

Force pumps and lift pumps that rely on atmospheric pressure for effectiveness have been almost entirely replaced by reciprocating pumps, centrifugal pumps, multistage turbine pumps, and submersible pumps, which do not depend on atmospheric pressure.

Two other important properties of liquids are (1) *viscosity*, or internal liquid friction, and (2) *volatility*, the measure of a liquid's readiness to assume the vapor state.

QUESTIONS AND EXERCISES

1. The specific gravity of a substance has the same numerical value as its density if the density is expressed in what units?

2. If a rock loses 1 lb weight when submerged 1 ft in water, how much will it lose when submerged 5 ft deep in water?

3. The term *fluid* is used to include what states of matter?

4. Will increasing the diameter of a capillary tube cause water to rise or fall in the tube?

5. An ice cube floats in a glass brimful of water. When the ice melts will the water overflow? Give reasons for your answer.

6. When a ship sails from the ocean into a freshwater river, does it sink deeper or ride higher out of the water? Give reasons for your answer.

7. A glass is brimful of water. What happens if a cube of ice is gently placed on the water surface? Explain.

8. A brass ball is weighed, first in kerosene (sp gr 0.82) and then in benzene (sp gr 0.90). In which will its apparent weight be the greatest? Explain.

9. Upon what factors does rate of liquid flow through a hole in the bottom of a tank depend?

10. Explain why kerosene will rise in the wick of a lantern.

11. Why are soap bubbles floating in air nearly spherical?

12. How is it possible in a hydraulic press to eventually obtain a greater linear displacement at the output piston than the linear displacement of one stroke at the input piston?

13. Explain how Bernoulli's equation is an example of the law of conservation of energy.

PROBLEMS

1. The height of water impounded behind two identical dams is the same. Dam A holds back a lake which stretches 1320 m from the dam. Dam B holds back a lake which stretches 5280 m from the dam. What is the ratio of the forces acting on the dams?

2. What is the force exerted on an area of 4.0 m² at a level in a liquid where the average pressure is 3 Pa?

3. If the specific gravity of brick is 1.8, what is its density in kilograms per cubic meter?

4. Convert a pressure of 1050 mb to torrs; to lb/in.²

5. What is the pressure (psi) in a city water main that is 120 ft below the surface of the water in the reservoir? (Assume water not flowing.)

6. What is the difference in water pressure in two offices on different floors of a building if one office is 20 m higher than the other? Answer in pascals.

7. Cotton from a gin is passed to a baler, where it is compressed by hydraulic presses into 500-lb bales. If the diameters of the pistons of the cotton press are 2 and 30 in., respectively, and the force required to bale the cotton is 12,000 lb, how much force must be applied on the small piston?

8. What air pressure will be necessary to lift a hydraulic auto hoist in a service station if the air acts against a cylinder 8 in. in diameter and is required to lift a 4000-lb car? (Assume the weight of the hoist to be 500 lb.)

9. A block of wood with dimensions 20 by 10 by 5 cm floats with its largest surface horizontal. If its specific gravity is 0.6, how deep will it sink in water?

10. A rectangular block of wood floats in water with four-tenths its volume out of water. What is the specific gravity of the wood?

11. An auto ferry is 30 ft long and 15 ft wide and has vertical sides. When six cars are driven on board, the barge sinks 8 in. deeper into water. How much do the six autos weigh?

12. With what velocity will water emerge from a hole in the bottom of a tank 6 m high if the tank is full?

13. A pressure gauge on a pipe indicates a hydrostatic water pressure of 34 lb/in.². When a valve is opened the gauge pressure drops to 29 lb/in.². What is the velocity of flow of the water in the pipe?

14. A cylindrical tank (7 m high) full of water develops a hole 3 m below the top. How far from the base of the tank will the escaping water strike the ground? (Neglect friction).

15. The water company of a residential district has drilled a new well and installed pumps to force the water into storage tanks. How high above ground must the storage tanks be placed to obtain a hydrostatic pressure head of 2.62×10^5 Pa on the water main at ground level?

16. A cylindrical tank 50 cm high is filled two-thirds with water and one-third with linseed oil. What is the pressure at the bottom of the tank? (Use Table 10.1).

17. A piece of metal weighs 41.0 lb in air. Completely submerged in water it weighs 25.4 lb, and submerged in an unknown liquid it weighs 28.2 lb. What is the density of the unknown liquid?

18. A solid was weighed in water, kerosene, and alcohol. Its loss of weight in water was 5.06 N, in kerosene 4.15 N, and in alcohol 4.00 N. What is the specific gravity of (a) kerosene and (b) alcohol?

19. An iron casting weighs 8.35 lb in air and 7.16 lb when submerged in water. (a) What is the volume of the casting? (b) What is the weight density of the casting?

20. With what speed will water escape from a hole 2.54 cm in diameter at the bottom of a water storage tank 11.0 m high? How many cubic meters will be discharged each second?

21. Water flows at a steady state in a horizontal irrigation pipe. The level of water in the standpipe is 15 ft above that in the irrigation pipe. If at a certain point in the horizontal pipe the friction head is 3 ft and the static pressure head is 4 ft, what is the velocity of flow? What volume in cubic feet of water per hour flows through the pipe if it has an inside diameter of 1 ft?

22. The water level in the intake tower behind a dam is 420 ft above the valve of a 6 ft diameter outlet pipe (called the *penstock*). (a) What is the hydrostatic pres-

sure head at center of the valve? (b) When the valve is opened, water flows into the penstock at the rate of 1600 ft³/sec. Find the velocity of flow through the penstock. (c) Find the velocity head and the pressure head (in feet of water) in the penstock when water is flowing.

23. Water flows at a velocity of 1.52 m/sec in a pipe with a 7.62-cm diameter. It then passes through a 5.08-cm pipe and finally through a 10.16-cm pipe. Find (a) the velocity in the last two pipes and (b) the velocity head in all three pipes.

24. A piece of metal has a mass of 0.287 kg. Submerged in water, it has an apparent weight of 1.69 N. What is the volume of the metal? What is its mass density?

25. Salt water (density 64 lb/ft³) flows through a tube of variable cross section and height as shown in Fig. 11.27. Find the velocity and pressure at b if $P_a = 24$ lb/in.², $v_a = 25$ ft/sec, $h_a = 32$ ft, $h_b = 22$ ft, and the cross-sectional area at b is one-half that at a. (Neglect frictional losses.)

12 Properties of Gases

In many respects, gases resemble liquids: both consist of many tiny molecules that move about in constant random motion. Both are capable of flowing and are therefore designated by the common term *fluid*. Each applies pressure on the surfaces within which it is confined, and each exerts upward buoyant forces in accordance with Archimedes' principle. The discharge velocities through an orifice can be determined for gases as they are determined for liquids. Gases, like liquids, conform to the shape of the containing vessel. Neither is able to exert shearing stresses, except those due to viscosity.

Gases, however, differ from liquids in two respects: gases are very compressible, whereas liquids are nearly incompressible. A liquid has a fixed volume, while a gas has an unlimited volume, determined only by the volume of the containing vessel.

While water is the most abundant liquid, air is the most common "gas." Actually, dry air is a relatively homogeneous mixture of gases consisting of about 77% nitrogen, 21% oxygen, and 1% argon. The remaining 1% includes small quantities of such gases as carbon dioxide, hydrogen, neon, krypton, and helium. Air from the frigid arctic wastes contains very nearly the same proportions of nitrogen, oxygen, and other gases as does air from the tropical deserts.

As is true for the water on the earth, we can no longer assume an unlimited supply of clean air in our environment. The atmosphere surrounding the earth is not boundless. Ninety-nine per-cent of the earth's atmosphere extends just 31 kilometers (19 mi) above the earth's surface. Today we are becoming more and more aware of the dangers involved in adding harmful pollutants to this limited air supply. Many nations are now making concerted efforts to avoid the hazards to health that may result from dumping smoke, coal dust, insecticides, herbicides, fungicides, exhaust fumes, radioactive wastes from nuclear reactions, and similar pollutants into the atmosphere.

12.1 Characteristics of Gases

We have seen in Chap. 9 that when matter exists in the gaseous state its molecules possess high velocities and such a great freedom of motion that any given mass of gas will completely fill any container, regardless of its shape. Gases obey the laws of Pascal and Archimedes. They are perfectly elastic, they can readily be compressed, and they provide buoyancy.

We have seen how the properties of liquids are used in machines, including the hydraulic press. Similarly, pneumatic machines use the properties of gases: air pressure, for example, is used in vacuum cleaners, pneumatic tools, air brakes, and jackhammers.

12.2 Variation of Volume of a Gas with Pressure – Boyle's Law

It can be proved experimentally that whenever the pressure on a gas is increased, the volume of the gas is decreased. On the other hand, a gas

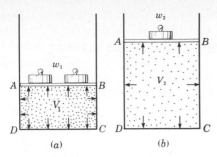

Fig. 12.1 Bombardment of molecules on piston AB supports the weight w.

which is allowed to expand to a greater volume will exert a lesser pressure on the container. The kinetic theory of matter offers a simple and adequate explanation of this pressure-volume relationship.

Let us assume that a certain mass of air is confined in a cylindrical container and that the pressure P_1 of the air is adequate to sustain a weight w_1 on the piston AB (Fig. 12.1a). The molecules of the air are continually bombarding each other, the sides of the vessel, and the bottom of the piston AB (of negligible weight) supporting the weight w_1. The impact of these molecules on AB develops a total force equal to the area of the piston times the pressure. Suppose the volume of the air under this condition to be V_1. When the weight w_1 is replaced by a weight w_2 equal to one-half of w_1, the pressure P_2 necessary to support w_2 is only half the original pressure. The piston AB is forced up until the volume V_2 of the confined gas is doubled (Fig. 12.1b) and the consequent total impact of the molecular bombardment on AB is halved.

This relationship between pressure and volume was first stated as a law in 1662 by Robert Boyle (1627–1691), after whom it is named.

Boyle's Law

When the temperature of a gas is kept constant, the volume of an enclosed mass of gas varies inversely as the absolute pressure upon it.

Boyle's law holds quite accurately for a wide range of pressures, but deviations from the law

must be considered when dealing with extreme pressures.

This relationship can be conveniently expressed as an equation:

$$\frac{V_1 \text{ (original volume)}}{V_2 \text{ (new volume)}}$$

$$= \frac{P_2 \text{ (new absolute pressure)}}{P_1 \text{ (original absolute pressure)}}$$

or $\qquad P_1 V_1 = P_2 V_2 \qquad\qquad (12.1)$

Absolute pressure will be explained in Sec. 12.16.

Illustrative Problem 12.1 The oxygen to be used in an oxy-acetylene welding outfit is stored in a cylinder 1.40 m long, and of 17.8-cm internal diameter. The gas is at an absolute pressure of 1.55×10^6 Pa. To what volume, in cubic meters, would the gas expand if released to atmospheric pressure? (Assume normal atmospheric pressure = 1.013×10^5 Pa.)

Solution Recall the formula for the volume of a right circular cylinder and substitute for known values.

$$V_1 = \pi r^2 h$$

$$= 3.14 \left(\frac{0.178 \text{ m}}{2}\right)^2 (1.40 \text{ m})$$

$$= 3.48 \times 10^{-2} \text{ m}^3$$

Now, it is given that

$$P_1 = 1.55 \times 10^6 \text{ Pa}$$

$$P_2 = 1.013 \times 10^5 \text{ Pa}$$

Next solve Eq. (12.1) for V_2 and substitute for known values:

$$V_2 = \frac{1.55 \times 10^6 \text{ Pa} \, (3.48 \times 10^{-2} \text{ m}^3)}{1.013 \times 10^5 \text{ Pa}}$$

$$= 0.532 \text{ m}^3 \qquad\qquad\qquad\qquad ans.$$

12.3 Variation of Volume of a Gas with Temperature

Charles' Law Boyle's law assumes conditions of constant temperature, and in actual industrial situations this is rarely the case. Temperature changes continually occur, and they affect the volume of a given mass of gas. Until we study the subject of heat and temperature in Chap. 13, the effect of adding heat to a gas will have to be treated superficially. It can be shown that if pressure is unchanged, an increase in temperature of a gas is accompanied by an increase in its volume.

If constant pressure is maintained, the volume of a gas is proportional to its absolute temperature.

("Absolute temperature" is more rigorously explained in Chap. 13.) This statement is known as *Charles' law* and may be given in algebraic form by

$$\frac{V_1 \text{ (original volume)}}{V_2 \text{ (new volume)}} = \frac{T_1 \text{ (original abs temp)}}{T_2 \text{ (new abs temp)}}$$
(12.2)

Illustrative Problem 12.2 A thin balloon is filled with 220 in.^{3} of air at a temperature of 280° abs. To what volume will the air expand if the temperature rises to 320° abs? (Neglect changing forces of constriction of balloon.)

Solution Original volume = 220 in.^{3}; new volume = V_2 in.^{3}; original abs temp = 280°; new abs temp = 320°. Use Eq. (12.2), and write

$$\frac{220 \text{ in.}^3}{V_2 \text{ in.}^3} = \frac{280°}{320°}$$

$$280V_2 = 220 \times 320$$

$$V_2 = 250 \text{ in.}^3 \qquad\qquad ans.$$

12.4   Variation of Pressure of a Gas with Temperature

Since an increase in the temperature of a gas will cause it to expand if the pressure is kept constant, it is reasonable to expect that if a certain mass of gas were heated in a closed container so that its volume had to stay constant, there would be a consequent increase in the gas pressure. Experiments have found this to be true.

The exact relationship between pressure, temperature, and volume of gases is developed at

greater length in Chap. 13. However, it should be noted that for every increase of 1°C in the temperature of a given mass of gas at constant pressure there is an accompanying volume increase equal to approximately $\frac{1}{273}$ of the gas volume at 0°C. For each decrease of 1°C the volume decreases by approximately $\frac{1}{273}$ of the volume at 0°C. Although this generalization breaks down at conditions of extremely low and extremely high temperatures, it is a satisfactory approximation for conditions normally encountered.

12.5   Vapors and Gases

A vapor is the gaseous form of any substance. The term *vapor*, however, is commonly used to describe only those gases which exist as liquids or solids at ordinary temperatures or pressures. Examples of vapors are water vapor, gasoline vapor, alcohol vapor, or naphthalene vapor. The ability of solids and liquids to change to a vapor state is essential to our everyday comfort and convenience—indeed to our very existence. The evaporation of water to the vapor state makes possible the formation of clouds, rain, and snow. The operation of the steam engine depends upon this same property. Before gasoline can effectively be used in an engine, it must first be converted to the vapor state. Perfumes, liquefied petroleum gas (LPG), refrigerants, and mothballs are only a few examples of substances whose effectiveness depends upon their being changed to the vapor state.

12.6   Density of Gases and Vapors

Just as different solids and liquids vary in density, so do gases and vapors. In determining the specific gravities of gases and vapors, the weights of equal volumes of the gases or vapors are compared with that of either oxygen or air. Table 12.1 lists the densities and specific gravities of a few of the more common gases and vapors.

$$\text{sp gr}_{gas} = \frac{\text{weight of a volume of gas}}{\text{weight of an equal volume of oxygen (or air)}}$$
(12.3)

Table 12.1 Density and specific gravity of gases and vapors

Substance	Density kg/m³ or gm/liter at 0°C and 760 mm	Density lb/ft³ at 32°F and 1 atm	Specific Gravity Air = 1.000	Specific Gravity Oxygen = 1.000
Acetylene	1.173	0.0732	0.9073	0.8208
Air	1.293	0.0807	1.0000	0.9047
Ammonia	0.7710	0.0481	0.5963	0.5395
Carbon dioxide	1.977	0.1234	1.529	1.383
Carbon monoxide	1.250	0.0781	0.9671	0.8750
Freon-12	5.39	0.337	4.21	3.77
Helium	0.1785	0.0111	0.1380	0.1249
Hydrogen	0.0899	0.0056	0.0695	0.0629
Isobutane	2.673	0.1669	2.067	1.870
Methyl chloride	2.3043	0.1434	1.7824	1.6123
Neon	0.9003	0.0562	0.6964	0.6300
Oxygen	1.429	0.0892	1.105	1.000
Sulfur dioxide	2.020	0.1261	1.562	1.414

12.7 Industrial Gases

There are hundreds of gases used commercially throughout the world today. Each gas has its own peculiar chemical and physical properties. These properties make many of them suitable for specific industrial purposes. Some of the more common gases and a few of the uses to which they have been put by industry are listed in Table 12.2.

Table 12.2 Some industrial gases and their uses

Gas	Use
Acetylene	High-temperature welding and cutting steel
Ammonia	Refrigerant, preparation of fertilizers, soap, and glass
Argon	Filler for electric-light bulbs and signs (blue color)
Carbon dioxide	Carbonated beverages, fire extinguishers, baking powders
Chlorine	Bleaching agents, germicides, anesthetics, solvents, chemical-warfare agents
Coal gas	Fuels
Coke-oven gas	Fuels
Natural gas	Fuels
Water gas	Fuels
Producer gas	Fuels
Freon (12, 21, 22, etc.)	Refrigerants, aerosol propellant for spray cans
Helium	Lighter-than-air craft, low-temperature physics research
Hydrogen	Balloons, conversion of cottonseed oil and peanut oil into solid edible fats, welding, fuel for rockets (in liquid form)
Methyl chloride	Refrigerant
Nitrogen	Filler for incandescent lamps, component of rocket fuel (liquid)
Oxygen	Welding and cutting, component of rocket fuel (in liquid form)
Sulfur dioxide	Bleaching agent, preservative, preparation of sulfuric acid (which enters into the manufacture of a multitude of products), refrigerant
Water vapor	Steam, for engines and turbines, heating plants, etc.

12.8 The Atmosphere

All creatures on the surface of the earth live at the bottom of a deep ocean of air. The total force (equal to the weight of the air) that this atmosphere exerts on the earth is almost incomprehensible. We gain a notion of this tremendous force when we realize that the total force exerted by the atmosphere on an acre of the earth's surface at sea level is 46,100 tons. Each square meter of surface is acted upon by a force equal to the weight of a column of air 1 m² in cross section, with an altitude equal to the height of the atmosphere. This is roughly 1.013×10^5 N. The force on each square inch of surface at sea level is roughly 14.7 lb. You may wonder why we are not conscious of these terrific forces or even crushed by them. Pascal's law offers a simple, adequate explanation. Air pressure is transmitted equally in all directions; consequently, though the external force on the human body due to air pressure is several tons, the body does not suffer, for an equal pressure on the inside of the body balances it.

The effect of atmospheric pressure can be effectively demonstrated by a simple experiment, diagramed in Fig. 12.2. Tie a thin sheet of rubber over the top end of a jar (Fig. 12.2a). Then seal (with grease) the mouth of the jar over a flat disk connected to a vacuum pump. As air is removed from the jar, the pressure inside the jar will be reduced. As a consequence, the outside (atmospheric) pressure will push the rubber down inside the jar as shown in Fig. 12.2b.

Houses with windows and doors closed in the path of a tornado frequently burst their walls outward. The pressure in the eye of a tornado is very low, and when the tornado passes by, the tremendous forces of atmospheric pressure within the house may easily push the walls out.

12.9 Variations in Pressure of the Atmosphere

Because air has weight, the force of gravity of the earth on the atmosphere causes about half the air in the atmosphere to lie within 3.5 mi (5.6 km) of the surface of the earth. Since the atmospheric pressure at any point is a measure of the weight of

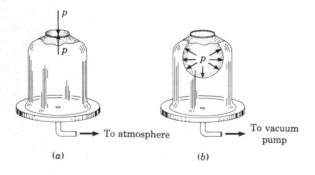

Fig. 12.2 Effect of atmospheric pressure on rubber diaphragm.

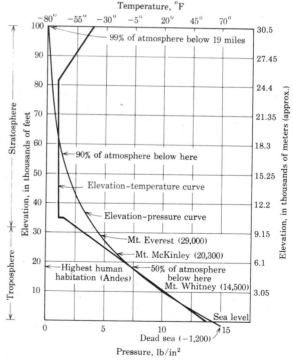

Fig. 12.3 Atmospheric temperatures, pressures, and densities diminish with increasing heights.

the column of air of unit cross-sectional area above that point, it is evident that the pressure will vary with elevation. The greatest atmospheric pressure is found at the lowest levels. Atmospheric pressure on the top of Mt. Everest (8882 m or 29,141 ft) is only about 2.8×10^4 Pa or 4 lb/in.^{2}. At 30 miles, the pressure has been calculated to be only about $\frac{1}{6000}$ that at sea level. (See Fig. 12.3 for variation of atmospheric pressure with height.)

12.10 The Barometer

Instruments which are employed to measure atmospheric pressure are called *barometers*. The earliest and still the simplest barometer is the *mercury barometer*. Evangelista Torricelli (1608–1647) filled a long glass tube, closed at one end, with mercury so as to permit no air in the tube. Then, keeping the open end sealed with his finger, he inverted the tube and placed it with the open end immersed in a vessel containing mercury (Fig. 12.4*a*). When he removed his finger, the mercury fell in the column until a definite height of mercury remained. He found that no matter what the length of the tube, as long as it had a length greater than 30 in. (76 cm), the mercury always leveled off at the same height. Torricelli reasoned that the pressure of the atmosphere on the free surface of the mercury in the vessel was transmitted (*Pascal's law*) to the mercury in the tube, thereby counterbalancing the downward pressure due to the weight of the column of mercury. At sea level the height of the mercury sustained was about 76 cm (30 in.).

If a similar experiment were conducted using water, one might expect (since mercury is about 13.6 times as heavy as water) the atmosphere at sea level to support a column of water 13.6×760 mm (30 in.), which is about 10.4 m (34 ft). Experiment reveals that this is the case. Pascal, when he read of Torricelli's experiment, reasoned that if the air were less dense at higher altitudes, the height of the mercury column supported on the top of a mountain should be appreciably less than it is at sea level. This theory was tested and proved by him.

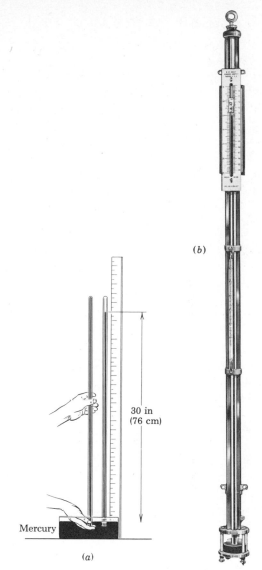

Fig. 12.4 (*a*) Torricelli's experiment for measuring atmospheric pressure. (*b*) Mercury barometer. (Central Scientific Co.)

The simple torricellian tube has been highly refined and perfected into the mercurial barometer (Fig. 12.4*b*). With it even the slightest changes in atmospheric pressure can be detected.

12.11 Atmospheric Pressure

Atmospheric pressure is usually expressed in terms of millimeters of mercury (mm Hg) or inches of mercury (in. Hg), not in Pascals or pounds per square inch. Pressure in millimeters or inches of mercury is read directly from the mercurial barometer, which usually has a vernier attachment for precision reading. The barometer represented in Fig. 12.4b is designed to measure pressures accurate to 0.01 in. Hg. Since a column of mercury 29.92 in. high with cross-sectional area of 1 in.^{2} weighs 14.7 lb, it follows that a pressure of 29.92 in. Hg is equivalent to 14.7 lb/in.^{2}.

Often a unit called the *bar* is employed for pressure. One *bar* is equivalent to 10^5 Pa. One *millibar* (mb) is equal to 10^2 Pa. The millibar is routinely used in meteorology. It can be shown that it is also equivalent to 100 N/m^2 (100 Pa).

Quite frequently—especially for large pressures—the unit used for pressure is the atmosphere (atm). The *atmosphere* represents the average atmospheric pressure at sea level on the earth.

One atmosphere (atm)

$$= 29.92 \text{ in. Hg} = 760 \text{ mm Hg}$$

The unit mm Hg is so frequently used that it has been given the name *torr*, in honor of Torricelli: 1 atm = 760 torrs.

Here are several equivalents of the *atmosphere* (pressure):

$$1 \text{ atm} = 14.7 \text{ lb/in.}^2$$
$$= 1.013 \times 10^5 \text{ N/m}^2$$
$$= 1.013 \times 10^5 \text{ Pa}$$
$$= 1013 \text{ millibars (mb)}$$
$$= 760 \text{ mm Hg}$$
$$= 760 \text{ torr}$$
$$= 29.92 \text{ in. Hg}$$
$$= 10.36 \text{ m of water}$$
$$= 34.0 \text{ ft of water}$$

12.12   The Aneroid Barometer

The mercury barometer is inconvenient in many cases because of its length and the use of a free

Fig. 12.5   The barograph is an aneroid barometer linked to a pen which traces changes in atmospheric pressure on a rotating record sheet. (Courtesy Taylor Instrument Co.)

liquid. For purposes where a more rugged instrument is desired, the mercurial barometer is replaced by the *aneroid barometer* (Fig. 12.5). It consists principally of one or more small disk-shaped, accordion-pleated metal boxes (called *sylphons*) which have been partially evacuated and sealed. As the pressure of the air changes, the sylphon moves slightly up or down. This small motion is multiplied manifold by a delicate system of levers and is communicated to a pointer which moves over a scale to correspond to the readings of a mercury barometer or to be read directly in pounds per square inch. Large aneroid barometers have been made so sensitive that they will show the difference in atmospheric pressure between the floor level and the top of a table. Mercury barometers have been made which are even more sensitive than the most sensitive aneroid.

12.13   The Altimeter

Aviators use a form of aneroid barometer called the *altimeter*, in which the pointer moves along a scale which is calibrated to read directly in feet above sea level. Table 12.3 shows how the pressure of the atmosphere will change as one ascends above sea level.

The barometer is not entirely satisfactory as an altimeter, because even at a single elevation the pressure of the atmosphere changes con-

Table 12.3 Standard atmospheric pressure at various heights above sea level

Altitude		Pressure			
Kilometers	Miles	Millibars	mm Hg (torr)	in. Hg	lb/in.²
0	0	1013	760	29.9	14.7
2	1.2	795	596	23.5	11.5
4	2.5	616	462	18.2	8.9
6	3.7	472	354	13.9	6.8
8	5.0	356	267	10.5	5.2
10	6.2	264	198	7.8	3.8
12	7.5	193	145	5.7	2.8
14	8.7	141	106	4.2	2.0
16	9.9	103	77	3.0	1.5
18	11.2	75	56	2.2	1.1
20	12.4	55	41	1.6	0.8

tinually because of changing weather conditions. It is imperative that the altimeter be constantly adjusted to the atmospheric pressure prevailing at the time of use. Also, the aviator is interested primarily in his height above the terrain below him, and not in his height above sea level. Modern altimeters, using radar or other electronic devices, measure height above terrain rather than height above sea level.

12.14 Weather Forecasting

Although the fluctuating barometric pressure due to changing weather conditions makes a barometer unreliable as an altimeter, it nevertheless renders it a valuable instrument in predicting weather conditions. Experience shows that a falling barometer often indicates an approaching storm, while a rising barometer usually indicates fair weather. These changes are a fair guide to the amateur forecaster in predicting weather conditions for short periods of time. However, the science of weather forecasting is a very complex one, and barometer readings alone cannot be relied on for accurate weather forecasting.

The United States government has established the *National Weather Service*, whose function it is to give weather reports and forecasts to all those concerned. Weather stations have been set up at strategic locations throughout the nation and at many places throughout the world. Weather satellites keep track of major storm movements. Simultaneous instrument readings at the many different places are radioed in to central stations, or obtained from orbiting satellites. Weather maps are prepared and distributed by video recorders and television. Wind velocities, temperatures, and relative humidities all play an important part in these forecasts. Figure 12.6 shows a typical weather map prepared by the National Weather Service. On the map can be found the low-pressure and high-pressure areas. The direction and velocities of winds are usually included. Generally speaking, the areas of low barometric pressure are the storm centers. These usually move in an easterly direction across the United States. Lines joining places of equal barometric pressure (expressed in millibars) on a weather map are called *isobars*.

12.15 The Bourdon Gauge

Various types of gauges are used in industry for measuring gas and steam pressure. One of the most common type is called the *Bourdon gauge*. This device (Fig. 12.7) consists essentially of a

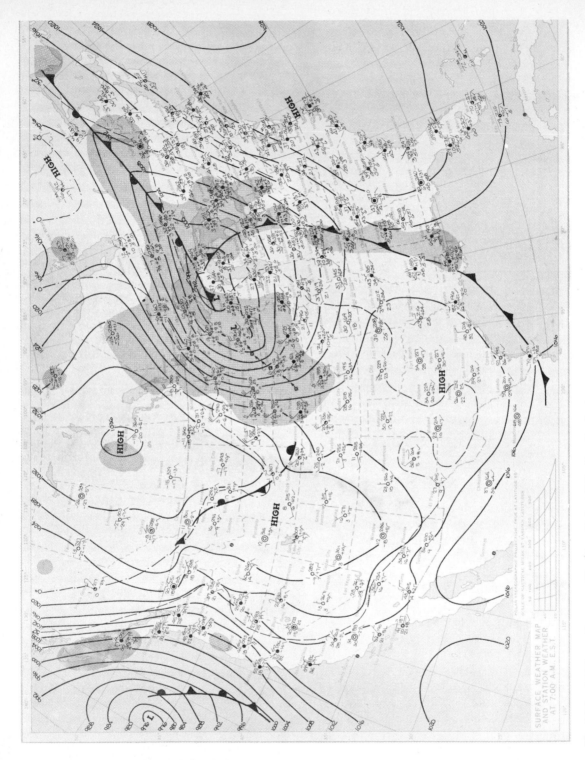

Fig. 12.6 A U.S. weather map. (Courtesy National Weather Service.)

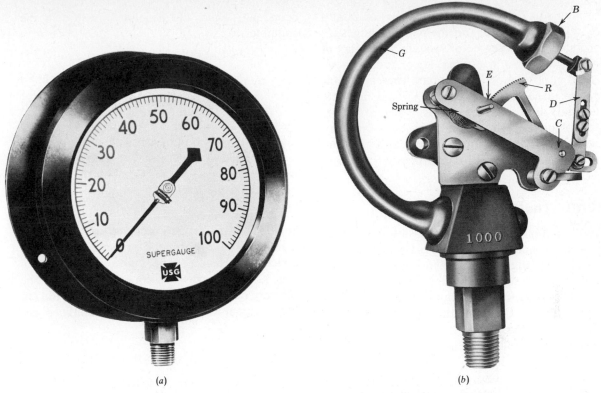

Fig. 12.7 (*a*) Bourdon gauge. (*b*) Pressure straightens out the curved tube, actuating a linkage which moves the pointer along a calibrated scale. (United Gauge Division, American Machine and Metals, Inc.)

hollow bronze tube *G* bent into a nearly complete ring and closed at one end *B*. When fluids enter the tube under pressure, the ring tends to straighten out. This motion at *B* is transmitted by a linkage *D* to a rack *R* (which is pivoted at *C*). The rack actuates a pinion attached to shaft *E*, to which a pointer is attached. By proper calibration, Bourdon gauges can be made to read pressures in pounds per square inch (psi), newtons per square meter (Pa), in. Hg, or mm Hg (torrs).

12.16   Gauge Pressure and Absolute Pressure

In practice we may be interested in two kinds of pressure readings. We may want to know the *difference* between the actual pressure being measured and the pressure of the atmosphere. This pressure is called *gauge pressure*. Or we may

wish to know the actual *total pressure* on the confined gas, including that of the atmosphere on the outside. This pressure is called the *absolute pressure*.

In predicting the behavior of gases under different conditions of temperature and pressure, it is essential to use *absolute pressures*. However, in most practical applications the pressure desired is the gauge pressure. The following relationship is used to determine either pressure if the other is known.

Absolute pressure = gauge pressure

+ atmospheric pressure

$$P_{abs} = P_g + P_{atm} \qquad (12.4)$$

12.17   Vacuum Gauge

When pressures less than 1 atm are to be measured, the instrument is calibrated to read from 0 to 14.7 lb/in.^{2} (1.013 × 10^5 **Pa**) or from 0 to 30 in. (76 cm) Hg, and the scale reading is interpreted as the degree of vacuum. A reading of 14.7 lb/in.^{2} or 30 in. (76 cm) Hg on such a gauge would indicate a perfect vacuum, whereas a reading of zero on either scale would indicate (normal) atmospheric pressure. Vacuum gauges find frequent application in the testing of reciprocating engines, vacuum pumps and vacuum systems, and refrigeration compressors and connecting lines.

12.18 The Manometer Gauge

Another common device for measuring the gauge pressure of a confined body of gas is the *open-tube manometer* (Fig. 12.8). The manometer is a U-shaped tube and contains a liquid of known density, usually mercury, but some contain water or oil. When both ends of the tube are open to the atmosphere (Fig. 12.8a), the level of the mercury will be the same in both arms of the manometer. (Why?)

In practice, one end of the manometer is connected to the container in which the gas pressure is to be measured; the other is open to the atmosphere (Fig. 12.8b). The mercury will rise in the open tube end until the pressures are equalized. The difference h between the two levels of mercury (or whatever fluid is in the tube) is a measure of the difference between the pressure in the chamber and the atmospheric pressure at the

open end. Thus, the absolute pressure in the container,

$$P_{abs} = P_{atm} + hD_{mercury} \qquad (12.5)$$

$$= P_{atm} + h\rho_{mercury}g \qquad (12.5')$$

and   Gauge pressure $= hD = h\rho g$ (12.6)

Illustrative Problem 12.3   An open-tube mercury manometer is used to measure the pressure inside a pressurized tank of gas. The difference in the mercury levels is 32 cm. What is the absolute pressure (in pascals) of the confined gas in the tank?

Solution The gauge pressure is 32 cm of mercury. The atmospheric pressure is 76 cm of mercury. Hence, the absolute pressure (Eq. 12.4) = 32 cm + 76 cm = 108 cm Hg. The pressure exerted by a column of mercury 108 cm high is equal to the pressure of the confined gas.

$$P_{abs} = \rho g h$$

$$= 13600 \text{ kg/m}^3(9.81 \text{ m/sec}^2)$$

$$\times (108 \text{ cm})(1 \text{ m/100 cm})$$

$$= 1.44 \times 10^5 \ \frac{\text{kg-m}}{\text{m}^2(\text{sec}^2)}\left(\frac{1 \text{ N}}{1 \text{ kg-m/sec}^2}\right)$$

$$= 1.44 \times 10^5 \text{ N/m}^2 \text{ (Pa)} \qquad ans.$$

The open-tube manometer is inconvenient for accurate measurements because barometric pressures are not constant and must therefore be measured frequently. The closed-tube manometer (Fig. 12.9) eliminates this problem by sealing one arm of the tube and highly evacuating the space above the mercury in that arm. With this device, the difference in the mercury columns is supported entirely by the pressure of the gas confined in the tank. Thus, the closed-tube manometer is used to measure directly the total (absolute) pressure of the gas. The pressure in pascals would be calculated from $P = h\rho g$.

Illustrative Problem 12.4   A closed-tube manometer is used to measure the pressure inside a tank of gas. The difference in the mercury levels is 32 cm. What is the gas pressure?

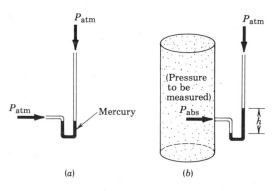

P_{atm} P_{atm}

P_{atm} Mercury   (Pressure to be measured) P_{abs}

(a)   (b)

Fig. 12.8 The open-tube manometer (not drawn to scale).

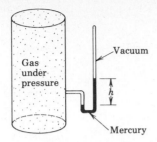

Fig. 12.9 The closed-tube manometer (not drawn to scale).

Solution The difference in the mercury levels is a direct measure of the pressure inside the tank, since the pressurized gas from the tank is the only factor contributing to the difference. Thus,

$$P = \rho g h$$
$$= 13600 \text{ kg/m}^3 (9.81 \text{ m/sec}^2)$$
$$\times (32 \text{ cm})(1 \text{ m}/100 \text{ cm})$$
$$= 4.27 \times 10^4 \text{ N/m}^2$$
$$= 4.27 \times 10^4 \text{ Pa} \qquad\qquad ans.$$

DEVICES OPERATED BY ATMOSPHERIC PRESSURE

12.19 The Lift Pump

The old-fashioned water pump (Fig. 12.10) used by our forefathers (and still in use in some parts of the United States) illustrates in a simple manner how atmospheric pressure is used to lift liquids. The principal parts of the pump are shown in Fig. 12.10. It consists of a cylinder *EF* into which a piston *CD* fits tightly. This piston is raised and lowered by the handle *RS* pivoted at *T*. The lower end of the cylinder is connected by a pipe to the source of the water *W*. Valves are placed at *A* in the piston and at *B* at the bottom of the cylinder joining the inlet pipe. When the piston is raised, the pressure of the air in space *ECD* of the cylinder decreases because of its increase in volume (*Boyle's law*), and this causes valve *A* to close and valve *B* to open. Since the atmospheric pressure *p* will then be greater than

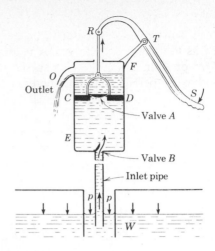

Fig. 12.10 The lift pump.

the pressure in *ECD*, water will be forced from *W* up the inlet pipe into the cylinder. On the downward stroke, valve *B* closes and valve *A* opens because of the increased pressure in *ECD*.

Repeated up-and-down strokes of the piston will result in atmospheric pressure forcing the water up until it passes through valve *A*. The water above *CD* is then lifted and flows out the spout *O* of the pump. But as the water above *CD* is lifted, more water is forced by atmospheric pressure from *W* into *ECD*. And so the process continues. It should be apparent that since the atmospheric pressure at sea level is approximately 34 ft of water, the most perfect pump of this type could not lift water more than 34 ft from the water level in the well. Usually the maximum height lifted is about 28 ft, since most such pumps do not attain a high degree of vacuum in *ECD*. However, liquids can be lifted many hundreds of feet *above the piston*, as long as the piston is not more than about 28 ft above the water level in the well or cistern.

12.20 The Siphon

The siphon is an interesting application of the laws of physics, in which water can be made to "run uphill," utilizing the force of atmospheric pressure. It is a bent tube of rubber, glass, metal,

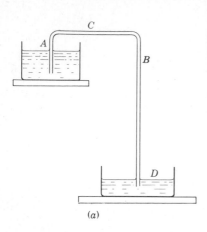

Fig. 12.11 (a) The siphon. (b) Irrigating by siphoning water from an open ditch. This method is preferred by many farmers over the more permanent "pipeline and stand" method. It gives a more uniform rate of flow into every crop row. It is cheaper to install, and the control of weeds near the ditch is no problem. (U.S. Department of Agriculture.)

(b)

or plastic which is used to carry water from one level over a small elevation to a lower level. To operate the siphon, one has only to fill the tube full of water, or other liquid, and then, holding closed both ends of the tube, lower it until one end is immersed in the vessel from which the liquid is to be drained and the other end is at a lower elevation. On opening the ends, liquid will flow from the higher to the lower elevation. The explanation of the siphon is quite simple. The atmospheric pressures at levels A and D (Fig. 12.11a) are essentially the same. Thus the atmosphere attempts to push water up the two arms with equal force. However, the downward force of the column of water in the shorter column at A, because of its weight, is less than that of the longer column of water at D; thus a greater resultant upward force acts on the shorter side A. Hence the water will move from the point of the greater force to that of the lesser—from A across C and down to D in the diagram.

The siphon will cease to function (1) when the elevation of D comes up to the elevation of A or (2) when the water level at A falls below the tube opening on the shorter side. It should be pointed out that the siphon will not work if the difference in elevation between A and C exceeds 34 ft.

12.21 Applications of Bernoulli's Principle to Gases

Bernoulli's principle (see Sec. 11.21) applies to all kinds of fluid motion. There are many interesting examples of its application associated with gases of which we shall cite a few.

Sports enthusiasts have for years argued pro and con on the question of whether a baseball pitcher can really throw a curve ball. Tests with research laboratory equipment and high-speed photography show that the ball really does curve a great deal. Anyone who is a beginner at golf

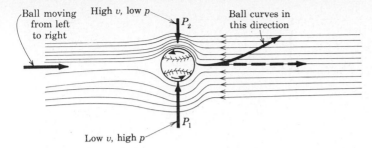

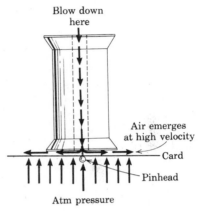

Fig. 12.12 A pitched spinning baseball travels in a curved path.

knows for sure that his "slice" or "hook" is not imaginary. Bernoulli's principle contributes to our understanding of the forces which cause a ball to curve.

Figure 12.12 shows a spinning ball thrown from left to right. This situation is equivalent to the ball standing still, but spinning, while the air moves from *right to left*. The spinning ball drags some air along with it, causing the air on one side of the ball to rush past the ball faster than the air on the other side. As a consequence, the pressure P_1 is greater than the pressure P_2. To produce a downward curve (a "drop ball"), the ball is given "overspin" around a horizontal axis. The "in" or "out" curve ball is achieved by spinning the ball on a vertical axis. Roughing the surface of the ball increases the effectiveness of the "curve." (Why?)

Put a pin through a light card and then place a spool over the pin as shown in Fig. 12.13. Next blow through the spool and try to blow the card off. The harder you blow, the tighter the card will be pressed to the spool by atmospheric pressure. Can you explain why the card will not fall from the spool?

When a moving airfoil is inclined at a slight angle with the horizontal, air will be deflected from the lower surface to create a reacting force which will tend to lift the foil. At the same time air rushes over the longer upper wing surface at a higher velocity than over the shorter (lower) wing surface. Thus (Bernoulli's principle) the pressure below the foil is greater than that above the foil (Fig. 12.14a). This difference in pressure provides about 85 percent of the lifting force which causes an airplane to rise and maintain altitude in flight.

Fig. 12.13 Livingroom demonstration of Bernoulli's principle.

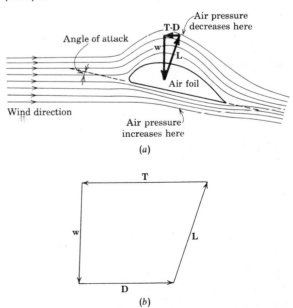

Fig. 12.14 (a) Lift produced on an airfoil; an example of Bernoulli's principle. (b) When the airfoil moves with constant velocity, **L** + **w** + **D** + **T** = 0.

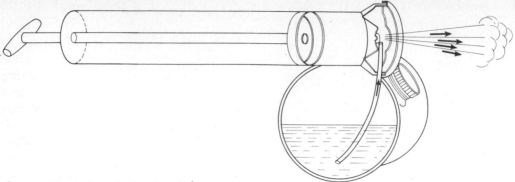

Fig. 12.15 Cross-sectional view of a hand garden sprayer.

When the plane flies with constant speed, the lift L, the weight of the plane w, the drag D due to air friction, and the thrust T due to the engine are in equilibrium. Vectors representing the forces form a closed polygon, as shown in Fig. 12.14b. Thus, $\mathbf{L} + \mathbf{W} + \mathbf{D} + \mathbf{T} = 0$.

The action of an atomizer (Fig. 12.15) or paint gun depends upon Bernoulli's principle. In the paint gun a stream of high-velocity air passes over the top of a vertical tube dipped in the paint in a jar. The high-velocity air creates a lower pressure at the top of the tube than the pressure at the surface of the paint in the jar. The unbalanced pressure causes a force which pushes the paint up the tube. As the liquid spills over the top of the tube, the airstream blows it out of the gun in a fine misty ("atomized") spray.

Gasoline undergoes atomizer action in the carburetor of an automobile and is quickly changed to a vapor state on mixing with the air before reaching the cylinders of the engine.

SUMMARY

The pressure of a gas is due to the molecular motion of the gas.

Boyle's law states: *The volume of a gas at constant temperature varies inversely as the absolute pressure on it.*

$$\frac{P_1}{P_2} = \frac{V_2}{V_1} \quad \text{or} \quad P_1 V_1 = P_2 V_2 \quad (12.1)$$

Charles' law states: *If constant pressure is maintained, the volume of a gas is proportional to its absolute temperature:*

$$\frac{V_1 \text{ (original volume)}}{V_2 \text{ (new volume)}} = \frac{T_1 \text{ (original abs temp)}}{T_2 \text{ (new abs temp)}} \quad (12.2)$$

Air has weight: 13 ft^3 at normal pressure and temperature weighs about 1 lb. The density of air is 0.081 lb/ft^3, at standard pressure and temperature (1 atm and 32°F) conditions. In metric units the density of air is 1.293 gm/liter at 1 atm and 0°C.

Atmospheric pressure is approximately equal to

> 760 mm Hg
> 760 torr
> 29.92 in. Hg
> 34 ft of water
> 14.7 lb/in.^{2}
> 1.013×10^5 N/m^2 (Pa)
> 1013 millibars (mb)

The barometer is used to measure air pressure. Barometric readings are of value in predicting weather. One form of altimeter is an aneroid barometer whose scale is calibrated in feet or meters above sea level.

The lift pump and the force pump both depend upon the pressure of the atmosphere for their operation.

Bourdon gauges and manometers are used to measure fluid (gas and liquid) pressures. *Absolute pressure = gauge pressure + atmospheric pressure.*

Practical applications of Bernoulli's principle involving gases include lift pump, siphon, atomizers, paint guns, and sprayers; the lift of an airplane wing; carburetors for automobiles; and the curving of balls used in various games.

QUESTIONS AND EXERCISES

1. Explain why the pressure in an automobile tire increases when the car has been traveling for a considerable distance.

2. A baseball is thrown so that it rotates with its top spinning directly away from the pitcher. Describe its probable motion.

3. What are the percentages of the two most abundant gases in air?

4. Why would a siphon fail to operate over a hill more than 34 ft high?

5. An object is weighed with great precision in the atmosphere. It is then weighed in a partial vacuum. Compare the two weights. Give reasons for your answer.

6. Two objects of different volumes have the same apparent weight when submerged in water. Compare their weights when weighed in air. Give reasons for your answer.

7. Discuss the advantages and disadvantages of the closed-type and the open-type manometers.

8. The system shown in Fig. 12.16 is filled with oil, closed by a piston P, and open to the atmosphere in pipes A and B. The diameter of pipe A = one-half the diameter of pipe B. The piston P is lowered 2 in. Compare how much the oil will rise in the two pipes?

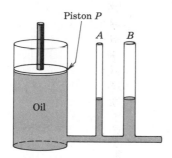

Fig. 12.16

9. Two manometers are used to measure gas pressures P_A and P_B (Fig. 12.17). h_A is greater than h_B. (a) Compare P_A and P_B. (b) What is the gauge pressure at A? (c) What is the absolute pressure at B?

10. Describe the situation which would cause a closed-tube manometer to appear as shown in the drawing of Fig. 12.18.

PROBLEMS

1. How much does a cubic yard of air at sea level and 32°F weigh? Assume barometric pressure of 29.92 in. Hg. (Use Table 12.2.)

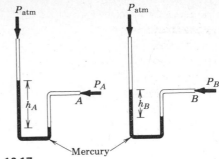

Fig. 12.17

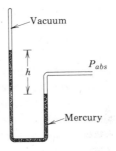

Fig. 12.18

2. Convert a barometric pressure of 720 mm Hg to an equivalent pressure in Psi units.

3. Find the greatest theoretical height that water can be raised over a siphon in the mountains where the barometer reads 22 in. Hg.

4. If a cylindrical can 8 cm in diameter and 10 cm tall were completely evacuated, what force would the atmosphere exert on each end of the can?

5. In the closed-tube manometer illustrated in Fig. 12.19, h = 4.5 in. Find P_{abs} in Psi.

6. In the open-tube manometer of Fig. 12.20, h = 25 mm, P_{atm} = 760 mm. Find P_{abs} in pascals.

7. In Fig. 12.11a, the difference in the elevations of A and D is 12 cm and the tube has a cross-sectional area

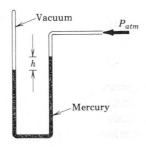

Fig. 12.19

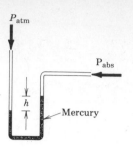

P_{atm}

P_{abs}

h

Mercury

Fig. 12.20

of 1.4 cm². Find (a) the velocity of efflux and (b) the quantity of water (in cubic meters per minute) transferred at D.

8. An open-tube manometer is used to measure the pressure of a tank of gas. The meter reads 37 cm Hg. Atmospheric pressure is 74 cm Hg. What is the absolute gas pressure (a) in centimeters of mercury; (b) in pounds per square inch; (c) in pascals?

9. A closed-tube manometer is used to measure the pressure of a gas system. Find the gas pressure (a) in pounds per square inch and (b) in newtons per square meter given that the difference of the mercury levels in the two arms is 12.5 cm.

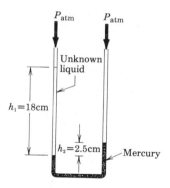

P_{atm} P_{atm}

Unknown liquid

$h_1 = 18$cm

$h_2 = 2.5$cm   Mercury

Fig. 12.21

10. An open U-tube (Fig. 12.21) is partially filled with mercury. Then another liquid of unknown density is poured in the left arm until the levels of the free liquid surfaces are stabilized as shown. Find the density of the unknown liquid.

11. An auto has a mass of 1930 kg and is supported by four identical tires. Each tire is filled with air at a gauge pressure of 1.655×10^5 Pa. Find the theoretical area of contact with the ground (in square centimeters) for each tire. Neglect the tensile stresses within the fabric of the tires.

12. A gas balloon, its trappings, and gondola weigh 3000 N. It is filled with 1500 m³ of helium at atmospheric pressure. What will be its maximum possible payload near the earth's surface?

13. The standard tank of oxygen used in industry has compressed into it gas which would occupy 244 ft³ at atmospheric pressure. The compressed gas has a gauge pressure (70°F) of 2200 lb/in.². Determine the inside volume of the tank.

14. The 150 gallon pressure tank of a home water-supply system is full of air at atmospheric pressure. 100 gal of water is then forced in, trapping and compressing all the air. What is the resultant water pressure delivered from the tank?

15. "Bottled gas" (butane) in a steel tank is under a gauge pressure of 125 lb/in.². After some of the gas has been used, the gauge pressure drops to 50 lb/in.². What fractional part of the original amount of gas still remains in the tank?

16. A compressed air tank has a capacity of 4.1 ft³ and is filled with air at normal atmospheric pressure. How many additional cubic feet of atmospheric air must be pumped into the tank in order to raise the gauge pressure to 90 lb/in.²?

17. A caisson used in constructing a tunnel under a river is lowered to the bed of the river 50 ft under water. Atmospheric pressure is 14.7 lb/in.², and the density of air is 0.0805 lb/ft³. (a) What is the pressure (due to water pressure head plus atmospheric pressure) at the bottom of the river? (b) When air is pumped to workers in the caisson at the bottom of the river, to what fraction of the original volume has the air been reduced? (c) What is the density of the air within the caisson? (Assume no temperature change.)

18. An air bubble released at the bottom of a lake will expand to three times its original volume by the time it reaches the surface of the lake. If atmospheric pressure is 76.0 cm Hg, how deep is the lake? (Assume no temperature change.)

19. The volume of a certain automobile tire is 1600 in.³. How much air in cubic inches at standard pressure must be forced in to raise the gauge pressure from 24 to 32 lb/in.²? (Assume no temperature change.)

20. A cylindrical diving bell, open at the bottom and filled with air at 1 atm pressure, has a diameter of 1 m and a height of 3 m. How high will the water rise in the bell when the bottom of the bell is immersed to a depth of 12 m? Assume atmospheric pressure of 1.013×10^5 N/m².

PART 4 Heat and Thermo-dynamics

Molten Pig Iron from a Blast Furnace being charged into an Open-Hearth Furnace at 2900°F. Heat energy is essential in nearly every phase of the metals industry. (Bethlehem Steel Corporation).

13 Temperature and the Effects of Heat

Heat is another form of energy. In earlier sections of the book *kinetic energy* and *potential energy* were the central focus of attention. Their relationships to mass, force, motion, and time were thoroughly discussed in the study of *mechanics*. We now turn to the study of heat, which is often referred to as *thermal energy*. Heat can be converted into mechanical energy by the use of suitable machines, and mechanical energy is continually being converted into heat as a result of friction. Heat can also be converted into electrical energy, and electrical energy into heat. Of all forms of energy, heat energy has been the most readily available during the past several centuries, in the form of *fossil fuels*—coal, petroleum, and natural gas. But these convenient sources of heat energy have been seriously depleted, and alternative sources of energy must now be found and developed.

Temperature is a measure of the hotness or coldness of a body or a substance. It is expressed in *degrees*, and is measured by instruments called *thermometers*. In this chapter we will explain temperature and its measurement, discuss the nature of heat, and examine some of the effects of heat on matter. In subsequent chapters heat transfer, change of state, heat engines, and refrigeration and air conditioning will be considered in detail.

TEMPERATURE AND ITS MEASUREMENT

13.1 Temperature

Our ideas of temperature are related to the sense of feel of the human body. If a substance feels *hot*, we say that its temperature is high. If it feels *cold*, it is said to have a low temperature. The terms *warm* and *cool* describe mild degrees of hotness and coldness, respectively. Bodily sensations are not very accurate indicators of temperature. You can do a simple experiment to demonstrate this fact. After a few minutes with one hand in hot water and the other in ice-cold water, place both hands in lukewarm water. The "cold" hand tells you that the lukewarm water is hot, and the "hot" hand tells you it is cold.

Try a second experiment. Place a block of wood or styrofoam and a block of copper or steel in the freezer and allow time for both of them to come to the freezer temperature. Remove them and touch one with each hand. Why does the metal block feel so much colder than the other, when actually they are at the same temperature?

These two experiments suggest that the sensation of temperature difference *depends on the rate at which heat is transferred to or from the body.* Lukewarm water transfers heat *into* the "cold" hand, and the "hot" hand transfers heat into the lukewarm water. In the second experiment, metal, being an excellent heat conductor, transfers heat away from the hand much faster than does wood or styrofoam, which are poor conductors of heat. Sensations of "hot" and "cold" are not very accurate indicators of temperature. Later, a definition of temperature will be proposed which relates temperature to the velocity (and kinetic energy) of moving molecules.

The accurate measurement of temperature depends on the use of *thermometers*, a number of which will now be described.

Fig. 13.1 Standard laboratory mercury-in-glass thermometer, Fahrenheit temperature scale.

13.2 The Liquid-in-glass Thermometer

In 1714, Gabriel Daniel Fahrenheit (1686–1736) designed the first successful liquid-in-glass thermometer and calibrated it in terms of a temperature scale (the *Fahrenheit scale*) which is still used throughout the English-speaking world for engineering and industrial purposes. The liquid-in-glass thermometer (see Fig. 13.1) consists of a glass bulb and narrow stem filled with either mercury or alcohol. The space above the liquid is evacuated. Changes in temperature cause the liquid to rise or fall in the narrow stem because of expansion or contraction of the liquid in the bulb. The expansion of the glass bulb is almost negligible when compared with that of either mercury or alcohol, and since the stem is sealed off, variations in atmospheric pressure have no effect on the readings. The stem bore must be uniform throughout if accuracy is expected at all readings. The sensitivity can be increased by making the bore smaller with respect to the volume of the bulb.

13.3 The Fahrenheit Temperature Scale

Fahrenheit was limited by the inability, in his day, to attain low temperatures by any other means than ice-salt mixtures. He experimented with various "zero points" and upper limits for his thermometer, using at one time or another such fixed points as the temperature of the human body, the boiling point of water, and the lowest temperature obtainable with an ice-salt mixture. As research methods improved, some of these "fixed points" were abandoned, and finally the thermal behavior of water was chosen to standardize the Fahrenheit thermometer.

The Fahrenheit scale is now defined as having its fixed points at the freezing point of water (32°F) and at the boiling point of water (212°F) at standard atmospheric pressure. The range between these fixed points is divided into 180 equal divisions, each corresponding to 1° of temperature. The scale is then extrapolated above 212° and below 32° as far as desired, subject to the temperature limitations of the liquid with which the particular thermometer is filled. The Fahrenheit scale is now used only for nonscientific purposes. A different scale, called the *Celsius scale*, is used where the SI-metric system has been adopted.

13.4 The Celsius Temperature Scale

Fahrenheit realized the convenience of setting the fixed points of his scale at 0 and 100, and started out to do so, but he had to abandon this plan. Some years later (about 1742) Anders Celsius (1701–1744) proposed the "centigrade" scale, with the freezing point of water to be pegged at 0°C and the boiling point of water at 100°C. Such a scale had the advantage of simplicity and ease of decimal calculation, and it soon became the standard throughout the world for scientific work. Although called the *centigrade scale* for over two centuries, it has recently been officially renamed the *Celsius scale*.

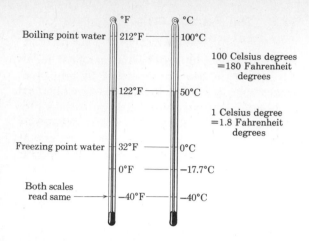

Fig. 13.2 Fahrenheit and Celsius temperature scales compared. The reference points for both are the freezing point and boiling point of water at 1 atm pressure.

The following rules for converting temperatures from one scale to another may be used, but they need not be memorized if the student visualizes the scales as shown in Fig. 13.2.

1. To convert Celsius to Fahrenheit, multiply the Celsius reading by 1.8 and add 32.

2. To convert Fahrenheit to Celsius, subtract 32 from the Fahrenheit reading and divide by 1.8.

In applying these rules the correct *algebraic sign* of the readings must be used. Stated as mathematical formulas the rules are:

$$F = 1.8C + 32 \qquad (13.1)$$

$$C = \frac{F - 32}{1.8} \qquad (13.2)$$

where F and C represent Fahrenheit and Celsius readings of the same actual temperature. Note that at $-40°$ both scales have the same reading.

13.5 Comparison of Fahrenheit and Celsius Scales

The two scales are compared in Fig. 13.2. Analysis reveals that 100 Celsius degrees are equivalent to 180 Fahrenheit degrees $(212 - 32 = 180)$. Therefore, 1 Celsius degree represents a temperature change equivalent to 1.8 Fahrenheit degrees. Also note that the freezing point of water is at 0°C and at 32°F. In comparing a temperature on one scale with that on another, the freezing point (fp) of water is always a reference point to be used.

Converting from one scale to the other, suppose it is desired to obtain the Fahrenheit reading which corresponds to a 50°C reading. Fifty Celsius degrees above the freezing point of water equal 50 × 1.8, or 90 Fahrenheit degrees above the freezing point. But freezing point on the Fahrenheit scale is 32°F, and we are 90°F above that. Adding 90 and 32, we obtain 122°F as the Fahrenheit equivalent of 50°C. As another example, what Celsius reading corresponds to a Fahrenheit reading of 20° below zero $(-20°F)$? Note that $-20°F$ is 52 Fahrenheit degrees below the freezing point of water; and 52 Fahrenheit degrees = 52/1.8, or 28.8 Celsius degrees below the freezing point of water. The Celsius reading is therefore $-28.8°C$.

13.6 Construction of Common Thermometers

Ordinary liquid-in-glass thermometers are made in many grades of quality for a wide range of purposes. The glass bulb is made as thin as possible consistent with sturdiness, because glass is a relatively poor conductor of heat and thickness of the bulb would cause poor sensitivity. The stem above the liquid is evacuated and sealed off, so there is no air pressure to act down on the liquid surface as it rises and falls because of changing temperature.

Mercury freezes at about $-38°F$, and consequently, for temperatures below this level, alcohol-filled thermometers are used. (Alcohol freezes at about $-200°F$.) Mercury is best for higher temperatures, since the boiling point of alcohol is about 171°F. For extremely high and extremely low temperatures, special types of thermometers, to be described later, must be used. There are many industrial, clinical, and home uses for liquid-in-glass thermometers. Figure 13.3 shows a type used by weather stations, and Fig. 13.4 two types of clinical thermometers to measure the human body temperature.

Fig. 13.3 Maximum and minimum thermometer for weather stations (alcohol-filled, range from −40 to 130°F). (Taylor Instrument Co.)

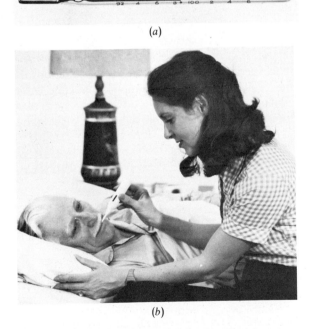

(a)

(b)

Fig. 13.4 Clinical thermometers are used by physicians and nurses to measure human-body temperature. The average "normal" temperature, taken orally, is 98.6°F. (a) Standard mercury-in-glass clinical thermometer. (b) A new digital-display, solid-state, electronic thermometer, battery powered. (Electromedics, Inc.).

13.7 Dial Thermometers

Many industrial thermometers register the temperature reading by means of a pointer on a dial. These thermometers do not contain any liquid. They may operate on the principle of thermal expansion, or as a result of the thermoelectric effect (see Sec. 13.9). Different metals expand at different rates (see Sec. 13.16), and some dial thermometers make use of this principle of un-

(a)

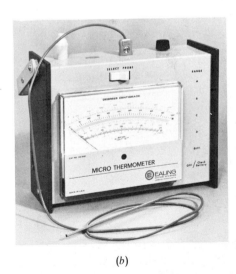

(b)

Fig. 13.5 (a) Dial thermometer for the Fahrenheit scale. The pointer rotates as a result of the differential expansion of a bimetallic element. (Central Scientific Co.). (b) Multi-range microthermometer for Celsius (centigrade) temperature readings. (The Ealing Corporation).

equal expansion. A *bimetallic element* (brass bonded to iron) bends one way on being heated, the opposite way on being cooled (see Sec. 13.15). This motion is transmitted by a suitable linkage to a pointer which moves across a calibrated scale (see Fig. 13.5a). Although dial thermometers are not as accurate as liquid-in-glass thermometers, their more rugged construction makes them ideal for many industrial purposes. Fig. 13.5b illustrates a thermoelectric thermometer of considerable precision.

13.8 Resistance Thermometers

Metals, which are usually good conductors of electricity, vary in the amount of *resistance* they offer to electric current. Also, any one metal shows a remarkable variation in *electric resistivity* with variations in temperature. The electric resistivity of pure metals decreases with falling temperature and increases with rising temperature. For some metals, notably platinum, this rise and fall in resistivity is such a direct and constant variation that the temperature can be very accurately determined by measuring the resistance, in *ohms*, of a given sample of the metal. A common form of such a resistance thermometer is shown in the sketch of Fig. 13.6. The electric leads from the thermometer are connected to a source of electric current and to a very sensitive *ohmmeter*, which measures the resistance of the platinum resistance coil at the temperature being measured. The resistance may be read and converted into a temperature reading from a

prepared table or graph, or the ohmmeter may be calibrated to read temperature directly. The quartz tube of the instrument and the platinum coil both have melting points of over 1650°C, and so the instrument is well suited for measuring reasonably high temperatures (up to 1350°C in actual practice) for heat-treating and annealing processes in the metallurgical industry.

13.9 Thermocouples

A very important discovery was made by T. J. Seebeck (1770–1831) about 1820. He found that if two different metals, such as copper and iron, are joined together to form a closed loop and one junction is kept at a different temperature from the other, an electric current can be observed in the closed loop. This discovery is known as the *thermoelectric effect*, or sometimes as the *Seebeck effect*. Figure 13.7 shows, in diagram form, the essentials of a temperature-measuring device based on the thermoelectric effect. It is called a *thermocouple*. In the diagram, T_h is the temperature of the hot junction and T_c the temperature of the cold junction. Careful experiments by Seebeck and others have shown that the electron flow is related in a predictable manner to the difference in temperature $(T_h - T_c)$ between the junctions. If the temperature of one junction is kept at a constant known value—say 0°C—the temperature of the other junction (which may be many

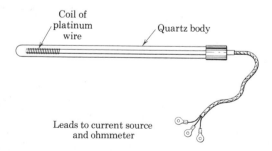

Fig. 13.6 Diagram of one form of platinum resistance thermometer.

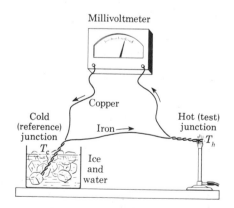

Fig. 13.7 Diagram of apparatus to demonstrate the *Seebeck effect*—the principle of the thermocouple.

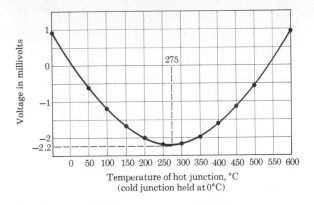

Fig. 13.8 Temperature-voltage graph for a copper-iron thermocouple. The curve shows the emf produced for a range of hot-junction temperatures, if the cold junction is kept at 0°C.

feet away—inside a furnace or refinery tank, for example) can be readily and accurately determined by noting the electric voltage produced, and then consulting a previously prepared temperature-voltage graph for the particular pair of metals being used. Figure 13.8 shows such a graph for a copper-iron thermocouple.

The thermocouple is convenient because of the small size of the test junction. It can be sealed in pipes, ducts, stacks, furnaces, cylinder heads, etc., wherever a temperature measurement is desired. Even a single junction is quite sensitive, and if additional sensitivity is desired, multiple junctions may be connected in series, resulting in an instrument of great sensitivity called a *thermopile.*

The copper-iron thermocouple is not much used in actual practice because of corrosion and the relatively low sensitivity and low melting point of these metals. Combinations of metals most often used for thermoelectric thermometers (called *thermels*) are iron and constantan, chromel and constantan, and platinum and platinum-rhodium alloy.

NOTE: Constantan is an alloy composed of 55 percent copper, 45 percent nickel; chromel is 90 percent nickel, 10 percent chromium. In addition to extreme sensitivity, the platinum-rhodium

combination has a high melting point and can be used for temperatures up to 1480°C.

The extreme sensitivity of thermopiles is difficult to appreciate without observing them in operation. As an example, however, a thermopile composed of just eight copper-iron thermocouples, series-connected, is capable of recording temperature differences of 0.005°C. The use of the more sensitive metal combinations together with an increase in the number of junctions will increase the sensitivity to almost any desired figure.

13.10 Optical Pyrometers

To measure temperatures beyond the range of resistance thermometers and thermocouples, *optical pyrometers* are often used. The brightness of an incandescent surface has been found to increase as the temperature increases. An optical pyrometer enables the operator to estimate rather accurately the temperatures of glowing surfaces by their brightness. The temperature (see Fig. 13.9*a*) is judged visually by comparing the brightness and color of the molten or glowing mass with the brightness and color of a heated filament (whose temperature is known) within the instrument. The filament is heated by a controlled electric current until it is judged by the operator to be of the same brightness as that of the incandescent body whose temperature is desired. The optical arrangement of the instrument is shown diagrammatically in Fig. 13.9*b*. The operator views the glowing filament against the background of the incandescent material and then adjusts the current which heats the filament until color and brightness of the one are judged to be the same as those of the other. The meter which measures the current supplied to the filament is calibrated to read in degrees of filament temperature rather than in units of electric current. Tungsten filaments are ordinarily used since the melting point of this metal is over 3300°C. Temperatures up to about 3000°C are measured directly. Above this range, the instrument can be modified by the introduction of wedge-type filters to bring the intensity and color

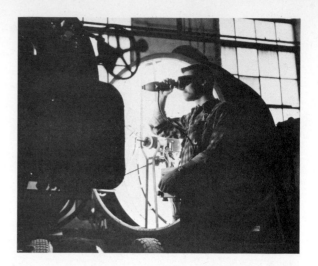

Fig. 13.9 The optical pyrometer. (*a*) The instrument in use, as a technician determines the temperature of molten steel. (*b*) Diagram of the construction details of an optical pyrometer. (Leeds and Northrup Co.)

(*a*)

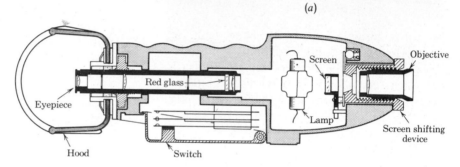

Optical pyrometer telescope

(*b*)

of the source down within the limits of the tungsten filament. A dial on the instrument reads corrected temperatures for the introduction of varying amounts of wedge thickness.

Optical pyrometers have their greatest use in estimating the temperature of molten metals in the iron and steel industry. They are also used by volcanologists in estimating the temperature within an active volcano.

Very low temperatures (below −250°C) require specialized thermometers, one form of which is shown in Fig. 13.10.

THE NATURE OF HEAT

Several hundred years ago scientists thought of heat as being a material substance which could be made to flow in and out of a body, with properties similar to those of a fluid. We still speak of heat "flowing" from one object to another. The French chemist Lavoisier (1743–1794) gave the name *caloric* to this hypothetical fluid, taking it from the Latin word for heat. From this same root comes our word *calorie* for a unit of heat.

13.11 Heat as a Form of Energy

Benjamin Thompson (1753–1814), born in colonial America but destined to spend most of his life in Europe, was the first to demonstrate that heat was not a fluid substance but a *form of energy*. While superintendent of a Bavarian ordnance works (about 1790), Thompson (known in Europe as Count Rumford) was in charge of boring brass cannon. Then, even as now, one of

Fig. 13.10 The ultrasonic thermometer. The measurement of low temperatures encountered in cryogenics research (see Sec. 17.10) can be accomplished by the use of ultra-high frequency sound waves. The tall Dewar flask (center) contains the low-temperature substance (liquid hydrogen or helium, for example) and also a quartz crystal which produces a high-frequency sound of known frequency. The speed of the sound through the cryogenic substance can be measured, and from the known speed, the temperature can be calculated. With such equipment, accurate temperature determinations are possible in the range from −250 to −270°C. (National Bureau of Standards.)

erated. He kept records on the change in temperature of known amounts of coolant water and also on the amount of mechanical work being done by the horses which were the factory "engines" of that day. He saw that as long as work was done by the drilling bit on the brass, heat was produced, and when work stopped, heat flow stopped. His conclusion was that heat was not a fluid, not *caloric*, not a substance at all, but a *form of energy* which appeared to flow into the metal as a result of mechanical work.

13.12 The Concept of Internal Energy

Any given quantity of matter possesses a certain amount of *internal energy*. This energy is in addition to whatever potential energy the body may have with respect to the earth or any other body. Internal energy is also in addition to any kinetic energy the body as a whole may have due to its motion.

The *internal energy* of a body or system is that energy which is the sum total of: (1) the translational kinetic energy of the body's or system's molecules in random motion; (2) the rotational or vibrational energy of its atoms; (3) the potential energy of action-at-a-distance forces between molecules, and of the binding forces between and among atoms and subatomic particles. The total amount of internal energy possessed by a body or system depends on its temperature, its mass, its physical state (i.e. solid, liquid, or gas), and on such other variables as pressure, volume, magnetic and electric properties, and chemical characteristics. Of all these variables, it is temperature that concerns us here, since it is temperature that is directly related to the translational kinetic energy of molecules. Temperature and heat are related in that they both are thought to be manifestations of molecular kinetic energy.

13.13 Temperature and Heat

The *kinetic theory of gases* was introduced and briefly discussed in Chap. 9. A further extension of that discussion is given here. Suppose a quantity of gas is enclosed in a container equipped with a pressure gauge and a thermometer. The

the basic problems of machine-shop work was to provide the proper coolant liquid to carry away the heat generated by machining operations. Thompson was struck by the fact that whenever work was done on the boring mill, heat was gen-

kinetic theory assumes that the gas molecules are in constant random motion in different directions at different speeds. For any given condition of pressure, volume, and temperature, we can conceive of some *average molecular speed* which causes the molecules to have a certain number of impacts per second against the walls of the container. The frequency and kinetic energy of these impacts are what cause the pressure, which is read on the gauge. If the container and enclosed gas are heated, what happens? The thermometer registers an increased temperature, and the gauge registers an increased pressure. Adding heat is therefore associated with a temperature increase. (Physical state (i.e. gas) is assumed to remain unchanged.) Since the number and total mass of the molecules is also unchanged (closed container) the increase in pressure can come only from more frequent impacts of the same number of molecules on the walls of the container. But more frequent impacts can result only from an increase in the *average molecular speed*. And molecular speed (velocity) is directly related to molecular kinetic energy.

The kinetic theory then, leads to a physical concept of temperature:

The (absolute) temperature T of a gas is proportional to the average translational kinetic energy of the molecules of the gas.

A "hot" gas has molecules with high average kinetic energy; a "cold" gas has molecules of relatively low kinetic energy. These same generalizations hold also for liquids and solids. (The meaning of *absolute* temperature will be explained in a later section.)

Consider two bodies or material systems in the same physical state (i.e. gas, liquid, or solid) at different temperatures T_1 and T_2 (T_1 hotter than T_2) (Fig. 13.11a). If the two systems are brought together (gases or liquids will mix, solids will attain good surface contact), it is observed that T_1 decreases and T_2 increases until some final temperature T_3 (intermediate between T_1 and T_2) is reached (Fig. 13.11b). Kinetic theory explains this phenomenon by supposing that the high-

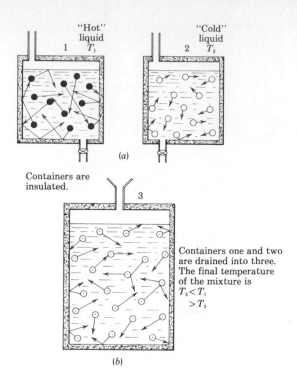

Fig. 13.11 Kinetic theory and thermal equilibrium. High-energy molecules transfer energy to low-energy molecules by collision.

energy molecules of the T_1 (hot) system transfer energy by collision to the low-energy molecules of the T_2 (cold) system, until finally the average translational KE of the molecules of both systems is the same. At this point the two systems are said to be in *thermal equilibrium*—their temperatures are equal.

For substances that do not undergo a change of state in the process (i.e. they do not boil or freeze or melt or condense), the kinetic theory leads to the conclusion that energy always flows from a hotter body to a colder body. This energy which flows, or is transferred, from a hot body to a cold body is *molecular kinetic energy*. We call it *heat energy*.

Heat energy is molecular kinetic energy which is transferred from one body or system to another as a result of temperature difference.

Going back to the idea of *internal energy*, we see that heat energy is that part of internal energy that is contributed by the translational kinetic energy of the moving molecules.

Heat energy may be added to a body or it may flow out of a body. A temperature difference between two bodies or systems always causes heat to flow. However, heat flow does not always result in a temperature change. Instead, heat flow may result in a change of state (melting, freezing, evaporating, condensing) with no change in temperature, or heat flow may cause work to be done (as in an engine cylinder) with no change in temperature. These relationships among internal energy, change of state (phase), and work constitute the science of *thermodynamics*, which we will study in later chapters.

SOME EFFECTS OF HEAT

13.14 Expansion Due to Heat

Steel rails are laid with gaps between the ends to allow for expansion due to temperature changes. Pistons are fitted in cylinders, and valve-rod clearances are adjusted to allow for expansion under operating conditions. Tanks full of a cool liquid will overflow as the liquid temperature rises. A toy balloon partially inflated in a cool room may expand to full size and burst if placed out in the hot sun. All these examples illustrate the general law that almost all substances (solids, liquids, and gases) expand on being heated and contract on being cooled.

Thermal expansion of a substance occurs in all directions simultaneously, but for convenience the discussion of expansion is treated under three separate headings:

1. *Linear expansion*, applicable to such objects as wires, cables, and rods, where change in length is the factor of primary importance (solids only)

2. *Area expansion*, applicable to flat sheets, where variation in thickness is immaterial (solids only)

3. *Volume expansion*, applicable to solids, liquids, and gases

13.15 Linear Expansion

Consider a metal rail 10 m long as it warms up from an early-morning temperature of 3°C to an afternoon temperature of 18°C. Each unit length of the rail becomes a bit longer as the temperature rises degree by degree. The amount of increase of each unit length for each degree rise in temperature is a property of the particular material, in this case, steel, and is called the *coefficient of linear expansion*.

The coefficient of linear expansion is defined as *the change in length per unit of original length per degree change in temperature*. As a mathematical formula,

$$\alpha = \frac{L_2 - L_1}{L_1(t_2 - t_1)} \qquad (13.3)$$

where α = coefficient of linear expansion

L_1 = length at some lower temperature t_1

L_2 = length at some higher temperature t_2

Or, if ΔL represents the change in length caused by a temperature change Δt,

$$\alpha = \frac{\Delta L}{L_1 \, \Delta t} \qquad (13.3')$$

Letting Δt stand for the change in temperature, and solving Eq. (13.3) for L_2, we obtain

$$L_2 = L_1(1 + \alpha \, \Delta t) \qquad (13.4)$$

a convenient formula for determining the new length if the original length and change in temperature are known.

The coefficient α has different values for different substances and varies somewhat over different temperature ranges even for the same substance. As might be expected, the values are very small, but nevertheless measurable and significant in almost every manufacturing and assembling operation. For steel, as an example, the coefficient is 0.0000061 ft/(ft)(F°). Note that

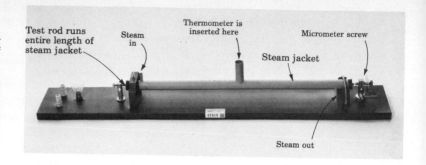

Fig. 13.12 Laboratory apparatus for determining the linear coefficient of expansion, α, of metals. The micrometer screw measures $L_2 - L_1$ (ΔL), and the thermometer provides readings t_2 and t_1, from which Δt can be determined. The micrometer screw is part of a low-voltage lamp circuit that is completed when the test rod expands and contacts the screw tip. (Central Scientific Co.)

this would also be 0.0000061 in./(in.)(F°), but that it would be 0.000011 in./in. or ft/ft per Celsius degree since the Celsius degree represents a change in temperature 1.8 times that of a Fahrenheit degree. Figure 13.12 shows an apparatus for the laboratory determination of α.

Values of α for some common substances, reasonably accurate for ordinary temperature ranges, are given in Table 13.1.

Illustrative Problem 13.1 A valve pushrod of mild steel measures 28.00 cm as the engine is assembled at a temperature of 20°C. What clearance should be allowed if the operating temperature is expected to be 100°C?

Solution We want to solve for the maximum elongation, or $L_2 - L_1$. Write Eq. (13.3) in the form

$$\Delta L = L_2 - L_1 = \alpha L_1(t_2 - t_1)$$

Table 13.1 Coefficients of linear expansion for some common materials

Material	Per °C	Per °F
Aluminum	24×10^{-6}	13×10^{-6}
Brass	18×10^{-6}	10×10^{-6}
Copper	17×10^{-6}	9.5×10^{-6}
Glass, ordinary	$7\text{-}9.5 \times 10^{-6}$	$4\text{-}5 \times 10^{-6}$
borosilicate (Pyrex)	3×10^{-6}	1.7×10^{-6}
Lead	30×10^{-6}	17×10^{-6}
Invar-steel alloy (average)	2.16×10^{-6}	1.2×10^{-6}
Iron (wrought)	12×10^{-6}	6.7×10^{-6}
Silver	20×10^{-6}	11×10^{-6}
Steel (mild)	12×10^{-6}	6.7×10^{-6}

Denoting elongation by ΔL and substituting numerical values (see Table 13.1),

$$\Delta L = [(12 \times 10^{-6})/°C] \times 28.00 \text{ cm} \times 80°C$$
$$= 0.0269 \text{ cm} \qquad ans.$$

Thermal Stresses Experience shows that tremendous forces are exerted by expansion. Buckling of pipes and girders, "freezing" of pistons in cylinders, and the force of explosions are all examples of expansion due to heat. As an example of the tremendous forces exerted by expanding metals consider the following problem.

Illustrative Problem 13.2 A steel I beam is 40.0 ft long at 20°F and has a cross-sectional area of 15.0 in.². (a) What is its increase in length when its temperature rises to 120°F? (b) What force would it exert against an object or an opposing force in the way of its expansion?

Solution
(a) From Eq. (13.3)

$$L_2 - L_1 = \Delta L = \alpha L_1(t_2 - t_1)$$

From Table 13.1

$$\alpha_{(mild\ steel)} = (6.7 \times 10^{-6})/°F$$

Substitution yields

$$\Delta L = [(6.7 \times 10^{-6})/°F] \times 40.0 \text{ ft} \times 100°F$$
$$= 2.68 \times 10^{-2} \text{ ft} \qquad ans.$$

(b) The force which the beam is capable of exerting as it expands is equal to the force required to

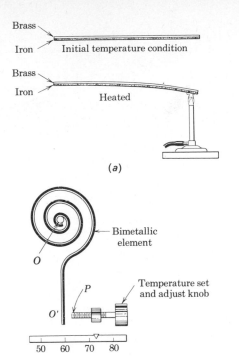

(a)

(b)

stretch it or compress it by the same amount as the expansion. Referring to Sec. 10.6 and using Young's modulus,

$$Y = \frac{\Delta F/A}{\Delta L/L} \quad \text{and} \quad \Delta F = \frac{YA\,\Delta L}{L}$$

From Table 10.2,

$$Y_{\text{(mild steel)}} = 25 \times 10^6 \text{ lb/in.}^2$$

Substitute numerical values:

$$\Delta F =$$

$$\frac{[(25 \times 10^6 \text{ lb)/in.}^2] \times 15 \text{ in.}^2 \times 2.68 \times 10^{-2} \text{ ft}}{40 \text{ ft}}$$

$$= 251{,}000 \text{ lb} = 126 \text{ tons} \qquad \textit{ans.}$$

Anyone who has seen the aftermath of a fire in a building where steel beams have buckled from the heat and where walls have been pushed out by the forces of expansion can appreciate the huge forces generated in the thermal expansion of metals.

13.16 Differential Expansion— Bimetallic Elements

If two strips of different metals, such as iron and brass, are riveted or bonded together at a certain initial temperature, as long as the temperature remains at that level the *compound bar* will remain flat and unbent. However, when the temperature changes, since the metals expand at different rates with great force, the bar will have to

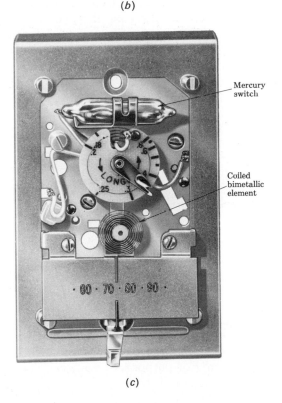

(c)

Fig. 13.13 (*a*) A bimetallic strip will bend when it undergoes a change in temperature, because the two metals of which it is made have different rates of expansion due to heat. (*b*) Schematic diagram of the essential components of a thermostat. *O* and *P* are connected to a low-voltage electric circuit. This circuit, when energized by contact between *O'* and *P*, actuates the burner of a furnace or boiler, or the refrigerating equipment and fan-coil unit of an air conditioner. The bimetallic unit (in the form of a helix to magnify the motion) is sensitive to the temperature of the surrounding air, thus opening and closing the contact points at *P*. Turning the temperature-set knob adjusts contact point *P*. (*c*) Air conditioning cooling thermostat with cover removed, showing the bimetallic element (coil) and mercury switch.

bend. For a given increase in temperature, brass expands more than iron and the compound bar, or *bimetallic element*, as it is called, will bend as shown in Fig. 13.13*a*.

Such bimetallic elements are the basis for dial thermometers and also form the working element in thermostats for heating and cooling controls. The essentials of a typical thermostat are diagramed in Fig. 13.13*b*. In actual practice the contact point *P* is usually in the form of a small magnet so that a "snap action" results. Thus the gap opens and closes with a positiveness which prevents the contact points from burning and pitting as a result of repeated arcing in the electric circuit. Heating thermostats close the electric circuit as temperature falls, while cooling thermostats are set to close the circuit as temperature rises. Temperature control of rooms to a tolerance of plus or minus 2°F can easily be achieved. Research laboratories, final assembly rooms for sensitive instruments, and electronic computer laboratories are just a few of the industrial facilities which need thermostatically controlled temperatures. (See Fig. 13.13*c*.)

The principle of differential expansion is used also in the construction of clocks and watches. Pendulums of clocks and balance wheels of watches expand and contract with temperature changes, and time-measurement errors result. Suitable compensation is effected by using steel and brass so combined and distributed in the pendulum or balance wheel that the expansion of the one works against, or *compensates* for, the expansion of the other.

Industrial Applications of Linear Expansion The effects of heat expansion must be reckoned with constantly throughout industry. In bridge and building construction, expansion joints must be provided for each major section. In oil-field piping and in the plumbing, refrigeration, and steam-fitting trades careful attention must be given to providing expansion joints at the right places (see Fig. 13.14). The importance of expansion to the fitting of engine parts has already been referred to. Surveyors take careful account of the effects of temperature variations on the steel tapes with which they measure the earth's surface.

Fig. 13.14 Expansion joints at a compressor station on a transcontinental gas pipeline. Hot gas coming from compressors causes the pipes to expand considerably. The large U-bends provide flexibility to prevent pipe breakage.

13.17 Area Expansion

When heated, a surface expands in both length and width, and consequently in area. The area coefficient of expansion of a material is defined as *the change in area per unit of original area per degree change in temperature.* Mathematically,

$$\alpha_A = \frac{A_2 - A_1}{A_1(t_2 - t_1)} \qquad (13.5)$$

where

α_A = area coefficient of expansion

$A_2 - A_1$ = change in area = ΔA

$t_2 - t_1$ = change in temperature = Δt

An alternative form is

$$\alpha_A = \frac{\Delta A}{A_1 \, \Delta t} \qquad (13.5')$$

The predicted new area A_2 is

$$A_2 = A_1(1 + \alpha_A \, \Delta t) \qquad (13.6)$$

Values of α_A are not usually listed in tables, since it turns out that α_A is in every case very approximately equal to 2α, where α is the *linear coefficient* of expansion. For purposes of ready calculation, Eq. (13.6) therefore becomes

$$A_2 = A_1(1 + 2\alpha \, \Delta t) \qquad (13.7)$$

The area of a hole cut in a sheet of metal expands just as if it were a sheet of the same material. In calculating the increase of area of the hole, the value of α to be used is that of the material in which the hole was cut.

Illustrative Problem 13.3 A round flat plate or disk of copper has a diameter of exactly 30 cm at a temperature of 15°C. What is its area when heated to 300°C?

Solution From Eq. (13.7),

$$A_2 = A_1(1 + 2\alpha \, \Delta t)$$

From Table 13.1,

$$\alpha_{(copper)} = (17 \times 10^{-6})/°C$$

Substituting,

$$A_2 = \pi \times (15 \text{ cm})^2$$
$$\times [1 + (2 \times 17 \times 10^{-6})/°C \times 285°C]$$
$$= 707 \text{ cm}^2 \times 1.00969 = 714 \text{ cm}^2 \qquad ans.$$

an increase of about 7 cm².

A 30-cm *hole* in a copper sheet subjected to the same temperature increase would have the same area increase.

13.18 Volume, or Cubical, Expansion—Solids

Volume coefficient of expansion β is defined as *the change in volume per unit of original volume per degree change in temperature*. As a formula,

$$\beta = \frac{V_2 - V_1}{V_1(t_2 - t_1)} \qquad (13.8)$$

Or, alternatively,

$$\beta = \frac{\Delta V}{V_1 \, \Delta t} \qquad (13.8')$$

Therefore

$$V_2 = V_1(1 + \beta \, \Delta t) \qquad (13.9)$$

Experimental results and mathematical analysis agree that the value of β very closely approximates 3α, for solids. We can therefore write

$$V_2 = V_1(1 + 3\alpha \, \Delta t) \qquad (13.10)$$

A cavity in a hollow object expands just as if it were a solid block of the same material. Thus a steel gasoline tank which holds 22 gal at 60°F will hold $22(1 + 3 \times 0.0000067 \times 50) = 22.02$ gal at 110°F. Note that α for the hollow space inside is the value of α for the steel of which the container is made.

Volume Expansion of Liquids Liquids follow the same law for volume expansion as solids, except that in general the coefficients are larger for liquids. β for liquids is determined by actual experiment. A list of values of β for some common liquids is given in Table 13.2.

In the experimental determination of values of β for liquids, it is of course necessary to have the liquid in some sort of container. The container will also change in volume during the experiment. An *apparent* value of β for the liquid will therefore be obtained. To find the true value of the expansion coefficient β_T, the coefficient for the container, $\beta_C = 3\alpha_C$, must be added to the apparent value β_A. As an equation,

$$\beta_T = \beta_A + \beta_C \qquad (13.11)$$

The values in Table 13.2 are *true* values.

Table 13.2 Volume coefficients of expansion for common liquids

Liquid	Per °C	Per °F
Alcohol, methyl	122×10^{-5}	68×10^{-5}
Gasoline	108×10^{-5}	60×10^{-5}
Glycerin	53×10^{-5}	29×10^{-5}
Mercury	18.2×10^{-5}	10.1×10^{-5}
Petroleum	89.9×10^{-5}	50×10^{-5}
Sulfuric acid	58×10^{-5}	32×10^{-5}
Turpentine	94×10^{-5}	52×10^{-5}
Water (at 20°C)	20.7×10^{-5}	11.5×10^{-5}

Illustrative Problem 13.4 A steel drum is filled to the brim with exactly 100 gal of turpentine at 40°F. How many gallons will overflow when the temperature rises to 120°F?

Solution Let V_{T_1} and V_{T_2} represent the initial and final volumes, respectively, of turpentine, and V_{D_1} and V_{D_2} the initial and final volumes of the drum. For the turpentine

$$V_{T_2} = V_{T_1}(1 + \beta_T \, \Delta t) = 100 \text{ gal}$$
$$\times [1 + (52 \times 10^{-5})/°F \times 80°F]$$
$$= 104.16 \text{ gal}$$

and the increase in volume of the turpentine is

$$\Delta V_T = V_{T_2} - V_{T_1} = 104.16 \text{ gal} - 100 \text{ gal}$$
$$= 4.16 \text{ gal}$$

For the steel drum,

$$V_{D_2} = V_{D_1}(1 + \beta_D \, \Delta t) = 100 \text{ gal}$$
$$\times [1 + (3 \times 6.7 \times 10^{-6})/°F \times 80°F]$$
$$= 100.16 \text{ gal}$$

and the increase in volume of the drum is

$$\Delta V_D = V_{D_2} - V_{D_1} = 100.16 \text{ gal} - 100 \text{ gal}$$
$$= 0.16 \text{ gal}$$

The overflow is

$$\Delta V_T - \Delta V_D = 4.16 \text{ gal} - 0.16 \text{ gal}$$
$$= 4 \text{ gal (approx.)}$$
ans.

13.19 The Unusual Expansion of Water

When water is cooled, its volume contracts steadily, as expected, until its temperature reaches 4°C. At this temperature, water has its greatest density. If cooled further, water *expands* slightly until the freezing point (0°C) is reached. As it freezes, it expands considerably. The bursting of water pipes in cold winter climates is a common result of this expansion due to freezing.

Table 13.3 gives the density of water at selected temperatures, and the curve of Fig. 13.15 shows graphically the peculiar expansion of water.

It is this peculiar behavior of water which causes lakes and oceans to freeze on top first, rather than at lower levels. Consider a body of water as winter sets in. The warmer water, being less dense, continually rises to the top, and a

Table 13.3 Density of water at various temperatures

Temperature, °C	Density, gm/cm³
0	0.99987
2	0.99997
4	1.0000
6	0.99997
10	0.99988
15	0.99912
20	0.99823
25	0.99705

steady circulation is maintained until the entire body of water reaches a temperature of 4°C. Below this temperature the *colder water is less dense* and stays on top, where it quickly freezes. Extremely cold weather is required to freeze a layer as much as a foot thick, since ice is a poor conductor of heat. What might the consequences be if water did not have this reversal in its expansion curve? What would happen to lakes and ponds in very cold climates?

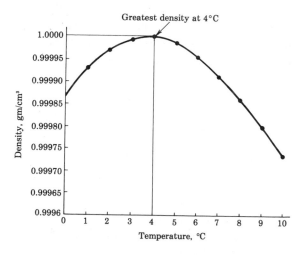

Fig. 13.15 The unique density-temperature curve for water. Water has its maximum density at 4°C.

THE LAWS OF GASES

13.20 The Thermal Behavior of Gases —Pressure-Volume-Temperature Relationships

In Chap. 12 some properties of gases were discussed in an elementary way, among them *Boyle's law*, which relates gas volume to changes in pressure. The present discussion will explore the pressure-volume-temperature relations of gases in greater detail.

The *condition* of a gas is determined by three factors—its *pressure*, its *temperature*, and its *volume*. Within reasonable limits of temperature and pressure, i.e., not close to the liquefaction points, *all gases* have the same pressure-volume-temperature $(P\text{-}V\text{-}T)$ behavior.

All of the so-called *gas laws* to be discussed below presume an "ideal gas." In practice, such a gas does not exist, but the behavior of real gases is close enough to that of the hypothesized ideal gas that the gas laws are extremely useful.

Gases at Constant Temperature; Boyle's Law A given mass of (an ideal) gas enclosed in a cylinder *at constant temperature* will occupy a volume which is inversely proportional to the *absolute pressure* (see Sec. 12.16) applied on the gas by a piston (see Fig. 13.16*a*). If a series of different pressures are applied (slowly), the corresponding volumes noted, and then absolute pressure is plotted against volume, a curve like that of

Fig. 13.16*b* results. Curves obtained by plotting constant temperature processes are called *isothermals*. Careful analysis of such curves reveals that for all combinations of pressure and volume, the areas of the rectangles of which P and V are sides are equal. In Fig. 13.16*b* for example, the area $P_1 A V_1 O$ equals the area $P_2 B V_2 O$. Since the area of a rectangle equals the product of length and width, for the present case

$$P_1 V_1 = P_2 V_2 = \text{constant} \qquad (13.12)$$

or

$$\frac{V_1}{V_2} = \frac{P_2}{P_1} \qquad (13.12')$$

which the student should recognize as a mathematical statement of Boyle's law.

Boyle's Law states:

The volume of a given mass of gas is inversely proportional to the absolute pressure if temperature is held constant.

Illustrative Problem 13.5 An air compressor takes atmospheric air $(P_1 = 14.7 \text{ lb/in.}^2 \text{ abs})$ and compresses it into a pressure tank whose volume is 20 ft^3. The initial pressure in the tank is 1 atm. After the temperature of the tank cools to room temperature, the gauge reads 600 lb/in.^{2}. What volume of atmospheric air was forced into the tank?

Solution Temperature is constant, and the problem is a pressure-volume, or Boyle's law, problem.

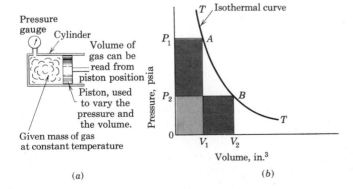

(a)

(b)

Volume, in.^{3}

Fig. 13.16 Pressure-volume $(P\text{-}V)$ relations of a gas at constant temperature. The curve is called an *isothermal*, meaning that the process occurs at constant temperature.

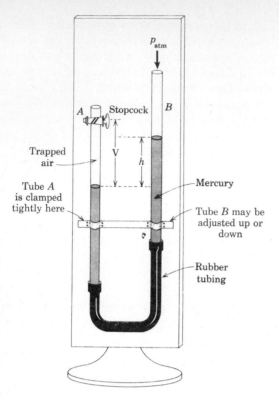

Fig. 13.17 Boyle's law apparatus, simple U-tube form.

form in Fig. 13.17. Two glass tubes, one open and the other capable of being tightly closed, are connected at the bottom with a U-tube loop of rubber hose. Tube A is rigidly mounted on a laboratory stand, and tube B can be adjusted vertically at the will of the experimenter. Mercury is poured into the open tube B, with the stopcock at A open, until the mercury rises about halfway up tube A. The stopcock is then tightly closed, trapping air in tube A at atmospheric pressure, and the apparatus is ready for use. Tube B is moved vertically with respect to tube A, and a series of readings of V (volume of trapped air) and h (difference in levels of mercury in the two tubes) is taken. Since the glass tubing is of uniform bore, the distance from the mercury level to the stopcock in tube A can be taken as the measure of volume of the trapped air, and the difference h in levels of mercury in the two tubes is the *gauge pressure* applied on the trapped air.

At each trial, the *absolute pressure* on the trapped air is equal to the pressure contributed by the mercury column of height h plus the atmospheric pressure:

P_{abs}(cm Hg)

$$= h \text{ (cm Hg)} + 76 \text{ cm Hg (atm press.)}$$

Proof of Boyle's law can be obtained from experimental data, by multiplying P for each trial by V for that trial. If all the PV products are essentially the same (that is, if $PV = $ a constant), Boyle's law has been verified (see Eq. 13.12). Results, even from such a crude apparatus, are usually in error by no more than 2 to 3 percent.

Gases at Constant Pressure; Charles' Law If a given mass of any gas is enclosed in a chamber fitted with a freely sliding piston, the *pressure will be constant*, for any slight increase or decrease of gas pressure will cause the (frictionless) piston to move in or out. If such a container of gas is heated (see Fig. 13.18a), the volume is found to increase with rising temperature in a manner depicted by the graph of Fig. 13.18b.

In a series of careful experiments the noted French physicist Jacques Charles (1746–1823)

$P_1 = 14.7$ lb/in.² abs
(in tank and atmospheric air)

$P_2 = $ gauge pressure + atmospheric pressure
$ = 600$ lb/in.² $+ 14.7$ lb/in.² $= 614.7$ lb/in.²

From Boyle's law, the total volume of air at pressure P_1 is given by

$$\frac{V_1}{V_2} = \frac{P_2}{P_1} \quad \text{or} \quad V_1 = \frac{V_2 P_2}{P_1}$$

Substituting,

$$V_1 = \frac{20 \text{ ft}^3 \times 614.7 \text{ lb/in.}^2}{14.7 \text{ lb/in.}^2} = 836 \text{ ft}^3$$

Volume of atmospheric air added is

$$836 \text{ ft}^3 - 20 \text{ ft}^3 = 816 \text{ ft}^3 \qquad ans.$$

Boyle's law can be verified in the laboratory with the simple apparatus shown in diagram

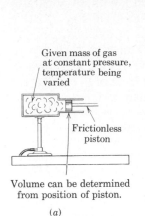

Given mass of gas at constant pressure, temperature being varied

Frictionless piston

Volume can be determined from position of piston.

(a)

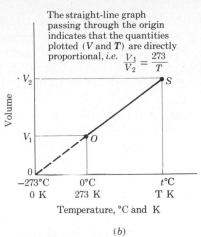

The straight-line graph passing through the origin indicates that the quantities plotted (V and T) are directly proportional, *i.e.* $\dfrac{V_1}{V_2} = \dfrac{273}{T}$

Volume

V_2 — — — — S

V_1 — — O

0

| −273°C | 0°C | t°C |
| 0 K | 273 K | T K |

Temperature, °C and K

(b)

Fig. 13.18 Volume-temperature (V-T) relations of a gas at constant pressure.

discovered that the volume of a gas varies with temperature (if the pressure is constant) in accordance with the relation

$$V_t = V_0(1 + \beta\,\Delta t) \qquad (13.13)$$

where β is the coefficient of volume expansion. Charles and subsequent workers found β *for all gases* to be $\frac{1}{273}$ per Celsius degree throughout normal temperature ranges (well above liquefaction). This discovery led to the following hypothesis: If any mass of gas at 0°C is cooled at constant pressure, its volume decreases by $\frac{1}{273}$ for each Celsius degree, and therefore at −273°C it should have zero volume! (Actually, all gases liquefy before reaching −273°C, and so the theory is not subject to experimental verification.)

Absolute Temperature This line of reasoning led to the proposal by Lord Kelvin (William Thomson, 1824–1907) of a third temperature scale, whose zero would be 273° below zero Celsius. This scale, of tremendous importance in all heat-engineering work, is referred to as the *Kelvin* or *absolute temperature scale*. A similar absolute scale developed for use with the Fahrenheit degree is called the *Rankine scale* for William J. M. Rankine (1820–1872), a Scottish engineer who proposed it. Figure 13.19 shows these absolute scales lined up with the Fahrenheit and Celsius scales for comparison. Any Kelvin

temperature is obtained by adding 273° to the Celsius temperature, and any Rankine temperature by adding 460° to the Fahrenheit temperature. *Kelvin temperatures use no degree symbol.* A temperature of 20°C is 293 kelvins, or 293 K.

$$T_K = t_C + 273$$

$$T_R = t_F + 460$$

Referring again to Fig. 13.18b suppose that the V and T values obtained in a range of temper-

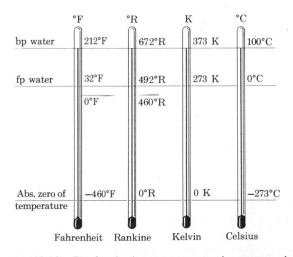

	°F	°R	K	°C
bp water	212°F	672°R	373 K	100°C
fp water	32°F	492°R	273 K	0°C
	0°F	460°R		
Abs. zero of temperature	−460°F	0°R	0 K	−273°C
	Fahrenheit	Rankine	Kelvin	Celsius

Fig. 13.19 The four basic temperature scales compared. This sketch is schematic only, since no liquid-in-glass thermometer could actually record such a range of temperature.

atures from 0°C to t°C plot to a straight line, as shown. Using the same scale and extending the "curve" downward and to the left (shown dashed), we see that it will hit the zero volume line at −273°C, or 0 K. This point is called the *absolute zero* of temperature. The straight-line graph tells us that *with pressure constant, the volume of a gas is directly proportional to the absolute temperature*. This relationship may be stated mathematically either by Eq. (13.13) or, more commonly, by the following:

$$\frac{V_1}{V_2} = \frac{T_1}{T_2} \qquad \text{Charles' law} \qquad (13.14)$$

where the capital T's indicate absolute temperatures, either Kelvin (K) or Rankine (R).

Charles' Law states:

For a given mass of gas, if pressure is held constant, the volume is directly proportional to the absolute temperature.

Gases at Constant Volume If a given mass of gas is held in a container of fixed volume while the temperature is varied, the *pressure-temperature relations at constant volume* can be obtained. (Actually, the volume of the container will vary slightly with temperature changes, but a correc-

tion factor can be applied.) With an apparatus like that shown schematically in Fig. 13.20a a series of pressure readings at different temperatures can be obtained. Data from such experiments, when plotted, give results as shown in Fig. 13.20b. Note that the locus of P-T points is a straight line, indicating that *at constant volume, the absolute pressure of a gas is proportional to the absolute temperature*. At constant volume:

$$\frac{P_1}{P_2} = \frac{T_1}{T_2} \qquad (13.15)$$

Illustrative Problem 13.6 A tank whose volume is assumed to remain constant is filled with carbon dioxide at a *gauge pressure* of 375 N/cm² when the temperature is 18°C. While being transported under a hot sun its temperature rises to 60°C. What is the new gauge pressure?

Solution Absolute pressures and temperatures must be used.

$$P_1 = 375 \text{ N/cm}^2 \text{ gauge} + 10.1 \text{ N/cm}^2 \text{ (1 atm)}$$

$$= 385.1 \text{ N/cm}^2 \text{ abs.}$$

$$T_1 = 18°C = 291 \text{ K} \qquad T_2 = 60°C = 333 \text{ K}$$

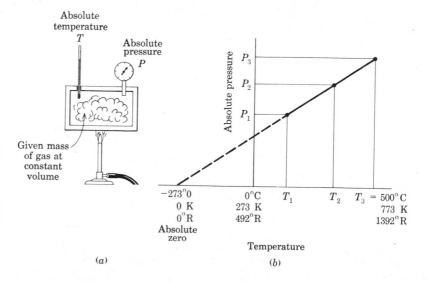

Fig. 13.20 Pressure-temperature (*P-T*) relations of a gas at constant volume. The path of *P-T* condition points is a straight line, indicating that $P_1/P_2 = T_1/T_2$. Note that extrapolating the "curve" to the left indicates that at −273°C (0 K or 0°R), the pressure reduces to zero. This condition would be true only for an "ideal" gas. All gases liquefy or solidify before reaching absolute zero.

Given mass of gas at constant volume

(a)

(b)

From Eq. (13.15),

$$P_2 = \frac{P_1 T_2}{T_1}$$

$$= \frac{385.1 \text{ N/cm}^2 \times 333 \text{ K}}{291 \text{ K}}$$

$$= 440.7 \text{ N/cm}^2 \text{ abs} = 430.6 \text{ N/cm}^2 \text{ gauge}$$

ans.

13.21   The General Gas Law (Ideal-gas Law)

We have found that *Boyle's law* describes the pressure-volume conditions of a gas held at constant temperature and that *Charles' law* governs the volume-temperature conditions of a gas maintained at constant pressure. However, these limitations, i.e., constant pressure or constant temperature, rarely occur in actual practice, since it is usually not possible to control either the pressure or the temperature of a gas very closely under actual conditions of commercial and industrial use.

Fig. 13.21   Butane tanks at an oil refinery. Industrial gases are usually stored under high pressure, so the volume can be minimized. (Standard Oil of New Jersey.)

The laws of Boyle and Charles are combined to form the *general gas law* (also called the *ideal-gas law*):

$$\frac{P_1 V_1}{T_1} = \frac{P_2 V_2}{T_2} = k \quad \text{(a constant)} \quad (13.16)$$

This equation relates *absolute* pressures, *absolute* temperatures, and volumes of a given mass of gas. The term *ideal* is used since real gases liquefy at low temperatures and at high pressures, and the "law" is therefore not subject to verification at these extremes.

An *ideal gas* is a theoretical concept. No such "perfect" gas actually exists. The concept involves a number of assumptions, of which the following are typical:

1.   A large number (billions) of molecules are in completely random motion within a finite (sizable) volume.

2.   No gravitational, magnetic, or other "action-at-a-distance" forces exist between molecules—only the forces of impact and reaction during collisions. This assumption is reasonably true only when the molecules are not too close together.

3.   Collisions between molecules occur very frequently, and they are perfectly elastic. Both KE and momentum are conserved at collision.

4.   The molecules are extremely small compared to the spacing between them.

It will be noted from Eq. (13.16) above, that when $T_1 = T_2$ (temperature constant), Boyle's law results:

$$P_1 V_1 = P_2 V_2 \qquad \text{or} \qquad PV = k$$

When $P_1 = P_2$ (pressure constant), Charles' law is obtained:

$$V_1/T_1 = V_2/T_2$$

Finally, a general statement of the ideal gas law is

$$PV/T = k \qquad (13.17)$$

Illustrative Problem 13.7   A quantity of acetylene gas is being held in a cylinder at 250 lb/in.² gauge pressure at 80°F. Another cylinder, previously

Table 13.4 Some fixed points on the International Practical Celsius Temperature Scale (IPCTS)

Standard	Temperature, °C
Normal boiling point of liquid hydrogen	−252.9
Normal boiling point of liquid oxygen	−182.96
Normal melting point of ice	0.00
Normal boiling point of water	100.00
Boiling point of liquid sulfur	444.60
Melting point of silver	961.9
Melting point of gold	1064

evacuated, of the same volume, is cross-connected to the original cylinder, the valve is opened, and the gas is allowed to equalize pressure between the two cylinders. If the temperature drops to 50°F, what will the new gauge pressure be?

Solution Equation (13.16) requires that absolute temperatures and absolute pressures be used.

$$P_1 = 250 \text{ lb/in.}^2 + 14.7 \text{ lb/in.}^2$$

$$= 264.7 \text{ lb/in.}^2 \text{ abs}$$

$$T_1 = 80°F + 460°F = 540°R$$

$$T_2 = 50°F + 460°F = 510°R$$

$$V_2 = 2V_1 \quad \text{or} \quad \frac{V_1}{V_2} = \frac{1}{2}$$

Substituting in Eq. (13.16) and solving for P_2 gives

$$P_2 = \frac{P_1 T_2}{T_1} \times \frac{V_1}{V_2}$$

$$= \frac{264.7 \text{ lb/in.}^2 \times 510°R}{540°R} \times \frac{1}{2}$$

$$= 125 \text{ lb/in.}^2 \text{ abs}$$

or

$$P_2 = 110.3 \text{ lb/in.}^2 \text{ gauge} \qquad ans.$$

13.22   International Standards for Temperature Measurement

The four temperature scales discussed above are very satisfactory for scientific and engineering work in normal temperature ranges. Based as they are on the boiling and freezing points of water and on the behavior of gases, they are not always useful for industrial purposes. Industry has a need for a more comprehensive scale for higher temperatures, utilizing certain of the practical industrial thermometers discussed in previous pages.

Such a practical temperature scale has been worked out and adopted by most nations of the world. It is called the International Practical Celsius Temperature Scale (IPCTS). A series of fixed points based on certain standard metals is the central idea of the scheme. Some of these fixed points are listed in Table 13.4.

It was agreed that certain thermometric methods would be used for appropriate temperature ranges. A few of these are given in Table 13.5.

Down to −260°C (13 K) a platinum resistance thermometer can be used with fair accuracy. Between this point and absolute zero, temperature-measuring methods have not been standardized, but low-temperature physicists are using methods based on the variations in magnetic susceptibility, with temperature, of certain crystalline salts (see Sec. 17.10 on *cryogenics*). Ultrasonic thermometers, which send very short, high-frequency vibrations through the substance

Table 13.5 Thermometric methods for stated temperature ranges

Range	Standard Method of Measurement
From −190 to 630°C	Platinum resistance thermometer
From 630°C to the gold melting point	Thermocouple of platinum and platinum-rhodium alloy
Above the gold point	Monochromatic optical pyrometer

in question, are also used. The smaller the wave velocity, the lower the temperature of the substance.

SUMMARY

Temperature is an indication of the hotness or coldness of a body or system. It is also a measure of average molecular kinetic energy.

The four basic temperature scales are (1) Fahrenheit, (2) Rankine, (3) Celsius, and (4) Kelvin.

Industry uses many and varied types of thermometers. In common use are liquid-in-glass thermometers, dial thermometers, resistance thermometers, thermocouples, and pyrometers.

Heat is defined as molecular kinetic energy which flows from one body or system to another as a result of temperature differences.

Expansion is one of the fundamental effects of heat. The linear coefficient of expansion of a substance is defined mathematically by the equation

$$\alpha = \frac{L_2 - L_1}{L_1(t_2 - t_1)} \quad \text{or} \quad \alpha = \frac{\Delta L}{L_1 \, \Delta t}$$

Areas expand when heated, and so do volumes. For solids, the area expansion coefficient $\alpha_A = 2\alpha$, and the volume expansion coefficient $\beta = 3\alpha$.

The principle of differential expansion is the basis for bimetallic elements and thermostats.

Water has very unusual expansion characteristics. Its greatest density occurs at 4°C. Between 4 and 0°C, water *expands as it is cooled*.

All gases expand and contract at the same rate, $\frac{1}{273}$ per Celsius degree. This fact leads to the concept of *absolute zero*: −273°C or zero K.

The condition of a gas is determined by three factors—pressure, volume, and temperature (P, V, T).

Boyle's law states that *with temperature constant, the volume of a gas is inversely proportional to the absolute pressure*, or

$$P_1 V_1 = P_2 V_2 \quad \text{or} \quad PV = \text{constant} = k$$

Charles' law states that *with pressure constant, the volume of a gas is directly proportional to the absolute temperature*, or

$$\frac{V_1}{V_2} = \frac{T_1}{T_2}$$

With volume held constant, the absolute pressure of a given mass of gas is directly proportional to the absolute temperature,

$$\frac{P_1}{P_2} = \frac{T_1}{T_2}$$

The general gas law (ideal-gas law) combines both Boyle's law and Charles' law, as follows:

$$\frac{P_1 V_1}{T_1} = \frac{P_2 V_2}{T_2} \quad \text{or} \quad \frac{PV}{T} = k$$

The general gas law, and in fact all the gas laws, are based on a set of assumptions about an *ideal* or *perfect* gas.

QUESTIONS AND EXERCISES

1. Diagram and explain the operation of (*a*) a dial thermometer, (*b*) a simple thermocouple.

2. If a service station owner takes delivery of gasoline from a tank truck on a hot summer day, and stores it in his cool underground tanks, does he lose or gain gallonage when he pumps it into his customers' cars?

3. On graph paper, plot a neat graph which will allow you to convert Fahrenheit temperatures to Celsius temperatures all the way from −50°F to 250°F. Use the vertical axis for Fahrenheit readings and the horizontal axis for Celsius readings. What happens at $-40°$?

4. Temperature readings from the freezing point of water down to zero degrees Fahrenheit are positive readings. When the same temperature range is observed with a Celsius thermometer, the readings are all negative. Explain.

5. Make use of a diagram or sketch, and explain why the top surface of a lake or pond freezes first as winter sets in. What might be the result if freezing occurred from the bottom up?

6. Explain the relationship between *heat* and *temperature*. Let your explanation show that you are familiar with the kinetic-molecular theory of heat energy.

7. Use a sketch to show how a bimetallic element works. If possible, examine a heating or cooling thermostat and diagram the basic elements which make it function.

8. Do some library reading about Benjamin Thompson (Count Rumford). Describe his experiments that led to the abandonment of the old *caloric theory* of heat.

9. The temperature of molten steel is to be measured inside an electric furnace. What methods and instruments would you recommend?

PROBLEMS

1. A Fahrenheit thermometer indicates a room temperature of 68°F. What Celsius temperature is this?

2. An influenza patient ran a temperature of 104.2°F. In three days of treatment the temperature was down to normal, 98.6°F. Express this range for a Celsius thermometer.

3. The temperature gauge on the instrument panel of a sports car registers 85°C. What Fahrenheit temperature is this?

4. The directions on an aerosol spray can say it is not to be stored at temperatures higher than 95°F. Express this in degrees Celsius.

5. On the radio you hear the temperature announced as $-12°C$. What Fahrenheit temperature is this?

6. A metal rod begins to glow red when heated to about 700°C. Express this temperature in degrees Rankine.

7. The boiling point of alcohol is 171°F. How many kelvins is this?

8. Steel rails are 40 ft long and are laid when the temperature is 40°F. What gap should be allowed between rails (in inches) if the maximum temperature expected is 130°F?

9. Steel cables on the suspension section of the San Francisco Bay Bridge are about 9000 ft long. If the temperature range is from 25 (winter) to 90°F (summer), what is the variation in cable length?

10. A metal rod 1.0 m long expands exactly 1.20 mm when heated from 10 to 100°C. Calculate the coefficient of linear expansion of the material.

11. A valve pushrod in an internal combustion engine is made of mild steel and is exactly 12.000 cm long at 20°C. If the clearance between the rod and the rocker arm is to be 0.038 mm at the engine operating temperature of 95°C, what must be the clearance at 20°C? (Answer to the nearest 0.001 mm.)

12. A brass hot-water pipe is 40 ft long when installed at a temperature of 50°F. By what length (in inches) will it expand when filled with hot water at 200°F?

13. A survey line is run on the desert (120°F) and measures exactly 18 mi. The 100-ft invar-steel alloy tape used was standardized at 60°F. By how many feet (too long or too short?) was the 18-mi measurement in error?

14. A Pyrex glass flask holds 1000 cm³ of methyl alcohol at 20°C. How much will overflow when the alcohol is heated to 70°C?

15. A 100-gal steel tank is filled with gasoline at 30°F. If the temperature rises to 100°F, how many gallons of gasoline will overflow?

16. The density of aluminum at 20°C is 2.7 gm/cm³. Calculate its density at 400°C, accurate to three significant figures.

17. The density of mercury at 0°C is 13.596 gm/cm³. Find its density at 100°C.

18. A certain mass of gas occupies 3.5 m³ at 150°C. If pressure remains constant, calculate the volume at 0°C.

19. A piston compresses gas in a cylinder. Before the compression, the volume of gas was 0.55 liter, pressure was 1 atm, and temperature was 80°C. At maximum compression, the volume is reduced to 0.08 liter, and the pressure is read from a gauge as 11×10^5 Pa. What is the temperature of the compressed gas in degrees Celsius?

20. A truck tire has a volume of 2.5 ft^3 when filled with air at a gauge pressure of 85 lb/in.^{2}. How many cubic feet of atmospheric air is required to fill this tire? (Assume no temperature change.)

21. An auto tire is inflated to a pressure of 26 lb/in.^{2} gauge on a cool morning when the temperature is 40°F. After a high-speed run the temperature of the tire is 150°F. Assuming no volume change, what is the new gauge pressure in lb/in.^{2}? In pascals?

22. Cold outside air at −20°C is drawn into a furnace heat exchanger and heated to a discharge temperature of 60°C. If pressure is assumed constant, by what factor is the volume multiplied?

23. An air pump with a piston whose diameter is 4 in. and whose stroke is 12 in. takes atmospheric air and compresses it into a tank whose volume is 5 ft^3. Assuming 100 percent volumetric efficiency of the pump, and no temperature change (neither assumption is true in actual practice), find the gauge pressure in the tank after 100 strokes of the pump. The tank has air at atmospheric pressure to start.

24. A refrigeration compressor takes in 20 in.^{3} of cold refrigerant " gas " at each stroke of the piston and compresses it to a volume of 9.0 in.^{3} (mass constant) before discharging it to the condenser. The "cold gas" enters the compressor cylinder at 50°F and 37 lb/in.^{2} gauge pressure. The "hot gas" leaves the compressor at 120 lb/in.^{2} gauge pressure. Find the Fahrenheit temperature of the hot gas.

25. A high-compression auto engine takes in 7.5 volumes of fuel-air mixture at 40°C and atmospheric pressure and compresses it to 1 volume at 245°C. Find the absolute pressure in the cylinder at the end of the compression stroke, in newtons per square meter (Pa).

26. The cross-sectional area of a steel rod is 1 in.^{2}. Find the least force that will prevent it from contracting as it cools from 125 to 20°F. (HINT: Review Young's modulus.)

14 Heat and Change of State

Three states of matter—solids, liquids, and gases—were defined in Chap. 9. The "state" of a substance depends on its molecular arrangement and also on its temperature and its internal energy at the time. Pressure is also a factor in determining the state of a substance.

Water undergoes changes of state which are easily observed. *Liquid* water (the "normal state" at most locations on earth) often changes to ice, snow, or hail—the *solid* state. And, although it is invisible, water vapor (the *gaseous* state) is continually forming as the heat of the sun evaporates water into the air. This moisture in the air is known as *humidity*. One of the most useful of all gases is high-pressure steam, which is water vapor that is generated in boilers to operate steam engines and turbines. Steam is just water vapor that has high heat content at a high temperature.

Solids can *melt* to liquids, and liquids can *freeze* to solids. Liquids *vaporize* to gases, and gases *condense* to liquids. In all of these changes of state there are heat gains or losses, temperature changes, and pressure and volume changes, either separately or in combination.

Many change-of-state processes occur naturally, but we are more concerned here with the changes of state that are made to take place for scientific, engineering, and industrial purposes. Common examples of forced changes of state are found in steam-generating plants, in chemical plants and petroleum refineries, in heating and refrigerating systems, in foundries, in steel making, and in the smelting and refining of ores.

Up to this point we have stressed the nature of heat as molecular kinetic energy which is trans-

Fig. 14.1 Molten (liquid) steel at nearly 3000 degrees Fahrenheit is poured from an electric furnace into a huge ladle. Heat and change of state are essential to the entire metals industry. (Bethlehem Steel Corporation.)

ferred (or flows) from one body or system to another as a result of temperature difference. Temperature measurement and some types of thermometers were explained, as were the principles and applications of expansion due to heat. Both Boyle's law and Charles' law were derived, leading to a discussion of the general (ideal) gas law. The present chapter will deal mainly with two concepts: (1) the measurement of heat, and (2) heat as a cause of *change of state*.

HEAT MEASUREMENT

Thermometers are useful instruments for measuring temperature, but a thermometer alone cannot measure the *amount* of heat energy that flows in or out of a substance. The total amount of heat required to warm up, or melt, or vaporize a substance, or the amount of heat a substance will give off as it cools (or as it condenses or solidifies) cannot be directly measured by a thermometer. What is needed is a unit of *heat measurement*, and a method of determining heat flow that involves temperature change as one of the factors in the calculation.

If two bodies at different temperatures are placed in contact, heat energy will flow from the hotter one to the cooler one. The temperature of the hotter one falls, and the temperature of the cooler one rises. The heat exchange will continue until thermal equilibrium (both bodies reach the same temperature) is attained. Heat energy always flows from the hotter object to the colder object.

14.1   Units of Heat Measurement

Since heat is a form of energy and not a material substance, it can only be measured in terms of the effect it has on some material substance. The substance chosen to standardize the heat unit is water. As usual, we define two basic units, one for the metric system, and one for the English (engineering) system.

The *basic unit of heat* in the metric system is that amount which will *raise the temperature of one*

gram of water one Celsius degree. This quantity is called a *calorie* (cal).

1 cal of heat raises the temperature of 1 gm of water 1 C°

Recently, to be compatible with SI-metric units, the kilogram-calorie, or *kilocalorie* (kcal), is increasingly being used in scientific and engineering work.

1 kcal (1000 cal) raises 1 kg of water 1 C°

The basic heat unit in the English (engineering) system is the *British thermal unit* (Btu). Its definition follows:

1 Btu is the amount of heat necessary to *raise the temperature of 1 lb of water 1 F°*

Since there are approximately 454 gm in the mass of a standard English pound, and since 1 Celsius degree is equivalent to 1.8 Fahrenheit degrees, relationships between the metric and English units of heat are as follows:

$$1 \text{ Btu} = \frac{454}{1.8} = 252 \text{ cal}$$

$$1 \text{ Btu} = 0.252 \text{ kcal}$$

$$1 \text{ kcal} = 3.97 \text{ Btu}$$

The phrase "mass of a standard English pound" is necessary for a rigorous definition, since the pound is a unit of weight or force. However, in industrial and commercial *practice* the pound-weight is used without the above distinction. We shall not repeat the "mass of a pound" phrase in these chapters on heat and heat engines.

Strictly speaking, the heat absorbing capacity of water is not a constant over a broad temperature range. As a result, the calorie is rigorously defined as the amount of heat required to raise the temperature of one gram of water from 14.5°C to 15.5°C. Similarly, a rigorous definition of the British thermal unit stipulates that it is the amount of heat required to raise the temperature of one pound of water from 63°F to 64°F. The variation with temperature is so small, however, that in engineering, industry, agriculture, and medicine, the value of the calorie and the British thermal unit are considered constant throughout the range of temperatures in ordinary use.

It should also be noted that the *kilocalorie* is called a *Calorie* (capitalized) by physicians and nutritionists. This measure of the energy content of foods is actually 1000 times as large as the metric *calorie*.

14.2 Measuring Heating Values of Fuels

On the basis of the definition of the unit heat quantity, the amount of heat supplied to a given mass of water can be determined as follows:

Let H = heat quantity
m = mass of water being heated
t_1 = initial temperature
t_2 = final temperature

The heat absorbed by the water is then

$$H = m(t_2 - t_1) \qquad (14.1)$$

If m is in grams and t_2 and t_1 in degrees Celsius, H will be in calories; or, if m is in kilograms and t_2 and t_1 in degrees Celsius, H will be in kilocalories.

For English-system heat measurements, Eq. (14.1) becomes

$$H = w(t_2 - t_1) \qquad (14.1')$$

where w is the *weight* in pounds on earth. The *slug*, as a unit of mass, is not used in heat calculations, probably because the Btu (based on the pound) has been and still is the basic unit of heat quantity in the English (engineering) system.

The *therm* (1 therm = 10^5 Btu) is often used as a unit to describe large quantities of heat.

In order to measure the heating value (or *heat of combustion*) of fuels, all the heat given off by the fuel must be absorbed by a given quantity of water. The container in which the water for heat-measurement experiments is heated is called a *calorimeter*. Fig. 14.2 shows the elements of a laboratory calorimeter suitable for student use. In actual fuel value tests a modification called a *continuous-flow calorimeter* is used, one form of which (for gaseous fuels) is diagramed in Fig. 14.3.

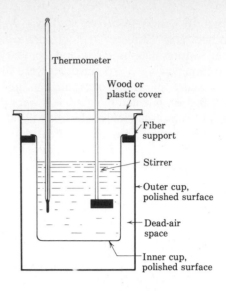

Fig. 14.2   Cross-sectional diagram of a simple laboratory calorimeter.

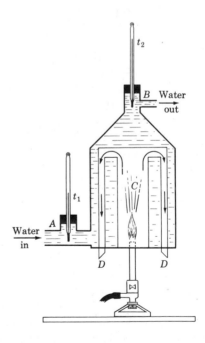

Fig. 14.3   One form of continuous-flow calorimeter for fuel-value determinations with gaseous fuels. Solid or liquid fuels would require a different kind of burner.

Table 14.1 Heating value, or heat of combustion, of some common fuels

Fuel	Heating Value		
	Btu/lb	*Btu/ft³*	*kcal/kg or cal/gm*
Coal (best grade)	14,000		7,800
Coke	13,000		7,200
Diesel oil	19,500		10,700
Fuel oil (varies)	19,000		10,800
Gasoline	20,000		11,300
Kerosene	19,800		11,000
TNT (for comparison)	6,520		3,600
Gases:			
Natural	24,000	1,100	13,300
Manufactured (varies)	10,000–15,000	450–700	5,500–8,300
Hydrogen	61,000	2,800	33,800
Wood	5,000–8,000		2,800–4,450

The hot gases of combustion go up through the central opening C, and then down through a number of flues D. Water flows in at A and out at B. The rate of water flow is carefully controlled in order to obtain a steady but rather small difference (10 to 20 F°) between water-in and water-out temperatures. The test is run for a definite time, during which the fuel consumed and the water flow through the calorimeter are carefully measured. From the amount of fuel used and from the number of Btu or kilocalories required to heat the measured amount of water to a fixed change in temperature (computed from Eq. 14.1), the fuel value can be determined. Before the actual test trial is made, the apparatus must operate long enough for a steady-state condition of water flow and temperature difference to be obtained.

Illustrative Problem 14.1 During a fuel-value test on natural gas, 4.62 ft³ of gas were consumed while 195 lb of water passed through the continuous-flow calorimeter. Throughout the test run the water-in temperature was 62°F, and the water-out temperature was 85.5°F. Assuming all the heat was absorbed by the water and that the burner used was 100 percent efficient, what is the fuel value of the gas in Btu per cubic foot?

Solution Substituting in Eq. (14.1'),

$$H = w(t_2 - t_1)$$

$$= 195 \text{ lb} \times 23.5 \text{ F}°$$

$$= 4580 \text{ Btu} \quad \text{(heat absorbed by the water)}$$

Dividing,

$$\text{fuel value} = \frac{H(\text{Btu})}{V(\text{ft}^3)} = \frac{4580 \text{ Btu}}{4.62 \text{ ft}^3}$$

$$= 990 \text{ Btu/ft}^3 \qquad ans.$$

The heating values (heat of combustion) of all common fuels have been determined by this and similar methods. Table 14.1 gives some average values.

14.3 Different Substances Absorb Heat at Different Rates

A slab of iron in the direct rays of the sun becomes unbearably hot to the touch while a pan of water alongside it merely becomes lukewarm. Many everyday experiences tell us that some substances heat up more rapidly than others. If equal masses of copper and water are heated over the same flame for equal periods of time, the temperature of the copper rises about 10 times as fast as the temperature of the water. To cause the same temperature change, the water would have to be heated 10 times as long. It is thus seen that water has far greater ability to absorb heat than copper does. We say it has greater *heat capacity*. The *heat capacity* of a substance is *the quantity of heat required to produce unit temperature change*. Water has the highest heat capacity of any common substance. By definition, the *thermal capacity of water* is 1 kcal/(kg)(C°), or 1 cal/(gm)(C°), or 1 Btu/(lb)(F°).

14.4 Specific Heat

The quantity of heat required to produce unit temperature change in unit mass of any substance is called specific heat. Since water is the chosen standard for defining heat units, the specific heat of water is 1. Since *water has the highest heat capacity of any common substance*, the specific heats of all other substances are less than 1. It is important to note that when the specific heat of water is said to be 1, this means 1 kcal/(kg)(C°), or 1 cal/(gm)(C°), or 1 Btu/(lb)(F°).

Specific heat may also be considered as *the ratio of the heat capacity of a substance to the heat capacity of an equal mass of water*. It may be looked upon as a measure of the "thermal inertia" of a substance.

In industrial heating processes it is desirable to know in advance how much heat a material will absorb before reaching a certain temperature. Knowing the specific heat of a substance, we can predict in advance how much heat will be required to raise its temperature by any given amount. If H is the heat required, m the mass of the substance, c the specific heat of the substance, and $t_2 - t_1$ the temperature change, we can write *the fundamental heat equation*

$$H = mc(t_2 - t_1) \quad \text{(SI-metric)} \qquad (14.2)$$

or

$$H = wc(t_2 - t_1) \quad \text{(English-engineering)} \quad (14.2')$$

If m is in grams and temperatures are in Celsius degrees, H will be in calories; and if m is in kilograms and temperatures are in Celsius degrees, H will be in kilocalories. For the English

system, remember to use weight in pounds, and Fahrenheit degrees. H will then be in Btu.

14.5 Measuring Specific Heat

The measurement of specific heats of substances is carried out in a calorimeter similar to that pictured in Fig. 14.2. Specific heats of liquids are determined by using the *method of mixtures*, in which a known mass of the test liquid at a known temperature is poured into a known mass of hot water whose exact temperature is also known. Any two liquids at different temperatures will, when mixed, soon come to some common intermediate temperature (thermal equilibrium). The heat given up by the hotter substance will warm up the cooler substance until the final temperature is the same throughout the mixture. *If there is no heat lost or gained in the process*, the heat units given up by the hot substance as it cools will be exactly equal to the heat units absorbed by the cold substance as it warms up. This *law of mixtures* results in the equation

$$m_h c_h(t_h - t_f) = m_c c_c(t_f - t_c) \qquad (14.3)$$

$$\frac{\text{Heat lost by}}{\text{hot substance}} = \frac{\text{heat gained by}}{\text{cold substance}}$$

where the subscript h refers to the hot substance, the subscript c refers to the cold substance, and t_f stands for the final temperature of the mixture.

As stressed above, Eq. (14.3) assumes no heat lost or gained in the mixing process. Of necessity, however, the mixing occurs in a container (the inner cup of the calorimeter), and it too will absorb heat from the hot substance and warm up to the final temperature of the mixture. (The dead-air space, fiber support ring, and polished surfaces tend to prevent heat transfer and radiation losses outside this inner cup.) We must therefore modify Eq. (14.3) to include the amount of heat absorbed by the calorimeter and its associated equipment. Writing m_{cal} and c_{cal} for the mass and specific heat, respectively, of the calorimeter's inner cup, we find that Eq. (14.3) becomes

$$m_h c_h(t_h - t_f) = m_c c_c(t_f - t_c) \quad + m_{cal} c_{cal}(t_f - t_c)$$

$$\frac{\text{Heat lost by}}{\text{hot substance}} = \frac{\text{heat gained by}}{\text{cold substance}} + \frac{\text{heat gained by}}{\text{calorimeter}}$$

which is usually rearranged and written in this form:

$$m_h c_h(t_h - t_f) = (m_c c_c + m_{cal} c_{cal})(t_f - t_c) \quad (14.4)$$

The quantity $m_{cal} c_{cal}$ is known as the *water equivalent* (WE) of the calorimeter, since this mass of water would have the same heat capacity as m mass units of the calorimeter material.

$$WE_{cal} = m_{cal} c_{cal} \quad (14.5)$$

Again, remember to use weight (w) for English-system heat exchange calculations.

14.6 Measuring the Specific Heat of Solids

Suppose it is desired to measure the specific heat of a certain metal sample whose mass is 125 gm. Arrange a calorimeter as shown in Fig. 14.4. Place a known amount (say 300 gm) of water in the calorimeter. Let its temperature (and that of the calorimeter cup) be 15.5°C. Heat the metal sample in an open steam bath for several minutes until it can be assumed to be at the steam temperature throughout. Record the temperature of the steam and also that of the water and calorimeter. Now lower the sample into the calorimeter, cover with the wooden lid and stir or shake gently until the temperature as indicated by the thermometer reaches a steady final value. If the steam temperature is 100°C, the final temperature is 19.5°C, and the mass of the aluminum ($c = 0.22$) calorimeter cup is 110 gm, we can substitute in Eq. (14.4) and obtain the specific heat of the metal sample as follows:

125 gm $\times c_h(100°C - 19.5°C)$

$\quad = (300$ gm $\times 1$ cal/gm C° $+ 110$ gm

$\qquad \times 0.22$ cal/gm C°$)(19.5°C - 15.5°C)$

from which

$\qquad c_h = 0.039$ cal/(gm)(C°)

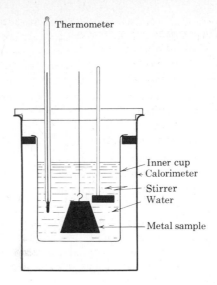

Fig. 14.4 A calorimeter fitted with items of equipment for measuring the specific heat of a solid.

Thermometer

Inner cup
Calorimeter

Stirrer
Water

Metal sample

Through carefully conducted experiments of this sort, specific heats for many common substances have been determined. Values for some common substances are given in Table 14.2.

Illustrative Problem 14.2 If 15 kg of steel ball bearings at 100°C are immersed in a bath of 25 kg of water at 20°C, assuming no losses of heat to or outside the container, what is the final temperature?

Solution Letting t_f represent the final temperature, the law of mixtures, Eq. (14.3), gives

$$m_h c_h(t_h - t_f) = m_c c_c(t_f - t_c)$$

Multiplying, we have

$$m_h c_h t_h - m_h c_h t_f = m_c c_c t_f - m_c c_c t_c$$

Collecting terms and factoring to facilitate solution gives

$$m_h c_h t_h + m_c c_c t_c = t_f(m_c c_c + m_h c_h)$$

and

$$t_f = \frac{m_h c_h t_h + m_c c_c t_c}{m_c c_c + m_h c_h}$$

Table 14.2 Specific heats of common substances

Substance	Specific Heat	Substance	Specific Heat
Air	0.24	Lead	0.03
Alcohol, ethyl	0.60	Mercury	0.033
Aluminum	0.22	Silver	0.056
Brass	0.091	Steam (100°C)	0.48
Copper	0.093	Stone (avg)	0.192
Earth (dry soil)	0.20	Tin	0.055
Glass	0.21	Water	1.00
Gold	0.032	Wood (avg)	0.42
Ice	0.50	Zinc	0.093
Iron (steel)	0.115		

NOTE: Specific heats have the same magnitudes in either metric or engineering units, kcal/(kg)(C°), cal/(gm)(C°), or Btu/(lb)(F°).

Substituting,

$$t_f = \frac{\begin{array}{l}15 \text{ kg} \times 0.115 \text{ kcal/(kg)(C°)} \times 100\text{C°} \\ + 25 \text{ kg} \times 1 \text{ kcal/(kg)(C°)} \times 20\text{C°}\end{array}}{\begin{array}{l}25 \text{ kg} \times 1 \text{ kcal/(kg)(C°)} + 15 \text{ kg} \\ \times 0.115 \text{ kcal/(kg)(C°)}\end{array}}$$

$$= \frac{672 \text{ kg C°}}{26.7 \text{ kg}} = 25.2°C \qquad ans.$$

Illustrative Problem 14.3 How much heat in Btu's is required to raise the temperature of a one-ton block of steel from 90°F to 800°F?

Solution Use the fundamental heat equation (14.2'), with the weight in pounds. The specific heat of steel (Table 14.2) is 0.115 Btu/(lb)(F°). Substituting values gives

$$H = 2000 \text{ lb} \times 0.115 \text{ Btu/(lb)(F°)}$$
$$\times (800°F - 90°F)$$
$$= 163,000 \text{ Btu} \qquad ans.$$

14.7 Water and Heat

Inspection of Table 14.2 shows how remarkably high the heat capacity of water is compared to other common substances. The amount of heat absorbed by a substance when it is heated is exactly equal to the amount it will give off as it cools through the same temperature range. Therefore large quantities of water can act as excellent storehouses for heat energy. In lakes and oceans, water even acts as a controlling factor, or brake, on seasonal temperature changes. In summer, large bodies of water absorb vast quantities of the sun's heat which would otherwise cause unbearably hot temperatures on the adjacent land; and in winter this heat in the water is slowly given up to the surrounding air, thus tending to moderate the winter weather. This accounts for the "even" climate of islands and narrow strips of land surrounded by the sea.

Water as a Heat-transfer Medium For the same reason (high thermal capacity), water is the most universally used substance for transferring heat in industry. Water is heated in boilers and pumped through pipes to radiators perhaps hundreds of feet away, there to give up heat as it warms homes, apartments, or shops. Cooled, it is pumped back to be heated again and the cycle continues, water serving as the carrier of heat, or the *heat-transfer medium.*

Besides being used to carry heat to spaces where heat is desired, water is used to carry heat away from spaces where it is not desired. Water circulates in the *jackets* around cylinders and combustion chambers of engines and picks up engine heat, preventing temperatures which would damage the engine parts. The hot water is pumped to a radiator and the heat is dissipated there to the surrounding air. Chilled water is cir-

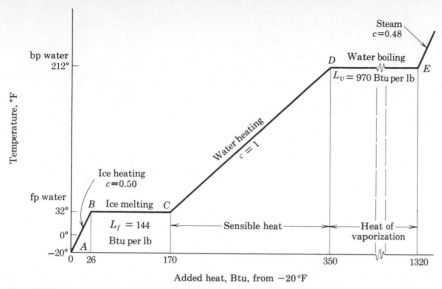

Fig. 14.5 Approximate temperature-heat (*T-H*) diagram for 1 lb of pure water at atmospheric pressure. Note the discontinuity or break in the horizontal (heat-content) scale (English-engineering units).

culated through air-conditioning cooling coils, and as warm air is forced over and through the chilled-water coil, the air gives up its heat to the water. As the water is warmed, the air is cooled, a heat-transfer process.

CHANGE OF STATE

The fact that all matter exists in one of three states—*solid*, *liquid*, or *gas*—has been discussed at some length. Whether a substance exists in the solid state, the liquid state, or the gaseous state depends on its temperature, its heat content, and on the pressure to which it is subjected. The temperature of a given mass of a substance depends on the heat content and on its thermal capacity. As an example let us investigate the effects of changing heat content on the temperature of water at standard atmospheric pressure (76 cm Hg or 14.7 lb/in.^{2}).

14.8 The Temperature-Heat (*T-H*) Diagram for Water—Pressure Constant

As an example, take 1 lb of ice at $-20°F$ and add heat to it slowly. Its temperature is found to rise

1 F° for each 0.5 Btu of heat added. (This means that the specific heat of ice is 0.5.) The line *AB* of the diagram of Fig. 14.5 shows how the temperature of ice rises as heat is added.

We see from the figure that 26 Btu of heat will raise the temperature of the pound of ice to 32°F, the melting point. Here it will be observed that the addition of more heat does not result in any further temperature increase but that the water begins to melt from solid to liquid form—a *change of state*. Careful measurements show that it takes 144 Btu to change (melt) 1 lb of ice at 32°F into liquid water at 32°F. Since this heat does not show up in increased temperature, we say it is hidden or *latent heat*. The process of melting is technically called *fusion*, and so we define *latent heat of fusion* of any substance as *the amount of heat required to melt unit mass of the substance without any change of temperature or pressure.*

Latent heat of fusion of water

$$L_f = 144 \text{ Btu/lb} \quad \text{(English system)}$$

The melting (fusion) process is shown by line *BC* of the diagram. By the time all the ice has melted (point *C* of the diagram) a total of 170 Btu (26 +

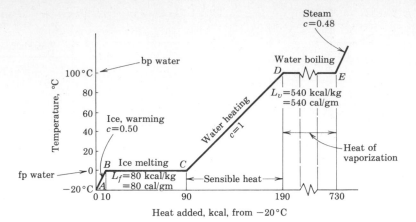

Fig. 14.6 Approximate temperature-heat (T-H) diagram for 1 kg of pure water at atmospheric pressure (SI-metric units).

144) of heat has been added since the process was started at $-20°F$.

The further addition of heat results in increased temperature at the rate of $1F°$ rise for each British thermal unit added, since the specific heat of liquid water is 1. (See line CD of the diagram.) After having absorbed another 180 Btu the water will be at *state point D*, at a temperature of 212°F, ready to boil. Total heat added (above $-20°F$) to this point is 350 Btu. The heat added up to this point is called *sensible heat*, since a change in heat added can be "sensed" as a temperature change. To boil away the entire pound of water requires the addition of 970 Btu of heat, and again this process occurs (line DE) *with no change in temperature*. Note that DE is a broken line, since the original scale cannot be maintained on the textbook page size. The process is technically called *vaporization*, and we define *latent heat of vaporization* of any substance as *the amount of heat required to vaporize unit mass of the substance without any change of temperature or pressure*.

Latent heat of vaporization of water

$$L_v = 970 \text{ Btu/lb} \quad \text{(English system)}$$

This brings us to *state point E* of the diagram with 1 lb of steam at 212°F. This steam is said to be *saturated* or *wet steam*. At point E 1320 Btu of heat has been added to the pound of water, measured from the starting point at $-20°F$.

The same process for one kilogram of water is diagramed in Fig. 14.6. Using it and the explanation above for the English system, you should be able to follow each step of the metric T-H diagram for water and understand what is happening in every stage of the process.

The *metric* values for *latent heat of fusion* and *latent heat of vaporization* are:

$$L_f = 80 \text{ kcal/kg} = 80 \text{ cal/gm} \quad \text{(metric)}$$

$$L_v = 540 \text{ kcal/kg} = 540 \text{ cal/gm} \quad \text{(metric)}$$

In a manner similar to that just described for water, but with markedly different ranges of temperature and heat content, most common substances will undergo changes of state as heat is supplied or taken away. Substances which are normally solids (metals, glass, etc.) can be melted and even vaporized if sufficient heat is added. Substances like air, oxygen, and nitrogen—normally gases—can be liquefied and finally frozen solid if enough heat is taken away from them. Table 14.3 gives melting points and boiling points for some common substances, and Table 14.4 the latent heats of fusion and vaporization of various substances.

Illustrative Problem 14.4 How much heat must be removed by a refrigerator to produce ice cubes whose temperature is 15°F from 20 lb of water whose initial temperature is 70°F?

Table 14.3 Approximate melting points and boiling points of some common substances at atmospheric pressure

Substance	Melting Point, °C	°F	Boiling Point, °C	°F
Air	(about) −215	−353	−195 to −185	−321 to −297
Ammonia	−75	−102	−33	−28
Carbon dioxide	−58	−70	−60	−76
Copper	1080	1980	2300	4170
Helium		−458		−452
Hydrogen	−258	−434	−253	−423
Iron (steel)	1530	2790	2980	5400
Lead	325	620	1620	2950
Mercury	−39	−38	356	675
Nitrogen		−346		−321
Oxygen	−218	−362	−182	−297
Platinum	1770	3220	4300	7770
Tin	232	450	2260	4100
Tungsten	3365	6090	5900	10,650

Table 14.4 Latent heats of fusion and vaporization of some common substances at atmospheric pressure

Substance	Latent Heat of Fusion L_f		Latent Heat of Vaporization L_v	
	kcal/kg	Btu/lb	kcal/kg	Btu/lb
Alcohol, ethyl	24.9	45	204	367
Ammonia	108	195	327	465
Lead	5.9	10.6	210	315
Mercury	2.8	5.0	71	128
Nitrogen	6.2	11	47.8	85
Oxygen	3.3	5.9	51	92
Water	80	144	540	970

Solution Heat removal occurs in three steps: (1) cooling the water to the freezing point, (2) removing the latent heat of fusion L_f, and (3) further cooling the ice from 32°F to 15°F.

Step 1: $H_1 = 20 \text{ lb} \times 1 \text{ Btu/(lb)(F°)}$
$$\times (70 - 32)°F \doteq 760 \text{ Btu}$$

Step 2: $H_2 = 20 \text{ lb} \times 144 \text{ Btu/lb} = 2880 \text{ Btu}$

Step 3: $H_3 = 20 \text{ lb} \times 0.50 \text{ Btu/(lb)(F°)}$
$$\times (32 - 15)°F = \underline{170 \text{ Btu}}$$

Total heat removed 3810 Btu
ans.

14.9 Melting and Freezing

Crystalline substances like ice and most metals melt at fairly definite temperatures. The melting point of these substances is generally at the same temperature as the freezing point. Many substances, on the other hand, are noncrystalline, or *amorphous*, like glass and the fats, and their melting is a gradual process which takes place not at a fixed temperature but over an appreciable temperature range.

The exact melting points and heats of fusion of crystalline substances are of great importance in the metallurgical industry. In metallurgy, almost every process depends on the ability to predict in advance what will happen when heat is added or taken away. The properties of toughness, hardness, malleability, tensile strength, and elasticity are determined by the molecular and crystalline structure of the finished metal. This molecular arrangement is, in turn, tailored to specifications by the metallurgist using his knowledge of the effects of heat on the molecular and crystalline structure of the particular metal.

Melting Is Actually a Cooling Process   When a substance such as ice melts without heat being intentionally supplied, the heat necessary to melt it is absorbed from its surroundings, with a consequent drop in temperature. Thus, *melting* can be thought of as a *cooling process*. For example, each pound of ice that melts in an icebox will absorb 144 Btu of heat from the food products in the space. If ice did not melt in an " icebox," it would do very little cooling. Conversely, *freezing* can be looked upon as a *heating process*, since as a liquid freezes, it gives up to its surroundings an amount of heat equal to its latent heat of fusion. When water freezes to ice, 144 Btu (latent heat of fusion) is given to the surroundings for each pound of ice frozen; or 80 kcal for each kilogram of ice frozen. Ranchers take advantage of this property of water by irrigating or spraying citrus trees and vineyards during freezing weather. The latent heat of fusion given off as the water freezes, "warms" the trees, vines, fruit, and the immediately surrounding air, by several degrees.

Change of Volume on Freezing   Most substances contract on freezing, resulting in a greater density of the solid. Water and some alloys are notable exceptions to this rule. Water expands markedly as it freezes, and ice has a specific gravity of only about 0.9. Ice therefore floats in water—a fact that is indeed fortunate, else the polar ice caps would spread well into what we now call the temperate zones.

Type metal (an alloy of lead, antimony, and copper) expands considerably on solidifying, a property which is essential for clear type with sharp, well-defined edges. So-called *gray iron* also expands slightly on solidifying, making it an ideal metal for making castings, since it expands against the mold and reproduces every detail which the design calls for.

Effect of Pressure on Freezing   Everything said about melting and freezing thus far has been based on the assumption that normal atmospheric pressure prevails. Every ice skater has observed that the pressure of his skate blades actually melts the ice under the blade. This forms a slippery film of water between the ice and the blade which facilitates skating. Part of the explanation for this melting is that *pressure causes it by a sudden lowering of the freezing point.* Carefully conducted experiments have shown that for liquids which expand on freezing, such as water, increasing the pressure lowers the freezing point; and for liquids which contract on freezing a pressure increase raises the freezing point. The effect is very slight, however, because it takes a pressure

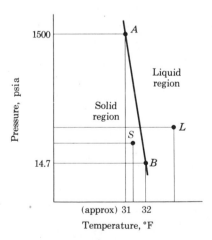

Fig. 14.7   Diagram showing the effect of pressure on the freezing point of water. A *P-T* curve (constant heat) for the solid-liquid boundary (not to scale). Pressure causes ice to melt at a lower temperature than it normally would.

of about 1500 lb/in.^{2} to lower the freezing point of water 1°F. An increase in pressure of 1 atm lowers the freezing point by only about 0.007°C.

The graph of Fig. 14.7 shows a pressure-temperature diagram for water in the vicinity of 32°F. The line *AB* represents a series of *P-T* points for which the solid and liquid phases are in equilibrium. *As long as no heat is added or taken away*, any point to the right of *AB*, for example, *L*, will represent a *P-T* situation for a *liquid*, and any point to the left of *AB*, for example, *S*, will represent a *P-T* situation for the *solid* phase.

We have been investigating some of the factors which cause a substance to undergo a change of state at the solid-liquid boundary. Let us now discuss change of state at the liquid-vapor boundary. We shall begin with an elementary analysis of the boiling of water.

14.10 What Makes Water Boil?

Fill a flask half full of water (Fig. 14.8), support it over a burner, and place a thermometer and a steam outlet in the stopper. As the water heats, the first effect noticed is that air which was dissolved in the water is driven out as small bubbles. Later, bubbles of steam will form at the bottom but as they rise to the cooler water near the top they condense and disappear. Finally, as the temperature of the entire mass of water reaches 100°C or 212°F (at standard sea-level pressure),

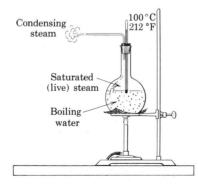

Fig. 14.8 The ordinary way to boil water is to add heat.

Condensing steam

100°C
212°F

Saturated
(live) steam

Boiling
water

bubbles of steam reach the surface and *boiling* is said to occur. The steam in the flask is invisible, but as it shoots out into the room, it condenses into tiny fog droplets, which can be seen and which are ordinarily called "steam" in everyday usage.

If the outlet valve is now *partly* closed (*careful!*) to create some back pressure in the flask, it will be noted that boiling stops until more heat is absorbed, the temperature rising perhaps 2 or 3°F (depending on the amount the pressure is increased) before boiling begins again.

As a final experiment, we now remove the burner, allow the water to cool down well below the boiling point (say to 140°F), and connect a vacuum pump to the flask. As the pressure is reduced on the surface, the water will burst into a vigorous "boil" even at this reduced temperature. (Pump very briefly, as moisture contaminates the vacuum pump oil rapidly.)

14.11 The True Nature of Boiling

We have seen that boiling depends not only on the temperature but on the pressure above the liquid. Heat energy added increases the temperature (see "sensible heat" portion of Figs. 14.5 and 14.6), which means an increased molecular speed of the liquid molecules. This causes some of the molecules of liquid to possess energies appropriate to the vapor state; i.e., bubbles of vapor are formed within the volume of the liquid. The formation of these bubbles is possible only when the pressure created by the vapor is equal to, or greater than, the pressure acting on the surface of the liquid. In accordance with these principles, *boiling point* is defined as *that temperature at which the vapor pressure within the liquid becomes equal to the pressure on the liquid surface* (see Fig. 14.9).

Vaporization of a liquid occurs at all temperatures at which the substance exists as a liquid. [Actually, there is direct vaporization of some solids: ice, naphthalene (mothballs), solid carbon dioxide (dry ice) are examples. This vaporization of solids is called *sublimation*. Ice kept

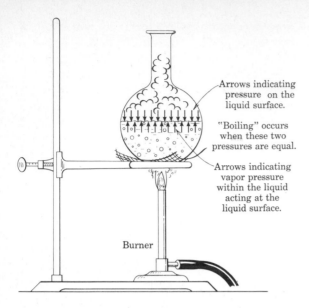

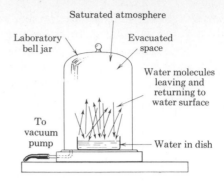

Fig. 14.10 Saturated vapor pressure—vapor and liquid in equilibrium. The space above the liquid is first evacuated so that it will contain only molecules of water vapor and no air. Liquid evaporates into this space until the pressure of the vapor above the liquid surface is equal to the internal vapor pressure of the liquid. This pressure is called the *saturated vapor pressure* for that temperature.

Fig. 14.9 Boiling point defined. The boiling point of a liquid is reached when the vapor pressure within the liquid becomes equal to the pressure directly above the surface of the liquid. Pressure equilibrium can result either from adding heat to the liquid and thus increasing its internal vapor pressure, or by reducing the pressure acting on the surface of the liquid.

for months at a temperature well below its melting point will slowly lose weight due to sublimation.] Even at extremely cold temperatures, some molecules possess energies sufficient to "fly off" from a liquid surface. The escape of these high-energy molecules into the atmosphere above the liquid creates a pressure called *vapor pressure*. If a liquid is placed in a container and the space above it is evacuated of air, molecules of the liquid will leave the surface (evaporate) and collect in the space above the liquid (see Fig. 14.10). Because of collisions with each other, some molecules fall back in. When the number leaving the surface is just equal to the number returning to the surface, an equilibrium condition is reached and the vapor above the liquid is said to be at its *saturated* vapor pressure for that temperature.

The higher the temperature of the liquid, the more high-energy molecules there are and the more will escape from the surface, creating a

higher vapor pressure. The value of the vapor pressure depends on the temperature, and *only on the temperature*, for any given liquid. Table 14.5 gives values of the saturated vapor pressure of some common liquids for a normal range of temperatures. Table 14.6 gives some properties of saturated steam for selected Fahrenheit temperatures.

14.12 The Kinetic-Molecular Hypothesis

Many phenomena of physics cannot be satisfactorily explained in the sense that *physical proof* is readily available. Sometimes *theories* or *hypotheses* are formulated which satisfactorily explain observed behavior. Change-of-state phenomena, for example, can best be explained by a pattern of reasoning known as the *kinetic-molecular hypothesis*. This hypothesis assumes that in any quantity of a substance there is a somewhat random distribution of molecular velocities (energies). For example, in a beaker of water whose temperature as read from a thermometer is 65°F, it might be assumed that the 65°F temperature is only an

Table 14.5 Saturated vapor pressure of some common liquids

Liquid	Temperature °C	°F	Pressure in. Hg	mm Hg (torrs)
Alcohol, ethyl	20	68	1.73	43.9
	60	122	8.74	223
	78.3	173	29.92 (1 atm)	760
Mercury	20	68	0.000047	0.0019
	100	212	0.0107	0.272
	357	675	29.92 (1 atm)	760
Water	0	32	0.18	4.57
	20	68	0.69	17.5
	60	122	3.64	92.5
	100	212	29.92 (1 atm)	760

Table 14.6 Properties of saturated steam at selected Fahrenheit temperatures

Absolute Pressure (Gauge + Atmospheric), lb/in.² abs (psia)	Temperature, °F	Heat Btu/lb above 32°F Sensible Heat (in Liquid)	Heat of Vaporization	Total Heat Content
1.0	102.2	70	1034	1104
14.7 (1 atm)	212 (bp)	180	970	1150
25	240	209	950	1159
50	281	250	923	1173
100	328	298	888	1186
250	401	376	824	1200
500	467	450	754	1204

average energy indication. The actual energies of the millions of molecules of water in the beaker are assumed to vary all the way from those with near zero energy to those with energies associated with a temperature of hundreds of degrees. It would be logical to assume, however, that most of the molecules would have speeds (energies) at any given time which would cluster around the *average* value and that speeds near either extreme would be much less common.

James Clerk Maxwell (1831-1879) was able to show that statistically, molecules do have a definite (though not entirely random) probability distribution of speeds. He assumed that in a random assembly of (ideal) gas molecules in motion, all energy values are theoretically possible. Under his assumptions, the distribution of molecular speeds can be represented by a curve resulting from a plot of the relative numbers of molecules at various speeds, against the speeds themselves. Figure 14.11 gives the general form of the curve of a *maxwellian distribution.*

If such a probability distribution of molecular speeds actually occurs, it is interesting to think about what would happen if the higher-energy molecules in a liquid could be drawn off in some way. Would the temperature of the remaining liquid change? Could a change of state result?

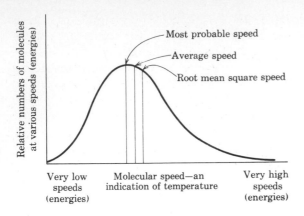

Fig. 14.11 The Maxwell-Boltzmann distribution curve of molecular velocities, for a gas at a specified temperature, T. The curve, which approximates that for a normal distribution but departs from it somewhat, illustrates one possible distribution of molecular speeds with respect to average speed. The average speed, in turn, is a measure of molecular kinetic energy, and therefore of the Kelvin temperature of the gas. (Not to any scale.)

The paragraphs that follow attempt to shed some light on these questions and to lend support to the kinetic-molecular hypothesis.

14.13 Evaporation and Condensation

Evaporation is a phenomenon that occurs at all temperatures, as pointed out above. Since it is the higher-energy molecules that leave the liquid, the average molecular energy (and speed) of the remaining molecules is lowered. This results in a drop in temperature of the remaining liquid. *Evaporation is therefore a cooling process.* There is direct, sensible evidence of this, since perspiration cools the body as moisture evaporates from the skin. Some factors which affect the rate of evaporation of liquids into the atmosphere are:

1. The *volatility* of the liquid, which is measured by the vapor pressure it will create at a given temperature. Note from Table 14.5 that alcohol has a vapor pressure about 2.5 times that of water at the same temperature. Alcohol is said to be more *volatile*. Thus, volatile liquids like alcohol, ether, and gasoline have a marked cooling effect

on the skin, since they evaporate quickly at normal temperatures.

2. The *temperature*. At higher temperatures liquids evaporate faster, since more molecules have energies sufficient to escape from the liquid surface. Processes requiring evaporation, e.g., crystallizing salts and sugars from solutions and making pure water from seawater, are carried on at high temperatures.

3. The *pressure above the liquid*. The lower the pressure above the liquid, the higher the evaporation rate. The crystallization processes just referred to are usually carried out under a partial vacuum.

4. The *surface area* of liquid exposed. Greater surface area gives more molecules a chance to escape without falling back into the liquid.

5. *Air currents* blowing over the surface. These aid evaporation by removing those molecules already evaporated, thus preventing an equilibrium state from being reached. The cooling effect of a breeze is thus explained, since the air movement causes faster evaporation.

6. The amount of *liquid vapor already present in the surrounding atmosphere*. Damp (high humidity) air will allow little evaporation of water, for example. Evaporation occurs readily into an atmosphere with low humidity. On the other hand, if the surrounding atmosphere is already saturated, for every molecule that leaves the liquid surface, one reenters from the saturated atmosphere above.

Condensation Saturated vapors condense to liquids at any given pressure if they lose an amount of heat equal to the latent heat of vaporization, L_v, at that pressure. For example (Table 14.4), at atmospheric pressure, one pound of saturated ammonia vapor will condense to saturated liquid ammonia if 465 Btu of heat is removed. Heat removal to cause a change of state is done commercially by a piece of equipment called a *condenser*, as we shall see in a later chapter. When vapors condense, the *latent heat* they possessed as vapors is transferred to and absorbed by the surroundings, where it shows up as *sensible heat*. This explains why, as clouds form and it begins to rain, the temperature of the atmosphere usually rises. *Condensation*, in this sense, *is a heating process*. *After* the rain, as water evaporates again, the temperature falls, confirming the observation made previously that *evaporation is a cooling process*.

14.14   Some Interesting Demonstrations

The definition of boiling as the condition which exists when the vapor pressure of the liquid equals the pressure above the liquid will explain the following interesting demonstrations, which at first may seem contrary to everyday experience.

1.   *Boiling by cooling.* First bring some water to a boil in an open flask. Take away the burner and *working rapidly*, insert a tight stopper, fitted with a thermometer, in the flask. (This arrangement must be "airtight.") Invert the flask and *quickly* pour cold water over it to prevent steam pressure from forcing the stopper out (see Fig. 14.12). As cold water is poured over the hot flask, the temperature drops but vigorous boiling continues. Cool the contents of the flask to about

150°F, and let stand. The boiling stops. Pour more cold water on; as the temperature drops further, boiling begins again!

Apparently we have boiled water by cooling it. The real explanation is left for the student to think through, keeping in mind the vapor-pressure explanation of boiling given above.

2.   *Freezing by boiling under reduced pressure.* Place a small quantity of water in a watch glass under a bell jar which can be evacuated by a high-capacity vacuum pump. Suspend a thermometer in the water in such a way that its scale can be read through the bell jar. Under the water dish place a crystallizing dish of *concentrated* sulfuric acid, which helps the pump by absorbing water vapor rapidly. Seal all joints with vacuum wax, for even the slightest leak will prevent the demonstration from being successful. The arrangement is diagramed in Fig. 14.13.

Turn on the pump and watch the vacuum gauge. At about 5 cm of mercury pressure, bubbles will begin to form in the water. These are air coming out of solution. At about 2 cm pressure, *real* boiling begins. As the pressure drops further, boiling continues, and the falling thermometer indicates that the water is being rapidly cooled. At about 7 mm pressure, boiling becomes intermittent, the bubbles appearing rather infrequently but bursting violently when they do appear. At about 5 mm pressure boiling almost ceases, and as the gauge drops toward 4 mm pressure, it will be observed that the thermometer

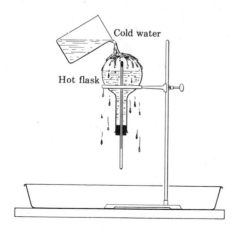

Cold water

Hot flask

Fig. 14.12   Boiling water by cooling it—a seeming paradox.

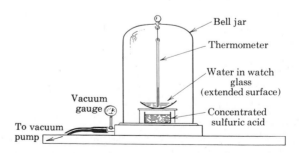

Bell jar

Thermometer

Water in watch glass (extended surface)

Vacuum gauge

Concentrated sulfuric acid

To vacuum pump

Fig. 14.13   Freezing water by boiling it without adding any heat. You have to see it to believe it. After you believe it, do you understand it?

registers 0°C. A few minutes' more pumping results in the sudden freezing of the water. Ice crystals form first around the edges; then needlelike crystals suddenly form across the entire dish. Watch closely or you may miss the sudden crystal formation.

We have boiled water, not by heating it but by reducing the pressure on it. We have cooled water by boiling it, and continued boiling has frozen it into ice! If you can explain these phenomena in terms of the kinetic-molecular theory you are beginning to understand change of state.

14.15 The Liquid-Vapor Boundary

Change of state at the liquid-vapor boundary can be portrayed by a P-T curve in much the way that Fig. 14.7 illustrates conditions at the solid-liquid boundary. The curve of Fig. 14.14 shows the P-T diagram (vapor-pressure curve) for water from about 230 to 32°F. The curve itself represents a series of P-T points for which the liquid and vapor phases are in equilibrium. Any point to the right of and below the curve, such as V, represents a P-T combination which will result in

vapor; and any point to the left of and above the curve, such as L, is a P-T situation for liquid. Only the curve for water is shown here. Different P-T conditions would apply to other substances, but the general nature of the curve is similar. Figure 14.15 shows how air at the liquid-vapor boundary is handled and stored.

14.16 The Triple Point

In the experiment illustrated by Fig. 14.13 (freezing by boiling under reduced pressure), water existed in a very unusual situation as the process reached its climax. Ice was forming, liquid water was still present, and vapor was being given off (boiling was occurring) all at the same time. The P-T condition for water at this point, if accurate

Fig. 14.15 Liquid air being stored in a Dewar flask. Liquid air boils at −190°C at atmospheric pressure. To liquefy air, it is first compressed to about 3000 lb/in.² pressure, then refrigerated to remove the heat of compression. It is then allowed to expand suddenly (adiabatic cooling—see Sec. 16.6 and Fig. 16.6). The sudden expansion causes enough additional cooling to liquefy some of the air. (Union Carbide Corp.)

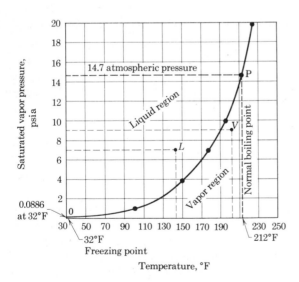

Fig. 14.14 The effect of pressure on the boiling point of water. A P-T curve for water at the liquid-vapor boundary.

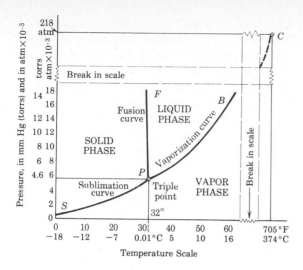

Fig. 14.16 The phase diagram for water. Note the *triple point*, a *P-T* condition where all three phases (states) can exist simultaneously. The *critical point, C,* the upper limit of the vaporization curve, is also shown. (Note the breaks in scale on both axes.)

pressure and temperature readings are taken, is found to be $P = 4.6$ mm Hg (torrs) = 0.18 in. Hg; $t = 0.01°C = 32.002°F$. This *condition point*, at which all three states of matter (solid, liquid, gas) momentarily exist simultaneously, is called the *triple point*. Many common substances exhibit triple-point behavior, but we shall consider only water in this discussion.

By choosing a somewhat different scale than that used in Figs. 14.7 and 14.14, it is possible to portray on a single diagram the relationships involved in the triple-point condition for water. In Fig. 14.16, *BP* is a section of the *vapor-pressure curve*, or boiling-point curve, for water. The *fusion curve FP*, the locus of state points in phase equilibrium between solid and liquid, is also drawn in, as is the *sublimation curve SP*, the locus of state points in phase equilibrium between solid and vapor. Pressures and temperatures shown on the diagram are only approximate. Note that at any temperature higher than about 0.01°C, no matter what the pressure, water will not exist as a solid, and that at any pressure less than about 4.6 torrs (0.006 atm), water will not exist as a liquid, no matter what the temperature.

The Critical Point There is an upper end point to the vapor-pressure curve of any substance. Called the *critical point*, it occurs for water at a pressure of 218 atm and a temperature of about 374°C. No substance can exist as a liquid at a temperature above its critical point. No matter how great a pressure may be exerted upon it, a substance will still be in gaseous form if its temperature exceeds the critical temperature for that substance.

14.17 The Importance of Change of State to Industry

Although much of the foregoing discussion of change of state has concerned itself with water, the student should not gather the impression that water is the only important substance in which industry uses change-of-state processes. Here are some other applications.

Refining Metals Most metals are mined in the form of ores, which contain many impurities, and frequently the metal is present in the form of an oxide. The basic procedure in obtaining the metal itself may be to heat the ore to the melting point, where impurities can be removed by chemical reduction, physical separation, or electrolytic means. Melting (a change of state) is the first step in the refining of many metals.

Refining Petroleum Petroleum is an exceedingly complex mixture of chemical compounds called hydrocarbons. The problem of the refiner is to separate out one useful compound from another, i.e., to separate butane, gasoline, naphtha, kerosene, distillate, diesel fuel, fuel oil, etc., from the crude oil. This separation is effected by heating under carefully controlled temperature and pressure conditions in petroleum stills called *fractionating towers* (see Fig. 14.17). The more volatile hydrocarbons undergo a change of state first and are drawn off. Then come the next most volatile, and so on. These vapors are condensed by cooling water and are drawn off at the proper point on the fractionating tower. Pressures are controlled as needed, by pressure pumps or vacuum pumps.

Fig. 14.17 A fractionating tower in a modern petroleum refinery. Pressure-temperature (*P-T*) conditions are carefully controlled over the entire height of the tower. Vaporized crude petroleum is separated into many components or *fractions* by such towers. Asphalt and heavy waxes remain at the bottom. Heavy oils come off next, then fuel oil, distillate, kerosene, and gasoline. Nearer the top the more volatile fractions (butane, propane, and methane) are collected. (Standard Oil Company of California.)

Heat Engines and Change of State Internal-combustion engines, though the fuel may be a liquid, receive their energy from the sudden expansion of hot gas in the cylinders. The fuel-air mixture must be supplied to the cylinder in the form of a vapor. *Carburetion* is the process of vaporizing the liquid fuel and mixing it in proper proportions with air. Steam engines, both reciprocating and turbine types, use water as the heat-transfer medium and cycle it back and forth from the liquid phase to the vapor phase (steam), using it over and over again. Heat engines are discussed in detail in a later chapter.

Change of State and Refrigeration The entire refrigeration and air-conditioning industry is based on the principles of change of state. The process of refrigeration consists in removing heat from some space or substance where its presence is not wanted and rejecting it to some space or substance where its presence is immaterial. The medium for this heat removal is an easily liquefied vapor called a *refrigerant*. Refrigeration and air conditioning are the subject of Chap. 17.

These few examples drawn at random from the metallurgical, mechanical-power, and refrigeration industries are illustrative of the extremely wide applications of change-of-state phenomena to industry.

SUMMARY

Heat is a form of energy. It is measured in terms of the effect it has on raising the temperature of water.

> 1 kcal raises 1 kg of water 1 C°
> 1 cal raises 1 gm of water 1 C°
> 1 Btu raises 1 lb of water 1 F°

The *heating value* or *heat of combustion*, of fuels is expressed in terms of Btu per pound (solid and liquid fuels) or Btu per cubic foot (gaseous fuels); and in kilocalories per kilogram or calories per gram.

The nutritionist's *Calorie*, a measure of the energy content of foods, is equivalent to 1000 cal.

The *heat capacity* of a substance is *the quantity of heat required to produce unit temperature change* of the substance.

Specific heat is the quantity of heat required to produce unit temperature change in unit mass of the substance. It is also defined as the ratio

$$\frac{\text{Heat capacity of a substance}}{\text{Heat capacity of an equal mass of water}}$$

The fundamental heat equation is

$$H = mc(t_2 - t_1) \qquad (14.2)$$

In English-engineering system calculations, weight w is used with Eq. (14.2) instead of mass m.

Water is the most important industrial heat-transfer medium.

Whether a substance exists as a *solid*, a *liquid*, or a *gas* depends on the *temperature and pressure conditions* to which it is subjected.

The *latent heat of fusion* L_f of a substance is defined as *the amount of heat required to melt unit mass of the substance from a solid to a liquid with no change in temperature or pressure.*

$$L_{f(\text{water})} = 80 \text{ kcal/kg or } 80 \text{ cal/gm or } 144 \text{ Btu/lb}$$

at 1 atm pressure

The *latent heat of vaporization* of a substance is defined as *the amount of heat required to boil away (vaporize) unit mass of a substance with no change in temperature or pressure.*

$$L_{v(\text{water})} = 540 \text{ kcal/kg or } 540 \text{ cal/gm or } 970 \text{ Btu/lb}$$

at 1 atm pressure

Boiling is technically defined as that condition which occurs when the vapor pressure of the liquid is equal to the pressure on the liquid surface.

The *kinetic-molecular hypothesis* and the concept of a Maxwellian-Boltzmann distribution of molecular velocities are ideas which lead to a better understanding of change-of-state processes.

Vaporization of liquids occurs at all temperatures. Vaporization is explained by the molecular theory—the faster-moving, higher-energy molecules escape the liquid boundary, resulting in a lower average energy of the remaining molecules. *Evaporation* is therefore a cooling process.

Boiling a liquid can be accomplished by the following methods: (1) raising its temperature, (2) lowering its pressure, (3) controlling both temperature and pressure simultaneously.

The *triple point* is a *P-T* condition at which a substance exists in all three states simultaneously. It is the intersection of the *vaporization curve*, the *fusion curve*, and the *sublimation curve*.

The *critical point* is a point at the upper end of the vaporization curve; no substance can exist as a liquid above its critical point temperature.

Change-of-state situations are basic to many industrial processes.

QUESTIONS AND EXERCISES

1. Explain what is meant by *heat capacity* and *specific heat.* How do they differ in meaning?

2. Why does a large body of water tend to modify the climate in its immediate vicinity?

3. Explain why *melting* is a cooling process and *freezing* is a heating process. Also why *evaporation* is a cooling process and *condensation* is a heating process.

4. Give a technical definition of *boiling.* Use it to explain why it is possible to "bring water to a boil" without adding any heat to it.

5. Define L_f and L_v. What conditions must be stated in each case?

6. Define *fractional distillation.* Do some library research and explain the approximate *P-T* conditions required in a fractionating tower to obtain distillate, kerosene, gasoline, and butane.

7. Under what conditions can heat be added to, or taken away from, a substance without any temperature change occurring?

8. In cold climates, in addition to heating the air in their houses, many people also humidify the air. What are some reasons for doing this?

9. If you want raw potatoes to cook as fast as possible in boiling water, would you turn up the flame and boil the water more vigorously? If so, explain why; if not, what could you do to hasten the cooking process in boiling water?

10. A manufacturer of iceboxes for camp and trailer use claims that the *ice compartment* is so well insulated that the ice will not melt for 10 days. Would this box be a good buy?

11. It is a common experience to boil (i.e. evaporate) water and other liquids by adding heat to them. Many liquids with low boiling points (e.g. ammonia, carbon dioxide) evaporate or " boil " if the pressure on them is lowered suddenly. Where does the heat to boil them come from?

PROBLEMS

1. How many Btu are required to raise the temperature of 1 gal of water from 32 to 212°F?

2. A slice of whole wheat bread is rated by dietitians at 100 Calories. If the bread were used as a fuel and converted 100 percent into heat, how many grams of water would it heat from 15 to 75°C?

3. How many kilocalories of heat are required to raise the temperature of 700 kg of copper from 18°C to its melting point at 1082°C?

4. It is desired to heat a steel ingot weighing 450 lb from 60 to 850°F. What is the minimum possible amount of heat (Btu) required?

5. Ten pounds of molten lead at 630°F are dropped through a sieve to form shot for shotgun shells. The shot cool to 600°F and solidify while falling. They land in 10 lb of water whose temperature is 50°F. What is the final temperature of the water?

6. An 80-kg man consumes 3000 kcal of food energy per day. If the same thermal energy were used to heat an 80-kg block of iron originally at 20°C, what would be the final temperature of the iron?

7. How many gallons of fuel oil (sp. gr. 0.76) are required to raise the temperature of 100,000 gal of water from 40°F to the boiling point?

8. During a test run with a continuous-flow calorimeter, 5 kg of high-grade coal were used. The water-in temperature was 12°C, the water-out temperature 60°C. A volume of 690 liters of water ran through during the test. The burner is known to be only 85 percent efficient. Calculate the fuel value of the coal in kilocalories per kilogram.

9. An oil burner for a home furnace must supply 40,000 kcal/hr to the furnace heat exchanger. The burner is 60 percent efficient, and the fuel oil density is 0.72 kg/liter. Find the consumption of fuel in liters per day (see Table 14.1).

10. Air at 32°F has a density of 0.081 lb/ft³. If a furnace must heat 5000 ft³/min of this air to a temperature of 140°F, using natural gas as the fuel, what will be the consumption of gas per day if the burner and heat-exchanger system is 65 percent efficient?

11. How much heat (Btu) is required to convert 1 gal of water at 50°F into saturated steam at 212°F?

12. How much heat is required to convert 500 kg of water at 10°C into saturated steam at 120°C?

13. Mercury in the laboratory is initially at 20°C. How much heat (kcal) must be removed from 10 kg of it to just freeze the entire quantity?

14. A 1-kg sample of an unknown metal is being tested for its specific heat. It is placed in a calorimeter whose water equivalent (WE) is known to be 50 gm.

Room temperature is 22°C. A mass of 0.45 kg of water at 95°C is poured into the calorimeter, covering the test sample. The final (equilibrium) temperature is 81°C. What is the specific heat of the sample, and (probably) what metal is it?

15. After 10 gal of water at 70°F is poured slowly over a 50-lb cake of ice at 32°F, the ice is dried and weighed. What is its weight? (Assume all the water is cooled to ice temperature.)

16. If 50 kg of ice at −10°C is placed in an icebox, how much heat (in kilocalories) will be removed from the contents of the box by the time all the ice has melted and drained out as a liquid at 0°C? (Assume no heat gains or losses from or to the outside of the box.)

17. A 100-gm aluminum calorimeter cup contains 200 gm of water at 20°C; 50 gm of ice at 0°C and 300 gm of iron at 100°C are placed in the cup, and the contents are stirred until a steady state is reached. Compute the mass of ice and water and the final (equilibrium) temperature.

18. A 200-kg steel forging is taken from a soaking pit at 900°C and immersed in a cooling tank containing 100 kg of water at 50°C. How much water is left when temperature equilibrium is reached?

19. Saturated steam is supplied to a radiator at 25 lb/in.² abs, and the condensed steam (water) leaves at 160°F and 1 atm pressure. How much heat is furnished to the room by each pound of steam? (Use Table 14.6.)

20. What is the end result of removing 1250 Btu from 1 lb of saturated steam at 212°F?

21. What capacity, in Btu per hour, must a refrigerating plant have to freeze twenty 300-lb cakes of ice in 4 hr and take their temperature down to 0°F? The water is at 60°F to start, and the cans in which the water is contained are of steel, each weighing 65 lb.

22. Calculate the cost of heating 50,000 gal of water from 60°F and converting it into saturated steam under 500 lb/in.² abs pressure if the fuel is natural gas costing $2.00 per thousand cubic feet. Assume the burner-boiler combination to be 65 percent efficient. (Use Table 14.6.)

23. A low-pressure steam engine is operating on 10,000 lb/hr of saturated steam, supplied from a boiler at 250 psia. (See Table 14.6.) The boiler "feed water" is supplied at 62°F. The boiler burners use commercial fuel oil, whose weight density is 6 lb/gal. The overall boiler/burner efficiency is 55 percent. Find the fuel oil demand in gallons per hour.

15 Heat Transfer

Heat is one of the most useful forms of energy. It is of great importance in industrial processes and in powering engines for air, ground, and sea transportation. We will study these applications in a subsequent chapter. Of equally great importance to millions of people is the fact that heat energy is necessary for health and comfort, and even to sustain life in the colder regions of the globe. In the tropics, the removal of heat from buildings by air conditioning makes life more pleasant and healthful for other millions. As conventional energy supplies become scarcer, the discovery—and development—of new fuel and energy sources is one of the greatest challenges ever faced by scientists, engineers, and technicians.

The nature of heat and some methods of measuring heat and temperature were explained in foregoing chapters. The relationship of heat to change of state was illustrated by a number of examples. In this chapter you will study how heat energy flows from one point to another in various substances. More often than not heat energy has to be used at some place well removed from where it is produced. This movement of heat is called *heat transfer*.

Heat energy always flows from a place or substance of high temperature to one of lower temperature. We say that "heat flows down a temperature hill." Its behavior in this respect is like that of water, which always flows downhill to lower elevations (see Fig. 15.1).

HEAT TRANSFER PROCESSES

Heat can be transferred, or moved, from its source by three different processes, all of which are common to everyday experience. These heat transfer methods are known as *conduction*, *convection*, and *radiation*.

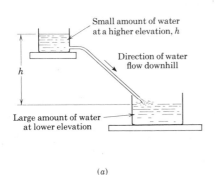

Small amount of water at a higher elevation, *h*

Direction of water flow downhill

h

Large amount of water at lower elevation

(*a*)

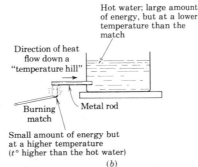

Hot water; large amount of energy, but at a lower temperature than the match

Direction of heat flow down a "temperature hill"

Burning match

Metal rod

Small amount of energy but at a higher temperature (*t*° higher than the hot water)

(*b*)

Fig. 15.1 Heat flow compared with water flow. (*a*) Water flows downhill to lower elevations regardless of the relative amounts of water at the two locations. (*b*) Heat flows "down a temperature hill" to a substance at a lower temperature, regardless of the *amount* or *quantity* of heat available at the two temperatures.

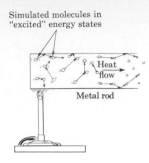

Simulated molecules in "excited" energy states

Heat flow

Metal rod

Fig. 15.2 A schematic explanation of heat conduction by molecular motion and molecular collisions.

15.1 Conduction

Conduction is the primary method of heat transfer in solids. When a metal rod is held with one end in a flame, the other end soon gets hot. A pan set on a stove conducts the heat through to the substance contained in the pan. Heat from engine cylinders passes through the cast-iron cylinder walls to the cooling liquid in the "jackets" surrounding them. No motion of the heated metal can be observed by the eye, but the kinetic–molecular hypothesis would predict that the hot molecules are in violent motion. These higher-energy molecules in contact with the flame collide with their neighbors, striking them hard blows and thus passing on energy to them. These in turn strike others, and heat energy is thus passed through the metal (see Fig. 15.2). The molecules themselves do not actually move along the rod; they vibrate more or less in place and pass the energy along, much as early fire-fighting companies passed water buckets along the line of men from the town pump to the blaze.

15.2 Thermal Conductivity

All solid materials conduct heat to some extent, some being excellent conductors, others very poor conductors. (Liquids and gases also conduct heat but very poorly compared to most solids. Recalling the molecular theory, can you suggest a possible reason for this?) Metals, in general, are

very good conductors; wood, glass, and brick are fair conductors; and porous substances with many tiny air pockets are poor conductors. It is an interesting fact that the heat conductivity of substances seems to parallel quite closely their electric conductivity. Good conductors of heat are generally good conductors of electricity, and conversely. Atoms of metals generally have "free electrons," and these play a part in both electrical conductivity and heat conductivity. The ability of a given substance to conduct heat is called its *thermal conductivity*, denoted by the letter k. The rate at which heat will flow through a slab or rod depends on a number of factors, as follows:

1. The *thermal conductivity k* of the material

2. The *cross-sectional area* of the material through which the heat is flowing

3. The *temperature difference* between the hot side and the cold side of the material

Figure 15.3 shows the factors involved in heat conduction through solids.

The total amount of heat conducted through a material also depends on the length of time the heat flows and on the thickness of the material. Careful experiments have verified the following equation for heat transfer:

$$H = \frac{kAT(t_2 - t_1)}{L} \qquad (15.1)$$

In SI-units, H is the heat in kilocalories conducted through a material whose cross-sectional area is A (in square meters) in time T (seconds), when the temperature difference is $(t_2 - t_1)$C°. The thickness of the material L is measured in meters.

In English engineering units, H is the heat (in Btu) conducted through a substance of cross-sectional area A (in square feet) in time T (hours) when the difference in temperature of the two sides is $(t_2 - t_1)$F°. L is the thickness of the material in inches.

The units of k depend, of course, on what system of measurement is being used with regard to the other factors in Eq. (15.1).

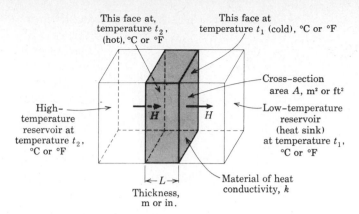

This face at,
temperature t_2,
(hot), °C or °F

This face at
temperature t_1 (cold), °C or °F

Cross–section
area A, m² or ft²

High–
temperature
reservoir at
temperature t_2,
°C or °F

Low–temperature
reservoir
(heat sink)
at temperature t_1,
°C or °F

H H

$\leftarrow L \rightarrow$
Thickness,
m or in.

Material of heat
conductivity, k

Fig. 15.3 Heat conduction through solids. The diagram shows the factors involved in the heat conduction equation (15.1).

In the mks (SI-metric) system

k has the units   kcal/(sec)(m²)(C°/m)

$$= \text{kcal/(sec)(m)(C°)}$$

In the engineering (English) system

k has the units   Btu/(hr)(ft²)(F°/in.)

Heat conductivities for all common substances have been determined by careful experiment. Table 15.1 lists values of k for a number of common materials and for some insulating materials.

    For walls, ceilings, floors, or any structure built up of *different materials*, "overall factors" are used instead of k. The *coefficient of transmission* (or *conductance*), U, in Btu/(hr)(ft²)(F°), for the wall or structure specified, can be found in engineering tables. *Resistance to heat flow, R*, is the reciprocal of conductance ($R = 1/U$). Tables of R are also readily available, for calculating heat gains and losses.

    Tables of U values and R values can be found in any standard air-conditioning text. Space does not allow their inclusion here.

Illustrative Problem 15.1 A plate-glass window in an office building measures 4 by 8 m and is 1.2 cm thick. When its outer surface is at a temperature of $-10°C$ and the inner surface at 0°C, how much heat is transferred through the glass in 1 hr?

Table 15.1 Heat conductivities of various materials (ordinary temperature ranges)

Material	k	
	$kcal/(sec)(m)(C°)$	$Btu/(hr)(ft^2)(F°/in.)$
Air, at rest	0.055×10^{-4}	0.168
Aluminum	5.0×10^{-2}	1480
Brick	1.7×10^{-4}	5.0
Concrete	4.1×10^{-4}	12.0
Copper	9.2×10^{-2}	2640
Cork board	0.1×10^{-4}	0.25
Window glass, plate	1.6×10^{-4}	5.5
Insulating board, fiber	0.14×10^{-4}	0.35
Iron, cast	1.1×10^{-2}	350
Kapok	0.08×10^{-4}	0.24
Fiberglass (loose batt)	0.09×10^{-4}	0.27
Sawdust	0.14×10^{-4}	0.41
Silver	10×10^{-2}	2880
Steel	1.1×10^{-2}	312
Water, liquid	1.4×10^{-4}	4.28
Water, ice	5.2×10^{-4}	15.6
Wood, oak	0.4×10^{-4}	1.10
pine	0.29×10^{-4}	0.84

Solution   Use Eq. (15.1):

$$H = \frac{kAT(t_2 - t_1)}{L}$$

Heat Transfer   **293**

From Table 15.1, for glass

$$k = 1.6 \times 10^{-4} \text{ kcal/(sec)(m)(°C)}$$

Substituting gives

$$H = \frac{1.6 \times 10^{-4} \text{ kcal/(sec)(m)(°C)} \times 32 \text{ m}^2 \times 3600 \text{ sec} \times 10°C}{1.2 \times 10^{-2} \text{ m} \times 1 \text{ hr}}$$

$$= 15,400 \text{ kcal/hr} \qquad ans.$$

Illustrative Problem 15.2 Compare the heat loss in Btu/hr from one wall of an apartment before it is insulated, with the heat loss after it has been insulated. The basic wall is 8 ft × 12 ft and consists of a 6-in. layer of brick. After it has been insulated, the wall has the heat transfer characteristics of 4 in. of fiberglass. Inside wall surface temperature is 50°F and outside wall surface temperature is −10°F. (See Table 15.1 for values of k.)

Solution

(*a*) Before the wall is insulated:

$$H = \frac{5 \text{ Btu/(hr)(ft}^2\text{)(°F/in.)} \times 72 \text{ ft}^2 \times 1 \text{ hr (60°F)}}{6 \text{ in.}}$$

$$= 3600 \text{ Btu/hr} \qquad ans.$$

(*b*) After the wall has been insulated:

$$H = \frac{0.27 \text{ Btu/(hr)(ft}^2\text{)(°F/in.)} \times 72 \text{ ft}^2 \times 1 \text{ hr (60°F)}}{4 \text{ in.}}$$

$$= 290 \text{ Btu/hr} \qquad ans.$$

Such a significant reduction in heat loss, with consequent savings on fuel and fuel costs, shows how important it is to insulate homes and commercial buildings properly.

15.3 Measuring Thermal Conductivity

In every phase of industrial and engineering work in which heat transfer is involved, it is of extreme importance to use the material which possesses the desired conductivity. Millions of dollars are saved annually by controlling heat losses (or gains) in industrial processes. Most test laboratories have equipment for determining the heat conductivity of materials. Figure 15.4 illustrates, in diagram form, a laboratory apparatus for determining k for metals.

A rod of the test material (HC in the diagram) has a steam jacket fitted at one end and a water jacket at the other. The entire assembly is enclosed in a heavily insulated case (not

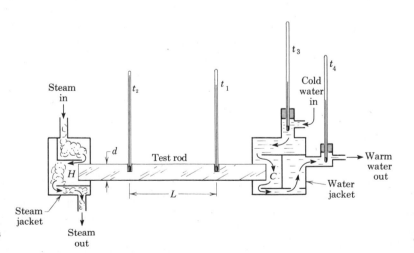

Fig. 15.4 Searle's apparatus for measuring the heat conductivity of metals and other solid materials.

diagramed) to prevent heat losses to the atmosphere. Live steam is supplied to the H end of the test rod, and the amount of heat transmitted to the C end is measured by the continuous-flow calorimeter arrangement provided. Cold water enters the jacket at the C end at a temperature t_3, is warmed by the heat transmitted by the test rod, and leaves the apparatus at temperature t_4. If a known mass of water m passes through the jacket, the total heat H which the water picks up from the test rod can be calculated from the basic heat equation $H = mc(t_4 - t_3)$, where $c = 1$ for water.

In a regular test run, the material to be tested is placed in the apparatus, the steam is turned in to the H end, and a small stream of water is circulated through the cold end. After 15 or 20 min, the water flow can be adjusted until a steady-state condition is obtained, i.e., thermometer readings t_2, t_1, t_3, and t_4 will stay at constant levels. When this condition is obtained, the *water out* is caught in a catch bucket for a measured time T and weighed. The four thermometer readings are recorded, and the length L and diameter d of the test rod are measured. From these data, k can be calculated, as in the following sample problem.

Illustrative Problem 15.3 A newly developed alloy is being tested for its heat conductivity by the apparatus and method described above. The following data are taken (English system):

$$L = 6 \text{ in.} \qquad t_3 = 60°\text{F}$$
$$d = 1 \text{ in.} \qquad t_4 = 68°\text{F}$$
$$t_2 = 185°\text{F} \qquad T = 25 \text{ min}$$
$$t_1 = 102°\text{F} \qquad w = 3.5 \text{ lb water}$$

Determine the heat conductivity of the alloy from these data.

Solution First calculate the heat supplied to the water at the C end. Substituting $H = wc(t_4 - t_3)$, (English system) gives

$$H = 3.5(68 - 60) = 28 \text{ Btu in 25 min}$$

Now solve Eq. (15.1) for k:

$$k = \frac{HL}{AT(t_2 - t_1)}$$

Substituting the values known, we obtain

$$k = \frac{28 \text{ Btu} \times 6 \text{ in.}}{\left(\dfrac{\pi \times 1^2}{144 \times 4}\right) \text{ft}^2 \times \frac{25}{60} \text{ hr} \times (185 - 102)°\text{F}}$$

$$= 890 \text{ Btu/(hr)(ft}^2\text{)(F}°/\text{in.)} \qquad ans.$$

Where strength is a factor, steel or iron is used in heat transfer equipment, even though its conductivity is not as high as that of some other metals. Boilers and furnaces are examples of equipment where steel is necessary. Copper is perhaps the best metal for heat transfer. It combines high heat conductivity ($k = 2640 \text{ Btu/(hr)}$ $\text{(ft}^2\text{)(F}°/\text{in.)}$, with reasonably good strength, resistance to corrosion, and relatively low price. As new developments on the energy front—solar energy, geothermal energy, and nuclear energy—come into full use, heat transfer will be a growth industry, and new metals and alloys with high values of k will be in great demand.

15.4 Convection

Heat is not *conducted* very well by liquids and gases, but heat transfer is easily accomplished by both liquids and gases by the process known as *convection*.

Warm winds carry heat from one place to another on the earth. Warm ocean currents carry heat from the tropics to warm what would otherwise be uninhabitable northern shores. Warm air and cold air (depending on the season) are blown (forced convection) through ducts and registers to heat and cool homes, offices, and factories. Hot water rises through pipes from boilers to remote radiators in other parts of the building or plant, carrying heat along with it. These are all examples of *heat transfer by convection.*

Convection Explained Note that in all the examples cited above the heat-transfer medium (water or air) moved from one place to another and *carried the heat with it.* Here we have the distinguishing characteristic of *convection—heat transfer by motion of the medium.* Conduction takes place by motion and collision of the *molecules* of the medium, whereas convection necessitates the *actual movement of the medium as a whole*. Recall the analogy of the early fire brigade: If instead of passing the buckets along the line, each man carried his own bucket from pump to blaze, this would be comparable to the manner in which heat is carried by convection.

15.5 Forced and Natural Convection

Some of the examples of convection cited above occur naturally while others depend upon mechanically operated blowers or pumps. Two simple experiments, one for liquids, one for gases, will serve to illustrate natural convection, frequently called a *convection current*.

Arrange a bent glass tube in the form shown in Fig. 15.5. Fill it with water in which is placed some sawdust or dye crystals. If the section *AB* is gently heated, a circulation begins almost immediately in the direction of the arrows.

Anyone sitting near a fireplace or a bonfire is aware almost equally of the warmth from the fire in front and the gentle but cold air movement coming from the rear. This air movement can be easily observed if someone in the room is smoking. The tobacco smoke drifts to the hearth, into the fireplace, and up the chimney.

These examples of natural convection show that convection currents are established in liquids

and gases. These convection currents are in turn caused by the expansion of the liquid or gas due to heat. The rapid expansion creates a quantity of liquid or gas of decreased density, and this lighter fluid rises, being pushed up by the surrounding heavier fluid. As hot gases rise in a chimney or over a bonfire, cold surrounding air flows in to take their place. This current continues as long as the fire burns, and in stoves or fireplaces it is called the *draft*. Draft is of considerable importance in the design of a fireplace, a furnace firebox, or an industrial chimney, since an adequate supply of oxygen to support combustion depends upon how good a draft is obtained. The design and location of the chimney are the most important factors in obtaining a good draft.

Many heating systems are based on natural convection, while others utilize forced convection.

Forced convection uses a moving medium (liquid or gas or air) to carry heat, but the motion of the medium is caused (or aided) by mechanical means. The cooling system of an automobile is a good example of a *convector* (it is mistakenly called a *radiator*), since the liquid (water or a coolant specified by the manufacturer) is circulated from the engine block to the heat exchanger ("radiator") and back again by means of a V-belt-driven water pump (see Fig. 15.6).

15.6 Radiation

The third method of heat transfer (*radiation*) cannot be explained either by the molecular theory or by familiar analogies such as that of the fire brigade. We can easily observe the process of radiation and measure its effects; we can create and use sources of radiant heat for many industrial processes; but the cause of radiant energy and the method by which it travels through space cannot be explained by common, everyday events.

We saw in the previous discussion that air currents flow from a room toward a burning fireplace. Yet, seated halfway across the room, we can "feel" the heat of the glowing embers on our

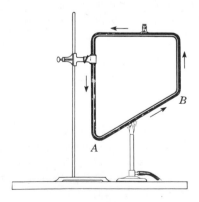

Fig. 15.5 Convection currents in water. Heat transfer by convection occurs in all fluids.

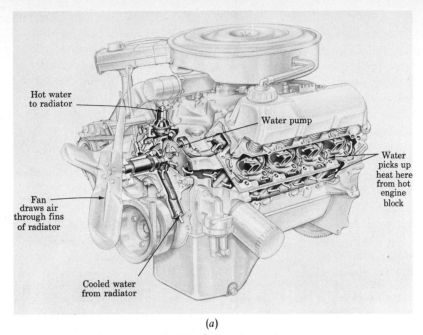

Hot water
to radiator

Water pump

Water
picks up
heat here
from hot
engine
block

Fan
draws air
through fins
of radiator

Cooled water
from radiator

(a)

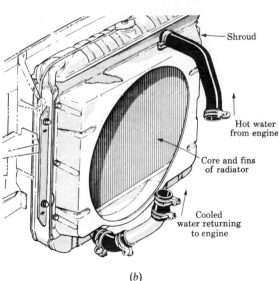

Shroud

Hot water
from engine

Core and fins
of radiator

Cooled
water returning
to engine

(b)

Fig. 15.6 Auto engine cooling system. (*a*) Heat from the engine cylinders is transmitted by *conduction* through the metal parts of the engine to the water in the passages or "jackets" surrounding the cylinders and cylinder heads. The water pump provides forced *convection* so that the hot water carries the engine heat to the *radiator*. (*b*) The radiator is actually a combination radiator, conductor and convector. It does radiate some heat, but most of it is transferred to the stream of air which passes through and around the coils and fins of highly conducting copper and aluminum of which the "radiator" is constructed. The fan maintains air flow through the radiator even if the auto is not moving. (Ford Motor Co.)

faces, shins, and outstretched hands. Quite obviously, this heat is not carried to us by the air, for the air is moving *away from us toward the fireplace*. Neither is it *conducted* to us by the air, for air is one of the poorest *conductors* of heat known.

As another example of radiation, consider the following: A steel bar, cherry-red from the forge, is held a few inches from and somewhat above the face. The heat is intense, but it cannot be due either to convection (hot air will rise from the bar) or to conduction. Now place a pane of transparent glass between the glowing bar and the face, and note the results: (1) You still see the cherry-red bar, and (2) despite the fact that glass is a very poor *conductor* of heat, you feel the burn-

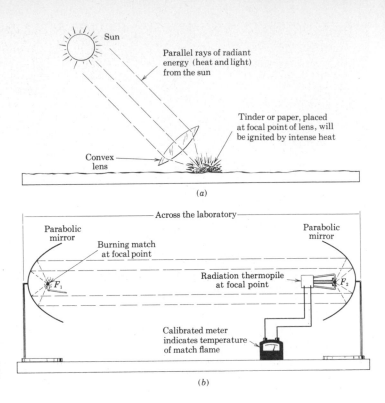

Parallel rays of radiant energy (heat and light) from the sun

Tinder or paper, placed at focal point of lens, will be ignited by intense heat

Convex lens

(a)

Across the laboratory

Parabolic mirror

Burning match at focal point

Parabolic mirror

Radiation thermopile at focal point

F_1

F_2

Calibrated meter indicates temperature of match flame

(b)

Fig. 15.7 Radiant heat, like light, can be brought to a focus by (a) lenses, and (b) mirrors.

ing almost as intensely as before. Now replace the pane of glass by a thin board, a sheet of paper, or anything else opaque to light. The immediate results noted are (1) that you no longer see the bar, since light cannot travel from it to your eyes, and (2) that you no longer feel its heat.

This simple experiment would seem to justify the conclusion that radiant heat is like light. The process of heat transfer by radiation seems to be similar to the passage of light through space. This conclusion is further borne out by the manner in which heat reaches us from the sun. Certainly convection and conduction are not involved, for some 93 million miles of nearly empty space separate us from the sun. When the sun sets or is temporarily obscured by a dense cloud, both its light and its heat diminish simultaneously.

Scientists have proposed the theory that empty space transmits heat energy and light energy by electromagnetic waves. The process is known as *radiation*. Radiant heat, like light, seems

to travel in straight lines. It is reflected and absorbed by the same surfaces which reflect and absorb light. It can be brought to a focus by lenses and mirrors in the same manner as light (see Fig. 15.7). Radiant heat travels with the same speed as light, 3×10^8 m/sec, or 186,000 mi/sec. Radiant heat and light both seem to be the *energy of wave motion*, and the difference between them is in the frequency and length of the waves. Heat energy for the most part is a longer-wavelength radiation than light energy. Heat waves include the region of the electromagnetic spectrum known as the "infrared." (See Fig. 21.14.) A rather conclusive proof of the supposition that radiant heat and light are essentially the same is given in Fig. 15.8. Though least understood, radiation is the most important of the three methods of heat transfer, for it is the only method by which heat energy reaches the earth from the sun. In full sunlight and perpendicular to the sun's rays, the earth receives about 19.4 kcal/(min)(m²) or

Fig. 15.8 Photography by radiant heat shows that light and radiant heat are very similar forms of energy. This photograph, with infrared film, was taken in total darkness with the irons set at full heat. Infrared photography is now being extensively used on the ground and from aircraft and satellites, to indicate excess "heat leakage" from buildings. More effective insulation can then be installed with consequent energy savings.

7.16 Btu/(min)(ft^2) in the form of radiant energy. This value is often referred to as the *solar constant*. (For more about radiation phenomena, see Chap. 21.)

SOLAR ENERGY

The possibilities for solar energy development can be appreciated by performing a simple calculation with the solar constant. In full sunlight, with the sun directly overhead, the amount of radiant heat striking the flat roof of a 1500-ft^2 house would be

$$H = 1500 \text{ ft}^2 \times 7.16 \text{ Btu/(min)(ft}^2)$$

$$\times \, 60 \text{ min/hr}$$

$$= 640,000 \text{ Btu/hr}$$

This is roughly four times the heat output rating of the furnace that would be needed to heat such a home on the coldest winter days. However, practical utilization of solar radiation as a reality is far less than the above theoretical value would indicate. For one thing, much of the incident heat is reflected. There is also the problem of storing the heat absorbed during periods of sunlight for use at night or during prolonged cloudy periods. Neither of these problems has been satisfactorily

solved as yet. Also, factors of geography and climate are such that in the regions needing the most heat during the winter months, the sun is not "directly overhead," and its radiant energy is far less than the given value of the solar constant. With slanting rays the solar constant is decreased by a factor equal to the cosine of the angle made with the perpendicular. Solar collector panels can be mechanized to "track" the sun, but this is an expensive installation. Solar heating apparatus in current use is sometimes unsightly and thus architecturally undesirable, and present costs are very high. However, dozens of different designs are under development, and in a few more years the state of the art will probably advance to the point where solar heating of homes, commercial buildings, and factories will be commonplace, rather than a novelty. Figure 15.9 shows one type of system that is presently on the market.

Solar heating systems are classified as *active* or *passive*. Skylights or windows with west and south exposures (northern hemisphere) are considered passive heat collectors. Passive systems usually have no mechanical parts or forced circulation.

Active solar systems feature *solar collectors*, roof-mounted and angled so as to receive maximum radiation from the winter sun (Fig. 15.9*b*).

(a)

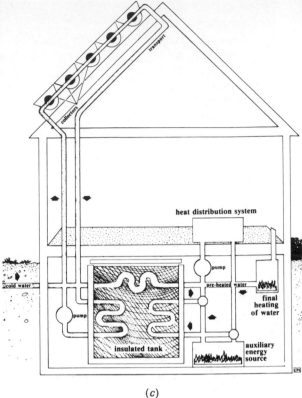

heat distribution system

pump

cold water

pre-heated water

final heating of water

pump

insulated tank

auxiliary energy source

(c)

(b)

Fig. 15.9 Solar heating. (a) Contemporary-style house, utilizing energy from roof-mounted solar panels. This home is an experimental solar house under test to determine its technical and economic feasibility. (b) Close-up of roof-mounted solar collector panels. (National Aeronautics and Space Administration). (c) Schematic diagram of the components that make up a typical solar heating system, as installed in colder climates where an auxiliary heat source will probably be needed.

Some installations allow for adjusting the collectors to provide for perpendicular incidence of the sun's rays at all seasons of the year. Active systems also utilize a heat-transfer medium (most often water), a heat energy storage unit (water tank, stones, or ceramic materials), a heat distribution system (pipes, ducts, pumps, blowers, controls, etc.), and in colder climates, an auxiliary, or "backup" system to augment or supplant the solar system when the sun's energy is insufficient (see Fig. 15.9c). The simplest form of roof collector is a flat, black-coated metal plate. The heat transfer fluid, usually water, flows in contact with this black metal absorber. The collectors are insulated so heat is not lost through the roof to the attic; and they are covered by sheets of glass that are infrared-opaque, thus preventing reradiation losses to the sky from the collector itself. This "trapping" of solar radiation is known as the "greenhouse effect" (see Fig. 15.13).

The heat energy absorbed by the black metal is transferred to the circulating water and carried to the storage tank (about 2000 gal capacity for a single-family residence). This tank is underground or is otherwise well insulated, and will retain large amounts of heat energy for several days.

Heat transfer to the rooms of the house may be by pipes carrying the hot water to radiation panels. Or, in another system, the storage tank is surrounded by loose stones or ceramic blocks, which absorb heat from the tank water, and air is blown through the hot stones and circulated to the rooms. New designs and equipment are continually appearing on the market and little purpose would be served by describing in detail any one system. The solar heating industry is now well established, and as the cost of oil and gas and coal continues to rise, solar heat will undoubtedly become an increasingly attractive option for homes and for commercial and institutional buildings. Solar *cooling* by absorption refrigeration (see Chap. 17) will also increase in popularity as cost ratios change in its favor

Energy from Photovoltaic Cells　When solar radiation strikes certain metallic elements and compounds, the radiant energy of the sun is converted directly into electric current. The theory of the *photoelectric effect* is dealt with in Chap. 33, and here we are concerned only with some possible practical applications for the present and future.

Silicon has proved to be a good material for photovoltaic (solar) cells. Cadmium/sulfur and gallium/arsenic compounds are also coming into favor. The efficiencies of solar cells are quite low, with silicon cells converting only about 10–12 percent of the sun's energy that falls on them into electric energy. Some newly developed gallium-arsenide cells are reported to be over 20 percent efficient, and 35-percent efficiency is hoped for in the future. The cost of solar-cell installations is very high at present. Compared to the approximate $400.00/kW cost for a coal- or oil-fired steam turbine-electric plant, solar-cell facilities range in cost from $5000/kW up to $50,000/kW, depending on location, size, and type of installation. Each solar cell is quite small (Fig. 15.10), on the order of one to two inches in diameter (they are not necessarily round), and nearly 7000 silicon cells are required to produce 1 kW of electrical output. Installations of massive size will be required to produce significant amounts of electric power. Using the silicon solar cells that are now the industry standard, a 100,000-kW plant (small, by steam plant standards) could require covering an area of over 100 acres with solar cells.

Solar-cell arrays are used to power satellites of all kinds (Skylab, for example, had an installation of nearly 150,000 cells, producing 22 kW), and they are frequently used to generate small amounts of power at remote locations on earth. Preliminary conceptual schemes for huge solar-powered satellite generating stations of perhaps 5000-MW capacity have been proposed by the National Aeronautics and Space Agency (NASA), but it may be many years before such a station is in operation. The sheer size and mass of such a plant (the solar cells and collection panels alone would have a mass of over 4.5 million kilograms) would involve hundreds of space missions, perhaps as many missions as are planned for all the activities of the Space Shuttle (see Fig. 7.21) for the decade of the 1980s, in order to get it assembled and operating in a suitable orbit. Furthermore, the problems of transmitting that amount of power through space and of receiving it at one or more ground stations are by no means simple of solution.

Solar energy is the hope of the present and a reality for the future, however, and despite the technical problems (and the economic and political problems that will inevitably follow), the promise of unlimited energy that is nonpolluting, renewable minute by minute, and that has zero "fuel" cost, suggests that research and development in the field will accelerate in the years ahead.

Absorbers and Radiators　Substances which absorb radiation are heated by it. This suggests that the absorbed radiation in some manner causes an increase in molecular and electronic motion within the body.

A completely satisfying explanation of electromagnetic phenomena like light and radiant heat is as yet unavailable; consequently scientists resort to theories and hypotheses. One theory is that light and radiant heat are composed of bundles of energy called *quanta* (singular *quantum*). The term *photons* is also used (see Chap. 33). According to the *quantum theory*, proposed by Max Planck (1858–1947) (see Chap. 33), the energy of each quantum is proportional to the frequency of the electromagnetic radiation which is the source of the energy. When light and/or radiant heat impinges on a surface such as the earth, the quanta impart their energies to the substance they strike. These "bundles" of energy increase the excitation of molecules and atoms on the surface, with consequent increases of both temperature and thermal energy. Transparent substances, through which radiation travels with little loss, are not appreciably heated, and neither are substances whose surfaces reflect most of the radiant energy falling upon them. In brief, then, before radiant energy in the form of waves or quanta can be converted into sensible heat, it must strike *and be absorbed* by some material substance.

All bodies radiate energy; the hotter the body the greater the amount of radiation. As a body radiates, its temperature falls, unless heat is supplied to keep the temperature up. All bodies also absorb radiant energy in varying amounts, and a continual exchange of energy is thus going on among bodies at different temperatures.

Gustav Kirchhoff (1824–1887) found that the emitting or radiating ability of a surface is

(a)

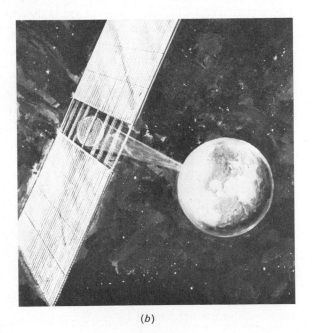

(b)

Fig. 15.10 (a) An array of photovoltaic (PV) cells used to provide electric power for an interstate expressway road sign. (Arizona Dept. of Transportation). (b) Artist's conception of a Satellite Solar Power Station (SSPS). The large rectangular structure, shown in two halves, would contain millions of photovoltaic cells. The circular device in the center represents a collector/microwave transmitter that beams the energy to earth. This particular concept envisions a hovering satellite, in a "geosynchronous" orbit over the earth's equator, whose dimensions are $3\frac{1}{2}$ miles by $2\frac{3}{4}$ miles. The microwave transmitter antenna alone would be 900 yards in diameter. (Grumman Aerospace Corp., courtesy of National Aeronautics and Space Administration).

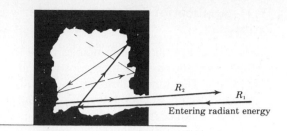

Fig. 15.11 Theoretical blackbody radiation. A cavity with rough dark-colored walls and only a small hole opening to the outside is a near-perfect absorber. Consider any ray of radiant energy, R_1, entering the cavity. As it strikes the rough cavity wall, both absorption and reflection occur at each incidence. After many such absorption-reflection events, the chance that any energy that originally entered from outside would be reflected back out through the entrance hole is virtually zero. The only radiation to the outside, such as ray R_2, is that which results from quanta whose source is heat energy *inside* the cavity. The combustion chamber of a boiler or the interior of a blast furnace for making steel are examples of actual devices that come close to blackbody radiation.

directly proportional to its absorbing ability. In other words, good radiators are also good absorbers, and *vice versa*.

A surface that could absorb all radiant energy that falls on it (a perfect absorber) would appear to be dead black. Such a surface would radiate, at any given temperature, better than any other surface. This "perfect" radiator is called an *ideal blackbody*, and we use the term "blackbody radiation" to describe this theoretical condition. Actually, there are no perfect absorbers. Black velvet and carbon black (soot) come closest, with absorption of about 97–98 percent of incident radiation. Polished, light-colored surfaces are poor radiators, and poor absorbers as well, since they reflect most of the radiant energy that falls on them.

The *ideal blackbody* cannot be attained in practice, and the best approach to it is a small hole leading into a cavity with rough, dark interior walls (Fig. 15.11). Such a cavity then radiates to the outside, through the hole, at a rate which depends on the absolute (Kelvin) temperature of the interior.

In 1879, Josef Stefan (1835–1893), an Austrian physicist, was able to show experimentally that the *rate of emission* of radiant energy from an ideal blackbody (cavity radiation) is proportional to the fourth power of the Kelvin temperature. Ludwig Boltzmann (1844–1906) subsequently confirmed Stefan's findings from theoretical calculations. The *Stefan-Boltzmann radiation law* is written

$$R = \sigma T^4 \qquad (15.2)$$

where R = the rate of total radiation emittance

T = the absolute temperature of the "blackbody" in kelvins

σ = the radiation constant

If R is in kcal/(sec)(m²),

$$\sigma = 1.36 \times 10^{-11} \text{ kcal/(sec)(m}^2)(\text{K}^4)$$

If R is in W/m², (W, watts)

$$\sigma = 5.67 \times 10^{-8} \text{ W/(sec)(m}^2)(\text{K}^4)$$

The critical effect of temperature on radiation is readily evident from Eq. (15.2), since doubling the Kelvin temperature increases the rate of radiation sixteenfold. As the temperature of a radiating body increases, the wavelength of the radiation becomes shorter. White-hot or blue-hot surfaces are thus much hotter than red-hot surfaces.

Illustrative Problem 15.4 The temperature of a cherry-red ingot from the soaking pit of a steel mill is 600°C. Assuming it to radiate like a blackbody, how much heat energy is radiated from all its surfaces per second if its dimensions are 0.5 by 1 by 3 m?

Solution Use Eq. (15.2):

$$R = \sigma T^4$$

$$\sigma = 1.36 \times 10^{-11} \text{ kcal/(sec)(m}^2)(\text{K}^4)$$

Substituting, and changing 600°C to 873 K,

$$R = 1.36 \times 10^{-11} \text{ kcal/(sec)}$$
$$\times (\text{m}^2)(\text{K}^4) \times (873 \text{ K})^4$$
$$= 7.88 \text{ kcal/(sec)(m}^2)$$

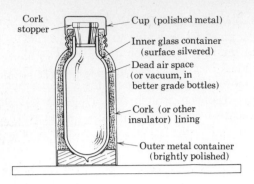

Cork stopper — Cup (polished metal)

— Inner glass container (surface silvered)

— Dead air space (or vacuum, in better grade bottles)

— Cork (or other insulator) lining

— Outer metal container (brightly polished)

Fig. 15.12 The vacuum (thermos) bottle is so constructed that heat losses and gains of all three types (conduction, convection, and radiation) are minimized.

Total surface area $= 10 \text{ m}^2$

Total heat radiated $= 78.8 \text{ kcal/sec}$ *ans.*

We take advantage of the known facts about radiation in many ways. For example, white clothing will keep us cooler in summer; in tropical climates houses are commonly painted white to reflect much of the sun's radiant energy. Boilers and steam and hot-water pipes are painted either white or aluminum color to reduce heat losses from radiation. Heating-system radiators should be (but for esthetic reasons seldom are) painted a dark color so they will readily radiate heat to the space surrounding them.

15.7 Applications of Heat-transfer Processes

Thermos bottles utilize the principles of all three methods of heat transfer in their construction (see Fig. 15.12). The inner container is usually of glass, which is a poor conductor. Its outer surface is silvered to reduce radiation losses. Between this surface and the outer container, the space is evacuated to prevent convection (and conduction) losses. The outer container is usually lined with cork or another heat insulator, and the outer surface is frequently of shiny metal to reduce the radiation losses further. If the bottle contains a cold liquid, the same principles apply, but heat gains are minimized instead of heat losses.

Greenhouses and cold frames used by gardeners for growing vegetables and flowers in unseasonable weather take advantage of the fact that radiation from hot bodies, e.g., the sun, is in general composed of short-wavelength high-frequency energy, while low-temperature radiation is of longer wavelength and lower frequency. Glass, although transparent to medium- and medium-short-wavelength radiation, is relatively opaque to extremely short (ultraviolet) and extremely long (infrared) radiation. The sun's heat rays pass through the glass roof and warm the interior. The plants and objects within radiate in turn, but their radiation is of longer wavelengths,

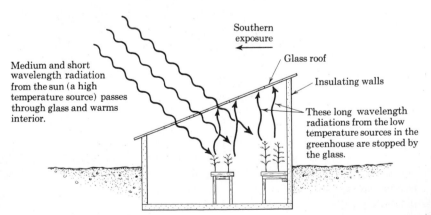

Southern exposure ←

Glass roof

Insulating walls

Medium and short wavelength radiation from the sun (a high temperature source) passes through glass and warms interior.

These long wavelength radiations from the low temperature sources in the greenhouse are stopped by the glass.

Fig. 15.13 The greenhouse or cold frame acts as a "trap" for radiant heat. So do some forms of solar heating systems for homes and commercial buildings.

Fig. 15.14 Heat-exchanger bundle from a petroleum refinery. Such heat exchangers are common in many industrial operations. They effect heat transfer mainly by conduction and convection. Radiation losses are kept to a minimum by painting the outer shell with a reflecting paint. (Standard Oil Co. of California.)

which is characteristic of low-temperature sources, and it is stopped by the glass. The greenhouse thus acts as a one-way valve for heat energy, and the structure itself becomes a "heat trap" (see Fig. 15.13). Crop production in commercial greenhouses is an important industry in north temperate-zone climates. Part of the necessary heat is solar heat "trapped" in this manner.

Many of the practical applications of heat transfer occur in, or are representative of, the heating, refrigeration, and air-conditioning industries. We shall examine some of these applications later in Chap. 17.

Heat exchange plays an essential role in many chemical-plant and refinery processes. Superheated steam is usually the heat-transfer medium, and the process of transferring the heat from the steam to the product being circulated is carried on in units of equipment called heat exchangers (see Fig. 15.14). The tubes of heat exchangers are ordinarily made of copper, because of its excellent conductivity.

SUMMARY

Heat transfer is the name given to all processes by which heat energy moves from one place to another.

There are three distinct methods of heat transfer: *conduction*, *convection*, and *radiation*.

Conduction is heat transfer by *molecular* motion or agitation within the substance (usually a solid) itself.

Convection is heat transfer by actual motion of the substance (liquid or gas) as a whole.

Radiation is heat transfer at the speed of light by waves or quanta. Radiation occurs without any material medium and is proportional to the fourth power of the Kelvin temperature. The Stefan-Boltzmann law of radiation is given by the equation $R = \sigma T^4$.

Heat will flow only from a hotter substance to a colder substance, and if there is no difference in temperature between two substances, no *net* heat flow will occur.

The heat conductivity k of a material is expressed in kcal/(sec)(m)(C°) or in Btu/(hr)(ft^2)(F°/in.).

Convection currents are of great importance in heating and cooling systems.

Rough, dark bodies are good radiators. Shiny, bright surfaces are poor radiators. The "ideal" radiator is the theoretical "blackbody."

The hotter the source, the higher the frequency of the radiation it emits.

Good radiators are in general good absorbers, and, conversely, good absorbers are good radiators.

Solar radiation as an energy source will be increasingly common in the future.

Water (liquid and vapor) is the universal medium for heat transfer.

Devices for transferring heat from a liquid or gas to some other liquid or gas, without mixing the heat carrier or the heat absorber, are called *heat exchangers*.

QUESTIONS AND EXERCISES

1. What is the determining factor in the direction of heat flow by conduction? convection? radiation?

2. Outdoors on a day when the temperature is well below freezing, one's bare hand will "freeze" to a metal tool handle. This does not occur with a wooden-handled tool. Explain.

3. Discuss in detail how the kinetic-molecular hypothesis explains the process of conduction in solids.

4. Trace every heat-transfer process involved in dissipating the heat from an operating automobile engine. Specify each material that takes part in the process and state whether conduction, convection, or radiation is involved at each stage of the process.

5. Pour a small quantity of water in a paper cup and then heat the bottom of the cup gently over an open flame until the water boils. Why does the paper not catch fire from the flame?

6. Why is it necessary for a fireplace or a furnace combustion chamber to have a proper chimney or flue for efficient operation? What is the purpose of the draft?

7. A pond will very quickly acquire a thin layer of ice when the air temperature falls below 32°F. Why is a much lower temperature required for several days before the ice layer thickens appreciably and is suitable for skating?

8. On a winter day the outside air temperature is −15°C and the air inside a house is +20°C. Would the outside and inside surface temperatures of a plate glass window be −15°C and +20°C respectively? Which surface (outside or inside) is likely to be closest to the temperature of the air immediately adjacent to it? Explain.

9. A house has a solar heating unit on the roof made up of many coils of copper pipe, encased in a glass-covered housing the inside of which is painted a dead black. Water is circulated in the pipes in a closed system which includes a pump and the "radiators" in the house. Trace every heat transfer process from the sun itself to a person sitting by one of the radiators who feels warmed by it.

10. At 40° N latitude on December 21st, what would be the approximate angle that the sun's rays make with the perpendicular? Using a diagram see if you can calculate the approximate value of the solar constant at this latitude at midday in bright sunlight.

PROBLEMS

1. "Four inches of fiberglass," usually in the form of batts or "blankets," is a common specification for insulating the walls of homes. How thick would the walls have to be to provide the same insulation effect if they were constructed of solid (*a*) pine, (*b*) brick, (*c*) concrete?

2. If one side of a 5 cm-thick corkboard wall is at a temperature of 30°C and the other at −10°C, how many kilocalories per hour will be conducted through 1 m² of the wall?

3. A glass window measures 8 by 5 ft and is $\frac{1}{4}$ in. thick. The outside glass surface temperature is +5°F, and the inside temperature of the glass is 40°F. How many Btu/hr are lost through this window?

4. A large household refrigerator has a cabinet which is equivalent to a box made of corkboard, 4 in. thick and 65 ft² in area. If the interior is to be maintained at 38°F while the kitchen temperature is 76°F, what is the conduction heat gain in Btu/hr into the cabinet? (Assume door is not opened during the hour of the test.)

5. A shoe store has a total wall surface 100 ft long by 14 ft high. The wall is of common brick, 8 in. thick. If the outdoor wall temperature is 10°F on a winter day and the inside wall temperature is being maintained at 54°F, find the heat loss in Btu/hr through the wall.

6. The ceiling of a building is insulated with 6 in. of fiberglass and measures 30 by 50 ft. If the inside ceiling temperature is 80°F and the attic temperature is 30°F, how many Btu/hr are lost through the ceiling?

7. A room for a computer laboratory is 30 by 40 by 9 ft high. The walls and ceiling are equivalent to 4 in. of fiberglass insulation. Assume no heat gains or losses through the floor. The computers dissipate heat energy into the space from their use of electric power at a rate of 8 kW. If a temperature of 70°F is to be maintained within the space against an outside wall temperature of 90°F and an attic temperature of 110°F, what is the required cooling capacity of an air conditioner in Btu/hr to balance the combined equipment and heat-transmission loads? (NOTE: 1 kW of electric energy dissipates 3400 Btu/hr of heat.)

8. An iron steam pipe has an outside diameter of 20 cm and is 0.8 cm thick. The steam temperature inside the pipe is 130°C, and the outside of the pipe is at 70°C. How many kcal/hr are lost by conduction from a 5-m section of this pipe? (HINT: Assume effective diameter of 19.2 cm.)

9. The inside dimensions of an icebox are 3 by 2 by 2 ft. The walls are of corkboard 4 in. thick. The inside temperature is 38°F, and the outside temperature is 95°F. How many pounds of ice are melted per 24-hr day by heat leakage through the walls?

10. A sample of metal is being tested to determine the value of k. With an apparatus like that of Fig. 15.4, the following data are recorded:

$$L = 5 \text{ in.} \qquad t_1 = 130°\text{F}$$

$$d = 2 \text{ in.} \qquad t_2 = 205°\text{F}$$

$$m = 15.3 \text{ lb water} \qquad t_3 = 50°\text{F}$$

$$T = 18 \text{ min} \qquad t_4 = 60°\text{F}$$

What is the value of k for this sample?

11. A furnace door has a round hole in it 6 cm in diameter. The interior of the furnace is at a temperature of 750°C. How many kilocalories will be radiated from this hole in 1 min? (Assume blackbody radiation.)

12. Find the rate of radiation of a blackbody whose temperature is 50°C and compare it with the rate from the same body when its temperature is raised to 450°C.

13. A steam radiator (painted black) is operating with a surface temperature of 210°F. Its effective surface area is 20 ft^2. If saturated steam enters at 240°F and leaves as hot water at 212°F, how many pounds of steam per hour must be supplied? [HINT: Convert problem conditions to SI-metric units and use Eq. (15.2). See Table 14.6 for steam properties.]

14. A copper kettle whose bottom surface is 0.5 cm thick and 50 cm in diameter rests on a burner which maintains the bottom surface of the kettle at 110°C. A steady-state heat flow occurs through the copper bottom into the vessel, where water is boiling at atmospheric pressure. The actual temperature of the inside surface of the kettle bottom is 105°C. How much water (in kilograms) boils away in 1 hr?

15. A layer of ice on a lake is 6 cm thick. If the top surface of the ice is at −10°C and the temperature of the water just below the ice layer is 0°C, at what rate does the ice become thicker? (Answer in centimeters per hour.)

16. A radiation pyrometer is used to measure the rate of energy emission from a hole of area 1 cm^2 in the wall of a furnace for making alloy steel. If the pyrometer reading is 140 W, what is the temperature (degrees Celsius) of the furnace interior? [HINT: Recall that $1 \text{ W} = 1 \text{ J/sec}$. (Given: 1 kcal = 4186 J.) Assume blackbody radiation.]

16 Thermodynamics —Heat Engines— Energy Resources

Of all forms of energy, heat energy has probably been the most important one over the past century. Heat energy generates electricity; it powers autos, trucks, trains, ships, aircraft, and spacecraft; and it produces most of the mechanical energy that industry needs.

Machines that utilize heat energy to produce energy of motion are called *heat engines*. The science of converting heat energy into motion is called *thermodynamics.*

The first successful heat engine was a crude steam-and-vacuum engine made in 1705 by Thomas Newcomen (1663–1729), an English blacksmith (see Fig. 16.2). Cumbersome and unreliable though it was, it could do the work of scores of men and many horses. Some 70 years later, James Watt, a Scotsman, improved Newcomen's engine by adding a flywheel and a slide valve. Other improvements followed rapidly and the reciprocating steam engine, by the late 1700s, was ready to power the industrial revolution.

Internal combustion engines and the steam turbine both came into use in the closing decades of the nineteenth century. These heat engines have been undergoing continual modification and improvement right up to the present day. In the last fifty years several new types of heat engines have been introduced and developed to a high level of performance. These include hot-gas turbines, turbojet engines, and rocket motors.

Fig. 16.1 Turbojet engines are good examples of modern heat engines. This model JT8D-209 engine has been designed for the 1980s to offer unusually low noise and air pollution levels, as well as an improved thrust-to-fuel usage ratio. It will deliver 18,500 pounds of thrust at full power. Except for controls, valves, and instrumentation, this engine's components are all in rotary motion. (Pratt and Whitney Aircraft Group of United Technologies.)

The present chapter will discuss all of these engines in some detail, but first it is necessary to develop some basic principles and explain the fundamental processes by which engines use heat to do mechanical work.

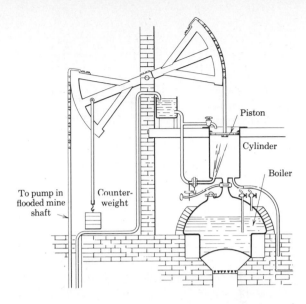

Fig. 16.2 Newcomen's steam-vacuum engine. Its reciprocating motion was used for operating pumps to control flooding in mines. In this engine the counterweight pulled the piston up. Steam was then let into the cylinder, filling it completely. Cold water was then sprayed over and into the cylinder, condensing the steam and creating a vacuum in the cylinder. Atmospheric pressure forced the piston down, pulling the pump plunger (in the mine shaft) up. The various controls and valves were hand-operated.

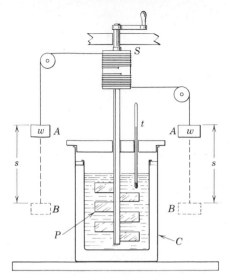

Fig. 16.3 Joule's apparatus (schematic) for determining the mechanical equivalent of heat.

BASIC PRINCIPLES

16.1 The Mechanical Equivalent of Heat

In an earlier chapter, the studies of Count Rumford on the nature of heat were mentioned. He found that heat could be produced indefinitely from mechanical work and suggested that heat was a form of energy. His studies, though carefully performed and well interpreted, did not go far enough to determine the quantitative relationship which exists between heat and mechanical work. It remained for James Prescott Joule (1818–1889), an English scientist, to determine the heat equivalent of mechanical work. Joule performed a famous experiment in 1843 with an apparatus (similar to that shown in Fig. 16.3)

which determined the equivalence between heat and work with considerable precision.

He fitted a calorimeter C with a paddle arrangement P which could be turned by falling weights w, unwinding cords from spindle S. As the weights move from A to B through a distance s, an amount of work $W = 2ws$ is done. This energy is expended by the paddle in churning the water in the calorimeter cup. Joule was successful in showing that as mechanical work was done on the water, its temperature increased. Knowing the mass of water in the calorimeter and its temperature rise, he calculated the amount of heat which resulted from a known amount of mechanical work. His findings, which have been verified by many subsequent workers, resulted in a determination of the *heat equivalent of mechanical work*. The *mechanical equivalent of heat* is defined as *the ratio of mechanical work done to the heat produced*. This ratio is known as *Joule's constant*, symbol J.

Joule's constant $\quad J = W/H \qquad (16.1)$

The presently accepted values for the constant J are

$J = 4.186$ joules/cal

or   4186 joules/kcal   (metric units)

or $J = 778$ ft-lb/Btu   (engineering units)

This is equivalent to saying that if an energy transformation could be effected without loss, 1 kcal would accomplish 4186 joules of work, and 1 Btu of heat would accomplish 778 ft-lb of mechanical work. The conversion of work into heat is continually going on, since friction is never absent when work is being done. Also, heat is continually being transformed into mechanical energy by natural processes and, in modern times, by man-made machines.

Note: *Joule's constant* is denoted by J, while the symbol for the energy unit *joule* is **J**.

Illustrative Problem 16.1   A sledgehammer, mass 2 kg, is moving at 25 m/sec when it strikes a 150-gm steel spike, driving it into a railroad tie. If 40 percent of the impact energy is converted into heat in the spike, calculate its rise in temperature

Solution

$$\text{KE of hammer} = \tfrac{1}{2}mv^2 = 1 \text{ kg} \times (25 \text{ m/sec})^2$$

$$= 625 \text{ kg-m}^2/\text{sec}^2$$

$$= 625 \text{ J}$$

(Review units of the joule, Sec. 5.2.) The energy which goes into heat is

$$W = 0.40 \times 625 \text{ J} = 250 \text{ J}$$

From Eq. (16.1), the heat produced is

$$H = \frac{W}{J} = \frac{250 \text{ J}}{4.186 \text{ J/cal}}$$

from which

$$H = 59.7 \text{ cal}$$

From Eq. (14.2)

$$H = mc(t_2 - t_1)$$

$$\text{Change in temperature} = t_2 - t_1 = \frac{H}{mc}$$

From Table 14.2, the specific heat c for steel is 0.115 cal/(gm)(C°). Substituting gives

$$t_2 - t_1 = \frac{59.7 \text{ cal}}{150 \text{ gm} \times 0.115 \text{ cal/(gm)(C°)}}$$

$$= 3.46\text{C}° \qquad\qquad\qquad ans.$$

16.2   Conservation in Energy Transformations

The conversion of mechanical energy into heat is easily accomplished—in fact, it is inescapable. This conversion is usually an undesirable process, and the heat produced is generally regarded as "wasted" for the purpose in mind. Friction is often the cause of "wasted" energy.

Converting heat to mechanical energy, on the other hand, is not so easily accomplished. This transformation requires a heat engine, and unfortunately, there is no heat engine which can carry out the transformation at anywhere near 100 percent efficiency. Some heat engines are more efficient than others, and reasons for this will be explored in the following pages. It is emphasized, however, that actually no energy is *lost* in either of these processes. The *law of conservation of energy* holds here, as in all other energy transformations. When producing heat from mechanical work, we can write

$$W \qquad = \qquad JH \qquad + \qquad W_w$$

$$\begin{array}{ccc} \text{Mechanical-} \\ \text{energy input} \end{array} = \begin{array}{c} \text{energy con-} \\ \text{verted to heat} \end{array} + \begin{array}{c} \text{"wasted"} \\ \text{work} \end{array}$$

$$(16.2)$$

and when work is obtained from heat, energy conservation is expressed by

$$H \qquad = \qquad \frac{W}{J} \qquad + \qquad H_w$$

$$\begin{array}{ccc} \text{Heat-energy} \\ \text{input} \end{array} = \begin{array}{c} \text{heat con-} \\ \text{verted to work} \end{array} + \begin{array}{c} \text{"wasted"} \\ \text{heat} \end{array}$$

$$(16.3)$$

There is a reasonable degree of certainty that it is impossible to create or destroy energy: Any increase in one form of energy will be accompanied by an equal decrease in some other form of energy. If electric energy is dissipated, for example, heat and/or mechanical energy will result. As nuclear energy is released by fission, thermal energy is the result; and when heat energy is applied to junctions of certain dissimilar metals, electric energy is produced. Countless experiments meticulously performed by the best scientists of the past 300 years lead to the following *general principle*:

Law of Conservation of Energy

Energy can neither be created nor destroyed, but it can be converted or transformed from one form to another.

The law of conservation of energy was perhaps the most comprehensive generalization in all of science prior to Einstein's special theory of relativity and his principle of mass-energy equivalence, $E = mc^2$ (see Chap. 33). Even so it is still possible to retain the principle of conservation of energy by broadening it to express the *principle of conservation of mass-energy*.

The following problem illustrates the conversion of heat to mechanical work.

Illustrative Problem 17.1 A compression re-bine generates 50,000 hp and has an overall efficiency of 22 percent. If the fuel oil used has a fuel value of 19,000 Btu/lb, calculate the fuel-oil consumption per hour.

Solution Determine the amount of mechanical work done in 1 hr, in foot-pounds, as follows:

$$W = 50,000 \text{ hp} \times \frac{33,000 \text{ ft-lb/min}}{1 \text{ hp}} \times \frac{60 \text{ min}}{1 \text{ hr}}$$

$$= 9.9 \times 10^{10} \text{ ft-lb/hr}$$

From Eq. (16.1),

$$H = \frac{W}{J} = \frac{9.9 \times 10^{10} \text{ ft-lb/hr}}{778 \text{ ft-lb/Btu}}$$

$$= 12.7 \times 10^7 \text{ Btu/hr}$$

which would be the heat input quantity required if the boiler-turbine system were 100 percent efficient. The oil *actually* consumed, however, amounts to

$$\frac{12.7 \times 10^7 \text{ Btu/hr}}{19,000 \text{ Btu/lb} \times 0.22} = 30,400 \text{ lb/hr} \quad ans.$$

Illustrating the reverse process, the conversion of mechanical work to heat, consider the following problem.

Illustrative Problem 16.3 A 1400-kg automobile traveling at 80 km/hr is brought to rest by application of the brakes. Assuming that one-half the total heat produced goes into heating the iron brake drums (four, each weighing 4 kg), what is their temperature rise?

Solution The kinetic energy of the auto is, by Eq. (5.4),

$$KE = \tfrac{1}{2}mv^2$$

$$= \frac{1400 \text{ kg}}{2}\left(80\,\frac{\text{km}}{\text{hr}} \times 1000\,\frac{\text{m}}{\text{km}} \times \frac{1}{3600\,\dfrac{\text{sec}}{\text{hr}}}\right)^2$$

$$= 34.6 \times 10^4 \text{ kg}\,\frac{\text{m}^2}{\text{sec}^2}$$

$$= 34.6 \times 10^4 \text{ N-m (or joules)}$$

[since 1 N is 1 (kg-m)/sec^2]. This amount of mechanical energy is converted entirely to heat, the amount of which is, from Eq. (16.1),

$$H = \frac{W}{J} = \frac{34.6 \times 10^4 \text{ J}}{4186 \text{ J/kcal}} = 82.6 \text{ kcal}$$

To determine how much the brake drums will be heated, note from Table 14.2 that the specific heat of iron is 0.115 kcal/(kg)(C°); and remembering that only one-half the total heat is assumed to go into heating the four brake drums, write the basic heat equation

$$H = mc(t_2 - t_1)$$

Solving for the temperature rise, we obtain

$$t_2 - t_1 = \frac{H}{mc} = \frac{\tfrac{1}{2} \times 82.6 \text{ kcal}}{4 \times 4 \text{ kg} \times 0.115 \text{ kcal/(kg)(C°)}}$$

Temp rise $= 22.4$ C° \qquad *ans.*

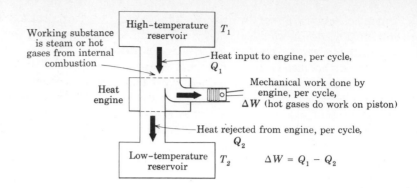

Fig. 16.4 Block diagram illustrating the first law of thermodynamics.

16.3   The Laws of Thermodynamics

These transformations of energy involve producing heat from motion or the reverse—motion from heat. Consequently, the name *thermodynamics* (literally, "motion from heat") is the technical term applied to this branch of physics. The science of thermodynamics is based on two fundamental laws, both of which have tremendous significance for industry and everyday living.

The First Law of Thermodynamics   This law is actually a restatement of the law of conservation of energy for the special case involving heat energy:

> *In any thermodynamic process, total energy remains constant—none is created or destroyed.*

The meaning of the first law of thermodynamics can be illustrated by a schematic diagram. In Fig. 16.4 a heat engine containing a working substance (usually a gas, such as steam, or the products of internal combustion) is shown operating between a high-temperature reservoir and a low-temperature reservoir. It is assumed that a supply Q_1 of heat energy is available at temperature T_1 from the high-temperature reservoir for each cycle. After an amount of work ΔW is done by the engine in that cycle, an amount of heat Q_2 is rejected to the low-temperature reservoir at temperature T_2. The law of conservation of energy, in this case the *first law of thermodynamics*, states that

$$\Delta W = Q_1 - Q_2 \qquad (16.4)$$

(First Law of Thermodynamics)

Or, stated another way, the mechanical work done by a heat engine in one cycle is equal to the difference between the heat energy taken in at the higher temperature and the heat energy rejected at the lower temperature. T_1 and T_2 are *absolute temperatures.*

The Second Law of Thermodynamics   This law has to do with the *availability* of heat energy for the purpose desired and states:

> *Heat will not flow "up a temperature hill" unless mechanical work is expended to force it to do so.*

A somewhat simplified statement of the second law is:

> *Mechanical energy cannot be obtained from any source of heat unless that source undergoes a drop in temperature.*

Thus, in an engine, a hot substance flows from a hot cylinder to a cooler exhaust and does mechanical work as the temperature drops. (A refrigerator, which is a heat engine in reverse, takes heat from a low-temperature source and carries it to a place of higher temperature, but *mechanical work must be done on the working substance by the compressor in the process.*) (See Chap. 17.)

The *first law* of thermodynamics tells us that energy is conserved in all processes *that actually* take place, but it does not tell us *whether or not a specific process will take place*. When a baseball's kinetic energy (at about 90 ft/sec velocity) is absorbed by a catcher's mitt, much of the energy of motion shows up as heat, which raises the temperature of the mitt. The first law is obeyed. But no one has ever observed the reverse process, which, *if it did happen*, would also be consistent with the first law. Sudden cooling of a catcher's mitt does not expel a baseball at 90 ft/sec. By the same token, a jar of hot water does not rise into the air gaining potential energy as it cools and freezes, even though there is sufficient energy in the water to supply the mechanical work needed for such a feat. An automobile is brought to a stop from high speed by the absorption of its kinetic energy by the braking system, with a consequent sudden heating of the brake drums and shoes. Luckily for us the reverse does not happen: we do not sharply accelerate in the auto as a result of the braking system's sudden surrender of heat energy with consequent frost formation on the brake drums.

These examples of what *does not happen* seem to say that there is some kind of natural law which stipulates the *direction* in which thermodynamic processes can take place. Analysis of the many processes theoretically possible for isolated systems under the first law reveals that only those which move in a direction *from an orderly state toward a more disorderly state* will actually occur. Steam or gas molecules in a confined chamber at high pressure represent an orderly arrangement of high-speed molecules, under control and available to do work. The exhaust steam or the spent gases from an internal-combustion engine and the "wasted heat" which heats up the engine parts—these are disorderly, more random, manifestations of thermal energy at lower temperatures, and we cannot put these lower levels of energy back through the engine and get useful work.

As another example, there is an incalculable amount of energy (measured above zero K) stored in the waters of the ocean, but its temperature is low compared to its surroundings, and up

to now we have not developed effective devices for tapping this vast energy source to operate practical engines. The heat energy is there, but it has been relatively *unavailable*, at least by past concepts of "engine" design. Pilot plants are presently in the design stage, however, to use ocean temperature gradients for energy production. One design will use ammonia as the working substance. Warm ocean layers near the surface will boil the ammonia (boiling point $-28°F$, at 1 atm), and the resulting vapor will drive a low-pressure "vapor engine." The ammonia vapor exhaust from the engine will be pumped down to cold layers of the ocean where it will be condensed back to a liquid and then be brought up for another cycle.

The second law of thermodynamics says that energy processes always move from order to disorder, and that heat flows of its own accord from high temperature to low temperature. Three men of science who devoted much of their lives to the study of heat and thermodynamics have made separate but equivalent statements of the second law, which are summarized here:

The Second Law of Thermodynamics

As formulated by Sadi Carnot (1796–1832):

It is impossible to construct an engine which will convert a given quantity of heat energy into an equivalent amount of work.

As formulated by Rudolph Clausius (1822–1888):

It is impossible for any self-operating device to take heat continuously from a reservoir at one temperature and deliver it to a reservoir at a higher temperature.

As formulated by Lord Kelvin:

As a result of natural processes, the world's supply of energy available to do work is continually decreasing.

16.4 Entropy

The thermodynamic property of a system which measures its degree of disorder is called *entropy*. Entropy may be thought of as a measure of the degree of energy degradation in any process. The second law of thermodynamics is sometimes called the *law of entropy*. Clausius defined entropy mathematically as follows: If ΔQ is a small reversible change in the heat content of a body, and if the change takes place at temperature T, then the *change in entropy* is

$$\Delta S = \frac{\Delta Q}{T} \qquad (16.5)$$

Change in entropy = heat gain (or loss) divided by absolute temperature

Natural phenomena always occur (in isolated systems) in such a direction that entropy (the state of disorder or *unavailability* of energy) *increases*. Only in an idealized reversible process can ΔS be zero. Referring to Fig. 16.4, we see that heat from a reservoir of hot gas (temperature T_1) is allowed to flow into a reservoir of "cold" gas (temperature T_2), with an engine in the path of the heat flow converting some of the heat to mechanical energy. Suppose no engine had been there. The high-temperature gas would then merely get cooler, and the low-temperature gas would get warmer until the mixture reached some equilibrium temperature. There is no decrease in *total energy* in this second process (First Law), but the opportunity to convert some of the heat in the "hot" reservoir to mechanical work has been lost, not temporarily, but forever (Second Law). The mixture will not, by itself, separate back into hot gas and cold gas. In this process, entropy will have increased in accordance with Eq. (16.5), and *availability* of energy for conversion to mechanical work will have decreased.

The natural condition throughout the universe seems to be one which tends toward "running down," and a final state at which the entire universe will finally come to the same temperature. Entropy will then be at a maximum, and theoretically all thermodynamic processes (including life processes, of course) would end. This so-called "heat death" of the universe is assumed to be many billions of years in the future, however.

16.5 Basic Principles of Heat Engines

A material substance must be employed in the transformation of heat to work accomplished by a heat engine. Ordinarily, a gas is heated and allowed to do work as it expands against a piston or in a turbine casing. In reciprocating steam engines, superheated steam expands and drives the piston down the cylinder. As the volume increases, the steam pressure decreases and its temperature drops. In steam turbines, high-pressure steam shoots through nozzles against blades on the turbine rotor, and the impulse and reaction forces whirl the rotor at high speed. In internal-combustion engines, hot gases are produced by high-temperature burning within the cylinder itself, and again the low-temperature exhaust indicates that mechanical work has been done. All heat engines which do work absorb heat at a high temperature, turn some of it into mechanical work, and exhaust or reject the rest at a lower temperature. In any case, $\Delta W = Q_1 - Q_2$.

Reciprocating engines must return the piston to the starting point. As the piston returns, the gas in the cylinder must be compressed. Work has to be done *by the piston* on the gas to accomplish this compression. The sequence of expansion and compression is called a *cycle*. During a cycle, work is done *by the gas* during expansion, and a smaller amount of work has to be done *on the gas* during compression. For an ideal engine, the difference between the amount of work done on the piston by the expanding gas and the amount of work done by the piston on the gas to compress it is the net work accomplished.

16.6 Cyclic Processes Involved in Heat Engines—Isothermals and Adiabatics

It will be recalled (Sec. 13.20) that an *isothermal process* is one in which the pressure and volume of a gas vary *without change in the temperature*. In order to keep temperature constant, heat must be allowed to flow in and out of the gas as its pressure and volume vary. Under these con-

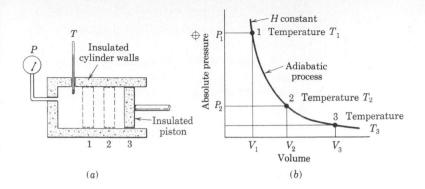

Fig. 16.5 A diagrammatic explanation of an adiabatic (constant-heat) process. (*a*) The piston-cylinder relationship. (*b*) A *P-V* diagram showing temperature changes, but indicating no heat flow.

ditions, Boyle's law of gases is obeyed (see Fig. 13.16).

If a gas is expanded or compressed (*P* and *V* changing) *without permitting heat to flow* from or to the gas, the temperature *will not stay constant*, and Boyle's law does not hold true. This kind of process, where there is no flow of heat into or out of the system, is known as a constant-heat, or *adiabatic*, process. Adiabatic expansion and compression are illustrated in Fig. 16.5.

An insulated cylinder (theoretically insulated so well that no heat is lost or gained during a complete cycle) is fitted with an insulated piston and a sensitive pressure gauge and thermometer (Fig. 16.5*a*). A scale along the cylinder enables the volume to be determined for any piston position. Let the cylinder contain a given mass of gas at temperature T_1 (degrees Rankine), pressure P_1 (pounds per square inch absolute, psia), and volume V_1 (cubic inches) when the piston is in position 1. This situation corresponds to point 1 on the graph of Fig. 16.5*b*. (NOTE: Heat-power engineering in the United States is still using the English-engineering system of units.)

As the piston is pushed to position 2, an adiabatic (constant-heat) expansion takes place, the pressure drops to P_2, and the temperature drops to a lower value T_2. Moving the piston to position 3 lowers the pressure and temperature still further and results in the plotted point marked 3 on the graph. A series of such points

would plot to the smooth curve shown. This curve is known as an *adiabatic*.

An *adiabatic process* for an ideal gas obeys the *general gas law* (Sec. 13.21), which says that the pressure of a given quantity of a gas is proportional to the absolute temperature, and inversely proportional to the volume. As an equation, this relation may be expressed as

$$\frac{PV}{T} = k \qquad (13.17)$$

Note the contrast with the equation for an *isothermal* (Eq. 13.12), in which the product of pressure and volume is a constant ($PV = k$).

Heat engines make use of processes which roughly approximate isothermals and adiabatics in the compression and expansion of gases. The two processes will now be compared and plotted with respect to the same *P-V* coordinates.

In Fig. 16.6 let *ic* indicate the direction of an isothermal compression (*V* decreases as *P* increases, temperature being held constant). The work of compression will produce heat, and the only way a constant temperature can be held is for the heat to be liberated as process *ic* occurs. *Isothermal compression is a process in which mechanical work is done on the gas and heat is given off, while temperature remains constant.*

On isothermal expansion (direction *ie*), the exact opposite is the case: mechanical work is done *by the gas* on the piston, and *heat must be supplied* if temperature is to remain constant.

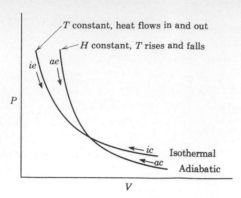

Fig. 16.6 Isothermal and adiabatic curves plotted with respect to the same set of *P-V* coordinate axes.

Referring to the adiabatic curve, we note that *ac* indicates the direction of adiabatic compression. As work is done *by the piston* on the gas to compress it, the heat equivalent of mechanical work evidences itself as an increase in temperature. *During adiabatic expansion (process ae), work is done by the gas on the piston and a drop in temperature occurs, without an exchange of heat.*

16.7 The Ideal Heat Engine—Carnot's Cycle

The French physicist Sadi Carnot, in 1824, proposed an ideal engine operating on a cycle (since called the *Carnot cycle*) which has the highest possible efficiency for the temperature limits within which it operates. His ideal engine was to operate on a cycle of four heat processes, in the following manner.

A theoretically frictionless one-cylinder engine (see Fig. 16.7) is assumed to have perfectly *nonconducting* walls, a *nonconducting* piston, and a *perfectly conducting* cylinder head. The cylinder is assumed to contain an ideal gas as the *working substance*. The box S_1 represents a source of heat from which heat at a high temperature T_1 (°R) can be obtained, and the box S_2 is a receiver of heat, or *sink*, to which heat at a low temperature T_2 can be rejected. The box I is an insulating stand. The four diagrams of Fig. 16.7 show the four stages in Carnot's ideal cycle, and the graph of Fig. 16.8 shows the resulting P-V-T conditions for a complete cycle.

An isothermal expansion is illustrated in Fig. 16.7a. At the start the cylinder is assumed to be standing on the heat source S_1, with the piston at A near the bottom of the cylinder. The gas in the cylinder is at temperature T_1 and it occupies a small volume at high pressure, as indicated by point A on the P-V diagram of Fig. 16.8. A quantity of heat Q_1 flows into the cylinder and the gas expands while the piston moves to B. The temperature stays constant at T_1 during this stage, since heat is flowing in from source S_1 to maintain the temperature at the same level. The isothermal expansion curve AB of Fig. 16.8 represents this stage of the cycle.

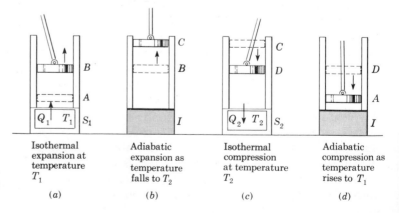

Fig. 16.7 Schematic diagram of the four processes of Carnot's ideal heat-engine cycle.

(a) Isothermal expansion at temperature T_1

(b) Adiabatic expansion as temperature falls to T_2

(c) Isothermal compression at temperature T_2

(d) Adiabatic compression as temperature rises to T_1

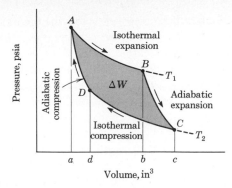

Fig. 16.8 The Carnot cycle represented by two isothermals and two adiabatics. The area enclosed by the four curves is the net work accomplished per cycle, ΔW.

Next the cylinder is placed on top of the insulator box I, and the gas is allowed to expand further, moving the piston to "top dead center" at C. During this second stage of the expansion no heat leaves or enters the cylinder and work is done on the piston as the temperature of the gas in the cylinder falls to a value T_2. The process is illustrated on the P-V diagram by the adiabatic BC.

The cylinder is now placed on the sink S_2, whose temperature is cold enough so that heat will flow into it from the cylinder. The piston is now thrust downward to D, and the heat equivalent of this mechanical work Q_2 flows into the sink. The temperature of the gas in the cylinder remains constant at T_2. This first phase of the compression is illustrated by the isothermal compression curve CD of the P-V diagram.

Finally, the cylinder is placed on the insulating stand once more, and the piston thrust downward to A. More work is thus done on the gas, and since no heat can escape, a rise in temperature is the result. The final P-V-T conditions are the same as those at the start, and the second stage of the compression is illustrated by the adiabatic DA on the P-V diagram.

During the complete cycle just described an amount of heat Q_1 flows into the cylinder at the high temperature T_1, and an amount Q_2 is exhausted at the lower temperature T_2. External work is done on the piston by the expanding gas during the expansion phase ABC, and some work is done on the gas by the piston during the compression CDA. The external work accomplished during expansion is represented by the area $ABCca$, and the energy given back during compression by the area $ADCca$. The net work ΔW accomplished in one complete cycle is given by $ABCca - ADCca$, which is the shaded area $ABCD$ within the closed curves. This is the same as writing $\Delta W = Q_1 - Q_2$.

16.8 Efficiency of the Ideal Engine

On the basis of these considerations, and using the concept of absolute temperature, Carnot was able to derive an expression for the ideal efficiency of any heat engine operating between the temperature limits T_1 and T_2.

The efficiency of *any machine* was given in Chap. 5:

$$\text{Eff} = \frac{\text{work output}}{\text{energy input}} \qquad (5.15)$$

For a heat engine

$$\text{Eff} = \frac{\text{work output}}{\text{mech. equiv. of heat input}} = \frac{\Delta W}{Q_1}$$

From the first law of thermodynamics

$$\Delta W = Q_1 - Q_2$$

Consequently

$$\text{Eff} = \frac{Q_1 - Q_2}{Q_1} = 1 - \frac{Q_2}{Q_1} \qquad (16.6)$$

In the ideal (Carnot) engine, the heat transferred to or from the working substance is directly proportional to the absolute temperature of the reservoir which supplies or receives the heat, or

$$\frac{Q_2}{Q_1} = \frac{T_2}{T_1}$$

Therefore Carnot efficiency (ideal engine) is

$$\text{Carnot eff} = 1 - \frac{T_2}{T_1} = \frac{T_1 - T_2}{T_1} \qquad (16.7)$$

where T_1 = absolute temperature of hot gas supplied to engine

T_2 = absolute temperature of (cooler) gas exhausted from engine

A real engine cannot possibly approach Carnot's theoretical cycle of operation, and actual operating engines are always much less efficient than Eq. (16.7) would indicate. Friction is always present, for one thing, and actual cycles only approximate the theoretical isothermal and adiabatic processes described above.

Equation (16.7) shows that the efficiency of a heat engine increases as the difference between T_1 and T_2 increases; or, an engine is the more efficient, the greater the difference between the temperature of the hot gas supplied (steam or gases from internal combustion) and the temperature of the exhausted gas. Exhaust temperature cannot be lower than that of the medium to which the working substance is exhausted (the atmosphere, or perhaps water). Since T_2 cannot be lowered at will by the engine designer or operator, the most feasible way to get higher efficiency is to make T_1 as high as possible. (T_2 *can be lowered* within limits by the use of condensers—see below.)

Illustrative Problem 16.4 A steam engine is supplied saturated steam at a pressure of 250 lb/in.^{2} abs and exhausts it into the atmosphere. What is the "ideal" Carnot efficiency of this engine?

Solution From Table 14.6 note that *saturated* steam at 250 lb/in.^{2} abs is at a temperature of 401°F, and at 14.7 lb/in.^{2} abs its temperature is 212°F. Changing these to Rankine temperatures and substituting in Eq. (16.7), we have

$$\text{Carnot eff} = 1 - \frac{672}{861} = 1 - 0.78 = 0.22 \text{ or } 22\%$$

ans.

Operating between these temperature limits, even the ideal engine could convert into mechanical energy no more than 22 percent of the heat energy supplied. Real engines, as pointed out

above, have efficiencies well below the theoretical Carnot efficiency. However, *real* engines operate on *superheated* steam rather than saturated steam, so that T_1 can be very high, and efficiency can therefore be increased somewhat.

STEAM ENGINES

Heat engines designed to use steam as the working substance are of two types, *reciprocating engines* and *turbines*. In both cases the complete steam plant involves a *boiler* in which to produce steam, the engine itself, and a *condenser* into which the exhaust steam is discharged to save both the water and the heat which still remain in the spent steam. Condensers can be operated in such a manner that they create a partial vacuum at the exhaust exit and lower the temperature T_2 appreciably. This increases efficiency.

16.9 Reciprocating Steam Engines

The *basic principles* of operation of the single-cylinder slide-valve reciprocating steam engine have not changed a great deal since the time of James Watt. The engine can be represented diagrammatically, as in Fig. 16.9. Steam enters the port L, which is opened at the proper instant by the slide valve S. The hot steam expands, pushing the piston P toward the right end of the cylinder

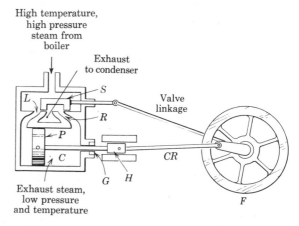

Fig. 16.9 Diagram of the slide-valve reciprocating steam engine.

C. As the piston moves to the right, the port *R* is also open, and the spent steam from the previous stroke exhausts to the condenser. The piston is joined to the crosshead *H* by the piston rod and thence to the crank of the flywheel by the connecting rod *CR*. The flywheel *F* is mounted on the main shaft of the engine, and so also is an eccentric, or cam, which operates the slide-valve linkage. The slide valve closes the intake port about one-third of the way through the stroke, and the stroke is completed by the expansion of the steam which is already in the cylinder. As the expanding steam does work on the piston, the steam temperature falls. At the end of the stroke, the slide valve interchanges the connections so that port *R* is now open to incoming live steam and port *L* is connected to the exhaust. Thus, a continual reciprocating motion of the piston is maintained, which is converted into rotary motion of the flywheel and shaft by the crankshaft and crank.

Such engines are frequently called *double-acting* engines, since work is done on the piston by the expanding steam from both ends of the cylinder. The piston rod enters the cylinder through a *packing gland G*, which must be carefully made and frequently inspected to ensure that steam does not leak out around the piston rod and that the packing is not tight enough to cause undue friction on the rod.

Superheated steam is supplied to such an engine from a *boiler* or *steam generator*. The spent steam exhausts to a *condenser* (usually water-cooled) which is operated at a sufficiently low temperature that steam condenses rapidly enough to create a partial vacuum at the exhaust valve ports. Both the partial vacuum and the low exhaust temperature tend to increase efficiency.

These processes are illustrated in simplified form by the *P-V* diagram of Fig. 16.10. This idealized cycle (the Rankine cycle) begins with liquid water coming from the condenser at point *A* on the diagram—low pressure and temperature. Line *AB* represents the water being pumped, under pressure, into the boiler. Line *BC* represents heating the water at constant pressure to

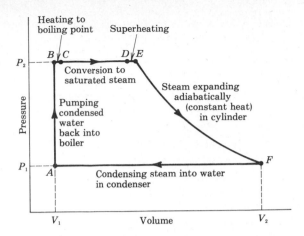

Fig. 16.10 The idealized steam cycle, or Rankine cycle, for a condenser-equipped reciprocating steam engine.

the boiling point; line *CD*, heating and evaporation to saturated steam; the line *DE*, further heating and expansion to superheat the steam. The superheated steam then enters the cylinder and expands adiabatically, doing work on the piston (*EF*). The line *FA* represents cooling and condensation at reduced pressure in the condenser, to complete the cycle. Water is the working substance.

The advantage of a condenser can be shown by recalculating Illustrative Problem 16.4, this time assuming that the steam exhausts from the cylinder at point *F* into a condenser at 110°F, instead of blowing to the atmosphere as saturated steam at 212°F. For the condenser-equipped engine (all other conditions as in Illustrative Problem 16.4),

$$\text{Carnot eff} = 1 - \frac{570}{861}$$

$$= 1 - 0.662$$

$$= 0.338, \quad \text{or 34 percent} \quad ans$$

The condenser thus increases the efficiency by over 10 percent.

It is again pointed out that real engines do not attain efficiencies predicted from the theoretical Carnot cycle.

Thermodynamics—Heat Engines—Energy Resources   **319**

16.10 Horsepower Developed by Reciprocating Steam Engines

We have seen from the analysis of the Rankine cycle that pressure is not constant during the stroke. However, an average or *mean effective pressure* (MEP) for the stroke can be determined. If MEP is multiplied by piston area, the average force is obtained. This force times the length of stroke gives the work done per stroke. Recalling that *power is work per unit time* and that 1 hp equals 33,000 ft-lb/min, we write

$$\underset{\text{(reciprocating engine)}}{\text{hp}} = \frac{PLAN}{33{,}000} \qquad (16.8)$$

where P = MEP, lb/in.^{2}
L = length of stroke, ft
A = piston area, in.^{2}
N = number of power strokes per minute

Illustrative Problem 16.5 Calculate the horsepower being developed by a double-acting reciprocating steam engine making 250 rev/min if piston diameter is 10 in., stroke is 22 in., and MEP is 240 lb/in.^{2}.

Solution Substitute in Eq. (16.8), being careful to use the units indicated and remembering that there are two power strokes per revolution in a double-acting engine.

$$\text{hp} = \frac{240 \text{ lb/in.}^2 \times \frac{22}{12} \text{ ft} \times \pi}{\times 25 \text{ in.}^2 \times 2 \times 250 \text{ rev/min}} \over (33{,}000 \text{ ft-lb/min})/1 \text{ hp}$$

$$= 523 \text{ hp} \qquad \qquad \textit{ans.}$$

16.11 The Steam Turbine

Since ancient times it has been known that continuous rotational motion from steam was possible. At some time between 150 B.C. and A.D. 250 (scholars disagree on the dates of his life) a Greek scientist of Alexandria, Hero (or Heron), made and demonstrated a toy steam engine operating as a *reaction turbine* (see Fig. 16.11). A simple laboratory demonstration will serve to

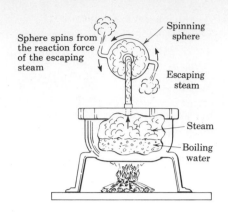

Fig. 16.11 Hero's engine. This first steam turbine ran on reaction principles. (General Electric Co.)

show the basic principles of another type of steam turbine—the *impulse turbine*. Boil water in a flask and allow the steam to issue at high velocity from a nozzle made from a bent glass tube. Let the steam jet strike a bladed wheel made of cardboard, as shown in Fig. 16.12. The heat contained by the steam makes for extremely high-speed, high-energy molecules, which hit the blades of the rotor. The loss in momentum of the steam molecules on impact with the vanes is imparted to the vanes of the rotor as *impulse*, in accordance with the impulse equation $Ft = mv_2 - mv_1$.

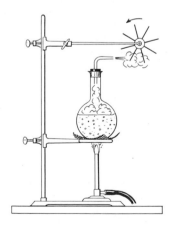

Fig. 16.12 Laboratory setup to illustrate an impulse turbine.

Table 16.1   Efficiencies of typical heat engines

Engine	Percent Efficiency
Reciprocating steam engine, noncondensing	6–8
Condenser-equipped	12–16
Steam turbine	Up to 35
Gasoline engine	Up to 28
Industrial gasoline engine (heavy)	30–35
Industrial engine (natural gas)	35
Hot-gas turbine	20
Diesel engine	30–38

An English engineer, Sir Charles Parsons (1854–1931), was the first to construct a commercially successful engine based on reaction principles (about 1884). Gustaf de Laval and Charles Curtis, beginning about 1896, made a series of improvements on Parsons' original design, the most important of which was to combine the impulse principle with the reaction principle in the same turbine. As a result of this and other refinements, the steam turbine is today among the most efficient of heat engines, as Table 16.1 shows.

16.12   Details of Turbine Construction

The essential parts of a steam turbine are the *rotor* and the *casing*. The rotor sections "mesh" with the casing sections (see Fig. 16.13), and each section is fitted with curved blades, or *buckets*. The blades on the casing sections are oppositely curved from those on the rotor sections, so that they will redirect the steam to strike the next set of rotor blades at the correct angle (see diagram of Fig. 16.14). The casing, of course, is steamtight so that no steam escapes until it has expanded through the entire set of "stages." Note that the rotor blades are curved in such a manner that the steam which strikes them has to undergo a partial reversal of direction. Consequently, in addition to the *action* of the impulse, there is *reaction* against the blades as the steam caroms off. Modern turbines, for this reason, have come to be called *impulse-reaction turbines*, and this single refinement of design increased the efficiency from a top of about 25 percent up to the present 35 percent.

Fig. 16.13   Steam turbine for generating electric energy, being assembled at the factory. The rotor has 16 sections or stages, each of which meshes with a section of the stator (housing). Steam at 1000 degrees F or more enters the first stage at pressures up to 3500 lb/in.² The diameter of the largest stage of the rotor may be 14 ft or more, resulting in very high linear speed of the blade tips. (General Electric Company.)

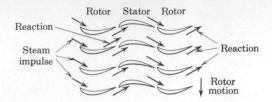

Fig. 16.14 Turbine bucket arrangement showing steam flow. Both impulse and reaction forces are involved in turning the rotor. The stator blades redirect the steam so that it will strike the next set of rotor blades at the proper angle.

16.13 Horsepower Developed by Steam Turbines

Theoretically the energy output of a steam turbine is the difference between the kinetic energy of the steam entering the turbine and that leaving it. If, in t sec, w lb of steam enters stage 1 at v_1 ft/sec and emerges from the final stage at v_2 ft/sec, then the reduction in kinetic energy is

$$\frac{wv_1^2}{2g} - \frac{wv_2^2}{2g}$$

Since this represents the energy absorbed by the turbine in t sec, the theoretical power output is (factoring to simplify):

$$P = \frac{w(v_1^2 - v_2^2)}{2gt} \text{ ft-lb/sec}$$

and the theoretical horsepower is

$$\text{hp}_{\text{(steam turbine)}} = \frac{w(v_1^2 - v_2^2) \text{ ft-lb/sec}}{2gt \times 550 \text{ ft-lb/sec-hp}} \quad (16.9)$$

16.14 Energy for Steam Plants

Currently, and for the foreseeable future, a major share of the electrical power requirements of the United States must come from steam-turbine generating plants. Natural gas and fuel oil have been the preferred energy sources for several decades but their impending scarcity and ever-rising costs are dictating an early change back to coal, which was the standard fuel 50 or more years ago. Relatively smaller amounts of electric power are

Fig. 16.15 A geothermal experimental facility in the Imperial Valley of California. The present capacity of this geothermal plant is only 10 megawatts, but geologists specializing in geothermal resources estimate that there are hundreds of geothermal reservoirs in the U.S. with a potential of extremely large amounts of electrical power. One plant in Sonoma County in California is already producing nearly 500 megawatts of power—enough to supply all the electricity needs of a city the size of San Francisco. (U.S. Department of Energy.)

produced by hydroelectric and gas-turbine plants (see Chap. 31). *Geothermal power*, using steam from high temperature zones deep in the earth, is an operating actuality, but suitable thermally-active zones are not numerous in the United States (Fig. 16.15). Nuclear *fission* reactors (see Chap. 34), using enriched uranium or plutonium "fuels," could probably supply up to 50 percent of the energy for steam plant operation by 1990, but recent political and environmental concerns, as well as a limited supply of uranium ore, make the future for nuclear fission energy rather uncertain. Thermonuclear *fusion* (Chap. 34) holds great promise for the distant future, but 20 or 30

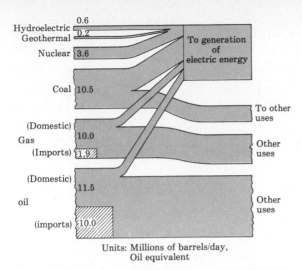

Units: Millions of barrels/day,
Oil equivalent

(*Source:* The National Petroleum Council and the U.S. Department of the Interior, as reported in *EXXON USA*, First Quarter, 1974.)

Fig. 16.16 Energy flow chart showing the 1980 expected energy flow from various sources into the production of electric energy. The estimates were made in the mid-1970s and the contribution from nuclear and coal-fired plants was probably over-estimated. (See Chapter 34 for an explanation of difficulties with nuclear plant expansion.)

years may pass before its promise becomes a reality.

All indications are that for the next 20 years at least, coal will have to be the primary fuel for steam-generating plants (see Fig. 16.16). There are serious problems also in mining, transporting, and burning coal within acceptable environmental standards. The solution to these and other energy-related problems will be a challenge to scientists, engineers, and technicians (and also for economists and politicians) for decades to come.

INTERNAL–COMBUSTION ENGINES

For steam engines of all types the *working substance* (steam) is heated in a boiler and carried to the cylinder by pipes. The products of combustion are wasted up the flue, or "stack," as it is called. In the internal-combustion engine, however, the burning takes place within the engine itself, and the products of combustion, i.e., the hot gases, are the actual working substance. The chemical energy of the fuel is thus converted into kinetic energy of a moving piston or turbine rotor. Internal-combustion engines may be classified as (1) *reciprocating*, (2) *turbine*, and (3) *jet*, or *rocket*. The reciprocating engine will be discussed first.

16.15 Reciprocating Engines

Such engines may have one or many cylinders, and there are at least four common arrangements of the cylinders. They are designed to burn many different fuels, but gasoline, diesel oil, and natural gas are the most common. Two different cycles of operation are possible, and many specialized types of engines are manufactured for widely differing demands. We shall begin by analyzing the operation of a simple one-cylinder, four stroke-cycle gasoline engine.

The Four-stroke Cycle The four-stroke cycle is called the *Otto cycle*, for the German engineer who built the first successful engine of this type. Figure 16.17 shows in diagrammatic form the basic components and cycle of operation of an engine operating on a four-stroke cycle. Air and gasoline vapor, previously mixed in the proper proportions by a carburetor, enter the cylinder through the intake valve I, shown open, in diagram (*a*). The exhaust valve E is closed. The piston P is starting down in the cylinder and is connected to the crankshaft CS by the connecting rod CR. Note that there is no crosshead or piston rod and that the cylinder is open to the crankcase at the bottom. This downward stroke with the intake valve open is called the *intake stroke*. Diagram (*b*) shows both valves closed and the piston starting upward. On this stroke the fuel-air mixture in the cylinder is compressed to about one-eighth its full-cylinder volume, depending on the design of the particular engine, and the sudden pressure increase causes an appreciable temperature rise. This stroke is the *compression stroke*. In diagram (*c*) a spark from the ignition system has jumped the gap of the spark plug S, and

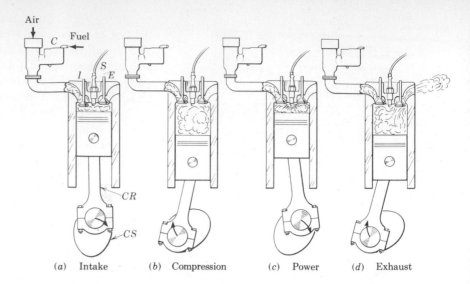

| (a) Intake | (b) Compression | (c) Power | (d) Exhaust |

Fig. 16.17 The four-stroke cycle of an internal-combustion engine. *C*, carburetor; *I*, intake valve; *E*, exhaust valve; *S*, spark plug; *P*, piston; *CR*, connecting rod; *CS*, crankshaft. (General Motors Corp.)

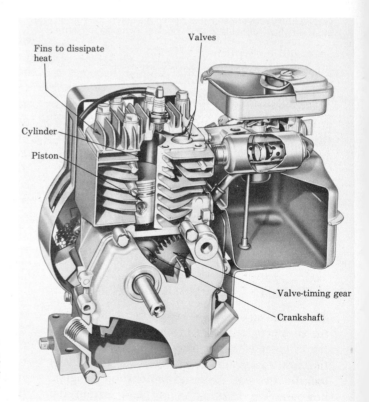

Fig. 16.18 Cutaway view of one-cylinder, air-cooled gasoline engine for four-cycle operation. Such engines are used by the millions to power small tractors and electric generating plants, and for other machines with low power requirements. (Briggs and Stratton Corp.)

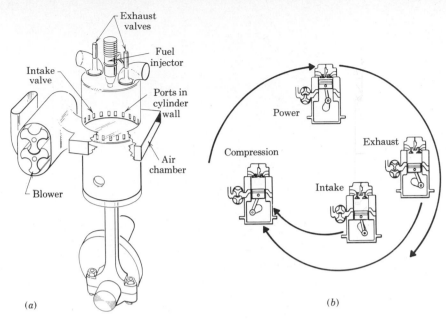

Fig. 16.19 The two-stroke cycle of a diesel engine.

combustion has begun. Modern fuels do not *explode* but burn progressively through most of the stroke. The rapidly expanding gases force the piston down on the *power stroke*. Both valves are closed. Diagram (*d*) shows the piston starting back up again after passing bottom dead center. The exhaust valve has opened, and the rising piston will force out most of the spent gases. This is the *exhaust stroke*, and, upon its completion, the cycle will begin again. The cycle is, then, *intake*, *compression*, *power*, *exhaust*. Only a part of the heat of combustion of the fuel goes into mechanical work. Much of it is "wasted" in heat conducted through the cylinder walls. The ideal Carnot efficiency is not approached by such engines. Fig. 16.18 shows a one-cylinder, air-cooled engine for four-cycle operation.

The Two-stroke Cycle　The four-stroke cycle has the disadvantage of producing only one power stroke per cylinder for every two revolutions of the engine shaft. This deficiency is serious where a lightweight engine is desired. To increase the horsepower-per-pound ratio, the two-stroke-

cycle engine has been developed. This cycle gives a power stroke every time the piston goes down, or one power stroke per revolution. The two-stroke cycle has only a *power stroke* and a *compression stroke*. In between these, the functions of *exhaust* and *intake* must be performed. There are several different methods in common use for achieving the two-stroke cycle. Two-stroke engines are used for outboard motors, motorcycles, small electric generators, chainsaws, pumps, truck refrigerators, and the like. Many diesel engines are two-cycle engines (see Fig. 16.19).

16.16　Engines for Transportation

Early in the development of the automobile, two- and four-cylinder engines were very common. Then for many years six-cylinder and eight-cylinder engines seemed the most satisfactory for the average motorist's needs. Current practice is moving rapidly toward small four- and six-cylinder engines again, in order to increase fuel economy in response to high fuel prices. Engines

Fig. 16.20 Typical four-stroke cycle automobile engine. This is a small, 2.3 liter, 4-cylinder engine for "subcompact" cars. (Ford Motor Co.)

in the "sub-compact" and "minicompact" cars range from 150 bhp down to 40 bhp. (See Fig. 16.20.) Large V-8 engines, developing up to 400 bhp still power a few big cars. Some autos are powered by diesel engines. Rotary engines power some imported autos.

Truck Engines Light-duty trucks are powered with automotive-type gasoline engines of somewhat slower speed and heavier construction. Heavy-duty trucks are usually powered by diesel engines. The diesel engine is described in detail in a later section.

16.17 Horsepower Standards for Automobile and Truck Engines

The Society of Automotive Engineers (SAE) has adopted certain standards for rating the horsepower of automobile and truck engines. These are a mean effective pressure of 67.2 lb/in.^{2} and an average piston speed of 1000 ft/min. The piston of a four-cycle engine travels four stroke lengths for each power stroke. If L is the length of stroke in feet and N the number of strokes per minute, we write $4LN = 1000$ ft/min, or

$L = 1000/4N$. Rewriting Eq. (16.8) for these SAE standards, we have

$$\text{(SAE) hp} \atop \text{(per cylinder)}$$

$$= \frac{PLAN}{33,000} = \frac{67.2 \times 1000/4N \times \pi D^2/4 \times N}{33,000}$$

which, when evaluated for an engine of n cylinders, gives

$$\underset{\text{(automotive-type engine)}}{\text{(SAE) hp}} = \frac{D^2 n}{2.5} \qquad (16.10)$$

where D is the cylinder bore in inches.

Illustrative Problem 16.6 Calculate the SAE horsepower of a four-cylinder auto engine whose cylinder bore is 3.75 in.

Solution Substituting in Eq. (16.10), we obtain

$$\text{hp} = \frac{(3.75)^2 \times 4}{2.5} = 22.5 \text{ hp} \qquad ans.$$

This seems a rather low value for the horsepower of a modern four-cylinder auto engine. The SAE formula, based as it is on a 1000-ft/min piston

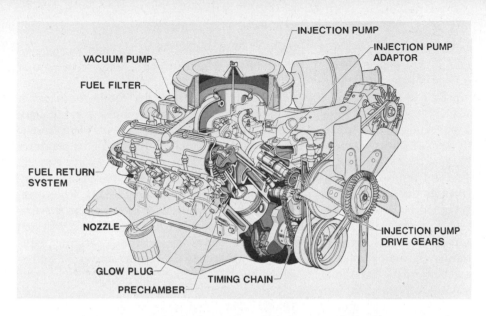

Fig. 16.21 Cutaway of diesel engine for automobile use, showing details of pistons, valves, and fuel injectors. (Oldsmobile Division, General Motors Corp.)

speed, is intended to give the horsepower output of the engine at a much slower road speed than 60 mph. For an engine of 4-in. stroke, for example, a piston speed of 1000 ft/min gives a rotational speed of only 1500 rev/min. SAE horsepower, then, is a conservative estimate of the engine's output at a slow speed while pulling a load. Brake horsepower, on the other hand, is the peak horsepower obtained when the engine is "wound up tight."

Automotive-type engines, when properly "tuned-up," may have efficiencies ranging from 25 to 30 percent.

16.18   Aviation Engines

Though jet and gas turbine engines (see Sec. 16.21) have replaced internal-combustion engines on nearly all large commercial and military aircraft, the gasoline-powered internal-combustion engine is still an important engine in the aviation industry. Many small 65- to 250-hp aviation engines are of the horizontal-opposed type. Larger engines are of *radial* design, with one or more "banks" of cylinders containing five, seven, or nine cylinders per bank.

Aircraft engines are essentially heavy-duty, low-speed, high-torque engines. Their design is for relatively constant-speed operation, usually around 2200 rev/min. All these design features must be obtained with a minimum of weight, for the aircraft engine must not only pull the plane forward through the air but lift the loaded airplane against the force of gravity.

16.19   The Diesel Engine

In 1892, Rudolf Diesel (1858–1913), a young German engineer, built a new type of internal-combustion engine based on the principle that heat of compression in a cylinder could flash the fuel and thus dispense with the spark plug and its associated components of the ignition system.

Early diesel engines were massive machines, sometimes weighing as much as 250 lb for each horsepower developed, and were thus suitable only for stationary installations. Improvements over the years have brought this ratio down to as low as 5 lb/hp in smaller, higher-speed engines.

Principles of Operation   In any internal-combustion engine, the higher the compression ratio, the greater the efficiency, provided the fuel used will burn slowly enough to prevent "knocking." The diesel engine uses compression ratios up to 20 : 1, instead of the 6 : 1 to 10 : 1 common for gasoline engines. The fuel is classified as *diesel oil*, a rather heavy type of distillate, which, though it has a high Btu content, burns slowly in the cylinder without "pinging." Air is forced into the cylinder (see Fig. 16.21) by atmospheric pressure or by a supercharger as the piston moves down on the intake stroke. On the upstroke, the air is compressed to less than one-sixteenth its full-cylinder volume, and this sudden compression raises the temperature in the cylinder to about 1000°F, which is well above the flash point of the fuel oil. At this instant, fuel is sprayed into the cylinder in a fine mist by the fuel injector which builds up pressures of over 2100 lb/in.^{2}. The oil begins to burn immediately and burns evenly throughout the stroke. This power stroke is followed by an exhaust stroke, and the four-stroke cycle is complete. Diesel engines are also built to operate on the two-stroke cycle (see Fig. 16.19). Two-cycle diesels *must* be supplied with a supercharger in order to get enough air in the cylinder for proper combustion.

Diesel-engine Applications Diesel engines are now made in many sizes, ranging from a one-cylinder model up to 16-cylinder V-type engines developing over 2000 hp. Railroad locomotives, trucks, buses, tractors, construction equipment, and farm machinery are all powered by diesel engines. Stationary diesels are used in small electric generating plants. The petroleum and mining industries use many sizes and types of diesels for their power requirements, running mine hoists, mine-ventilating equipment, petroleum drilling rigs, and pumping units. Diesels have extensive marine applications, both as main propulsion engines and as auxiliaries. Diesel engines are slowly increasing in popularity in automobiles also and will probably continue to do so.

16.20   Gas Turbines

Another type of heat engine, now used in many industrial applications, is the gas turbine. Its design and principles of operation are similar to those of the steam turbine. There are two major differences: (1) the working substance is the mixture of hot gases from the combustion, rather than steam; and (2) the combustion takes place within the engine itself. The fuel may be kerosene or petroleum distillate. Operating temperatures are very high, 2000°F and more in some engines. These extreme temperatures require the development of special heat-resisting alloys. Rotational speeds are also high, and gear reduction is necessary. A few locomotives are now being powered by gas turbines. Some electric generating plants use gas turbines as the "prime mover." "Turbo-prop" aircraft also use these engines.

16.21   Jet Engines

The most recently developed heat engine is the *turbojet engine*, designed for aircraft use. Its combustion chamber and general design are somewhat similar to the gas turbine. There is a turbine rotor which is driven by the hot gases, and its sole function is to drive a rotary air compressor which is required to force in enough air for combustion (see Fig. 16.1). Forward thrust is obtained not from a propeller but from the reaction force of the hot gases escaping through the "jet" to the rear at fantastic speeds. The discussions of Chaps. 4 and 5, which explained the basic principle of jet propulsion should be recalled here. The diameter and shape of the orifice through which the hot gases escape are matters of tremendous importance in jet-engine design. It is only with these engines that airplane speeds in excess of 600 mi/hr have been attained. Speeds in the 2000-mi/hr range are common for some military aircraft, and supersonic commercial transport aircraft (SSTs) capable of 1900 mi/hr are now operating on transoceanic flights.

16.22   Rocket Motors

All the internal-combustion engines so far discussed have had one feature in common. In order to sustain combustion, oxygen must be obtained from the atmosphere. An engine with self-sustaining combustion, with built-in oxidizers, is

Fig. 16.22 A Tartar antiaircraft missile just fired from a deck launcher on a Navy ship. These missiles have solid-propellant rocket engines. (Official U.S. Navy photo.)

called a *rocket motor*. Though successful rocket motors are of relatively recent development (40 years), man has experimented with rockets for several centuries. The principle of flight used by a rocket is jet propulsion, as with the jet engines discussed above. There are no moving parts to the rocket motor proper, and it is the fuel which makes the difference between rockets and jet engines. Rocket fuel may be in either liquid or solid form, but in either case the rocket *carries its own oxygen* or other oxidizing agent. Solid propellant fuels are shaped to burn evenly and progressively, and their chemical composition is such that just the right amount of oxygen is present to assure good combustion. Liquid-fueled rockets may use liquid hydrogen or alcohol as the fuel, and the oxygen supply is usually liquid oxygen (LOX). The proper mixture of fuel and oxygen is regulated by a complex valve system.

Guided-missile research and space-vehicle development are proceeding rapidly in military and space-agency laboratories, and many improvements in rocket propulsion are reported each year.

Most of the large rocket engines for space exploration operate on liquid fuels. To date, greater thrusts have been obtained with liquid than with solid propellants. However, the difficulties involved in handling liquid fuels and oxidizers have spurred the development of solid-fuel rockets. Antiaircraft rockets (Fig. 16.22) and submarine-fired ballistic missiles are examples of currently produced solid-propellant rocket engines.

Rocket Motors for Spacecraft   The huge rocket engines used for the Apollo space program (moon exploration) were liquid-fueled and developed thrusts of over 7 million pounds. New rocket motors with thrusts of 10 million pounds and more are under development.

The Saturn 5 vehicle (Fig. 16.23) was the workhorse of the Apollo program. Its first-stage booster is 138 ft long and 33 ft in diameter. Its motor consists of five liquid-fueled engines each developing 1.5 million pounds of thrust. It weighs nearly 4.5 million pounds when fully loaded. The second stage is 82 ft long and also has five engines, each developing 200,000 pounds of thrust. On top of the S-2 stage is the S-4B stage, which is 58 ft long and has a 200,000-lb-thrust rocket motor. It was able to place 240,000 lb in an earth orbit and 90,000 lb in the vicinity of the moon.

Fig. 16.23 The mammoth rocket engine for the Saturn 5 booster. This is the engine for the first-stage vehicle. Each of the five jets delivers 1.5 million pounds of thrust. This was the "lift-off" engine for the Apollo series of flights to the moon in the late 1960s and early 1970s.

Fig. 16.24 Artist's concept of Space Shuttle in flight. Note the three main engines which constitute the booster and propulsion power plant. The Space Shuttle was test flown and landed on a regular-length runway several times in 1977–1978. Actual take-off and space flight under its own power is anticipated in 1980–81. Present plans call for 570 shuttle missions into space during the first 11 years of operation. (Rockwell International, Space Division, and National Aeronautics and Space Administration.)

Engines for the Space Shuttle The U.S. space program for the 1980s and beyond features plans for large orbiting satellites, of which Skylab was a prototype. Commuting from earth to such satellites will be accomplished by space shuttles. Unlike the Apollo (moon) space vehicle, in which the return to earth was made by parachuting into the ocean, the space shuttle will "fly" to a landing at more or less conventional airports. Engines for the space shuttle have been under development for several years. The main engine is reported to have a thrust of about 375,000 lb (1,608,000 N) at sea level. Three such engines in tandem "boost" the shuttle on its way to port at the satellite. Figure 16.24 shows the space shuttle in flight.

The shuttle is designed to provide more or less routine access to space, with return to earth under aircraft-flight conditions. When the space shuttle is operational, large and complex satellites can be constructed in orbit and serviced on a regular basis. Continuously manned space stations can be operated, with the "work force" being shuttled back and forth to earth on a planned schedule.

16.23 Thermodynamics and the Energy Crisis

The case could be made that since the earth is an isolated system, there is just as much energy in and on the earth today as there ever was. But of all forms of energy, heat energy is probably the

Fig. 16.25 Sixty-ton magnet built at the Argonne National Laboratory for experimental research in generating electricity by magnetohydrodynamics (MHD). MHD generators produce electricity directly, as a hot ionized gas from the combustion of fossil fuels moves at high speed in a magnetic field. MHD generators are not presently operational on a commercial basis in the United States, but they show promise for the 1980s. (U.S. Department of Energy.)

of engine weight. In contrast, nuclear power, which is one option for the future, involves massive, extremely ponderous equipment suitable only for stationary installations, or at best, propulsion plants for large ships.

As a matter of fact, though nuclear and other forms of energy to be discussed below promise abundant power for *stationary* electric generating plants, there does not appear to be at the present time any ready solution to the long-term problem of *mobile* energy (autos, trucks, farm machinery, aircraft, and small ships) when the earth's store of fossil fuels is finally exhausted. Consequently, these nonrenewable energy resources must be conserved for mobile energy uses. The energy conservation effort should have two thrusts: first, existing forms of heat engines should be made as efficient as possible; and second, new forms of heat engines that can operate from renewable, lower-temperature heat sources must be devised.

Magnetohydrodynamics (MHD) A highly efficient hot-gas engine is now under development in the United States and in Russia, based on the concept of *magnetohydrodynamics*. Conventional fuels (oil, coal dust) are burned in a special combustion chamber "supercharged" with hot compressed air. The resulting hot gases at about 5000°F are highly ionized (i.e. large numbers of electrons have been freed from their parent atoms). To further improve electrical conductivity of the stream of hot gas, finely divided metal particles are injected into the combustion chamber. The hot gas, with its many free electrons, streams at high velocity through a strong magnetic field (Fig. 16.25). Electrodes are placed at proper points to collect the electric charges and lead them out onto wires as an electric current. MHD generators in pilot-plant operations have already attained overall efficiencies of nearly 60 percent. In contrast, the best steam-turbine electric plants achieve only about 42 percent efficiency, and internal-combustion-engine electric plants, only about 35 percent. However, even with such an appreciable increase in efficiency, it is questionable whether our dwindling supplies of

most essential to modern societies; and the application of heat energy to the performance of useful work is rigidly governed by the two laws of thermodynamics. High-temperature sources of heat are required to operate heat engines, and a low-temperature reservoir or "sink" must be available to receive the exhaust. But entropy throughout the universe is always increasing, which is just another way of saying that the temperature difference between "high-temperature reservoirs" and "low-temperature reservoirs" is continually decreasing. Fossil fuels have, for centuries, been the readily available means of creating high-temperature reservoirs. But petroleum, coal, and natural gas are being depleted at an alarming rate, and fossil-fuel thermodynamics, with its large temperature differentials and high pressures, may be supplanted in the future by a kind of low-order thermodynamics, featuring devices that provide useful energy from larger *quantities* of heat at *lower temperature differentials*. The era of the small, lightweight, high-speed engine may be passing. Today's typical auto engines produce about one horsepower per pound

fossil fuels should be even partially committed to stationary-plant production of electric power.

A major research effort is now being directed to the development of engines and devices to use *renewable* sources of energy. The sun is the ultimate renewable source, and solar energy development (see p. 299) is moving ahead rapidly on many fronts. Besides the energy available from direct solar radiation, there are many energy sources awaiting further development, which represent stored energy from the sun. A discussior of some of these follows.

Hydroelectric (*water*) *power* is a renewable energy resource, and greatly increased development of hydropower is necessary. Installations for hydroelectric power generation are discussed in Chap. 31.

Bioconversion is another name for "energy farms." Certain kinds of high-energy plants (for example, trees and shrubs containing flammable oils and resins) can yield 200 million Btu's of heat energy per acre per year as a result of controlled combustion. Pilot plant operation of energy farms is already beginning. Also, there is renewed interest in the use of forest-product wastes and combustible wastes of all kinds to provide fuel for power plants. Several public utilities are already making use of these wastes for fuel.

Ocean tides and waves represent renewable sources of energy of incalculable quantity. Engineers have attempted for over a century to construct ponderous "engines" to use these tremendous sources of energy. At locations in estuaries, bays, and river outlets, where the tidal variation results in differences in level of 20 feet or more, water turbines can be installed in a huge dam to generate electric power. With suitable gates and valves, the turbines can operate on both the incoming and outgoing tides. An installation on the northern coast of France is presently the largest operating tidal-power project. It has 24 turbines and generates a total of 240 MW at maximum output.

Experimental projects to extract energy from ocean waves and from ocean currents like the Gulf Stream have been proposed, and a few prototype models of equipment have been tried, but as yet none has been commercially successful.

Ocean thermal currents have been studied for a long time and their effects on continental climates, on marine life, and on the world's storm systems are well known. The ocean surface is usually much warmer than the water at great depths, and in certain locations in the Caribbean Sea and near Hawaii, the surface temperature rarely drops below 82°F and at the same time cold currents moving southward along the ocean bottom from the Arctic may result in a temperature as low as 42°F directly below. This 40°F difference is not enough to operate a conventional heat engine with any acceptable level of efficiency. (The Carnot efficiency of an ideal engine operating between these temperature limits would be only 7 percent.)

Recently, however, development has begun on a Rankine-cycle heat engine (see Fig. 16.10) using ammonia as the working substance, and operating on a temperature difference of 40°F. (This engine was mentioned on p. 313.) Warm water at the surface heats and boils the ammonia and the sensible heat loss of the warm water becomes latent heat in the ammonia vapor. The "hot" (about 80°F) ammonia vapor at a pressure of 150 lb/in.^{2} abs. is used to drive a low-pressure turbine turning an electric generator. The ammonia vapor exhausted from the turbine is cooled and condensed by cold (42°F) water pumped up from the 2000-ft deep cold layer, and the ammonia liquid is then ready for another cycle. Figure 16.26 shows a nonoperational scale model, or mockup of an ocean thermal-energy power plant. The first closed-cycle Ocean Thermal Energy Conversion (OTEC) plant is scheduled to begin test operations in mid-1980 off the coast of the "Big-Island" of Hawaii.

Many technical and economic problems will have to be overcome before commercially useful amounts of electric energy can be obtained from the ocean. Principal among these are: The problem of sea-water corrosion, the high cost of building and maintaining power plants at sea, and the

Fig. 16.26 A scale model (not operational) of a free-floating ocean thermal power plant. Small temperature differences of ocean layers at various depths will be used to drive a heat engine. This design envisions a structure 340 feet across and 17 stories high. It would supply 100 MW of electric power, enough for a city of 60,000 population. (U.S. Department of Energy and TRW Systems Group).

provision of the vertical stability needed for rotating machinery. The potential for ocean energy is so vast, however, that it offers one of the most enticing avenues for research and development. There is enough energy in ocean tides, waves, and thermal currents to provide all of man's energy needs for the foreseeable future, if technical and economic problems can be solved. Three factors stand out: it is renewable on a daily basis as the sun shines; it is pollution free; and, there is no "fuel" cost. Estimates of the rate of development tend to be quite conservative—of the order of two to five percent of the nation's electrical energy production by the year 2025.

Wind energy has been used for centuries, but only recently has there been a serious engineering effort to evaluate wind-machine design and potential locations for wind-powered electric energy generating plants. NASA scientists have predicted that wind machines could supply as much as 10 percent of the nation's electricity by the year 1990. Figure 16.27 illustrates one form of wind turbine presently under development.

One final energy project is worthy of mention. *Bacterial fermentation* of sewage, garbage, and other organic wastes produces methane gas (CH_4), which is an excellent fuel for heating, firing steam boilers, and running gas engines. Several such plants are already in operation, and the results are good enough to call for a great deal of further research and development.

SUMMARY

Heat engines convert heat into mechanical energy.

The mechanical equivalent of heat (Joule's constant) is 4.186×10^3 J/kcal, or 778 ft-lb/Btu. It is defined by the equation

$$J = \frac{W}{H}$$

The term *thermodynamics* is applied to the science that deals with the relationships between mechanical energy and heat.

The first law of thermodynamics states that *in any thermodynamic process, the total energy remains constant—none is created or destroyed.*

Fig. 16.27 A vertical-axis wind turbine under test at Sandia Laboratories in Albuquerque, New Mexico. The 15-foot blades, when driven by a 20 mile per hour wind, produce about three horsepower. (U.S. Department of Energy).

The second law of thermodynamics states that *mechanical energy cannot be obtained from any source of heat unless that source undergoes a drop in temperature.*

An alternative statement of the second law is: *As a result of natural processes, the world's supply of energy available to do work is continually decreasing.*

The thermodynamic property of a system which measures its degree of disorder, or unavailability of heat energy, is called *entropy.*

An *isothermal process* occurs at constant temperature.

An *adiabatic process* occurs at constant heat.

The basic principle of heat engines is that a hot gas (called the *working substance*) is allowed to expand and do work. As it expands, its temperature drops, and its loss of heat energy equals the amount of mechanical energy expended ($\Delta W = Q_1 - Q_2$).

The ideal efficiency of a heat engine is given by

$$\underset{\text{(ideal engine)}}{\text{Carnot eff}} = 1 - \frac{T_2}{T_1} = \frac{T_1 - T_2}{T_1} \qquad (16.7)$$

Steam engines are of two types, *reciprocating* and *turbine*. The horsepower developed by a double-acting reciprocating steam engine is given by the formula

$$\text{hp} = \frac{PLAN}{33,000}$$

Modern steam turbines make use of both impulse and reaction principles.

Internal-combustion engines have many different cylinder arrangements and they operate on two basic cycles—the four-stroke cycle and the two-stroke cycle.

The SAE-rated horsepower developed by an automotive-type engine is given by the formula

$$\text{(SAE) hp} = \frac{D^2 n}{2.5}$$

SAE horsepower is a conservative estimate of the engine's power while pulling a load at a slow speed.

The diesel engine is an oil-burning engine with a compression ratio so high (up to $20:1$) that the heat of compression is sufficient to fire the fuel. Diesel engines are in general more efficient than gasoline engines.

Newer heat engines include the gas turbine, the turbojet, and the rocket motor. Except for the gas turbine, the application of these engines is currently limited to aircraft and to boosters for military and space-exploration purposes.

Thermodynamics, as an engineering science, will be heavily involved in developing and improving machines for the future that will operate from renewable sources of energy.

QUESTIONS AND EXERCISES

1. What do you think were some of the most difficult problems encountered by Joule in his attempts to measure the mechanical equivalent of heat? Recall the approximate date of his work.

2. Exactly what does a heat engine accomplish? What is meant by the term *working substance*? Why is it necessary? How do heat engines illustrate the law of conservation of energy (the First Law of Thermodynamics)?

3. The First Law of Thermodynamics tells us that energy is conserved in all heat-to-work and work-to-heat processes *that actually take place*. It does not tell us whether or not a specific process *will take place*. In general, what kinds of processes will occur and what kinds will not? Suggest several unique and interesting consequences for everyday life if the Second Law could be "repealed." (See Sec. 16.3.)

4. A refrigerator seems to violate the second law of thermodynamics, in that it takes heat away from bodies already quite cold and rejects this heat to bodies (or water or air) which may be very warm. How can this process be consistent with the second law?

5. Starting with the definition of work, $W = Fs$, show that the amount of work done by a gas expanding at constant pressure p pushing a piston through a swept volume V is $W = pV$.

6. An inventor claims to have built an engine which will operate between 800 and 400 K and which will give a work output of 2500 J for every kilocalorie of heat input. He wants financing from you to build and market his engine. Would you invest? Explain your answer with a mathematical analysis.

7. Make a list of relative advantages and disadvantages for steam turbines vs. reciprocating steam engines.

8. Diagram and explain in detail the four-stroke operating cycle of an automotive-type gasoline engine.

9. Compare the advantages and disadvantages of diesel engines and gasoline engines for automotive and truck use, and do the same for diesel engines and steam turbines for marine use.

10. Use diagrams and explain briefly the principle of operation of (*a*) the gas turbine, (*b*) the turbojet engine, (*c*) the rocket motor.

11. Recalling Joule's constant and Carnot's ideal cycle and its maximum efficiency, show that a high-compression engine will be more efficient than a low-compression engine, assuming that all other conditions are equal.

12. The oceans of the world contain incalculably great amounts of heat energy. Why is it so difficult to invent and perfect a machine to utilize this renewable source of heat? If such machines are eventually put into operation, what could you predict about their efficiency? Why?

13. The same sample of hot gas expands in a cylinder from a volume V_1 to a volume V_2, in two separate expansions. First it is expanded *isothermally*, then *adiabatically*. In which expansion does the gas perform more work? Which process is more efficient? Explain your answers.

PROBLEMS

1. A coal-fired steam plant burns 0.5 kg/hr of coal per kW of electric power produced. Calculate the overall efficiency of the burner-boiler steam-turbine generator system.

2. A steam-turbine generator plant burns 8300 gal/hr of fuel oil and delivers 50,000 kW of electric power. What is the overall efficiency of this plant? Take specific gravity of fuel oil as 0.76.

3. A 1-hp motor drives a paddle in 10 gal of water for 1 hr. If the device is 100 percent efficient (no heat lost), how much of a temperature rise in Fahrenheit degrees will occur?

4. What is the maximum possible (Carnot) efficiency of an engine which is supplied heat at 350°C and rejects heat at 110°C?

5. Calculate the ideal, or Carnot, efficiency of a steam turbine if saturated steam is supplied to stage 1 at 2000 lb/in.^{2} abs (900°F) and exhausts to the condenser at 1 lb/in.^{2} abs (see Table 14.6).

6. While resting, the average adult produces heat at a rate of about 65 kcal/hr (basal metabolism rate). How many watts of power is this?

7. How many Btu are required to melt a 1-lb ice cube (32°F) and then heat the water to the boiling point? If an equivalent amount of work were done on the ice cube against the force of gravity, how high (feet) would it be raised?

8. How much work in foot-pounds will a four-cycle gasoline engine accomplish from each gallon of gasoline, if 25 percent efficiency is assumed? (See Table 14.1.)

9. An engine is assumed to be operating at Carnot efficiency between the temperatures of 800 and 450 K, and it rejects 1 kcal of thermal energy during each cycle. How much useful mechanical work (joules) does it accomplish during each cycle?

10. A Carnot (ideal) engine has an efficiency of 45 percent. If hot gas enters the cylinder at 650°C, what is the exhaust temperature?

11. A hot-gas turbine uses kerosene as fuel. It is rated at 15,000 hp and operates a 10-MW electric generator at full load. The engine burns kerosene at 4000 lb/hr. Find the overall efficiency of the engine-generator system.

12. While an engine is idling, it expends energy at the rate of 1.5 hp. If 25 percent of this energy goes into heat development by friction of the moving parts, how many Btu/min are dissipated due to friction?

13. A jet aircraft uses 75,000 lb of kerosene for a five-hour flight. How many foot-pounds of energy are expended? If the engines provide an average of 40,000 hp during the flight, what is their efficiency?

14. An "ideal" Carnot engine operates between reservoirs at 400°C and 20°C. (*a*) Find its theoretical efficiency. (*b*) If the work output per cycle is 9750 **J**, how many kilocalories of heat must the hot reservoir supply for each cycle? (*c*) How much heat does the low-temperature reservoir absorb per cycle?

15. A single-cylinder double-acting steam engine has a cylinder diameter of 14 in. and a stroke of 28 in. The MEP is 315 lb/in.^{2}, and the engine speed 260 rev/min. Find its horsepower output.

16. A steam turbine uses 80,000 lb of steam per hour. The steam velocity entering stage 1 is 1650 ft/sec, and the exit velocity from the final stage is 300 ft/sec. Calculate the theoretical horsepower output of this turbine.

17. A truck engine burns 7.5 gal/hr of diesel fuel. On a dynamometer test it shows 110 hp output at 2400 rev/min. What is its efficiency at this speed? Specific gravity of diesel fuel is 0.72.

18. A certain eight-cylinder V-type auto engine is said by the manufacturer's sales literature to develop 300 bhp. Its specifications are $3\frac{7}{8}$ in. bore, $4\frac{1}{4}$ in. stroke, compression ratio 8 : 1. What is its SAE horsepower? Why is this so much lower than the manufacturer's rating?

19. Compute the velocity with which an ice cube (at 0°C) would have to strike against a concrete wall in order to melt it to water at 0°C. Assume all the kinetic energy is converted to heat and that all the heat generated goes into melting the ice cube.

20. A Carnot engine whose hot reservoir is at 380 K uses 320 cal in useful work per cycle and rejects 250 cal to the cold reservoir. (*a*) What is the efficiency of this cycle? (*b*) Find the temperature of the cold reservoir.

21. The temperature of steam entering a steam turbine is 1350°R. It exits into the condenser under a partial vacuum at 665°R. Find the theoretical maximum efficiency of the turbine.

22. An "ideal" engine operates from a hot reservoir at 1340°R. It uses 80 Btu per cycle in useful work and rejects 180 Btu to the cold reservoir. (*a*) Find its efficiency. (*b*) What is the temperature of the cold reservoir?

17 Refrigeration and Air Conditioning

Refrigeration is an important industry in most regions of the world and absolutely essential in advanced industrial nations. The processing, freezing, and preservation of foods, and the cooling of beverages is necessary to the health and well-being of much of the world's population. Refrigerating machines make use of thermodynamic principles and "reversed" heat engines to remove heat from products and freeze them if desired.

Air conditioning is the process of controlling the temperature, humidity, and cleanliness of air for human comfort. Year-round air conditioning systems heat and humidify air in winter; cool and dehumidify it in summer; and filter out odors and impurities the year around. Air conditioning is a major industry in the United States and throughout most of the world. Besides human comfort, many scientific and industrial processes require a controlled atmosphere.

Although refrigeration and summer air conditioning have many factors in common, we treat them in separate sections in the present chapter. Summer air conditioning makes use of refrigerating equipment to cool and dehumidify air. Winter air conditioning involves the use of furnaces, boilers, or solar heaters to provide heat. Moisture is added to room air by humidifiers, in well-designed systems.

At present, in the United States, the refrigeration and air conditioning industries are still based on the English (engineering) system of units. This chapter will, therefore, not use SI-metric units.

REFRIGERATION

Refrigeration is defined as the process of removing heat from a space or a product in order to lower its temperature to some specified value. People sometimes speak loosely of refrigeration as the "production of cold," as if *heat* and *cold* were two separate conditions capable of independent existence. But cold is merely the *absence* of heat. Some heat is present in all substances whose temperature is above absolute zero, and refrigeration is the process of removing some of this heat and dissipating it to some other substance or space where its presence will be unobjectionable. *Heat removal* and *heat dissipation* are the two processes which are always involved in refrigeration. The inaccurate concept of "cold production" should not be allowed to confuse the issue.

17.1 The Unit of Refrigeration Capacity

Since refrigeration is a process of heat removal, refrigeration machines are rated as to capacity in terms of the quantity of heat which they are capable of removing from a given space or product in unit time. The standard unit of refrigeration capacity is the *Btu per hour*. A much larger unit is the *ton of refrigeration*. This is the quantity of

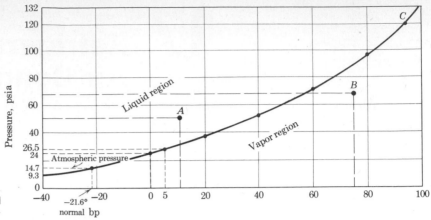

Fig. 17.1 The pressure-temperature (P-T) curve (saturated vapor-pressure curve) for Freon-12.

heat removal required to freeze 1 ton of ice from water which is already at 32°F, in a 24-hr day. The latent heat of fusion of ice is $L_f = 144$ Btu/lb, and therefore the ton of refrigeration is the equivalent of 144×2000, or 288,000 Btu of heat removal per day. It is more common to rate the ton of refrigeration on an hourly basis, and the ton of refrigeration capacity is practically defined as

$$1 \text{ ton of refrigeration} = 12,000 \text{ Btu/hr}$$

This amount of heat removal will freeze about 10 gal/hr of water (at 32°F) into ice.

17.2 Basic Thermodynamics of Refrigeration

It was emphasized in the previous chapter that in any thermodynamic process total energy is conserved (first law). Also, heat energy can accomplish mechanical work only when the circumstances are such that the heat can "flow down a temperature hill" (second law). As examples of the *second law of thermodynamics* we studied in detail several heat engines which perform mechanical work as hot, high-pressure gas *does work on a piston*, and low-pressure gas at reduced temperature is exhausted.

The refrigeration process at first glance seems to violate the second law of thermodynamics, for a refrigerating machine takes heat from a low-temperature source and rejects it to a space or substance at a higher temperature. A refrigerator makes heat *run up a "temperature hill"!* The explanation is, of course, that work must be done on the refrigerant gas by the piston in a refrigerating machine. A heat engine *accomplishes* work as heat flows "downhill." A refrigerating machine *must have work done on it* to make heat flow "uphill." In this sense, then, a refrigerating machine is a reversed heat engine. Ordinarily the work is done on the refrigerant by a compressor which is driven by an electric motor. The compressor may be either a piston-in-cylinder, reciprocating type; or a turbo-compressor with rotating blades or vanes.

17.3 Mechanical Refrigeration

Almost all refrigeration (including the manufacture of ice) is today based upon the use of certain low-boiling-point substances called *refrigerants* as heat-transfer media. *Ammonia* (bp $-28°F$ at 1 atm pressure), *methyl chloride* ($-10.6°F$), *sulfur dioxide* (13.8°F), *Freon-12* ($-21.6°F$), *Freon-22* ($-41.4°F$), and *Carrene-1* (105.2°F), are examples of commonly used refrigerants.

These substances are all vapors (gases) at normal temperatures and atmospheric pressure but are easily liquefied by an increase in pressure or a drop in temperature. The curve of Fig. 17.1 illustrates the pressure-temperature behavior of a typical refrigerant. The diagram shows the pressure-temperature (P-T) curve for saturated Freon-12 vapor for a range of temperatures typical of a refrigerating machine using this gas as a refrigerant. The curve can be considered as the path of all pressure-temperature condition points which represent equilibrium between the liquid and the vapor state. Any P-T combination which would result in a point plotted to the left of and above the curve (such as A) would result in the liquid phase. A P-T combination resulting in any point B, to the right and below the curve, indicates a vapor condition.

Each refrigerant has its own pressure-temperature curve, and the choice of a refrigerant depends to a great extent on how it reacts to changes in temperature and pressure. The boiling point of the refrigerant must be low enough to attain the desired temperature without having to pull a high vacuum on it. On the other hand, it should not require excessively high pressures to liquefy the refrigerant at the temperature of the surrounding air or that of the cooling water available. Note from the curve of Fig. 17.1 that Freon-12 satisfies both these conditions. The liquid phase boils at −21.6°F without any vacuum. Pulling a slight vacuum will lower the boiling point still further and make colder temperatures possible. At the other end of the curve note that gaseous Freon-12 will begin condensing to a liquid at 132 lb/in.^{2} abs pressure at the temperature of a hot summer day, 100°F. If cool water were used to condense it, the condenser pressure could be correspondingly less.

NOTE: In mechanical engineering the notation lb/in.^{2} abs is commonly written psia, and we shall use this notation in this chapter. The student will recall that absolute pressure equals gauge pressure plus atmospheric (14.7 lb/in.^{2}) pressure.

17.4 Refrigerating Systems

Two basic, but fundamentally different, systems account for almost all mechanical refrigeration. They are (1) *compression refrigeration systems* and (2) *absorption refrigeration systems*. Compression systems account for more than 80 per-cent of installed tonnage. The theory of absorption refrigeration will be briefly discussed, and an elementary explanation of *thermoelectric refrigeration* will also be given.

COMPRESSION REFRIGERATION SYSTEMS

Refrigerating machines take many forms, depending on the uses for which they are designed. The relative location of the component parts of a commercial refrigeration system will differ, for example, from the arrangement of components in a household refrigerator. The following discussion considers some basic principles common to all compression refrigerating systems.

17.5 Compression-system Components

Figure 17.2 is a schematic diagram showing the basic components of a compression refrigerating system using Freon-12 as the refrigerant.

In such a system, a cycle consists of alternate *compression*, *liquefaction*, *expansion*, and *evaporation* of the refrigerant. Every component of the system must aid in removing heat from the space to be cooled and rejecting it to a place where its presence is not objectionable. The first heat transfer occurs as the space and products being cooled give up heat to the refrigerant through the walls of a copper coil called the *evaporator*. This heat boils the liquid Freon contained therein and is "carried away" by the gaseous Freon as latent heat of evaporation. Evaporators for refrigeration machines are of many different sizes, shapes, and designs, depending on the cooling job to be done. The evaporator is located within the space to be cooled and may be at a considerable distance from the other components of the system. It may be a simple coil of pipe, as in the diagram of Fig. 17.2, or it may take the form of a blower-equipped and "finned" evaporator as shown in Fig. 17.3.

The *compressor* is a pump for removing vapor from the evaporator on the *suction stroke* and compressing it on the *compression stroke*.

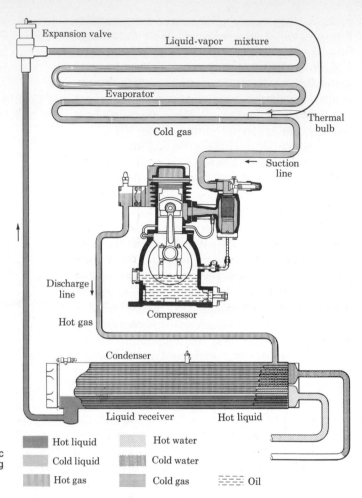

Expansion valve

Liquid-vapor mixture

Evaporator

Cold gas

Thermal bulb

Suction line

Discharge line

Hot gas

Compressor

Condenser

Liquid receiver

Hot liquid

Hot liquid

Cold liquid

Hot gas

Hot water

Cold water

Cold gas

Oil

Fig. 17.2 Schematic diagram of the basic components of a compression refrigerating system. (Carrier Corp.)

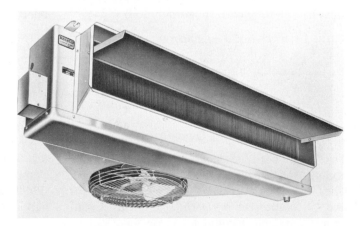

Fig. 17.3 Example of blower-equipped evaporator intended for suspension from ceiling in walk-in boxes and cold-storage rooms. (Recold Corp.)

Vapor "suction" from the evaporator is essential to keep the evaporator pressure (and therefore the temperature) low (see the curve of Fig. 17.1). Vapor compression is necessary to boost the gas temperature high enough (superheated) to permit heat rejection from the hot vapor along a descending temperature gradient into the cooling water or air.

The final heat rejection to cooling water or air is effected by the *condenser*. The typical water-cooled condenser is a steel shell containing a coil of copper pipe in which the refrigerant vapor circulates while being cooled by the water flowing through the shell. Air-cooled condensers are often used, and they look somewhat like the radiator of an automobile. In air-cooled condensers the refrigerant gas circulates in coiled copper tubing while air is blown across the "fins" and around the tubes by a fan. "Fin-and-tube" construction gives a large surface area for heat transfer.

Cooling the superheated refrigerant gas in the condenser while it is at the high pressure built up by the compressor brings the gas to a *P-T* condition representing saturation, i.e., to some point such as *C* (120 psia, 95°F) on the curve of Fig. 17.1. Further cooling results in condensation, and the warm liquid refrigerant collects in the *liquid receiver*, ready for another cycle.

Liquid refrigerant is supplied to the evaporator by the *expansion valve*, a device which automatically regulates the quantity of refrigerant supplied. The expansion valve is usually operated by a bellows arrangement containing a gas, the pressure of which is controlled by the evaporator temperature. When evaporator temperature is low, the gas pressure in the bellows is low, allowing the valve to close. As evaporator temperature rises, gas pressure in the bellows increases, opening the expansion valve, and liquid refrigerant is supplied to the evaporator. Figure 17.4 shows a typical expansion valve in a cross-section. (There are, of course, other methods of controlling refrigerant flow, e.g., float valves, and capillaries; see any standard text on refrigeration.)

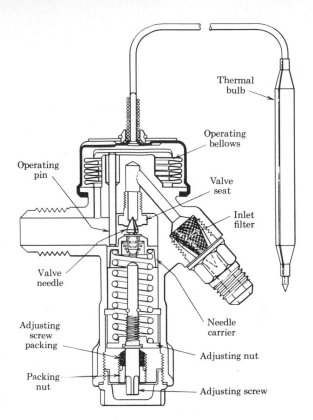

Fig. 17.4 Typical refrigeration expansion valve, cutaway view. The thermal bulb (right) is strapped to the suction line between the evaporator and condenser as shown in Fig. 17.2. Gas pressure in the thermal bulb actuates the bellows arrangement of the valve and controls the flow of refrigerant to the evaporator. (Frigidaire Division, General Motors Corp.)

17.6 Analysis of a Compression Cycle

A step-by-step analysis of a compression cycle, using Fig. 17.5 for reference, will be instructive. Consider the evaporator, located within the space to be cooled, being supplied with liquid Freon-12 from the expansion valve. The temperature to be maintained in the evaporator is determined by the temperature desired in the cooled space (for example, 15°F in an ice-cream cabinet). An evaporator temperature of 5°F, which is common for many systems, is shown. This temperature is produced when the compressor maintains a pres-

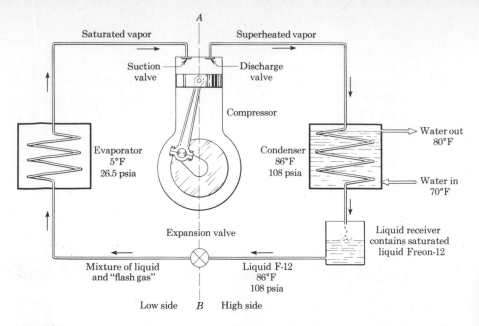

Fig. 17.5 Diagram illustrating the operating cycle of a Freon-12 refrigerating machine equipped with a water-cooled condenser.

sure in the evaporator corresponding to a 5°F boiling point, namely, 26.5 psia for Freon-12 (see Fig. 17.1).

Heat from the space and products being cooled boils the refrigerant in the evaporator, and saturated vapor is formed. This saturated vapor usually picks up more heat before leaving the evaporator and is therefore somewhat super-heated before it is drawn off by the suction stroke of the compressor piston.

The components thus far discussed, those to the left of the dashed line AB of Fig. 17.5, are said to constitute the "low side" of the system, since both temperature and pressure of the refrigerant gas are low.

On the compression stroke of the compressor piston, the vapor is compressed and the work done by the piston *on the gas* increases the gas temperature to a point where heat rejection to the cooling water in the condenser is possible. The standard Freon-12 cycle under discussion operates at a high-side pressure of 108 psia, which gives a condenser temperature of 86°F. Water is

ordinarily supplied to the condenser from a *water cooling tower*. The temperature of the water coming into the condenser is usually around 70°F, and an automatic water-regulator valve controls the flow of water to the condenser so that a 10°F rise through the condenser is maintained. As the cooling water absorbs the latent heat of the vapor and also the heat of compression, the vapor begins to condense. It collects in the receiver as a warm liquid still under the high-side pressure of 108 psia.

In the case of a system using an air-cooled condenser where the surrounding air might be at a temperature of 90°F or more, the operating pressure would have to be much higher. Reference to Fig. 17.1 shows that a condenser temperature of 100°F would require a pressure in excess of 132 psia before condensation would begin. A different refrigerant would, of course, have a different set of *P-T* operating conditions. Compressors, condensers, evaporators and all other components must be carefully matched as to capacity and operating conditions.

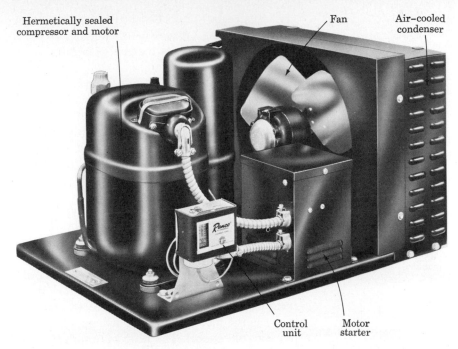

Fig. 17.6 A typical air-cooled refrigerated condensing unit, with components labeled. (Tecumseh Products Company)

This elementary analysis from a schematic, standard-cycle point of view serves to illustrate several basic concepts of a compression refrigeration system:

1. The net effect is *heat transfer*—from the space or product being cooled, along a descending temperature gradient to the refrigerant in the evaporator; and after compression, from the hot gas along a second descending temperature gradient to the condenser cooling medium.

2. The purpose of the compressor is to work as a *reversed heat engine*. Work is done *by the piston on a low-pressure, low-temperature gas* to raise its temperature to a point high enough *to make heat rejection possible*. This is just the reverse of a conventional heat engine in which a *high-pressure, high-temperature gas does work on a piston* and low-temperature gas at reduced pressure is exhausted.

3. The expansion valve may be regarded as the balance system between the high side and the low

side, furnishing refrigerant at an automatically controlled rate to meet the demands of the product- and space-cooling loads.

Figure 17.6 shows an air-cooled refrigerating unit with components labelled.

17.7 Thermodynamics of Compression Refrigeration Systems

A refrigerator is a heat engine which absorbs heat at a low temperature and rejects it at a higher temperature. This process at first thought seems at variance with the *second law of thermodynamics*. More careful consideration reveals, however, that the *availability-of-energy* criterion demanded by the second law is met by the fact that work is done *on the refrigerant gas* (working substance) by an external source of power. Heat can be made to "flow up a temperature hill" if it is "pushed up" by an external source of energy.

A compression refrigerating machine, then, acts as a reversed heat engine. We can use the

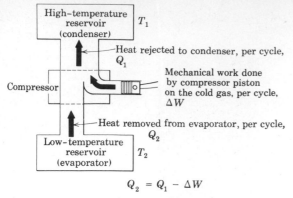

$$Q_2 = Q_1 - \Delta W$$

Refrigerating effect (per cycle) = heat rejected to condenser
− heat equivalent of
compressor work

Fig. 17.7 Block diagram illustrating the first law of thermodynamics for a refrigerating machine.

analysis applied to the *ideal* heat engine (see Chapter 16) and apply the same analysis to an *idealized* refrigerator. Carnot's theoretical treatment, applied to a compression refrigeration system (see Fig. 17.7), follows: Let

Q_1 = heat rejected in one cycle to a high-temperature reservoir at temperature T_1
Q_2 = heat absorbed in one cycle from a low-temperature reservoir at temperature T_2 (Q_2 is the *refrigerating effect*)
ΔW = work done in one cycle *on the refrigerant gas* by an *external* source of energy

Recalling Eq. (16.4) and keeping in mind that the refrigerator is acting as a reversed heat engine, the work ΔW, done in one cycle on the refrigerant gas, is

$$\Delta W = Q_1 - Q_2 \qquad (17.1)$$

(First law of thermodynamics for a refrigerating machine)

Stated another way,

$$Q_2 = Q_1 - \Delta W \qquad (17.1')$$

which says that the refrigerating effect per cycle, Q_2, is equal to the heat rejected to the condenser minus the heat equivalent of the mechanical work done by the compressor.

Figure 17.7 illustrates the energy balance for a refrigerating machine. Compare it with Fig. 16.4, which shows the energy balance for a heat engine.

It appears that Eqs. (17.1) and (16.4) are identical in form, but it must be remembered that the term ΔW in Eq. (16.4) represents the *work output of the engine* in one cycle, while in Eq. (17.1), ΔW is the *work which must be done on the refrigerator* in one cycle, by some external source of power.

Efficiency of a Refrigerator—Coefficient of Performance In evaluating the efficiency of a refrigerating machine we compare the amount of *heat removed* from the evaporator per cycle with the amount of external work which must be done *on the refrigerator*. In this case, the *output* is the heat extracted from the evaporator Q_2, and the *input* is the mechanical work ΔW supplied from an external source. The ratio of output to input for a refrigerator is not called efficiency; the more descriptive term *coefficient of performance* (c.o.p.) is used.

$$\text{c.o.p.}_{\text{(refrigerator)}} = \frac{Q_2}{\Delta W} = \frac{Q_2}{Q_1 - Q_2} \qquad (17.2)$$

In a refrigerator, as in a heat engine (see Sec. 16.8), heat transfer to or from the working substance in an *ideal* machine is directly proportional to the absolute temperature T at which the heat flow occurs. In mathematical terms,

$$\frac{Q_2}{Q_1} = \frac{T_2}{T_1} \qquad \text{or} \qquad Q_2 = \frac{Q_1 T_2}{T_1} \qquad (17.3)$$

Substituting the value of Q_2 from Eq. (17.3) in Eq. (17.2) and simplifying gives

$$\text{c.o.p.}_{\text{(ideal refrigerator)}} = \frac{T_2}{T_1 - T_2} \qquad (17.4)$$

where

T_2 = absolute temperature of evaporator (cold reservoir), °R or K
T_1 = absolute temperature of hot gas as it enters condenser (hot reservoir)

Analysis of Eq. (17.4) indicates that the colder the required evaporator temperature (smaller value of T_2), the lower the coefficient of performance will be. The more nearly equal the temperatures of the evaporator and the condenser are, the higher the c.o.p. will be. But the essential purpose of a refrigerating machine is to produce low evaporator temperatures, and consequently the value of ΔW must be high and c.o.p.'s must be relatively low in order to obtain a low evaporator temperature.

Furthermore, heat transfer in the condenser from the hot gas to the cooling water (or air) is greater if T_1 is high compared to the temperature of the condenser cooling medium. Consequently, when engineers design for *overall performance* of a refrigerating system, it is not sufficient merely to attain a high value of the c.o.p., without giving attention also to other operating requirements and conditions. All factors (necessary evaporator temperature, cost of electric power, availability of suitable condenser-cooling media, space and noise considerations, etc.) must be taken into account by the designers of the components of a refrigerating system.

Unlike the *efficiencies* of heat engines, the *coefficients of performance* of refrigerating machines are ordinarily *greater than unity*. Consider the following illustrative problem.

Illustrative Problem 17.1 A compression refrigeration system is operating between 10°F evaporating and 130°F condensing. Find its coefficient of performance.

Solution

$$T_2 = 460 + 10 = 470°R$$

$$T_1 = 460 + 130 = 590°R$$

From Eq. (17.4)

$$\text{c.o.p.} = \frac{470}{590 - 470} = 3.92 \qquad ans.$$

For an *ideal* refrigerator operating between the temperatures given, the above result indicates that the *net refrigerating effect* is 3.92 times as great as the heat equivalent of the mechanical

Fig. 17.8 A 50-ton refrigerating plant for a cold-storage warehouse. The evaporator is not shown, since the evaporator-blower unit(s) would be in the cold spaces. Note the large number of valves, controls, and pumps. (York Division, Borg-Warner Corp., and Allen-Bradley Co.)

work done on the refrigerator by the external source of energy. In simpler terms, for the system described, 3.92 Btu of heat will be removed from the evaporator for every 778 ft-lb of work (the mechanical equivalent of 1.0 Btu) done *on the compressor* by the driving motor. From the *ideal cycle*, c.o.p.'s of from 2.0 to 7.0 are possible. Actual refrigerators do not give coefficients as high as those computed from Eq. (17.4), however, because there are many losses which the ideal cycle does not take into account. When the *actual net refrigerating effect* is measured by suitable instruments and compared to the *actual energy input* to the compressor from the driving motor (converted to heat units, using Joule's constant), the coefficient of performance is calculated from

$$\underset{\text{(actual)}}{\text{c.o.p.}} = \frac{\text{net refrigerating effect}}{\text{heat equiv. of mech. work}} \quad (17.5)$$

Coefficients of performance of refrigerating machines under actual operating conditions range from 1.5 to 4.0 depending on the application, and on the state of maintenance and repair of the equipment.

Figure 17.8 illustrates the arrangement of components for a typical cold-storage refrigerating plant.

ABSORPTION REFRIGERATION SYSTEMS

At first thought it seems strange that heat can be removed by doing work on a gas, since the work done on the gas will raise its temperature rather than lower it. It seems even more strange that we can *remove heat with a flame*. The *absorption refrigerator* removes heat (refrigerates), and in the process the refrigerant, in water solution, is heated by a gas flame. A brief explanation of this system follows.

17.8 Description of an Absorption System

The principles of absorption refrigeration were first discovered by Michael Faraday more than a hundred years ago. A detailed technical discussion of absorption refrigeration will not be given here, but the basic principle will be explained with the aid of the diagram of Fig. 17.9. Both the explanation and the diagram are much simplified. There are no moving (rotating) parts to such systems, but the valve-and-control arrangement is, in many cases, exceedingly complex.

There is an *evaporator*, as with any other refrigerator. It is shown at *A* in the diagram. Ammonia boils in the evaporator and absorbs heat from the refrigerator cabinet and its contents. The evaporator coils contain some hydrogen, and the presence of this gas speeds up the evaporation of the ammonia. This resulting mixture of ammonia and hydrogen is heavier than either gas alone and flows by gravity down a pipe to the *absorber B*. Here the ammonia from the ammonia-hydrogen mixture is absorbed into solution in water, and the light hydrogen is free to return to the evaporator. The heat picked up in the evaporator is now in the ammonia-water mixture, and this mixture drains to the *generator*

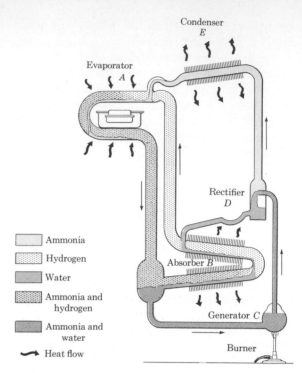

Fig. 17.9 Schematic diagram of an absorption refrigeration system. (Frigidaire Division, General Motors Corp.)

C. Heat from a small gas flame causes the liquid to boil. A percolator-like arrangement is used so that the generator pressure will force the vapors up into the *rectifier D*. Since ammonia evaporates at a much lower temperature than water, practically all the vapor which rises to the rectifier is ammonia. The little water which does go up is separated in the rectifier and drains back through the absorber. The warm ammonia gas rises from the rectifier to the *condenser E*, where it is cooled (either by water or by air, depending on the size and type of the unit) and liquefied. The final heat rejection to the condenser includes the heat picked up in the evaporator and the heat added by the flame to drive the ammonia out of water solution. Finally the liquid ammonia flows by gravity back to the evaporator, where it is ready to repeat the cycle.

For a more detailed discussion of absorption refrigeration systems, the student may consult

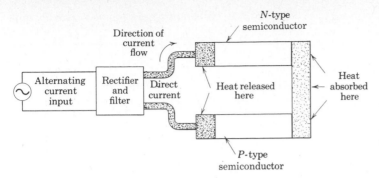

Fig. 17.10 A Peltier couple, diagram form. Both heating and cooling are accomplished by the passage of direct-current electricity through the junctions of two dissimilar semiconductors. At the present time, overall efficiency is lower than that for conventional heating and cooling equipment.

any standard textbook on refrigeration. Absorption systems are now being applied to large-tonnage operations with excellent results. Solar energy can also be used to activate absorption refrigeration machines.

17.9   Thermoelectric Refrigeration

In Sec. 13.9 the *thermoelectric effect*, discovered by T. J. Seebeck in 1820, was discussed. It will be recalled that the process is one of converting heat energy directly into electric energy. A difference in temperature between two junctions of a loop made up of two wires of dissimilar materials causes electrons to flow in the loop. The closed loop with its two junctions is known as a *thermocouple*. (See Fig. 13.7).

In 1834 the French physicist Jean Peltier (1785–1845) discovered the opposite effect—that a forced flow of direct-current (dc) electricity in such a couple produces a temperature difference between the two junctions, one junction being heated and the other cooled. Recent research developments have indicated the feasibility of refrigeration on a commercially successful basis by means of the *Peltier effect*.

Peltier effect refrigeration can be thought of as using electrons as the "working substance" instead of the usual refrigerant. Each electron is assumed to carry with it a certain fraction of the internal energy contained in the entire material. As the electrons move in response to the dc voltage applied to the Peltier couple, they carry internal energy away from one junction and deliver it to the other junction where it is dissipated.

Successful Peltier couples make use of a class of materials known as *semiconductors* (see Chap. 32). Bismuth telluride is one of the materials now being used in several prototype models of thermoelectric refrigerators. Thermoelectric refrigeration is still somewhat in the development stage, not in full mass production. Small-capacity units have been used in space vehicles, in hospitals, in submarines and in some recreational vehicles. Figure 17.10 shows the principle of the Peltier effect in diagram form. To achieve greater cooling capacity many such "couples" are combined to form a *thermoelectric module*.

17.10   Cryogenics

There has been a great deal of interest in the past three decades in the physics of extremely low temperatures. The term *cryogenics* (from the Greek, meaning "icy cold") has been applied to the study of physical phenomena at temperatures below that of liquid air (86 K, or 130°R). Liquid air is considered to be at the "warm end" of the cryogenic region, and *low-temperature physics* usually implies temperatures below 10 K.

Many unusual things happen at very low temperatures, a few of which will be cited here as examples. At about 20 K and below many electric conductors become *superconductors*, i.e., their electric resistance drops to near zero. Below about 2.2 K liquid helium behaves in a most peculiar manner. Its viscosity reduces to zero (*superfluidity*), and it climbs right up the side of a glass tube or container. If an empty test tube, previously cooled to about 2 K, is pushed partly

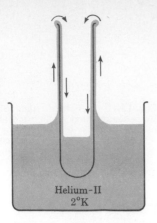

Fig. 17.11 Illustration of "creep" at cryogenic temperatures. "Superfluid" helium II climbing the side of a test tube at a temperature of 2 K.

Table 17.1 Boiling points of some cryogenic and common substances

| Substance | Boiling Point | | | |
	°F	°R	°C	K
Water	212	672	100	373
Ether	95	555	35	308
Ammonia	−28	432	−33	240
Methane	−259	201	−161	112
Oxygen	−297	163	−183	90
Nitrogen	−320	140	−196	77
Air	−330	130	−187	86
Hydrogen	−423	37	−253	20.3
Helium IV	−452	8	−269	4.2
Helium III	−454	6	−270	3.2

below the surface of superfluid helium-II, the helium will climb the outside of the test tube, go right over the top, and down the inside wall until the test tube is filled to the same level inside as the helium level is on the outside (see Fig. 17.11). This phenomenon is aptly called *creep.*

Living tissue can be preserved and metabolic processes held in abeyance for long periods of time at cryogenic temperatures. Cryogenic surgery (destruction or excision of selected tissues or cells by freezing with liquid nitrogen) is now successful in some types of cancer therapy and in brain surgery. And there is a popular dream (it is indeed *only a dream at present*) of being able to freeze a human being and suspend all life processes so that aging would not occur. The "dream" is to be thawed out, still young and vigorous, perhaps a century later.

Cryogenic Liquids A number of substances which are gases at normal temperatures exist as liquids in the cryogenic region, e.g., air, carbon dioxide, methane, oxygen, nitrogen, hydrogen, and helium. The boiling temperatures of some cryogenic liquids are listed in Table 17.1, with a few more common substances also listed for comparison.

Liquid oxygen (LOX) is used in large quantities as an oxidizing agent for liquid-fuel rocket motors. Liquid hydrogen serves as the fuel in some rockets and is also manufactured in large quantities. Methane in liquid form is the well-known *liquefied petroleum gas* (LPG) sold all over the world in pressure tanks for fuel. Liquid air is, of course, the intermediate stage in obtaining pure oxygen and pure nitrogen. Liquid helium is not of great importance *industrially*, but it is of extreme value in low-temperature laboratory research, since the lowest temperatures obtainable are reached by using liquid helium as the cooling medium. A temperature of 0.71 K has been reached by pulling a vacuum on liquid helium and thus cooling it adiabatically through its own evaporation.

Methods in Low-temperature Research Low-temperature physics research has been going on since late in the nineteenth century, when in 1898, James Dewar (1842–1923), for whom the Dewar flask is named (see Fig. 14.15), liquefied hydrogen and reached 15 K. In 1908 a Dutch physicist, H. Kamerlingh Onnes (1853–1926) liquefied helium (4.2 K) and went on to reach a low of 0.7 K. In 1935 a temperature of 0.003 K was reached, and during the past 40 years researchers have been able to make further reductions, with the lowest temperature reported now being of the order of

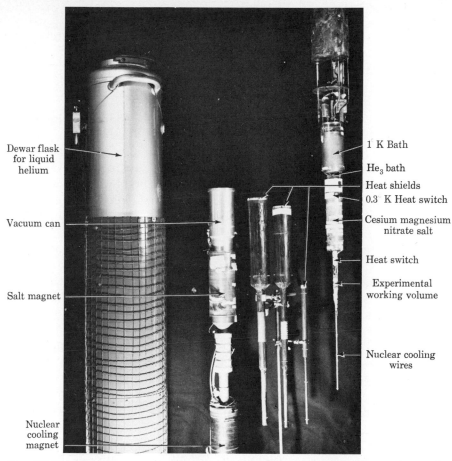

Dewar flask for liquid helium

Vacuum can

Salt magnet

Nuclear cooling magnet

1 K Bath

He₃ bath

Heat shields

0.3 K Heat switch

Cesium magnesium nitrate salt

Heat switch

Experimental working volume

Nuclear cooling wires

Fig. 17.12 Research apparatus for reaching extremely low temperatures by nuclear demagnetization methods. (Department of Physics, University of California, San Diego.)

10^{-6} K, as a result of nuclear adiabatic demagnetization.

The methods in most common use for attaining extremely low temperatures are:

1. Adiabatic expansion of a gas (see Sec. 16.6) while it does external work. This method is often used as a starting point to get the temperature of a gas down close to the liquefaction point.

2. Sudden expansion of a gas through a throttling valve, the *Joule-Thomson effect*. Most cryogenic substances are obtained in liquid form by this method.

3. Rapid evaporation of the cryogenic liquid under vacuum.

4. Demagnetization of a paramagnetic salt already at a very low temperature.

5. Nuclear adiabatic demagnetization methods.

By suitable combinations of methods 1, 2, and 3, helium can be liquefied, and temperatures as low as 0.7 K can be reached. These methods can be operated on a more or less continuous basis, so that low-temperature researchers now have "refrigerators" capable of maintaining 0.7 K for extended periods. The last two methods mentioned above have been, until very recently, "one-shot" experiments. In the late 1960s physicists in San Diego, California; in England; and in Helsinki and Moscow have developed "dilution refrigerators" and "nuclear demagnetization refrigerators" capable of *maintaining* temperatures as low as 1 millidegree. Starting from this base temperature of 0.001 K and using nuclear adiabatic demagnetization on a "one-shot" basis, temperatures of 0.000001 K have been attained. (See Fig. 17.12.)

AIR CONDITIONING

Seasonal extremes of heat and cold in most parts of the United States are greater than people want to endure. Furthermore, humidity, dust, and smoke conditions in many areas are detrimental to health and comfort. Winter heating of homes and places of work has been common for centuries, but it is only fairly recently that humidity controls have been added to heating systems. Cooling and dehumidification of moist summer air are fairly recent innovations in the realm of human comfort. Also of comparatively recent origin are the demands of many industrial processes for accurate temperature and humidity control and for dust-free manufacturing areas.

17.11   What Is Air Conditioning?

Air conditioning may be classified under two basic headings: (1) *winter air conditioning* and (2) *summer air conditioning*. As might be expected, *heating* is the most important aspect of the winter phase, and *cooling* that of the summer phase. Besides *temperature control*, both winter and summer air-conditioning systems should provide other essentials to human comfort, namely, *humidity control*, *control of dust, smoke, odors, and noise*, and *proper air circulation*, including a supply of fresh outside air for ventilation.

17.12   The Atmosphere and Human Comfort

Air is a mixture of several gases, together with water vapor, dust, smoke, bacteria, and other particles. Pure dry air has approximately the following composition:

Component	Percent
Nitrogen	78
Oxygen	21
Carbon dioxide	0.04
Other gases	0.96

Oxygen is the vital element in the atmosphere since it is essential to life. The nitrogen serves to dilute the oxygen. Carbon dioxide is essential to plant life and also exercises a regulatory function on human breathing. The other gases have no known effect on human health or comfort. Impurities such as smoke, dust, and bacteria are frequently harmful.

Water vapor is the most variable constituent of air. The amount of water vapor present in air ranges all the way from a mere trace to as much as 0.03 lb of water in a pound of air. Air is said to be *saturated* with water vapor when it contains as much water as it possibly can hold at that temperature—when any drop in temperature causes condensation. The higher the temperature, the greater the amount of water vapor that can be mixed with air before saturation is reached. *Humidity* is the term used to describe the amount of water vapor in the air.

Atmospheric factors which affect human comfort are the air temperature, the amount of water vapor in the air, the relative cleanliness of the air, and its freedom from odors, dust, and smoke. Air which is too dry irritates the nasal passages and chaps lips and skin, while air containing too much water vapor feels dank and "heavy," slows the normal perspiration rate, and aggravates certain respiratory difficulties.

17.13   Humidity

The amount of water vapor contained in the air is measured in two ways. *Specific humidity* (*W*) gives *the actual weight of moisture contained in any given amount of air. Relative humidity* (RH) is the *ratio of the amount of water vapor actually present in the atmosphere to that amount which would be present if the air were saturated* at that temperature. Specific humidity is often expressed in *grains* of water per pound of dry air, the *grain* (gr) being a unit of weight equal to 1/7000 lb. Some tables and charts give specific humidity in *pounds of water vapor per pound of dry air*. Relative humidity is expressed as a percentage of saturation. At saturation the relative humidity (RH) is 100 percent, while absolutely dry air has a relative humidity of 0 percent. Table 17.2 shows the maximum amount of water vapor that 1 lb of air can contain at various Fahrenheit temperatures.

Table 17.2 Water vapor contained in saturated air at various temperatures

Temperature, °F	Weight of Water Vapor, gr/lb of Air
50	53.5
60	77.3
70	110.5
80	155.8
90	217.6
100	301.3
110	415.0
120	569.0
130	780.0

It has been found by experiment that human comfort is much more closely related to *relative humidity* than to *specific humidity*. This is reasonable, for it is the nearness to saturation which interferes with normal bodily cooling processes by evaporation of moisture from the skin. Most people feel comfortable when the indoor condition is in the range 70–72°F with RH at 40–50 percent (winter); and at 74–78°F, 50 percent RH (summer). The actual amount of water vapor which air can hold is determined solely by the temperature, and the temperature is an important factor in determining *relative humidity*, also. Note from Table 17.2 that at 80°F, 156 gr of water vapor per pound of air results in saturation, or 100 percent relative humidity. At 110°F, however, this same amount (156 gr) of water

vapor would give a RH of only $\frac{156}{415} = 0.376$ or 37.6 percent RH.

17.14 Measuring Relative Humidity

The most convenient means for measuring relative humidity is the *sling psychrometer*. Figure 17.13 shows a common type. The instrument consists of two identical thermometers mounted on a light frame which can be whirled in the air. One thermometer, the *wet bulb* (WB), is covered with a wick which is saturated with water before taking a reading. The other thermometer, the *dry bulb* (DB), has no wick. As the instrument is whirled or "slung" through the air, evaporation from the wet wick occurs, cooling the bulb of the wet-bulb thermometer. Its temperature reading falls below that of the dry-bulb thermometer, and

Fig. 17.13 Sling psychrometer for relative-humidity measurements. (Carrier Corp.)

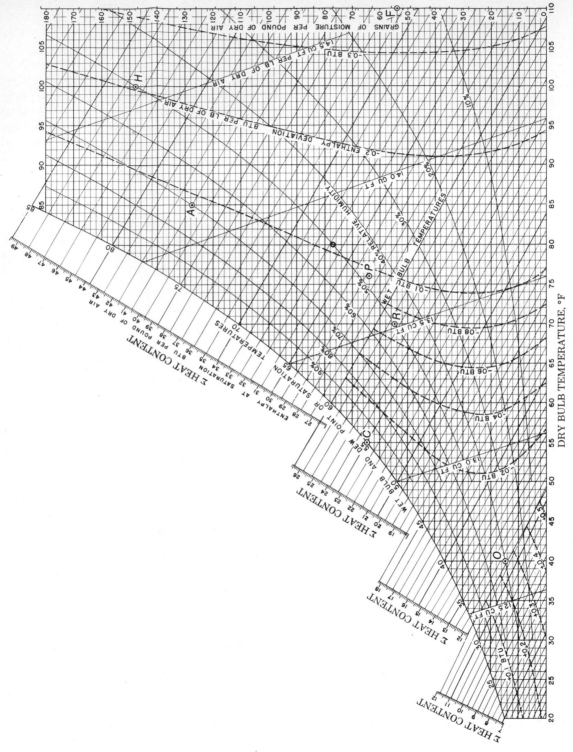

Fig. 17.14 Psychrometric chart (normal temperatures) for comfort air-conditioning use. The Σ heat-content scale gives values of the total heat content in air (sensible and latent), measured from an arbitrary base of 0°F, dry air. (Carrier Corp.)

the difference between the two readings is called the *wet-bulb depression*. Since the cooling effect on the "wet bulb" depends on the evaporation rate from the wick, and the evaporation, in turn, depends on the degree of saturation of the surrounding air, it follows that the wet-bulb depression is a measure of relative humidity.

The properties of air are ordinarily presented in graphical form by means of a *psychrometric chart*. Many air-conditioning problems can be successfully solved by reference to such a chart. A portion of a psychrometric chart for normal temperatures is shown in Fig. 17.14.

Illustrative Problem 17.2 The RH of the air in an auditorium is being determined with a sling psychrometer. The instrument gives these readings: dry bulb, 85°F; wet bulb, 77°F. What is the RH in percent?

Solution Enter the psychrometric chart (Fig. 17.14) at 85° (dry bulb). Follow the vertical line upward until it intersects the 77° (wet bulb) line slanting downward and to the right. Note the intersection (point *A* on the chart) is on the 70 percent RH line.

$$70\% \text{ RH} \qquad \qquad ans.$$

WINTER AIR-CONDITIONING SYSTEMS

There are two basic types of systems for winter air conditioning: (1) *steam and hot-water systems* and (2) *warm-air systems*. Each has several variations, and we shall treat only the more typical arrangements in a very elementary way. Some are true air-conditioning systems, and others do only part of the job.

17.15 Steam and Hot-water Systems

The typical steam (or hot-water) heating system for homes and factories comprises a furnace and boiler usually located in a basement or separate boiler room, with radiators located in the several spaces to be heated. Figure 17.15 shows a typical hot-water system in diagram form. Coal, oil, or gas may be used as the fuel. The hot water (or steam) circulates to the radiators in pipes, and return lines are provided to take the water back to the boiler. Each radiator may have a valve which is thermostatically controlled, so that the temperature in each space can be controlled independently. Automatic valves control fuel supply to gas and oil burners, and automatic stokers keep coal burners at the proper level of operation. Types of controls vary depending on whether it is a steam system or a hot-water system, but the basic principles of operation are quite similar. Base board *radiators* or fan-coil unit *convectors* are used in the rooms to be heated.

Steam and hot-water heating systems are dependable and usually quiet in operation, and temperature can be effectively controlled. They usually have no provision for humidifying the air in the heated spaces, and for two reasons (no ventilation and no humidity control) steam and hot-water radiator systems are not true airconditioning systems but merely *heating systems*. The tendency of these systems is to produce a poorly ventilated and extremely dry condition which is not conducive to optimum human comfort. Console-type humidifiers can, of course, be provided in each space to be heated.

17.16 Warm-air Systems

There are several different types of warm-air systems, but the *forced-air system* is the most common. Figure 17.16 shows a typical, small forced-air furnace. The furnace may be in a basement, but can be located at any level since the air circulation is forced by the blower and does not depend on convection currents. It may be gas-, oil-, or coal-fired. It includes a large air-heating chamber, called the *heat exchanger*, through which air is circulated in close contact with the hot iron *firebox*. Fresh outside air is pulled into the system at all times, the ratio of fresh to recirculated air being controlled by damper settings. In the furnace bonnet, there is often a spray-type humidifier, which may be automatically controlled from *humidistats* in the conditioned

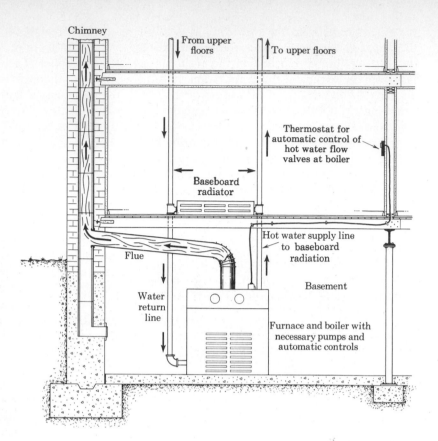

Fig. 17.15 Schematic diagram of a hot-water heating system for a home, very much simplified. (See page 353.)

Labels in figure: Chimney — From upper floors — To upper floors — Thermostat for automatic control of hot water flow valves at boiler — Baseboard radiator — Hot water supply line to baseboard radiation — Flue — Basement — Water return line — Furnace and boiler with necessary pumps and automatic controls

spaces. Relative humidity can be controlled within limits of ± 10 percent. *Filters* for dust control are installed in the path of the return air and fresh air.

The conditioned air is circulated through insulated sheet-metal *ducts* to the rooms in the system. *Registers* discharge the conditioned air into the rooms with a minimum of draft and air noise. Return registers and ducts bring the cold air (near the floors) back to the furnace for reheating. *Thermostats* in the heated spaces control the airflow to the space, or sometimes the burner itself. A *fan-limit switch* delays the start of the blower motor until furnace bonnet temperature has risen to a set value (usually 180°), thus eliminating cold-air blasts to the rooms when the system first turns on. Safety controls called *duct stats* may be installed at critical points in the ducts and will automatically shut down the furnace if duct temperatures rise to dangerous levels.

Electrically Heated Systems In areas where electric energy is competitive in cost, electric heat is very common. One typical electric-heat installation is similar in most respects to the forced-air system described above. The furnace unit is fitted with electric strip (resistance) heaters instead of facilities for the burning of a fuel. A blower forces the air through the strip heaters and circulates it to the rooms through ducts and registers in the same manner as described above. Thermostats and electric controls cycle the operation of the strip heaters and the blowers to regulate room temperatures. Another system features electric-resistance baseboard heaters in each room.

As an approximation, the heat output from electric strip (resistance) heaters can be calculated from the following:

$$\underset{\text{(electric power)}}{1 \text{ kW}} = \underset{\text{(heat)}}{3400 \text{ Btu/hr}}$$

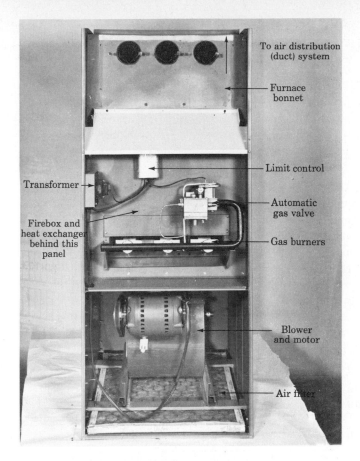

To air distribution (duct) system

Furnace bonnet

Limit control

Transformer

Automatic gas valve

Firebox and heat exchanger behind this panel

Gas burners

Blower and motor

Air filter

Fig. 17.16 Gas-fired, forced-air furnace with front panel removed, showing essential operating parts and controls. Air distribution system not shown. (York Division, Borg-Warner Corp.)

Humidifiers and filters are installed in electric heat systems in the same manner as in other forced-air systems.

The following illustrative problem will illustrate the use of the psychrometric chart in the solution of heating problems.

Illustrative Problem 17.3 A forced-air furnace handles 1000 ft³/min (or cfm, the term used in the air-conditioning industry) of all outside air (40°F dry bulb and 40 percent relative humidity). The air is heated and humidified and leaves the furnace at 110°F (dry bulb). This air is supplied to the conditioned space, where a condition of 70°F and 50 percent relative humidity is maintained. (*a*) How many gallons of water per hour must the humidifier use to give the desired room condition? (*b*) What is the output of the furnace in Btu per hour?

Solution

(*a*) Locate the outside-air condition on the psychrometric chart (Fig. 17.14, point *O*). Note that the specific humidity (scale at right of chart) is 14 gr of water per pound of air. Now locate the desired room condition on the chart (point *R*), and note that the specific humidity is now 54 gr of water per pound of air. The humidifier must have added 54 − 14, or 40, gr of water per pound of air. Now note the lines running steeply up and to the left on the chart. These are lines of *specific volume*, which give the number of cubic feet occupied by 1 lb of air. We may estimate the *specific volume* of the outside air as about 12.6 ft³/lb. Since we are handling 1000 ft³/min,

$$\text{Air quantity} = \frac{1000 \text{ ft}^3/\text{min}}{12.6 \text{ ft}^3/\text{lb}} = 79.4 \text{ lb/min}$$

Multiplying yields

$$79.4 \text{ lb/min} \times 40 \text{ gr/lb} = 3170 \text{ gr/min}$$

$$\text{Gal water/hr} = \frac{3170 \text{ gr/min} \times 60 \text{ min}}{7000 \text{ gr/lb} \times 8.33 \text{ lb/gal}}$$

$$= 3.27 \text{ gal/hr} \qquad ans.$$

(*b*) Since no moisture is gained or lost after leaving the humidifier (in the furnace), the air leaving the furnace has the same *specific humidity* (54 gr/lb) as the air in the room. The condition of the air leaving the furnace is therefore shown by the point *F* on the chart. To determine the heat required to get the air from condition *O* (outside air) to condition *F* (air leaving furnace), locate the Σ Heat Content scale at the left of the chart. This scale gives values of the heat added (sensible and latent) per pound of air to bring it to the *condition point* being considered (in this case, condition point *F*) from an arbitrary base reference point of 0°F, zero humidity. The calculation is as follows:

Air leaving furnace (point *F*)	35.3 Btu/lb
Air entering furnace (point *O*)	11.8 Btu/lb
Heat added by furnace to each pound of air	23.5 Btu/lb

From part (*a*) above, the air quantity is 79.4 lb/min. Multiplying the air quantity by time and heat added per pound, we obtain

Furnace-heat output

$$= 79.4 \text{ lb/min} \times 23.5 \text{ Btu/lb} \times 60 \text{ min/hr}$$

$$= 112,000 \text{ Btu/hr} \qquad ans.$$

SUMMER AIR-CONDITIONING SYSTEMS

It will be recalled that the basic problem of a summer air-conditioning system is to cool and dehumidify the air. Both cooling the air and dehumidifying it are problems of *heat removal*. To cool hot dry air, we remove its *sensible* heat, which is that heat whose effect is noted as an increase in temperature, so called because the body "senses" the fact that the air is hotter. (*Latent heat*, on the other hand, causes no dry-bulb temperature increase but indicates the presence of more water vapor in the air.) This sensible heat is removed in accordance with the basic heat equation $H = wc(t_2 - t_1)$, where c is the specific heat of dry air (0.24), w is the weight of air being cooled, and $t_2 - t_1$ the temperature drop desired. In dehumidifying air, we remove the latent heat of vaporization L_v of the water vapor in the air. The vapor then condenses and drains off, carrying latent heat away with it.

17.17 Classification of Heat Gains

In summer weather the air in buildings gains heat from several sources. A cooling system must be designed and sized to dissipate all these heat gains. The main sources of heat gains are the following:

1. Heat gained by *conduction* through walls, ceilings, and floors due to the difference between outside and inside temperatures.

2. *Radiant-heat energy* which enters through windows and other glass areas.

3. Heat gains from lights, stoves, equipment, and machinery *in the building* [3.4 Btu/(hr)(W), or 2540 Btu/(hr)(hp) of electric motors in operation].

4. Heat gained from *people* (400 to 1000 Btu/hr per person, depending on the degree of muscular activity).

5. *Sensible heat* in the outside air which enters through cracks, windows, and doors, or is taken in for ventilation.

6. *Latent heat* in the moisture of ventilation air, or given off by equipment in the space. This latent-heat load is very small (less than 5 percent of the total) in buildings which contain few people and which are situated in dry climates. On the other hand, in crowded theaters or restaurants located in humid climates the latent-heat load may be well over one-half the total load.

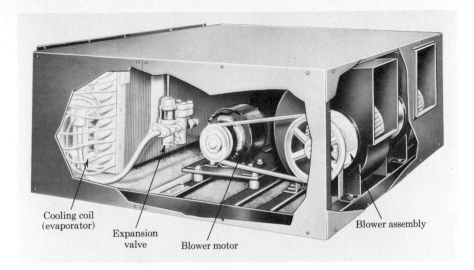

Cooling coil
(evaporator)

Expansion
valve

Blower motor

Blower assembly

Fig. 17.17 Summer air conditioning unit, or fan-coil unit, for installation in a duct serving one "zone" of an air conditioning system. The blower and its motor are at the right and the finned cooling coil is at the left. (Frigidaire Division, General Motors Corp.)

17.18 Basic Components of Summer Air-conditioning Systems

The student should keep in mind that it is not a *building* that is being cooled and dehumidified— it is the *air* in the building. The basic components of an air-conditioning system must operate in such a manner that the air in the building will be passed through the air conditioner, where both sensible heat and latent heat can be removed. A certain proportion of ventilation air is also cooled and dehumidified and mixed with the recirculated air. The conditioned air is then delivered to the occupied spaces through the duct-and-register system.

Many different arrangements of equipment for summer air conditioning are possible. Any system, however, must contain the following basic components:

1. The *air-conditioning unit* itself, consisting of the cooling and dehumidifying coil, the blower, and filters. This unit is ordinarily housed in a sheet-metal casing and is installed in the duct system in such a way that it can handle both the recirculated air and the ventilation air from the

outside. Figure 17.17 shows a common type of air-conditioning unit.

2. The *refrigerating equipment*, which may be at a remote location from the air-conditioning unit. Several air-conditioning units may operate from a single refrigerating plant.

3. The *duct-and-register system*, properly sized and insulated to deliver conditioned air to the spaces without objectionable noise or drafts.

4. The *control system*, which cycles the refrigerating equipment, controls the air-conditioning unit, and operates mixing dampers to control airflow.

17.19 The Air-conditioning Unit

The heart of the air-conditioning unit is the cooling and dehumidifying apparatus. This may be in the form of a *direct-expansion cooling coil* with extended, finned surface; or a chilled water spray; or a chilled-water coil with finned surface. The present trend for all but the largest installations is toward the direct-expansion cooling coil. The

Fig. 17.18 Centrifugal compressor and water chiller for a large air-conditioning system. The chilled water is pumped to air-conditioning units (also called fan-coil units) in or near the spaces to be cooled. (Carrier Corp.)

Labels in figure: Shell-and-tube condenser; Turbo-compressor; Speed increaser; Synchronous motor; Discharge line; Suction line; Water chiller evaporator

direct-expansion coil (see Fig. 17.17) is actually the *evaporator* of the refrigerating system and is made up of copper tubing coiled several tubes thick, with extended-surface fins for greater heat transfer. The coil is mounted in the air-conditioning-unit casing in such a way that all the air (fresh and recirculated) must pass through it. Some of the moisture in the air may condense out on the cold coil and fall into a drain below, the amount of condensate depending on the coil temperature and the condition of the entering air. This condensed water removes latent heat from the air. Contact with the metal surface of the coil removes much of the sensible heat from the air. The expansion valve supplies refrigerant to the coil, and the compressor maintains a low pressure in the coil which will give a coil temperature of about 36°F.

The blower is ordinarily of the "squirrel-cage" type (see Fig. 17.17), mounted in sleeve bearings and V-belt-driven by an electric motor. It is common practice to handle about 400 ft³/min of air per ton of refrigerating capacity.

Filters may be either metal honeycomb-type cells coated with oil (these are cleanable for reuse) or disposable-type filters made of fiberglass or honey-combed cardboard impregnated with oil. An interesting type of filter is the electrostatic-precipitation type, which utilizes a high-voltage electrostatic charge to attract dust and smoke particles to the charged plates, thus removing the particles from the airstream.

17.20 The Refrigerating Equipment

Refrigerating equipment for air-conditioning systems varies widely with size and type of system. If a chilled-water type of air-conditioning unit is used, the evaporator is located in the water-chilling apparatus in the equipment room, and centrifugal compressors may be used (see Fig. 17.18). For direct-expansion air-conditioning coils, Freon-12 or Freon-22 are the refrigerants most often used, and the tendency is to use multi-cylinder compressors of V, W, or radial design. These may operate at from 600 to 1750 rev/min.

There are some basic differences between compressors designed for air-conditioning work and those designed for most other refrigerating applications. The standard working cycle is different, of course, since the evaporator temperature must not fall below 32°F for air-conditioning systems. Control design and control settings are therefore different from those on low-temperature applications for ice-making, cold storage, and freezing plants.

17.21 Ducts and Registers

The ducts which supply conditioned air to the spaces and the ducts which return warm air to the air conditioner are ordinarily made of galvanized iron, although sheet aluminum and fiberboard are also used. The supply ducts are usually wrapped with fiberglass or other insulating material to prevent heat gains from warming up the chilled air before it arrives in the conditioned space, and also to prevent cold ducts from condensing moisture on their surfaces and dripping. Ducts frequently run through attics, and since attic temperatures may rise to 140°F on a hot summer day, insulation is very important. Ducts should be sized so that air velocities within them do not exceed around 1000 ft/min. Higher velocities will usually result in an undesirable noise level as well as unacceptable pressure drops and fan-power requirements. The proper sizing of ducts is a rather technical engineering problem, since each air-conditioning system varies in its air-supply demands. The basic equation for duct sizing is

$$Q = Av \qquad (17.6)$$

where Q = quantity of air to be supplied, ft^3/min
  A = cross-sectional area of the duct, ft^2
  v = air velocity, ft/min

Ordinarily the quantity of air to be supplied should result in a complete air change for the conditioned space about every 6 to 10 min. The amount of outside (ventilation) air to be brought into the system depends on the nature of the spaces to be cooled and the number of people in them. From 10 to 30 ft^3/min should be supplied per person, so that in theaters or crowded restaurants the system will often have to operate on 100 percent outside air. Cooking odors, as in restaurants, and heavy smoking will also increase the outside-air requirements. Registers should be so sized and placed that objec-

tionable drafts will not result, and velocities must be low enough so that air whistle is not evident.

17.22 Year-round Air-conditioning Units

Many homes, theaters, restaurants, and commercial establishments are equipped with "year-round" air-conditioning systems with automatic controls to cycle both the winter equipment and the summer equipment. An additional control refinement can be provided to cycle automatically from one system to another. In some areas, a small amount of heat is required in the early morning of the same day in which cooling might be required during the warm afternoon. Many industrial processes require close temperature control regardless of the ups and downs of outdoor weather, and both the summer and winter equipment are required. Figure 17.19 shows a typical year-round air conditioner for a home.

Illustrative Problem 17.4 An air conditioner handles 2000 ft^3/min of all outside air for a hospital operating room. The outside condition is 100°F dry bulb and 50 percent relative humidity. The air off the cooling coil is 55°F, saturated (wet bulb = dry bulb). (*a*) How many gallons of water per hour are condensed and drip off the cooling coil? (*b*) How much cooling (tons of refrigeration) is the unit doing? (*c*) Assuming no further latent gains or losses, what is the relative humidity in the operating room if the dry-bulb temperature is 76°F? (*d*) The duct to deliver the air to the operating room is limited to a 12-in. vertical dimension. If air velocity in the duct is to be 600 ft/min, what should the horizontal dimensions of the duct be?

Solution

(*a*) Locate the outside-air condition (point H on the psychrometric chart, Fig. 17.14) and the air-off-the-coil condition (point C). Note that the moisture condensed from each pound of air is

$$148 \text{ gr (at } H) - 65 \text{ gr (at } C) = 83 \text{ gr/lb}$$

Assume a specific volume (SpV) of 14.6 ft^3/lb. (See SpV lines on psychrometric chart). The weight of

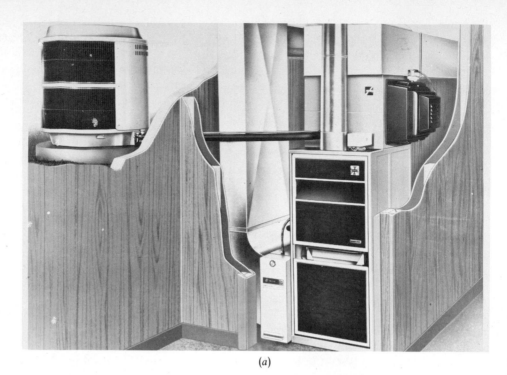

(a)

Fig. 17.19 (a) Phantom view of a year-round air-conditioning unit for a small residence. This is a "split system," with the condensing unit located outdoors to minimize noise and vibration in the house, and the cooling coil located on top of the furnace in the basement. The furnace blower circulates warm, humidified air in winter, and cool, dehumidified air in summer, through the same duct system. (b) Cutaway view of condensing unit for "split system" air conditioner, showing compressor (center) surrounded by finned-coil condenser assembly. Fan and motor are at top center. (Both photos courtesy of Carrier Corporation.)

(b)

air being handled is then

$$\frac{2000 \text{ ft}^3/\text{min}}{14.6 \text{ ft}^3/\text{lb}} = 137 \text{ lb/min}$$

Therefore the moisture condensed per hour is

$$\frac{83 \text{ gr/lb} \times 137 \text{ lb/min} \times 60 \text{ min/hr}}{7000 \text{ gr/lb} \times 8.33 \text{ lb/gal}} = 11.7 \text{ gal/hr}$$

ans.

(*b*) The total heat content of the incoming air (at *H*) is 47.5 Btu/lb, and that of the air which is leaving the air conditioner (at *C*) is 23.2 Btu/lb:

$$47.5 - 23.2 = 24.3 \text{ Btu of cooling per pound}$$
of air

From part (*a*) we found approximately 137 lb/min of air passing through the coil.

Cooling output

$$= \frac{24.3 \text{ Btu/lb} \times 137 \text{ lb/min} \times 60 \text{ min/hr}}{12,000 \text{ Btu/hr}}$$

$$= 16.6 \text{ tons of refrigeration} \qquad ans.$$

(*c*) The *specific* humidity in the operating room is the same as that of the air leaving the cooling coil (65 gr/lb of air), since we assumed that no moisture is added or lost. Locate the 65 gr/lb specific-humidity point on the right side of the chart and go horizontally to the left until the 76°F dry-bulb line (vertical) is reached. This point is marked *P* on the chart and gives

$$\text{Relative humidity} = 48\% \text{ (approx)} \qquad ans.$$

(*d*) Since a velocity of 600 ft/min is not to be exceeded, with an air quantity of 2000 ft³/min, use Eq. (17.6) and solve it for the area, *A*.

$$Q = Av, \quad \text{and} \quad A = Q/v$$

Substituting values

$$A = \frac{2000 \text{ ft}^3/\text{min}}{500 \text{ ft/min}} = 4.0 \text{ ft}^2$$

Since the height of the duct is limited to 12 in. (1 ft), the width is

$$4 \text{ ft} \qquad ans.$$

17.23 The Heat Pump

Thermodynamics is a fascinating science, full of seeming paradoxes. One of these is the refrigerator itself, which at first thought seems to defy the second law of thermodynamics. Another is the absorption refrigerating machine, which operates by *adding heat* from a flame to the refrigerant. A third is the *heat pump*, a compression refrigeration air conditioner that both heats and cools air, cycling automatically from cooling to heating on demand from a room thermostat.

A commercial heat pump for air conditioning operates like any air-cooled compression refrigeration air conditioner when summer cooling is desired. Heat is absorbed from the room air by the *cooling coil* (evaporator) and is rejected (along with the heat of compressor work) to outdoor air by the condenser (Fig. 17.19b). The heat pump has a complex system of valves and controls which allows the evaporator and the condenser to exchange functions. When the machine is switched to the *heating cycle*, the outdoor coil (Figs. 17.20 and 17.21) becomes the evaporator. It must operate at a temperature well below that of the cold winter air, so that heat flow *from the winter air into the very cold refrigerant* will occur. This heat, together with *the heat equivalent of compressor work*, is rejected to room air by the indoor coil, now operating as the condenser at a high temperature. The usual blower-and-duct system forces air over the indoor coil to be heated, and circulates the warm air to the conditioned spaces.

Heat pumps have rather interesting thermodynamic properties. The coefficient of performance for a heat pump on the winter cycle is greater than that for a refrigerator operating between the same temperature limits, because in the heat pump, the heat equivalent of compressor work is *added* to the useful heat output of the machine. In contrast, for summer cooling the heat pump is on a regular refrigeration cycle, and the heat of compressor work is a loss factor which must be removed from the refrigerant by the condenser and rejected to the outside air.

The theoretical (i.e. Carnot) coefficient of performance of a heat pump on winter cycle is

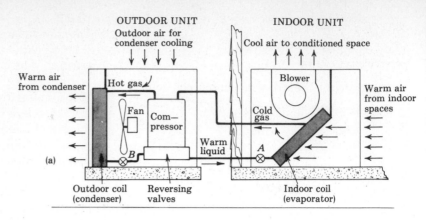

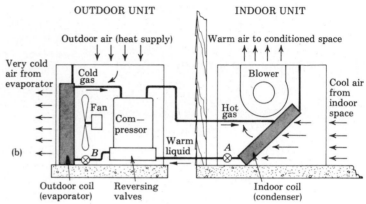

Fig. 17.20 The heat pump. (*a*) Schematic diagram of cooling cycle. (*b*) Schematic diagram of heating cycle.

given by

$$\text{c.o.p.}_{\substack{\text{(heat pump)}\\\text{Carnot}}} = \frac{T_1}{T_1 - T_2} \qquad (17.7)$$

where

T_1 = absolute temperature of the heating (indoor) coil (the condenser on winter cycle).

T_2 = absolute temperature of the evaporator (outdoor) coil.

Heat pumps are quite effective heating devices in climates where mild winters are the rule. In very cold climates however, their efficiencies drop off rapidly, as the following calculations show.

Illustrative Problem 17.5 A heat pump is set to produce an indoor coil temperature of 120°F.

What is its *theoretical coefficient of performance* when the outdoor coil operates at (*a*) 30°F? (*b*) −10°F?

Solution

(*a*) $\quad \text{c.o.p.}_{\text{(Carnot)}} = \dfrac{580°\text{R}}{(580 - 490)°\text{R}}$

$\qquad\qquad = 6.45 \qquad\qquad\qquad ans.$

(*b*) $\quad \text{c.o.p.}_{\text{(Carnot)}} = \dfrac{580°\text{R}}{(580 - 450)°\text{R}}$

$\qquad\qquad = 4.45 \qquad\qquad\qquad ans.$

From these idealized-cycle Carnot coefficients, the decrease in heating effectiveness in cold climates is readily apparent. The *actual coefficient of performance* of a heat pump on winter cycle is given by

Fig. 17.21 A single package, high efficiency heat pump for year-round residential and light commercial air conditioning use. Panels have been removed to show the essential components. (York Division of Borg-Warner Corp.)

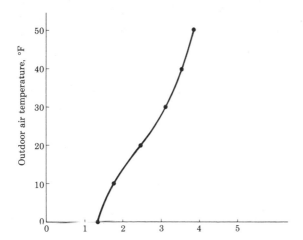

Fig. 17.22 Actual c.o.p., air-to-air heat pumps, at various outdoor temperatures.

$$\underset{\text{(actual)}}{\text{c.o.p}} = \frac{\text{heat obtained from condenser (indoor coil)}}{\text{heat equivalent of electric energy input to compressor motor}} \quad (17.8)$$

Using this indicator of *actual* performance, and substituting data obtained under operating conditions, the coefficients of performance for

heat pumps on winter cycle range from a high of around 4.0 in very mild climates down to about 1.5 in cold winter conditions. Figure 17.22 gives an indication of the typical performance of commercial heat pumps operating on an air-to-air cycle. It is common practice to supplement the heat output of heat pumps with electric strip (resistance) heaters for use whenever the outdoor temperature drops much below 20°F.

17.24 Conserving Energy in Air Conditioning

In this era of critical shortages and high costs of energy, every possible step must be taken to conserve electricity and fuels. Instead of simply installing more cooling capacity or a larger furnace, attention should be given to increasing the efficiency of the equipment. Even more important is the need to insulate all spaces to be air conditioned in order to reduce significantly the heat gains and losses.

Human comfort, too, is somewhat flexible. Long accustomed to 72°F as a comfortable indoor winter temperature, perhaps we will have to "dial down" to 68°F. And, by the opposite token, a room need not be at 72°F to be comfortable in summer when the outdoor temperature is 100°F. Actually 80°F DB, 50% RH indoors is quite "comfortable" when the outside temperature is 100°F. With more than a little justification, a critic has referred to air conditioning as "a system which enables people to keep their homes colder in the summer than they are in the winter."

Proper insulation and weatherstripping are extremely important to conserve energy. Ceiling insulation is especially critical, since attics are very cold in winter and very hot in summer. At least six inches of insulation should be standard for ceilings, and a minimum of three inches in exterior walls. In cold climates, windows should be double-glazed to reduce heat conduction, or storm windows should be installed in the winter. All doors and windows should be weatherstripped to minimize heat gains and losses due to convection. In the summer, depending on the orientation of windows (west and south), very large heat gains can occur as a result of solar radiation. These can be reduced by roof overhang, awnings, shades or plantings of trees and shrubbery.

The calculation of heating and cooling loads due to convection and radiation is quite complex and will not be discussed here. Consult any standard text on air conditioning for this information. We present here a simplified problem involving heat gains and losses from *conduction* only, to illustrate the importance of good insulation.

Illustrative Problem 17.6 A family residence has 2000 ft² of ceiling area, 1400 ft² of exterior wall area, and 600 ft² of window glass. A full basement is under the floor, but it is assumed that there will be no heat losses to or gains from the basement. The following additional conditions apply:

Winter temperatures:
  outside walls 0°F;
  inside walls and ceiling, 60°F; attic 10°F
Summer temperatures:
  outside walls, 95°F;
  inside walls and ceiling, 80°F; attic 105°F

	Before insulation and double glazing	After
Wall conduction factor, U_{wall}	0.28 Btu/hr(ft²)(F°)	0.08
Ceiling conduction factor, U_{ceil}	0.75 Btu/hr(ft²)(°F)	0.05
Window conduction factor, U_{win}	0.74 Btu/hr(ft²)(°F)	0.60

Calculate the conduction heat loss in winter and the heat gain in summer.

Solution

1. Winter heating (conduction loss)

(*a*) Before insulation and double glazing:

$$H_{wall} = 1400 \text{ ft}^2 \times 0.28 \text{ Btu/hr(ft}^2)(F°)$$
$$\times 60°F = 23,500 \text{ Btu/hr}$$
$$H_{ceil} = 2000 \text{ ft}^2 \times 0.74 \text{ Btu/hr (ft}^2)(F°)$$
$$\times 50°F = 74,000 \text{ Btu/hr}$$
$$H_{win} = 600 \text{ ft}^2 \times 0.75 \text{ Btu/hr(ft}^2)(F°)$$
$$\times 60°F = 27,000 \text{ Btu/hr}$$

Winter heat loss before insulation

124,500 Btu/hr *ans.*

(*b*) After insulation and double glazing:

$$H_{wall} = 1400 \text{ ft}^2 \times 0.08 \text{ Btu/hr(ft}^2)(F°)$$
$$\times 60°F = 6,700 \text{ Btu/hr}$$
$$H_{ceil} = 2000 \text{ ft}^2 \times 0.05 \text{ Btu/hr(ft}^2)(F°)$$
$$\times 50°F = 5,000 \text{ Btu/hr}$$
$$H_{win} = 600 \text{ ft}^2 \times 0.60 \text{ Btu/hr(ft}^2)(F°)$$
$$\times 60°F = 21,600 \text{ Btu/hr}$$

Winter heat loss after insulation

33,300 Btu/hr *ans.*

2. Summer cooling A similar set of calculations is left for the student. The results are:

(*a*) Conduction heat gain before insulation
73,500 Btu/hr *ans.*

(*b*) Conduction heat gain after insulation
11,200 Btu/hr *ans.*

Even though these calculations neglect radiation and convection loads, and do not deal at all with *latent heat* (moisture) loads, the importance of proper insulation should be apparent. For a home under these "before" and "after" conditions, averaged for different family sizes and life styles and for current fuel and electricity rates, the annual savings on energy bills could easily run to $250.00 for the heating season and $140.00 for the cooling season.

SUMMARY

Refrigeration is a *heat-removal* process, not a "cold-production" process.

The unit of refrigeration capacity is the ton of refrigeration. One ton of refrigeration equals 12,000 Btu/hr.

A refrigerating machine is a reversed heat engine in the sense that work is done *on* the refrigerant so that heat can "flow *up* a temperature hill."

Refrigerants are low-boiling-point substances which are used as media *for heat transfer through a change of state.* They are vapors (gases) at normal temperatures and atmospheric pressure but are easily liquefied by an increase of pressure or a drop in temperature.

Some commonly used refrigerants are *ammonia, Freon-12, Freon-22, methyl chloride, sulfur dioxide,* and *Carrene.*

There are two basic refrigeration systems: compression systems and absorption systems.

The basic components of a compression refrigeration system are evaporator, compressor, condenser, receiver, and controls.

Compressors for refrigerating plants may be of reciprocating, centrifugal, or rotary design.

Refrigeration, without any moving parts, can be accomplished by the *Peltier effect*, using copper and semiconductors combined to form a thermoelectric couple.

The actual *coefficient of performance* of a refrigerating machine is given by the expression

$$\underset{\text{(actual)}}{\text{c.o.p.}} = \frac{\text{heat absorbed in evaporator per cycle}}{\text{heat equiv. of mech. work per cycle}}$$

$$= \frac{Q_2}{\Delta W}$$

In terms of operating temperatures, if T_2 is the evaporator temperature, and T_1 is the condenser temperature, the theoretical (Carnot) coefficient of performance is

$$\underset{\text{(Carnot)}}{\text{c.o.p.}} = \frac{T_2}{T_1 - T_2}$$

True air conditioning involves all of the following: temperature control, humidity control, ventilation, and the elimination of dust, smoke, odors, noise, and drafts.

Specific humidity is the actual water vapor content of the air, expressed in grains per pound of air (7000 gr = 1 lb).

Relative humidity is a ratio:

$$\text{RH} = \frac{\text{amount of water vapor present}}{\text{amount present at saturation}}$$

It is expressed as a percentage, and it is a better indicator of human comfort than specific humidity.

Psychrometric charts give the properties of air essential to the air-conditioning problem.

Human comfort is related to both temperature and relative humidity.

The summer air-conditioning problem is to cool and dehumidify the air, while the winter problem is to heat and humidify the air.

Heat pumps cool and heat air with the same refrigeration equipment. Provision is made to exchange functions between the evaporator and the condenser. In winter, heat is absorbed from outside air into the refrigerant, and then rejected (along with the heat equivalent of compressor work) to inside air.

QUESTIONS AND EXERCISES

1. Explain how it is possible for a "warm" liquid refrigerant in the condenser to enter the evaporator and immediately begin absorbing heat from products that are much colder than the condenser temperature. [HINT: What happens to the "warm" liquid as it goes through the expansion valve from high pressure (condenser) to low pressure (evaporator)?]

2. Compressing a refrigerant gas adds the heat equivalent of mechanical work to the gas. This would seem to be the wrong thing to do since the refrigerant later in the evaporator must be very cold to absorb heat from the product being cooled. Explain this seeming contradiction.

3. Starting with the basic definition of a ton of refrigeration, show that 1 ton of refrigerating capacity equals 12,000 Btu/hr.

4. Explain in detail how both the first and second laws of thermodynamics apply to the compression refrigerating cycle.

5. Diagram a compression refrigerating system and discuss in detail the function of each of the basic components? What is included in the *high side?* The *low side?*

6. What are the essential purposes of the compressor in a compression refrigerating system? Explain in terms of the second law of thermodynamics.

7. Does the fact that refrigerating machines have coefficients of performance greater than unity mean that such machines do not obey the first law of thermodynamics? (The net refrigerating effect per cycle may be 3 or 4 times the thermal equivalent of compressor work.) Explain in detail how it is possible to get more cooling out of a machine than the heat equivalent of mechanical work (input) to the machine.

8. What process on the "low side" in the refrigeration cycle is an adiabatic process? What does it accomplish?

9. Hold a small can of (liquid) Freon in your hands. It is at room temperature. Open the valve and let some escape into a beaker. There it will start to boil at −21.6°F. Explain the sudden temperature drop. Why does a mere release of pressure produce this effect? What name is given to this cooling process?

10. Why is humidity an important factor in human comfort? Why is relative humidity of more importance than specific humidity?

11. List the relative advantages and disadvantages of steam or hot-water vs. forced-air heating systems.

12. Explain what is meant by *sensible heat; latent heat.* List the types of heat gains present in a crowded lunchroom (summer) and classify them as sensible or latent.

13. What are the four basic components which any air-conditioning system must have?

14. Explain how it is possible to extract heat energy from cold outdoor air at 10°F and use it to heat indoor spaces with warm air at 90°F. For a given amount of electric energy used, why will a heat pump provide more heat than an electric resistance heater?

PROBLEMS

1. A 100-lb cake of ice at 32°F is placed in a refrigerator so well insulated that gains and losses to the exterior are negligible. In exactly 8 hr heat from the products being cooled has melted all the ice and the 32°F water has drained off. What was the cooling capacity of the ice in tons of refrigeration?

2. An "ideal" refrigerator is operating at a rate of exactly one ton of refrigeration (net refrigerating effect). Tests show that in one hour, 15,000 Btu are rejected to the condenser. How much mechanical work (foot-pounds) does the compressor do on the refrigerant gas in one hour?

3. An ice plant manufactures 100 tons of ice daily (24 hr) from supply water whose temperature is 65°F. The ice cakes are at a temperature of 10°F. What is the capacity of the refrigeration plant in tons of refrigeration? (Assume no losses.)

4. An ice plant freezes 100 rectangular cakes per 24-hr day, each weighing 300 lb. The starting point is water at 65°F, and the ending point is ice at 10°F. Each cake is frozen in a stainless steel can whose weight is 75 lb. Assuming no losses, what must be the refrigerating capacity (tons of refrigeration) of the plant?

5. A 500-lb batch of ice-cream mix whose initial temperature is 70°F is to be frozen (freezing point is 28°F) and then subcooled to 10°F in 20 min. The specific heat of ice-cream mix is 0.78 before freezing and 0.45 after freezing. Its latent heat of fusion L_f is 126 Btu/lb. What refrigeration capacity (tons) must the freezer have?

6. A total of 8000 lb of warm sides of beef (initial temperature is 80°F) is placed in a freezer for cooling and freezing. It is desired to reduce the temperature of the beef to the freezing point (28°F), then freeze it, and further subcool the entire load to 15°F. Specific heat of beef before freezing is 0.75; after freezing, 0.40. The latent heat of fusion of beef is 100 Btu/lb. The entire process is to be completed in 24 hr. Neglecting any other losses or loads, calculate the refrigeration capacity required, in tons of refrigeration.

7. An actual refrigerator is operating between 5°F (evaporator) and 95°F (condenser). It is found to have a coefficient of performance of 3.2. If a Carnot refrigerator operated under these conditions, what would be its c.o.p.?

8. An inventor has submitted plans and specifications for a new compressor-operated, air-cooled refrigerator. It is to operate between 10°F and 120°F. He claims that his design will produce one ton of refrigeration with exactly one horsepower of mechanical input. Evaluate this design thoroughly. Show mathematically whether it is feasible or not.

9. Calculate (from Table 17.2) the approximate relative humidity if the specific humidity is found to be 128 gr/lb of air when the temperature is 105°F.

10. From the psychrometric chart determine the relative humidity RH and specific humidity W for each of the following combinations of dry-bulb and wet-bulb temperatures: (a) 105° DB, 69° WB, (b) 95° DB, 72° WB, (c) 92° DB, 86° WB, (d) 40° DB, 35° WB.

11. A blower is handling 7500 ft³/min (cfm) of outside air at 35°F dry-bulb, 60 percent RH. How many pounds of air per hour are being circulated? (See the specific volume lines on the psychrometric chart.)

12. An air-conditioning unit takes in 5000 ft³/min of all outside air (100°F dry bulb and 50 percent RH). The air coming off the cooling coil is at 54°F, saturated. What minimum capacity refrigeration unit (tons) is required?

13. A winter heating system takes in 2000 ft³/min of cold, dry air (30°F, 80 percent RH) and heats it to 105°F dry bulb. If it is desired to add moisture to the air off the furnace with a humidifier, how many gallons per hour would be required to bring the 105°F air up to a relative humidity of 40 percent?

14. If a rectangular duct is to carry 10,000 ft³/min at a velocity of 800 ft/min, what vertical dimension must it have if its horizontal dimension is limited to 38 in.?

15. A round duct has a diameter of 18 in. What flow of air (cubic feet per minute) can this duct deliver at a velocity of 600 ft/min?

16. A rectangular duct measures 34 by 20 in. If it must deliver 4000 ft³/min, what velocity will be required?

17. A rectangular duct whose inside dimensions are 14 by 18 in. carries chilled air whose condition is 50°F, saturated. If the air velocity is 500 ft/min, how many pounds of air per minute flow past a given point in the duct?

18. An air conditioner is handling 2000 ft³/min of all outside air (95° DB and 60 percent RH). The air coming off the cooling coil is saturated at 55°F (DB = WB). (a) How many pounds of water are condensed per minute? (Assume the average specific volume of the air-water-vapor mixture is 14 ft³/lb.) (b) What must be the capacity of the cooling coil in tons of refrigeration? (See Σ Heat Content scale, left of chart, Fig. 17.14.) (c) This air is discharged into the conditioned space where heat gains (all sensible heat) result in a dry-bulb temperature of 80°F. What is the relative humidity in the room?

19. A furnace blower is handling 3000 cfm of outside air at 35°F DB and 40 percent RH. The furnace heats the air and also provides heat for the humidifier (located in the top of the furnace) to evaporate moisture into the air so that the air-off-the-furnace condition is 110°F, 10 percent RH. What is the heat output of the furnace in Btu/hr? (All information needed is on the psychrometric chart.)

20. A small retail store has the following construction details:

		U, Btu/hr(ft²) F°
Rear wall	600 ft² 8-in. brick	0.32
Side wall	1200 ft² 8-in. brick	0.32
Front	200 ft² 8-in. brick	0.32
	400 ft² triple plate glass	0.50
Ceiling	5000 ft² attic above	0.08
		(4 in. fiberglass)

Additional data are:

Inside wall temperature	60°F
Inside glass temperature	50°F
Outdoor wall and glass	10°F
Attic temperature	20°F

Heat losses to or gains from the adjoining store and the basement are assumed zero.

Calculate the conduction heat loss in Btu per hour.

21. A commercial air-to-air heat pump is operating on an outdoor air temperature of 30°F. (a) From the curve of Fig. 17.22, what is its approximate actual c.o.p.? (b) If the total heating requirement of your house at that outdoor temperature is 140,000 Btu/hr, what must be the kilowatt input to the compressor motor to balance that heating load on a steady-state basis?

PART 5 Wave Motion and Sound

Ocean waves possess vast amounts of unharnessed energy.

18 Simple Harmonic Motion

The resultant motion of a body when acted upon by a constant force or torque was considered in Chaps. 4 and 8. Such constantly accelerated motion was described by equations for the position and velocity of the body as a function of time. Formulas relating linear (and angular) displacements and uniform linear (and angular) velocities were developed.

We shall now study another basic kind of motion. In this kind of motion the body is acted upon by a force which is not constant but varies during the motion. In general, when the forces acting on a body are not constant, the results are complicated types of nonuniformly accelerated motion which are often quite difficult to analyze.

However, there is one common and important type of nonuniformly accelerated motion that can be analyzed quite easily. That type of motion is called *periodic* or *harmonic* motion.

Closely related to periodic motion is wave motion. In subsequent chapters we shall study how periodic motion and its related wave motion can be used to describe sound, light, alternating current, and radio phenomena.

18.1 Periodic Motion

Any motion which repeats itself in equal intervals of time is called *periodic motion*. Many phenomena in every day experiences are examples of periodic motion—a swinging pendulum, the turning balance wheel of a watch, the pistons in an automobile engine, a vibrating string of a violin, a vibrating weight on the end of a coiled spring, and vibrating air columns in musical instruments.

Periodic motion is set up when objects are stretched, compressed, or bent from their normal positions and then released from the applied stresses. The object moves back and forth about a fixed point and along a fixed path, executing over and over a sequence of motions in which the object returns to the same position with the same velocity after a definite period of time (friction neglected).

18.2 Elastic Potential Energy

Suppose we attach a mass m at the end of a spring (Fig. 18.1). When we stretch the spring, it resists being lengthened. If we release the stretching force, the spring returns to its original length (zero gravitational and frictional forces are assumed).

When the spring is stretched by a force F, the magnitude of the force required is proportional to the elongation s (see Sec. 10.5). Thus,

$$F = ks \qquad \text{(Hooke's law)} \qquad (18.1)$$

where k is a proportionality constant that depends upon the units used for F and s.

In like manner, if we compress the spring, it resists being shortened and will return to its original length upon removal of the compressional force. In each case, the restoring force is always equal in magnitude, but opposite in direction, to the deforming force. Hence the restoring force F is $-ks$; that is it acts in a direction opposite to

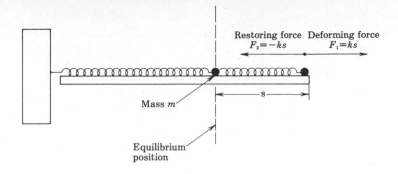

Restoring force $F_2 = -ks$ Deforming force $F_1 = ks$

Mass m

s

Equilibrium position

Fig. 18.1 Within limits of perfect elasticity, the stretching force F is proportional to the elongation s (Hooke's law).

the deforming force and toward the neutral position O.

The work done in stretching (or compressing) the spring is the product of the force F and the distance through which it acts. In these instances, we will need to use the average force $\bar{F}$ to find the work:

$$\bar{F} = \frac{F_{\text{initial}} + F_{\text{final}}}{2}$$

$$= \frac{0 + ks}{2}$$

$$= \tfrac{1}{2}ks$$

Thus the work done on the spring is

$$W = \tfrac{1}{2}ks(s) = \tfrac{1}{2}ks^2 \qquad (18.2)$$

This work is stored in the spring as *elastic potential* energy.

When the restraining forces are released, the potential energy $\tfrac{1}{2}ks^2$ is converted into kinetic energy of the mass (or into work done on some other object). If we can neglect the mass of the spring (assumed small compared to m) and friction losses, we can readily determine the velocity of m as it passes the equilibrium point O. At the equilibrium position, $s = 0$, and all the potential energy of the spring will have been converted to the kinetic energy of the mass m:

$$\tfrac{1}{2}mv_{\text{max}}^2 = \tfrac{1}{2}ks^2$$

or $\qquad v_{\text{max}} = \sqrt{(k/m)}\,s \qquad (18.3)$

Illustrative Problem 18.1 The spring shown in Fig. 18.1 has a spring constant k for elongation of 15 N/m, and $m = 0.75$ kg. The spring is pulled by a force which will produce an elongation of 50 cm and is then released. What will be the speed of the mass when it passes through its equilibrium position?

Solution At the equilibrium position the velocity of m will be at a maximum.

$$v_{\text{max}} = \sqrt{\frac{k}{m}}\,s$$

$$= \sqrt{\frac{15 \text{ N/m}}{0.75 \text{ kg}}}\,(50 \text{ cm})$$

$$= \sqrt{20\frac{\text{N/m}}{\text{kg}}\frac{(1 \text{ kg-m/sec}^2)}{1 \text{ N}}}\,(0.50 \text{ m})$$

$$= 4.47 \text{ sec}^{-1}(0.50 \text{ m})$$

$$= 2.2 \text{ m/sec} \qquad\qquad\qquad ans.$$

18.3 Simple Harmonic Motion

Consider a mass suspended from a steel spring attached to a fixed support (Fig. 18.2a). At the equilibrium position the only force acting on the spring is the weight $w = mg$ of the object. This force is balanced by the upward tension T in the spring.

If the mass is pulled down, the tension on the spring is increased. Before the mass (Fig. 18.2b) is released, the tension is greater than the weight w.

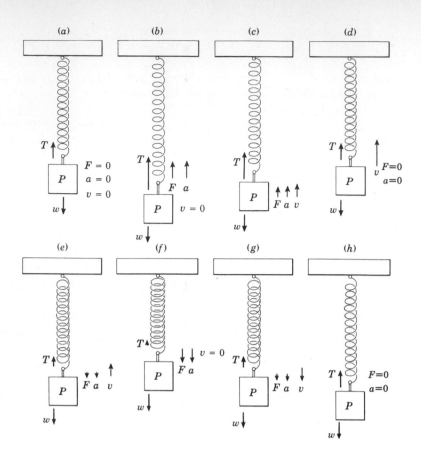

Fig. 18.2 Motion of mass hung from a spring.

The resultant force $(F = T - w)$ will cause the object to be pulled upward, with an initial velocity $v = 0$ and an acceleration that is directed toward the equilibrium position. If the spring is stretched s units, and if k is the spring constant $(F = -ks)$, the acceleration will be proportional to s but opposite in direction.

As the mass moves upward toward the equilibrium position (Fig. 18.2c), the tension T decreases (according to Hooke's law), but T will still be greater than w, and the unbalanced force F will continue to accelerate the body but at a decreased rate. Even so, the velocity will still be increasing.

When the object has returned to its equilibrium position (Fig. 18.2d), the net force F is zero. Hence, the acceleration will be zero, but the velocity will be maximum. The momentum of the moving mass will carry it upward past the original position.

As it moves above the neutral position (Fig. 18.2e) the weight w will be greater than the tension T, yielding a resultant force F directed downward. This force will cause an acceleration downward which decreases the upward velocity.

The mass will continue to rise until it has zero velocity, at a point which is as high above the equilibrium position as the original release point was below the neutral position at the instant the mass was released (Fig. 18.2f). At the upper limit the tension on the spring will be quite small. The direction of motion will change (maximum acceleration), and the force of gravity $(w = mg)$ will cause the body to accelerate downward.

As the body moves downward (Fig. 18.2g), tensile forces in the spring will once again increase. The resultant decreased net downward force acting on the object will lessen the acceleration, but the velocity increases downward because acceleration in that direction still exists.

When the body once again reaches the equilibrium position, both the resultant force and the acceleration will be zero, but the velocity will have reached a maximum in a downward direction (Fig. 18.2h).

Once more the momentum of the moving mass will cause the body to continue downward with an acceleration directed toward the equilibrium position. When it has reached the maximum downward displacement, the body will start to repeat the same vibratory motion. Such a motion is termed simple harmonic motion. *Simple harmonic motion (SHM) is motion described by a particle about an equilibrium position such that the acceleration is directed toward the equilibrium position and is proportional to the displacement.*

18.4 Period, Frequency, and Amplitude

When an object P (Fig. 18.2) undergoes a regularly repeating motion, the time between successive passages of P in the same direction through any point is called the *period T* of the motion. The *frequency f* of the motion is the number of cycles, or complete vibrations, the body makes in unit time. Frequency is the reciprocal of period:

$$f \text{(cycles/sec)} = \frac{1}{T \text{(sec/cycle)}} \qquad (18.4)$$

The standard unit of frequency is the hertz (Hz), named after the German physicist Heinrich Hertz (1857–1894).

$$\text{One hertz} = 1 \text{ cycle/sec.}$$

Multiples of cycles/second (Hz) are used for high frequencies. Thus,

1 kilocycle/sec (kc/sec)

$$= 10^3 \text{ cycles/sec} = 1 \text{ kilohertz (kHz)}$$

1 megacycle/sec (Mc/sec)

$$= 10^6 \text{ cycles/sec} = 1 \text{ megahertz (MHz)}$$

The distance of the body from the equilibrium position at a given instant is called its displacement s. The *amplitude A* is the maximum displacement of the vibrating object.

If there is no appreciable dissipation of energy for the system of Fig. 18.1 or Fig. 18.2, the sum of the potential energy and the kinetic energy is a constant at every instant during the oscillation. The total energy is equal to the initial energy given the system. Thus

$$E_p + E_k = \text{constant} = \tfrac{1}{2}kA^2$$

or $\qquad \tfrac{1}{2}ks^2 + \tfrac{1}{2}mv^2 = \tfrac{1}{2}kA^2 \qquad (18.5)$

Illustrative Problem 18.2 A 4.0-kg body vibrates in SHM with an amplitude of 5.0 cm. Find the speed of the body (a) at the midpoint of the vibration and (b) at a point 2.0 cm from the midpoint. Assume $k = 0.75$ N/m.

Solution

(a) At the midpoint of the vibration, $s = 0$. Then

$$\tfrac{1}{2}mv^2 = \tfrac{1}{2}kA^2 \qquad \text{or} \qquad v^2 = \frac{kA^2}{m}$$

$$v^2 = \frac{0.75 \text{ N/m } (0.05 \text{ m})^2}{4 \text{ kg}}$$

$$= 4.69 \times 10^{-4} \frac{\text{N-m}}{\text{kg}} \left(\frac{1 \text{ kg-m/sec}^2}{1 \text{ N}} \right)$$

$$= 4.69 \times 10^{-4} \text{ m}^2/\text{sec}^2$$

$$v_{\text{max}} = 0.022 \text{ m/sec} \qquad \text{or} \qquad 2.2 \text{ cm/sec} \quad ans.$$

(b) At $s = 2.0$ cm,

$$\tfrac{1}{2}ks^2 + \tfrac{1}{2}mv^2 = \tfrac{1}{2}kA^2$$

$$\tfrac{1}{2}(0.75 \text{ N/m})(0.02 \text{ m})^2 + \tfrac{1}{2}(4 \text{ kg})v^2$$

$$= \tfrac{1}{2}(0.75 \text{ N/m})(0.05 \text{ m})^2$$

$$(2 \text{ kg})v^2 = 7.875 \times 10^{-4} \text{ N-m}$$

$$v^2 = \frac{7.875 \times 10^{-4} \text{ N-m} \left(\frac{1 \text{ kg-m/sec}^2}{1 \text{ N}} \right)}{2 \text{ kg}}$$

$$= 3.9375 \times 10^{-4} \text{ m}^2/\text{sec}^2$$

$$v = 1.98 \times 10^{-2} \text{ m/sec}^2 \approx 2.0 \text{ cm/sec} \quad ans.$$

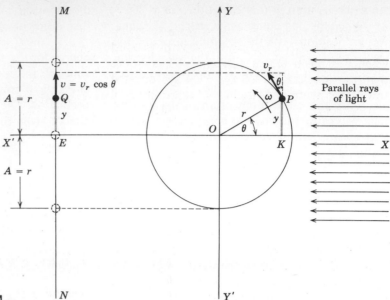

Fig. 18.3 Circle of reference for SHM.

To develop equations relating displacement, amplitude, frequency, and acceleration of bodies in SHM we shall show how SHM can be related to uniform circular motion (UCM).

18.5 Simple Harmonic Motion and Uniform Circular Motion

Consider a particle P moving with uniform linear speed v_r and uniform angular speed ω around a circle, called the *reference circle*, of radius r with center at O (Fig. 18.3). Imagine rays of light parallel to the X axis producing a moving shadow of P on a line MN lying in the XY plane and perpendicular to the X axis at E. Let Q represent the shadow of P on line MN. As P moves with uniform circular motion, point Q will oscillate back and forth along MN with SHM in the same manner as the swinging mass of Fig. 18.2.

The displacement EQ of point Q is

$$y = r \sin \theta \qquad (18.6)$$

The velocity of Q is the component of the *instantaneous* velocity v_r of point P along a line parallel to YY', or

$$v = v_r \cos \theta \qquad (18.7)$$

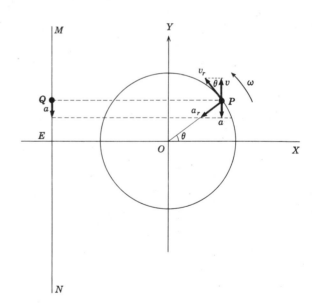

Fig. 18.4 Velocity and acceleration for UCM and SHM.

The centripetal acceleration a_r (see Fig. 18.4) of particle P toward the center of the circle is v^2/r or $\omega^2 r$. The acceleration of Q is the component of the centripetal acceleration of P along a line parallel to the Y axis and may be written

$$a = -a_r \sin \theta = -\omega^2 r \sin \theta = -\omega^2 y \quad (18.8)$$

The negative sign is introduced since ω is constant and positive and a is directed downward when y is positive, and vice versa. Thus we see that the acceleration of Q is always directed toward the neutral position E and is proportional to the displacement y, as required in SHM.

18.6 Period of Simple Harmonic Motion

It is usual to express the period T for a particle with SHM in terms of the acceleration and displacement of the particle. Consider a mass m suspended from a spring undergoing SHM (see Fig. 18.5). The period T of mass m will be the same as that of a similar mass P in uniform circular motion (UCM) of the same amplitude. The angular velocity ω is

$$\omega = 2\pi f = \frac{2\pi}{T} \quad (18.9)$$

From Eq. (18.8)

$$a = -\omega^2 y = -\frac{4\pi^2}{T^2} y \quad (18.10)$$

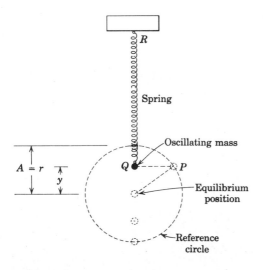

Fig. 18.5 Mass at end of spring executes SHM.

Solving first for T^2 and then for T, we get

$$T^2 = -4\pi^2 \frac{y}{a}$$

$$T = 2\pi\sqrt{-y/a} \quad (18.11)$$

where the acceleration is that of the body when it is displaced y units from the equilibrium position. It is evident that the term under the radical sign will always be positive since the displacement of a body with SHM will always be of opposite sign to the acceleration.

The period can also be expressed in terms of the spring constant (or other measure of elasticity) and the mass of the body.

From Hooke's law we have the relation $F = -ky$, where F is the restoring force, y is the displacement, and k is the (spring constant) measure of elasticity. From Newton's second law we know also that the force F to accelerate the mass m is equal to ma. Equating the two expressions for F,

$$ma = -ky \qquad \text{or} \qquad -\frac{y}{a} = \frac{m}{k}$$

and substituting m/k for $-y/a$ in Eq. (18.11), we get

$$T = 2\pi\sqrt{m/k} \quad (18.12)$$

The student is urged to verify that consistent units to use for T, m, and k in Eq. (18.12) are seconds, kilograms, and newtons per meter; or seconds, slugs, and pounds per foot. For example,

$$T \text{ sec} = 2\pi\sqrt{\frac{m \text{ slugs} \times [(1 \text{ lb-sec}^2/\text{ft})/1 \text{ slug}]}{k \text{ lb/ft}}}$$

Note that increasing the mass increases the period, while increasing the stiffness (elasticity) of a spring decreases the period. It should also be evident, since amplitude A does not appear in Eq. (18.12), that the period does not depend upon the amplitude of motion.

Illustrative Problem 18.3 A 0.100-kg mass is hung from a spring. It is noted that when an additional 0.400-kg mass is added to the spring, the spring elongates 80.0 cm. If the 0.500-kg mass is

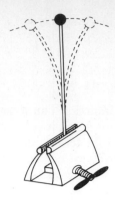

Fig. 18.6 Illustrative Problem 18.4.

then set vibrating, what should the period of oscillation be? What is the frequency?

Solution

$$m = 0.500 \text{ kg} \quad F = -ky \quad \text{or} \quad k = -F/y$$

$$k = -\frac{-0.400 \text{ kg}}{80.0 \text{ cm}} \times \frac{9.81 \text{ N}}{1 \text{ kg}} \times \frac{100 \text{ cm}}{1 \text{ m}}$$

$$= 4.90 \text{ N/m}$$

By Eq. (18.12),

$$T = 2\pi\sqrt{m/k}$$

$$= 2\pi\sqrt{0.500 \text{ kg}/(4.90 \text{ N/m})}$$

$$= 2.01 \text{ sec} \qquad \qquad ans.$$

$$f = 1/T = 1/2.01 \text{ sec}$$

$$= 0.50 \text{ cycles/sec}$$

$$= 0.50 \text{ Hz} \qquad \qquad ans.$$

Illustrative Problem 18.4 A 2.8-lb ball is fastened to a metal strip clamped tightly at one end (Fig. 18.6). A force of 3.4 lb will pull the ball 6 in. to one side. Find the period of vibration of the weighted strip. Assume SHM.

Solution

$$k = -\frac{F}{y} = \frac{3.4 \text{ lb}}{6 \text{ in.}} \times \frac{12 \text{ in.}}{1 \text{ ft}}$$

$$= 6.8 \text{ lb/ft}$$

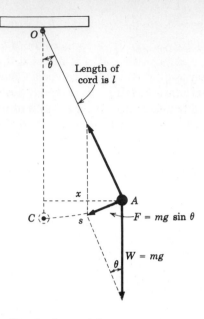

Fig. 18.7 The simple pendulum.

$$m = \frac{w}{g} = \frac{2.8 \text{ lb}}{32 \text{ ft/sec}^2}$$

$$= 0.0875 \frac{\text{lb-sec}^2}{\text{ft}} \text{ (slug)}$$

$$T = 2\pi\sqrt{m/k}$$

$$= 2\pi\sqrt{0.0875 \text{ slug}/(6.8 \text{ lb/ft})}$$

$$= 0.71 \text{ sec} \qquad \qquad ans.$$

18.7   The Simple Pendulum

The *simple pendulum* consists of a concentrated mass at the end of a cord of negligible weight. The motion of the simple pendulum closely approximates simple harmonic motion. Consider a small, relatively heavy, bob at the end of a very light rod or string. If such a bob of mass m is displaced from its equilibrium position as shown in Fig. 18.7, two forces act upon the bob: the attraction of gravity mg vertically downward and the pull P of the string acting toward O.

The restoring force is the component of the weight perpendicular to the string. This component is

$$-mg \sin \theta = -mg \sin \angle COA$$

The negative sign is used to indicate that the force F is directed toward the neutral position C. (Angles measured in a counterclockwise direction are given positive signs, and those measured in a clockwise direction are given negative signs.) The displacement along the arc is $s = l\theta$ (if θ is in radians). We see that the force is proportional to $\sin \theta$ while the displacement is proportional to θ. However, if θ is small, $\sin \theta$ may be replaced by θ (in radians) without serious error. Hence, to the degree of this approximation, the restoring force F is directly proportional to the displacement, and we can apply the equations of SHM to the motion of the pendulum:

$$F = ma$$

$$= -mg \sin \theta = -mg \frac{x}{l} = -mg \frac{s}{l} \quad \text{(approx)}$$

Equating the two forces, we get

$$-mg \frac{s}{l} = ma$$

and
$$-s/a = l/g$$

Substitute in Eq. (18.11):

$$T = 2\pi\sqrt{l/g} \quad \text{period of a pendulum} \quad (18.13)$$

The period of a pendulum is seen to depend only upon the length of the pendulum and the acceleration due to gravity at the location of the pendulum. The period does not depend on the mass of the bob and is independent of the amplitude so long as the amplitudes are small (remember that for small θ's, $\theta \approx \sin \theta$). It can be shown that an angle of 10° will produce an error of about 0.2 percent if Eq. (18.13) is used.

Equation (18.13) can be applied in determining the value of g. By using a pendulum of fixed length and measuring its period of vibration, the value of g at a particular location can readily be determined.

Illustrative Problem 18.5 A simple pendulum has a length of 4.05 m and executes 20 complete vibrations in 80.8 sec. What is the value of the acceleration of gravity g where the pendulum is located?

Solution Solve Eq. (18.13) for g:

$$T = 2\pi\sqrt{l/g}$$

$$T^2 = 4\pi^2(l/g)$$

$$g = 4\pi^2 l/T^2$$

Substituting $l = 4.05$ m and $T = 80.8/20 = 4.04$ sec in the equation for g, we get

$$g = \frac{4\pi^2 \times 4.05 \text{ m}}{(4.04 \text{ sec})^2} = 9.79 \text{ m/sec}^2 \quad ans.$$

18.8 Simple Harmonic Motion and the Sine Curve

In Sec. 18.5 it was determined that the displacement of a particle undergoing simple harmonic motion is a function of the sine of θ, where θ is the angular displacement of the related particle undergoing uniform circular motion; that is, $y = A \sin \theta$.

If T is the period of the motion and t is the time for the particle of Fig. 18.3 to move from E to Q, we know from Eq. (7.4) that

$$\theta = \omega t$$

But

$$\omega = \frac{2\pi}{T}$$

Then
$$\theta = \frac{2\pi}{T} t$$

and
$$y = A \sin \theta = A \sin \frac{2\pi}{T} t \qquad (18.14)$$

The graph of Eq. (18.14) is called a *sine curve*.

The relation between SHM and the sine curve can readily be demonstrated in the laboratory. Hang a funnel filled with fine sand, as shown in Fig. 18.8 and place a dark-colored cardboard

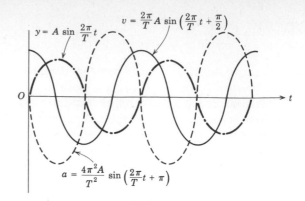

Fig. 18.9 Graphs of displacement, velocity, and acceleration of SHM as a function of time.

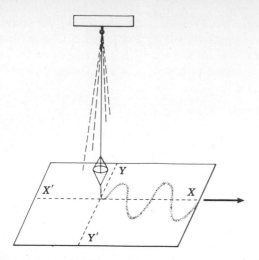

Fig. 18.8 Sand falling from a swinging funnel forms a sine curve on a cardboard moved with constant speed.

under it. Set the funnel swinging along YY'. Then move the cardboard slowly with constant speed at right angles to the motion of the pendulum. The falling sand will trace the sine curve.

In the study of trigonometric functions it can be shown that $\cos \theta = \sin(\theta + \pi/2)$ and $-\sin \theta = \sin(\theta + \pi)$. Using these relations, we can alter Eqs. (18.7) and (18.8) to express the velocity v and acceleration a of a body describing SHM as sine functions of time t.

From Eq. (7.2) we know that v_r, the tangential velocity of particle P, is equal to ωr or ωA. Hence

$$v_r = \omega A = \frac{2\pi}{T} A$$

and Eq. (18.7) becomes

$$v = v_r \cos \theta$$
$$= v_r \sin\left(\theta + \frac{\pi}{2}\right)$$

or $\qquad v = \frac{2\pi}{T} A \sin\left(\frac{2\pi}{T} t + \frac{\pi}{2}\right) \qquad (18.15)$

Equation (7.5′) states that $a_r = r\omega^2$. But $r = A$ and $r\omega^2 = A\omega^2$. Hence Eq. (18.8) becomes

$$a = -a_r \sin \theta$$
$$= a_r \sin(\theta + \pi)$$
$$= A\left(\frac{2\pi}{T}\right)^2 \sin\left(\frac{2\pi}{T} t + \pi\right)$$

or $\qquad a = \frac{4\pi^2}{T^2} A \sin\left(\frac{2\pi}{T} t + \pi\right) \qquad (18.16)$

The graphs of Eqs. (18.14), (18.15), and (18.16) are shown in Fig. 18.9. Thus, we see that the velocity v reaches its maximum and minimum $90°$ ($\pi/2$ rad) ahead of the displacement y. We say that the velocity *leads* the displacement by $90°$. Similarly, note that the acceleration leads the velocity by $90°$ and the displacement by $180°$. The displacement, velocity, and acceleration can each be represented by sine (or cosine) curves of the same frequency but differing in *phase*. Two variables that are represented by sine waves are said to be *in phase* if they reach maxima, zeros, and minima at the same instant.

NOTE: Recall that $\sec^{-1}$ is read "per second."

Illustrative Problem 18.6 A wave is represented (metric units) by the equation $y = 8 \sin 0.400\pi t$, where y is measured in centimeters and t in seconds. Find (*a*) the amplitude, (*b*) the period, and (*c*) the frequency of the wave. What are the (*d*) displacement, (*e*) velocity, and (*f*) acceleration of the wave particle when $t = 0.625$ sec?

Solution From Eq. (18.14),

$$y = A \sin \frac{2\pi}{T} t$$

(a) Compare $y = 8 \sin 0.400\pi t$ with Eq. (18.14) to get

$$A = 8 \text{ cm} \qquad ans.$$

$$\frac{2\pi}{T} = 0.400\pi \text{ sec}^{-1}$$

Thus,

(b) $$T = \frac{2\pi}{0.400\pi \text{ sec}^{-1}} = 5 \text{ sec} \qquad ans.$$

(c) $$f = \frac{1}{T} = \frac{1}{5 \text{ sec}} = 0.2 \text{ sec}^{-1}$$

$$= 0.2 \text{ Hz} \qquad ans.$$

When $t = 0.625$ sec,

$$\frac{2\pi}{T} t = (0.400\pi \text{ sec}^{-1})(0.625 \text{ sec})$$

$$= 0.25\pi = \pi/4$$

(d) Substituting known values in Eqs. (18.14), (18.15), and (18.16) gives

$$y = A \sin \frac{2\pi}{T} t$$

$$= (8 \text{ cm})\left(\sin \frac{\pi}{4} \right)$$

$$= (8 \text{ cm})(0.707)$$

$$= 5.66 \text{ cm} \qquad ans.$$

(e) $$v = \frac{2\pi}{T} A \sin\left(\frac{2\pi}{T} t + \frac{\pi}{2} \right)$$

$$= (0.400\pi \text{ sec}^{-1})(8 \text{ cm})\sin\left(\frac{\pi}{4} + \frac{\pi}{2} \right)$$

$$= (3.20\pi \text{ cm/sec})\sin \tfrac{3}{4}\pi$$

$$= (3.20\pi \text{ cm/sec})(0.707)$$

$$= 7.10 \text{ cm/sec} \qquad ans.$$

(f) $$a = \frac{4\pi^2}{T^2} A \sin\left(\frac{\pi}{4} + \pi \right)$$

$$= (0.400\pi \text{ sec}^{-1})^2(8 \text{ cm})\sin \tfrac{5}{4}\pi$$

$$= (0.160\pi^2 \text{ sec}^{-2})(8 \text{ cm})(-0.707)$$

$$= -8.93 \text{ cm/sec}^2 \qquad ans.$$

18.9 Complex Periodic Vibrations Can Be Resolved into Simple Harmonic Motions

If we hang a pendulum B from the bob A of a first swinging pendulum (Fig. 18.10a), the bob B of the second pendulum will undergo a complex vibration which will be the resultant of the simpler motions of the individual pendulums. In like manner, we might support a mass C from a spring and then from mass C hang a second weighted spring D (Fig. 18.10b). When both masses are set in motion, the motion of the lower mass will be complex. Suppose the period of the spring with mass C is 2 sec and the period of the spring with mass D is 1 sec. The dashed black line of Fig. 18.11 is the graph of the motion of C with only the top spring-mass combination. The solid black line represents the graph of the motion of D

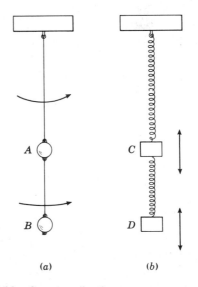

(a)   (b)

Fig. 18.10 Complex vibrations.

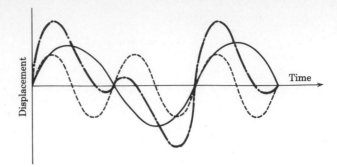

Fig. 18.11 Complex periodic motions are combinations of several SHMs.

with only the bottom spring-mass combination. The dashed-dotted line is the graph of the complicated motion of *D* when the two are combined and set into motion.

If can be shown that any periodic complex motion is the vector sum of a number of simple harmonic motions. We shall make use of this fact in our study of sound waves and electric currents in later chapters.

18.10 Forced Vibrations and Resonance

Suppose a heavy ball is mounted on a flat steel spring and clamped in a vise, as shown in Fig. 18.12. Assume a *natural frequency* of vibration of the system to be 2 vib/sec. Next imagine some means of applying a to-and-fro force with a frequency of 5 vib/sec to the bottom of the spring. As a consequence, the system will be forced to vibrate at 5 vib/sec. During part of each cycle the system will oppose the driving force. Hence the amplitude of vibration will be quite small. The system will be undergoing a *forced vibration*.

Now suppose the frequency of the to-and-fro motion is gradually slowed until it is equal to the natural frequency of the system, 2 vib/sec. When this happens, the amplitude of vibration becomes relatively large and the system is said to be in *resonance* with the driving force.

If the frequency of the driving force is reduced still more, the amplitude of vibration of the spring-ball system will diminish until it is again quite small.

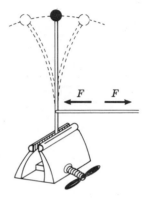

Fig. 18.12 Forced vibrations.

The principle of resonance can be demonstrated by a simple experiment illustrated in Fig. 18.13. Hang three pendulums on the same light rod. Let the lengths of pendulums *B* and *C* be fixed and that of *A* be variable. Next set pendulum *A* to swinging. If no two pendulum lengths are equal, pendulums *B* and *C* will jiggle a little, but they will not swing appreciably.

Now if the length of the swinging pendulum *A* is shortened to equal that of pendulum *B*, the motion of pendulum *A* will be transferred to *B* in greater measure and *B* will start to swing with an appreciable amplitude while *C* continues to jiggle slightly. As the length of pendulum *A* is shortened more until it equals that of pendulum *C*, *C* will start to swing and the amplitude of *B* will diminish until it is negligible.

The ability of a small periodic driving external force of proper frequency to build up a very

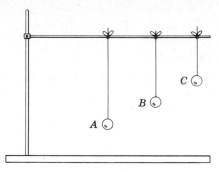

Fig. 18.13 Illustration of resonance.

large amplitude of motion in another system of the same natural frequency is evident when the child in the swing "pumps" at the right time to increase the amplitude of motion. The rattle of a jalopy, the vibrations set up in dishes, a swinging chandelier, a tuned radio circuit, and the vibrations of the sounding box of a violin are all examples of a resonant response to external vibrations.

18.11   Types of Waves

One of the most important means of transmitting energy from one point to another is wave motion. Many kinds of waves commonly occur in nature. One of the most familiar is *water waves*. We are able to hear when our senses respond to *sound waves*. *Electromagnetic waves* transmit radio signals, light, and x-rays. These types of waves have some qualities in common and some that are distinct and different.

Water Waves   When one watches the waves in a body of water from a distance, it appears that the water itself is traveling in the direction of the wave; but on observing an object floating on the water surface it becomes apparent that the water does not actually travel with the wave. The floating object can be seen to follow an elliptical or circular path, returning to its original position with each complete vibration. In like manner, the water molecules move in orbit about their original positions when a wave passes by (see Fig. 18.14).

It is apparent that what moves forward as successive waves pass over the water is not the water itself, but the energy which makes the masses of molecules of water rise and fall. This energy is supplied by the agent that creates the disturbance in the water which produces the waves.

Transverse Waves   If an ordinary rope, fastened at one end, is stretched across the room and given a sudden up-and-down motion at the free end, a wave will start at the free end and run the length of the rope. Each portion of the rope will move up and down perpendicular (*transverse*) to the direction of the wave motion. If the up-and-down motion of the hand is continuous, a series of waves travels down the rope (Fig. 18.15). These waves are called *transverse waves*.

In order to study the process by which a transverse wave advances let us refer to the mechanical model of Fig. 18.16. A long *elastic*

Fig. 18.15   Transverse waves in a rope.

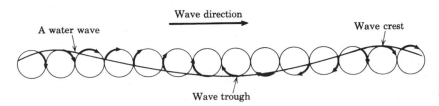

Wave direction

A water wave

Wave crest

Wave trough

Fig. 18.14 Water molecules move in circular or elliptical orbits about their original positions as the wave passes by.

Simple Harmonic Motion   **381**

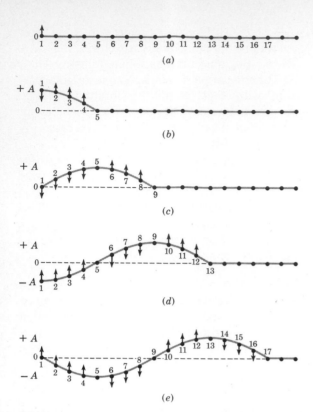

until $+A$ is reached and then beginning down-ward movements because of the elastic forces of the cord to the left.

When particle 1 returns to its original posi-tion (moving in a downward direction), the cord will have the form indicated in Fig. 18.16c. As 1 continues downward, the others follow, but always lag in motion a bit behind the particles to the left.

When particle 1 has gone through a com-plete vibration the particles are distributed as shown in Fig. 18.16e. Particle 1 is now ready to start its second vibration. As can be observed from the figure, the wave has advanced to particle 17. As the particles continue their up-and-down motion, the wave advances to the right.

Light waves, radio waves, heat waves, and x-rays are examples of transverse waves.

Longitudinal Waves A similar model can be used to illustrate a different type of waveform. Instead of having mass 1 move with a motion at right angles to the elastic cord, let it move back and forth along the direction of the cord (Fig. 18.17).

Motion of mass 1 toward the right will compress that portion of the cord between 1 and 2. Hence mass 2 will be forced to the right. As 2 moves to the right, that portion of the cord be-tween 2 and 3 will be compressed, and so on. In each case the moving particles will follow the pat-tern of those to their left but will always lag a little because of their inertia. When mass 1 reaches its maximum displacement to the right, the particles appear as illustrated in Fig. 18.17b. At this instant, mass 1 begins its motion to the left, while particles 2, 3, 4, etc., continue their motion to the right until the elastic forces to the left overcome the forces of inertia to the right. The motion of each particle, after it has been displaced to the right as far as particle 1 was displaced from its neutral position, will reverse and follow the motion of the particle to its left.

As particle 1 moves back to the left, particle 2 follows it but lags behind because of its inertia; thus a state of tension will exist in the cord be-

Fig. 18.16 Motion of particles in transverse wave motion.

cord, with markers, 1, 2, 3, 4, etc., equally spaced along its length to represent the vibrating par-ticles, is attached to a rigid support on one end and is free to oscillate at the other.

If particle 1 is given an upward motion at right angles to the cord, particle 2 will follow because of the elasticity of the cord, but it will lag behind particle 1 because of its inertia. As particle 2 rises, particle 3 will follow but will lag behind particle 2. Particle 4 will lag behind 3, and so on. Let us assume that when 1 reaches its highest position $(+A)$, 5 is just beginning to be raised, and at that instant 1 starts in downward motion. Particle 2 will continue in an upward motion, because of its inertia, until it reaches the height $+A$; it will then begin to descend, while particles 3, 4, 5, . . . follow the same pattern, moving upward because of elasticity and inertia

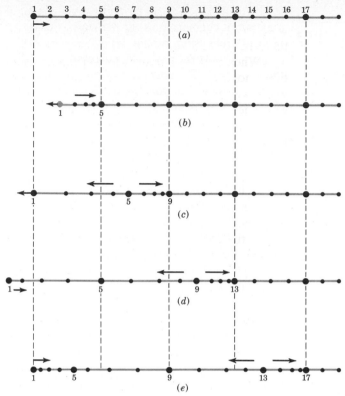

Fig. 18.17 Motion of particles in longitudinal wave motion.

tween 1 and 2, and the particles will begin to separate.

When 1, moving to the left, has reached its initial position (Fig. 18.17c), 9 will just begin its motion to the right.

When 1 has reached its maximum displacement to the left (Fig. 18.17d), particles 1 and 9 will reach their maximum separation. At this instant, 1 will begin to move to the right, 9 will change direction to the left, and the wave motion will have reached 13, compressing it to the right.

As particle 1 continues to the right, the state of tension in the section between 1 and 2 is reduced. This reduces the tension between 2 and 3, and so on. When 1 returns to its original position moving to the right, it then begins to repeat its vibrations and the wavefront will have moved to particle 17. Particles 1 and 17 are then said to be in the same *phase*.

Waves in which the particles of the medium vibrate in the direction of the motion of the wave are called *longitudinal waves*. Sound waves are of this type.

18.12   Condensation and Rarefaction

We have just seen that in longitudinal waves the particles of the transmitting medium vibrate to and fro in the direction of the propagation. Particles of the medium are alternately in a state of compression and tension. When these waves are set up, some of the particles are crowded together with their neighbors during one instant to form a *condensation*; an instant later they are drawn apart to form a *rarefaction*. The lines drawn in Fig. 18.18 represent layers or particles of air alternately crowded and separated as sound waves move through air.

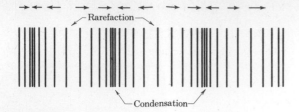

Fig. 18.18 Condensation and rarefaction in a sound wave.

It should be observed that in all types of wave motion the actual particles which form the wave do not travel very far from their normal position. The particles are displaced only a little from their normal positions and oscillate about that point.

18.13 Curves Representing Transverse and Longitudinal Waves

It is common practice to represent a longitudinal wave graphically by a curve similar to that of a transverse wave. Let us imagine a vibrating tuning fork imparting a series of pulses to an air column. Successive condensations and rarefactions of the air particles are indicated by C and R in Fig. 18.19a. The wave is shown traveling from the fork to the ear, but the air molecules themselves oscillate only a very little distance to and fro from their normal positions; actually, air molecules vibrate only about 1 mm even for painfully loud sounds. Figure 18.19b represents graphically the pressures on the air particles. Points on the line, such as A, B, C, indicate points of normal pressure. The distances above or below the horizontal line are a measure of the degree of compression or rarefaction on the particle.

The distance between successive points of maximum condensation or between successive points of maximum rarefaction or, in general, the distance between any two successive particles that are in the same condition of vibration (phase) is called a *wavelength*. Wavelength (see Fig. 18.20) is denoted by the Greek letter λ (lambda). A complete wave includes a condensation followed by a rarefaction. The wavelength can be determined by measuring the distance between two corresponding points in any two adjacent waves, for example, *AC*, *BD*, or *EF*. The maximum displacement of any particle from its neutral position is called *amplitude* of the vibration. The distance a in Fig. 18.20 represents graphically the *amplitude* of the wave. Peaks of the waveform are called *crests*, and the valleys are called *troughs*.

18.14 Speed of Waves

The distance which a crest (or trough) of a wave travels in 1 sec is called its *speed*. The number of crests passing a given point in 1 sec is called the *frequency*. The frequency of a wave is equal to the number of vibrations per second of the wave. Hence the total distance the wave will advance in 1 sec (*speed*) *is equal to the product of the frequency times the wavelength*. As an equation

$$v = f\lambda$$

$$Speed = \text{frequency} \times \text{wavelength} \quad (18.17)$$

This is the basic equation of wave motion. It holds for any kind of wave motion, whether sound, light, heat, radio, or water wave.

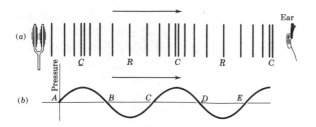

Fig. 18.19 Curve graphically representing a sound wave.

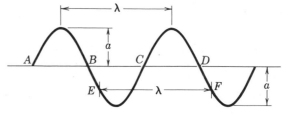

Fig. 18.20 Amplitude and length of wave.

Illustrative Problem 18.7 Waves roll up on a beach at the rate of five waves every 10 sec. If the distance between the crests of the waves is 7.50 m, with what speed are the waves moving?

Solution A rate of five waves every 10 sec is equivalent to a frequency of $\frac{1}{2}$ sec^{-1}. Substituting the known values for f and λ in Eq. (18.17), we obtain

$$v = f\lambda$$

$$= \tfrac{1}{2} \text{ sec}^{-1} \times 7.5 \text{ m}$$

$$= 3.75 \text{ m/sec} \qquad \qquad ans.$$

Illustrative Problem 18.8 A sound wave which travels 1100 ft/sec has a frequency of 256 Hz. What is the length of the wave?

Solution Solving Eq. (18.17) for λ and substituting known values, we get

$$\text{Wavelength} = \frac{\text{speed}}{\text{frequency}}$$

$$\lambda = \frac{v}{f}$$

$$= \frac{1100 \text{ ft/sec}}{256 \text{ vib/sec}}$$

$$= 4.30 \text{ ft/vib} \qquad \qquad ans.$$

This is the wavelength of middle C of the musical scale when $v = 1100$ ft/sec.

SUMMARY

Simple harmonic motion (SHM) is motion about an equilibrium position such that its acceleration is directed toward the equilibrium position and is proportional to the displacement.

In SHM the sum of the kinetic energy and the potential energy is constant and equal to the initial energy supplied to the vibrating system.

The frequency of periodic motion is the reciprocal of the period.

$$f = 1/T \qquad \qquad (18.4)$$

The period of SHM is found by

$$T = 2\pi\sqrt{-y/a} \qquad \qquad (18.11)$$

where y is the displacement and a is the acceleration of the particle.

The period is also found by

$$T = 2\pi\sqrt{m/k} \qquad \qquad (18.12)$$

where m is the mass and k is the measure of elasticity of the material undergoing SHM.

The period of a simple pendulum is solved by a formula

$$T = 2\pi\sqrt{l/g} \qquad \qquad (18.13)$$

where l is the length of the pendulum and g is the acceleration of gravity.

The graphs of the displacement, velocity, and acceleration of a particle in SHM are each sine curves, where the velocity leads the displacement by 90° and the acceleration leads the velocity by 90°:

$$y = A \sin \frac{2\pi}{T} t \qquad \qquad (18.14)$$

$$v = \frac{2\pi}{T} A \sin\left(\frac{2\pi}{T} t + \frac{\pi}{2}\right) \qquad (18.15)$$

$$a = \frac{4\pi^2}{T^2} A \sin\left(\frac{2\pi}{T} t + \pi\right) \qquad (18.16)$$

Two sine curves are *in phase* if they reach maxima, zeros, and minima at the same time.

Every periodic complex motion is the vector sum of simple harmonic motions.

When a periodic driving force is impressed upon a system whose natural frequency of vibration is the same as that of the driving force, *resonance* occurs.

A great deal of energy around us is transmitted by waves, e.g., heat, light, radio, sound, and water waves.

The distance between two consecutive particles in phase is the *wavelength* λ.

The number of vibrations per second of a particle undergoing periodic motion is called its frequency f.

The speed v of a wave equals the product of its frequency and its wavelength ($v = f\lambda$).

QUESTIONS AND EXERCISES

1. In SHM, when the acceleration is greatest what is true about the (a) displacement and (b) the velocity?

2. In SHM, when the velocity is greatest, what is true about (a) displacement and (b) acceleration?

3. What is the numerical value for the product of the frequency and the period of periodic motion?

4. Which, if any, of the following will be doubled by doubling the amplitude of a swinging pendulum: (a) the period, (b) the frequency, (c) the maximum velocity?

5. A pendulum clock is found to run too fast so that it gains time. In trying to adjust the clock, should you make the length of the pendulum (a) longer or (b) shorter? Give the reasoning behind your answer.

6. How would the period of a pendulum be affected if the pendulum were moved from sea level to the top of a mountain?

7. When a particle undergoes SHM, its displacement is always directly proportional to which of the following: (a) mass, (b) frequency, (c) acceleration, (d) velocity?

8. Which of the following properties of waves is independent of the others: (a) frequency, (b) wavelength, (c) amplitude, (d) velocity?

9. Which of the following will always change when two SHM waves are added: (a) phase, (b) wavelength, (c) amplitude, (d) frequency?

10. What is transferred along the direction of wave motion?

11. What is the primary difference between longitudinal and transverse waves?

12. When a rock is dropped in a large pool of water, wave crests spread out as circles with the center at the point of impact. Explain why the water wave crests die down as the wave continues to radiate from the center.

PROBLEMS

1. What is the period of vibration of a wave if its frequency is 400 Hz?

2. A train of waves moves along a cable with a speed of 12 m/sec. The wavelength is equal to 3 cm. What are the frequency and period of the source of the wave vibration?

3. The velocity of radio waves is 3.0×10^8 m/sec. What is the wavelength of the waves from a transmitter broadcasting on a frequency of 830 kHz?

4. A radio transmitter uses a carrier wave frequency of 1200 kHz. The speed of the waves is $3.00 \times$ 10^8 m/sec. At what wavelength must a radio be tuned to receive the broadcast signal?

5. Doubling both the length and mass of the bob of a pendulum will multiply the period of the pendulum by what numerical factor?

6. Doubling both the mass and the spring constant of a vibrating system will multiply the frequency of vibration by what numerical factor?

7. A 1-lb mass on the end of a spring describes SHM with a period of 2 sec and amplitude of 3 in. Find (a) the maximum velocity of the mass and (b) the maximum acceleration of the mass. (HINT: Relate the SHM to a corresponding UCM.)

8. An object executes SHM with a period of 0.55 sec and an amplitude of 2.5 cm. Calculate (a) its maximum velocity and (b) its maximum acceleration.

9. The maximum velocity of an object moving with SHM is 0.45 m/sec. Its period of vibration is 0.8 sec. Calculate (a) its amplitude and (b) its maximum acceleration.

10. An object moving with SHM has an amplitude of 5 cm and a period of 0.04 sec. Find (a) its maximum velocity and (b) its maximum acceleration.

11. An object, at the end of a spring, vibrates with SHM. The shadow of the object is projected on a screen by a beam of parallel light. The length of the shadow's path from one end to the other is 6 cm. The time required for the shadow to travel this distance is 2 sec. Find (a) the frequency of the SHM and (b) the velocity of the shadow at the center of its paths.

12. A bob weighing 2.5 lb is hung from the end of a spring whose Hooke's law constant is 0.65 lb/in. The bob is set to oscillating with an amplitude of 5 in. What is the speed in feet per second of the bob when it passes through its equilibrium position?

13. A body describing SHM has a maximum velocity of 4 m/sec and a maximum acceleration of 16π m/sec². Find (a) the period and (b) the amplitude of motion.

14. A 60-gm mass when suspended from a spring will stretch it a distance of 12 cm. If the 60-gm mass is now replaced by a 100-gm mass and then set to vibrating up and down, what will be the resultant frequency of vibration of the system?

15. A body whose mass is 0.30 kg is suspended from a spring whose elasticity constant is 0.15 N/m. It is set to vibrating with an amplitude of 2.0 cm. Find the speed of the mass when it passes through the midpoint of its vibration.

16. Determine the speed of the mass of Prob. 15 when it is 1.2 cm from the midpoint of its vibration.

17. A mass attached to a spring undergoes SHM with an amplitude of 6 cm. When the mass is 3 cm from the equilibrium position, what is the ratio of the kinetic energy to the potential energy of the system?

18. When a mass of 40 gm is suspended from a spring, the spring stretches 1.0 cm. If an additional 160-gm mass is suspended from the spring, what will be the natural frequency of the up-and-down motion of the system? (Assume Hooke's law holds.)

19. A 12-lb weight when suspended from a spring will temporarily (the elastic limit is not exceeded) stretch it a distance of 4 in. If the 12-lb weight is replaced by a new weight and the system set to vibrating up and down, the period of vibration will be 1.5 sec. How heavy is the new weight?

20. A 100-gm mass is suspended from a spring. The mass is pulled down 2 cm by a force of 0.3 N and then released. Find (*a*) the frequency of the motion and (*b*) the maximum velocity.

21. A simple pendulum is to have a period of 1 sec. How long, in inches, should it be if $g = 31.5$ ft/sec^2 at that location?

22. A pendulum has a length of 4 m and executes 50 complete vibrations in 210 sec. Find the acceleration of gravity at the location of the pendulum.

23. A pendulum has a length of 29.50 in. Find, accurate to four significant figures, its period at a location where $g = 32.16$ ft/sec^2.

24. A clock pendulum has a period of 1 sec where $g = 980$ cm/sec^2. How many seconds will it gain in 24 hr if it is taken to an altitude where $g = 976$ cm/sec^2?

25. A body describing SHM has an amplitude of 6.00 cm and a period of 0.300 sec. Find its velocity and acceleration when the displacement is (*a*) 6.00 cm, (*b*) 3.00 cm, and (*c*) 0 cm.

19 Sound Waves

A person, on hearing a sudden sound, usually turns in the direction from which the sound seems to come, expecting to see the thing that causes the sound. What causes sound? Why are some sounds different from others? Why are some sounds pleasant to hear while others are not? What media transmit sound from the source to the ear? How is the sound transmitted? In this chapter we seek out the answers to these questions.

The word "sound" usually means different things to the physiologist than it does to the physicist. For the physiologist sound is an auditory sensation transmitted by the ear and interpreted by the brain. He uses such terms as pitch, loudness, and tone quality. The physicist is more interested in the production and transmission of the disturbance which gives rise to the auditory sensation. He describes sound in terms of frequency, intensity (energy), and overtones. Both of these aspects of sound will be discussed in this chapter.

19.1 Source of Sound

If any sound is traced to its source, we find that it originates in some material that is in vibratory motion. There are many familiar examples that even a casual observer will recognize as a vibrating body emitting sound, such as a ringing bell, the clatter resulting from objects dropped on the floor, a tuning fork, the vibrating head of a drum,

the warning noise of a rattlesnake, and the busy roar of traffic. With some sounds the vibration of the source cannot be readily observed, but if investigated closely, the vibration can be verified. The sound of the human voice originates with vibrating vocal chords; the notes from the clarinet and the saxophone start with a vibrating reed; the piano wire and the string of the violin vibrate as they give off their musical tones; the auto horn uses a vibrating diaphragm; and pipe organs make use of vibrating air columns.

19.2 Transmission of Sound

In order for a sound to be heard, three conditions must be met: (1) there must be a source of wave energy consisting of a vibrating body; (2) there must be a material medium for transmitting the sound; (3) there must be a device for receiving the sound, such as the ear. If any of these three conditions is not met, no sound will be heard.

The material medium for sound transmission can be in solid, liquid, or gaseous form. Tapping on the water pipes in the basement can be clearly heard in a room several floors above. In the past miners trapped deep underground have communicated to workers on the surface by tapping on solid rock formations in the mine. A mechanic occasionally listens to the sounds of a running engine by placing one end of a rod against the engine and the other against the ear.

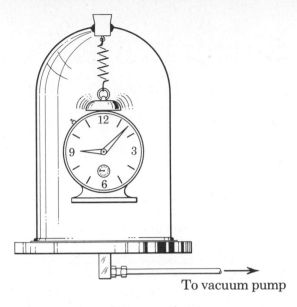

Fig. 19.1 Sound is not transmitted in a vacuum.

We are all familiar with the tales of Indians listening to the rails for the approach of trains along the western plains. Two rocks clapped together under water can be distinctly heard at a considerable distance. Sound transmitted by water is utilized by many ships at sea.

Experimental proof that sound will not travel in a vacuum can be obtained by placing an alarm clock under a bell jar to which is attached a vacuum pump (Fig. 19.1). The alarm bell may be clearly heard when air is present in the jar. As the air is gradually evacuated from the jar, the sounds become fainter and fainter. If enough air is removed, the sound can hardly be heard. As air is once more permitted to enter the jar, the sound regains its original intensity.

19.3 Speed of Sound

When lightning is seen from a distance, observers wait a considerable time before they hear the clap of thunder. When a band is followed by a long line of men marching in step to its music, the marching men are seen to be progressively more out of step as their distance from the band increases.

These and many more examples can be cited to illustrate the relatively slow speed of sound in air. Light on the other hand, travels at such a prodigious speed that over short distances its passage can be considered as instantaneous. Consequently, the speed of sound can be determined by dividing the distance between a gun and an observer by the time that elapses between the *observation* of the flash of the gun and *hearing* the report of the gun. By measuring the velocity, first in one direction and then in an opposite direction, errors due to the velocity and direction of the wind can be minimized by averaging.

The accepted values for the speed of sound in air are 331.4 m/sec at 0°C or 1087 ft/sec at 32°F. For small differences in temperature, it can be shown that the speed of sound in air increases about 0.61 m/sec per degree Celsius increase in temperature. Written as an equation,

$$V_s = 331.4 + 0.61t \qquad (19.1)$$

where V_s = speed, m/sec at t°C, and where t may have either + or − values.

For each degree Fahrenheit increase in temperature, the speed of sound in air increases about 1.1 ft/sec. As an equation,

$$V_s = 1087 + 1.1(t - 32) \qquad (19.2)$$

where V_s = speed, ft/sec at t°F. (t can be either greater or smaller than 32.)

This change in speed due to change in temperature may cause pipe organs and other wind instruments to be appreciably out of tune if the temperature change is extreme.

It is important to realize that in speaking of the speed of sound we are referring to the speed with which energy of vibration is passed from molecule to molecule and not to the velocity of the individual molecules.

Illustrative Problem 19.1 The flash of lightning from an electrical storm is seen in the sky 7 sec before the sound of the thunder is heard. How far from the observer is the lightning if the temperature is 15°C?

Solution The speed of sound in air at 15°C is 331.4 + 0.61(15) or 341 m/sec. Distance = speed × time, or

$$s = vt$$

Substituting, we have

$$s = 341 \text{ m/sec} \times 7 \text{ sec}$$

$$= 2390 \text{ m} \qquad \qquad \text{ans.}$$

Sound waves are transmitted by gases, liquids, and solids. In general, the speed of sound is greater in liquids than in gases, and greater still in solids than in liquids.

The speed of sound in a fluid medium is given by the formula

$$v = \sqrt{B/\rho} \qquad \qquad (19.3)$$

where B is the bulk modulus of the fluid and ρ is its density.

For a solid rod or wire speed of the sound wave is given by the formula

$$v = \sqrt{Y/\rho} \qquad \qquad (19.4)$$

where Y is Young's modulus for the solid material and ρ is its density.

Illustrative Problem 19.2 Compute the speed of sound in sea water. Assume a bulk modulus of $2.25 \times 10^9 \text{ N/m}^2$ for the sea water.

Solution The density of sea water is 1025 kg/m³ (from Table 10.1).

We have

$$v = \sqrt{\frac{B}{\rho}}$$

$$= \sqrt{\frac{2.25 \times 10^9 \text{ N/m}^2}{1025 \text{ kg/m}^3} \times \frac{1 \text{ kg-m/sec}^2}{1 \text{ N}}}$$

$$= 1480 \text{ m/sec} \qquad \qquad \text{ans.}$$

Illustrative Problem 19.3 Compute the time required for sound waves to travel the length of a 1.5-km rod made of drawn steel.

Solution The density of steel is 7830 kg/m³ (Table 10.1) and Young's modulus is 20.0×10^{10} N/m² (Table 10.2). Then

$$v = \sqrt{\frac{Y}{\rho}} = \sqrt{\frac{20.0 \times 10^{10} \text{ N/m}^2}{7830 \text{ kg/m}^3}}$$

$$= 5.05 \times 10^3 \text{ m/sec}$$

Substituting in Eq. (3.1') yields

$$t = \frac{s}{v} = \frac{1.5 \text{ km} \times (1000 \text{ m/1 km})}{5.05 \times 10^3 \text{ m/sec}} = 0.30 \text{ sec}$$

$$\text{ans.}$$

Table 19.1 gives the speed of sound in some common media. It is emphasized that neither the frequency nor the intensity of vibration of the sound source determines the sound speed in a given medium. The medium itself and the temperature determine the speed of sound.

Table 19.1 Speed of sound in various media

Medium	Velocity	
	m/sec	ft/sec
Aluminum	5,104	16,740
Brass	3,500	11,480
Glass	5,010–6,010	16,410–19,690
Iron	4,720–5,210	15,480–17,390
Steel	5,103	16,730
Brick	3,660	11,980
Granite	6,010	19,685
Oak	3,850	12,620
Pine	3,320	10,900
Alcohol	1,240	4,070
Water (4°C)	1,420	4,700
Air, dry (32°F)	331.4	1,087

19.4 Speed of Sound and Speed of Projectiles

When a high-speed projectile whose speed is greater than the speed of sound in the air passes through air, it forms a *bow wave* of compression which precedes it. The same effect can be seen when a boat moves with considerable speed on the surface of water. The bow wave can be seen fanning out from the front of the boat in an ever lengthening V.

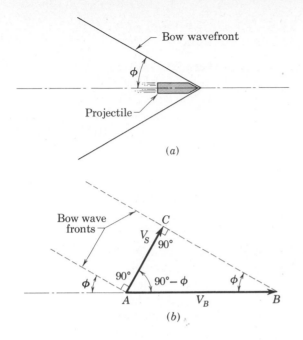

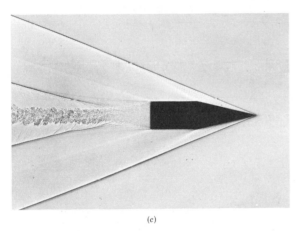

Fig. 19.2 (*a*) (*b*) Measuring the speed of a projectile. (*c*) Shadowgram of a cone-cylinder 20-mm projectile fired from a gun. The speed of the projectile can be determined from such pictures. (Official U.S. Navy photograph.)

The bow wave of compression from a high-speed projectile can be photographed by ultraspeed photography. Such a picture, called a *shadowgram*, is seen in Fig. 19.2*c*.

Let us represent the conditions shown in such a photograph by the diagram of Fig. 19.2*a*. Let the angle ϕ be the angle between the bow-wave front and the axis of the projectile. The wavefront represents a condensation traveling out from the projectile, and the wavefront is perpendicular to the direction of the propagation of the sound itself. The vector diagram of Fig. 19.2*b* shows V_s representing the velocity of the sound in a direction perpendicular to the bow-wave front. The value of V_s is known if the air temperature is known. The length of vector $\mathbf{V}_B$, representing the velocity of the projectile, is obtained by erecting a perpendicular to $\mathbf{V}_s$ at C, which will cut $\mathbf{V}_B$ at B. From the figure,

$$V_s = V_B \sin \phi$$

from which $\qquad V_B = V_s/\sin \phi \qquad\qquad (19.5)$

Illustrative Problem 19.4 Ballistic test photographs to determine the speed of a projectile showed a bow-wave angle $\phi = 24°$. The air temperature was 23°C. What was the speed of the projectile?

Solution The speed of sound at the time was

$$V_s = 331.4 + 0.61(23) = 345.4 \text{ m/sec}$$

From Eq. (19.5),

$$V_B = V_s/\sin \phi$$

we obtain

$$V_B = \frac{345.4}{\sin 24°} = \frac{345.4}{0.41} = 840 \text{ m/sec} \qquad ans.$$

19.5 Supersonic Speeds—Shock Waves

In this era of jet aircraft and high-altitude rockets, man-made objects moving through air at speeds in excess of the speed of sound are commonplace. Because the flow of air around such high-speed objects is a complex phenomenon, the study of *supersonic* speeds has a major place in research laboratory programs throughout the nation.

When an object moves through air with a speed greater than that of sound at the same altitude, the particles of air are unable to get out of the way fast enough, causing a tremendous compression of air in front of the object. This highly compressed air followed by a considerable rar-

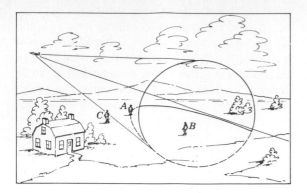

Fig. 19.3 Sonic shock wave produced when plane travels with supersonic speeds. *B* has heard the boom, *A* is experiencing it, and *C* will soon hear it.

efaction forms a series of conic waves similar to those of Fig. 19.3.

If you are in the line of travel of such a shock wave, the great pressure change from condensation to rarefaction as the wave passes you causes a loud boom. One reason the sonic boom startles a person is that the approach of a supersonic plane is absolutely silent, since it travels ahead of the noise it makes. Some booms that reach the earth can be quite destructive in terms of window breakage, as well as being annoying for their loud sound.

Mach Number Speeds which approach or exceed the speed of sound are described by the *Mach number*. This quantity is the ratio of speed of the body to the speed of sound at that place and time:

$$\text{Mach number} = \frac{\text{speed of body}}{\text{speed of sound}} \quad (19.6)$$

Since the speed of sound varies appreciably with the temperature, the Mach number speed of a plane or missile depends on whether it is flying at sea level or at a high altitude, and on the temperature of the air at the time. For sea-level flight under normal daytime temperature conditions a speed of Mach 1.00 would be about 344 m/sec (1130 ft/sec), or 1240 km/hr (770 mph).

19.6 Characteristics of Musical Sounds

The difference between a musical sound and noise is sometimes not clearly defined. Most of the common sounds that we hear may be classified as noises. The clanking streetcar, the slamming door, the boom of thunder, the barking dog, the cheering crowd, the clacking rivet hammer are all examples of noises. Musical sounds, by contrast, are more pleasing to the ear and are characterized by a definite periodicity of vibration.

19.7 Pitch of a Musical Sound

The *pitch* of a musical sound depends upon the frequency of vibration of the source. We think of pitch in terms of the "highness" or "lowness" of the tone as interpreted by the ears and brain. The relation between the pitch of a tone and its frequency of vibration can be experimentally demonstrated by the use of Savart's wheel (Fig. 19.4). This is a wheel with pointed teeth equally distributed on its circumference and mounted on a shaft so that it can be rotated at

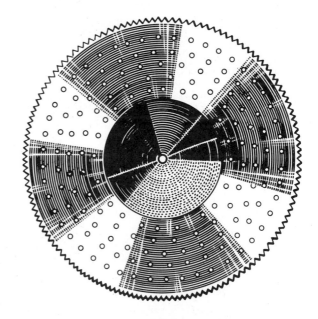

Fig. 19.4 Savart's wheel. (Central Scientific Co.)

varying speeds. When the wheel is rotating, a card is pressed against the teeth. One transverse vibration will be imparted to the card every time a tooth strikes it. The card in turn sends out a longitudinal wave throughout the surrounding air medium. When the frequency is about 20 or 30 vib/sec, a recognizable musical tone of low pitch can be heard. Increasing the speed of the wheel will increase the vibrations per second of the card and consequently the frequency of the forced vibrations of the air about it. This will be recognized as a higher pitch.

Thus it can be demonstrated that a sound has a definite pitch when the source is vibrating at a fixed frequency. Increasing the frequency of vibration of the source causes a rise in pitch, while decreasing the frequency will cause the pitch to fall.

If the teeth of the wheel were irregularly spaced, the sound would have no definite pitch; only a *noise* would be heard. We can represent the difference between a pure musical tone and a noise graphically (Fig. 19.5). A pure musical tone is distinguished by a regularity of the condensation-rarefaction pattern, while a noise is represented by a very irregular and haphazard curve.

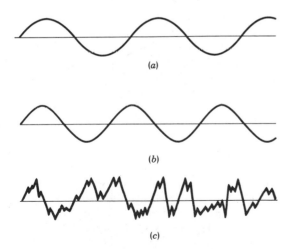

(a)

(b)

(c)

Fig. 19.5 Curves representing (a) pure musical tone, (b) pure musical tone of higher pitch, (c) noise.

19.8 Audible Frequencies

The range of frequencies audible to the human ear varies with the individual, but the average normal ear is sensitive to frequencies lying in the range from 20 to 20,000 Hz. Generally speaking, the range decreases with the age of the individual. The upper limit is usually a bit higher for children and for women than it is for men. The range of audible frequencies depends to some extent on the loudness, or intensity, of the sound.

Figure 19.6 shows graphically how the upper and lower limits of audible sound intensity vary for the normal ear as a function of frequency. Intensities below the lower line are generally insufficient to produce any sensation of hearing to the ear. Intensities near the upper line are painfully loud to the ear; frequencies in excess of about 20,000 Hz are not heard by the normal ear.

It will be noticed that the frequency scale is not linear (it is logarithmic; see Sec. 19.9). A tremendous difference in the sensitivity of the normal ear for different frequencies is apparent from a study of the graph. The ear has its greatest sensitivity for low-intensity levels in the frequency range 2000 to 4000 Hz. The intensity which produces the sensation of pain does not vary a great deal with the frequency. Various other interesting observations can be made. The intensity necessary for hearing near the high- and low-frequency limits of audibility is many times as great as that necessary near the region of greatest sensitivity. For example, the normal ear may hear a sound whose frequency is 3000 Hz when its intensity is only about 10^{-16} W/m^2; but the intensity of a note whose frequency is 30 Hz will have to be in the neighborhood of 10^{-8} W/m^2 (100 million times the energy) to be audible.

Most commercial reproducing systems, such as the telephone, the phonograph, the radio, tape recordings, and public-address systems, reproduce faithfully over a frequency range considerably less than the hearing range of the normal ear. For example, a good radio receiver will cover a range of from 40 to about 7000 Hz. Since this does not cover the range of frequency

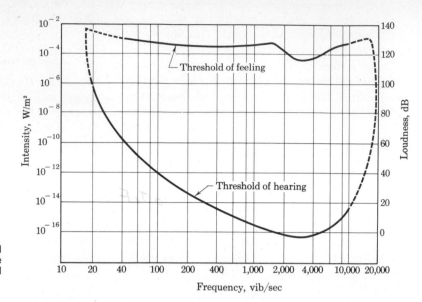

Fig. 19.6 Limits of audible sound intensities. The audible range for the normal ear lies in the region enclosed by the two curves.

of most orchestral music, some of the quality of fine music is lost in reproduction. Modern high-fidelity reproducing equipment (tapes and disk recordings) covers the range from 50 to 16,000 Hz.

Sounds with frequencies above 20,000 Hz are generally beyond the range of the human ear. They are called *ultrasonic*. In the next chapter we shall study some of the technical uses to which ultrasonic waves have been put. Experiments conducted with animals reveal that some animals can hear sounds with frequencies as high as 40,000 Hz. Many insects, such as the cricket, are able to produce and hear ultrasonic frequencies.

19.9 Loudness and Intensity of Sound—the Decibel

The harder we hit a drum with a padded hammer, the "louder" the sound emitted. Just as it is diffi-cult to describe anything as being "twice as red," "twice as cold," or "twice as sweet," so it is that loudness is a characteristic of sound that is difficult to describe in exact terms. Such expres-sions as a "pitch twice as high," or "a tone twice as loud" are quite meaningless. The impressions of sounds are sensory and depend on the

physiology of the ear and the interpretation within the brain. However, there is a relationship between the sensation of loudness interpreted by the ear and certain physical properties of the sound wave reaching the ear. The sensation of loudness depends upon two things: (1) the ampli-tude of the vibrating source and (2) the distance of the source from the ear.

There is an approximate law which states that the magnitude of the sensation of loudness is proportional to the logarithm of the sound intensity, or

$$L = \log \frac{I}{I_0} \qquad (19.7)$$

where
L = magnitude of the sensation of loudness
log = logarithm to base 10
I = intensity of the sound in W/m^2
I_0 = intensity at the threshold of hearing $(10^{-12}\ \text{W/m}^2)$

This law is called the *Weber-Fechner law*, after the two scientists who did the basic research leading to its discovery. It is a fair approximation for more or less pure tones throughout a normal range of frequencies.

The unit of loudness is called the *bel* in honor of Alexander Graham Bell (1847–1922),

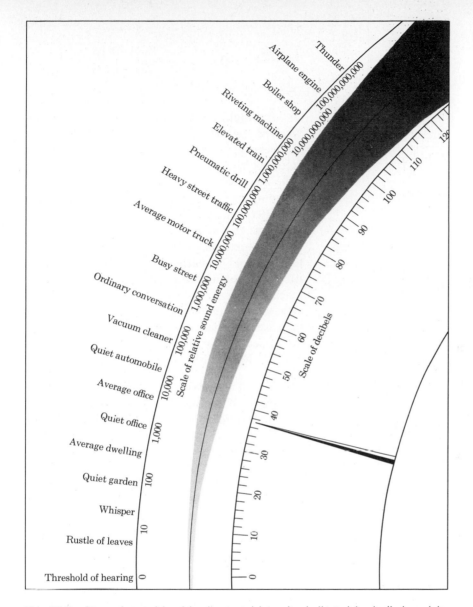

Fig. 19.7 Chart of sound-level loudness and intensity, indicated in decibels and in units of relative sound energy. (Electronics.)

inventor of the telephone. The definition of the bel is quite technical, but for our purposes we can say that each time the power delivered to the ear is multiplied by 10, the loudness sensation that results is increased 1 bel, or 10 decibels (dB). Because the bel is too large for ordinary purposes of loudness measurement, and because tests have shown that the average ear can distinguish sound-intensity differences equal to about one-tenth of the bel, the *decibel* ($\frac{1}{10}$ bel) is usually used. Intensity levels, in decibels, for some common sounds can be found in Fig. 19.7.

$$L \text{ (bels)} = \log \frac{I_1}{I_0} \qquad (19.8)$$

$$l \text{ (dB)} = 10 \log \frac{I_1}{I_0} \qquad (19.9)$$

Illustrative Problem 19.5 Find the magnitude of loudness in decibels at the threshold of hearing.

Solution In this case $I_1 = I_0$ in Eq. (19.9). Thus

$$l \text{ (dB)} = 10 \log \frac{I_1}{I_0} = 10 \log \frac{I_0}{I_0}$$

$$= 10 \log 1$$

$$= 0 \text{ dB} \qquad \qquad ans.$$

Illustrative Problem 19.6 Street noises entering through an open window register a reading of 68 dB on a sound meter on the window sill. When the double window is closed, the meter registers 42 dB. What percentage of the sound is prevented from entering the room?

Solution Use Eq. (19.9):

$$68 \text{ (dB)} = 10 \log \frac{I_1(\text{W/m}^2)}{I_0(\text{W/m}^2)}$$

$$\log \frac{I_1}{I_0} = 6.8$$

Use Table 4 of the Appendix to get

$$I_1/I_0 = 6.31 \times 10^6$$

Then

$$I_1 = 6.31 \times 10^6 \, I_0 \text{ W/m}^2$$

In like manner,

$$42 \text{ (dB)} = 10 \log \frac{I_2(\text{W/m}^2)}{I_0(\text{W/m}^2)}$$

and

$$I_2/I_0 = \text{antilog } 4.2$$

$$= 1.59 \times 10^4$$

$$I_2 = 1.59 \times 10^4 \, I_0$$

Then

$$\frac{I_1 - I_2}{I_1} \times 100 = \frac{(6.31 - 1.59)(10^4)\text{W/m}^2}{6.31 \times 10^6 \text{ W/m}^2}$$

$$\times 100 = 99.7\% \qquad ans.$$

It should be emphasized that as the power received by the ear is *multiplied* by 10, the loudness scale is increased (*added*) by 10 dB. Hence it should be clear that the wave-power scale increases very much faster than the corresponding decibel scale. For example, from Figs. 19.6 and 19.7, the student should be able to recognize that when the loudness level increases from the extremely faint sounds that border on the threshold of hearing (0 dB) to the point where it becomes painful to the ear (120 dB) the power reaching the ear to give these sensations is increased 10^{12} times. The loudness of a sound will increase by 5 bels (50 dB) when the intensity is multiplied by 10^5. Note that 10^5 is 100,000 and that log 100,000 is 5.

Effect of Distance from Source on Loudness Loudness of sounds decreases with distance from the source. Sound disturbances send waves which spread uniformly in all directions in space, forming concentric spheres of condensation and rarefaction. The intensity of disturbance is inversely proportional to the surface area of the wave. Since the value for the surface area of a sphere is $4\pi r^2$, sound intensity is inversely proportional to the square of the distance from the source (radius) to the wavefront. Thus at a point 5 m from a source of sound the intensity of the sound will be one-twenty-fifth $(\frac{1}{5})^2$ its value at a distance of 1 m. The intensity of a sound falls off rapidly with increasing distance from the source. This falling off, known as *attenuation*, follows the inverse-square law, as will be shown.

Illustrative Problem 19.7 Find (a) the acoustic energy level and (b) the sound intensity level at a point 30 m from a source of sound if the sound source radiates acoustic energy uniformly in all directions at the rate of 2.0 W.

Solution

(a)

$$I = \frac{P}{A} = \frac{2.0 \text{ W}}{4\pi(30 \text{ m})^2} = 1.8 \times 10^{-4} \text{ W/m}^2$$

ans.

(b)

$$L = 10 \log \frac{I}{I_0} = 10 \log \frac{1.8 \times 10^{-4} \text{ W/m}^2}{(10^{-12} \text{ W/m}^2)}$$

$$= 10 \log (1.8 \times 10^8)$$

$$= 10 [\log 1.8 + 8 \log 10]$$

$$= 10 [0.255 + 8]$$

$$= 83 \text{ dB} \qquad ans.$$

Consider a source of sound at P (Fig. 19.8). At a distance s_1 the original sound energy is dissipated over a total area of $4\pi s_1^2$, and the sound intensity (energy per unit of area) could be expressed as

$$I_1 = E/4\pi s_1^2$$

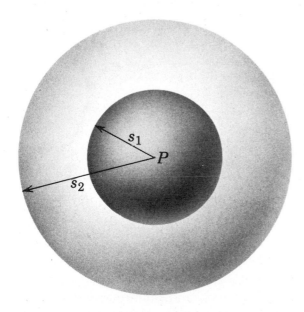

Fig. 19.8 Sound waves travel out from a source in alternate compression-rarefaction waves, which are concentric spheres.

When the sound wave arrives at the distance s_2, the original energy is dissipated over an area $4\pi s_2^2$, and

$$I_2 = E/4\pi s_2^2$$

Dividing, we obtain the formula which relates the intensity of sound and distance from the source,

$$I_1/I_2 = s_2^2/s_1^2 \qquad (19.10)$$

where I_1 and I_2 are intensities of the sound at respective distances s_1 and s_2. This formula is strictly true, however, only if the distances involved are large compared to the size of the source of vibration, if no sound is reflected to the listener, and if no absorbing material is between the source and listener.

Illustrative Problem 19.8 If the loudspeaker of a public-address system can produce loudness rated at 90 dB at a distance of 1.22 m from the speaker, how far can the sound travel and still give a loudness at the listener's ear of 40 dB (equivalent in intensity to that of ordinary conversation at a 0.91-m distance)? Consider the speaker as a point source of sound.

Solution The difference in loudness of the sound at the two locations is equal to $90 - 40$ or 50 dB (5 bels). Thus the intensity of the sound 1.22 m from the source is 10^5 times the intensity at the desired distance.

$$I_1 = 10^5 I_2$$

$$s_1 = 1.22 \text{ m}$$

Using Eq. (19.6),

$$s_2^2/s_1^2 = I_1/I_2$$

$$s_2^2/1.22^2 = 10^5 I_2/I_2$$

$$s_2^2 = 1.22^2 \times 10^5$$

$$s_2 = 386 \text{ m} \qquad ans.$$

The loudness level interpreted by the ear depends considerably also on the frequency (pitch) of the sound. This is true because the ear is

Fig. 19.9 Acoustical engineers working in an anechoic chamber are testing a car for possible noise and source of vibration. (Courtesy General Motors.)

not equally sensitive for all frequencies (see Fig. 19.6).

Engineers are now able to measure sound intensities quite accurately with electronic devices which determine the wave-energy levels (see Fig. 19.9).

As sound spreads out from a source its intensity decreases and the wave is said to be *attenuated*. When the sound intensity is increased the sound is said to be *amplified*. In communication and audio systems the gain in decibels is given by the expression

$$l_2 \text{ (dB)} - l_1 \text{ (dB)} = 10 \log\left(\frac{P_{out}}{P_{in}}\right) \quad (19.11)$$

where P_{out} is the output power and P_{in} is the input power of the system.

Illustrative Problem 19.9 Determine the gain in decibels in an audio amplifier if it has an output power of 9.0 W when the input power is 1.5 mW.

Solution Use Eq. (19.11):

$$l_2 \text{ (dB)} - l_2 \text{ (dB)} = 10 \log \frac{P_{out}}{P_{in}}$$

$$= 10 \log \frac{9 \text{ W}}{1.5 \times 10^{-3} \text{ W}}$$

$$= 10 \log (6 \times 10^3)$$

$$= 10(\log 6 + 3 \log 10)$$

$$= 10(0.7782 + 3)$$

$$= 38 \text{ dB} \qquad ans.$$

19.10 Quality of Sound

We all know that two people singing the same note do not sound exactly the same. The difference in the sound is due to the *quality* of the note. It is the difference in the quality of tone of a violin that distinguishes a good instrument from a poor one.

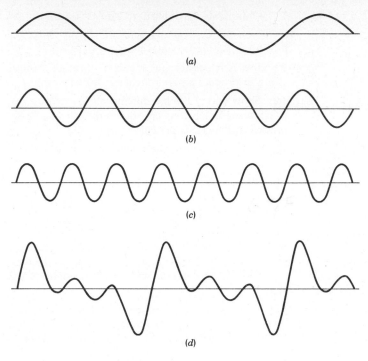

Fig. 19.10 Resolving the fundamental tone and overtones into one blended musical sound.

In order to understand the reasons for differences in quality, it must be realized that musical sounds do not usually result from single vibrations of a given frequency. When a tone is the result of a single frequency of vibration it is called a *pure* tone. However, most musical sounds are a blending of vibrations of different frequencies in a particular way. The different frequencies occur when the sounding body vibrates in parts at the same time that it is vibrating as a whole. For example, the string of a musical instrument may vibrate as a whole, but usually the whole is also vibrating in two or more equal parts, giving off frequencies which are exact multiples of the frequency of vibration of the whole. The result is a complex vibration which gives it a characteristic quality.

When a body vibrates as a whole, it produces its lowest tone, called the *fundamental* tone. Sounds due to vibration of segments of the whole are called *overtones* or *harmonics*.

The quality of a musical sound depends upon the number, the frequencies, and the relative intensity of the various overtones that blend with the fundamental. The rich tone of a violin may contain 10 or more overtones. In some instruments, the overtones are relatively strong, while in others they may be weak or missing completely. Figure 19.10 illustrates the blending of the fundamental (*a*) and the first two overtones (*b* and *c*) into the complex musical sound (*d*) which results. The pitch that is most easily recognized as a musical sound is that produced by the fundamental while the quality of the sound depends on the number and relative amplitude of its harmonics.

SUMMARY

A great deal of the energy around us is transmitted by waves, e.g., heat, light, radio, sound, and water waves. The source of every sound is a vibrating body.

Sound waves will not travel in a vacuum but can be transmitted through any material medium that possesses elasticity.

In any fluid medium sound waves travel from their source in spherical wavefronts unless reflected.

Sound waves are longitudinal in nature, i.e., the particles vibrate in a direction parallel to the direction of the wave.

Two particles that are in the same condition of condensation or rarefaction are said to be in the same *phase*.

The accepted value for the speed of sound in air is 331.4 m/sec at 0°C (1087 ft/sec at 32°F). This speed increases about 0.61 m/sec for each increase of 1C° (1.1 ft/sec/F°).

The speed of sound in a fluid medium can be found by the formula

$$v = \sqrt{B/\rho} \qquad (19.3)$$

The speed of sound in a solid rod or wire is found by

$$v = \sqrt{Y/\rho} \qquad (19.4)$$

The ratio of the speed of a body to the speed of sound at that place is called *Mach number*.

Musical sounds are characterized by their *pitch*, *loudness*, and *quality*. Pitch is determined mostly by the frequency of the source: the greater the frequency, the higher the pitch, and conversely. Loudness depends on the amplitude of vibration and on the distance from the source. Quality of a tone is determined by the number of ways in which the source is vibrating at the same time, e.g., in 1, 2, 3, 4, or more parts.

Loudness is usually expressed in units called *decibels*. The intensity of a sound is inversely proportional to the square of the distance from the source.

QUESTIONS AND EXERCISES

1. What determines the pitch of a sound?

2. Give a fundamental difference between a sound wave and a light wave.

3. What is meant by "Mach number"?

4. Why is it impossible for a person to hear a gunshot 25 ft away on the surface of the moon?

5. Give an illustration which would show that the speed of sound is independent of frequency.

6. Distinguish between intensity and loudness of a sound.

7. What name is applied to sound vibrations (a) below the audio range, (b) above the audio range?

8. What is a fundamental tone?

9. How is it possible for a single vibrating musical string to produce several pitches at the same time?

10. Show that $\sqrt{Y/\rho}$ has dimensions for velocity if Y is expressed in newtons per square meter and ρ is expressed in kilograms per cubic meter.

11. What is meant by the statement, "the sound level is 30 dB"?

12. Name the three characteristics of a musical sound. Tell what physical property of the wave determines the characteristics as heard by a person.

13. The speed of sound in a given medium depends upon which of the following? (a) the frequency of the sound wave, (b) the loudness of the sound, (c) the temperature of the medium, (d) the wavelength of the sound.

14. Which of the following are not correlated: (a) pitch and frequency, (b) quality and waveform, (c) loudness and intensity, (d) loudness and resonance?

15. Tell how to estimate the distance to a bolt of lightning.

16. Two waves traveling in the same direction meet at a time when the first wave has an amplitude of A and the second wave has an amplitude of B. Which of the following will represent the amplitude of the combined wave at that instant: (a) $A - B$, (b) $A + B$, (c) something between A and B, (d) something between $A - B$ and $A + B$.

PROBLEMS

1. Approximately how many times as much energy is there in the sounds of normal conversation than there is in a whisper? (Use Fig. 19.7.)

2. If a person has lost 20 dB of hearing intensity, to what intensity must a sound be magnified to give the individual the same effect that a 30-dB sound has on the normal ear?

3. Find the time required for sound to travel 3 km in air at 30°C.

4. A clap of thunder is heard 8 sec after the lightning flash is seen. How far away was the lightning if the temperature is 25°C?

5. At what air temperature is the speed of sound 1137 ft/sec?

6. Sounds with frequencies of 20 and 20,000 Hz represent the usual range of audibility the human ear can hear. What are the wavelengths of these sounds? (Assume the velocity of sound to be 1100 ft/sec.)

7. A depth-sounding device emits a signal of 38,000 Hz in water whose average temperature is 4°C. The impulse is reflected from the ocean bed and returned to the sounding device 1.62 sec after the signal has been emitted. (a) What is the depth of the water? (b) What is the wavelength of the signal in water? (See Table 19.1.)

8. Steam can be seen coming from a factory whistle on a day when the temperature is −5°C, and 6 sec later the sound is heard. How far away is the whistle?

9. A workman strikes the steel rail of a railroad track with a sledgehammer. The sound of the blow reaches an observer through the rail and through the air. The difference in time for the sound to reach the observer is 3 sec. How far, in meters, is the observer from the workman? (Use Table 19.1.)

10. In foggy weather a lighthouse keeper sends sound signals simultaneously under water (4°C) and through the air (15°C). A vessel receives the signals 2.15 sec apart. How far, in kilometers, is the vessel from the lighthouse?

11. A blow struck on a steel rail was heard through the rail in 0.2 sec and through the air in 2.9 sec. If the temperature was 54°F, (a) how far from the observer was the blow struck and (b) what was the speed of sound in the rail?

12. A marksman fires at a target with a rifle. The bullet travels at the average rate of 615 m/sec, and 4 sec after he fires the gun, he hears the bullet strike the target. How far away is the target if the temperature is 25°C?

13. Compute the speed of sound in a column of mercury. (Use Tables 10.1 and 10.4.)

14. At 0°C and 760 mm Hg, the density of oxygen is 1.43 kg/m^3 and the bulk modulus is 1.418×10^5 N/m^2. Find the speed of sound in oxygen under those conditions.

15. A ship tracking a submarine sends pulsed sound waves which are reflected by the submarine. The reflected waves are detected 4.26 sec after emission by the tracking ship. How far is the submarine from the ship? (Assume sea water has a bulk modulus of 0.225×10^{10} N/m^2.)

16. How long will it take for sound waves to travel 450 yd in a copper tube?

17. What is the intensity in watts per square centimeter of a sound which has an intensity of 50 dB?

18. What is the intensity level in decibels of a sound which has an intensity of 10^{-8} W/m^2?

19. If the angle between the bow-wave front and the axis of the projectile in Fig. 19.2c is 21° and the air temperature at the time of the ballistic test was 15°C, what was the speed of the projectile?

20. Two sounds of the same frequency have intensities of 10^{-7} and 10^{-10} μW/cm^2. How many decibels louder is the first sound than the second?

21. A tuning fork vibrates at a rate of 440 sec^{-1}. Find the wavelength of the sound in air at 83°F.

22. A jet plane travels 650 mi/hr near the earth's surface. The temperature of the atmosphere is 80°F. What is the plane's Mach number?

23. A jet plane is flying at a speed of Mach 2 at an altitude of 6000 ft. How far from an observer will the plane be when the shock waves hit him?

24. Sound energy is radiated uniformly in all directions from a source at the rate of 40 W. (a) What is the sound intensity in watts per square meter at a point 35 m from the source? (b) What is the intensity level, in decibels, at that point? (HINT: Use Eq. 19.7.)

25. What is the intensity level, in decibels, for Prob. 24 if there is a 15 percent absorption of acoustic energy in the 35-m path?

20 Technical Applications of Sound Waves

Today industries which deal with the production and reproduction of sound are among the largest in the world. The science of sound is basic to many occupations. The sound technician and sound engineer are essential in the maintenance of our standards of living.

The term *acoustics* is used to denote the science of the production, transmission, reception, and effects of sound. *Ultrasonics* is a branch of acoustics dealing with periodic waves with frequencies above the audible range. Although ultrasonic waves were known to scientists for many years, their practical application in science, industry, and medicine is of a rather recent date.

Sounds are present almost everywhere in our environment: some are pleasing, others are quite irritating. The control of unwanted sounds (noise) is an important consideration in our everyday lives. Yet it is only since the beginning of the nineteenth century that much attention has been given to proper acoustical design in the construction of buildings and houses. The beginning study of acoustics can be traced almost entirely to the efforts of W. C. Sabine (1868–1919) of Harvard University. Today it is possible for architects to meet exacting acoustical specifications while at the same time providing specific conditions of air conditioning and lighting.

In the previous chapter we studied the three characteristic properties of sound waves by means of which we are able to distinguish one sound from another, i.e., *pitch*, *loudness*, and *quality*. There are other properties of sound waves which must now be examined in order to understand technical methods of sound control— properties like reflection, refraction, interference, and absorption.

20.1 Reflection of Sound Waves

We are all familiar with the rebounding of a rubber ball when it is thrown against a wall. In much the same manner sound waves are reflected from various surfaces. Thunder, for example, is due to successive reflections of sound from cloud to cloud or from cloud to land surface. Certain buildings, such as the Statuary Hall at the United States Capitol in Washington, D.C., and the Mormon Tabernacle in Salt Lake City, are so constructed that sounds will be reflected from their walls and ceilings to give a "whispering gallery" effect. Because of reflected sounds, a pin dropped in a certain place in the Tabernacle can be clearly heard at the other end of the room. In Statuary Hall it is possible to stand at a certain spot in the building and distinctly hear a whisper uttered by someone at certain spots on the other side of the room, although the whisper may not be heard at points in between.

A simple experiment can be performed to illustrate the phenomenon of sound reflection.

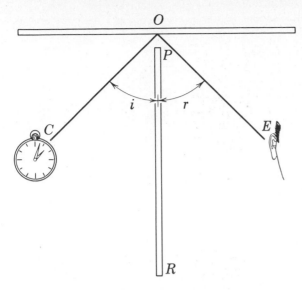

Fig. 20.1 The clock is heard when angle i = angle r.

Place a clock C near a hard wall surface (Fig. 20.1) and separate the hearer and the clock by a panel PR. As the hearer E moves about the room, he will discover that it is difficult to hear the clock on the other side of the panel for all positions except those that lie in a given direction. It can be shown that the direction of best audibility is the one which makes angle COR equal to angle EOR, which illustrates the following law:

The Law of Reflected Waves

The angle of reflection r is equal to the angle of incidence i:

$$\text{Angle } r = \text{angle } i$$

A sound that is reflected from a surface to the listener's ear is called an *echo*. The human ear is able to distinguish two sounds as being distinctly separate only if they reach the ear at least $\frac{1}{10}$ sec apart. If the two sounds reach the ear with a shorter time interval between them, they will blend together and give the impression of a single sound. The distance sound must travel to be audible as an echo can readily be determined. Since sound travels at the rate of about 330 m/sec (1100 ft/sec), or 33 m (110 ft) each 0.1 sec, the distance the reflected sound must travel to be distinguished from the original sound must be 33 m (110 ft) more than the distance the original sound travels directly to the ear. Echoes are commonly heard in long corridors and in deep canyons. The depth of a well can be determined by making a sound at the top of the well and noting the time required for the sound to reach the bottom and be reflected back to the ear.

Megaphones are shaped to concentrate the sound waves in a desired direction. In various musical instruments, such as the trumpet, the shape of the horn is such as to direct the sound waves in a desired fashion. Proper directing or focusing of sound waves is important in large auditoriums and amphitheaters. Many of the older auditoriums and churches were provided with curved sounding boards placed behind the speaker. The advent of the modern public-address system has made this practice less common. Many large amphitheaters, however, are still provided with curved bowls behind the stage to direct the sound, by reflection, to the audience (see Fig. 20.2).

20.2   Refraction of Sound Waves

Whenever a sound wave passes from one medium into another of different density or elasticity, its speed is changed. This change may alter the direction of propagation of the wave. The bending of the direction of a wave due to a change of speed is called *refraction*. All sound waves traveling in the open air are refracted to some extent. This is true because the air is seldom at rest and varies throughout in temperature and density. These variations in temperature and density cause changes of speed, with consequent refraction of the sound wave. This phenomenon explains why sound usually does not appear to carry so far in the daytime as at night. The sun in the daytime warms the surface of the earth and the air immediately adjacent to the surface. Hence the air next to the surface is warmer than at higher levels. Since the speed of sound increases with an increase in temperature, the part of the wavefront next to the earth's surface travels faster than that

Fig. 20.2 The Sidney Myer Music Bowl, Melbourne, Australia. The bowl consists of an excavated fixed seating area of 2000 seats, which is covered almost entirely by a large suspended canopy opening up at the rear to a sloping lawn seating some 20,000 people with unobstructed view of the stage. Beyond this area, sound is projected to 30,000 or more picnicking and strolling listeners. (Bolt, Beranek, and Newman, Inc.)

at higher levels. The effect is a bending upward of the wavefront away from the observer on the ground (Fig. 20.3a). At night the reverse is true. The ground surface cools faster than the air above. Hence the layers of air next to the ground are the colder ones, and the sound waves are refracted downward (Fig. 20.3b). Often in the evening one can hear the sound of voices or music distinctly, even though they may be at a considerable distance from the listener.

Wind may also be a factor causing refraction of sound. The speed of wind is usually slower next to the surface of the earth than at higher levels, because of greater air friction at the surface. Hence if the sound travels in a direction against

that of the wind, the top part of the wave is slowed more than the bottom; the result is a bending upward of the wavefront (Fig. 20.4a). Conversely, if the wave is traveling with the wind, the faster upper layers of wind bend the wavefront downward (Fig. 20.4b). Sound is thus seen to stay longer on the surface if it travels with the wind than if it travels against it.

20.3 The Doppler Effect

Many people have observed the phenomenon of the Doppler effect at one time or another without realizing the cause. As an auto (or diesel locomotive) goes by at high speed sounding its horn, the

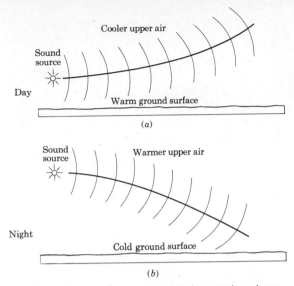

Fig. 20.3 Sound waves are refracted upward or downward according to whether the layers next to the ground are the warmer or the colder.

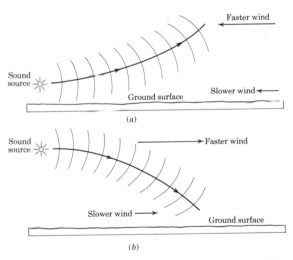

Fig. 20.4 Sound waves are refracted upward traveling against the wind and downward traveling with the wind.

pitch of the tone drops appreciably as the source of the sound passes and recedes from the observer.

This change of pitch is of course due to an apparent change in frequency, or a *frequency shift*. The principle underlying the phenomenon was developed by Christian Doppler (1803–1853). The principle states that the frequency of a wave motion as determined by an observer differs from the frequency of the source whenever there is relative motion of the source or observer. For sound, the pitch change (due to frequency shift) is quite marked, even when the relative speed of source and observer is fairly low. The diagram of Fig. 20.5 will serve to clarify the wave relationships in the Doppler effect. Let v_s be the speed of the source of the sound waves and v be the speed of sound. Let O_1 represent an observer who is being approached by the police car with sounding siren and O_2 one from whom the police car is receding.

The siren is moving away from the waves traveling to O_2 and toward the waves traveling to O_1. The waves behind the horn are therefore stretched out, and those in front are compressed or crowded together. The actual speed of the sound waves is the same in all directions, but the net effect of the motion of the source is that observer O_1 receives more waves per second than O_3 and O_4 do, and observer O_2 receives fewer per second. The pitch at O_1 is higher than the actual pitch (at O_3 and O_4), and the pitch at O_2 is lower than the actual pitch.

Source Moving If a source of sound of frequency f_s moves with speed v_s, the frequency f_0 of the tone heard by a stationary observer may be obtained from

$$f_0 = f_s \frac{v}{v \pm v_s} \qquad (20.1)$$

where v is the speed of sound in air at that place. If the source is approaching the observer, the algebraic sign in the denominator is minus; if receding, the sign is plus.

Observer Moving If an observer moves with a speed v_0 with respect to a stationary source of sound of frequency f_s, the frequency of the tone heard by the moving observer is

$$f_0 = f_s \frac{v \pm v_0}{v} \qquad (20.2)$$

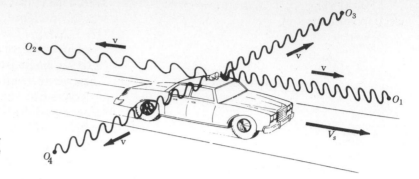

Fig. 20.5 A fast-moving police car sounding its horn illustrates the Doppler effect.

where the plus sign is applied when the observer is approaching the source and the minus sign when the observer recedes from the source.

Equations (20.1) and (20.2) can be combined into one that includes all four cases. It is

$$f_0 = f_s\left(\frac{v \pm v_0}{v \mp v_s}\right) \qquad (20.3)$$

The upper sign is to be used if the velocity is one of approach. The lower sign is used if the velocity is one of receding motion.

Doppler's principle, though an interesting phenomenon in sound, has more practical applications in the fields of light and radar. Frequency shifts in light waves received from distant stars tell us whether the star is approaching our solar system or receding from it. The same principle is used by highway patrolmen to detect "speeders."

An observed frequency shift in a returning radar wave reflected from a target airplane tells whether the target is approaching or receding, and the application of timed pulsing to this "doppler" gives the *range rate*, the actual rate at which the range to the target is closing or opening.

Applied to supersonic sounds, Doppler's principle is used in determining the range rate of enemy submarines and in the operation of sonar.

Illustrative Problem 20.1 A factory whistle is known to have a frequency of 480 Hz. An auto is approaching the source at 100 km/hr when the air temperature is 32°C. What is the pitch heard by the auto's occupants?

Solution Since this is a case of observer moving toward a stationary sound, Eq. (20.2) is used, with the plus sign.

$$v = 331.4 + 0.61(32) = 350.9 \text{ m/sec}$$

$$v_0 = 100 \text{ km/hr} = 27.77 \text{ m/sec}$$

Substituting known values in Eq. 20.2 gives

$$f_0 = 480 \text{ Hz} \frac{(350.9 + 27.8) \text{ m/sec}}{350.9 \text{ m/sec}}$$

$$= 518 \text{ Hz} \qquad\qquad ans.$$

Illustrative Problem 20.2 An ambulance, traveling 88 km/hr, is moving toward an approaching car which is traveling at a speed of 50 km/hr. The air temperature is 24°C. If the siren is emitting a frequency of 1020 Hz, what is the frequency heard by the driver of the car? (Assume no wind.)

Solution Use Eq. (20.3) with $f_s = 1020$ Hz, $v_s = 88$ km/hr, and $v_0 = 50$ km/hr. The velocity of the sound in air is

$$v = 331.4 + 0.61(32) = 350.9 \text{ m/sec}$$

$$v_s = 88 \frac{\text{km}}{\text{hr}}\left(\frac{1 \text{ hr}}{3600 \text{ sec}}\right)\left(\frac{1000 \text{ m}}{1 \text{ km}}\right) = 24.4 \text{ m/sec}$$

$$v_0 = 50 \text{ km/hr} = 13.9 \text{ m/sec}$$

Substituting known converted values and using proper signs, we get

$$f_0 = f_s\left(\frac{v + v_0}{v - v_s}\right)$$

$$= 1020\left(\frac{350.9 + 13.9}{350.9 - 24.4}\right)$$

$$= 1140 \, \text{Hz} \qquad \qquad \textit{ans.}$$

20.4 Locating Earthquakes

The approximate locations of earthquakes can be determined by the use of an instrument called a *seismograph*. When an earthquake occurs, the vibration source produces two distinct types of waves: longitudinal and transverse. These two types of waves travel with different velocities through the earth's crust. Hence it is possible to determine the distance to the source by recording with the seismograph the time interval between the arrival of the two types of waves. By recording the character and time pattern of the vibrations with seismographs at several points on the earth's surface, it is possible to locate fairly accurately the center (called *epicenter*) of the disturbance.

A study of earthquakes has given us our most reliable information about the makeup of the earth's interior. Engineers, by studying the character of the shocks, are now able to construct dams, canals, pipelines, and buildings which will resist or withstand them.

20.5 Prospecting for Oil

Seismographs are also used in geophysical prospecting for petroleum by determining underground structures. Waves set off by the explosion of a buried charge of dynamite or TNT travel through the earth and are partially refracted and reflected to the surface through adjacent layers of formation of different densities (Fig. 20.6). The returning seismic waves are picked up by sets of sound detectors, called *geophones*, arrayed on the ground at equal intervals, usually 12 on each side of the explosion site. The geophones convert the

mechanical earth vibrations into weak electrical impulses. The impulses are amplified and recorded on a rapidly moving photographic paper on which are time lines usually spaced at 0.01 sec intervals. The instant of the firing of the explosive charge is also recorded on the paper.

The first impulses recorded by the geophones are results of the refracted waves. Impulses from the reflected waves will follow. The time for the reflected waves to appear will depend on their velocity within each of the layers penetrated and on the depths of the reflecting surfaces. The time intervals which elapse between the firing of the charge and the arrival of the reflected waves can be used to determine the depths of subterranean structures that may trap the oil and/or gas. Impulses are timed with an accuracy of 0.001 sec. This produces an accuracy in depth determination of 1.0 to 1.5 m (3 to 5 ft) for every 300 m (1000 ft) of depth.

20.6 Ultrasonics

The science and technology of high frequency sound vibrations are called *ultrasonics*. The normal human ear can hear vibrations of approximately 20,000 Hz. Sound waves with vibrations greater than 20,000 Hz (beyond the limit of audibility) are ultrasonic. Ultrasonic waves have been developed that vibrate at approximately 20 billion hertz. Ultrasonic waves can be generated from mechanical (such as whistles and sirens), electromagnetic, or thermal sources. Devices which convert magnetic energy or heat energy into acoustic or ultrasonic energy are called *transducers*.

The applications of ultrasonic vibrations are many. And the potential is great for more practical applications to be developed in the near future. We cite a few current uses of ultrasonics.

Sonar Although audible sound waves travel considerably farther in water than in air, the range in water is still quite limited. Ultrasonic waves can pass through many miles of water before the intensity drops to half its original value. The device for using ultrasonic waves for

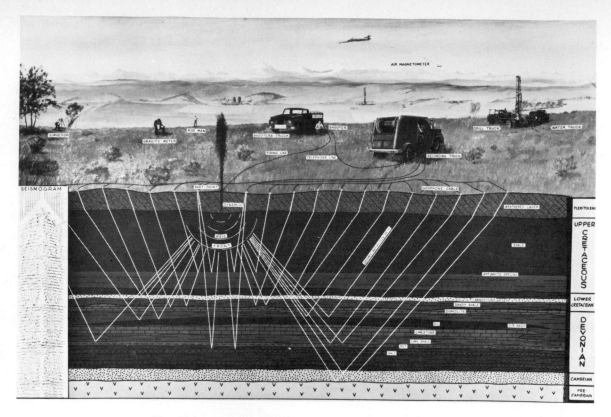

Fig. 20.6 A geophysical crew in the field using three primary geophysical exploration methods: seismic, gravity, and magnetic. In seismic exploration, maps of subsurface formations are obtained by setting off explosive charges near the surface. These create sound waves which are refracted and reflected from formations of composition different from that of the formation above it. By timing the return of the shock waves the geologist is able to draw accurate profiles of the subterranean strata. (Courtesy Society of Exploration Geophysicists.)

underwater observation, communication, and navigation is called *sonar*. The word is an abbreviation for Sound Navigation and Ranging. The frequencies most often used in sonar systems range from 20,000 to 50,000 Hz.

Ultrasonic beams can be made that are narrow and highly directional. Thus, when an ultrasonic signal is sent through the water as a narrow beam, any obstacle in its path will reflect the beam back to a receiver. The receiver, then, converts the reflected signal into an audible sound, to a pattern on an oscilloscope, or to a record on a roll of chart paper. Since the opera-

tion is silent, it is valuable in antisubmarine warfare.

By superposing or modulating audible speech frequencies on ultrasonic waves of about 40,000 Hz, submarines are now able to talk from ship to ship while under water. The method is similar to the modulation of carrier radio waves which will be described in Chap. 32.

Oceanography Ultrasonic waves are used in exploration of the ocean and lake floors, underwater distance measurements, and fishing. The operation is called *depth sounding* and is similar to that used in sonar ranging. The instrument used to detect the reflected waves is known as an *echo sounder*. The sounder on the oceanographic vessel functions automatically. A signal is periodically directed vertically downward into the water, and the time interval between the transmission of the signal and the reception of its echo is a measure of the water depth. The method of depth sounding is also of value in locating submerged icebergs, derelicts, or other obstructions in the channels of ship travel. Similar equipment allows nuclear powered submarines to cruise safely just below the ceiling of the ice cap of the polar regions.

Illustrative Problem 20.3 A crew on a ship in the U.S. Coast and Geodetic Survey, doing depth-sounding work off the Alaskan coast, found that an interval of 2.90 sec elapsed between the sending of a wave pulse to the ocean bottom and the return of the reflected wave. If the speed of sound in the water was 1460 m/sec, what was the ocean depth?

Solution Using the formula $v = s/t$, solve for s and substitute the known values for v and t:

$$s = vt$$

$$= 1460 \text{ m/sec} \times 2.90 \text{ sec}$$

$$= 4234 \text{ m} \qquad \text{distance traveled by sound}$$

Since the distance to the ocean floor is one-half the distance the sound travels, we get the ocean depth

$$D_{ocean} = \tfrac{1}{2}(4234 \text{ m})$$

$$= 2120 \text{ m} \qquad \qquad ans.$$

Nondestructive Testing The sonar echo reflection method is widely employed by industry in the inspection and detection of internal flaws or holes in many materials. By the proper adjustment of the transmission frequency of the waves, solid materials up to 20 feet thick can be penetrated by ultrasonic waves. Because of differences in densities and elastic properties, the holes, cracks, and impurities found in solid materials will reflect the ultrasonic waves. Ultrasonic waves, thus, can detect flaws in objects such as welds, castings, forgings, and machine parts without destroying the object.

Ultrasonic Cleaning One of the largest commercial applications of ultrasonic energy is that of ultrasonic cleaning. The process is used in the cleaning and degreasing of parts and assemblies where extreme cleanliness is required or where laborious hand-cleaning is to be avoided. It is used especially in cleaning small complicated parts used in the automotive, aircraft, and electronics industries. It is also widely used to clean optical, surgical, and other precision instruments. The object to be cleaned is subjected to high-intensity ultrasonic waves in a cleaning fluid. The waves tear the adhering foreign particles from the surface (Fig. 20.7).

Medicine The penetrating power of ultrasonic waves makes them valuable in certain types of diagnostic work in medicine. Transmission, reflection, and refraction of ultrasonic waves help the doctor locate solid particles, organs, and growths (e.g. gallstones, kidney stones, cysts, and tumors) in the human body.

Ultrasonic waves cause an oscillating movement between the cells and various tissues of the human body. As a result, "ultrasonic heat" is generated in the body which can be used in the clinical treatment of arthritis, bursitis, lumbago, and similar ailments.

Other Uses Ultrasonic drills are used in machining very hard materials. Since the drill does not rotate, it can produce holes of any shape.

Ultrasonic welding is used in bonding such metals as aluminum and tin. The intense vibratory action of the ultrasonic welding head removes the oxide scale from the aluminum, thus eliminating the need for fluxes.

20.7 Absorption of Sound

A sound wave will continue to recede from its source until it is converted into some other form of energy. When a sound wave passes through a given material, some of the sound energy is absorbed and converted into heat energy. That is, as

Fig. 20.7 (*a*) Ultrasonic vapor degreasing takes place in a two-compartment enclosure containing a boiling solvent sump and an ultrasonic cleaning sump. A typical three-step cleaning cycle is shown for degreasing a ball bearing. It consists of (*b*) initial vapor degreasing; (*c*) ultrasonic immersion cleaning and (*d*) final vapor rinsing. (Courtesy of Branson Cleaning Equipment Co.)

the sound-wave energy strikes the absorbing material, it increases the motion of its molecules. This increase in molecular motion appears as added heat energy (see Secs. 13.12 and 13.13). Porous materials are effective sound absorbers because they contain many pockets of air whose molecules can readily be set into increased motion.

The greater the conversion to heat, the greater the *absorption coefficient*. The absorption coefficient of a given material is the fraction of the sound energy that it will absorb at each reflection or transmission. Some materials have low absorption coefficients, and sound waves pass through them or are reflected from them with little loss of energy. Other materials, such as sponge rubber, rugs, draperies, pressed plant fibers, and porous felt, are good absorbing materials and are used commercially for such purposes.

20.8 Acoustics of Buildings

Materials with a high coefficient of absorption are of importance for the acoustical treatment of rooms and auditoriums. An auditorium is said to have good acoustics when speech can be heard almost equally well throughout the space, without troublesome echoes and reverberations. The podium and stage should be so designed that speech sounds are projected out into the audience and not "lost" backstage. Multiple echoes from the ceiling and walls of the room should not be entirely absent or the room will be acoustically "dead," as if the speaker were addressing a crowd in the open air. On the other hand, if multiple echoes (called *reverberations*) persist too long, the echoes from previous syllables uttered by the speaker will arrive at the listener's ear just in time to interfere with the hearing of the next syllable. Music, too, can be adversely affected by excessive reverberation.

Interference is another factor that must be considered in designing auditoriums. Interference will cause variations in intensity—loud spots in some places, and dead spots in others. Interference can be minimized by a proper choice of the dimensions and shape of the auditorium and by

having "clean" lines, free from pillars, overhangs, and unnecessary architectural embellishments.

The time it takes for an average sound intensity in a room to diminish to one-millionth of its original value at the instant the sound source is turned off is called the *reverberation time*. The reverberation time depends directly on the volume of the room and inversely on the absorption of the sound by the various surfaces in the room.

The best reverberation time for sound in a room depends on its intended use. For satisfactory reception of speech, a relatively short reverberation time is essential. This is necessary because successive utterances would overlap because of a long reverberation time. For moderately sized auditoriums (with volumes of between 2×10^5 ft^3 and 5×10^5 ft^3) the best reverberation time should be about 1 sec. In larger auditoriums, no speaker's voice can be heard without the use of a public address system, regardless of how short the reverberation time. For moderately sized concert halls the optimum reverberation time is about 1.7 sec.

As has been noted, Sabine made some of the earliest studies of acoustics for auditoriums, and although many researchers have since refined his techniques and supplemented his findings, his basic formula for reverberation time is still used. It is

$$t = 0.05 \frac{V}{a} \tag{20.4}$$

where

t = reverberation time, sec

V = volume of room, ft^3

a = absorbing power of all subjects, objects, and materials in room

Unit absorbing power is defined as that of a unit area of a perfectly absorbing material, such as an open window. The English unit, called the *sabin*, uses the square foot for the unit area. The *metric sabin* is equivalent to that of one square meter of perfectly absorbing material. In the following discussion we will use English units.

The *absorption coefficient* of a material is a number, always less than 1.00, which compares the absorbing power of 1 ft^2 of that material or surface to the absorbing power of 1 ft^2 of open window. The symbol β is used for absorption coefficient. Total absorbing power of all the surfaces and materials in a room is equal to the sum of all the separate areas A, multiplied by their respective absorption coefficients.

$$a = \beta_1 A_1 + \beta_2 A_2 + \beta_3 A_3 + \cdots$$

Table 20.1 gives the absorption coefficients of certain common surfaces and acoustical materials.

Table 20.1   Sound absorption

Absorption coefficient	β
Open window	1.00
Acousti-Celotex	0.82
Acoustic plaster	0.25
Carpet, heavy	0.40
Chalkboard	0.06
Draperies, cotton	0.50
Felt, heavy	0.70
Glass	0.025
Linoleum	0.03
Marble	0.01
Plaster wall, smooth	0.03
Wood wall, painted	0.04
Wood wall, floor, or paneling (unpainted)	0.08

Absorbing power	Sabins
Audience, each person	4.0
Auditorium chairs, wood, each	0.3
Upholstered, each, empty	2.0
Occupied	1.5

Illustrative Problem 20.4   Compute the reverberation time of a small auditorium 70 by 110 by 25 ft high. The following are the areas, in square feet, of the building surfaces: plaster, 10,500; wood panel, 6000; linoleum, 6700; heavy carpet, 1000; heavy draperies, 500; Acousti-Celotex, 1000. There are 500 people (seated) and 300 empty upholstered chairs in the auditorium.

Solution   The calculation of absorbing power a is carried out in tabular form.

Plaster, 10,500 × 0.03	=	315 sabins
Wood panel, 6000 × 0.08	=	480
Linoleum, 6700 × 0.03	=	200
Heavy carpet, 1000 × 0.40	=	400
Heavy draperies 500 × 0.70	=	350
Acousti-Celotex, 1000 × 0.82	=	820
500 people, 500 × 4.0	=	2000
500 chairs, occupied, 500 × 1.5	=	750
300 chairs, empty, 300 × 2.0	=	600

$$a = 5915 \text{ sabins}$$

From Eq. (20.3),

$$t = 0.05 \times \frac{70 \times 110 \times 25}{5915} = 1.63 \text{ sec} \quad ans.$$

In addition to acoustical problems in auditoriums, the science of acoustics is applied to public-address and hi-fi equipment, to the design of recording studios, and to the problems of suppressing the unwanted sounds of everyday life.

20.9   Interference of Sound Waves—Beats

Two wave motions being propagated through a medium, each producing its own disturbance in the medium, will interfere with each other if they are *out of phase* and will support each other if they are *in phase*. The combined action of both waves may be shown schematically as in Fig. 20.8. A transverse sinusoidal waveform is used, since the effect can be more clearly presented than with a longitudinal waveform.

Two waves of the same frequency, in phase, and moving in the same direction reinforce or support each other as shown in Fig. 20.8a. The two original waves P and Q have a resultant R, which is in phase with P and Q, and which has an amplitude equal to the sum of their amplitudes. Each point of the R wave is obtained by merely adding *algebraically* the ordinates of the corresponding points of the P and Q waves.

Figure 20.8b shows two waves of equal magnitude, completely out of phase with each

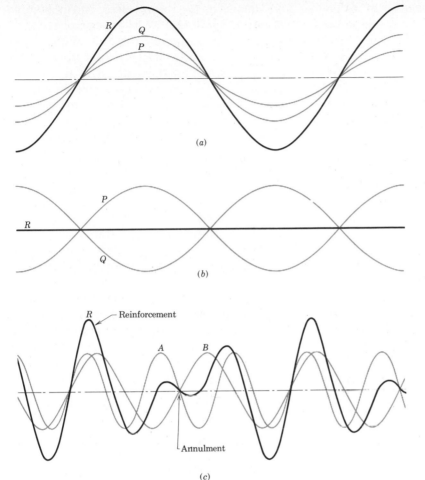

Fig. 20.8 Interference of sound waves. (*a*) Reinforcement of two waves in phase. (*b*) Annulment (destructive interference) of two waves of equal amplitude which are completely out of phase. (*c*) Interference between two sound waves slightly out of phase, producing beats.

other. The result is complete destructive interference, indicated by the resultant *R*, which has zero amplitude at all points. If *P* and *Q* were sound waves, each audible when sounded alone, the result when sounded together would be silence.

An interesting case of alternate interference and reinforcement occurs when the two sound waves are approximately of the same amplitude but differ in frequency. This situation is illustrated in Fig. 20.8*c*. The resultant waveform *R* indicates an alternate increase and attenuation (weakening) of the sound, an observation that can readily be checked by the ear if two tuning forks or gongs of slightly different frequency are struck simultaneously. These alternating periods of increasing and decreasing volume of sound are

called *beats*, and the number of times per second the sound builds to a maximum and dies away again is known as the *beat frequency*. An analysis of the waveforms of Fig. 20.8*c* shows that the beat frequency is equal to the difference in frequency between the two component sounds causing the beats.

The phenomenon of beats thus offers a ready method of tuning a string, horn, or other instrument against a standard. First the two tones are matched as closely as possible by ear. Then while the test string or instrument and the standard are sounded together, the musician listens for beats. The faster the beat frequency, the farther out of tune his instrument is. He makes adjustments until the beat frequency is reduced to one or fewer beats per second, the accuracy of his final tuning depending on the demands of the particular situation.

20.10 Sound Reproduction

By utilizing the properties of sound, engineers have made it possible to record and preserve famous voices, great music, and the sounds of important events. All sound-recording devices obey the same general principles. All sounds are vibrations. To reproduce the sounds, these vibrations must be recorded without distortion on some sort of disk or tape. It then remains for the impressions on the record to be reconverted to the identical sound-wave patterns produced by the original source.

Modern disc recording instruments convert mechanical sound waves into a series of fluctuating electric currents. These electric currents are strengthened by an electronic amplifier. The amplified current actuates an electromagnetic "head" with an attached cutting tool. The cutting tool cuts a "wavy" groove in a master plate. This plate is used to make a *matrix*, or male die, from which all other records are stamped or pressed. The record copies are made from blank plates of warm, thermosetting vinyl plastic, each pressing being a replica of the original master.

Sound is taken from the record by a process the reverse of the one which put it on the die. Figure 20.9 represents a diagrammatic outline of the process. As the record revolves on a turntable, the needle is held in the groove by a pickup arm and vibrates as the record turns. The motion of the needle is transmitted to crystals of rochelle salt or to electromagnets. These vibrations cause the crystal or electromagnet to set up a feeble electric current that fluctuates with the forced vibrations of the needle. This feeble current is amplified many thousands of times by an electronic amplifier. The amplified currents actuate one or more diaphragms in the loudspeaker electromagnetically.

By modulating both edges of the master disk groove, stereophonic sound signals can be etched in a single groove and picked up separately by a special stereophonic cartridge. Two electrical signals are picked up by a single transducer. These signals are amplified separately and are reproduced by two or more loudspeakers. When recording for stereophonic sound of a large sound source, e.g., a large orchestra, microphones are placed near each end of the ensemble. When these recordings are played back through a two-speaker system, the listener experiences an angular separation of sounds which approximates that which he would experience at the live concert.

The familiar tape recorder impresses the sound waves on a thin plastic tape in the form of patterns of magnetized ferric oxide or chromium dioxide particles. The magnetic patterns on the tape correspond to the wave motion of the original sound. The tape can be played back immediately, or the sound can be wiped out by demagnetization and the tape used over and over again. Both records and tapes can be used as components of high-fidelity and stereophonic sound systems.

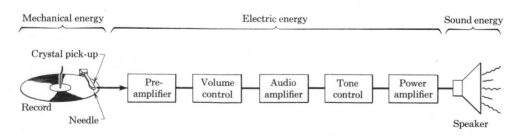

Fig. 20.9 The loudspeaker diaphragms set up the mechanical vibrations in the air which we hear as sound.

20.11 Sound on Film

The sound which accompanies a motion picture is sometimes recorded, by electromechanical devices, as a small variable-density (or variable-area) sound track on the edge of the film. The sound waves are converted into a small strip on the side of the film with variations in film density corresponding to the variations in the voice currents. Light that is allowed to shine through the sound track will then have variations corresponding to the voice or music recorded on the film. These variations in light energy are converted into corresponding changes in electric energy by a photoelectric cell. These are amplified and converted into sound by the loudspeaker. Dialogue for the film may be recorded as the pictures are being taken. However, other sounds are usually not recorded on the film at the same time the pictures are taken. They are recorded in the studio later and then synchronized to assure exact coordination with the action on the film.

With the introduction of stereophonic sound reproduction, magnetic sound tracks are fast replacing optical tracks. It is now possible to place five or six magnetic sound tracks in the space formerly occupied by one optical track.

20.12 Resonance

A body that vibrates or oscillates unhampered by any other body or system is said to be executing *free vibrations*. The frequency of its vibration is said to be its *natural frequency*. Thus a child's swing, a pendulum, and a violin string have their natural frequencies. When a body is compelled to vibrate with a frequency other than its natural frequency, it is said to be executing *forced vibrations*. The sounding boards of the piano and the violin are forced to vibrate at all the frequencies emitted by the natural frequencies of the strings.

If two bodies having the same natural frequencies are set next to each other and one is set into vibration, the wave energy from it reaches the second body and sets it into *sympathetic* vibration. The phenomenon of the transferring of energy from one body to another of the same natural frequency is called *resonance*. All wind musical instruments are based on this phenomenon. The air columns of the instrument resonate with the vibrating reed or lip. Pieces of chinaware or glassware on a shelf will occasionally be set into resonant vibration when a certain note is struck on the piano.

Resonance plays a very important role in mechanical and electrical systems as well as in sound. Soldiers marching across a bridge will break step to avoid a regular rhythm of marching feet that might set the bridge structure into violent and dangerous vibrations if the rhythm corresponded to the natural frequency of the structure. The child pumping in a swing and the man rocking a boat employ resonance. Vibrations can be set up mechanically with a small force if the frequency of the applied force is set to equal the natural frequency of the body. We are all familiar with the appearance of annoying rattles or sounds in an automobile at certain running speeds and their disappearance with a slight speed change.

In heavy machinery and structures it is important that the distribution of mass be such that the natural frequencies of the various parts will not be the same as the possible frequencies of its moving parts. Otherwise the vibrations of the smaller moving parts create a condition of resonance with heavier components and may build up a very large amplitude of motion, with serious consequences.

SUMMARY

Sound waves striking a hard surface are reflected, and *the angle of reflection equals the angle of incidence.*

Sound waves reflected to the ear are called *echoes*.

Refraction is a bending of sound waves due to their passing from one medium into a second medium which causes a change of speed.

Change of pitch due to frequency shift is known as the *Doppler effect*.

$$f_0 = f_s \frac{v}{v \pm v_s} \quad \text{(source moving)} \quad (20.1)$$

$$f_0 = f_s \frac{v \pm v_0}{v} \quad \text{(observer moving)} \quad (20.2)$$

Sound ranging, seismography, and depth sounding depend on an understanding and utilization of the speed of sound and on the reflection of sound.

Ultrasonics or supersonics are longitudinal waves having frequencies above the audible range. They are used in sonic ranging, navigating, and in submarine-detection devices known as *sonar*.

Inelastic materials of loose, porous texture are used to absorb or deaden sound.

The science of sound control in rooms is called *acoustics*.

In making sound recordings, the sound is recorded as wavy grooves on disks, magnetic fields on a tape, or variable density on film, to correspond to the wave motion of the original sound. In reproducing the sounds the recording process is reversed.

Resonance applies to the response of a body which has a natural frequency of its own to impulses from a body vibrating with the same frequency.

QUESTIONS, EXERCISES, AND PROBLEMS

1. Which of the following remains unchanged when a sound wave passes from one medium into another of different density: (a) speed, (b) wavelength, (c) amplitude, (d) frequency?

2. Which of the following does not determine the quality of a musical sound: (a) the fundamental tone, (b) the number of overtones, (c) the frequency of the overtones, (d) the intensity of the overtones?

3. Explain how a pure note of high frequency and intensity is capable of shattering a fine glass tumbler.

4. Explain how the sounding box of a violin amplifies the sound produced by the vibrating string.

5. Will the apparent pitch be the same whether the sound source approaches the listener with a given speed or the listener approaches the sound source with the same speed? Explain your reasoning.

6. How many beats will be heard each second when a string with a frequency of 327 sec^{-1} is plucked simultaneously with another string whose frequency is 324 sec^{-1}?

7. With a depth-sounding apparatus, it is found that 3.4 sec elapses between the sending of a sound and the receiving of its echo. If the velocity of sound in sea water is 4820 ft/sec, what is the ocean depth?

8. A gun is fired near a cliff, and 3 sec later the man who fired the gun hears the echo. If the temperature is 25°C, how far from the gunner is the cliff?

9. How far away (minimum distance) from you must a wall be in order to give you a distinct echo if the temperature is 23°C?

10. A particular groove in a phonograph record moves past the needle at a speed of 0.24 m/sec. The sound produced has a frequency of 440 vib/sec. What is the wavelength of the wavy indentations in the record groove?

11. A stone is dropped down a water well, and 5.7 sec later the splash is heard. How deep is the well? (Use 1100 ft/sec for the speed of sound and 32 ft/sec^2 for the acceleration of gravity. You will need to know how to solve quadratic equations to work this problem.)

12. A locomotive approaches a railway crossing at a speed of 105 km/hr. The engineer blasts the horn, which has a frequency of 320 Hz. What is the frequency of the sound heard by someone at the crossing if the temperature is 18°C?

13. The engineer of a diesel train approaching a stationary observer at 70 mi/hr sounds an air horn. The tone emitted has a true frequency of 328 vib/sec. If the air temperature at the time is 60°F, what is the pitch of the tone heard by the observer?

14. What is the frequency heard by a listener who is leaving (at a speed of 25 m/sec) a stationary sound source emitting a frequency of 980 Hz if the temperature is 24°C?

15. A sonar signal of frequency 30,000 Hz is sent out from the transducer of a surface vessel tracking a submarine. The "ping" returns in 5.25 sec and has an apparent frequency of 30,300 Hz. The speed of sound in seawater at the time is 4850 ft/sec. Find (a) the range to the submarine, in yards, and (b) the rate at which the submarine is closing the range, in knots. (A knot is a nautical mile per hour.)

16. A classroom has the dimensions 36 by 45 by 12 ft. There are 220 ft^2 of glass, 250 ft^2 of wood paneling, 300 ft^2 of chalkboard, 1000 ft^2 of smooth plaster, and 175 ft^2 of painted wood wainscot. The ceiling is covered with acoustic plaster, and the floor with linoleum tile. There are 70 tablet armchairs (wood), 50 of which are occupied by students. Find the reverberation time in seconds.

17. It is desired to reduce the reverberation time of the classroom of Prob. 16 to exactly 0.8 sec. It is decided that Acousti-celotex tile will be used. How many square feet of the ceiling will have to be covered?

PART 6 Light and Optics

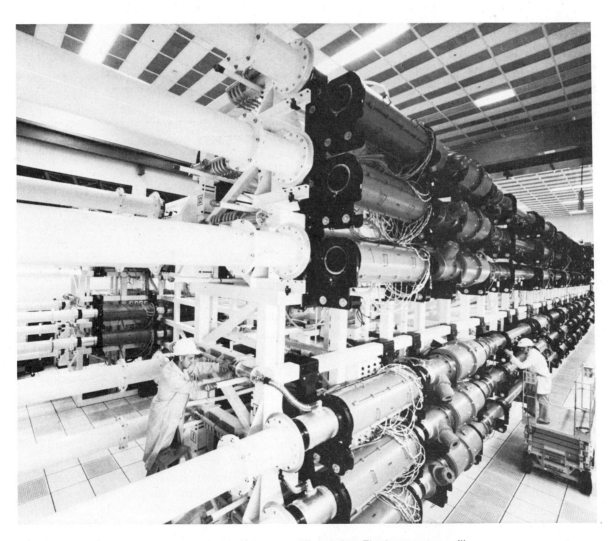

View of the north bank of six of the 20 Shiva laser amplifier chains. The laser system will develop more than 30 trillion watts of electrical power in less than one billionth of a second to a tiny fusion target the size of a grain of sand (Lawrence Livermore Laboratory)

21 The Nature of Light and Illumination

We see by means of light. Plants grow by the process of photosynthesis. Without light there can be no photosynthesis. Because of light we gain information about ourselves and our environment, even to the far reaches of space.

But what is light? How is it transmitted? How can light travel through empty space? What causes the sensation of color?

From earliest recorded times people have sought answers to these questions. They have pried into the mysteries of light and speculated about its exact nature and its relationship to human vision. Today much is known about light and how it behaves. We know that light is a form of energy. Laws have been developed which will predict its behavior, but there is still considerable concern about the mechanism by which light energy is transmitted from a source to an object.

In this and the next two chapters we will study some of the properties of and laws developed for this form of wave energy called light.

21.1 Light and Vision

In order to understand the nature of light it is necessary to recognize the relationship between light and vision. An object is *visible* when light coming from the object enters the human eye.

A body is visible either because it is luminous or because it is illuminated. A *luminous* body is one that produces and emits its own light energy. Luminous bodies include the sun, stars, incandescent lamps, and fires. Nonluminous bodies, such as the moon, the planets, and most common objects about us, are visible because they reflect light from a luminous body. When such bodies reflect light from a luminous source, they are said to be *illuminated*.

Air, glass, and water readily transmit light and are said to be *transparent*. Other substances transmit some light but scatter or diffuse the light so that objects seen through them are not clearly visible or identifiable. Such substances are called *translucent*. Frosted light bulbs and parchment lampshades are made of translucent material. Substances which do not transmit light at all are said to be *opaque*.

21.2 Early Theories of Light

Energy can be transmitted from one place to another by two means: (1) by the actual motion of matter from one point to the other or (2) by a wave disturbance traveling through the intervening medium. Both these methods of transmitting light energy have been advanced by scientists in the past. The *corpuscular theory of light* was accepted for many centuries by leading scientists. Sir Isaac Newton was perhaps its most famous champion. He assumed light to be a stream of small particles or *corpuscles* originating from the light source and moving in straight lines through space. These small corpuscles, in passing through the eye, were thought to stimulate the sensory nerves and produce the sense of vision. Christian Huygens (1629–1695), on the other hand, conceived of light as a form of wave motion transmitted through a vague, intangible medium which he called the *ether*. Each of these two seemingly incompatible theories has had its staunch supporters, and the relative merits of the two theories have been debated for three centuries.

21.3 Two Newer Theories for Light

In the latter part of the nineteenth century most scientists supported the *wave theory*, as a result of the *electromagnetic wave* studies of James Clerk Maxwell. He pictured light as transverse waves of progressively changing electric and magnetic field intensities. These fields were presumed to act at right angles to each other and to the direction of propagation. The electromagnetic wave theory satisfactorily explains such phenomena as color, interference, polarization, and varying wavelengths in different media but fails to explain the easily observed *photoelectric effect* (Sec. 33.7) and certain other phenomena.

Evidence began to accumulate early in the twentieth century that when light interacts with matter, it behaves as if its energy were in the form of small particles. This theory, known as the *quantum theory*, was first advanced by Max Planck (1858–1947) and occupies an important place in the study of light. The small particles of energy are called *photons* (see Sec. 33.7).

And so it is that we do not have a single, consistent theory of the nature of light. Our present view must be a combination of the two theories; we think of light as being dual in character. In considering the *propagation* of light, we find that the wave theory generally gives correct solutions, while the quantum theory (Chap. 33) is best fitted to the explanation of the effects of light interacting with matter.

The lack of a single theory need not handicap us in the study of applied physics, for we are primarily interested in studying what light *does* and how it *behaves* under varying circumstances. Perhaps research scientists will eventually develop one theory compatible with all light phenomena. As a matter of fact, in describing most optical phenomena by the wave theory, we need not consider the quantum concept. For the purposes of this chapter we shall consider the properties of the light wave to be the same as for other transverse waves, as discussed in Chap. 19. We shall employ the quantum theory only in discussing the photoelectric effect.

21.4 Wavefronts and Rays

Light from a point source S (see Fig. 21.1) spreads out in a succession of spherical wavefronts. As they recede from the point source, the successive spherical wavefronts become more and more like planes. Imaginary lines perpendicular to the wavefronts, called *rays*, are used to show the direction in which the waves are moving. We shall find it convenient to use rays in describing many of the effects of light.

Common experience has shown that when light enters a darkened room through a small hole, it travels in a straight line. The surveyor and the navigator, as they use their instruments, assume light to be traveling in straight lines. A line of light propagation is termed a *ray*. This rectilinear propagation of light can be quite clearly illustrated by placing an incandescent lamp on one side of a cardboard which has a small hole punched in it and placing a white screen on the other side (Fig. 21.2). Light will come from the lamp, pass through the hole, and form an inverted image of the bulb on the screen. Points on the image will lie on the straight lines joining the hole and the corresponding points on the object. Quite clear photographs can be taken with a *pinhole camera*, which employs this principle. A hole is punched in a closed box, and light traveling from the object is allowed to pass through the hole onto the photographic film at the back of the box. The hole must be made small in order to ensure a sharp image.

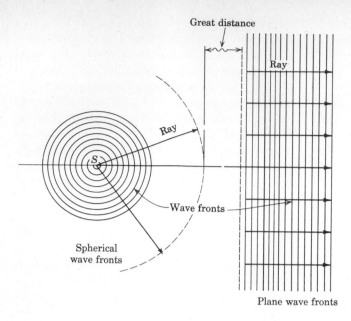

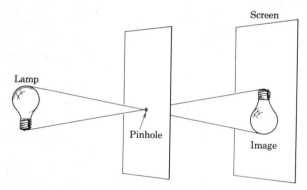

Fig. 21.1 Spherical wavefronts from a point source appear to be plane wavefronts a great distance from the source, just as the earth's surface *appears* to be flat.

Fig. 21.2 Rectilinear propagation of light.

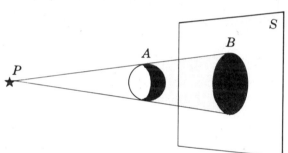

Fig. 21.3 A sharp shadow is cast by an opaque object if the light source is small.

21.5 Shadows

Straight-line propagation of light from a point source P can be illustrated by showing that the light will cast a sharp shadow B of an opaque object A on a screen S (see Fig. 21.3). The fact that the shadow is sharp shows that the light from P does not bend appreciably in traveling from A to B.

If the light source is not small, the outer boundary of the shadow will not be sharp (see

Fig. 21.4). Points on the screen outside circle B are fully illuminated by P. The points of the screen inside circle C receive almost no light. The part of the screen between C and the outer boundary of B will be partially illuminated by P. This outer shadow will gradually shade off from complete shadow at C to complete illumination outside of B.

The part from which all the rays of light are excluded (total shadow) is called the *umbra*; the partially shadowed region is called the *penumbra*.

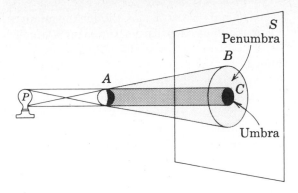

Fig. 21.4 Shadows cast by a light source larger than a point.

Eclipses are caused by shadows of heavenly bodies. An eclipse of the sun occurs when the moon appears directly between the sun and the earth (Fig. 21.5). When this happens, the moon's shadow falls upon the earth, forming an umbra and a penumbra. If the earth is sufficiently near the moon and passes in the umbra, the eclipse is *total*. If, however, the earth passes through the penumbra, the eclipse is said to be *partial* or *annular*.

An eclipse of the moon occurs when the moon passes through the shadow of the earth.

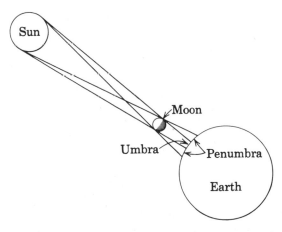

Fig. 21.5 Eclipse of the sun (distances are severely distorted).

21.6 Speed of Light

In 1926 to 1929, A. A. Michelson (1852–1931) measured the speed of light over a distance of about 44 miles between southern California's Mt. San Antonio and Mt. Wilson and return. The distance was measured by triangulation to an accuracy of about 1 in. As a result of his measurements, the velocity of light has now been determined as one of the most accurately known physical constants. The essential features of Michelson's method are illustrated in Fig. 21.6. Light from an arc lamp S was focused by a lens L onto face 1 of an octagonal mirror M at an angle of 45° and reflected to a distant concave mirror D and thence to a plane mirror m. From m the beam was reflected back to D, which returned it to face 3 of M. From there it was reflected into a telescope T. The mirror was then set in rotation and its speed gradually increased until face 2 moved to the point where it would just catch the beam of reflected light from D. It would then reflect the light to the telescope T. The speed of rotation of M was very accurately determined. Hence, the time for the light to travel the distance of $2MD$ was equal to the time for the mirror to make $\frac{1}{8}$ rev. The average of nearly 3000 such measurements gave the speed of light as 186,284 mi/sec or 2.99796×10^8 m/sec.

Michelson conducted similar experiments using an evacuated tube 1 mile long. He found that the speed of light is slightly greater in a vacuum than in air. Many other scientists have conducted extensive experiments measuring the speed of light in various media. In recent years electromagnetic waves much longer than light waves have been used (see Sec. 21.14). By measuring the frequencies and wavelengths of these *microwaves*, scientists have confirmed that in free space the speeds of waves in all parts of the electromagnetic spectrum have the same value.

In 1972 a research team led by Kenneth M. Evenson of the U.S. Bureau of Standards measured the speed of light, c, using a helium-neon laser light source. They arrived at a value of 299,792,456.2 m/sec, with an uncertainty of

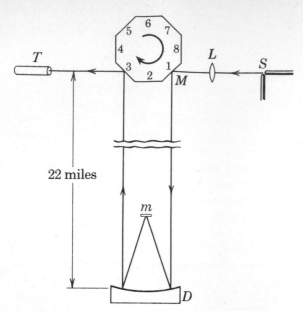

Fig. 21.6 Determining the speed of light by Michelson's method.

1.1 m/sec. This is equivalent to a relative uncertainty of only 3.5 parts per billion. For most practical purposes we can take the speed of light in air or in a vacuum to be 186,000 mi/sec or 3.00×10^8 m/sec.

21.7 Speed of Light in Other Media

Direct but independent measurements by Michelson and Jean Foucault (1819–1868) indicated that light travels more slowly in water than in a vacuum. The speed of light in water was found to be about three-fourths its speed in air. Indirect methods, which we shall discuss later, show that the speed of light in any medium is less than that in a vacuum. It also can be shown that the speed *of light in a vacuum is independent of* (1) *the speed of the light source or observer or* (2) *the brightness or color of the source.*

21.8 Luminous Intensity of Light

The science of light measurement, called *photometry*, is concerned with three quantities: the *luminous intensity I* of the source, the *luminous flux F* or flow of light from a source, and the

illumination E on a surface. Although the ultimate concern in light-intensity measurement is usually the amount of light actually falling on a given surface, we must also consider the intensity of the source. The intensity of illumination on a surface varies directly with the intensity of the light source. Since there are so many different types of lamps in common use, it is important that we have a means of measuring their luminous intensities. This is usually done by comparing the unknown source with a standard source of known intensity. The instrument used for the comparison is called a *photometer.*

For years the unit of luminous intensity, the *candle*, was based on the early use of candles as standard sources of light. One *candle* was originally defined as the quantity of light given out by a certain candle that burned whale oil at the rate of 120 grains/hr.

Since not all whale oils have the same fuel value and because it is difficult to work with a flickering flame, scientists have sought more highly reproducible sources of luminous intensity. In 1948, a *new international candle*, called the *candela*, was adopted by action of the International Committee on Weights and Measures. The *candle*, or candela (cd), is now defined as one-sixtieth the luminous intensity of 1 cm² of a blackbody radiator (see Sec. 15.6) maintained at the temperature of freezing platinum, 2045K (see Fig. 21.7).

Incandescent lamps, operated with a specified amount of electric-power consumption, are standardized on the incandescent platinum cavity, and these standard lamps are used, in turn, to rate other lamps. The intensity of a light source is often called its *candlepower* (cp). Thus, a lamp with an intensity of 60 candles (candelas) is a 60-cp lamp. The choice of words here is unfortunate, since luminous intensity is not really power.

21.9 The Photometer

A photometer capable of fair precision was devised by Robert Bunsen (1811–1899), a German chemist and inventor. It consists essen-

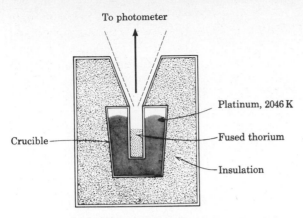

Fig. 21.7 Platinum at 2045 K provides 60 international candles (candelas).

tially of a flat screen of white paper with a central, translucent grease spot which transmits some light (Fig. 21.8). The grease-spot screen is mounted in a box with openings on three sides so that light entering from X and S will strike the grease spot. The intensity of illumination on the two sides of the grease spot can be viewed from the third open side of the box by means of tilted mirrors. If the illumination on one side of the screen is more intense than that on the other, the grease spot when viewed in the mirror on that side will appear darker (in contrast) than when viewed in the mirror of the other side. Conversely, it will appear lighter if viewed from the side of weaker illumination. If the two sides are illuminated equally, the grease spot will appear to be the same neutral shade of gray when viewed in either mirror. If the two light sources and the photometer box are mounted on an optical bench as shown in Fig. 21.9, the distances d_x and d_s can be varied until the two sides of the screen appear equally bright. When they are equally bright, the luminous intensity of the unknown source can be found from the equation

$$\frac{\text{Intensity of } X}{\text{Intensity of } S} = \frac{(\text{distance of } X \text{ from screen})^2}{(\text{distance of } S \text{ from screen})^2}$$

$$I_x/I_s = d_x^2/d_s^2 \tag{21.1}$$

This is known as the *photometric* equation.

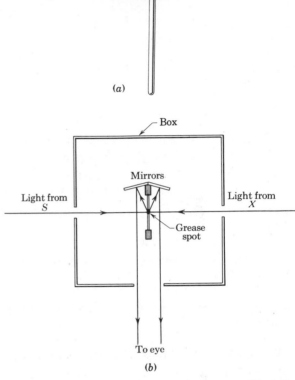

(a)

Fig. 21.8 Bunsen grease-spot photometer. (Central Scientific Co.)

Illustrative Problem 21.1 Equal illumination appeared on a photometer screen which was placed 80 cm from a 40-cd standard lamp and 120 cm from a second lamp. What was the luminous intensity of the second lamp?

Solution Substituting in Eq. (21.1) gives

$$\frac{I_x}{40 \text{ cd}} = \left(\frac{120 \text{ cm}}{80 \text{ cm}}\right)^2$$

$$I_x = 90 \text{ cd} \qquad \qquad ans.$$

Many other forms of photometers have been devised, but they all depend upon an adjustment which will make the illumination on two surfaces

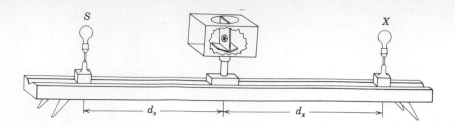

Fig. 21.9 Photometer used to compare lamp *X* of unknown intensity with standard lamp *S*.

appear the same. In all the discussion on photometric measurement thus far we have assumed that both the unknown and the known light sources have the same color. If they do not have the same color, the methods described cannot be used in determining equal illumination on a screen.

21.10 Luminous Flux

Not all the energy radiated from a source of light is capable of producing a visual sensation. *Luminous flux* is the total radiant energy per unit time that is effective in producing the sensation of sight. The unit of luminous flux is the *lumen* (abbreviated lm). Think of an idealized 1-cd point source of light at the center of a hollow sphere of radius 1 m (or 1 ft) sending out energy at a constant rate in all directions (Fig. 21.10). If we next imagine a hole with area of 1 m² (or 1 ft²) cut from the sphere, 1 lm of flux will be radiated through the hole by a 1-cd source. The *lumen* is

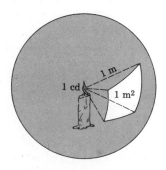

Fig. 21.10 One lumen of light flux through 1 m² at a distance of 1 m from a 1-cd source.

the luminous flux which falls on a unit surface all points of which are at a unit distance from a point source of 1 cd. The solid angle subtended at the center of a sphere by a portion of the surface whose area is equal to the square of the radius is called a *steradian* (sr). Thus, a light source emitting one candela uniformly in all directions will emit one lumen over a solid angle of one steradian.

Since the formula for the surface area *A* of a sphere is $A = 4\pi r^2$, we know that there are 4π unit areas in the surface of a unit sphere. Hence the total luminous flux emitted by a 1-cd source is 4π lm.

Light sources are usually rated in terms of total flux emitted. If the intensity of a light source is *I* cd and the total flux it emits is *F* lm, we know that

$$F = 4\pi I \qquad (21.2)$$

The common incandescent light bulbs in use today give a little more than 1 cd/W of electric power used. For example, a 100-W bulb has a luminous intensity of about 125 cd. Thus the 100-W bulb will emit a total flux of about $4\pi \times 125 = 1580$ lm. Since luminous flux represents a *flow* of light energy, the same flow will occur through a 1-m² area at a distance of 1 m from a 1-cd source as will flow through a 1-ft² area at a distance of 1 ft from the same source.

21.11 Illumination

When light strikes a surface, we say the surface is illuminated. *Illumination* (or *illuminance*) on a surface is defined as the luminous flux per unit area of that surface. If the surface is uniformly illuminated,

$$E = F/A \qquad (21.3)$$

where
F = luminous flux, lumens
A = area, square meters or square feet
E = illumination, luxes or footcandles

The two common units of illumination are the lumen per square meter (called *lux*) and the lumen per square foot (called *footcandle*, ft-c). One lux is equivalent (approx.) to 10.76 ft-c. Approximate values of the illumination from various sources are listed in Table 21.1.

Table 21.1 Approximate illumination from some light sources

	Illumination	
Source	Luxes	Footcandles
Starlight	0.00003	0.0003
Full moonlight	0.25	0.023
Clear sky	11,000	1000
Snow in sunlight	32,000	3000
Sun overhead	100,000	10,000

To summarize:

Property	Unit	Definition
Luminous intensity (I)	candela (cd)	
Luminous flux (F)	lumen (lm)	cd/sr
Illumination (E)	lux (lx) or footcandle (ft-c)	lm/m^2 or lm/ft^2

Let us imagine a point source of light S emitting light energy uniformly in all directions, and consider two spheres of radii d_1 and d_2 having S as a common center (Fig. 21.11). Unless some of the light energy is absorbed, the amount of luminous flux F_1 passing per second through the spherical surface whose area is $4\pi d_1^2$ must equal the luminous flux F_2 through the surface whose area is $4\pi d_2^2$. Hence if we let E_1 and E_2 equal the

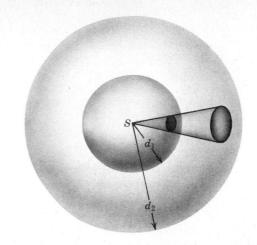

Fig. 21.11 Light travels from its source in concentric spheres.

intensity of illumination at d_1 and d_2, respectively, it must be true that

$$F_1 = F_2$$

and, from Eq. (21.3),

$$E_1(4\pi d_1^2) = E_2(4\pi d_2^2)$$

Dividing both sides of the equation by $4\pi E_2 d_1^2$, we get

$$E_1/E_2 = d_2^2/d_1^2 \qquad (21.4)$$

This is an example of the *inverse-square law*:

The illumination on a surface from a point source of light varies inversely as the square of the distance from the surface to the source.

(This law, it will be recalled, is also true for the intensity of sound). It should be emphasized that Eq. (21.4) is true only if the illuminated surface is at right angles to the rays of light and if the source of light is a point source.

The illumination E on a surface can be found if the distance d between the surface and the light source and the intensity I of the light source are known. Thus

$$E = F/A = 4\pi I/4\pi d^2 = I/d^2 \qquad (21.5)$$

If d is measured in meters and I in candelas, E will be expressed in luxes; if d is measured in feet and I in candelas, E will be expressed in footcandles.

Illustrative Problem 21.2 What is the illumination, in footcandles, on a surface placed 5 ft directly below a 160-cd lamp if all light from reflecting surfaces is blanked out?

Solution Substituting known values in Eq. (21.5), we get

$$\text{Illumination } E = I/d^2$$

$$= 160 \text{ candelas}/(5 \text{ ft})^2$$

$$= 6.4 \text{ ft-c} \qquad ans.$$

If the rays of light from the source strike a surface with an angle of incidence θ (measured from the normal to the surface) the intensity of illumination on the surface (see Fig. 21.12) is

$$E = \frac{I \cos \theta}{d^2} \qquad (21.6)$$

Illustrative Problem 21.3 A sheet of paper on a table is located 4.00 m directly below a 1650-lm lamp (Fig. 21.12). (a) What illumination is provided on the paper? (b) What is the illumination on the paper if it is moved 3 m horizontally from the original position?

Solution

(a) $\qquad E_1 = I/d^2 \qquad$ and $\qquad F = 4\pi I$

Solve for I:

$$I = F/4\pi$$

Then

$$E_1 = F/4\pi d^2 = 1650 \text{ lm}/4\pi(4.00 \text{ m})^2$$

$$= 8.21 \text{ lm/m}^2 = 8.21 \text{ luxes} \qquad ans.$$

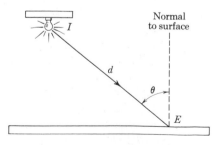

Fig. 21.12 Intensity of illumination on a surface at an angle θ to the incident rays of light.

(b) $\qquad d^2 = (4 \text{ m})^2 + (3 \text{ m})^2 = 25 \text{ m}^2$

$$d = 5 \text{ m}$$

$$\cos \theta = 4 \text{ m}/5 \text{ m} = 0.800$$

$$E_2 = I \cos \theta/d^2$$

$$= \frac{(F/4\pi) \cos \theta}{d^2}$$

$$= \frac{(1650 \text{ lm})(0.800)}{4\pi(5 \text{ m})^2}$$

$$= 4.20 \text{ lm/m}^2 = 4.20 \text{ luxes} \qquad ans.$$

21.12 Proper Illumination

The amount and kind of illumination needed to produce optimum light conditions in a room vary greatly with the function or use of the room. Our discussion to this point has considered only light coming *directly* from only one or two light sources. In practice, however, the illumination of a room generally comes from several incandescent lamps and possibly from diffused daylight or light from adjoining rooms, reflected from the walls, ceilings, furnishings, and objects in the room. So the problem of *computing* the actual intensity of illumination on any given surface in a room is quite difficult, if not impossible.

In actual practice, computations of illumination intensity are rarely used. The intensity is obtained by direct measurement using instruments known as *lux meters* and *footcandle meters* (Fig. 21.13). The most common type of this instrument makes use of the *photoelectric effect* (see Sec. 33.7) and is calibrated to read light intensities directly in luxes or footcandles. Photographic *exposure meters* are specialized footcandle meters.

Today we are growing more conscious of the need for proper illumination. Statistics show that 1 of every 5 persons under 20 years of age has defective vision. The ratio increases rapidly with age until at age of 60 about 4 out of 5 have defective vision.

Industrial plants find a marked improvement in the efficiency and output of employees working under proper illumination. A definite decrease in eyestrain, accidents, and poor work-

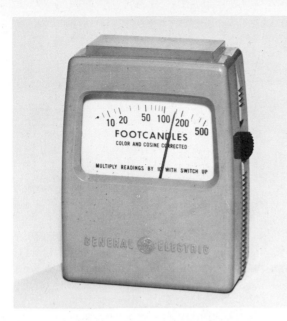

Fig. 21.13 Pocket-sized light meter. Weighing only 3 oz, it is a convenient device for people concerned with proper industrial lighting. It can also be used to indicate the brightness of surfaces and to determine relative transmission and reflection factors of materials. With the multiplying switch up, the operator can measure the higher light levels characteristic of plant-growth chambers, spotlight beams, and daylight. (General Electric Co.)

Table 21.2 Optimum illumination requirements

Situation	Luxes	Footcandles
Hallways	54–110	5–10
Library, reading rooms	320–540	30–50
Kitchen	160–270	15–25
Office, classroom, laboratory	650–1100	60–100
Industrial assembly plant	430–1100	40–100
Stores (except jewelry stores)	220–540	20–50
Hospital operating rooms	540–11,000	500–1000
Night baseball	320–430	30–40
Continued close work (drafting, etc.)	540–1100	50–100
Jewelry manufacturing	1100–3200	100–300

manship follows the installation of proper lighting. Good lighting consists of not only enough light but light of the proper kind. Glare and strong shadows should be avoided. Too much light is as undesirable as not enough. Proper illumination should provide uniform, diffused, and shadowless light. Indirect-lighting systems, employing reflection from walls and ceilings, generally give the best type of illumination.

Table 21.2 suggests optimum values of intensity of illumination for various situations.

21.13 Color and the Length of Light Waves

Most of us have been impressed at times with beautiful displays of color produced by nature. Just as we found certain aspects of sound to have a dual meaning, so does the word color have two distinctly different meanings. To the average person, it means the various sensations registered by the brain when light strikes the eye. It is measured by the psychological and physiological effects of light waves on the individual. The physicist is more interested in what causes these sensations. Analysis of light has shown that the *different color sensations are mainly due to the varying wavelengths of light to which the eye is sensitive.* Of all the colors of the rainbow, it has been found that red light has the longest wavelength, orange the next longest, and so on down through the yellow, green, blue, indigo, and finally violet with the shortest wavelength. Light from the sun is a combination of many wavelengths which produces the effect on the eye usually described as white light. Scientists have found it possible by the use of prisms (discussed in Chap. 23) and gratings to analyze sunlight or light from any other source into its constituent colors. They have been able to measure the wavelengths and, with suitable instruments, to distinguish thousands of variations in colors each with a different wavelength (or mixtures of different wavelengths). The unaided human eye can, at best, distinguish only about 40 distinctly different colors. The colors we see around us are mostly variations in the intensities and amounts of these fundamental hues.

To avoid long decimal expressions, scientists use a special unit to express the wavelength of light called the *angstrom* (Å). It is equal to 10^{-10} m. Hence the wavelengths given in Table

Table 21.3 Approximate wavelengths and frequencies of certain colors of light—the visible spectrum

Color	Wavelengths			Frequency, Hz
	m	in.	Å	
Red	67×10^{-8}	0.000028	6700	4.48×10^{14}
Orange	62×10^{-8}	0.000024	6200	4.83×10^{14}
Yellow	57×10^{-8}	0.000022	5700	5.25×10^{14}
Green	52×10^{-8}	0.000020	5200	5.76×10^{14}
Blue	47×10^{-8}	0.000018	4700	6.39×10^{14}
Violet	41×10^{-8}	0.000016	4100	7.32×10^{14}

21.3 will vary from 6900 Å (690 nanometers) for red to 4100 Å (410 nanometers) for violet.

The spread of light waves in the order of their wavelengths, from the deepest red to the deepest violet, is called the *visible spectrum* (Fig. 23.9). The wavelengths and frequencies of the central portion of each of the colored regions in the visible spectrum are given approximately in Table 21.3.

21.14 The Electromagnetic Spectrum

It should be mentioned here that the *visible spectrum* constitutes only a very small portion of the whole range of electromagnetic waves which are similar in nature to light, but which are *invisible*. By far the larger portion of the electromagnetic spectrum is composed of waves that are too long or too short to affect the sensory nerves of the eye. They are detected by other means.

The electromagnetic spectrum is continuous and has been measured for wavelengths of about 5,000,000 m to about 0.000000000000032 m. This range includes the types of radiation known as radio, infrared, visible spectrum, ultraviolet, x-rays, gamma radiations, and cosmic rays. A chart of the entire electromagnetic spectrum is given in Fig. 21.14. But even though the chart in the figure has limits (as now known), it is

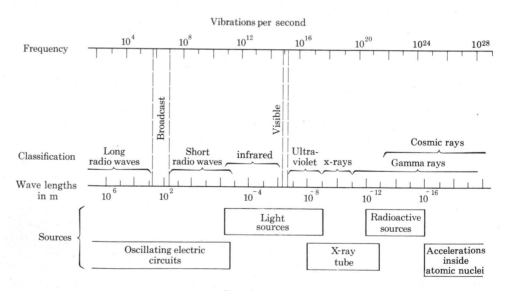

Fig. 21.14 Chart of the electromagnetic spectrum.

probably safe to assume that still longer or shorter waves exist and will some day be discovered.

The waves indicated in Fig. 21.14 are all alike in nature and they all travel through space with the same speed, the velocity of light. They differ only in their frequency and wavelength. It can be shown that the energy transmitted by any of the waves increases as its wavelength decreases (frequency increases).

21.15 Colors of Objects

When white light strikes a colored *transparent* object, e.g., a piece of blue glass, only wavelengths in the blue region of the spectrum are transmitted. All or almost all the other colors are absorbed. Light passing through a transparent color filter owes its transmitted color to a process of *selective absorption*. For example, a blue filter placed in the path of a beam of white light will absorb the red, orange, and some of the yellow colors (Fig. 21.15a). If a yellow filter is placed in the path of white light, it will absorb the blue, violet, and some of the red and green colors (Fig. 21.15b). When both are inserted in the path of the beam, only green and a little yellow or blue are transmitted. To the eye the transmitted beam will appear bright green.

Opaque objects, such as a piece of cloth, have characteristic colors because we see only the light that is *reflected* from them. For example, if white light shines on a piece of red cloth, it appears red because the dye in the cloth absorbs most of the wavelengths of the incident light except those near the red region of the visible spectrum. The red wavelengths are then *reflected*, giving a red appearance to the cloth. If an object reflects all the light it receives, we say it is *white*. If an object absorbs all the colors it receives, we say it is *black*. Since light is a form of energy, when some of it is absorbed, it must be converted to another form. It is converted largely to heat. This is one reasoning why dark clothing is worn in cold climates and light-colored or white suits in the tropics.

Many interesting effects can be obtained by using artificial light of one predominant color on fabrics or pigments of other colors. For example,

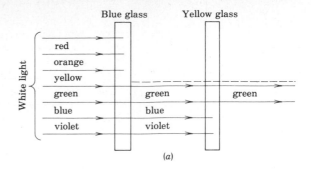

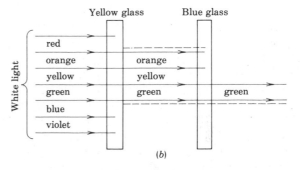

Fig. 21.15 Absorption and transmission of white light by blue and yellow filters.

a pure blue necktie will appear black when viewed under a red or yellow light source. This is because blue pigments reflect little or no yellow or red light. A car which appears blue in the daytime will appear black under yellow sodium street lamps.

When white light shines on a piece of red glass, the glass looks red, whether it is seen by reflected light or by transmitted light. A few materials, however, appear to be one color by reflected light and a different color by transmitted light. A thin piece of gold, for example, looks yellow-orange by reflected light, but if it is thin enough, it will appear blue-green by transmitted light. Lubricating oil may appear greenish by reflected light and reddish by transmitted light.

21.16 Complementary Colors

We have seen how colors may be subtracted from a beam of light when the light is reflected or

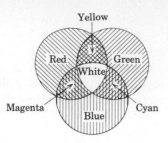

Fig. 21.16 Overlapping of the primary colors red, green, and blue produces white light. Any two primary colors combine to give the complement of the third. The complementary colors are cyan, magenta, and yellow.

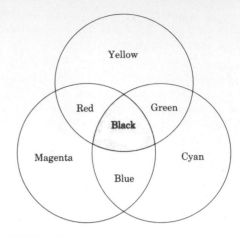

Fig. 21.17 Primary pigments cyan, magenta, and yellow, when mixed in proper proportions, appear as black. Combined in pairs, two primary pigments give the complement of the third pigment, forming red, green, and blue colors.

passes through a filter. The reflecting surface and the filter remove certain frequencies from the light.

It would seem reasonable to suppose that simple colors can be combined to form new colors. Experiments with beams of different colored lights have shown that most colors and hues can be produced by adding varying amounts of three different colors. The three colors which can most effectively be used to produce other colors and hues are red, green, and blue. Consequently, these have been called the *primary colors*.

When these primary colors are projected on a white screen so they overlap as shown in Fig. 21.16, additive mixtures of the primary colors will be produced. You will note that the addition of the red and green colors will produce yellow; that red and blue colors produce magenta; and green and blue colors produce cyan. By varying the intensities of the red, green, and blue lights, combinations can be produced which to the human eye will match every color and hue of the spectrum.

Observe that proper combinations of the primary colors will produce what to the eye appears white. Any two colors which combine to form white light are said to be *complementary*. The complement of red is cyan; of green is magenta; of blue is yellow. However, many other pairs of colors can be complementary.

21.17 Subtraction of Colors When Pigments Are Mixed

Most of us are familiar with the fact that if we mix yellow and blue pigments, we get a green mixture. This is because the yellow pigment absorbs most of the wavelengths of the visible spectrum except those which produce a sensation of yellow and some of the waves on either side of yellow, i.e., orange and green (see Table 21.3). In like manner, blue pigments absorb most wavelengths except blue, some green, and some indigo. The only color that is not completely absorbed by one or the other of the pigments is green. Hence green is reflected from the mixture. Similar explanations can be given for the resultant color of mixtures of all kinds of dyes and pigments.

When pigments are mixed, each one subtracts certain colors from the incident light and the resulting color depends upon the light waves that are not absorbed. The *primary pigments* are the complements of the primary colors. They are cyan, magenta, and yellow. When the three primary pigments are mixed in proper proportions, all wavelengths of the spectrum will be subtracted and the mixture will appear black (see Fig. 21.17).

Color mixing and color analyzing has become an important science. Many industries depend on a thorough knowledge and skillful application of this science. Consider the variety of colors that are now available and used for clothing, automobile finishes, wallpaper, floor coverings, and paints for interior and exterior decorating.

21.18 Commercial Light Sources

Before a body can emit light waves, its atoms and molecules must be set in violent motion. We learned in Chap. 13 that heating an object increases its molecular and atomic motion. If iron is heated to about 1000°F, it begins to glow with a dull red color. As the temperature of the iron is raised, the light appears yellow and finally bluish. Thus we see that the wavelengths of light emitted by a body shift from the longer red waves to the shorter blue waves as the temperature is raised. Astronomers have deduced the surface temperature of the sun to be about 5500°C (10,000°F), by such color analyses.

The carbon arc used in searchlights develops a temperature near that of the sun's surface. Incandescent electric lights (tungsten filaments) reach temperatures of around 2800°C (5000°F). If air were in the lamp at this temperature, the filament in the lamp would burn up (oxidize) almost instantly. In order to avoid this rapid oxidation of the filament, incandescent lamps made many years ago were quite highly evacuated. However, in a vacuum and at 2800°C the tungsten filament evaporated quite rapidly. It is desirable that lamps operate at a temperature as high as possible in order that more of the electromagnetic waves emitted will be radiated as visible light and less as the longer invisible heat (infrared) rays. Today most common incandescent lamps are filled with inert gas (usually a mixture of argon and nitrogen). The presence of the gas slows down the rate of evaporation and permits higher operating temperatures.

Despite the many modern improvements in incandescent lamps, only about 15 percent of the electric energy consumed is radiated as visible light, the rest being dissipated as heat. Most of the visible light radiated from the common incandescent lamp lies in the red and blue portions of the spectrum, and the human eye is relatively insensitive to these wavelengths. The colors to which our eyes are most sensitive lie in the yellow-green region. Hence the *visual efficiency* of the common light bulb is considerably below 15 percent. The yellow sodium lamp has a visual efficiency considerably above that of the tungsten-filament incandescent lamp, since a very large portion of its radiation is in the green and yellow region.

The *efficiency* of electric lamps may be expressed in terms of candelas *per watt*. Ordinary tungsten-filament lamps have efficiencies of a little more than 1 cd/W. Fluorescent lights yield about 4 cd/W.

Many lamp manufacturers list the efficiencies of their lamps in *lumens per watt*, rather than in candelas per watt. The efficiency of ordinary tungsten-filament lamps is of the order of 8 to 15 lm/W.

21.19 Penetration of Light Waves

Many people have enjoyed the beautiful blue of a midday sky and the gorgeous shades of red and yellow that frequently accompany a sunrise or sunset. These phenomena can be explained in terms of the *scattering* of the sun's rays. When white light passes through the atmosphere, the short-wavelength blue rays are scattered in all directions by the particles of the atmosphere, while the longer-wavelength red rays tend to continue straight through. The blue rays which are scattered by particles of air, water, and dust in the atmosphere are what we see when we look up in the sky. At sunrise and sunset we see only the longer red and yellow rays which have passed a long distance through the atmosphere. The shorter blue and green rays have been scattered and dissipated.

Red aircraft-beacon lights are used because of their greater penetrability in the fog and smoke of "heavy" weather. In like manner, yellow sodium lights are used on many bridges where foggy weather persists. It should be made clear, however, that merely covering a source of white light by a red or amber filter does not necessarily increase the penetrability of its rays, since by eliminating the shorter rays the intensity of the beam will be decreased.

21.20 Uses of Infrared Rays

Infrared rays have an even greater penetrating power than red and yellow rays. Using camera film which is sensitive to infrared rays, one can take pictures of objects which are completely invisible to the eye. Infrared photography is employed in mapping from aircraft and satellites, since these rays can penetrate fog and haze and give a much sharper picture.

Considerable use is made of infrared rays in medical practice. These rays tend to penetrate the skin and warm the tissue of patients suffering

Fig. 21.18 If any painted surface on a new car has been scratched or otherwise damaged during final assembly operations, portions of the automobile are repainted. The car then moves through a bake oven, using infrared lamps, which rapidly dries the repainted area to a hard finish. (Cadillac Division, General Motors Corp.)

from muscular aches and pains. In industry, freshly painted autos and refrigerators are today dried under large batteries of infrared lamps (Fig. 21.18). The penetrating rays afford a more uniform drying of many layers of paint at the same time. Many foods are dehydrated by using these rays—with a resultant saving of time as well as a saving of many of the vitamins in the food. Food warmers and ovens make use of infrared rays.

Although infrared rays are sometimes termed *heat rays*, the rays themselves are not warm. They *produce* heat by stimulating an increased motion of molecules in the objects they strike.

21.21 Properties of Ultraviolet Radiation

Ultraviolet radiation is scattered by the atmosphere even more readily than the visible blue rays. Hence very few ultraviolet rays reach the earth from the sun. It is fortunate that most of these rays are scattered and absorbed before reaching the earth, for if ultraviolet radiation from the sun reached the earth undiminished, it would prove disastrous to most life on the globe. These rays can produce serious burns of which "sunburn" is a painful example. Many mountain climbers have belatedly come to realize this fact. Since there is less atmosphere to penetrate at high altitudes, more ultraviolet rays reach the earth at high altitudes than at sea level. Hence sunburning takes place more rapidly on mountain peaks than at lower elevations.

Artificially produced ultraviolet rays are used in hospitals and other sickrooms to kill germs that may be in the air (Fig. 21.19).

The sun's ultraviolet radiation, acting on the human body, aids in the formation of vitamin D in the body, which helps to make healthy teeth and bone. Some food products are purposely irradiated with ultraviolet rays until the desired amount of vitamin D is developed in the food.

Carbon arcs and mercury arcs, which emit strong ultraviolet radiations, are used in sunlamps. Window glass is almost perfectly transpar-

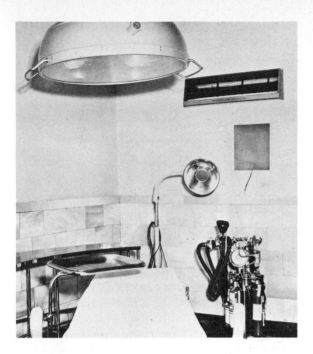

Fig. 21.19 Germicidal ultraviolet lamps in hospital operating rooms reduce the hazard of infection during surgery. (General Electric Co.)

ent to visible light, but it does not transmit the ultraviolet light. Therefore it is almost impossible to get a sunburn from sunlight that has come through a closed window. Pure quartz glass, which will transmit ultraviolet rays, is used in sunlamps. It is important that the eyes be shielded whenever sunlamps are used. Likewise, since the radiation from a welding arc contains a dangerous amount of ultraviolet, the welder's eyes must be protected by a mask with a suitable filter.

21.22 Fluorescent Lighting

When light falls on certain chemical compounds, the radiant energy is absorbed and then immediately reemitted, with an appreciable part of the new radiation being of longer wavelengths than those absorbed. Such materials are called fluorescent.

The presence of certain mineral ores can be recognized by the characteristic glow (*fluor-*

escence) they give off in the presence of ultraviolet radiation. Police often are able to detect forged documents by testing the ink for ultraviolet fluorescence, since different kinds of inks are distinguished by their fluorescence. Gems, such as diamonds, are often recognized by their ultraviolet fluorescence. The dentist is able to distinguish a dead tooth from a live one, because only the live tooth will fluoresce in the presence of ultraviolet light.

The ability of ultraviolet radiation to produce fluorescence is used extensively in fluorescent lamps used for lighting. The commercial fluorescent lamp consists of a glass tube 2 ft or more long with an electrode sealed in each end. The flow of electric current takes place through mercury vapor, which emits a predominance of ultraviolet radiation. The inside of the tube is coated with material which fluoresces when struck by the ultraviolet radiation. The fluorescent materials are made of different powders according to the color of light desired. By combining various powders, an almost exact match for daylight is obtained. Since the invisible ultraviolet light, which is ordinarily absorbed by the glass, is converted into useful radiations, the efficiency (candela per watt) of such a lamp is 3 to 4 times that of tungsten lamps. Fluorescent lamps give light that is free from glare while operating at relatively low temperatures.

Not much fluorescent lighting is used in homes, except in utility areas, as kitchens, and in workshops. This is partly due to the larger sizes of the fluorescent lamps. However, factory lighting is provided almost universally by ceiling-mounted fluorescent fixtures which give a high level of illumination. One of the disadvantages of fluorescent lamps is that they have a residual 120-Hz flicker. Fluorescent lamps near the end of their useful life tend to have an annoying flicker.

Fluorescence is not limited to ultraviolet radiations. Examples of fluorescence due to other than ultraviolet radiations are: (1) fluorescent paints, now used extensively for highway markers, advertising signs, etc.; (2) the screens of cathode-ray oscilloscopes, radar receivers, and television tubes; and (3) fluoroscopes used by radiologists to make bodily structures visible by means of x-rays.

21.23 X-rays

X-rays have wavelengths between $\frac{1}{100}$ and $\frac{1}{100,000}$ that of visible light. The method for producing them will be discussed in Chaps. 32 and 33. The most familiar, and among the earliest, application of x-rays was diagnosis and therapy in medicine. The rays are used for the detection of bone fractures, dental cavities, foreign objects in the body, and the presence of cancer cells. In therapeutic treatment x-rays are used to inhibit the growth of malignant tumors.

In industry, x-ray radiographs permit nondestructive inspection for flaws in castings, welds, and engineering structures.

21.24 Lasers

Ordinary light sources are incoherent; i.e., they produce waves of light energy of many different wavelengths and with random phase relationships. As a result, there is little reinforcement among them, and their energy is soon diffused and lost. The *laser* is a device that produces an intense, highly concentrated, and almost parallel beam of light. The word laser is an acronym made up of the first letters of the words: Light Amplification of Stimulated Emitted Radiation. A discussion of the theoretical principles involved in laser production requires familiarity with the energy levels of the electrons in the structure of various atoms. Such an account is given in Chap. 33 but we shall discuss here some properties and uses of lasers.

Lasers can use a solid, a liquid, or a gas as the amplification medium. One common laser consists of a long thin chamber or single large crystal of ruby (aluminum oxide) with flat parallel ends highly polished and silvered—one end silvered completely to reflect as perfectly as possible, and the other silvered only partially so that a fraction of the light can escape (Fig. 21.20). The atomic lattice structure of the ruby absorbs light of a certain frequency and can hold the ab-

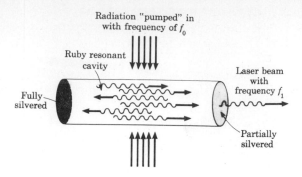

Fig. 21.20 Coherent stimulation of light waves in a solid-state laser.

sorbed energy for a period of time. The chamber is surrounded by a flash tube containing xenon or some other suitable gas. The gas ionizes and glows when a pulse of high-voltage electricity passes through it. When light of frequency f_0 is "pumped" into the laser, the ruby atoms are stimulated to emit radiation at the characteristic frequency f_1. This light with frequency f_1 bounces back and forth between the silvered ends. As the waves travel back and forth, they stimulate other emissions, thus amplifying the original beam. These waves will all be in phase, and by introducing other emissions they give rise to a chain reaction. The standing waves grow to enormous intensity until the light energy bursts ("lazes spontaneously") from the slightly transparent end of the tube. All this happens within a few microseconds. Activating the flash tube continuously gives a steady stream of coherent light.

Lasers produce an intense, narrow light beam of a very pure single color (monochromatic light). Laser beams can be intense enough to vaporize the hardest and most heat-resistant materials. This property is useful in drilling holes in diamonds for wire-drawing dies and to destroy tumors or to weld a detached retina in modern eye surgery. It is used to perform microsurgery on cells of the body.

The helium-neon laser provides an ideal narrow straight line for all kinds of alignment purposes. Such a beam diverges by less than one part in a thousand for distances approaching infinity. An even smaller divergence of a laser beam directed to the surface of the moon, 240,000 mi distant, has produced a spot of light only a few feet in diameter. Laser beams are widely used to guide machines for drilling tunnels and to align jigs employed in producing large jet aircraft.

Laser beams directed vertically downward from an airplane in flight can be used to determine the plane's height above ground or to map fine details of the land below.

Laser beams can be used for communications. Today one laser beam can carry the outlet from all existing radio stations, and one beam can carry as many as seven television programs at one time. However, the beams can be blocked by fog or rain. So, to be effective under all weather conditions, the beams would need to be enclosed in protecting pipes or conducted along optic fibers. Laser beams, because of their directionality and use of small amounts of power, do find uses in long-distance communication through outer space.

Most lasers in use today are safe when reasonably precaution is applied. Most lasers are of low power and do not cause injury if carefully used. However, one should never look directly down a laser beam or along a reflected beam. Helium-neon lasers are now in the physics laboratories of many colleges and technical institutes.

21.25 Masers

The maser is a device which produces radiation similar to that produced by the laser but in the radio (microwave) part of the electromagnetic spectrum. The word *maser* is an acronym made up of the first letters of the words: Microwave Amplification of Stimulated Emission of Radiation. Like laser beams, maser beams have been produced using solids, liquids, and gases as the amplification media. Actually, the maser was invented first, but the laser has proved the more useful.

Masers have been used to amplify faint signals from distant satellites. The atomic hydrogen maser is one of the most accurate standards of time in existence. A clock based on the hydrogen-maser frequency would not have an error greater than one second per 100,000 years.

SUMMARY

For most purposes light can be considered as being transverse waves emanating in all directions from the source. However, some light phenomena can be explained only by assuming light to be small particles of energy, called *photons*.

Lines drawn in the direction a light wave travels are called *rays*.

A body is visible either because it is luminous or because it is illuminated.

Fig. 21.21 Industrial laser in operation. The CO_2 laser emits a fine beam of light in the infrared region at 10.6μ (micrometer) wavelength. The coherent light can be focused to a spot the size of a fraction of a millimeter, providing very high densities of millions of watts per centimeter. It can be used in cutting, drilling, and welding a wide variety of materials. The laser machining principle depends on materials being highly absorptive at the $10.6\text{-}\mu$ wavelength. (Courtesy Laser Division, Coherent, Inc.)

The Nature of Light and Illumination   **435**

The speed of light has been measured by several methods and today has an accepted value of 2.997924×10^8 m/sec or 186,284 mi/sec (in a vacuum). In material media, the speed of light is less than in a vacuum.

The speed of light in a vacuum is independent of the speed of the light source, the brightness, or color of the source.

Light from the sun is a mixture of a vast number of different wavelengths which, when combined, give the sensation we call *white*.

Difference in color is due to difference in wavelengths reaching the eye. The visible spectrum ranges from the red, with a mean wavelength of about 67×10^{-8} m, to violet, with a mean wavelength of about 41×10^{-8} m.

Most sources of visible radiation also simultaneously emit invisible radiations.

The *candela* or *candle* is a unit of luminous intensity. The instrument for comparing the luminous intensities of light sources is called a *photometer*. The equation used in comparing different light-source intensities is

$$I_x/I_s = d_x^2/d_s^2$$

The total flow of light energy or flux from a source is measured in *lumens*. The *lumen* is the luminous flux which falls on a unit surface all points of which are a unit distance from a point source of one candela.

Inverse-square law: *The intensity of illumination from a point source of light varies inversely as the square of the distance from the source:*

$$E_1/E_2 = d_2^2/d_1^2$$

The unit of *intensity of illumination* on a surface is the lux or footcandle.

$$E \text{ (luxes)} = \frac{F \text{ (lm)}}{A \text{ (m}^2)}$$

$$E \text{ (ft-c)} = \frac{F \text{ (lm)}}{A \text{ (ft}^2)} \qquad (21.3)$$

$$E \text{ (luxes)} = \frac{I \text{ (cd)}}{d^2 \text{ (m}^2)} \qquad (21.5)$$

For nonperpendicular incidence

$$E = \frac{I \cos \theta}{d^2} \qquad (21.6)$$

Transparent objects transmit color through *selective absorption*.

Opaque objects exhibit color through *selective reflection*.

Colors are *additive* when colored *lights* are mixed.

The *primary colors* are red, green, and blue.

Two colors which combine to produce white light are called *complementary colors*.

Colors are *subtracted* when *pigments* are mixed.

The *primary pigments* are the complements of the primary colors, i.e., cyan, magenta, and yellow.

Proper mixing of the primary pigments produces a mixture that appears black.

The process by which a body absorbs radiant energy and then reemits waves of longer wavelengths than those received is called *fluorescence*.

Infrared radiation has wavelengths greater than those of the visible spectrum.

Ultraviolet radiation has wavelengths shorter than those of the visible spectrum.

The *laser* produces an intense, highly concentrated, monochromatic, and almost parallel beam of light.

QUESTIONS AND EXERCISES

1. List four ways in which light waves differ from sound waves. Then list four ways in which light waves are similar to sound waves.

2. The following pairs of colors (light) are mixed additively. Name the resultant colors: (*a*) blue and green, (*b*) red and green, (*c*) blue and yellow, (*d*) red and blue, (*e*) magenta and green.

3. The following pairs of colors (pigments) are mixed subtractively. Name the resultant colors: (*a*) yellow and cyan, (*b*) yellow and blue, (*c*) cyan and magenta, (*d*) red and cyan, (*e*) magenta and yellow.

4. (*a*) What color added to red will give white? (*b*) What color subtracted from yellow will give black?

5. What color is the complement of (*a*) red, (*b*) yellow, (*c*) green, (*d*) cyan, (*e*) blue, (*f*) magenta?

6. Distinguish between the meaning of the terms *light* and *illumination*.

7. Does the atmosphere scatter more red rays or more blue rays from the sun? Give reasons for your answer.

8. How effective will placing a yellow filter over the headlights of a car be in producing increased visibility in foggy weather? Why?

9. Six lamps, each emitting one of the following colors—violet, blue, green, yellow, orange, and red—are viewed through a red glass filter. No other light is present. Describe the "color" you would see as you view each lamp through the filter.

10. Which of the following does *not* properly pair with the unit of measurement: (*a*) luminous intensity—candela; (*b*) intensity of illumination—lux; (*c*) luminous flux—footcandle; (*d*) luminous flux—lumen?

11. Which of the following is the process by which a substance absorbs energy and then reemits electromagnetic waves longer than those received: (*a*) fluorescence, (*b*) luminescence, (*c*) incandescence, (*d*) coherence?

12. A hot iron can be photographed in a dark room only if which one of the following is true: (*a*) the lens aperture is opened wide; (*b*) a long exposure time is used; (*c*) infrared sensitive film is used; (*d*) the iron is painted with a luminous substance?

PROBLEMS

1. If 3 sec is the proper time of exposure in printing a photograph using an incandescent lamp 70 cm from the printing frame, what length of exposure would be required in printing a picture from the same negative at a distance of 2 m from the same light source?

2. If 20 luxes is correct for reading, how far from your reading material should a 64-cd light be placed in order to give proper illumination?

3. A screen is 5 m from an 80-cd source of light, and the screen surface makes an angle of 60° with the line drawn from the source to the center of the screen. What is the intensity of illumination on the screen?

4. What would be the intensity of illumination on a magazine held 5 ft from a 40-cd lamp? What would be the intensity of illumination 10 ft from the lamp? (Assume no reflection from the walls.)

5. A photometer has a standard 32-cd lamp at one end and a lamp of unknown intensity at the other. The two sides of a screen appear equally illuminated when the screen is 3.5 ft from the standard lamp and 5.2 ft from the unknown. What is the luminous intensity of the unknown lamp?

6. A standard 24-cd lamp at a distance of 40 cm from a screen gives the same illumination as a lamp of unknown intensity placed 120 cm from the screen. What is the intensity of the unknown lamp?

7. A 24-cd lamp is placed 0.36 m from a screen. How far from the screen must a 64-cd lamp be placed for the illumination on the screen to be 3 times as great as it was with the first lamp?

8. A football stadium is illuminated at night from eight towers, each with a bank of twenty-four 1000-W lamps with luminous efficiencies of 32 lm/W. One-third of the luminous flux reaches the playing field, which has dimensions of 120 by 75 m. Find the average intensity of illumination on the field.

9. A camera is set with its shutter speed at $\frac{1}{500}$ sec. How far will light travel during the time of exposure?

10. The North Star (Polaris) is 1.60×10^{15} mi from the earth. How many years elapse before light emitted from the star reaches the earth?

11. The efficiency of a 300-W tungsten lamp is rated at 12 lm/W. What illumination in footcandles will be given by such a lamp on a surface 5 m from the lamp?

12. At what angle must a surface be inclined so that light falling on it from an 80-cd luminous intensity source at a distance of 4 ft will give an illumination of 1.21 ft-c?

13. In measuring the speed of light by Michelson's method, a rotating mirror having 32 sides was used. The rotating mirror was 5.000 mi from the stationary mirror. What minimum speed, in revolutions per second, was required of the rotating mirror to catch the reflected beam on the very next mirror surface?

14. If four lighted candles are placed at one end of a meterstick and one identical candle at the other end, where must a screen be placed on the stick in order to get equal illumination from both sources? (HINT: Let x = the distance in centimeters of the one candle to the screen and $100 - x$ = the distance in centimeters of the four candles to the screen.)

22 Reflection and Refraction of Light

For thousands of years it was possible to study only what could be seen with unaided human eyes. The only mirrors in existence were those provided by natural lakes and pools. Stars and planets were seen only as shining points of light, and nothing was known of the countless microscopic phenomena that existed.

Then a few centuries ago, with the discovery of the properties of reflection and refraction of light, new vistas were opened up. With mirrors and lenses in use, the scope of human vision increased tremendously. In this and the next chapter we shall study the principles of reflection and refraction and see how they are applied in such optical devices as mirrors, eyeglasses, projectors, enlargers, binoculars, microscopes, and the telescope.

22.1 Graphical Representation of Light

According to the wave theory of light, light energy is transmitted through space from a point source as a series of spherical waves with the source as center. These spheres increase in size as the light spreads from the source. All points on the same crest of a wave are said to be in the same *phase* of vibration. The crest of a wave is followed by a trough whose radius is always $\frac{1}{2}$ wavelength less than that of the preceding crest. Figure 22.1 illustrates a graphical representation of light waves. The curved surfaces are the *wavefronts*. If the point source of light is a great distance away

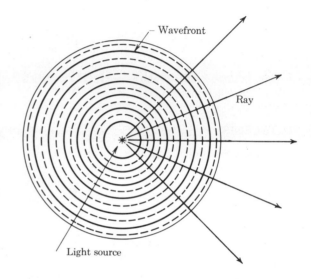

Fig. 22.1 Light waves and light rays.

and we observe only a small portion of the wavefront, for most practical purposes it may be considered as lying in a plane; hence it is called a *plane wave*.

The straight lines of Fig. 22.1 indicate the direction of advance of small sections of the wavefront. These lines are called *rays*. A small bundle of these rays is called a *beam*. When the wavefronts are plane, the rays are parallel and we speak of a beam of parallel rays or a parallel beam (Fig. 22.2). We can describe the propagation of light either in terms of its wavefronts or in terms of its rays. But it is important to recognize

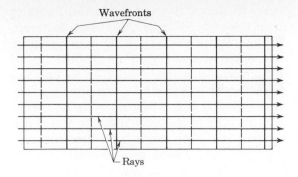

Wavefronts

Rays

Fig. 22.2 Parallel beam of light with plane wavefronts.

that rays and wavefronts are only imaginary lines in these diagrams. They are used to describe the characteristics and the laws of light, but it is actually the *motions* of the waves themselves that are really significant.

REFLECTION OF LIGHT

22.2 Regular and Irregular Reflection

Common experience shows that when light strikes most objects, some of the light is reflected from the object. Indeed, it is by this reflected light that we are able to see the object. If the surface is rough, e.g., a newspaper, light striking it is reflected in many directions. This kind of reflection (Fig. 22.3a) is called *diffuse reflection*. If, however, the surface is shiny and polished so that it is very smooth, the reflection is *regular* (= *specular*) (Fig. 22.3b), and the reflecting surface is said to be a *mirror*. The reproduction of an object in the mirror is called the *image* of the object.

22.3 Laws of Reflection

The laws governing the direction of beams of light which are regularly reflected are the same for any smooth surface. When a beam of light enters a small opening in a dark room, the path of

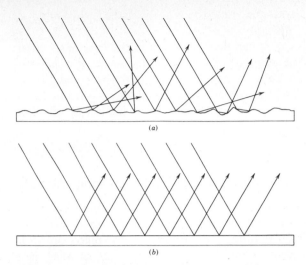

Fig. 22.3 Reflection from (*a*) rough surface and (*b*) smooth surface.

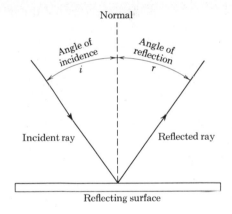

Fig. 22.4 Angle of incidence equals angle of reflection.

the beam can be clearly seen by the illuminated dust particles in the room. If this beam is allowed to strike a plane mirror (Fig. 22.4), the directions of the *incident* and *reflected* rays can easily be seen. If a perpendicular (normal) is drawn to the reflecting surface at the point where the incident ray strikes the surface, the angle between the normal and the incident ray is called the *angle of incidence*, while the angle between the normal and the reflected ray is called the *angle of reflection.*

It can be shown that in the case of specular, i.e. regular, reflection (1) *the angle of incidence i is equal to the angle of reflection r and* (2) *the incident ray, the normal, and the reflected ray lie in the same plane.* Algebraically the law of reflection can be stated by

$$\text{Angle } i = \text{angle } r \qquad (22.1)$$

22.4 Images Formed by Plane Mirrors

In order to see the image of an object in a plane mirror, it is necessary that the object be either self-luminous or luminous by reflection from another light source. In Fig. 22.5, let MM' be a plane mirror. Each point on an object, represented by the arrow AB, may be considered as a point source of light, sending out rays in all directions. One of these rays coming from each point of the object will strike the surface of the mirror at just the right spot to be reflected to the eye E. For example, a light ray coming from point A

strikes the mirror at P and is reflected, according to the law *angle i = angle r*, and comes to the eye. The light *appears* to be coming from a point A' back of the mirror. Light from B is reflected by the mirror as if coming from B', lying on a line perpendicular to the mirror and the same distance behind it as B is in front of the mirror. By applying rules of geometry, it can be proved that *an image in a plane mirror is the same size as its object and that the image appears as far behind the mirror as the object is in front of it.*

The image of an object in a plane mirror appears laterally reversed with respect to the reflecting plane. For example, the reflection of the printed page of a book is reversed from right to left, and as you look at yourself in a mirror, what appears to be your right hand is actually your left.

An interesting experiment that can be performed in the laboratory is illustrated in Fig. 22.6. A lighted candle O is placed in front of a plane mirror (only partially silvered) MM' and shielded from the eye by a box B. A glass of water G is placed a distance behind the "mirror" approximately equal to the distance between the candle and the mirror. Since the observer at E will see the candle by *reflected* light and the water glass *through* the mirror (remember it is only partially silvered) by transmitted light, the reflected image I of the candle will be seen as though it were burning in the glass of water. The image the viewer sees is not a real image but a *virtual* image.

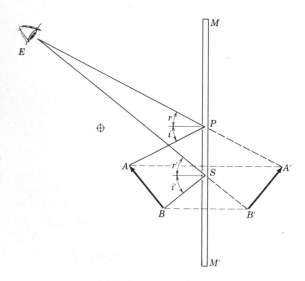

Fig. 22.5 Image formed by a plane mirror. The image appears to be as far behind the mirror as the object actually is in front of it.

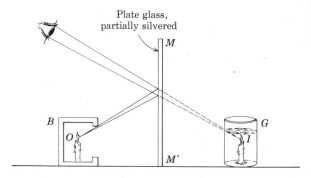

Fig. 22.6 The phantom candle.

A virtual image is formed by rays which seem to radiate from the image but which actually come from another source.

This experiment illustrates how ghostlike figures can be made to appear and move about a room or stage. A real person or an object below or above the stage can be reflected by a large plate of glass or a plane mirror at the front of the stage and made to appear on the stage. Proper draperies and lighting can make the optical illusion very effective.

22.5 The Optical Lever

In many scientific instruments small rotational displacements are magnified by attaching a very small plane mirror to the rotating part. Let us consider an incident ray of light from a source S striking the mirror MM' at O, and reflected as OR (Fig. 22.7). If we rotate the mirror through an angle ϕ, the normal ON is rotated through an equal angle. The new angle of incidence will be $i' = i + \phi$. By the law of reflection, the new angle of reflection r' also equals $i + \phi$. We see then that the increased deviation of the reflected ray

$$\text{Angle } ROR' = \text{angle } SOR' - \text{angle } SOR$$

or $\qquad \angle ROR' = 2(i + \phi) - 2i = 2\phi$

Hence such an arrangement, called an *optical lever*, will turn a reflected ray of light through *twice* the angle through which the mirror is turned. By shining the reflected ray on a scale several feet from the mirror, extremely small angular displacements can be measured. Optical levers of this type are used in magnifying the motions of indicators in several types of measuring instruments, such as the sextant.

22.6 Spherical Mirrors

Numerous applications are made today of curved mirrors. Most of these are spherical; i.e., they are small portions of the surface of a sphere. Spherical mirrors are classified as *concave* or *convex*, depending on whether the reflecting mirror is an inner or an outer segment of a sphere (Fig. 22.8, for example, is a diagram of a concave spherical mirror). The *center of curvature C* of the mirror is the center of the sphere. The *radius of curvature R* of the mirror is the radius of the sphere. The radius is taken as *positive* for concave mirrors and *negative* for convex mirrors. The *vertex* of the mirror is the central point V of the mirror, while a line XX' passing through the center of curvature and the vertex is called the *principal axis* of the

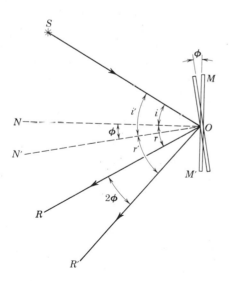

Fig. 22.7 Radiation of a reflected ray (the optical lever).

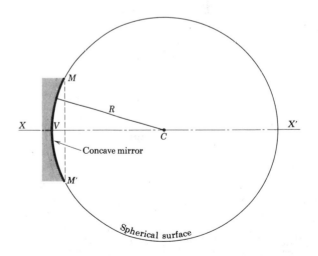

Fig. 22.8 Spherical mirror.

Fig. 22.9   Reflection from a concave mirror.

mirror. The *aperture* of the mirror is the diameter *MM'* of the small circle which is the base of the spherical segment.

Most spherical mirrors are comparatively flat; i.e., the size of the aperture is small in comparison with the radius of curvature. With a mirror of small aperture, rays of light parallel to the principal axis will strike the mirror and be so reflected as to pass through, or very close to, a single point *F*, called the *principal focus* of the mirror (Fig. 22.9). The distance *FV* from the principal focus to the mirror is called the *focal length f*.

The ability of a concave mirror to gather parallel rays from the sun and focus them at a point is used in one type of solar furnace. Temperatures at the focus as high as 5000°C have been produced in this way.*

It can be shown geometrically that *for apertures which are small compared to the radius of curvature*, the principal focus *F* lies approximately halfway between the vertex and the center of curvature of the mirror.

In the diagram of Fig. 22.10 let *PVM* be a *concave* spherical mirror whose center of curvature is at *C*. The principal axis intersects the

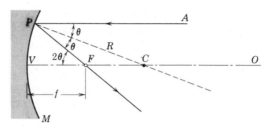

Fig. 22.10   Proof that $R = 2f$ for a concave spherical mirror.

mirror at *V*, the vertex. The *aperture PM* is *assumed to be small compared to R*. Any ray of light coming in to the mirror parallel to the optical axis (such as *AP*) will be reflected through the focal point *F*. The angle of incidence, $\angle APC$, equals the angle of reflection, $\angle CPF$. These two angles are both designated as θ in the diagram. But, from the geometry of parallel lines, $\angle APC = \angle PCV = \theta$ and $\angle APF = 2\theta$. Since the arc *PV* is small compared to *R*, *PV* may be assumed to be a straight line perpendicular to *OV* at *V*. Therefore

$$PV = CV \tan \theta$$

$$= FV \tan 2\theta$$

When *PV* is small, $\tan \theta = \theta$ (approximately). Consequently,

$$CV = 2FV$$

---

*   Although the solar energy received at the earth's surface is immense (approx. 430 Btu/hr/ft² or 1160 kcal/hr/m²), to effectively harness it is quite difficult.

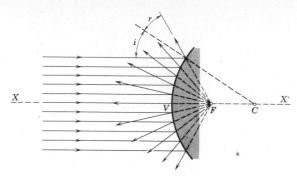

Fig. 22.11 Reflection from a concave mirror.

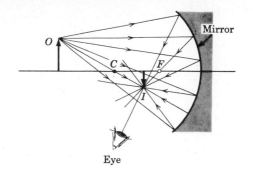

Fig. 22.13 All rays from O that strike the mirror contribute to the image.

and
$$R = 2f \qquad (22.2)$$

Parallel rays which strike a *convex* mirror will diverge after reflection as though they originated at a common focal point F behind the mirror (Fig. 22.11).

22.7 Images Formed by Spherical Mirrors—Real and Virtual Images

Let us consider a concave mirror (Fig. 22.12). Any ray OA parallel to the axis is reflected back through the focal point F, while the ray OC passing through the center of curvature C will be perpendicular to the mirror and will be reflected back upon itself. After reflection these rays coming from O will intersect at I, forming a *real image* of O at point I. A real image is formed by converging rays which *actually pass through it*. A real image can be projected on a screen; i.e., a

reproduction of the object O will be cast upon a screen placed at I. Even though we have been using two specific rays to locate the image, it must not be inferred that only these two are effective in forming the image. All other rays from point O which strike the mirror contribute to the image (see Fig. 22.13); the larger the aperture of the mirror, the brighter the image will be, although a large aperture causes the image to be less sharply focused.

A similar procedure is followed in locating the image in a convex mirror (Fig. 22.14). From the figure it is apparent that the rays, after reflection, will appear as if they came from behind the mirror at I. For all positions of O the image will be smaller than the object. An image that is formed by reflected rays that only appear to converge is called a *virtual image*. Such an image cannot be projected on a screen as a real image can.

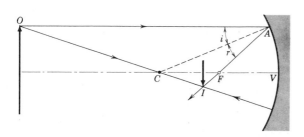

Fig. 22.12 Location of image formed by a concave mirror.

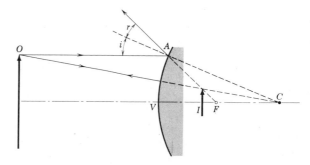

Fig. 22.14 Virtual image formed by a convex mirror.

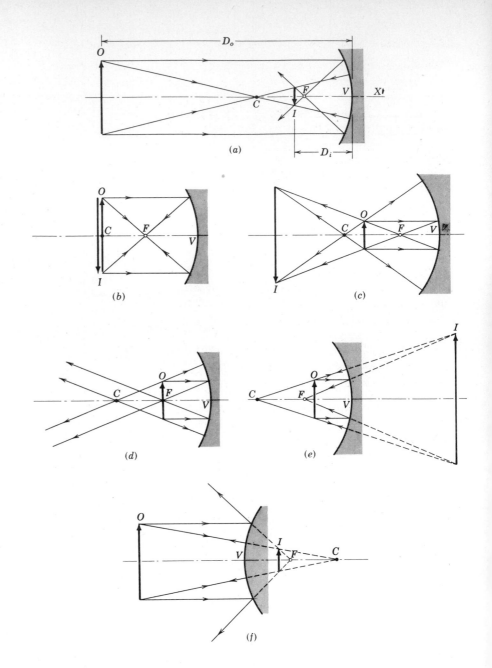

Fig. 22.15 Images formed by concave and convex mirrors. (*a*) Image is real, inverted, smaller than object. (*b*) Image is real, inverted, same size as object. (*c*) Image is real, inverted, larger than object. (*d*) No image, reflected rays are parallel. (*e*) Image is virtual, erect, larger than object. (*f*) Image is virtual, erect, smaller than object.

22.8 Real and Virtual Images

Whether an image will be real or virtual depends not only on whether the mirror is concave or convex but also on the position of the object with respect to the mirror. Six cases may occur, as represented in Fig. 22.15a to 22.15f.

Concave-Mirror Images

1. When the object lies at a distance from the vertex greater than the radius of the mirror, the image is located between the center and the focus. It is real, inverted, and smaller than the object (Fig. 22.15a). Reflecting telescopes use such an arrangement.
2. When the object is at the exact center of curvature, the image is also at the center. It is real, inverted, and the same size as the object (Fig. 22.15b).
3. When the object is located between the center and the focus, the image lies beyond the center. It is real, inverted, and larger than the object (Fig. 22.15c). Floodlights have the light source between the center and the focus of a concave reflector in order to spread the rays.
4. When the object is at the focus, the reflected rays are parallel. Hence there is no image formed by the mirror (Fig. 22.15d). Searchlights, car headlights, and flashlights have the light source near the focus of a concave mirror to use this property. The common slide projector of the amateur photographer has a lamp placed at the focal point of a concave mirror in order that the light energy will not be dissipated by spreading out in all directions. Lenses in front of the lamp direct the rays through the slide. A similar arrangement is used in a photographic enlarger.
5. When the object is between the focus and the vertex, the reflected rays do not actually meet but appear to meet behind the mirror. Thus the image is virtual, erect, and enlarged (Fig. 22.15e). Shaving mirrors and dental mirrors use this arrangement to get virtual magnified images.

Convex-Mirror Images

6. No matter where the object is located with respect to a convex mirror, the image is always virtual, erect, and smaller than the object (Fig. 22.15f). Ornamental convex mirrors are sometimes hung in living rooms. In stores they are used to protect against shoplifters because they make it possible to see the entire room in miniature.

22.9 The Mirror Equation

The relation between the distance of object D_o, the distance of image D_i, and the focal length f, for any spherical mirror, is given by the formula

$$\frac{1}{D_o} + \frac{1}{D_i} = \frac{1}{f} = \frac{2}{R} \qquad (22.3)$$

If any two of these quantities are known, the third can be computed. In order to make the equation applicable to both concave and convex mirrors, distances back of the mirror must be considered negative. Thus D_i and f for convex mirrors will always have negative values. A negative value for D_i with a concave mirror will therefore also signify that the image is behind the mirror and virtual.

Illustrative Problem 22.1 An object is placed 120 cm in front of a concave mirror whose focal length is 30 cm. Find the location of the image.

Solution Since the mirror is concave, f is positive. The image will be real, since D_o is greater than f; hence D_o is taken positive and D_i should solve to be positive. Substitute known values in Eq. (22.3) and obtain

$$\frac{1}{120} + \frac{1}{D_i} = \frac{1}{30}$$

Multiply each term by $120D_i$ to clear fractions.

$$D_i + 120 = 4D_i$$

Transpose, collect terms, and solve.

$$3D_i = 120$$

$$D_i = 40 \text{ cm} \qquad \textit{ans.}$$

Illustrative Problem 22.2 An object is placed 36 in. in front of a convex spherical mirror of 45-in. radius. Locate the image.

Solution In this problem take f negative, since the mirror is convex and the image is virtual and solve for D_i (which should be negative). The focal length is equal to one-half the radius of curvature, or $\frac{45}{2}$ in. Substituting known values in Eq. (22.3), we get

$$\frac{1}{D_o} + \frac{1}{D_i} = \frac{1}{f}$$

$$\frac{1}{36} + \frac{1}{D_i} = \frac{-2}{45}$$

Multiplying each term by the lowest common denominator $180D_i$, transposing, and solving for D_i, we get

$$5D_i + 180 = -8D_i$$

$$13D_i = -180$$

$$D_i = -13.8 \text{ in. (behind the mirror)} \quad ans.$$

22.10 Size of the Image

By using any of the diagrams of Fig. 22.15a through 22.15f it can be readily shown that the size of the image I is to the size of the object O as the image distance D_i is to the object distance D_o, neglecting signs. The *magnification* is expressed as the ratio of the size of the image to the size of the object. Thus for any spherical mirror this relationship can be expressed algebraically as

$$\text{Magnification} = \frac{I}{O} = \frac{D_i}{D_o} \qquad (22.4)$$

Illustrative Problem 22.3   An object 8 cm tall is placed 40 cm from a concave mirror of focal length 60 cm. Locate and describe the image, and find its size.

Solution   Substitute known values in Eq. (22.3), and solve for the image distance.

$$\frac{1}{D_o} + \frac{1}{D_i} = \frac{1}{f}$$

$$\frac{1}{40} + \frac{1}{D_i} = \frac{1}{60}$$

$$3D_i + 120 = 2D_i$$

$$D_i = -120 \text{ cm} \qquad ans.$$

The minus sign indicates a virtual image behind the mirror. Substituting the value for D_i and other known values in Eq. (22.4), we can determine the magnification.

$$\frac{I}{O} = \frac{D_i}{D_o}$$

$$\frac{I}{8} = \frac{120}{40}$$

$$I = 24 \text{ cm} \qquad ans.$$

Hence the image is virtual, situated 120 cm behind the mirror, and 24 cm tall.

22.11 Uses of Concave and Spherical Mirrors

Concave mirrors are used on *compound microscopes* to reflect light from a window or lamp upon the object to be examined. The *ophthalmoscope* is a concave mirror with a small hole in the center. The instrument enables the physician to reflect light from a lamp into the patient's eye, nose, or throat while looking through the hole into the cavity thus illuminated. Convex mirrors are frequently used as rearview mirrors on trucks and automobiles, since they form small erect images of a large area behind the vehicle. Most mirrors are silvered (or aluminized) on the back surface because the coating quickly tarnishes if the mirror is front-surfaced with silver. Mirrors for scientific work are often front-surfaced and then carefully protected from tarnishing agents.

22.12 Spherical Aberration with Mirrors

It should be reemphasized that the foregoing discussion on spherical mirrors and their images applies only to mirrors whose apertures are small compared to their radii of curvature and for objects on or near the principal axis of the mirror. As the aperture becomes relatively larger, the images become blurred and imperfect. The effect of a large aperture on parallel rays can be illustrated by Fig. 22.16. Here it will be noticed that only those rays close to the principal axis converge in the immediate vicinity of the focal point. The farther the incident ray is from the axis the more distant from the focal point (and the closer to the vertex) does its reflection pass. This

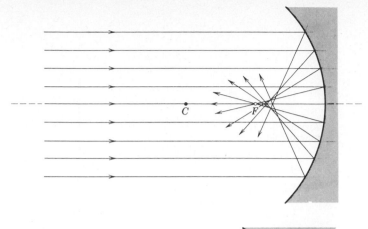

Fig. 22.16 Spherical aberration produced by a mirror with a large aperture.

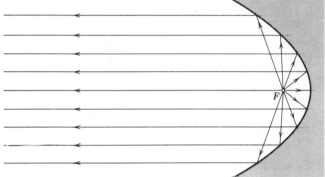

Fig. 22.17 Reflection from a parabolic mirror.

effect is called *spherical aberration*. If we use a parabolic mirror, however, the parallel rays from a distant object will all be focused at a common focal point. Conversely, light radiating from a point source at the focal point *F* of a parabolic mirror is reflected as parallel rays (Fig. 22.17). Parabolic mirrors are used for both purposes— for forming sharp images of distant objects and for reflecting intense beams of parallel rays. Auto headlights and military and display searchlights are equipped with parabolic mirrors. Reflecting-type astronomical telescopes (see Fig. 23.20) have concave parabolic mirrors. Since it is difficult and expensive to grind parabolic mirrors, spherical mirrors with small apertures are sometimes used on commercially produced, inexpensive equipment.

REFRACTION OF LIGHT

One of the most interesting and useful properties of light is its bending as it passes obliquely from one substance into another. Were it not for this bending, we could not have lenses, magnifiers, telescopes, binoculars, microscopes, eyeglasses, cameras, and the many other instruments which extend the usefulness and power of our eyes. This bending of the path of light is called *refraction*. Refraction is a matter of common observance. For example, a stick which is partially submerged in clear water will often appear sharply bent where it enters the water; the water in a swimming pool will appear less deep than it actually is. A coin placed on the bottom of an empty cup where it is just out of sight behind the rim will

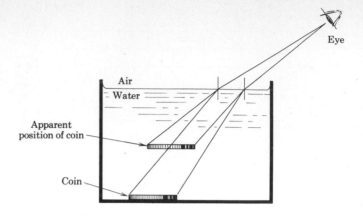

Fig. 22.18 The coin is visible because light is bent in passing from water to air.

appear to be lifted into view and closer to the observer when the cup is filled with water (Fig. 22.18).

22.13 Refraction of a Plane Wave at a Plane Surface

Let us consider one of the simplest cases of refraction, that of a plane wave of light passing obliquely from air into water (Fig. 22.19). The beam of light in passing from air into water from an incident direction IO is bent in the direction OR in such a way as to make the angle of refraction r' less than the angle of incidence i.

Refraction of light is caused by a change in the speed of light as it leaves one medium and enters another. The speed of light in water is about three-fourths that in air. Let us consider the wavefront AC of a bundle of parallel rays represented by AO, BD, and CE. The rays will advance in the air with a constant speed, making OE parallel to AC.

When ray IO enters the water, its speed is reduced. This ray will travel a distance OT in the water, while the ray BD will move in air a greater distance DS ($OT \approx \frac{3}{4}DS$). In like manner, while these two rays travel equal distances TK and SL in water, ray CE will travel the greater distance PM in air. The new wavefront will be represented by the straight line KM. The effect will be a bending of the beam of rays toward the lines that are

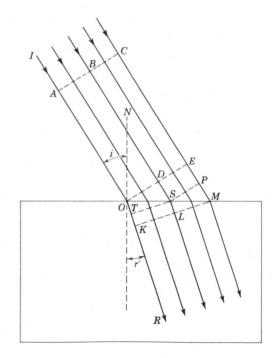

Fig. 22.19 Refraction of a plane wave passing from air into water.

perpendicular (called normals) to the water surface and drawn through the points at which the rays enter the water.

Thus, when a ray of light passes from one material into another in which the speed is less, it is bent *toward the normal*. In like manner, it can

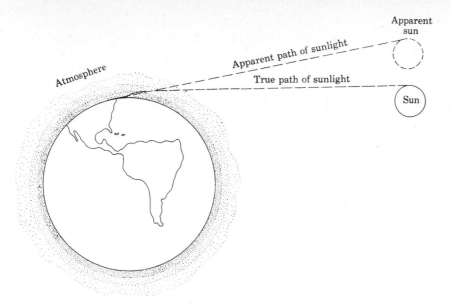

Fig. 22.20　Refraction of sun's rays at sunset (exaggerated).

be shown that when the speed of light is greater in the second medium, the ray is bent *away from the normal*. As a matter of fact, if we could consider a ray in Fig. 22.19 as moving in the direction *RO* as an incident ray, we would find that the direction of refraction would be *OI*. That is, the angles *i* and *r'* in the figure can be interchanged, if the direction of the ray is reversed.

22.14　Refraction Caused by Earth's Atmosphere

Light from the sun can be refracted by the earth's atmosphere. Whenever you watch a sunset, you will see the sun several minutes after it actually sinks below the horizon (see Fig. 22.20) because the earth's atmosphere bends the rays of the sun. Light from stars seen at night moves through varying masses of air, some hotter, some cooler, some more dense and some less dense. The light is bent from side to side as it moves from one mass to another. These masses of air are in motion and give the starlight its twinkling effect. Astronomers get their best photographs on cold,

still nights or by sending instruments aloft in high-altitude balloons or spacecraft. Space stations outside the region of atmospheric disturbances show great promise for the future for making detailed and precise astronomical observations.

Objects viewed through the air above a pavement on a hot summer day have a wavy appearance because the convection air currents of varying densities cause uneven refraction of light. Often a still hot layer of air above a paved surface or the floor of the desert refracts the sky light to the observer at ground level. The surface thus seems to be covered with a layer of water. These optical illusions are called *mirages*.

22.15　Index of Refraction

If Fig. 22.21, we mark off equal distances *IO* and *OR* on the incident and refracted rays and draw perpendiculars *IA* and *RB* to the normal *NN'*, we can show that no matter what the angle of incidence, the segment *IA* is always a definite number of times segment *RB*. If the light travels from air to water, *IA* will equal $\frac{4}{3}RB$ or

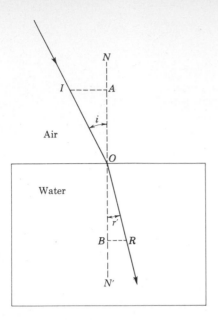

Fig. 22.21 Measurements for computing index of refraction.

$IA/RB = \frac{4}{3}$. Similarly, for light traveling from any material to another, we shall get $IA/RB = \mu$, which is a constant for each pair of materials. Since $IO = RO$, we can divide numerator and denominator by equals and get

$$\frac{IA/IO}{RB/RO} = \frac{\sin i}{\sin r'} = \mu \qquad (22.5)$$

This simple relationship, now known as *Snell's law*, was discovered by the Dutch physicist Willebrord Snellius (Snell van Royen) about 1620:

Snell's Law

The ratio of the sine of the angle of incidence to the sine of the angle of refraction is a constant, for any two given media.

By application of the laws of geometry, it can be shown that IA and RB (Fig. 22.21) are proportional, respectively, to the speed of light in the air v_1 and the speed of light in water v_2. Hence Eq. (22.5) can be written

$$\frac{\sin i}{\sin r'} = \frac{v_1}{v_2} = \mu_{1,2} \qquad (22.6)$$

where the order of the subscripts of μ indicates that the light is traveling from the first medium to the second. This constant μ is called the *index of refraction*. It is independent of the angle of incidence and depends on the characteristics of the two transmission media *and upon the wavelength of the light transmitted.*

The *absolute refractive index* of a substance is its index with respect to a vacuum. However, this has practically the same value as the *relative index* of that substance against air. Thus if we let the speed of light in a particular substance be v_s, in a vacuum c, and in air v_a, the index of refraction μ of the substance is

$$\mu = \frac{c}{v_s} \approx \frac{v_a}{v_s} \qquad (22.7)$$

In passing through two transparent media, the greater the difference between the index of refraction of the second medium and that of the first, the greater the amount of refraction. Thus, light passing from air into glass will be refracted more than if it passes from air into water.

The index of refraction of a substance can be measured by passing a narrow beam of light into it as suggested by Fig. 22.21, measuring the angles of incidence and refraction, and applying Eq. (22.5). A yellow light source of wavelength 5893 Å is used as a standard in determining indices of refraction.

Illustrative Problem 22.4 A narrow beam of light strikes a plane glass surface at an angle of incidence of 52°. The refracted ray makes an angle of 31° with the normal to the surface (Fig. 22.22). What is the index of refraction of the glass?

Solution Substitute the known angles (Fig. 22.22) for i and r' in Eq. (22.5) and solve the trigonometric equation:

$$\mu = \frac{\sin i}{\sin r'}$$

$$= \frac{\sin 52°}{\sin 31°} = \frac{0.788}{0.515}$$

$$= 1.53 \qquad\qquad ans.$$

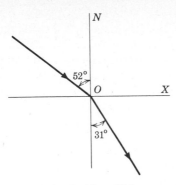

Fig. 22.22 Illustrative Problem 22.4.

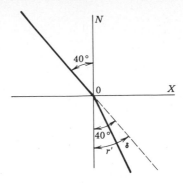

Fig. 22.23 Illustrative Problem 22.5.

Table 22.1 Index of refraction μ, for various substances, for light of wavelength 5893 Å

Substance	Index
Calcite	1.486–1.658
Diamond	2.417
Glass, ordinary crown	1.517
Light flint	1.575
Dense flint	1.656
Ice at −8°C	1.31
Quartz, fused	1.458
Carbon disulfide at 20°C	1.625
Carbon tetrachloride at 20°C	1.461
Ethyl alcohol at 20°C	1.360
Water at 0°C	1.334
At 20°C	1.333
At 40°C	1.331
At 60°C	1.327
Air	1.000293
Carbon dioxide	1.000450
Ethyl ether	1.001521

Table 22.1 gives values of the index of refraction μ for a number of common substances.

Illustrative Problem 22.5 Light, in air, is incident at an angle of 40° on the surface of a light flint-glass plate. Through what angle is the light deviated from its incident direction by refraction?

Solution From Table 22.1 the index of refraction of flint glass is 1.575. Solve for r' in Eq. (22.5). Substituting known quantities, we then have

$$\sin r' = \sin i/\mu$$
$$= \sin 40°/1.575 = 0.643/1.575$$
$$= 0.408$$
$$r' = 24°$$

The deviation is the difference between angles i and r' (see Fig. 22.23).

$$\text{Deviation} = 40° - 24°$$
$$= 16° \qquad\qquad ans.$$

22.16 Total Reflection

Let us consider light in a medium of high refractive index and directed toward one of lower index, e.g., light traveling from water into air (Fig. 22.24). Light traveling from a point O along OA is mostly refracted into the air along AR' in such a direction that $(\sin i)/(\sin r') = 1/\mu$ (since the light rays are passing from a medium of lower light speed to one of higher light speed), but a small part is reflected along AR. As the angle of incidence at the water-air surface increases, the angle of refraction in the less dense medium increases until we finally come to ray OC, which is refracted along CT' and just grazes the surface ($r' = 90°$). The angle i_c for which this occurs is called the *critical angle*. Any ray, such as OD, for which the angle of incidence is greater than i_c, does not emerge in air at all but is *totally reflected* along some line DW—just as if it had fallen on a highly polished metal surface at D. The critical

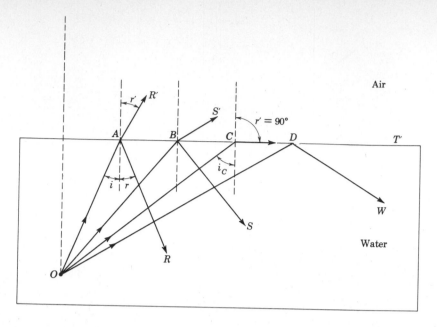

Fig. 22.24 Critical angle and total reflection.

angle must not be exceeded if the ray is to pass out of the denser medium. If we substitute 90° for r' in the formula $(\sin i)/(\sin r') = 1/\mu$, we get the value for the critical angle:

$$\sin i_c = \frac{1}{\mu} \qquad (22.8)$$

Illustrative Problem 22.6 What is the critical angle for crown glass and air?

Solution From Table 22.1, the index of refraction for crown glass is 1.517. Substituting this value in Eq. (22.8) and solving for i_c, we get

$$\sin i_c = 1/\mu$$

$$= 1/1.517 = 0.659$$

$$i_c = 41°14' \qquad \textit{ans.}$$

Total reflection is employed in optical instruments such as telescopes, microscopes, periscopes, prism binoculars, and spectroscopes. A *right-angle prism* with polished sides is used in optical instruments where a nearly perfect reflector is necessary. Figure 22.25 shows how a crown-glass prism serves as a reflector. When ray AB strikes the side of RT of such a prism at right

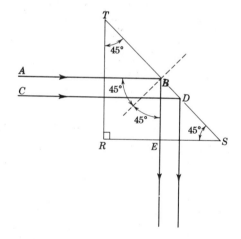

Fig. 22.25 Total reflection of light by a right-angle prism.

angles, it suffers no refraction but passes on through the glass to B on the side ST, where its angle of incidence is 45°. Since the critical angle for crown glass is less than 45°, the ray does not emerge from the glass but is totally reflected in the direction BE. It is not refracted at E since it strikes the surface RS at right angles. Prisms are

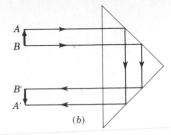

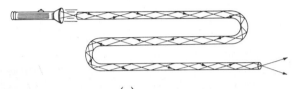

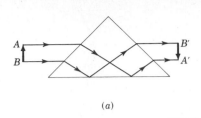

(a)

(b)

Fig. 22.26 Reflections by right-angle prisms.

Fig. 22.27 Prism binocular. The double total reflections give an optical path length between lenses considerably greater than the overall length of the instrument. (Bausch & Lomb Co.)

used in this manner in periscopes and range finders.

Figure 22.26 illustrates two other ways a right-angle prism is used. In Fig. 22.26a, we see how a prism can be used to invert the image. Projection lanterns frequently use prisms in this manner. When the prism is rotated to the position of Fig. 22.26b, it will reverse the beam and invert the image. Two such prisms are used for each eye in *prism binoculars*—one to reverse the image *up* and *down*, one to reverse it *left* to *right* (Fig. 22.27).

Light can be "piped" from one point to another by using total internal reflection of light in transparent rods of certain plastic and glass fibers (Fig. 22.28). On entering one end of a

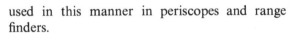

(a)

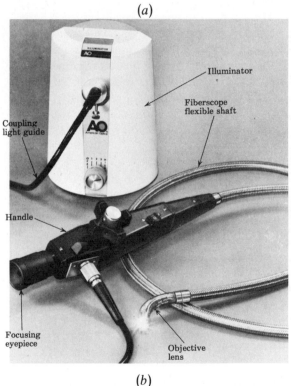

(b)

Fig. 22.28 Fiber optics. (a) The bundle of tapered fibers transmits light around curves by total internal reflection. (b) The fiberscope is used to produce images from hard-to-reach areas. For endoscopy, specialized instruments have been developed to view various internal body cavities such as the esophagus, stomach, bronchi, and the large bowel. Focus of the objective lens is remotely controlled by a knob on the eyepiece. Fiberscopes are also used for industrial inspection of internal or remote areas hidden from direct view. (Courtesy American Optical Co.)

Reflection and Refraction of Light **453**

curved rod or fiber, light undergoes repeated total internal reflection at the boundary of the rod and will follow its contour, emerging at the other end. Images can be transferred from inaccessible places along such a bundle of fibers.

A whole new field of *fiber optics* has developed in recent years. Fiber-optics techniques have made possible useful devices for transmitting and transforming luminous images. Physicians and surgeons use various instruments which employ fiber optics in observing internal body cavities. A typical ray may undergo 48,000 reflections in a flexible 7-ft length of fiber rod. Telephone companies use lasers and fiber optics to transmit signals (Sec. 32.22).

Total reflection is also used in enhancing the brilliance of cut diamonds. By taking advantage of the fact that the critical angle of diamond is only 24°, the skilled artisan will cut the facets of a diamond in such a way that most of the light entering it will be totally reflected many times before emerging through the outer surfaces. The effect will be to give the diamond added sparkle.

LENSES

A lens is any transparent medium, usually glass, with smooth spherical surfaces. We see because images are formed by lenses of the human eye. Defective eye lenses can often be compensated for by the proper use of other lenses made of glass or plastic. By the use of combinations of lenses, scientists delve into the realm of astronomical research on the one hand and into the secrets of extremely small microscopic life on the other.

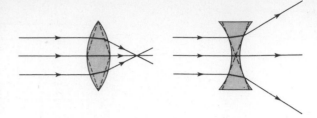

Fig. 22.30 Lenses act like prisms.

22.17 Types of Lenses

Lenses are divided into two classes: *converging lenses*, which are thicker at the center than at the edge, and *diverging lenses*, which are thinner at the center than at the edge. Typical lens forms are shown in Fig. 22.29. Special names associated with each of the six types are the following: (*a*) double-convex, (*b*) plano-convex, (*c*) concavo-convex, (*d*) double-concave, (*e*) plano-concave, and (*f*) convexo-concave.

By referring to Fig. 22.30 it can be seen that a double-convex lens is in cross section essentially equivalent to two triangular prisms placed base to base. Light rays entering the prisms are bent toward the bases, causing them to converge. The double-concave lens can be represented in cross section as essentially equivalent to two similar prisms but apex to apex. Light rays falling on these prisms will in like manner be bent toward the bases, causing them to diverge.

Figure 22.31 illustrates converging and diverging lenses from the standpoint of changing wavefronts. We have learned that light travels more slowly in glass than in air and that light

Fig. 22.29 Types of converging and diverging lenses: (*a*) double convex, (*b*) plano-convex, (*c*) concavo-convex, (*d*) double concave, (*e*) plano-concave, (*f*) convexo-concave.

Converging

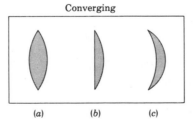

(*a*)   (*b*)   (*c*)

Diverging

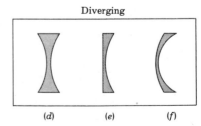

(*d*)   (*e*)   (*f*)

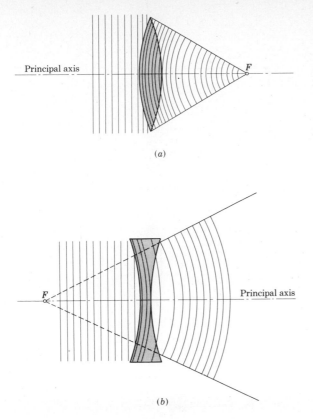

(a)

(b)

Fig. 22.31 Action of converging and diverging lenses on a plane wavefront. (*a*) Converging lens; (*b*) diverging lens.

coming from a distant object will be nearly in the form of plane waves. In the convex lens (Fig. 22.31*a*), the wavefront passing through the thicker center is retarded more than at other portions of the lens, the least retardation being at the edges. The result is a changed wavefront

which will converge to a focus at *F*. In the concave lens (Fig. 22.31*b*), the reverse is true: that part of the wavefront passing through the center is retarded least, and the greatest retardation occurs at the outside edge. Hence the plane wavefront will be altered so that the emerging waves diverge as if they were coming from a point *F* on the same side of the lens as the actual source.

The *principal axis* of a lens is the straight line passing through the centers of curvatures of the surfaces of the lens. (It must be remembered that a plane is a circle with an infinite radius.) The point at which rays parallel to the principal axis converge after passing through the lens is called the *principal focus* of the lens (see Fig. 22.32).

Parallel rays striking a lens from either side of the lens will converge or appear to converge at a given point. Hence every lens has two such *focal points F* and *F'* equidistant from the lens. The distance from the center of the lens to either focal point is called the *focal length f* of the lens. The *aperture* of a lens is the diameter of the uncovered part of the lens.

22.18 Formation of Images by Lenses

When rays from a light source *converge* after passing through a lens, an image of the source can be formed and viewed on a screen. Such an image is called a *real* image. If the rays *diverge* after passing through a lens, a real image cannot be formed on a screen, but the image can be seen by looking through the lens. The image viewed by looking through such a lens is called a *virtual* image.

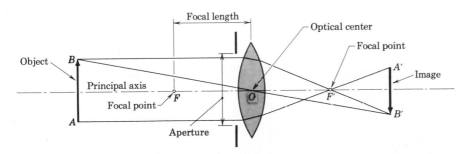

Fig. 22.32 Diagram illustrating terminology used with lenses.

22.19 Determination of Image Location and Size by Means of Rays

It is possible by the application of a few simple geometric constructions to determine the position and size of images formed by lenses. The method is similar to that used for spherical mirrors. If we trace any two rays from a point on the object to their intersection after refraction, we shall have the position of the corresponding point on the image. The two rays generally selected are (1) a ray from the object parallel to the principal axis and (2) a ray from the object through the center of the lens. The first ray will, after refraction, pass through the focal point, while the second will pass directly through the lens without deviation.

Figure 22.33 illustrates four cases of image formation by converging lenses.

1. In Fig. 22.33a, the object AB, if a considerable distance from the lens, will appear as a small, real, inverted image $A'B'$ at a relatively short distance from the lens. The eye lens and the camera lens form this type of image.

2. When the object is placed at a distance equal to twice the focal distance ($D_o = 2f$), the image of the object is real, inverted, and the same size as the object and is the same distance from the lens (Fig. 22.33b).

3. If the object is placed between $2f$ and f, the image will appear beyond $2f$ on the other side, as an enlarged, real, and inverted image (Fig. 22.33c). This effect is used by film and slide projectors, and photographic enlargers. It should be evident from the diagram why the operator of the projector must put the slides upside down and inverted in the machine.

4. If the object is placed between the lens and its principal focus (Fig. 22.33d), the rays will not intersect on the opposite side of the lens but will appear to intersect on the same side of the lens. The eye can see their *apparent* intersection as a virtual, erect image larger than the object. Ordinary magnifying glasses and the eyepieces on microscopes and telescopes are used in this manner.

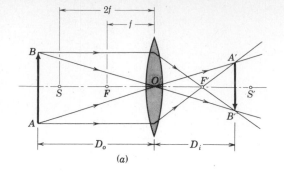

(a)

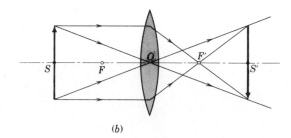

(b)

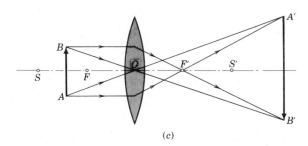

(c)

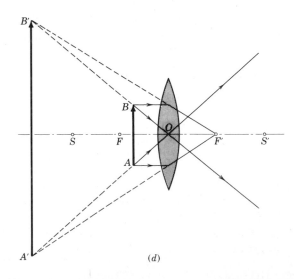

(d)

Fig. 22.33 Images formed by convex lenses: (a) Image is real, inverted, smaller than the object. (b) Image is real, inverted, same size as the object. (c) Image is real, inverted, larger than the object. (d) Image is virtual, erect, larger than the object.

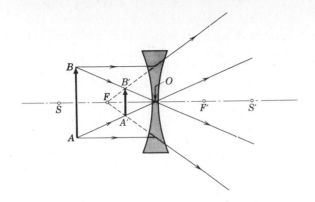

Fig. 22.34 Image formed by concave lens is virtual, erect, and smaller than the object.

The ray diagram of Fig. 22.34 indicates that no matter where the object location with reference to a concave (diverging) lens, the image will appear as a virtual, erect, small image, nearer to the lens than the object. An important use of concave lenses is in eyeglasses for the correction of nearsighted vision. They are also used as eyepieces in telescopes, field glasses, and binoculars.

22.20   The Lens Equation

It can be shown that the relation between the distance of the object from a lens D_o, the distance of its image from the lens D_i, and the focal length f is given by the same equation as the one previously given for mirrors.

$$\frac{1}{D_o} + \frac{1}{D_i} = \frac{1}{f} \qquad (22.3)$$

In order to make Eq. (22.3) applicable to all types of lenses, it is necessary to adopt the following *convention of algebraic signs:*

1.   Take f positive for converging lenses and negative for diverging lenses.

2.   Take D_o positive when the object is real and negative when the object is virtual. (In optical instruments the image formed by one lens often serves as the virtual object for another lens.)

3.   Take D_i positive when the image is real and negative when the image is virtual.

Illustrative Problem 22.7   A double-concave lens has a focal length of 60 cm. Where does it form an image of an object placed 75 cm from it?

Solution   Since the object is real and the lens is diverging, take D_o as positive and f as negative in accordance with the above convention of signs. Equation (22.3) then becomes

$$\frac{1}{75} + \frac{1}{D_i} = \frac{1}{-60}$$

Multiplying through by the lowest common denominator, transposing, and solving for D_i yields

$$4D_i + 300 = -5D_i$$

$$9D_i = -300$$

$$D_i = -33\tfrac{1}{3} \text{ cm} \qquad ans.$$

The minus sign indicates that the image is virtual, and it is therefore on the same side of the lens as the object.

Illustrative Problem 22.8   An object placed 100 cm from a lens will form an image that can be focused on a screen placed 22.3 cm on the other side of the lens. What kind of lens is it? Is the image real or virtual? Find the focal length of the lens.

Solution   Since the image is formed on the opposite side of the lens, the lens must be a converging one and the image is real (see Fig. 22.33a). Hence take positive values for D_o and D_i. We should get a positive answer for f. Substitute known values for D_o and D_i in Eq. (22.3) and solve for f:

$$1/D_0 + 1/D_i = 1/f$$

$$1/100 + 1/22.3 = 1/f$$

$$0.01 + 0.0448 = 1/f$$

$$f = 1/0.0548$$

$$= 18.2 \text{ cm} \qquad ans.$$

22.21   Size of Image

By comparing similar triangles AOB and $A'OB'$ in Figs. 22.33 and 22.34 made by the rays AA' and BB' passing through the center of the lens O, it is seen that

$$\frac{I}{O} = \frac{\text{Size of image}}{\text{Size of object}} = \frac{\text{image distance}}{\text{object distance}} = \frac{D_i}{D_o} \tag{22.9}$$

The absolute value of this ratio, as in the case of curved mirrors, is called *magnification*.

Illustrative Problem 22.9 Determine the size of the image of an object 12.5 cm tall with the arrangement of Illustrative Problem 22.8 above.

Solution

Magnification = D_i/D_o from Eq. (22.9)

$$= 22.3/100 = 0.223$$

Size of image = $0.223 \times 12.5 = 2.79$ cm high

ans.

22.22 The Lens Maker's Formula

In order to manufacture lenses which will refract light by exactly the right amounts to meet specifications of optical equipment, the lens maker must deal with the following variables:

The refractive index of the glass to be used
The radii of curvature to be used for the lens surfaces
The desired focal length for the lens

These variables are all related in the *lens maker's formula*

$$\frac{1}{f} = (\mu - 1)\left(\frac{1}{R_1} + \frac{1}{R_2}\right) \tag{22.10}$$

where

R_1, R_2 = radii of curvature of the two lens surfaces
f = focal length of lens
μ = index of refraction of glass to be used

A fourth factor, the *aperture*, determines the light-gathering power of the lens and consequently the brightness of the image. It should be pointed out, however, that increasing the aperture in order to increase brightness of the image will also increase the spherical aberration and therefore reduce the sharpness of focus of the image. (See Sec. 22.24 and Fig. 22.35).

Illustrative Problem 22.10 A manufacturer produces a quantity of double-convex lenses of 30-in. focal length from optical glass whose refractive index is 1.52. Both faces are to have the same radius of curvature. Calculate the radius to be used.

Solution Since the effect of both surfaces of a double-convex lens is to *converge* the rays, the algebraic sign of both R_1 and R_2 will be positive. Since, for this problem, $R_1 = R_2$, Eq. (22.10) becomes

$$\frac{1}{f} = (\mu - 1)\frac{2}{R}$$

Solving for R, we have

$$R = 2f(\mu - 1)$$

Substituting values, we get

$$R = 2 \times 30 \text{ in. } (1.52 - 1)$$

$$= R_1 = R_2 = 31.2 \text{ in.} \qquad ans.$$

22.23 Power of a Lens

Opticians in dealing with optical instruments employ a term called the power of a lens or of a combination of lenses. The power P of a lens, measured in units called *diopters*, is the reciprocal of its focal length f measured in meters.

$$P \text{ (diopters)} = \frac{1}{f \text{(meters)}} \tag{22.11}$$

Thus, the shorter the focal length, the greater the power of a lens. Converging lenses are said to have positive $(+)$ powers and diverging lenses negative $(-)$ powers.

The chief advantage of expressing lens power in diopters is that *for thin lenses* (if the thickness of the lens is small compared to its diameter) *in contact, the power in diopters of such a combination of lenses is equal to the sum of the powers of the separate lenses.*

Thus

$$1/f = 1/f_1 + 1/f_2 \tag{22.12}$$

where f_1 and f_2 are the focal lengths of the two

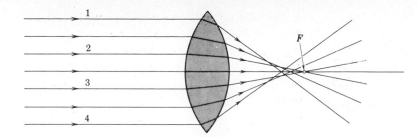

Fig. 22.35 Spherical aberration, a defect of lenses. Not all rays are focused at one point. Light rays that strike the outer portion of the lens are refracted more than the rays that fall on the central portion.

thin lenses in contact and f is the focal length of the combination.

Illustrative Problem 22.11 A convex lens of focal length 0.25 m is combined with a concave lens of -1.25 m focal length. Find the lens power of the combination. What would be the focal length of an equivalent single lens?

Solution The converging lens will have a power of 1/0.25 or $+4$ diopters, while the diverging lens will have a power of $1/(-1.25)$ or -0.80 diopters. Hence the power of the combination is

$$P = +4 - 0.80 = +3.20 \text{ diopters} \qquad ans.$$

To solve for the focal length of a single equivalent lens, we solve for f in Eq. (22.11):

$$f = 1/P = 1/3.20$$
$$= 0.31 \text{ m} = 31 \text{ cm} \qquad ans.$$

22.24 Spherical Aberration with Lenses

In the geometric construction of images formed by spherical lenses (Fig. 22.33), it was assumed that all rays parallel to the principal axis converge at a common point. This is not exactly true. Actually, if the surfaces of the lens in Fig. 22.35 are spherical, rays such as 1 and 4 which strike the lens near the edges are focused nearer the lens than rays such as 2 and 3 which strike the lens near the focal point. This imperfect focusing of spherical lenses is called *spherical aberration*. The effect of this aberration is an indistinct and distorted image. This defect can be partly overcome by special grinding of the outer thin portion of the lens (called *aspherizing*), or by using a diaphragm with a circular aperture in front of the

lens which will eliminate the outer rays. This reduction of aperture is called *stopping down*. Decreasing the effective aperture will result in a sharper image but at the same time will diminish the brightness of the image. Generally, in practice, combinations of lenses are used instead of single lenses. These combinations are so designed that any imperfection in one lens is compensated for or nullified by the optical properties of another lens.

SUMMARY

Light waves falling on a smooth surface may be reflected. If reflected, they will obey the laws of reflection: (1) *the angle of incidence = the angle of reflection*, and (2) *the incident ray, the reflected ray, and the normal to the mirror at the point of incidence all lie in the same plane.*

When light is reflected from a smooth plane surface to the eye, the eye sees an *image* of the source. This image appears to be as far behind the mirror as the object is in front of it; it is laterally reversed and virtual.

Rough surfaces scatter light in all directions. Such light is said to be *diffused*, and it is by such light that we see most objects.

Concave mirrors produce *inverted* and *real images* that lie between the focus and the center of curvature if the object is outside the focus; and they will produce *erect* and *virtual* images behind the mirror if the object is inside the focus.

Convex mirrors will always produce *erect, virtual* images which are *smaller* than the object and appear to be located behind the mirror.

The mirror equation is

$$\frac{1}{D_o} + \frac{1}{D_i} = \frac{1}{f} \qquad (22.3)$$

The magnification equation is

$$\text{Magnification} = \frac{I}{O} = \frac{D_i}{D_o} \qquad (22.4)$$

These two equations are also applicable to lenses, with proper attention to algebraic signs.

Refraction is the change of direction of light as it passes obliquely into a transparent medium of different optical density. When light enters a denser substance, it is refracted toward the normal obliquely, and it is refracted away from the normal in passing obliquely into a rarer medium. Refraction is caused by the change of speed of light as it enters a new medium.

The amount of bending depends on the index of refraction of a substance:

Index of refraction

$$\mu = \frac{\text{sine of angle of incidence}}{\text{sine of angle of refraction}} = \frac{\sin i}{\sin r'} \qquad (22.5)$$

$$= \frac{\text{speed of light in air (or vacuum)}}{\text{speed of light in other substance}} = \frac{c}{v_s} \qquad (22.7)$$

The *critical angle* is the largest angle of incidence that will permit light to be refracted along the surface instead of being reflected. Total reflection occurs when light tends to pass from a denser to a rarer medium at an angle that is greater than the critical angle.

$$\sin i_c = \frac{1}{\mu} \qquad (22.8)$$

A prism bends light toward the thick edge.

Concave mirrors and convex lenses have the same optical properties, while convex mirrors and concave lenses have the same optical properties.

The lens maker's formula is

$$\frac{1}{f} = (\mu - 1)\left(\frac{1}{R_1} + \frac{1}{R_2}\right) \qquad (22.10)$$

The power of a lens is measured in diopters:

$$P \text{ (diopters)} = \frac{1}{f\text{(meters)}} \qquad (22.11)$$

for two thin lenses in contact:

$$1/f = 1/f_1 + 1/f_2 \qquad (22.12)$$

QUESTIONS AND EXERCISES

1. What will the image of the letter F look like when reflected from a plane mirror?

2. Why can a virtual image be seen with the eye but cannot be projected on a screen?

3. Is it possible to get an image formed by a lens that is virtual and smaller than the object? If so, how?

4. Under what conditions will the image formed by a mirror be smaller than the object? Is the image real or virtual?

5. What causes spherical aberration in an image formed by a lens? How can the aberration be decreased?

6. Which of the following will increase the sharpness of a photograph? (a) Use a longer exposure time. (b) Use a smaller diaphragm opening. (c) Use flood lights. (d) None of these.

7. Why is there no spherical aberration when a parabolic mirror is used?

8. Describe the path of a ray of light which passes (obliquely) through a rectangular plate of glass. Illustrate with a diagram.

9. In which of the following is the optical lever used? (a) Telescope. (b) Microscope. (c) To magnify linear displacements. (d) To magnify angular displacements.

10. A person swimming under the surface of water looks diagonally up at the diving board. Does the diving board appear higher or lower than it actually is?

11. The flexible fiberglass rods used in medical instruments are exceedingly delicate. How would a broken fiber within the rod affect the produced image?

12. Which of the following will be true when two parallel rays of light pass through a combination of thin lenses with respective focal lengths of 30 cm and −120 cm? (a) The rays diverge. (b) The rays will converge. (c) The focal length of the combination is −90 cm. (d) The power of the combination is $(\frac{1}{30} - \frac{1}{120})$ diopters.

PROBLEMS

1. A lens which has a focal length of −25 cm will have what power, expressed in diopters?

2. What is the critical angle for rays passing from fused quartz to air?

3. Calculate the critical angle for rays passing from a piece of dense flint glass to air.

4. What is the approximate speed of light in water at 20°C? (HINT: Use Table 22.1 and Eq. 22.7)

5. A woman 1.8 m tall faces a vertical plane mirror. What is the minimum height of the mirror in which she can see her full-length image?

6. Two concave mirrors use the sun to heat objects placed at their foci. The mirrors have equal focal lengths, but the first has a 2-m aperture while the second has a 3-m aperture. Which will produce more intense heat (same heat source)? How much more?

7. Find the focal length of a combination of two thin lenses placed in contact if their focal lengths are 30 cm and −15 cm.

8. A ray of light strikes the surface of carbon disulfide at an angle of 50°. By what angle is the ray deviated in passing into the liquid?

9. An incandescent bulb placed 20 in. in front of a concave mirror has its image formed 50 in. in front of the mirror. Determine the radius of curvature of the mirror.

10. A concave spherical mirror whose radius of curvature is 75 cm is used as a shaving mirror. If the mirror is held 30 cm from the face, how far behind the mirror does the image appear to be? What is the magnification?

11. An object 4 in. high is placed 20 in. in front of a concave spherical mirror. The image is sharp on a screen 100 in. away from the mirror. Find the radius of curvature of the mirror and the size of the image.

12. A convex mirror has a focal length of 30 cm. An object 10 cm high is placed 20 cm in front of the mirror. Calculate the position and the height of the image formed.

13. A dentist uses a concave mirror whose radius of curvature is 6 cm. What will be the apparent magnification of a tooth when the mirror is held 2 cm from the tooth?

14. A concave lens has a focal length of 20 cm. Determine the image distance when the object is (*a*) 40 cm from the lens and (*b*) 10 cm from the lens.

15. How large is the image in each case of Prob. 14 if the object in each case is 2.5 cm high?

16. A convex lens has a focal length of 20 cm. Determine the image distance when the object is (*a*) 40 cm from the lens and (*b*) 10 cm from the lens.

17. How large is the image in each case of Prob. 16 if the object in each case is 3 cm high?

18. A farsighted person sees distant objects clearly (rays assumed parallel) with eyeglasses whose lenses have a power of 1.35 diopters. What power lenses should be prescribed to enable him to read without strain when the printed page is 40 cm away?

19. Light strikes a plane of ordinary crown glass at an angle of incidence of 30°. What is (*a*) the angle of reflection? (*b*) the angle of refraction?

20. A double-convex lens is used in a projector for colored slides. If the slides are placed in the carrier 20 cm behind the center of the lens and the image is sharp on a screen 9 m away on the other side of the lens, what focal-length lens must be used? If the slide is 5 by 5 cm, what will be the size of the image on the screen?

21. An object is 4 in. high. What type of mirror would have to be used to obtain a real image 1 in. high 100 in. from the object? What is the focal length of the mirror which will do this? (HINT: Let $D_i = x$ and $D_o = 100 + x$ in Eq. (22.4); solve for D_i and D_o and then use Eq. (22.3).)

22. A plano-convex lens is to be made from glass whose refractive index is 1.57. A focal length of 75 cm is desired. What must be the radius of curvature of the convex face of the lens?

23. In aerial mapping a camera uses a lens with a 50-in. focal length. How high must the airplane fly in order to photograph a strip of terrain 1 mi long so that its image will fit exactly on the filmstrip, which is 10 in. long?

24. A ray of monochromatic light strikes the first face of a prism at an angle of 30° with the normal to that face. The vertex angle of the prism is 50°, and the refractive index of the prism material is 1.55. Find the total angle through which the ray is bent in passing through the prism and emerging into the air again.

23 Dispersion, Polarized Light, and Optical Instruments

In Chaps. 21 and 22 light has been studied as a succession of waves with many frequencies. Imaginary lines, called rays, were used to represent graphically what happens when light waves fall on reflecting and refracting surfaces. Rules for the formation of real and virtual images have been developed.

In this chapter we apply this knowledge to study the operation of such common optical instruments as the human eye, the camera, and magnifiers, projectors, telescopes, microscopes, and spectroscopes.

23.1 Dispersion by a Prism

In a vacuum all colors of light travel with the same speed, 3×10^8 m/sec or 186,000 mi/sec (approx.). In transparent media, such as water or glass, they travel at different and considerably lower speeds. In air the speeds of the various colors are very nearly the same as their speeds in a vacuum.

Light in passing from air to a glass prism is refracted toward the normal because of the reduced speed of light in glass (see Sec. 22.13). However, light consisting of shorter wavelengths travels more slowly in glass than light consisting of longer wavelengths, which results in the spreading out of the incident rays of light into the various constituent wavelengths (colors). Thus, when a narrow beam of light travels through the prism, the shorter (violet) wavelengths are refracted the most, while the longer (red) wavelengths are refracted the least (Fig. 23.1). The separation of light into its component colors is called *dispersion*. The angular spread ϕ of all the colors is called the *angle*

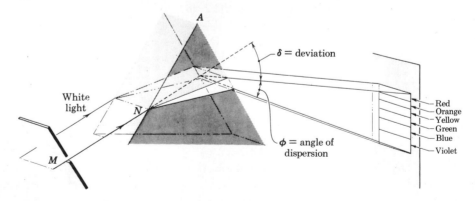

Fig. 23.1 The hues in the spectrum blend imperceptibly into one another, but they can be grouped into six principal colors.

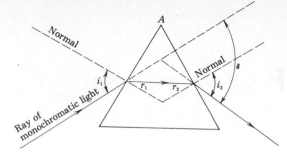

Fig. 23.2 Angle of minimum deviation occurs when $i_1 = i_2$.

of dispersion. The band of colors produced by dispersion is called a *spectrum.* Figure 23.9 will show the spectrum of the sun, for example. Spectra are used to study the structure of atoms and molecules, for chemical analyses, and for the study of the composition of the stars in our universe.

The angle δ (delta) between the entering ray and the emerging ray of a given color is termed the *angle of deviation* for that color. Deviations for a given color will depend upon the index of refraction of the prism, the angle of incidence of the ray, and the size of the (apex) angle A of the prism. The angle of deviation for a given color will have its smallest value when the angle of incidence i_1 at the first surface separating air and glass, and the angle of refraction i_2 at the second surface separating glass and air are equal (Fig. 23.2). It can be proved that, when the deviation is a minimum,

the index of refraction μ is equal to

$$\mu = \frac{\sin[(A + \delta)/2]}{\sin A/2} \qquad (23.1)$$

Equation (23.1) also holds reasonably well when the deviation is *near* the minimum.

The refractive indices for various wavelengths of light for some common substances are given in Table 23.1. Note the relatively high indices of refraction for diamond and the low indices for ice. Media that produce large deviations also produce large dispersions but the two properties are not proportional to each other.

Illustrative Problem 23.1 In a glass prism with angle $A = 60°$, the angle of minimum deviation for light of blue color is found to be 40°. What is the index of refraction of the prism?

Solution Substitute known values in Eq. (23.1):

$$\mu = \frac{\sin[(A + \delta)/2]}{\sin(A/2)}$$

$$- \frac{\sin[(60° + 40°)/2]}{\sin(60°/2)}$$

$$= \frac{0.766}{0.500}$$

$$= 1.53 \qquad\qquad ans.$$

Illustrative Problem 23.2 Find the angle of dispersion from the red rays of the spectrum to

Table 23.1 Refractive indices of various wavelengths of light for some transparent media

	Color and Wavelength					
Substance	*Red* 6700 Å	*Orange* 6200 Å	*Yellow* 5700 Å	*Green* 5200 Å	*Blue* 4700 Å	*Violet* 4100 Å
Crown glass	1.515	1.522	1.523	1.526	1.532	1.538
Diamond	2.410	2.415	2.417	2.426	2.444	2.458
Flint glass	1.624	1.626	1.627	1.632	1.640	1.651
Ice	1.306	1.308	1.309	1.311	1.314	1.317

the blue rays produced by a flint-glass prism if the refracting angle is 60°.

Solution From Table 23.1, we find $\mu_r = 1.624$, $\mu_b = 1.640$. Solve Eq. (23.1) for $\sin \frac{1}{2}(A + \delta_r)$, to get

$$\sin \tfrac{1}{2}(A + \delta_r) = \mu \sin A/2$$

$$\sin \tfrac{1}{2}(60 + \delta_r) = 1.624 \sin \tfrac{1}{2}(60°)$$

$$= 0.812$$

Referring to Appendix III for the angle whose sine is 0.812, we get

$$\tfrac{1}{2}(60° + \delta_r) = 54.3°$$

and

$$60° + \delta_r = 108.6°$$

Then

$$\delta_r = 48.6°$$

In like manner,

$$\sin \tfrac{1}{2}(60° + \delta_b) = 1.640 \sin 30°$$

$$= 0.820$$

$$\tfrac{1}{2}(60° + \delta_b) = 55.1°$$

$$\delta_b = 50.2°$$

$$\phi = \delta_b - \delta_r = 1.6° \quad \text{(see Fig. 23.1)} \quad ans.$$

The question may well be raised as to why the refracted beams of light in lenses studied in Chap. 22 were not spread out into spectra. The answer lies in the fact that the light beams used in Chap. 22 were considered to be much wider than those which will produce spectra. Actually,

dispersion does occur with lenses, but, because of the width of the beam, the colors of one part of the beam overlap the colors of the other parts of the beam. The result is a predominantly white color. At the edges of the lenses there is evidence of dispersion, but it is so dim compared to the bright central image that it is usually unnoticed.

23.2 Prism Combinations

If we position two identical prisms of the same material in close proximity, as in Fig. 23.3, each prism will produce an equal but opposite deviation and dispersion. The first prism will spread the colors out while the second prism bends them back together. The result is that a white spot appears on a screen near the second prism.

A similar effect can be produced by using a crown-glass prism combined with a flint-glass prism with a smaller central angle, as shown in Fig. 23.4, but since the deviations produced by the two prisms are not the same, the *emerging* white light will not be parallel to the *incident* light. Since the deviation produced by the second prism is opposite to that produced by the first prism, the net deviation of the combination is the difference between the two. A prism that deviates light without dispersion is called an *achromatic* prism.

If the angle of the flint glass is increased until the deviation it produces is equal to that produced by a crown-glass prism and they are placed together as shown in Fig. 23.5, the white light is dispersed into a spectrum but its direction is unchanged.

Fig. 23.3 The dispersion produced by the first prism is nullified by the second, and a white spot is observed on the screen.

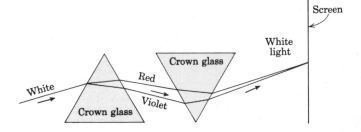

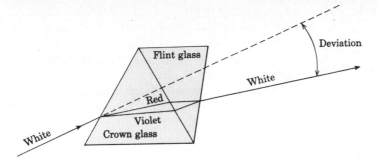

Fig. 23.4 Deviation without dispersion. The two prisms transmit white light with a net deviation equal to the difference between the two deviations produced by the separate prisms.

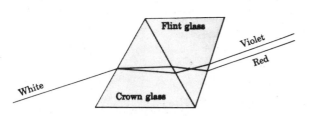

Fig. 23.5 A prism combination which produces dispersion without deviation.

23.3 Chromatic Aberration

When white light passes through a single converging lens, it is dispersed in much the same way that it is when passing through a prism. Since violet light is refracted most, it comes to a focus nearer the lens than red light does (Fig. 23.6). For this reason, the image formed by a single lens

with white light is not sharply defined. For example, the images formed by a pair of cheap binoculars often are surrounded by a faint halo of color, usually red or blue. The focusing of light of different colors at different locations is called *chromatic aberration*.

The presence of chromatic aberration in the early telescopes led Newton to invent the reflecting telescope (see Fig. 23.20) in which all wavelengths are reflected to the same focal point.

To correct a lens for chromatic aberration a converging lens of crown glass is combined with a diverging lens of flint glass, as illustrated in Fig. 23.7.

With this arrangement the dispersion of white light into its spectrum caused by one of the lenses is partly canceled by the other. Lenses made in this way are called *achromatic lenses*, but because the dispersion and deviation of the lens for each color are not exactly proportional, the

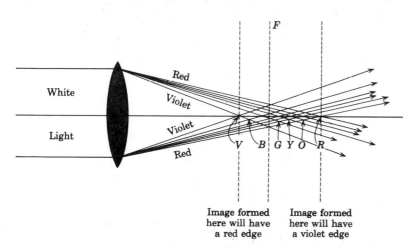

Image formed here will have a red edge

Image formed here will have a violet edge

Fig. 23.6 Diagram illustrating chromatic aberration (exaggerated).

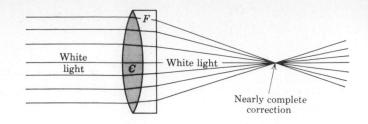

Fig. 23.7 Achromatic lens combination of a double convex crown glass lens and a plano-concave flint glass lens.

correction is not perfect. The best achromatic lenses are made of more than two components.

Illustrative Problem 23.3 A double convex crown-glass lens whose radii of curvature are 12 cm for each surface is combined with a flint-glass lens in contact with the crown-glass lens at the surface of the 12-cm radius (see Fig. 23.7). What should be the radius of the second surface of the flint glass to make the combination achromatic for red and blue light?

Solution Use the lens formula. For the crown-glass lens, $\mu_r = 1.515$, $\mu_b = 1.532$, $R_1 = R_2 = 12$ cm.

$$\frac{1}{f_r} = (\mu_r - 1)\left(\frac{1}{R_1} + \frac{1}{R_2}\right)$$

$$= (1.515 - 1)\left(\frac{1}{12 \text{ cm}} + \frac{1}{12 \text{ cm}}\right)$$

$$= 0.08583 \text{ cm}^{-1}$$

$$\frac{1}{f_b} = (1.532 - 1)\left(\frac{2}{12}\right) = 0.08867 \text{ cm}^{-1}$$

The difference of the powers of the first lens for blue and red light is

$$(0.08867 - 0.08583) \text{ cm}^{-1} = 0.00284 \text{ cm}^{-1}$$

For the flint glass (diverging) lens, $\mu_b = 1.640$, $\mu_r = 1.624$, $R_1 = -12$ cm.

$$\frac{1}{f_b} - \frac{1}{f_r} = [(\mu_b - 1) - (\mu_r - 1)]\left[\frac{1}{R_1} + \frac{1}{R_2}\right]$$

$$= (1.640 - 1.624)\left(\frac{1}{-12 \text{ cm}} + \frac{1}{R_2}\right)$$

$$= 0.016\left(\frac{1}{-12 \text{ cm}} + \frac{1}{R_2}\right)$$

To produce an achromatic lens, the differences between the crown-glass and flint-glass lenses must be equal in magnitude and opposite in sign. Thus

$$0.016\left(\frac{1}{-12 \text{ cm}} + \frac{1}{R_2}\right) = -0.00284 \text{ cm}^{-1}$$

and

$$\frac{1}{R_2} = \frac{-0.00284}{0.016} \text{ cm}^{-1} + \tfrac{1}{12} \text{ cm}$$

$$= -0.0942 \text{ cm}^{-1}$$

$$R_2 = -10.6 \text{ cm} \quad \text{(concave surface)} \quad ans.$$

23.4 The Spectroscope

The principle of dispersion is used extensively in the prism spectroscope (Fig. 23.8), which is used to produce and examine a spectrum. It consists essentially of four main parts: a *collimator*, which has a small adjustable slit *S* at one end, placed at the principal focus of a convex lens L_1 at the other end; a *prism*; a *circular scale* with a vernier; and a *telescope*, whose objective lens L_2 forms real images of the slit *S*. Since the slit is at the principal focus of the collimator, the rays are made parallel before they reach the prism. The prism refracts and disperses the various colors of the light source. The spectrum, as it emerges from the prism, is magnified by the lenses of the telescope until each wavelength of the source of light produces an image of the slit at the focal point of the objective lens. In this manner all the constituent colors which come from the source of light can be separated and analyzed. If the telescope is replaced by a camera so that a photograph of the spectrum can be made, the instrument is called a *spectrograph*.

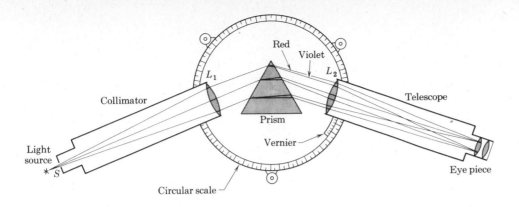

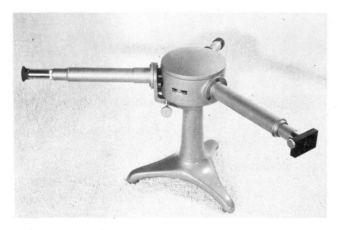

Fig. 23.8 A prism spectroscope. (Gaertner Scientific Corp.)

A spectroscope shows that different sources of light produce spectra that vary greatly. Each light source produces a unique spectrum which is characteristic of it. Spectra can be classified into five groups:

1. Continuous emission spectra
2. Bright-line spectra
3. Continuous absorption spectra
4. Line absorption spectra
5. Band spectra

23.5 Continuous Emission Spectra

When a spectrum is produced by light from a hot glowing solid, a glowing liquid, or glowing gas under high pressure, a continuous band of color from red to violet is observed. The spectrum of an incandescent lamp and that of a carbon arc are of this type and are known as *continuous spectra*. The intensity of the colors depends upon the temperature and upon the hot body itself. While the spectrum of tungsten will be brightest in the red and orange region, others may be brightest in some other region.

As the temperature of a given body increases, its spectrum will first be brightest in the red and orange region, with the brightness gradually shifting toward the blue and violet end. A heated iron poker will first glow red, then change color to orange, yellow, and on to white. It will appear white when all the colors are emitted.

A study of the spectrum of a star will reveal its temperature even though it may be thousands of light-years away.

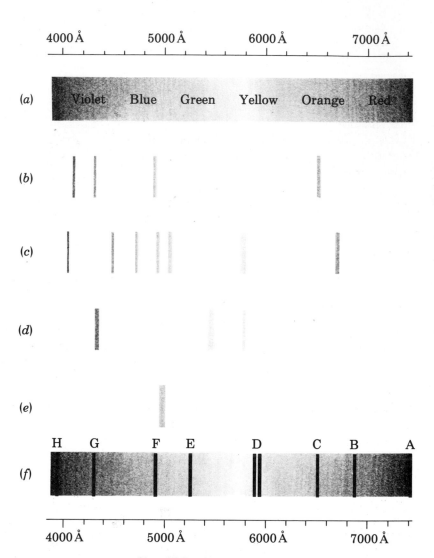

Fig. 23.9 Continuous, emission, and line spectra. (*a*) Continuous emission spectrum. (*b*) Bright line spectrum for hydrogen. (*c*) Bright line spectrum for helium. (*d*) Bright line spectrum for mercury. (*e*) Bright line spectrum for sodium. (*f*) Line absorption in the sun's spectrum (Fraunhofer lines).

23.6 Bright-line Spectra

When vaporized under moderate or low pressure, each of the chemical elements produces a spectrum that differs from that emitted by any other element. When the narrow slit of a spectrograph is illuminated by atoms of glowing vapors, a number of bright lines appear on the photographic plate in place of a continuous spectrum. Bright-line spectra of various elements can be seen in Fig. 23.9. Since each element has a unique bright-line spectrum, the spectrum becomes the "fingerprint" of the element. One way of identifying the elements present in a bit of material is to vaporize some of it and study the spectrum of the light emitted.

Bright-line spectroscopic analyses are used in criminology, archeology, medicine, and engineering. The fundamental metric unit of length, the meter, is defined as 1,650,763.73 wavelengths of a particular line in the spectrum of krypton (see p. 23).

The name *line spectra* is derived from the fact that a narrow slit in the collimator is used and its image is a line. If a small circular opening had been used in the collimator, a disk image would appear in place of each line.

23.7 Continuous Absorption Spectra

When white light is passed through various transparent solids or liquids and then examined in a spectroscope, some of the colors present in the spectrum of white light are often found to be missing. A common example occurs when white light is allowed to pass through a colored glass and then to the prism. The missing colors may cover a wide band of wavelengths. For example, a piece of red glass absorbs all visible light except the red, and a magenta-colored glass absorbs the central part of the spectrum.

23.8 Line Absorption Spectra

When white light is passed through a solid or liquid, we get an absorption spectrum consisting of one or more black bands in an otherwise continuous spectrum. However, when white light is passed through a gas or vapor, we obtain a line absorption spectrum consisting of individual black lines interspersed throughout the continuous spectrum.

The composition of the sun has been studied by examining its spectrum. Since the main body of the sun is very hot and consists of incandescent gases, it should produce a continuous spectrum. However, if the spectrum of the sun is examined in the spectroscope, it will be found that certain wavelengths are missing in the form of black lines distributed throughout the spectrum.

In the early part of the nineteenth century, Joseph Fraunhofer (1787–1826), a Bavarian optician, counted several hundred dark lines in the sun's spectrum and labeled eight of the most prominent lines by the first letters of the alphabet. The strongest of these are called *Fraunhofer lines* (see Fig. 23.9).

Scientists concluded that the Fraunhofer lines indicate that the sun is surrounded by an atmosphere of cooler gases that absorb some of the wavelengths coming from the main body of the sun. A gas absorbs the very same wavelengths of light that it emits in its own spectrum when it is incandescent. Hence, the dark absorption lines tell us which gases are doing the absorbing. For example, two Fraunhofer lines, called the D-lines, appear in the yellow part of the sun's spectrum exactly where two lines of the emission spectrum of heated metallic sodium vapors are normally found. This would indicate that sodium vapor is present in the sun's atmosphere. As a matter of interest, in 1868 scientists were able to identify in the sun an element unknown at the time. They called it *helium*. More than 25 years later the element was first discovered on earth.

23.9 Band Spectra

All the line spectra discussed thus far arise from single free atoms of a gas. Molecules of two or more atoms also give rise to spectrum lines. A *band spectrum* is produced by glowing molecules of a gas. The bands turn out to be a series of many bright lines very close together. A chemical compound can be identified by studying its band spectrum.

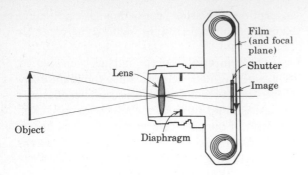

Fig. 23.10 Essential features of a simple camera.

OPTICAL INSTRUMENTS

The principles of light studied this far find wide applications in a variety of optical instruments. The number of such instruments is much too large to discuss in an introductory physics text. We shall, however, describe the more important features of cameras, spectacles, projectors, telescopes, binoculars, and microscopes.

23.10 The Camera

The simplest photographic camera employs a single lens unit, as illustrated in Fig. 23.10. It is essentially a lightproof box with a converging lens system at one end and a light-sensitive film at the opposite end. When a shutter is opened for a fraction of a second, the film receives a real, inverted image formed by the lens. An adjustable diaphragm controls the aperture and permits the intensity of the light entering the camera to be varied to suit the shutter speed and the film used. The amount of light reaching the film depends upon the time the shutter is open and the effective area of the lens aperture (controlled by the diaphragm). The lens aperture (effective area) is given as a fraction of the focal length. An $f/8$ lens setting means that the diameter of the aperture is one-eighth of the focal length of the lens. Thus, the f-number of a lens is equal to its focal length F divided by its effective diameter D:

$$f\text{-number} = F/D \tag{23.2}$$

The amount of light reaching the film is proportional to the square of the diameter of the lens

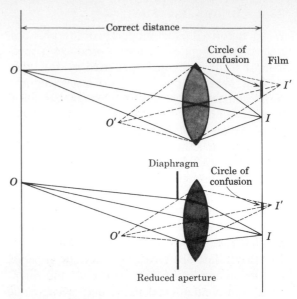

Fig. 23.11 Blurring of image is lessened by reducing the effective aperture.

opening and is inversely proportional to the square of the focal length of the lens. Thus an $f/8$ lens will let four times as much light reach the film as an $f/16$ lens and only one-fourth as much light as an $f/4$ lens. We say that an $f/4$ lens is "4 times as fast" as an $f/8$ lens.

Illustrative Problem 23.4 A photographer has set his camera lens with an $f/8$ stop and a correct film-exposure time of $\frac{1}{60}$ sec. What exposure time should he use to get the same amount of light exposure if he sets the diaphragm of the lens for $f/5.6$?

Solution The amount of light passing through the lens is proportional to the square of the diameter of the aperture. The time of exposure is directly proportional to the amount of light received on the film. Thus,

$$\frac{A_1}{A_2} = \frac{d_1^2}{d_2^2} = \frac{(8.0)^2}{(5.6)^2} \approx \frac{2}{1}$$

Or twice the amount of light is available through $f/5.6$ as was available through $f/8$.

Let t_1 be the original $\frac{1}{60}$-sec exposure time, and t_2 the new required exposure time. Then

$$\frac{t_2}{t_1} = \frac{A_2}{A_1} = \frac{1}{2}$$

and $\qquad t_2 = \dfrac{t_1}{2} = \dfrac{1}{60 \times 2} = \tfrac{1}{120}$ sec $\qquad$ *ans.*

When the lens of a camera is focused for a certain distance, only objects at that distance from the camera will be in sharp focus on the film. Points at other distances from the camera will focus on the sensitive film as a small blurred circle called a *circle of confusion*. The farther an object is from the accurately focused distance, the larger the circle of confusion will be. The range of distances of an object from the camera lens that will produce acceptably sharp images on the focal plane of the film is called the *depth of field*.

The size of the circle of confusion can be decreased by reducing the effective aperture of the lens (see Fig. 23.11), but it should be apparent that increased sharpness of the image is gained at a cost of decreased illumination of the image on the film.

Cheaper cameras use only a single converging lens, which means that the camera will be subject to all the discrepancies and aberrations discussed in Chap. 22 for converging lenses. More expensive cameras correct much of the spherical and chromatic aberrations by using three to five lenses.

23.11　The Eye

The human eye is like a camera. Optically it consists of a light-tight enclosure having an elaborate lens system at one end and a light-sensitive "film" of nerve fibers at the other. The eyeball

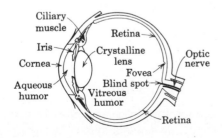

Fig. 23.12　Cross section of human eye.

(Fig. 23.12) is a nearly spherical opaque chamber of about 1.5 cm diameter filled with a transparent semifluid substance, called the *vitreous humor*. Light enters through a rather firm transparent tissue called the *cornea* into a clear fluid known as the *aqueous humor*. Behind the cornea is a circular diaphragm, the *iris*, with a central hole called the *pupil*. The iris has muscles which can shrink or dilate the diameter of the pupil as the light intensity increases or decreases. The iris contains the pigment noticeable as the color of the eye. Immediately behind the iris is the *crystalline lens* and the vitreous humor. The lens, which is biconvex, is composed of microscopic glassy fibers which can change shape by sliding over each other under the control of the *ciliary muscle*. In the interior of the back wall of the eye is a light-sensitive surface, the *retina*. Attached to the retina is the *optic nerve*, which carries information from the retina to the brain.

The refracting media of the eye are the cornea, the aqueous humor, lens, and the vitreous humor. The indices of refraction of the various transparent portions of the eye range from 1.34 to 1.44. The effect of all the refractions is to form an image of external objects on the retina, as discussed in Chap. 22. The principal bending of light occurs at the cornea, since its radius of curvature is much less than the remainder of the eyeball. The cornea provides roughly two-thirds of the refraction required to focus the image of a distant object on the retina. When a normal eye is viewing intermediate and distant objects, it is relaxed, and the lens has its thinnest shape, with a converging power of about 20 diopters (see Sec. 22.23). To bring the image of a near object in focus on the retina, the ciliary muscles contract and thicken the lens so that the radii of curvature of the lens decrease. Thus, the focal length of the lens is decreased and the image is focused on the retina. The ability of the eye to adjust its focal length is called *accommodation*.

The ability of the eye to bring the image of an object upon the retina (accommodation) decreases as a person gets older. In general, the 60-year old person cannot read a book held at a convenient distance of about 40 cm (16 in.).

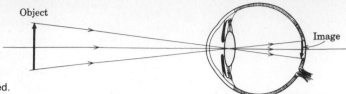

Fig. 23.13 Retinal images are real and inverted.

Thus, the extra power required must be provided by convex lenses in front of the eye (see Sec. 23.12).

The normal eye is most relaxed when it is focused for parallel rays, i.e., for distant objects, but an object to be studied in detail must often be brought closer to the eye. The reason, as we discussed in Chap. 22, is that the closer the object to the eye, the larger the image formed on the retina. For normal eyes, vision is most distinct when they are focused on objects about 25 cm (10 in.) away. Prolonged observations at distances of 25 cm or less result in considerable fatigue and eyestrain.

It is interesting to note that the image on the retina is upside-down (see Fig. 23.13). It is the brain that interprets the image as being erect.

23.12 Defects of Vision

The defective eye may be nearsighted, farsighted, or astigmatic. If the image formed by a distant object falls in front of the retina, as shown in Fig. 23.14a, the eye is *nearsighted*, or *myopic*. This may occur because the eyeball is too long, because the lens has too short a focal length, or for some other reason. In such cases, the ciliary muscle lacks the power to reduce the curvature of the eye lens sufficiently to focus the rays from a distant object on the retina. Only objects close to the eye can be focused sharply on the retina; all others appear blurred.

Nearsighted, or myopic, eyes can be corrected by using a suitable diverging lens (see Fig. 23.14a).

The eyes of a *farsighted*, or *hyperopic*, person form images behind the retina, as shown in Fig. 23.14b. This defect arises because the eyeball is too short or the lens has too long a focal length. As an object is moved farther away from the eye,

the image moves nearer the retina. Hence, the farsighted person can see distant objects clearly, while images of nearer objects appear blurred.

Farsighted vision can be corrected by using a suitable converging lens in front of the eye. As an object is brought still closer to the hyperopic eye, the same eye will require the use of a converging lens of still greater power. Bifocal spectacles have lenses with different focal lengths in the upper and lower halves of the glasses. Trifocal spectacles are quite common today.

Another common defect of the eye is *astigmatism*, which occurs when at least one of the refracting surfaces (cornea or lens) of the eye is not spherical. Since such an eye has varying focal lengths in different planes, the image formed may be distinct in one direction and blurred in another (see Fig. 23.15). Astigmatism can be corrected by glasses that have greater curvature in the plane in which the cornea or lens has less curvature.

Today millions of people are wearing contact lenses made of various new soft and pliable plastics. These lenses are worn to correct such conditions as nearsightedness, farsightedness and astigmatism. Athletes wear them in sports in which eyeglasses are unsafe.

No matter which kind of defect the eye may have, it should be apparent that it will see more clearly in bright light, even without glasses, because as the pupil of the eye becomes smaller automatically, due to the bright light, the size of circles of confusion become smaller (see Sec. 23.10).

23.13 The Projector

The projector for slides and films consists optically of the *illuminating system* and the *projection lens* (Fig. 23.16). The illuminating system

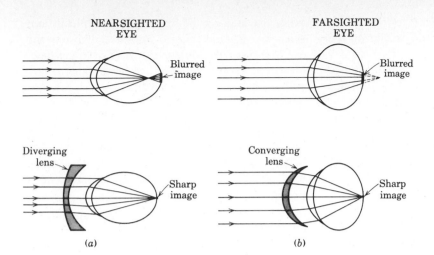

NEARSIGHTED EYE

FARSIGHTED EYE

Blurred image

Blurred image

Diverging lens

Converging lens

Sharp image

Sharp image

(a)

(b)

Fig. 23.14 Eyeglasses correct defects of vision. (a) Myopia (nearsightedness) and (b) hyperopia (farsightedness).

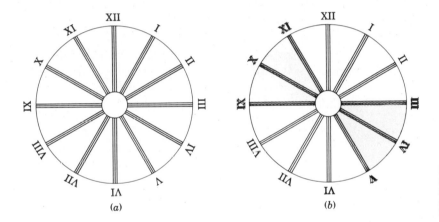

(a)

(b)

Fig. 23.15 Astigmatism. (a) Each set of parallels appears equally sharp to the normal eye. (b) To the astigmatic eye, the sets of radial lines differ in sharpness.

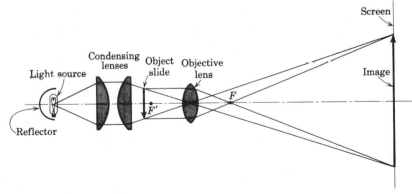

Screen

Condensing lenses

Object slide

Objective lens

Light source

Image

F

F'

Reflector

Fig. 23.16 Optical system of a projector for colored slides.

consists of a light source and a condensing pair of lenses which collects light from the source and concentrates it upon the film or slide. The slide is placed slightly beyond the focal length of the

Dispersion, Polarized Light, and Optical Instruments   **473** objective lens, which placement forms an enlarged inverted image of the slide or film on the screen.

Since all the light rays forming the image must come from the object (the slide or film), the object must be strongly illuminated if a bright image is to be projected. This is accomplished by having a powerful source of light with a reflector behind it and condensing lenses in front of the object to direct the rays through the slide.

In order to get an erect image on the screen, the slide must be placed in the projector upside-down, with left and right interchanged.

Motion-picture projection is essentially the same as slide projection, but mechanical means must be provided to change from one picture, or *frame*, to another many times a second. A rate of 18 frames per second is sufficient to produce a satisfactory illusion of motion. "Slow motion" effects are obtained by running the film in the camera at faster rates when the movie is being filmed.

23.14 The Simple Magnifier

If you looked at the print on this page from across a room, you probably would not be able to read it because the lenses of your eyes would form on the retina a very small image of the distant object. As the book is brought nearer and nearer the eye, the image gets larger and larger, until you can read the print. For the normal eye the print is most easily read at a distance of about 25 cm (10 in.) from the eye. If it were possible for the normal eye to accommodate to much nearer dis-

tances by making the lens of the eye thicker and thicker, you could bring an object very close to the eye and have the object enormously magnified. However, unless you are very nearsighted, you will see a blurred image if the object is brought closer than about 10 cm.

If a magnifying glass of short focal length is now placed before the eye, as shown in Fig. 23.17, the object can be brought much closer to the eye and can be observed in detail. If the object to be examined is placed a little nearer than one focal length to the lens, then an enlarged, erect, virtual image of the object can be seen. The magnifying glass is also called a simple microscope.

The linear magnification M of a simple microscope is the ratio of the image size I to the object size O:

$$M = I/O \qquad (23.3)$$

Using the properties of similar triangles, we can prove that $I/O = D_i/D_o$, where D_o is the object distance and D_i is the image distance. Then

$$M = D_i/D_o \qquad (23.4)$$

If we use Eq. (22.3) to solve for D_i/D_o, we get

$$\frac{1}{D_o} + \frac{1}{D_i} = \frac{1}{f}$$

Multiplying each term by D_i yields

$$\frac{D_i}{D_o} + 1 = \frac{D_i}{f}$$

or

$$\frac{D_i}{D_o} = \frac{D_i}{f} - 1 \qquad (23.5)$$

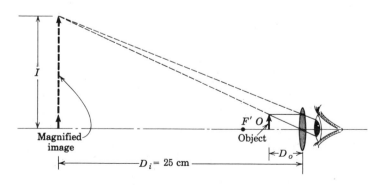

Fig. 23.17 A simple magnifier.

Using $D_i = -25$ cm (or -10 in.), the distance of most distinct vision for the normal eye, Eq. (23.5), becomes

$$\frac{D_i}{D_o} = \frac{-25 \text{ cm}}{f \text{ cm}} - 1 \qquad (23.6)$$

or

$$\frac{D_i}{D_o} = \frac{-10 \text{ in.}}{f \text{ in.}} - 1 \qquad (23.7)$$

Illustrative Problem 23.5   A converging lens has a focal length of 6.0 cm. If it is to be used as a simple magnifier, how far from the object should the lens be placed to produce a virtual image 25 cm from the eye? What is the magnification?

Solution   Solve Eq. (23.5) for D_o:

$$\frac{D_i}{D_o} = \frac{D_i}{f} - 1$$

$$= \frac{D_i - f}{f}$$

Inverting each term, we get

$$\frac{D_o}{D_i} = \frac{f}{D_i - f}$$

Multiplying through by D_i gives

$$D_o = \frac{D_i f}{D_i - f}$$

Substitute $D_i = -25$ cm and $f = 6.0$ cm in the above equation:

$$D_o = \frac{-25 \text{ cm} \times 6.0 \text{ cm}}{-25 \text{ cm} - 6.0 \text{ cm}} = 4.8 \text{ cm} \qquad ans.$$

$$M = \frac{D_i}{D_o} = \frac{-25 \text{ cm}}{4.8 \text{ cm}} = -5.2 \qquad ans.$$

The minus sign merely indicates that the image is virtual.

23.15   The Compound Microscope

When a simple magnifier is used to its best advantage, the object is placed just inside the focus of the lens. This will form an image at about 25 cm from the lens. The magnification M

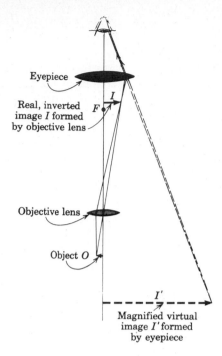

Fig. 23.18   Optical system (not drawn to scale) of the compound microscope.

produced is the ratio of the image distance, 25 cm, to the object distance from the lens, which is approximately equal to f, the focal length of the lens. Hence, the magnifying power of a simple microscope is equal to $25/f$ (if f is given in centimeters) or $10/f$ (if f is given in inches). Thus, a converging lens with a focal length of 5 cm can magnify an object $\frac{25}{5}$ or 5 times, and a converging lens with a focal length of 2.5 in. will have a magnifying power of 4.

To achieve a magnification greater than that obtainable with a single converging lens a compound microscope is used. It consists of a group of lenses acting as a single converging lens of very short focal length, called the *objective* lens, and a converging lens of moderately short focal length, called the *eyepiece*. Figure 23.18 shows the lens and ray diagram of a compound microscope.

The object is placed just beyond the principal focus of the objective lens; this position produces a real, somewhat magnified image I of the object. The image is then magnified again by the eyepiece, which acts as a simple microscope. The eyepiece produces an enlarged virtual image I' of the real image I.

The magnification M of a compound microscope is equal to the product of the magnifying power M_o of its objective and the magnifying power M_e of its eyepiece.

$$M = M_o M_e \qquad (23.8)$$

Illustrative Problem 23.6 A compound microscope has an objective lens of 7.50 mm focal length and an eyepiece of 30.0 mm focal length. What is the magnifying power of the microscope if the object is in sharp focus when it is 8.00 mm from the objective?

Solution Consider the objective lens alone and use Eq. (22.3):

$$\frac{1}{D_o} + \frac{1}{D_i} = \frac{1}{f}$$

$$\frac{1}{8.00 \text{ mm}} + \frac{1}{D_i} = \frac{1}{7.50 \text{ mm}}$$

$$D_i = 120 \text{ mm}$$

and

$$M_o = \frac{120 \text{ mm}}{8 \text{ mm}} = 15$$

Next consider the eyepiece alone. Let $D_i = -250$ mm, the distance for most distinct vision:

$$\frac{1}{D_o'} + \frac{1}{-250 \text{ mm}} = \frac{1}{30.0 \text{ mm}}$$

$$D_o' = 26.8 \text{ mm}$$

and

$$M_e = \frac{-250 \text{ mm}}{26.8 \text{ mm}} = -9.33$$

The total magnification is equal to

$$M = M_o M_e = (15)(-9.33)$$

$$= -140 \qquad \textit{ans.}$$

Again, the minus sign merely indicates a virtual image.

Most high-quality microscopes have a set of interchangeable eyepieces and/or a turret carrying three objectives, each of a different magnifying power. Turning the turret gives a variety of magnifying powers. Microscopes with magnifying powers of a little more than 2000 are quite common today. This is near the upper limit of magnification for optical microscopes.

For greater magnification, an instrument called the *electron microscope*, which employs electromagnetic lenses, can be used. Magnifications as high as 100,000 have been achieved with these instruments. (See Sec. 33.11).

23.16 Refracting Telescopes

The principle of the refracting telescope is similar to that of the compound microscope. The telescope, too, consists of an objective lens system and an eyepiece. The objective lens in this instrument, however, is a large converging lens of long focal length. A diagram of a small telescope is shown in Fig. 23.19. Parallel light rays from a point on a far-distant object enter the objective, and an image of this point is formed at I. In similar manner, parallel sets of rays from other points of the object form point images in the focal plane (through F_o) of the objective. Thus, if the distant object is an arrow pointing upward, the image will be a real and inverted arrow.

The eyepiece of the telescope magnifies the image formed by the objective. The eyepiece, which has a short focal length, is moved until the image I is just inside its focal plane, a plane through F_e. The eyepiece, used as a simple magnifier, leaves the final image I' inverted. For normal observations, the focal planes through F_o and F_e are the same, and light emerges from the eyepiece in "parallel" rays. Hence, the image I' will appear to be far off, "at infinity."

The magnifying power of a refracting telescope is defined as the ratio of the angle subtended at the eye by the final image I' and the angle subtended by the object itself. By applying the plane geometry of Fig. 23.19, it can be shown that the magnifying power is equal to the ratio of the focal lengths of the two lenses. Thus,

$$\text{Magnification } M = \frac{f_o}{f_e} \qquad (23.9)$$

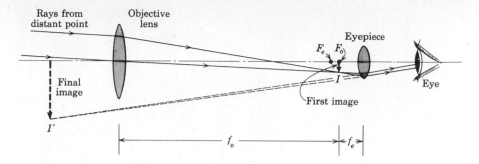

Fig. 23.19 The optical system of a refracting telescope.

where

$$f_o = \text{focal length of the objective}$$

$$f_e = \text{focal length of the eyepiece}$$

Illustrative Problem 23.7 The objective lens of a small telescope has a focal length of 120 cm, and the focal length of the eyepiece is 2 cm. Find the magnifying power of the telescope for distant objects.

Solution By Eq. (23.9),

$$M = f_o/f_e$$

Hence $\qquad M = 120 \text{ cm}/2 \text{ cm} = 60$

The magnification is 60. *ans.*

The image formed by the telescope just described is inverted. For the astronomer this poses no problem, but to observe objects on the earth the magnified image should be erect. Several methods are used to right the inverted image in a telescope. One is to use a third converging lens between the objective and the eyepiece in such a way that the image is again inverted before it is viewed in the eyepiece. Another is to insert an erecting prism in front of the eyepiece, as is done in prism binoculars. In the binoculars the image, after total reflection in two right-angle prisms, is restored to the upright position (see Fig. 22.27).

23.17 Reflecting Telescope

Most of the very large astronomical telescopes in the world are *reflecting telescopes*, in which the

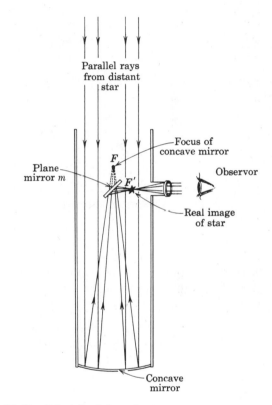

Fig. 23.20 Principle of the reflecting telescope.

main optical part is a large (parabolic) concave mirror. Reflecting telescopes have several advantages over refracting ones.

1. Large light-gathering devices are needed in both instruments. A large glass lens tends to distort under its own weight since it must be supported at its thin outer edges. Only one surface of a mirror need be ground, and the outer sup-

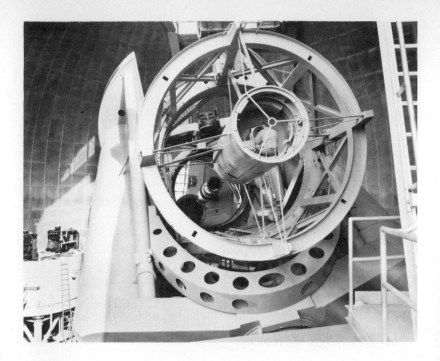

Fig. 23.21 The 200-in. Hale telescope, showing observer in the prime-focus cage, and reflecting surface of 200-in. mirror. (Photograph from the Hale Observatories.)

porting edges of the mirror are thicker, permitting a far more substantial support for the mirror than for the lens.

2. The reflecting telescope has the large, heavy optical part (the mirror) at the bottom of the instrument, while the refracting telescope has its large lens near the top of the instrument. This results in increased stability for the reflecting telescope.

The reflecting telescope consists essentially of a tube at the bottom of which is a concave mirror of long focal length (Fig. 23.20).

Parallel rays from a star enter the tube and are brought to a focus at F, where images can be viewed or photographed. Often a small plane mirror m reflects the convergent rays to a focus at F', out of the path of the incoming light. The small mirror removes only a small percent of the light from the star.

The Hale reflecting telescope on Mount Palomar in southern California has a concave mirror 200 in. in diameter. The observer or photographer sits inside the telescope at the focus F of the mirror and looks down into it (see Fig. 23.21). The largest reflecting telescope has a diameter of 6.007 m (236.5 in.) and is located in the Soviet Union.

POLARIZATION OF LIGHT

Perhaps one of the strongest indications that light behaves like a transverse wave is the phenomenon of *polarization*, which, until a few decades ago, was almost unknown outside scientific laboratories. Improved methods of polarizing light in use today mean that this laboratory curiosity has developed into an extremely practical technique in many industries. In order to understand what is meant by polarization, let us graphically represent an ordinary beam of light as composed of many transverse waves whose vibrations are along straight lines perpendicular to the direction of propagation. A beam of light consists of millions of such waves, each with its own plane of vibration. These vibrations occur in all planes around the axis of propagation (Fig. 23.22).

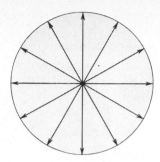

Fig. 23.22 Front view (schematic) of a beam of light showing waves vibrating in all planes.

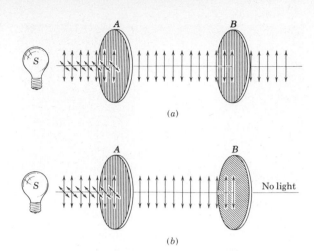

Fig. 23.23 Production and detection of polarized light.

23.18 Production of Polarized Light

By suitable interaction with matter, it is possible to eliminate components of the waves in all but one given plane or in parallel planes of vibration. Such a beam is said to be *plane-polarized*. There are several ways in which light can be polarized. When light is reflected from a transparent surface at a given angle, called the *polarizing angle*, the reflected light is polarized. Crystals such as quartz, Iceland spar, and tourmaline, when cut and arranged properly, are effective in polarizing light. The scattering of light by small particles of dust and smoke produces partially polarized light. However, each of these methods is difficult to control, is too expensive, absorbs too much light, or is limited to too small a light beam. Most of these limitations were removed with the introduction of the commercial polarizing film called *Polaroid* (not to be confused with the photographic process and camera of the same name). This film comes in thin sheets in which are embedded millions of ultramicroscopic needlelike crystals (sulfate of iodoquinine), which have all been made to align in one direction during the process of manufacture. When ordinary light passes through such a film, nearly all the components of the light vibrating in one direction are transmitted by the sheet, while nearly all the other components of the light are absorbed. Such sheets will transmit about 40 percent of the energy of the incident beam as polarized light

vibrating in parallel planes; the remainder of the light energy is absorbed.

23.19 Detection of Polarized Light

To the unaided human eye, polarized light will appear no different from unpolarized light. However, the presence of polarized light can be easily detected by the use of a second sheet of Polaroid. Let us set up a simple demonstration (Fig. 23.23) involving a source of light *S* and two polarizers *A* and *B*. Unpolarized light from *S* will be polarized by the *polarizer A*. If the polarizing axis of the second polarizer *B*, called the *analyzer*, is parallel to that of the polarizer *A* (Fig. 23.23a), the light beam polarized by the first sheet will pass through the second sheet without appreciable loss in intensity. If the analyzer is rotated through 90° about an axis parallel to the beam (Fig. 23.23b), the crystals of the analyzer will absorb the polarized light from *A*, and almost no light will be transmitted through it. If the analyzer is rotated another 90°, the polarized light from *A* will again be transmitted through *B*.

23.20 Applications of Polarized Light

The commercial production of improved and larger sheets of Polaroid has opened new indus-

Fig. 23.24 Strain pattern in a plastic model of meshing gears detected by polarized light. (Polaroid Corp.)

(a)

(b)

Fig. 23.25 (a) Picture taken without a filter. (b) Picture taken with polarizing filter.

trial possibilities. Polaroid glasses for motorists are available. A considerable amount of the glare of daylight driving comes from sunlight that is partially polarized as it is reflected from roads and other surfaces. By properly aligning the Polaroid lenses in the glasses, a great deal of this glare can be eliminated by the absorption of some of the reflected rays.

Industry has used polarized light in analyzing strains set up in transparent models of complex engineering structures (Fig. 23.24). Certain transparent materials (such as glass or Celluloid) possess polarizing qualities when they are subjected to mechanical strains. With proper polarizing equipment, these regions of strain in the models can be detected. In this manner it can be determined whether excessive strains might occur at critical points in the real structure.

Windows made of two polarizing panes, one of which can be rotated, could provide for curtainless windows and permit the variation at will of the intensity of light passing through the window. Photographers seeking to produce effective pictures of clouds in the sky commonly use Polaroid filters (Fig. 23.25) to eliminate some of the glare of the sky (which is composed of partially polarized light). For good color photography, Polaroid filters have an obvious advantage over the common yellow filter in that they do not cut out any part of the color spectrum.

23.21 Rotation of Plane of Polarization

Certain materials, such as sugar and cellophane, have the ability to rotate the plane of polarization while transmitting the polarized light. They are called optically active substances. This can be demonstrated by arranging an apparatus as shown in Fig. 23.26, in which light from a source S is passed through a polarizer P_1, then through a tube T with glass ends containing a sugar solution, and finally through an analyzer P_2 to the eye E. With the tube T removed, we turn the analyzer about an axis joining S and E until no light passes through P_2. If the tube with the sugar

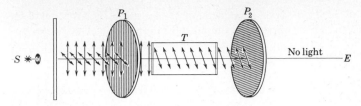

Fig. 23.26 Illustration of the principle of the saccharimeter.

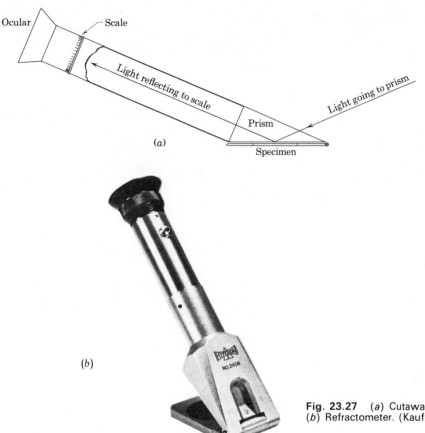

(a)

(b)

Fig. 23.27 (a) Cutaway diagram of juice refractometer. (b) Refractometer. (Kaufmann & Jost.)

solution is then placed between the two Polaroids, some light will once more be transmitted. This is true because the plane of vibration of the polarized light from P_1 has been rotated around the direction SE. The light can once more be extinguished by rotating the analyzer. The amount of rotation required for a given light source is proportional to the length of the tube T and proportional to the concentration of the solution. By adopting a standard length of the tube T, one can very readily determine the concentration of an optically active substance. This method is used quite extensively in determining the amount of sugar present in syrups. The instrument is called a *saccharimeter*. Grape growers use such a device to determine in the field the sugar content of their grapes (Fig. 23.27), and thus decide on the proper time to harvest them.

SUMMARY

White light, which is a mixture of a vast number of waves of different lengths, can be spread out into its constituent colors by passing through a prism. The process, called *dispersion*, is due to the unequal refraction of waves of different lengths when they pass obliquely into or out of a transparent medium. Short waves are refracted more than the longer waves.

When deviation is minimum for a prism, the index of refraction for the prism material is found by

$$\mu = \frac{\sin(A + \delta)/2}{\sin A/2} \qquad (23.1)$$

The spectroscope is an instrument used to study the constituent wavelengths of a given source of light.

Achromatic lenses minimizes chromatic aberration.

Spectra can be classified into five types: (1) continuous emission spectra, (2) bright-line spectra, (3) continuous absorption spectra, (4) line absorption spectra, (5) band spectra. Elements can be identified by examining their bright-line spectra.

The prominent lines in the absorption spectra of the sun are called *Fraunhofer lines*.

Chemical compounds can be identified by examining band spectra.

The *f*-number of a lens is the ratio of its focal length to its effective diameter.

$$f\text{-number} = F/D \qquad (23.2)$$

Common defects of the eye are nearsightedness, farsightedness, and astigmatism. Each defect can be corrected by using suitable spectacles.

The linear magnification M of a simple converging lens is equal to the ratio of the image size I to the object size O.

$$M = \frac{I}{O} \qquad (23.3)$$

Also $\qquad M = \dfrac{D_i \,(\text{image distance})}{D_o \,(\text{object distance})} \qquad (23.4)$

For most distinct vision the magnification of a simple magnifier equals

$$M = \frac{D_i}{D_o} = \frac{-25 \text{ cm}}{f \text{ cm}} - 1 \qquad (23.6)$$

or $\qquad M = \dfrac{-10 \text{ in.}}{f \text{ in.}} - 1 \qquad (23.7)$

Total magnification M of a microscope is equal to the product of the magnification of the objective M_o and the magnification of the eyepiece M_e.

$$M = M_o M_e \qquad (23.8)$$

The magnifying power of a refracting telescope is equal to the ratio of the focal length of the objective f_o and the focal length of the eyepiece f_e:

$$M = f_o/f_e \qquad (23.9)$$

The largest telescopes in the world are reflecting telescopes.

Polarization phenomena are among the strongest evidences that light is a set of transverse waves.

QUESTIONS AND EXERCISES

1. Will an increase in the temperature of a hot metal shift the brightness of its spectrum from the blue to the red region or from the red to the blue region? Give reasons for your answer.

2. Give two ways in which the speed of light in a vacuum differs from that in a transparent solid.

3. Distinguish between spherical aberration and chromatic aberration. Describe methods for minimizing each.

4. Given the magnifying powers of the objective lens and the eyepiece, how is the magnification of a compound microscope determined?

5. What is the principle on which the saccharimeter works?

6. Which of the following will decrease blurring of an image in a camera? (*a*) Decrease *f*-stop of the lens. (*b*) Decrease the aperture diameter. (*c*) Increase the time of exposure. (*d*) Decrease the time of exposure.

7. Which of the following methods of inserting a slide in a common projector will produce an erect image? (*a*) Upside down with right and left not interchanged. (*b*) Upside down with right and left interchanged. (*c*) Right side up with right and left interchanged. (*d*) Right side up with right and left not interchanged.

8. For which of the following is an achromatic lens designed? (*a*) To produce deviation without dispersion. (*b*) To produce dispersion without deviation. (*c*) To eliminate spherical aberration. (*d*) To produce a continuous spectrum.

9. The images on the retina of the eye are inverted. Then don't things appear upside down to the observer?

10. Do some research of the sextant and then describe how it is used.

11. Study library references on Huygens's Principle and then show how it can be applied to explain refraction of light.

12. Look up the meaning of diffraction and then describe what happens when light is diffracted.

PROBLEMS

1. A converging reading glass has a focal length of 8.0 cm. What is the magnification if the lens is to produce an image 25 cm from the eye?

2. The objective lens of a telescope has a focal length of 180 cm. The eyepiece has a focal length of 6.0 cm. What is the magnification of the telescope?

3. The objective lens of a refracting telescope has a focal length of 45 in. What should the focal length of the eyepiece be to produce a telescope with a magnification of 25?

4. The lens aperture for a portrait camera is 2.5 cm when the *f*-stop is 2.8. What is the focal length of the lens?

5. The lens of a camera has an effective diameter of 32 mm and a focal length of 180 mm. What is the *f*-number?

6. The objective lens of a telescope has a focal length of 150 cm and the eyepiece has a focal length of 2.5 cm. Find the magnification of the telescope.

7. A camera lens has a focal length of 40 mm. Calculate the image distances for the following object distances: (*a*) infinity, (*b*) 5 m, (*c*) 1 m, (*d*) 0.4 m.

8. An object is placed 4.0 cm from a concave lens of focal length -16.0 cm. Find (*a*) the image distance and (*b*) the magnification.

9. Where should the object be with respect to a $+6.0$-cm-focal length magnifier in order to produce a magnification of 8?

10. A double-convex lens, having a 7.5-cm focal length, is used as a simple microscope. If the lens is held close to the eye, where should the object be located in order to form an image 25 cm from the eye? What will be the magnification?

11. The lens of a slide projector has a focal length of 15 cm. It is set 6 m from a screen. If the height of the slide is 22 mm, what is the height of the image on the screen?

12. A slide projector uses 22- by 32-mm slides and projects the slides upon a screen at a distance of 12 m from the projector. What must the focal length of the lens of the projector be to produce images with dimensions 1.54 by 2.24 m?

13. A nearsighted person's greatest range of clear vision equals 80 cm. (*a*) What power spectacles must he wear to see distant objects clearly? (*b*) What would be the greatest distance he could see clearly if he wore spectacles with a power of -0.60 diopters?

14. A farsighted person cannot form sharp images of objects closer than 60.0 cm. (*a*) What power spectacles will enable him to see clearly objects 20.0 cm distant? (*b*) If he is fitted with lenses of 5.5-diopter power, what is the nearest distance which allows him to observe objects clearly?

15. A nearsighted person can see objects clearly only if they are not more than 75 cm distant. (*a*) What is the power of the weakest lens that will permit him to see distant ($D_o = \infty$) objects distinctly? (*b*) If he wears a pair of -1.25-diopter glasses, what will be his greatest distance of clear vision?

16. A microscope has an objective lens of focal length 1.25 cm and an eyepiece of focal length 3.60 cm. A specimen is placed 1.50 cm from the objective. What is the magnification if the virtual image formed by the eyepiece is 25 cm from the eye?

17. A small object is placed 15 cm in front of a converging lens of 10 cm focal length. A diverging lens of focal length -8 cm is placed 25 cm beyond the converging lens (see Fig. 23.28). Find the position, nature, and magnification of the image formed by the combination of lenses.

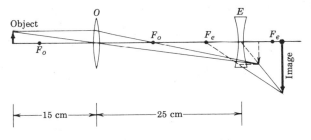

Fig. 23.28 Figure for Problem 23.17.

18. A 58° prism produces an angle of minimum deviation of 48° for a given color. Calculate the refractive index for that color.

19. A 62° prism has refractive indices of 1.572 for orange light and 1.597 for violet light. Find the angle of dispersion from the violet to the orange refracted rays.

20. A crown-glass lens with radii of $+10$ cm and $+8$ cm is combined with a flint-glass lens with one surface radius of -8 cm. What must be the radius of the second surface of the flint-glass lens to produce a lens that is achromatic for violet and red light?

PART 7 Electricity and Magnetism

Stator for a 930,000 kW generator (alternator) for a nuclear power plant. (Westinghouse Electric Corp)

24 Electrostatics

Uses for electricity are familiar to all of us, our high standard of living is in large measure attributable to the many applications of electricity.

Most of the common applications of electric energy involve electric charges in motion. However, we will begin our discussion of electricity by studying *electrostatics*, the science of stationary electric charges. Historically, electric charges at rest were discovered long before the discovery of electric charges in motion. The principles that govern electric charges at rest underlie those for current electricity.

Today we find an increasing number of devices that are based upon the principles of electrostatics. Among them are high-energy atom-smashing accelerators, electrostatic precipitators for removing dust and smoke particles from the air, electric copiers, and devices for reducing static charges on rolls of printing paper.

24.1 Electric Charges

The discovery of electric charges is credited to Thales of ancient Greece (about 600 B.C.). He observed that amber, when rubbed with a cloth, would pick up small fibrous materials such as straw or feathers. We can easily perform like experiments to illustrate electrification. First rub a hard-rubber rod briskly with cat's fur. Then place the rod near small pieces of paper, dry grass, or aluminum foil. The particles will "jump" toward the rod and cling to it (Fig. 24.1). The experiment can be repeated with a glass rod rubbed with silk. The same effect will be noted.

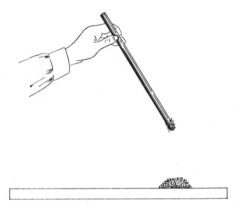

Fig. 24.1 Small particles of paper and straw will be attracted to a rod that has been electrified.

Electrical phenomena, such as lightning and the northern lights (aurora borealis), were known for thousands of years. At the beginning of the seventeenth century William Gilbert (1540–1603) of England announced that many substances could be electrified by friction or contact. Gilbert called these substances "electrics," which comes from the Greek word *elektron*, meaning amber. Our modern terms *electron* and *electricity* are derived from this Greek word.

Static electric charges frequently develop when leather belts travel over iron pulleys. Severe electric shocks can occur when paper in a printing press comes off the rollers. Such charges are also developed when hair is combed on a dry day with a hard-rubber comb.

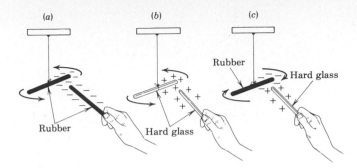

24.2 Positive and Negative Charges

There are two kinds of electric charge. To illustrate this, rub a rubber rod with cat's fur. Suspend it with a silk thread. Next bring the tip of a similarly charged rubber rod near one end of the fixed rod. The two charged rods repel each other as shown in Fig. 24.2*a*. Repeat the experiment, using glass rods rubbed with silk. These rods, too, will repel each other (Fig. 24.2*b*). When a charged glass rod is brought near the tip of a charged rubber rod however, the rods are attracted to each other (Fig. 24.2*b*). Evidently, the kind of charge on the rubber rod must be different from that on the glass rod.

Scientists have conducted similar experiments with many kinds of materials but they have never found more than the two kinds of charge. Benjamin Franklin (1706–1790) arbitrarily introduced the terms *negative* and *positive* to designate these two kinds of electric charge. We apply the term *negative* to substances that behave like the electrified rubber rod and the term *positive* to those that behave like the electrified glass rod. Thus, we can deduce the first law of interaction between electrified bodies:

Like electric charges repel each other, and unlike charges attract each other.

24.3 Electrical Structure of Matter

In Chap. 9 we learned that matter is made up of molecules. Molecules, in turn, are built of more elemental particles, called *atoms*. Atoms, in general, have quite complex structures. We will use a simplified, idealized structure to explain the electrical nature of matter.

Basically, each atom has a small, dense, positively charged mass at its center, called the *nucleus*. This nucleus is surrounded by larger and much lighter negatively charged particles called *electrons*. The nucleus consists of one or more *protons*. Each proton has a single unit of positive charge. The nucleus (except that of hydrogen) also contains one or more *neutrons*. A neutron has a mass very nearly that of a proton. It has no charge. The electrons can be idealized as revolving about the positive nuclei, moving in elliptic orbits much as planets revolve around the sun. The proton and neutron have masses about 1840 times that of an electron. An electron has a rest mass of 9.11×10^{-31} kg.

Each electron carries a single negative charge. The proton carries a single positive charge of the same magnitude. The atom is mostly empty space. Almost all the mass of the atom resides in the small, but extremely dense, central nucleus. The weight density of a proton is of the order of 8 billion tons per cubic inch! In the planetary model, the electrons revolve about the nucleus at tremendous speeds. The centripetal forces needed to keep the electrons in their orbits are provided by the electrical attraction forces between the negative electrons and the positive nucleus (Fig. 24.3).

Normally each atom exists in a neutral or uncharged state. It has an equal number of electrons and protons. The outermost electrons are less strongly bound to the atom than the inner ones are, and that is why it is the outer electrons

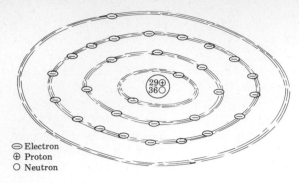

⊖ Electron
⊕ Proton
○ Neutron

Fig. 24.3 Planetary model of a copper atom.

that take part in chemical reactions between atoms. They are responsible for accumulation of charges on bodies. Atoms of different substances have varying degrees of affinity or attraction for outer electrons. (See Chap. 33 for further discussions on atomic structure.) For example, when hard rubber is rubbed with fur, the atoms of rubber have a greater attraction for electrons than the atoms of the fur do; hence they attract some of the nearby electrons from the atoms of cat's fur. Since the rubber atoms now have a greater number of (negative) electrons than (positive) protons, the rubber rod will have acquired a net negative charge. The atoms of the fur, having lost electrons, will have acquired a resultant positive charge.

A negatively charged body has an excess of electrons; a positively charged body has a deficiency of electrons.

When a glass rod is rubbed with silk cloth, the glass will have a positive charge and the silk an equal negative charge. This happens because the silk fibers have a greater affinity for electrons than the glass does. When two bodies are rubbed together to produce a given charge on one of the bodies, *an equal but opposite charge* will be left on the other. This concept can be expressed as the *law of conservation of charge.* In solid materials, only some of the outer electrons may have freedom to move. Other electrons are firmly fixed in their orbits (unless they are bombarded by highly energized charged particles).

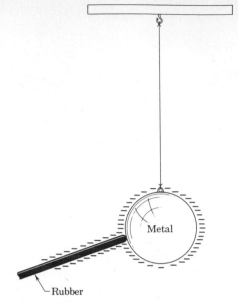

Metal

Rubber

Fig. 24.4 The metal sphere is charged by contact with the charged rubber rod.

24.4 Conductors and Insulators

Some substances can be charged by simply bringing a charged body in contact with them. For example, if a neutral metal sphere is suspended by a silk string and touched with a negatively charged rubber rod (Fig. 24.4), it can be shown that the sphere will acquire a negative charge. Some of the surplus electrons from the rubber will have left the rod and attached themselves to the sphere. If the uncharged sphere had been touched with a positively charged glass rod, the sphere would have become positively charged because some of its electrons would have been attracted away by the positively charged glass rod. This manner of charging a body is called *conduction.* Some substances will conduct electricity quite readily, while others will not. The metal in Fig. 24.4 will conduct the electricity quite readily, while the silk will not.

Those substances in which electrical charge flows quite freely are called *conductors.* Materials in which charges are not conducted freely are termed *insulators.* Insulators of porcelain or glass are used to prevent leakage of electricity from

Table 24.1 Insulators and conductors

Insulators	Poor Conductors	Conductors
Hard rubber	Dry wood	Metals
Dry air	Paper	The earth
Paraffin	Oil	Moist materials
Porcelain	Distilled water	Water solutions
Sulfur		of salts
Sealing wax		The human
Glass		body
Dry silk		
Bakelite and		
similar		
plastics		

wires and electric equipment. There is no sharp line dividing conductors and insulators. Table 24.1 gives a few common substances arranged according to their insulating ability.

(a)

Substances which are the most easily charged by friction are all classified as insulators. When an electric charge is developed on an insulator, it remains localized and does not flow freely throughout its length and leak away.

We will discuss a third type of conductor in Chap. 32 when we consider transistors. Such materials are called *semiconductors*. Silicon and germanium are examples of semiconductors. Although they have few electrons available for moving electric charges, the conductivity can be greatly increased by the addition of certain impurities or by changes in temperature.

24.5 The Electroscope

An instrument which can be used to detect the presence of an electric charge is called an *electroscope*. A common and very sensitive type is the *gold-leaf* electroscope. The essential features of the instrument are shown in Fig. 24.5a. It consists of a metal (copper or brass) rod that has a brass knob or plate attached at one end and one or two strips of aluminum or gold foil at the other. This rod passes vertically through the center of an in-

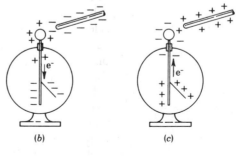

(b)   (c)

Fig. 24.5 Goldleaf electroscope.

sulating ring of sulfur or plastic at the top of the instrument. It is housed in a metal case with glass windows front and back.

When a charged rod is brought near the knob of the electroscope, the leaves will repel each other and diverge. When the charged rod is removed, the leaves will come together again. It will be noticed that *either a positive or a negative charge will cause the leaves to diverge.* When a negative charge is placed near, but not touching, the knob of the electroscope, some of the electrons (represented by e^- in the sketch) in the knob are repelled into the leaves. This makes both leaves predominantly negative (Fig. 24.5*b*) so that they repel each other. If a positive charge is placed near the knob, it will attract some of the electrons from the leaves to the knob. The result will be a predominant positive charge on the leaves, which again will cause them to repel each other (Fig. 24.5*c*). The greater the charge, the greater the divergence of the leaves; also, the closer the charging rod, the greater the divergence.

The electroscope (or any insulated conductor) can be charged by conduction. Touch the knob of an uncharged electroscope with a negatively charged rod (Fig. 24.6*a*). Electrons from the rod will enter the knob of the electroscope. In addition, the excess electrons left on parts of the rod that are not in contact with the knob will repel electrons down to the leaves. The leaves will diverge.

Remove the rod and with it the force of repulsion. The leaves will collapse slightly, but not completely. A residual charge of a somewhat lower density will be left on the electroscope.

The electroscope can be charged *positively* by using a positively charged glass rod in a similar manner (Fig. 24.6*b*).

Charge Acquired by Conduction

An insulated conductor charged by conduction acquires a residual charge of the same sign as that of the body touching it.

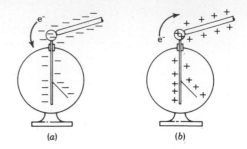

Fig. 24.6   Charging an electroscope by conduction.

24.6   Electrostatic Induction

We have considered the electrification of insulated bodies by friction and by conduction. There is still another way of placing a charge on a body. Figure 24.7 shows three brass spheres, each insulated by a glass support. Assume that sphere *A* has a positive charge, while spheres *B* and *C* are neutral or uncharged. If sphere *B* is brought near *A* (diagram *a*), the free electrons in *B* will be attracted to the surface nearest *A*. This will leave the one side of *B* charged negatively. The other will have an equal positive charge on it. Now bring sphere *C* in contact with *B* (diagram *b*). Some of the electrons of *C* will be attracted by the positive side of sphere *B* and will pass on to sphere *B*. When the two spheres are separated (diagram *c*), *B* will have a net negative charge, and sphere *C* will have an equal net positive charge. Now remove sphere *A*. The charges on *B* and *C* (negative and positive, respectively), since the spheres are conductors, will distribute themselves uniformly as shown in diagram *d*. Spheres *B* and *C* are said to have been charged by *induction*. The presence of sphere *A* in close proximity has *induced* a negative charge on *B* and a positive charge on *C*.

24.7   Electrical Ground and Charging by Induction

In Fig. 24.7 charges were induced on two spheres, but a charge can also be induced on a single sphere. This can be done by using the earth as a reservoir for charge.

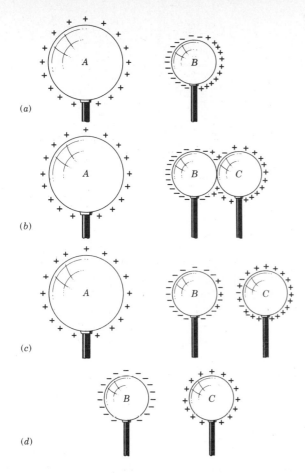

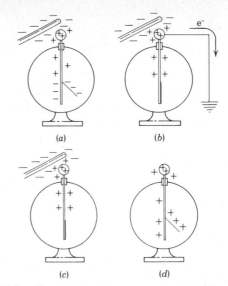

Fig. 24.8 Charging an electroscope by induction.

(a)   *(b)*

(c)   *(d)*

Fig. 24.7 Spheres *B* and *C* are charged by induction.

The earth is a large massive object and hence contains an extremely large number of atoms. Electrons can flow from or to the earth without appreciably changing the earth's relatively uncharged state. When a charged object is connected to the earth by a conductor, it is said to be grounded. *Grounding* is represented schematically by the symbol:

Charged objects when grounded to the earth will cause electrons to flow between the earth and the object in a direction which tends to neutralize the charge.

When a negatively charged rod is brought near an uncharged electroscope, electrons in the knob will be repelled to the leaves of the electroscope. The excess electrons will cause the leaves to diverge (Fig. 24.8*a*), and a deficiency of electrons will be left on the knob of the electroscope. Upon grounding the electroscope, free electrons will escape to the ground, and the leaves will converge (Fig. 24.8*b*). When the ground is removed, the positive charge on the knob will be held by the mutual attraction with the negatively charged rod (Fig. 24.8*c*), and the leaves will remain converged. Removal of the rod will then permit an even distribution of the charges in the electron-deficient electroscope (Fig. 24.8*d*), and the leaves will once more diverge. The residual charge on the electroscope will be positive.

The electroscope may also be given a net negative charge by a similar process using a positively charged object in place of the negatively charged rod.

Charge Acquired by Induction

The induced charge on a single object (or in using a ground) has a charge opposite in sign to that of the charging body.

Fig. 24.9 A proof plane.

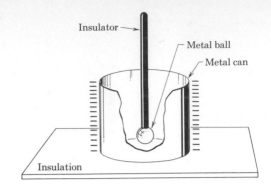

Fig. 24.11 Electric charges exist only on the outside of the metal can.

24.8 Distribution of Charge on Conductors

The distribution of charges on conductors of various shapes can be determined by means of a small metal disk or ball mounted on an insulating handle (Fig. 24.9) and known as a *proof plane*. Consider metallic surfaces of different shapes, each with a negative charge on them (Fig. 24.10). If we touch the proof plane to the surface of the sphere in several places and test the intensity of the charge by touching the proof plane to an electroscope, the charge on the sphere will be found to be uniformly distributed. When the same test is applied to surfaces like those shown in diagrams (*b*) and (*c*), it will be found that the charge density is greatest in the regions of greatest curvature. If the curvature is sharp enough, the surface density of the charge may be so great that the charge will actually leak off the sharp surface into the air. Such a leakage, called a *corona discharge*, is often observed in nature and in experiments conducted in electrical laboratories. It is used in electrostatic

precipitators and copiers. The air in the vicinity of corona discharges is said to be *ionized*.

On testing the intensity of charge at various places on a *hollow* metal can which has had a charge placed on it (Fig. 24.11), it will be found that *the charge exists entirely on the outer surface*. An electroscope placed in a metal cage which has a large charge on it will show that no charge exists on the inside. This suggests that a safe place in case of an electrical storm might well be inside an automobile. Even though lightning might possibly hit the car the charge is not likely to reach the occupants inside. (It is very unlikely that an auto or an airplane would be struck in the first place, for neither is grounded, and lightning takes the easiest path to the ground.) Protecting a region from electric fields in this manner is called *electrostatic shielding*.

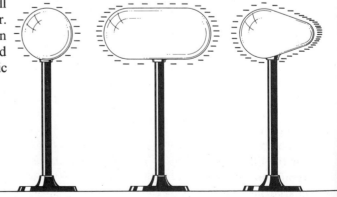

Fig. 24.10 Distribution of electric charge on various metallic surfaces.

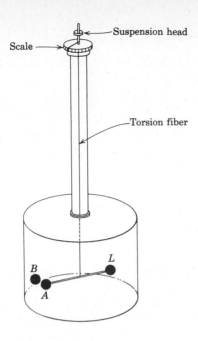

Scale

Suspension head

Torsion fiber

L

B

A

Fig. 24.12 Torsion balance measures forces of repulsion and attraction between charged spheres.

Shielding has numerous applications in radio and electronics equipment generally.

24.9 Coulomb's Law

The first recorded experimental investigations of the laws governing the forces of electrical attractions and repulsions between charged bodies was accomplished by the French physicist Charles Coulomb (1736–1806). He used a torsion balance (Fig. 24.12) which consisted essentially of a metal sphere A mounted on one end of a light rod and balanced by a load L at the other. The rod was suspended by a torsion fiber in which the restoring torque was proportional to the angle of twist.

Coulomb first charged the ball A. He then moved a second, equally charged, ball B near it. By turning the suspension head through an angle he was able to restore the system to its initial position. Knowing the modulus of twist for the fiber, he was able to compute the force of repulsion between the two charges. When the charge on ball B was doubled, the force between A

and B was doubled. Halving the charge on B halved the mutual force of repulsion. Similar tests were made with unlike charges and varying distances r between the two charges. The results of his experiments led Coulomb to establish the inverse-square law of electrostatic interaction.

Coulomb's Law

The force between two point charges is directly proportional to the product of their charges and inversely proportional to the square of the distance between them.

Coulomb's law can be expressed algebraically as

$$F = k\frac{Q_1 Q_2}{r^2} \qquad (24.1)$$

where

F = magnitude of force between two charges
Q_1, Q_2 = magnitudes of charges
r = distance between charges
k = constant of proportionality

The value of k depends upon the nature of the medium and the units used for the several quantities. Equation (24.1) assumes that the dimensions of the charged objects are small (point charges, actually) compared with the distance r between the bodies.

The unit of charge used in the SI system is called the *coulomb* (C). The formal definition of the coulomb will be given in Sec. 27.20, after we have studied electric currents. At present, we can give its value in terms of the electron charge:

$$1 \text{ C} = 6.242 \times 10^{18} \text{ electrons}$$

Therefore,

$$1 \text{ electron charge} = 1.602 \times 10^{-19} \text{ C}$$

Similarly, the charge on a proton is 1.602×10^{-19} C. The sign of Q is + for a proton, − for an electron.

If Q_1 and Q_2 are measured in coulombs, and if the distance r is measured in meters and the force in newtons, the value of k for free space (vacuum) in Eq. (24.1) has been found by experiment to be

$$k_0 = 8.988 \times 10^9 \text{ N-m}^2/\text{C}^2$$

For most calculations involving charges in vacuum or in air, k may be rounded off to 9.0×10^9. The value of k for air is only 0.06 percent less than that for a vacuum. Note that $8.988 \times 10^9 = 10^{-7} c^2$, where c is the speed of light in free space in meters per second.

One coulomb is that point charge which repels a like charge 1 m away in a vacuum with a force of approximately 9.0×10^9 N.

The force expressed by Coulomb's law is a *vector* quantity. Equation (24.1) gives only the *magnitude* of the force. The *direction* of the force is along the line joining Q_1 and Q_2. The plus and minus signs affixed to the charges have no mathematical meaning in Eq. (24.1). A positive value for F indicates a repulsive force between like charges; a negative value of F indicates an attractive force between unlike charges.

Illustrative Problem 24.1 Two insulated small objects have charges of 1.0 C and -2.0 C and are 50 cm apart. What will be the electrostatic force between them?

Solution Given:

$$Q_1 = 1.0 \text{ C} \qquad r = 50 \text{ cm} = 0.50 \text{ m}$$

$$Q_2 = -2.0 \text{ C} \qquad k = 9.0 \times 10^9 \text{ N-m}^2/\text{C}^2$$

To find F, use Eq. (24.1):

$$F = k\frac{Q_1 Q_2}{r^2}$$

$$= 9.0 \times 10^9 \text{ N-m}^2/\text{C}^2$$

$$\times \frac{1.0 \text{ C} \times (-2.0 \text{ C})}{(0.50 \text{ m})^2}$$

$$= -7.2 \times 10^{10} \text{ N} \qquad \qquad ans.$$

The negative sign indicates an attractive force between the charges.

The answer to Illus. Prob. 24.1 is more than 16 billion pounds of force! Obviously, the coulomb is a very large unit of charge, much too large for most electrostatic situations. The high-

Fig. 24.13 Illustrative Problem 24.2.

est charges that can be produced on bodies seldom contain more than a very small fraction of a coulomb. Frequently it is more convenient to work with the unit called the microcoulomb (μC):

$$1 \ \mu\text{C} = 10^{-6} \text{ C}$$

When two or more forces act on a charge, the resultant force is the vector sum of the separate forces.

Illustrative Problem 24.2 A charge of $-10 \ \mu$C is located 30.0 cm from a charge of $+2.5 \ \mu$C. A charge of $-15 \ \mu$C is placed on the line joining the two charges. It is 20 cm from the charge of $-10 \ \mu$C (Fig. 24.13). What is the force on the $-15 \ \mu$C-charge?

Solution The force F_1 on Q_2 due to Q_1 will be

$$F_1 = \frac{kQ_1 Q_2}{r^2} = 9.0 \times 10^9 \frac{\text{N-m}^2}{\text{C}^2}$$

$$\times \frac{(-10 \times 10^{-6} \text{ C}) \times (-15 \times 10^{-6} \text{ C})}{(0.20 \text{ m})^2}$$

$$= 33.75 \text{ N} \ (Q_2 \text{ is repelled to the right})$$

The force F_2 on Q_2 due to Q_3 will be

$$F_2 = 9.0 \times 10^9 \frac{\text{N-m}^2}{\text{C}^2}$$

$$\times \frac{(-15 \times 10^{-6} \text{ C}) \times (2.5 \times 10^{-6} \text{ C})}{(0.10 \text{ m})^2}$$

$$= -37.5 \text{ N} \ (Q_2 \text{ is attracted to the right})$$

$$F = F_1 + F_2$$

$$= 33.75 \text{ N} + 37.5 \text{ N}$$

$$= 71 \text{ N directed to the right} \qquad ans.$$

Illustrative Problem 24.3 Charges A, B, and C of $+12$, -16, and $+20 \ \mu$C, respectively, are arranged as shown in Fig. 24.14. Find the magni-

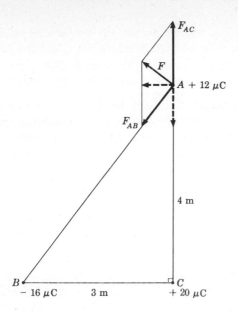

F_{AC}

F

$A + 12\ \mu C$

F_{AB}

4 m

B C

$-16\ \mu C$ 3 m $+20\ \mu C$

Fig. 24.14 Illustrative Problem 24.3.

tude of the force on charge A. Angle C is a right angle.

Solution Given

$$BC = 3 \text{ m} \qquad AC = 4 \text{ m}$$

$$Q_A = +12\ \mu C \qquad Q_B = -16\ \mu C$$

$$Q_C = +20\ \mu C \qquad \angle C = \text{right angle}$$

Use the Pythagorean theorem to find

$$AB = \sqrt{(3 \text{ m})^2 + (4 \text{ m})^2} = 5 \text{ m}$$

$$F_{AB} = 9.0 \times 10^9\ \frac{\text{N-m}^2}{\text{C}^2}$$

$$\times\ \frac{(12 \times 10^{-6} \text{ C})(-16 \times 10^{-6} \text{ C})}{(5 \text{ m})^2}$$

$$= -6.91 \times 10^{-2} \text{ N}$$

$$F_{AC} = 9.0 \times 10^9\ \frac{\text{N-m}^2}{\text{C}^2}$$

$$\times\ \frac{(12 \times 10^{-6} \text{ C})(20 \times 10^{-6} \text{ C})}{(5 \text{ m}^2)}$$

$$= 8.64 \times 10^{-2} \text{ N}$$

$$(F_{AB})_x = \tfrac{3}{5}F_{AB} = -4.15 \times 10^{-2} \text{ N}$$

$$(F_{AB})_y = \tfrac{4}{5}F_{AB} = -5.53 \times 10^{-2} \text{ N}$$

$$F_x = (F_{AB})_x = -4.15 \times 10^{-2} \text{ N}$$

$$F_y = F_{AC} + (F_{AB})_y = 8.64 \times 10^{-2} \text{ N}$$

$$-5.53 \times 10^{-2} \text{ N}$$

$$= 0.0311 \text{ N}$$

$$F = \sqrt{(F_x)^2 + (F_y)^2}$$

$$= \sqrt{(0.0415 \text{ N})^2 + (0.0311 \text{ N})^2}$$

$$= 0.052 \text{ N} \qquad \qquad ans.$$

Scientists have found it convenient to introduce a new constant ϵ_0 (epsilon) where $\epsilon_0 = 8.854 \times 10^{-12}\ \text{C}^2/\text{N-m}$ and is called the *permittivity of free space* (a vacuum). The constant k_0 and the constant ϵ_0 are related by the equation

$$k_0 = \frac{1}{4\pi\epsilon_0} \qquad (24.2)$$

While replacing k_0 by $1/4\pi\epsilon_0$ seems only to complicate Eq. (24.1), it will lead to simpler expressions for capacitances to be studied later in this chapter. An alternative expression for Coulomb's law is:

$$F = \frac{1}{4\pi\epsilon_0} \times \frac{Q_1 Q_2}{r^2} \qquad (24.3)$$

for free space.

24.10 The Electric Field

The notion of forces acting across empty space has puzzled scientists for many years. The effect of gravitational forces acting on bodies in space was studied in Chap. 4. In the study of electrostatics and magnetism we are again faced with examples of forces "acting at a distance".

Like charges repel each other, and unlike charges attract each other. These forces occur even though the charged objects do not touch. Electric charges must have an effect on the region surrounding them which is somehow altered because of the presence of the charge. This region of influence is called an *electric field*. When a charged object is placed in an electric field, the force it experiences is ascribed to the field itself rather than to any direct action of the field-producing charges. The field acts as an intermediary for the transmission of electrostatic forces.

An electric field can be detected by placing in it a *test body*, such as a small charged body suspended by a silk thread, and noting the effect of the field on it.

24.11 Electric-Field Intensity

The electric field **E** has magnitude and direction. Thus it is a vector quantity. The *electric-field intensity* (or *strength*) **E** at any point is defined as the force per unit positive charge acting on a charged body at that point:

$$\mathbf{E} = \frac{\mathbf{F}}{Q} \qquad (24.4)$$

The SI unit of field intensity is the *newton per coulomb* (**N/C**).

Direction of Electric Field

The direction of the electric field at a point is defined as the direction of the force on a positive charge placed at that point. Note the electric field and the force will be in opposite directions on a negative charge.

It is important that a small test charge be used, since a large test charge could of itself change the very electric field that is being measured.

The *magnitude E* of the intensity of the electric field at a distance r from an isolated point charge Q can be found by using Eqs. (24.3) and (24.4). If we let the test charge be q, Eq. (24.3) becomes

$$F = \frac{1}{4\pi\epsilon_0} \times \frac{qQ}{r^2}$$

Substituting this value for F in Eq. (24.4), gives

$$E = \frac{(1/4\pi\epsilon_0)(qQ/r^2)}{q} = \frac{1}{4\pi\epsilon_0} \times \frac{Q}{r^2} \quad (24.5)$$

Electric fields set up by several charges superpose to form a single net field. The vector specifying the field intensity **E** at any point in such a field is the vector sum of the fields due to each charge taken separately.

Illustrative Problem 24.4 Find the magnitude of the electric field intensity at A in Fig. 24.14 due to the charges at B and C.

Solution In Problem 24.3 we found the magnitude of the force acting on the charge of $+12\ \mu C$ at A to be 0.052 N. Then by Eq. (24.4),

$$E = \frac{F}{Q} = \frac{0.052\ \text{N}}{+12\ \mu C} = 4330\ \text{N/C} \qquad ans.$$

Illustrative Problem 24.5 In Fig. 24.15 the charge of body A is $+12\ \mu C$ and the charge of body B is $-8\ \mu C$. Find the electric field intensity at point P. Assume free space.

Solution The field intensity at P due to the $+12\ \mu C$ charge has a magnitude of

$$E_{12} = \frac{1}{4\pi\epsilon_0} \times \frac{Q_{12}}{r^2}$$

$$= 9.0 \times 10^9\ \text{N-m}^2/\text{C}^2 \times \frac{12 \times 10^{-6}\ \text{C}}{(0.20\ \text{m})^2}$$

$$= 27 \times 10^5\ \text{N/C}$$

Similarly,

$$E_{-8} = 9.0 \times 10^9\ \text{N-m}^2/\text{C}^2 \times \frac{8 \times 10^{-6}\ \text{C}}{(0.40\ \text{m})^2}$$

$$= 4.5 \times 10^5\ \text{N/C}$$

Consider a positive unit charge at P. The fields due to both charges will be to the right. Therefore the magnitude of the total field intensity E at P

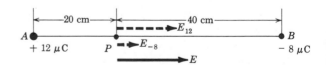

Fig. 24.15 Illustrative Problem 24.5.

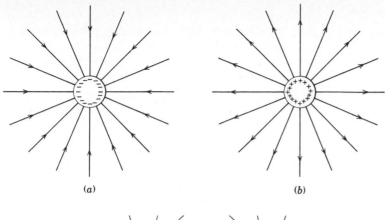

(a)   (b)

Fig. 24.16 Lines of force around a single charged sphere.

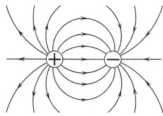

(a)

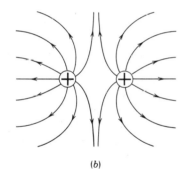

(b)

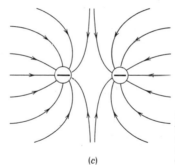

(c)

Fig. 24.17 The nature of lines of force and the electric field in the region of charged bodies: (a) two unlike charges, (b) like positive charges, (c) like negative charges.

will be $27 \times 10^5 + 4.5 \times 10^5 = 31.5 \times 10^5$ **N/C**. It is directed toward the right. *ans.*

24.12 Electric Lines of Force

The electric field near one or more charged bodies can be represented by drawing "lines of force" as an aid in visualizing the field. The relationship between the (imaginary) lines of force and the electric field intensity vector is this: (1) the direction of the line of force at any point is that in which a positive charge would move if

placed at that point, and (2) the lines of force are drawn so that the number of lines per unit cross-sectional area is proportional to the magnitude of **E**.

Figure 24.16a shows the lines of force near a negatively charged sphere, and Fig. 24.16b shows the lines of force near a positively charged sphere. The lines of force lie along radii of the spheres. Note how the magnitude E of the electric field intensity is not constant but decreases with increasing distance from the charge. This is shown

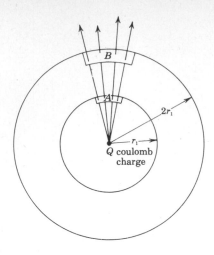

Fig. 24.18 The electric field intensity is proportional to the number of lines per area.

$$N = DA \qquad (24.6)$$

We will assign a constant ϵ_0 to the ratio of D to E:

$$\epsilon_0 = D/E \qquad (24.7)$$

The constant ϵ_0 is called the *permittivity of empty space* (see Sec. 24.9).

The magnitude of the electric field intensity near the point charge Q is

$$E = \frac{Q}{4\pi\epsilon_0 r^2} \qquad (24.5)$$

Recall the formula for the surface area of a sphere, $A = 4\pi r^2$. From Eq. (24.7), we get

$$D = \epsilon_0 E \qquad (24.7')$$

Then

$$N = DA = \epsilon_0 EA$$

$$= \epsilon_0 \left(\frac{Q}{4\pi\epsilon_0 r^2} \right) 4\pi r^2$$

and

$$N = Q \qquad (24.8)$$

We derived Eq. (24.8) for a point charge and a spherical surface. However, it holds for any collection of charges Q surrounded by a closed surface of any shape. A generalized statement of Eq. (24.8) is known as

Gauss's law

The total number of electric lines of force crossing a closed surface in an outward direction is numerically equal to the total positive charge enclosed within that surface.

24.14 Potential Energy in an Electric Field

In Chap. 4 we studied the work done moving a mass against forces in a gravitational field. This work was expressed as an increase in gravitational potential energy (GPE). Often problems in mechanics were simplified by considering the potential energies of masses. This will also be found true for forces in electric fields.

Work must be done in moving a mass against a gravitational field. In like manner, work

by the lines being spaced farther apart at greater distances. From symmetry, we know that the magnitude E is the same for all points that are a given distance from the center of the charge.

Figure 24.17 shows the lines of force around two equal like charges and two equal unlike charges. It should be emphasized that the actual lines do not exist but are introduced merely as an aid to understanding the concept of a field of force.

24.13 Lines of Force and Field Intensity

Consider the field (lines of force) directed radially outward from a positive point charge Q (Fig. 24.18). Let N equal the number of lines drawn to represent the field. Next imagine a spherical surface surrounding the point charge at a distance r from the charge. Whatever number of lines of force we choose to represent the electric field around a point charge, the same number must pass through *any* spherical surface around the charge.

We will define electric *flux density* D as the number of lines per unit area crossing a surface at right angles to the direction of the field. Let A be the surface area of the sphere. Then

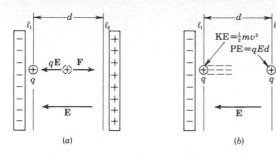

Fig. 24.19 The gain in potential energy = qEd when a positive charge $+q$ is moved distance d against a field E. When the restraining force is released, the charge will gain kinetic energy = $\frac{1}{2}mv = qEd$.

must be done in moving an electric charge in an electric field. This work can be expressed in terms of energy.

Consider a positive charge $+q$ resting on line l_1 in a uniform field E, as shown in Fig. 24.19a. The electric force on the charge is qE

$$\mathbf{F} = q\mathbf{E} \qquad (24.4')$$

An external force $\mathbf{F}$ equal to $q\mathbf{E}$ is required to move the charge against the field $\mathbf{E}$. The work to move the charge to line l_2, a distance d, against the field, is qEd. This work can be expressed as an increase in the potential energy of the charge with respect to line l_1. Thus the potential energy of the charge at line l_2 is

$$PE = qEd \qquad (24.9)$$

When the charge q is then released $(F = 0)$, the electric field will perform work on the charge. The field can do so because the work is stored in the field. As the charge is accelerated back toward line l_1, the potential energy decreases and the charged particle gains an equivalent kinetic energy. If the mass of the charged particle is m, the kinetic energy of the charged mass will be

$$KE = \tfrac{1}{2}mv^2 = qEd \qquad (24.10)$$

when it returns to its initial position (Fig. 24.19b).

The electric potential energy when $+q$ is at l_2 is independent of the path taken to reach l_2. This result is similar to that of a mass being moved in a gravitational force field.

Gravitational forces are always forces of attraction. A mass moved against the gravitational field always increases the potential energy. Moving an electric charge against an electric field does not always result in a gain of potential energy because there are two kinds of charges.

If we replace the charge $+q$ of Fig. 24.19 by a negative charge $-q$, no external force will be needed to move the charge against $\mathbf{E}$. The work will be done *by* the field. Therefore, energy leaves the field, and there will be a lower potential energy at l_2 than at l_1. Since we arbitrarily defined the electric field in terms of a positive charge, we have the rule:

Moving a positive charge against an electric field increases the potential energy. Moving a negative charge against an electric field decreases the potential energy.

24.15 Potential Difference

In dealing with gravitational potential energy we are often concerned with differences in potential energies. For falling bodies, differences in potential energies were equated to gains in kinetic energies. In like manner, in this discussion we are concerned with differences in electrical potential energies.

The *electric potential difference V* between two points *A* and *B* is defined *as the work W done per unit positive charge q to move the charge from A to B*

$$V = V_A - V_B = W/q \qquad (24.11)$$

The unit for electrical potential difference in the SI system is the *volt*.

Definition of the Volt
The potential difference between two points is one volt (V) if it requires one joule of external work to move one coulomb of charge from one point to the other.

$$1 \text{ volt (V)} = 1 \text{ joule/coulomb (1 J/C)}$$

$$W = qV \qquad (24.11')$$

A "twelve-volt" battery is one that has a potential difference of 12 V between its terminals. The two electric wires connected to a household outlet plug have an effective difference of potential of 110 to 120 V. On the average, 110 to 120 J of work must be expended for each coulomb of electricity which is transferred through an appliance connected between the wires.

Illustrative Problem 24.6 An electric iron is plugged into a 120-V outlet plug. The iron is rated at 1200 W. How many coulombs will be transferred through the iron in 1 hr?

Solution The work done on the iron is

$$\text{Work} = \text{power} \times \text{time} \quad \text{(from Eq. 5.12),}$$

$$= 1200 \text{ W} \left(\frac{1 \text{ J/sec}}{1 \text{ watt}} \right)$$

$$\times (1 \text{ hr}) \left(\frac{3600 \text{ sec}}{1 \text{ hr}} \right)$$

$$= 4{,}320{,}000 \text{ J}$$

Solve Eq. (24.11') for q. Substitute values for W and V.

$$q = \frac{W}{V} = \frac{4{,}320{,}000 \text{ J}}{120 \text{ V}} \times \frac{1 \text{ V}}{1 \text{ J/C}}$$

$$= 36{,}000 \text{ C} \qquad\qquad ans.$$

Small potential differences are often expressed in *millivolts* or *microvolts*.

$$1 \text{ millivolt } (1 \text{ mV}) = 10^{-3} \text{ V}$$

$$1 \text{ microvolt } (1 \text{ } \mu\text{V}) = 10^{-6} \text{ V}$$

Large potential differences are expressed in *kilovolts* or *megavolts*.

$$1 \text{ kilovolt } (1 \text{ kV}) = 10^{3} \text{ V}$$

$$1 \text{ megavolt } (1 \text{ MV}) = 10^{6} \text{ V}$$

Illustrative Problem 24.7 An alpha particle moves in a particle accelerator through the field where the potential difference is 2.5 MV. The mass of the particle is 4.69×10^{-27} kg. An alpha particle has a positive charge of 3.204×10^{-19} C. (*a*) What is the energy of the particle when it reaches its final speed? (*b*) What is that speed?

Solution

(*a*) Substitute known values in Eq. (24.11').

$$W = qV$$

$$= (3.204 \times 10^{-19} \text{ C})(2.5 \times 10^{6} \text{ V}) \left(\frac{1 \text{ J/C}}{1 \text{ V}} \right)$$

$$= 8.01 \times 10^{-13} \text{ J}$$

The kinetic energy of the particle is equal to the work done on it, namely, 8.01×10^{-13} J. *ans.*

(*b*) $\text{KE} = \frac{1}{2}mv^2 = W$

or

$$v = \sqrt{\frac{2W}{m}} = \sqrt{\frac{2(8.01 \times 10^{-13} \text{ J})}{4.69 \times 10^{-27} \text{ kg}}}$$

Remembering that 1 J = 1 N-m and 1 N = 1 kg-m/sec², we have

$$v = \sqrt{\frac{16.02 \times 10^{-13} \text{ kg-m/sec}^2 \times \text{m}}{4.69 \times 10^{-27} \text{ kg}}}$$

$$= \sqrt{3.414 \times 10^{14} \text{ m}^2/\text{sec}^2}$$

$$= 1.85 \times 10^{7} \text{ m/sec} \qquad ans.$$

24.16 Potential Difference in a Uniform Electric Field

Consider two large plates A and B separated by a distance d (Fig. 24.20). Let plate A bear a charge $+Q$ and plate B a charge $-Q$. The electric field **E** between the two plates is constant in both magnitude and direction. Such a field is said to be *uniform*. A charge $+q$ placed in the region between the plates will experience a force $\mathbf{F} = q\mathbf{E}$ (Eq. 22.4'). The work W done by the field **E** in moving the $+q$ charge a distance d from A to B is $Fd = qEd$.

The potential difference V between plates A and B is W/q (Eq. 24.11). Thus

$$V = V_A - V_B = \frac{W}{q} = \frac{qEd}{q}$$

If we divide both numerator and denominator by q, we get

$$V = Ed \qquad\qquad (24.12)$$

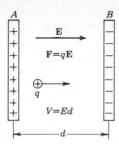

Fig. 24.20 Potential difference between two oppositely charged plates is equal to *Ed*.

The potential difference between two oppositely charged plates is equal to the product of the electric field intensity and the distance between the plates.

Potential difference, like work, is a scalar quantity. The defining equation for potential difference (Eq. 24.11) assumes q to be a *positive* charge. Thus, a positive potential difference means that the electric potential energy of the charge is greater at A than at B. A negative potential difference means that the electric potential energy at A is less than that at B. If V is positive, the positive charge at A will tend to return to B. If V is negative, the positive charge at A will tend to move farther away from B.

Equation (24.12) can be solved for E.

$$E = V/d \qquad (24.13)$$

The SI metric unit for E in Eq. (24.13) is the *volt per meter*. It is left as an exercise for the student to prove that the volt per meter is equivalent to the *newton per coulomb*.

Illustrative Problem 24.8 Two large plates are 1.5 cm apart. The potential difference between the plates is 20,000 V. Determine the field intensity E between the plates.

Solution Use Eq. (24.13) and substitute given values:

$$E = \frac{V}{d} = \frac{2 \times 10^4 \text{ V}}{1.5 \times 10^{-2} \text{ m}}$$

$$= 1.3 \times 10^6 \text{ V/m} \qquad ans.$$

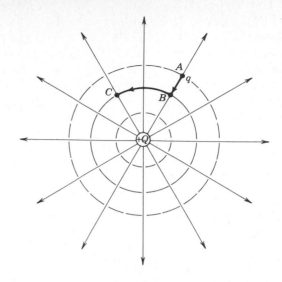

Fig. 24.21 Motion of charge q in a field due to charge $+Q$.

24.17 Potential Difference in a Nonuniform Field

We have been considering the potential differences in a *uniform* electric field. We will next consider a case where the field is not uniform.

Consider the field around an isolated charge $+Q$ (Fig. 24.21). The field is directed radially away from the charge. The field intensity varies inversely as the square of the distance from the center of the charge. At points A and B, the field intensities are

$$E_A = kQ/r_A^2 \qquad E_B = kQ/r_B^2$$

where r_A and r_B are respective distances from the charge to A and B.

The forces on a test charge q at A and B are

$$F_A = qE_A = kqQ/r_A^2 \qquad F_B = qE_B = kqQ/r_B^2$$

It can be shown that the average of forces F_A and F_B can be approximated by

$$F = kqQ/r_A r_B$$

Then, the work done by an external agent in moving the test charge the distance $r_A - r_B$ is

$$W_{A \to B} = \frac{kqQ}{r_A r_B}(r_A - r_B)$$

$$= kqQ\left(\frac{1}{r_B} - \frac{1}{r_A}\right) \qquad (24.14)$$

Using the defining equation (24.11) for potential difference and Eq. (23.14), we get

$$V = V_A - V_B = \frac{W_{A \to B}}{q}$$

$$= kQ\left(\frac{1}{r_B} - \frac{1}{r_A}\right) \qquad (24.15)$$

Potential differences are independent of the path followed in moving the test charge. Thus the potential difference between two points at distances r_A and r_B from a point charge depends only on the magnitude of the charge and the distances of the points from the charge.

24.18 Electric Potential

The *potential at a point* is defined as the work done per unit charge in moving a positive charge from some zero potential to the point. This work results in an increase in the electric potential energy. Let us choose infinity as the reference point for zero potential.

Referring to the past example (Fig. 24.21), we can choose A at a very large (actually an infinite) distance from Q. We will arbitrarily assign zero electric potential V_A at this infinite distance. This then permits us to define the electric potential at a point. Putting $V_A = 0$, $r_B = r$, and $r_A = \infty$ leads to the equation

$$V = V_B - V_A = kQ\left(\frac{1}{r} - \frac{1}{\infty}\right)$$

$$= \frac{kQ}{r} \qquad (24.16)$$

or

$$V = \frac{1}{4\pi\epsilon_0}\frac{Q}{r} \qquad (24.17)$$

In the SI system V is in volts, Q in coulombs, and r in meters. The value of k is 9.0×10^9 N-m²/C². The value of ϵ_0 is 8.9×10^{-12} C²/N-m² (see Sec. 24.9). The units are consistent because

$$1\ V = 1\frac{J}{C} = 1\frac{N\text{-}m}{C}$$

The student must keep in mind that choosing infinity as the reference point for zero potential is purely arbitrary. Any other reference position could well have been chosen for zero potential. In subsequent chapters, however, **potential differences** will generally be our fundamental concern.

Illustrative Problem 24.9 An electric potential of 150 V exists at all points 12 cm from an isolated positive charge. What is the magnitude of the charge?

Solution Solve Eq. (24.17) for Q. Then substitute known values for V and r. Remember 1 N/C $= 1$ V/m.

$$Q = 4\pi\epsilon_0 r V$$

$$= 4\pi(8.9 \times 10^{-12}\ \text{C}^2/\text{N-m}^2)$$

$$\times (0.12\ \text{m})(150\ \text{V})$$

$$= 2.0 \times 10^{-9}\left(\frac{\text{C}^2}{\text{N}}\right)(\text{V/m})\left(\frac{1\ \text{N/C}}{1\ \text{V/m}}\right)$$

$$= 2.0 \times 10^{-9}\ \text{C} \qquad \qquad ans.$$

Illustrative Problem 24.10 The diameter of the nucleus of the copper atom (see Fig. 24.3) is roughly 6×10^{-13} cm. What is the electric potential at the surface of the copper nucleus?

Solution The surface of the nucleus is assumed spherical. The charge on the nucleus for external purposes behaves as if it were a point charge at the center. Recall that the positive charge on a proton is 1.6×10^{-19} C (see Sec. 24.9). There are 29 protons in the nucleus. Substitute known values in Eq. (24.16).

$$V = k\frac{Q}{r}$$

$$= \frac{(9.0 \times 10^9\ \text{N-m}^2/\text{C}^2)(29)(1.6 \times 10^{-19}\ \text{C})}{3 \times 10^{-15}\ \text{m}}$$

$$= 1.4 \times 10^7\frac{\text{N-m}}{\text{C}}\left(\frac{1\ \text{V/m}}{1\ \text{N/C}}\right) = 1.4 \times 10^7\ \text{V}$$

$$= 14\ \text{MV} \qquad \qquad ans.$$

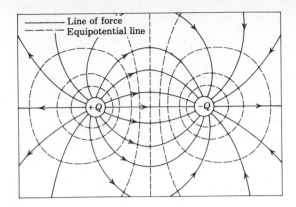

Fig. 24.22 Lines of force and equipotential lines in the field formed by two oppositely charged bodies.

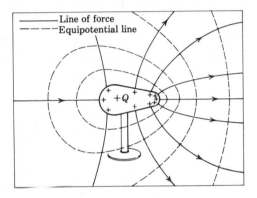

Fig. 24.23 Lines of force and equipotential lines in a nonuniform field.

24.19 Equipotential Surfaces

If all the points which have the same potential in an electric field near a charged object are joined, an *equipotential line* or *surface* within the field is formed. No work is done when a test charge is moved along an equipotential surface in an electric field. Thus, in Fig. 24.21, no work would be done in moving the charge from *B* to *C*.

Equipotential lines (dashed lines) and lines of force (solid lines) are illustrated in Fig. 24.22 and Fig. 24.23. Note that the electric lines of force are normal to equipotential surfaces.

The *electric potential gradient* is the rate of change of potential with distance along a line of force. If ΔV is a potential change that corresponds to a change Δr along a line of force, the potential gradient E_r is equal to $-\Delta V/\Delta r$. The electric intensity **E** is the negative of the potential gradient.

24.20 Electric Potential Referred to the Earth

In Chap. 4 we used sea level as an arbitrary choice for the zero of gravitational potential. Heights above sea level were used in determining potential energies. In other problems we used the difference of level between two points, neither of which was at sea level.

In Sec. 24.18 zero potential is assumed to be at infinity. It is frequently convenient to take the electric potential of the earth as zero. The earth can be considered as an (almost) infinite reservoir of positive and negative charges.

If a charged body connected to the earth by a metallic conductor of electricity allows electrons to flow to that body from the ground, the body is at a *positive potential* (see Fig. 24.24a). Conversely, if the connection to the ground allows electrons to flow from the body into the ground, the body is at a *negative potential* (see Fig. 24.24b). If the negative terminal of a 12-V battery is connected to the ground, it will be at 0-V potential, and the potential at the positive terminal will be +12 V. In practice, the negative terminal is connected to the frame of an auto-

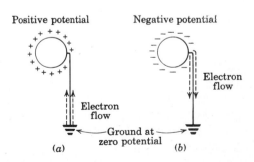

Fig. 24.24 Direction of flow of electrons when a positively charged body and a negatively charged body are grounded.

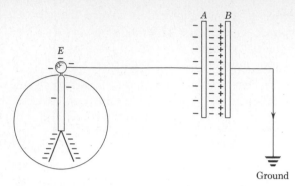

Fig. 24.25 The principle of the capacitor.

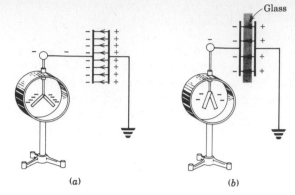

Fig. 24.26 Insertion of a glass slab between the plates of a charged capacitor lowers the potential difference between the plates.

mobile and the frame is regarded as being at zero potential. In radios and television receivers the chassis is usually regarded as being at zero potential. Electric tools and appliances are grounded for safety, so there is no potential difference between them and ground.

24.21 The Capacitor

A device that possesses the ability to store electrons and hold an electric charge is called a *capacitor* (or *condenser*). Almost any insulated body can hold a limited electric charge. The greater the surface area the greater the charge that can be stored. In many industrial applications of electricity it is necessary to use capacitors whose electron-storing ability is tremendous. Let us consider a simple experiment to illustrate the principle of a capacitor. Connect a metallic plate *A* to the knob of an electroscope *E* (Fig. 24.25). If we charge the plate *A* (with a negative charge in the figure), the leaves of the electroscope will diverge. As we increase the charge on *A*, the leaves will diverge farther. Now let us bring up a second plate *B*, similar to *A* except that it is "grounded" to the earth. As we bring *B* near *A*, the leaves of the electroscope will begin to collapse. The nearer we bring *B* (without touching) to *A* the more the leaves will collapse. Thus when *B* is held near *A*, a considerably larger charge can be placed on *A* before the same deflection of the leaves occurs. By moving the second plate near *A*, we have increased the

electron-holding capacity of the *E-A* system. The plates *A* and *B*, separated by air, constitute a *capacitor*.

Repeating the entire experiment with plates of larger area will show that the capacity of a capacitor increases with plate-surface area.

24.22 Capacitance

The capacitance of a capacitor is a measure of its ability to store up electric charge. We have discussed two ways to increase capacitance: (1) the area of the plates can be increased; (2) the plates can be brought closer together. A third way to increase the capacitance would be to insert a thin slab of nonconducting materials (other than air) between the plates. These materials are called *dielectrics*. Common dielectrics are air, glass, mica, oil, and waxed paper.

The effectiveness of the dielectric in increasing the ability of a capacitor to hold a charge can be illustrated in the laboratory. Figure 24.26a shows a parallel-plate capacitor with an electroscope connected across the plates. If a charge is placed on the capacitor, a difference of potential between the plates will be indicated by the diverging leaves of the electroscope. If a slab of glass or plastic is inserted between the plates (Fig. 24.26b), the leaves will partially collapse, indicating a smaller potential difference between the plates. When the slab is removed, the original potential difference will be indicated by the electroscope.

If the glass plate is kept between the plates, a greater charge must be put on the plates to produce the potential difference exhibited with air as the dielectric medium. The *capacitance* of a capacitor is defined as the *ratio of the charge on either plate to the potential difference between the two plates.*

$$C = \frac{Q}{V} \qquad (24.18)$$

where Q = charge on either plate, coulombs
V = potential difference between the conducting plates, volts
C = capacitance of capacitor, coulombs/volt

The Unit of Capacitance

The *coulomb per volt* unit has been given a special name, the *farad* (F). Thus *a capacitor whose capacitance is one farad will hold a charge of one coulomb if a difference of one volt is applied between the plates.*

In practice, the farad proves to be extremely large. For example, a capacitor of 1 F capacity, with two parallel plates 1 mm apart in air would require each plate to have an area of over 40 sq. mi!

Practical capacitors are usually calibrated in microfarads and picofarads:

$$1 \text{ microfarad } (\mu F) = 10^{-6} \text{ F}$$
$$1 \text{ picofarad } (pF) = 10^{-12} \text{ F}$$

Illustrative Problem 24.11 What is the capacitance of a two-plate capacitor that holds a charge of $7.5 \times 10^{-4} \ \mu C$ when the potential difference between the plates is 400 V?

Solution Using Eq. (24.18), we have

$$C = Q/V$$
$$= \frac{7.5 \times 10^{-10} \text{ C}}{400 \text{ V}}$$
$$= 1.875 \times 10^{-12} \text{ F}$$
$$= 1.875 \text{ pF} \qquad\qquad ans.$$

24.23   Dielectrics

In Fig. 24.3 the atom is pictured as consisting of a small, but extremely dense, positive nucleus surrounded by electrons whirling in orbits around the nucleus. When a conductor is placed in an electric field, some of the free electrons move from atom to neighboring atom in the conductor.

Materials which make good dielectrics are generally those whose atoms have few outer orbit electrons free to migrate. Atoms of dielectrics, as atoms of all substances, are electrically neutral. Normally the centers of their orbital electrons also coincide with the center of the nucleus. However, in an electric field, the electrons and protons are somewhat displaced from their normal positions. An electric field will cause the electrons to shift slightly against the field. The protons will shift slightly in the direction of the field.

Figure 24.27 illustrates schematically what happens when a dielectric slab is placed between the charged plates of a capacitor. The atom of the dielectric becomes an *induced electric dipole* with one end positively charged and the other negatively charged.

In the electric field the tiny dipoles tend to line up with their positive ends pointing toward

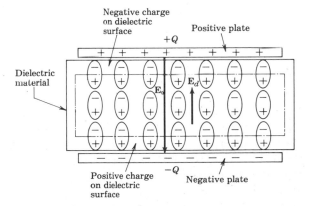

Fig. 24.27 The surfaces of the dielectric slab in an electric field produce an electric field E_d which opposes the external field E_o due to the charged plates.

the negative plate. When the dielectric is removed from the field, the atoms quickly cease to be aligned as dipoles. This is due to the constant state of thermal agitation or internal energy of the molecules.

The degree of alignment of the dipoles depends upon the intensity of the electric field, the temperature, and the nature of the dielectric. The greater the field intensity, the greater the alignment. As the temperature is decreased, the alignment will increase. In Fig. 24.27, note that all the positive and negative charges in the interior of the dielectric (within the dashed lines) neutralize each other. There are left, however, layers of positive and negative charges on opposite surfaces of the dielectric. These give the dielectric a positive charge on one surface and a negative bound charge on the other.

These surface charges produce an electric field E_d in the dielectric slab which opposes the external field E_0 of the capacitor. The effect is to weaken the original field. The net electric field E is the vector sum of the two fields E_0 and E_d. Thus, E will be a weaker field in the direction of E_0.

$$\mathbf{E} = \mathbf{E}_0 + \mathbf{E}_d \qquad (24.19)$$

The net effect of the dielectric in the charged capacitor is to lower the potential gradient of the electric field (see Eq. 24.13). Since $C = Q/V$ for the capacitor, we know that lowering V for a constant Q results in an increase in C. Or, in other words, if a dielectric is placed between the plates of a capacitor, Q will be larger for the same V. *The presence of the dielectric increases the capacitance of the capacitor.*

24.24 Dielectric Constant

The ratio of the capacitance of a capacitor with the dielectric between the plates to that with plates separated by a vacuum is the *dielectric constant K* of the material.

$$K = \frac{C}{C_0} \qquad (24.20)$$

The relation between the three factors which determine the capacitance of a parallel-plate capacitor can be expressed by the formula

$$C = \epsilon \frac{A}{d} \qquad (24.21)$$

where
A = area of either parallel plate, square meters
d = distance between plates, meters
ϵ = constant of seperating medium
C = capacitance, farads

The constant ϵ is called the *permittivity of the separating medium.* The value of ϵ is found by

$$\epsilon = \epsilon_0 K \qquad (24.22)$$

where $\epsilon_0 = 8.85 \times 10^{-12}$ C^2/(N)(m^2) (see Sec. 24.9) and K is the *dielectric constant.* Recall that ϵ_0 is the *permittivity of free space.*

Values of the dielectric constant for a few substances are given in Table 24.2. Note that K is dimensionless since it equals the quotient of like quantities.

Table 24.2 Dielectric constants K of some common electric materials

Material	K
Air (1 atm)	1.00059
Ammonia, liquid at $-34°C$	22
Glass	5–10
Mica	3–6
Paraffin wax at 20°C	2.1–2.5
Polyethylene at 20°C	2.25–2.30
Porcelain	6.0–8.0
Rubber	2.5–3.0
Sulfur at 20°C	4.0
Transformer oil at 20°C	2.24
Vacuum	1.00000
Water at 20°C	80

Illustrative Problem 24.12 Two rectangular sheets of copper foil 16 by 20 cm are separated by a thin layer of paraffin wax 0.2 mm thick. Calculate the capacitance if the dielectric constant for the wax is 2.4.

Solution Substitute $K = 2.4$, $d = 2 \times 10^{-4}$ m, and $A = 0.16$ by 0.20 m in Eqs. (24.21) and (24.22):

$$C = K\epsilon_0 \frac{A}{d} = 2.4 \times 8.85 \times 10^{-12} \text{ C}^2$$

$$\div (\text{N})(\text{m}^2) \times \frac{0.032 \text{ m}^2}{2 \times 10^{-4} \text{ m}}$$

$$= 3.4 \times 10^{-9} \text{ C}^2/(\text{N})(\text{m})$$

$$\times \frac{1 \text{ N-m/V}}{\text{C}} \times \frac{1 \text{ F}}{1 \text{ C/V}}$$

$$= 3.4 \times 10^{-9} \text{ F} = 3400 \text{ pF} \qquad \textit{ans.}$$

There is a limit to how much charge can be placed on a given capacitor. When sufficiently large charges are placed on a capacitor, the electrons break loose from the parent atoms of the dielectric. The electric field stress applied will make the dielectric temporarily conductive. Breakdown occurs. The breakdown consists in the passage of a spark from plate to plate through the dielectric. In dielectrics such as air, oil, and those used in electrolytic capacitors the breakdown is only temporary. The capacitive properties are restored after the spark disappears. But, in glass, mica, and other solid dielectrics, the dielectrics are punctured or shattered when the breakdown occurs. After breakdown the dielectrics have only the strength of an equal thickness of air. The *dielectric strength* of a dielectric is the field intensity, measured in megavolts per

Table 24.3 Approximate dielectric strengths

Material	Dielectric Strength (MV/m *to breakdown*)
Air	3.0
Bakelite	10–30
Glass	20–60
Mica	50–220
Paraffin, solid	25–45
Paraffined paper	30–50
Porcelain	5.5
Rubber	15–50
Transformer oil	5–15

meter (MV/m), required to breakdown the dielectric. Table 24.3 lists approximate dielectric strengths for some common dielectrics.

Illustrative Problem 24.13 The plates of a parallel-plate capacitor are separated by paraffined paper 0.015 mm thick. What voltage applied to the capacitor will cause the capacitor to break down? Assume the dielectric strength of the paper to be 35 MV/m.

Solution Let the answer be x volts.

$$35 \text{ MV/m} = 35 \times 10^6 \frac{\text{V}}{\text{m}} \times \frac{1 \text{ m}}{1000 \text{ mm}}$$

$$= 35 \times 10^3 \frac{\text{V}}{\text{mm}}$$

Then

$$x/0.015 \text{ mm} = 35 \times 10^3 \text{ V/mm}$$

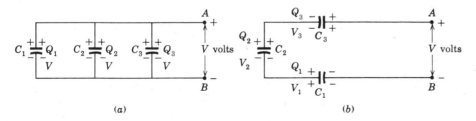

Fig. 24.28 Capacitors connected in (*a*) parallel and (*b*) series.

and

$$x = 35 \times 10^3 \text{ V/mm} \times 0.015 \text{ mm}$$

$$= 525 \text{ V} \qquad\qquad ans.$$

24.25 Combination of Capacitors

Capacitors are connected in *parallel* when one plate of each capacitor is connected to a common conductor while the remaining plates are connected to a second conductor (see Fig. 24.28*a*). (The symbol $\dashv\vdash$ or $\dashv\vdash$ is used to indicate a fixed capacitor in the diagram.) Essentially this system is equivalent to a single capacitor with two larger plates each equal in area to the sum of one set of three smaller areas. If the capacitors are charged, they must have the same potential difference across their plates. The charge on each capacitor will be

$$Q_1 = C_1 V \qquad Q_2 = C_2 V \qquad Q_3 = C_3 V$$

The total charge Q_T of the system must be the sum of the separate charges on the three capacitors.

$$Q_T = Q_1 + Q_2 + Q_3$$

Then $\qquad Q_T = C_1 V + C_2 V + C_3 V$

But the total charge is equivalent to the product of the total capacitance C_T and the common potential difference V:

$$Q_T = C_T V$$

Substituting, we obtain

$$C_T V = C_1 V + C_2 V + C_3 V$$

and dividing through by V yields

$$C_T = C_1 + C_2 + C_3 \qquad (24.23)$$

Capacitors in Parallel

When capacitors are connected in parallel their total capacitance is the sum of the individual capacitances.

When capacitors are connected in *series*, as shown in Fig. 24.28*b*, a positive charge on one plate of C_3 induces a negative charge on the adja-

cent plate. The electrons will be attracted away from the plate of C_2 that is connected to C_3. A positive charge of the same magnitude will be left on the other plate of C_2. This plate will in turn draw electrons from a plate of C_1, leaving it positive. All three capacitors will have equal charges. Thus,

$$Q_T = Q_1 = Q_2 = Q_3$$

The total difference in potential V_T across the series combination must equal the sum of the separate potential differences across each capacitor:

$$V_T = V_1 + V_2 + V_3$$

But

$$V_T = \frac{Q_T}{C_T} \qquad V_1 = \frac{Q_1}{C_1} \qquad V_2 = \frac{Q_2}{C_2} \qquad V_3 = \frac{Q_3}{C_3}$$

Substituting, we have

$$\frac{Q_T}{C_T} = \frac{Q_T}{C_1} + \frac{Q_T}{C_2} + \frac{Q_T}{C_3}$$

and dividing by Q_T, we obtain

$$\frac{1}{C_T} = \frac{1}{C_1} + \frac{1}{C_2} + \frac{1}{C_3} \qquad (24.24)$$

Capacitors in Series

When capacitors are connected in series the reciprocal of their total capacitance is equal to the sum of the reciprocals of their individual capacitances.

Illustrative Problem 24.14 Three capacitors having capacitances of 2.0, 3.0, and 5.0 μF are connected in parallel to a 12-V source. (*a*) Find the charge on each capacitor. (*b*) Find the total charge of the combination. (*c*) If the capacitors are discharged and connected in series to the 12-V source, find the total charge of the combination.

Solution

(*a*) $\qquad V_T = V_1 = V_2 = V_3 = 12 \text{ V}$

Fig. 24.29 Common multiplate capacitor.

Fig. 24.30 Capacitors of different sizes and shapes. *Back row* (left to right): oil-filled 10,000-V capacitor; 2000-V capacitor; liquid electrolyte capacitor; paper 2000-V capacitor. *Second row:* liquid electrolyte, mica-molded capacitor; mica-variable capacitor. *Front:* tantalum (solid electrolyte) capacitor. This capacitor has the same capacitance as the large oil-filled capacitor in the back row but at a much lower voltage.

Then

$$Q_1 = C_1 V_1 = 2.0 \ \mu F \times 12 \ V = 24 \ \mu C$$

$$Q_2 = C_2 V_2 = 3.0 \ \mu F \times 12 \ V = 36 \ \mu C$$

$$Q_3 = C_3 V_3 = 5.0 \ \mu F \times 12 \ V = 60 \ \mu C \quad ans.$$

(*b*) $\qquad Q_T = Q_1 + Q_2 + Q_3 = 120 \ \mu C \qquad ans.$

(*c*) $\qquad \dfrac{1}{C_T} = \dfrac{1}{C_1} + \dfrac{1}{C_2} + \dfrac{1}{C_3}$

$$= \frac{1}{2.0 \ \mu F} + \frac{1}{3.0 \ \mu F} + \frac{1}{5.0 \ \mu F}$$

$$C_T = \tfrac{30}{31} \ \mu F$$

$$V_T = \frac{Q_T}{C_T} \quad \text{or} \quad Q_T = V_T C_T$$

Fig. 24.31 Variable capacitors. On the left is a single-section capacitor with air as the dielectric. In the middle are two capacitors of different sizes mounted on the same shaft. In front of the two capacitors can be seen two small mica-variable capacitors. At the right is a gear-driven capacitor (really three separate capacitors mounted on the same shaft). To the right of the top two capacitors are two small variable air capacitors, called *trimmers*, used to adjust for minor variations in the larger capacitors and to bring the two up to the third capacitance.

$$Q_T = 12 \ V \times \tfrac{30}{31} \ \mu F \times \frac{1 \ F}{10^6 \ \mu F}$$

$$= 1.2 \times 10^{-6} \ C$$

$$= 1.2 \ \mu C \qquad\qquad ans.$$

24.26 Commercial Capacitors

Most capacitors are modifications of the parallel-plate capacitor. Increased capacitance in a capacitor can be achieved by using two sets of plates connected in parallel (idealized in Fig. 24.29). Such a capacitor is called a *multiple-plate capacitor*.

Commercial capacitors vary considerably in shape and size (Fig. 24.30 and Fig. 24.31). They are usually classified in terms of the dielectric used.

The *variable-air capacitor* consists of two sets of parallel aluminum plates that are separated by air. One set is fixed while the other is mounted on a shaft between the fixed plates (Fig. 24.31). Only the overlapping areas contribute to the capacitance, which can be varied by turning the shaft. The capacitance will be maximum when the movable plates are completely meshed within the fixed plates. This type of capacitor is used in tuning radio receiving sets and in many other types of electronic equipment where tuning one circuit to the frequency of another is necessary.

Paper capacitors are constructed by rolling two long strips of metal foil with a long thin strip of a paper dielectric. The paper is first treated with wax or oil. The rolled strips are then sealed in a metal container. Paper capacitors are used in radios and the ignition systems of automobiles.

The *plastic-film capacitor* is similar to the paper capacitor. The dielectric is one of the newer plastics on the market, such as teflon, mylar, or polystyrene.

The sheets of metal foil in a *mica capacitor* are separated by sheets of mica dielectric. The entire capacitor is sealed in a metal or bakelite case.

In the *electrolytic capacitor* a dielectric layer is formed by chemical action on the metal plates of the capacitor. The slight space between the layers is filled with an electrolyte in liquid or paste form. The electrolyte constitutes one of the plates. Electrolytic capacitors usually have relatively large capacitances.

SUMMARY

A neutral body contains equal amounts of positive and negative electricity.

Like charges repel each other; unlike charges attract each other.

To charge a body positively, some of its electrons must be removed; to charge it negatively, electrons must be added.

Whenever an electrostatic charge is developed in one body, an equal but opposite charge is developed in some other body.

Substances in which electrons are free to move are termed *conductors*; substances in which electrons are not free to move are termed *insulators*.

Bodies can be charged by conduction and by induction, as well as by friction.

An electroscope is a device for determining the presence and intensity of an electric charge.

Coulomb's law expresses the magnitude of the force of attraction or repulsion between two charges Q_1 and Q_2:

$$F = k \frac{Q_1 Q_2}{r^2} \tag{24.1}$$

where F is in newtons, k is in N-m^2/C^2, Q_1 and Q_2 are in coulombs, and r is in meters;

$$k = 1/4\pi\epsilon \tag{24.2}$$

where ϵ is called *permittivity* of the dielectric.

The electric field intensity **E** at any point is defined as the force per unit charge acting on the body at that point:

$$\mathbf{E} = \frac{\mathbf{F}}{Q} \tag{24.4}$$

The potential difference between two points A and B is equal to the work required to bring a unit positive charge from A to B.

The potential difference between two points is *one volt* if it requires one joule of external work to move a one-coulomb charge from one point to the other:

$$\text{Volts} = \frac{\text{joules}}{\text{coulombs}}$$

The potential difference between two oppositely charged plates is equal to the product of the field intensity and the distance between the plates

$$V = Ed \tag{24.12}$$

Electric potential gradient is the rate of change of potential with distance along a line of force.

A capacitor is a device for storing an electric charge.

The capacitance of a capacitor is proportional to the area of the plates and inversely proportional to the distance between them. The nature of the insulator (dielectric) between the plates also affects the capacitance.

The capacitance of a capacitor is the ratio of the charge on either plate to the potential difference between the plates.

$$C \text{ (farads)} = \frac{Q \text{ (coulombs)}}{V \text{ (volts)}} \tag{24.18}$$

$$1 \ \mu\text{F} = 10^{-6} \text{ F}$$

$$1 \ \text{pF} = 10^{-12} \text{ F}$$

For capacitors in parallel, the total capacitance is equal to the sum of the capacitances:

$$C_T = C_1 + C_2 + C_3 \qquad (24.23)$$

For capacitors in series, the reciprocal of the total capacitance is equal to the sum of the reciprocals of the capacitances:

$$\frac{1}{C_T} = \frac{1}{C_1} + \frac{1}{C_2} + \frac{1}{C_3} \qquad (24.24)$$

QUESTIONS AND EXERCISES

1. When a small neutral pith ball suspended by a silk thread is brought near a negatively charged sphere, the ball is first attracted to the sphere until contact is made; then the ball is repelled. Explain in terms of the electron theory the reason for the action.

2. An electroscope is charged by induction from a positively charged body. (a) What charge is left on the electroscope? (b) Explain how the induced charge is produced.

3. How would you determine the sign of the charge obtained on sulfur when it is rubbed with leather?

4. Do some research on lightning rods. Describe how they are installed and why they may reduce the chances of a building being struck by lightning in a storm.

5. Two small hollow metal spheres are suspended next to each other by silk strings. Describe what you think will happen if they are both touched with a positively charged glass rod. Tell why.

6. Where would be the safest place for you to be if you were caught in an electrical storm in some wooded hills? Why?

7. Discuss the relative masses of the electron, proton and the neutron? What are their relative charges?

8. Which of the following is not a measure of electric field intensity: newton/coulomb, coulomb/meter, volt/meter, farad/coulomb?

9. When a charged rod is brought near bits of paper, the paper will first cling to the rod. Soon thereafter the paper will drop from the rod. Why?

10. When a negatively charged rod is brought near a positively charged electroscope, the leaves collapse. As the rod is brought still closer (but not touching), the leaves again diverge. Give reasons for this behavior.

PROBLEMS

1. Find the force between charges of $+120\ \mu C$ and $-25\ \mu C$ located 50 cm apart in air.

2. Two charges attract each other with a force of 18×10^{-6} N when they are 40 cm apart. Find the force between them when their separation is 60 cm.

3. Find the force on an electron placed in an electric field whose intensity is 5.25×10^4 N/C.

4. A 50-μF capacitor is connected to a 12-V battery. What is the charge on each plate of the capacitor?

5. What is the electric potential 15 cm from a point charge of 2.5 μC?

6. An electrostatic generator accelerates protons through a potential difference of 9.00×10^6 V. What is the energy of a proton when it reaches its maximum velocity?

7. The potential difference between the two terminals of a battery is 12 V. How much work is required to transfer 15 C of electricity from one terminal to the other?

8. Four equal positive charges $+Q_1$ are placed at the four corners of a square which has sides of length s. What is the net force on a negative charge Q_2 placed at the center of the square?

9. A parallel-plate capacitor is made of two metal sheets 10×10 cm, 1.5 mm apart. The plates are connected to the terminals of a 32-V battery. What is the electric field intensity between the plates in newtons/coulomb?

10. Two charges of $+60$ and $-20\ \mu C$ are 40 cm apart. A third charge of $+30\ \mu C$ is located halfway between the two charges. (a) What is the magnitude and direction of the force on the 30-μC charge? (b) What is the field intensity at the midway point?

11. Two charges of 40 and 30 μC are 50 cm apart. Find the electric field strength at a point 20 cm from the smaller charge on a line connecting the two charges.

12. What is the potential difference between two points if 450 J of work is required to move 75 C from one point to the other?

13. The field intensity between two deflecting plates of an electrostatic-deflection television tube is 50,000 N/C. (a) What will be the force on an electron passing between the two plates? (b) What acceleration does the electron experience? Assume a mass of 9.11×10^{-31} kg for the electron.

14. A charge of 5.0×10^{-9} C is placed in a uniform field of intensity 75,000 N/C. (a) How much work is done if the charge is moved 0.40 m in the direction of the field? (b) How much work is done if the charge is moved 0.40 m perpendicular to the field?

15. Three capacitors 4, 8, and 12 μF, respectively, are connected in series. Find the capacitance of the combination.

16. Three capacitors 3, 6, and 12 μF, respectively, are connected in series to a 360-V battery. Find (a) the capacitance of the series combination, (b) the charge on each capacitor, and (c) the voltage across each capacitor.

17. Given three capacitors connected as shown in Fig. 24.32. Find the total capacitance if $C_1 = 2.0$ μF, $C_2 = 3.0$ μF, and $C_3 = 5.0$ μF.

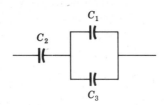

Fig. 24.32 Problem 24.17.

18. It is desired to combine 1.0-μF capacitors to make a combination with a capacity of 0.75 μF. How should they be connected to use as few 1.0-μF capacitors as possible?

19. Point charges of $+0.24$ μC and -0.18 μC are 60 cm apart in a vacuum. Point A is midway between the charges. Point B is 20 cm from the 0.24 μC charge and 40 cm from the -0.18 μC charge. What is the potential difference between A and B?

20. Four equal 30 μC positive charges are placed at the four corners of a square. The length of each side of the square is 300 cm. Find the force on one of the charges.

21. Electrons are accelerated in a uniform electric field whose field intensity is 1.8×10^{-4} N/C for a distance of 2.4 cm. If the electrons were initially at rest, find (a) their final speed, and (b) the time taken to travel the 2.4 cm. Assume an electron mass of 9.109×10^{-31} kg. [HINT: You will need to use Eqs. (24.4), (4.17), (4.6), and (4.3) in that order.]

22. An electron is accelerated between two plates that are 1.5 cm apart. One of the plates is grounded. The other has a potential of $+240$ V with respect to the earth. What is (a) the field intensity between the plates and (b) the force acting on the electron? (c) If the electron starts at rest at one of the plates, what is the speed with which it strikes the other plate? (d) Calculate the gravitational force at the earth's surface on the electron. Compare the answers to (b) and (d).

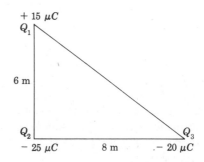

Fig. 24.33 Problem 24.23

23. Three point charges $Q_1 = +15$ μC, $Q_2 = -25$ μC, and $Q_3 = -20$ μC are located at the vertices of a right triangle as in Fig. 24.33. What will be the magnitude of the force on Q_1?

25 Basic Electric Circuits

In Chap. 24, we studied electric forces, fields, and energies as they relate to static electric charges. Static electricity played an important part in the early understanding of electricity. But electric charges remained quite unproductive until there was developed a means for continuous movement of the charges. A flow of electric charges constitutes an electric current. Wherever there is an electric current there is a source of electric energy, such as a dry cell, storage battery, or generator. We will study some of the sources of electric energy in Chap. 26. The battery or generator continually supplies the electric energy being carried by the current. The electric charges flow through a *closed electric circuit*.

In this chapter, we discuss elements that make up an electric circuit and develop laws governing such circuits. In developing the circuit theory in this and the next three chapters, we are primarily concerned with *direct currents*. A direct current (dc) is a unidirectional current. It will be assumed to have a constant magnitude. *Alternating current* (ac), in which the direction of the current keeps reversing, will be discussed in Chap. 29.

25.1 Electric Current

A metallic conductor, such as a copper wire, consists of a tremendously large number of atoms, each having electrons from outer orbits that are free to move through the material. In the absence of an electric field the rate at which electrons pass a point in the wire from left to right is the same as the rate at which they pass the point from right to left. The *net* rate is zero.

If the ends of the wire are connected to the terminals of a battery or generator, an electric field or potential gradient will be set up within the wire. This potential gradient will cause a continuous motion of electrons through the metallic conductor. If the terminals of the battery or generator are reversed, the motion of the electrons will be reversed. We therefore speak of the sources of electric energy as having fixed polarity. One terminal is called *positive* and the other *negative*.

Benjamin Franklin, one of the early investigators of electrical phenomena, assumed that an electric current was an "electric fluid" which flows from the positive terminal through the external circuit and back to the negative terminal. The early investigators had no knowledge of the concept of the electron. Thus developed the convention, still in use today, that "direction of current" is from a positively charged body to a negative (or less positively) charged body. Today it is known that the actual *direction of electron flow is from negative to positive* (Fig. 25.1).

As far as the *external* effects are concerned, it makes no difference if we consider hypothetical positive charges moving in one direction or real negative charges (electrons) moving in the opposite direction. The two ideas are equivalent. In this and subsequent chapters, we will adopt the *conventional direction* of current. We will even

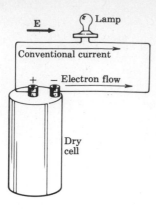

Fig. 25.1 In a metallic conductor, the conventional current is in the direction of the electric field. Electron flow is in the direction opposite that of the conventional current.

speak of the current in a metal conductor as though it consisted of motion of positive charges.

In later chapters we will discuss the motion of atoms (or groups of atoms) in solutions or gases that have lost or gained negative charges. These charged atoms are called *ions*. Those atoms that gain electrons are negative ions and those that lose electrons are positive ions. Positive ions do truly move in the direction of the conventional current in ionic conduction in solutions and gases. We will also study how positive " holes " in semiconductors move in the direction of the conventional current.

The American Institute of Electrical Engineers and The Institute of Radio Engineers use conventional current direction in their standards and in their publications.

> *The conventional direction of current is the same as the direction in which a positive charge will move.*

Thus our conventional current direction will agree with the concepts developed in the chapter on electrostatics. You will remember that electric field, potential energy, and potential difference were defined in terms of positive charges. We will learn later, also, that the conventional direction will result in loss in electric energy " down a potential gradient " when it flows through the external circuit from + to −.

In the cases where we will need to consider actual electron movement (as in transistors, electron tubes), we will speak of *electron flow* and not current.

25.2 The Magnitude of Electric Current—the Ampere

An electric current is a flow of charge. The electron is an extremely small electric charge. Consequently, a larger unit of electric charge or quantity of electricity is used for practical measurements. This unit, the *coulomb*, equals approximately 6.3×10^{18} (6.3 billion billion) electrons. The magnitude of an electric current in a circuit is the *time rate of flow of electric charge*. Current is represented by the letter I (or i). The SI-metric unit of current is the *coulomb per second* and is called the *ampere* (A), named after the French physicist André Ampère. Thus the formula relating amperes I, coulombs Q, and seconds t is:

$$I = \frac{Q}{t} \qquad (25.1)$$

We speak of a 75-W incandescent lamp as using about 0.62 A of current. This means that about 0.62 C of electricity passes through the lamp each second. Small currents are often expressed in milliamperes (mA) or microamperes (μA) where

$$1 \text{ mA} = 10^{-3} \text{ A}$$

$$1 \text{ } \mu\text{A} = 10^{-6} \text{ A}$$

Illustrative Problem 25.1 What quantity of electricity passes through the filament of an incandescent lamp in 3 hr if it draws 0.41 A?

Solution

$$Q = I \times t$$
$$= 0.41 \text{ A} \times 3 \text{ hr} \times 60 \text{ min/hr} \times 60 \text{ sec/min}$$
$$= 4400 \text{ C} \qquad\qquad\qquad ans.$$

In actual practice the term *coulomb* is seldom used, for we are less interested in the quantity of charge supplied than in the *electric energy* available. The ampere has another (*official*) definition. That definition will be given in Chap. 25.

25.3 Hydraulic Analogy of a Circuit

When a wire is connected to the terminals of a continuous source of potential difference, electrons will flow through the wire from the negative to the positive terminal. The electrons flowing through the wire are repeatedly deflected or stopped as they strike neighboring atoms. The kinetic energy lost by the electrons in these collisions is gained by the atoms. The flow of electrons through the conductor will generate heat in the wire and raise its temperature. Thus, the electron motion is not a constant accelerated motion from one end of the wire to the other. Rather there is a slow drift flow (about 0.2 mm/sec average speed) of electrons. However, the *changes* in flow rate are propagated with a speed approaching the speed of light. The propagation is similar to that which occurs when a water valve connected to a full hose of water is opened or closed. Almost as soon as the water flow is altered at the valve, an equal change occurs at the other end of the hose. Yet each molecule of water is likely to move very little in the interval of time it takes for the change to occur at the open end.

We can illustrate the flow of electricity by comparing it to water flow. The water pipe corresponds to the electric conductor. The amount of water per second (gallons per second) flowing past a given point in the pipe corresponds to the electric current (coulombs per second). When one ampere flows in a wire it means that 6.3×10^{18} electrons per second enter one end of the wire and an equal number of electrons per second leave the wire at the other end. Chances are that *very few* are the same electrons.

25.4 Difference of Potential—the Volt

Let us carry our hydraulic analogy further. Water will not move in pipes unless the pressure behind it is greater than the pressure in front of it. Similarly, electrons move along a wire only if there is a difference of electric pressure (called *potential*

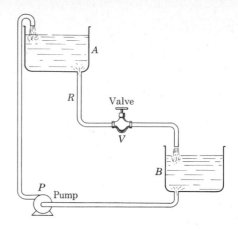

Fig. 25.2 Hydraulic analogy of an electric current.

difference) along the conductor. This difference is produced by a battery or a generator, which acts as a pressure pump to produce a potential difference, just as a water pump (Fig. 25.2) builds up water pressure. Points along the pipe R are at different pressures; hence, when the valve V is open, water will flow. The pump P maintains the pressure by lifting water to reservoir A. The unit difference of mechanical pressure is in newtons per square meter or in pounds per square inch. In a similar manner (see Fig. 25.3) a generator G builds up electric potential difference so that electric charges will flow in the circuit when the switch K is closed. The unit of electric pressure is the *volt*. The common dry cell develops a voltage of 1.5 V. Most household circuits are supplied at 120 or 220 V. Industrial circuits are typically supplied at 220 or 440 V.

A source of electric current, such as a battery or a generator, produces a potential difference across its terminals (battery) or windings (generator). This *potential difference across a source* is known as *electromotive force* (emf). It is the zero-current potential difference supplied by the source. Electromotive force is also measured in volts, the same as potential difference. We speak of the emf (pronounced "ee-em-eff") of a circuit as the total voltage applied to the circuit, or as the total potential drop across the entire circuit. The

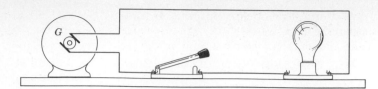

Fig. 25.3 Simple electric circuit.

term *potential drop* may be used to describe the voltage drop across any portion of a circuit. Thus we see that though emf and potential drop are measured in the same units (volts), they are not at all synonymous terms. Hereafter, *emf* or $\mathscr{E}$ will be used to designate the potential difference that exists across a battery, generator, or other *source* of electrical energy that is not connected to an external circuit. On the other hand, V will designate the potential difference between any two points of a closed electric circuit. An electromotive force $\mathscr{E}$ causes differences of potential V to exist between points in a closed circuit. The concept of emf will be covered in greater detail in Chap. 26.

25.5 The Unit of Resistance Is the Ohm

If there were no resistance to a current in a conductor, no potential difference would be needed along its length. But just as the stream of water flowing in the pipe of Fig. 25.2 is retarded by frictional resistances, so various materials offer resistances to electric currents. The degree to which a substance offers resistance to an electric current divides the substances into two broad classes: *conductors* and *nonconductors* or *insulators*. All conductors, except the superconductors, offer some resistance to the flow of electricity. Superconductors are a group of metals, alloys, and compounds that, when cooled to temperatures below 4.15 K, offer essentially zero resistance to the passage of an electric current.

George Simon Ohm (1787–1854), a German scientist, studied the relationship between potential differences applied to the ends of a metal wire and a current applied through it. He applied increasing potential differences to the ends of a metal wire whose temperature was kept constant.

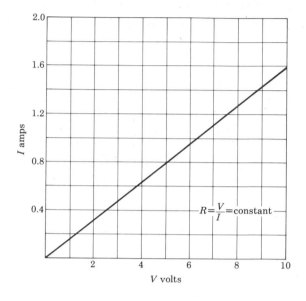

Fig. 25.4 For a metal conductor at constant temperature, V/I is a constant.

He then measured the current in the wire and discovered that the current increased in direct proportion to the difference of potential. The potential difference-current graph for a given wire at a fixed temperature is a straight line (see Fig. 25.4).

The constant ratio of potential difference to current is called the *resistance* of the conductor. The relationship between potential difference and current in a metal conductor is known as *Ohm's law*. In equation form

$$\frac{\text{Potential difference}}{\text{Current}} = \text{resistance} \quad \text{(a constant)}$$

$$\frac{V}{I} = R \qquad (25.2)$$

When V is expressed in volts and I is expressed in amperes, R is expressed in the unit called *ohm* (Ω). Thus, the ohm has the dimension volt/ampere.

Unit of Electrical Resistance
The ohm is the resistance of a conductor such that a potential difference of one volt will maintain a current of one ampere in the conductor.

Large resistances are usually expressed in megohms (MΩ). Small resistances are expressed in micro-ohms ($\mu\Omega$).

$$1 \text{ megohm} = 10^6 \ \Omega$$

$$1 \text{ micro-ohm} = 10^{-6} \ \Omega$$

Illustrative Problem 25.2 A 1000-W toaster when plugged into a 110-V outlet draws 12 A. What is the operating resistance of the toaster?

Solution

$$V = 120 \text{ V}, I = 12 \text{ A}$$

$$R = V/I$$

$$= \frac{120 \text{ V}}{12 \text{ A}} = 10 \ \Omega \qquad \qquad ans.$$

The electrical resistance of a metallic conductor depends on four factors: (1) the kind of material, (2) the length, (3) the cross-sectional area, and (4) the temperature of the wire. As a general rule, pure metals offer smaller resistances, while alloys offer considerably greater resistances to current flow. A conductor 2 ft long offers twice as much resistance as a similar conductor 1 ft long. Also, the larger the cross-sectional area of a conductor of a given material and of a fixed length, the lower its resistance. This is true because there are more electrons free to move in the larger conductor. The resistance of most metal conductors increases with an increase in temperature.

25.6 Wire Measurements

Since most common wires in the United States still have diameters which are fractions of an inch, it is more convenient to express these diameters in terms of a unit called the *mil*. A mil is equal to 0.001 in. Thus a wire whose diameter is 0.072 in. is said to have a diameter of 72 mils.

Since most electric wires are circular in cross section, it is simpler to specify the area in *circular mils* than in square units. A *circular mil* (cmil) is defined as *the area of a circle whose diameter is 1 mil*. Using circular units eliminates the use of π. Area in circular mils equals the square of the diameter in mils. The area of a circle, $A = \pi d^2/4$, is proportional to the diameter squared. By definition, the area whose diameter is 1 mil is 1 cmil. Then the area of a circle whose diameter is d mils is d^2 cmils.

Illustrative Problem 25.3 What is the area (a) in circular mils and (b) in square inches of a wire 0.032 in. in diameter?

Solution

(a) $\quad$ Diameter $= 0.032$ in. $\left(\dfrac{1 \text{ mil}}{0.001 \text{ in.}} \right)$

$$= 32 \text{ mils}$$

Then $\quad$ Area $= (32 \text{ mil})^2 \left(\dfrac{1 \text{ cmil}}{1 \text{ mil}^2} \right)$

$$= 1024 \text{ cmils} \qquad \qquad ans.$$

(b) $\quad$ Area $= \dfrac{\pi d^2}{4}$ in conventional square units

$$= \frac{3.14(0.032 \text{ in.})^2}{4}$$

$$= 0.000804 \text{ in.}^2 \qquad \qquad ans.$$

25.7 Resistivity

The practical (English system) unit of length for wires is the foot. The resistances of most conductors are generally standardized in terms of the resistance of a wire having an area of 1 cmil and a length of 1 ft. Such a unit wire is called the *circular mil/foot*. The *resistivity* of a substance is equal numerically to the resistance of 1 cmil/ft of the substance. The unit of resistivity is the *ohm-circular mil per foot*. In the metric system the resistivity of a material is numerically equal to the resistance of a piece of the material 1 m long and 1 m^2 in cross-sectional area. The unit of resistivity in the SI-metric system is called the *ohm-meter* (ohm-meter^2/meter = ohm-meter). The symbol ρ is used to denote resistivity units. To obtain ρ in ohm-mils per foot, multiply ρ in ohm-meters by 6.02×10^8; to obtain ρ in ohm-meters, multiply ρ in ohm-mils per foot by 0.166×10^8. Table 25.1 lists resistivities of some common materials.

Table 25.1 shows that the pure metals have the lower resistivities. Copper and aluminum are generally used for electric-transmission purposes. Silver is an even better conductor of electricity than copper, but it is not used extensively because of its high cost. Gold, however, is used frequently especially for electric contact. Because of its low resistivity it doesn't oxidize (burn) readily and is quite ductile. Tungsten is used almost exclusively for filaments of incandescent lamps. Alloys have relatively higher resistances than the pure metals. They are used in heating units because they have high resistances and because they do not oxidize (burn) readily at high temperatures. The alloys listed are quite hard and durable.

25.8 Calculation of the Electric Resistance of a Conductor

The resistance of a conductor of uniform cross section is directly proportional to the resistivity and the length and inversely proportional to the cross-sectional area of the conductor. In the English system

$$R = \rho \frac{L}{d^2} \qquad (25.3)$$

where

R = resistance, ohms
ρ = resistivity, ohm-cmils/foot
L = length, foot
d^2 = cross-sectional area of conductor, cmils

In the SI system

$$R = \rho \frac{L}{A} \qquad (25.4)$$

Table 25.1 Resistivities of common conductors at 20°C

	Resistivity ρ	
	ohm-meter	*ohm-circular mils/foot*
Elements:		
Aluminum, commercial	2.8×10^{-8}	16.9
Copper, commercial	1.7×10^{-8}	10.3
Gold	2.4×10^{-8}	14
Iron	$12–14 \times 10^{-8}$	72–84
Lead	22×10^{-8}	132
Mercury	96×10^{-8}	575
Silver	1.6×10^{-8}	9.7
Tungsten	5.5×10^{-8}	33
Zinc	6.1×10^{-8}	36.7
Alloys:		
Brass	$6.3–8.6 \times 10^{-8}$	38–52
Chromel	$104–109 \times 10^{-8}$	625–655
Constantan (Cu 60%, Ni 40%)	49×10^{-8}	295
German silver	33×10^{-8}	199
Manganin (Cu 84%, Mn 12%, Ni 4%)	43×10^{-8}	260
Nichrome (Ni 60%, Cr 12%, Mn 2%, Fe 26%)	110×10^{-8}	660

If R is expressed in ohms, A in square centimeters, and L in centimeters, then ρ is in ohm-cm^2/cm, or *ohm-centimeters*. This unit is usually more convenient than the ohm-meter. The value of ρ in ohm-meters is multiplied by 10^2 to get ρ in ohm-cm.

Illustrative Problem 25.4 Find the resistance of 2000 ft of commercial copper wire having a diameter of 0.102 in.

Solution The cross-sectional area of a wire whose diameter is 0.102 in. is

$$d^2 = (102 \text{ mils})^2 = 10{,}404 \text{ cmils}$$

From Table 25.1, ρ (for commercial copper) = 10.3 Ω-cmils/ft. Substituting in Eq. (25.3) gives

$$R = 10.3 \ \Omega\text{-cmils/ft} \times \frac{2000 \text{ ft}}{10{,}404 \text{ cmils}}$$

$$= 1.98 \ \Omega \qquad\qquad\qquad ans.$$

Illustrative Problem 25.5 How much nichrome wire of diameter 0.081 cm will have to be used in order to build a resistor of 5 Ω (20°C)?

Solution Solving for L in Eq. (25.4), we have

$$L = R\frac{A}{\rho}$$

From Table 25.1,

$$\rho = 110 \times 10^{-8} \ \Omega\text{-m}$$

Substituting yields

$$L = \frac{5 \ \Omega \times (\pi/4)(8.1 \times 10^{-4} \text{ m})^2}{110 \times 10^{-8} \ \Omega\text{-m}}$$

$$= 2.34 \text{ m} \qquad\qquad\qquad ans.$$

Illustrative Problem 25.6 Find the resistivity of a wire made of an alloy of copper, zinc, and nickel whose mean diameter, measured by a micrometer caliper, is 0.072 in. The wire is 100 ft long and has a total resistance of 3.88 Ω.

Solution Solve Eq. (25.3) for ρ to get

$$\rho = R\frac{d^2}{L}$$

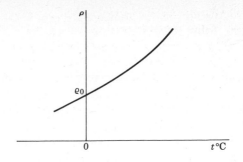

Fig. 25.5 Variation of resistivity of a metal with temperature.

$$d^2 = (72 \text{ mils})^2 = 5184 \text{ cmils}$$

Then

$$\rho = 3.88 \ \Omega \times \frac{5184 \text{ cmils}}{100 \text{ ft}}$$

$$= 201 \ \Omega\text{-cmils/ft} \qquad\qquad ans.$$

25.9 Temperature and Resistivity

The resistivity of most pure metals increases with a rise in temperature. In nonmetallic conductors the resistivity decreases as the temperature rises. Most alloys not only have relatively high resistivities, but their resistivities are much less affected by temperature changes.

The relationship between resistivity and temperature for a typical metallic conductor is shown in Fig. 25.5. For temperatures which are not too great, the graph is approximately a straight line. Thus we say that resistivity varies directly with the temperature. We can express this relationship by the equation

$$\rho = \rho_0(1 + \alpha t) \qquad\qquad (25.5)$$

where ρ_0 is the resistivity at 0°C and ρ is the resistivity at t°C. The quantity α (Greek letter *alpha*) is called the *temperature coefficient of resistivity*. The temperature coefficient of resistivity is the fractional increase in resistivity per degree increase in temperature. Its units are deg^{-1} or "reciprocal degrees." Table 25.2 lists temperature coefficients of resistivity for several common materials.

Table 25.2 Temperature coefficients of resistivity

Material	α (°C^{-1})
Metals:	
Aluminum	0.0039
Carbon	−0.0005
Copper, annealed	0.0039
Gold	0.0034
Iron	0.0050
Lead	0.0043
Mercury	0.0089
Platinum	0.0030
Tin	0.0042
Tungsten	0.0045
Zinc	0.0020
Alloys:	
Brass	0.0020
Constantin	0.00003
Nichrome	0.0004

Since the resistance of a given conductor is proportional to its resistivity, and its length and cross sectional area change little with temperature changes, its resistance at any temperature t may be written

$$R_t = R_0(1 + \alpha t) \qquad (25.6)$$

where R_0 is the resistance of the conductor at 0°C and R_t is the resistance at t°C.

Illustrative Problem 25.7 The resistance of a coil of copper wire at 20°C is 2.4 Ω. What is the resistance of the coil at 50°C?

Solution Solve Eq. (25.6) for R_0 and substitute values for R_t, α, and t.

$$R_0 = \frac{R_t}{1 + \alpha t}$$

$$= \frac{2.4\ \Omega}{1 + (0.0039/°C)(20°C)}$$

$$= 2.23\ \Omega \quad \text{(the resistance of the coil at 0°C)}$$

Then at 50°C,

$$R_{50} = R_0(1 + \alpha t)$$
$$= 2.23\ \Omega(1 + 0.0039\ °C^{-1} \times 50°C)$$
$$= 2.7\ \Omega \qquad\qquad ans.$$

Illustrative Problem 25.8 The resistance of a tungsten filament at 20°C is 2.5 Ω. What is the temperature of the filament if its resistance is 17.5 Ω? Assume α is constant over the temperature range.

Solution From Table 25.2, we find $\alpha = 0.0045$ °C^{-1}.

$$R_0 = \frac{R_t}{1 + \alpha t} = \frac{2.5\ \Omega}{1 + 0.0045°C^{-1} \times 20°C}$$

$$= 2.29\ \Omega \text{ at } 0°C$$

Solve Eq. (25.6) for t to get

$$t = \frac{R_t - R_0}{\alpha R_0}$$

$$= \frac{17.5\ \Omega - 2.5\ \Omega}{(0.0045°C^{-1})(2.5°C)}$$

$$= 1300°C \qquad\qquad ans.$$

25.10 Wire Gauge

Copper wire is commonly manufactured in accordance with standard specifications. For convenience, wire sizes are indicated by gauge numbers. In the United States the commonly used gauge is the *Brown & Sharpe* (B & S), often called the *American wire gauge* (AWG). The largest commercial-size round wire has an AWG number of 0000. The next smaller size is No. 000, then No. 00, then No. 0. The sizes then vary in whole-number values ranging from No. 1 to 40, which is the smallest size. Standard copper wire sizes and their gauge numbers plus other valuable information will be found in Appendix V. This table will simplify wire computations. It reveals that the wires grow smaller as the numbers increase and that the ratio of the areas of any two consecutive standard sizes is about 1.26. It also shows that every third gauge number halves the area of cross section and thus doubles the resistance.

In checking the table, you will note that the resistance of a copper wire 1000 ft long and 0.1 in. in diameter is about 1 Ω. This will be convenient to remember in estimating resistances of other copper wires. For example, you should be able to determine readily that the resistance of a copper wire 4000 ft long and 0.2 in. in diameter will also be 1 Ω.

25.11 Ohm's Law

Ohm's law (Eq. 25.2) can be rewritten in the forms

$$I = \frac{V}{R} \qquad (25.2')$$

and

$$V = IR \qquad (25.2'')$$

Equation (25.2') expresses the fact that the current passing through a metal conductor is directly proportional to the potential difference and inversely proportional to the resistance of the conductor (when the temperature is kept constant).

Equation (25.2'') is used to find the voltage V required to cause a current I to flow in a conductor whose resistance is R.

Two limitations to Ohm's law should be noted: (1) the law is true only if external conditions, e.g., the temperature of the conductor, are maintained constant; (2) the law is strictly valid only for ordinary metallic conductors. The current is not proportional to the potential difference, except over limited ranges, when electricity is conducted in liquids, gases, or solid-state devices.

Illustrative Problem 25.9 How much current will an electric heater draw from a 120-V line if the resistance of the heater (when hot) is 26.7 Ω?

Solution Substitute known values in Eq. (25.2') and solve:

$$I = \frac{V}{R}$$

$$= \frac{120 \text{ V}}{26.7 \text{ Ω}}$$

$$= 4.5 \text{ A} \qquad ans.$$

It is essential to distinguish carefully between the application of Ohm's law to a part of a circuit and to a complete circuit. Equation 25.2 expresses the relationship between the resistance R, the current I, and the potential difference V for a *part of a closed circuit*. When a *complete circuit* is to be considered, one must take into account all the potential differences (which are due to the emfs $\mathscr{E}$ of the sources of energy) and all the resistances R_t in the circuit. In equation form, Ohm's law, then, becomes

$$\frac{\text{Net emf}}{\text{Current}} = \text{total resistance}$$

$$\frac{\mathscr{E}}{I} = R_t \qquad (25.7)$$

25.12 Measurement of Current and Voltage

The two instruments most commonly used to measure the current and voltage in a given circuit are the *ammeter* and the *voltmeter* (Fig. 25.6). They are both direct-reading instruments; i.e., by reading the deflection of a pointer on each instrument when the current is flowing, one can read directly the amperage and the voltage. The ammeter has an extremely low resistance. Thus, when an ammeter is placed *in a circuit*, the total resistance of the circuit is not appreciably changed. On the other hand, the voltmeter has a very high internal resistance. When it is connected *across a circuit*, the total current flow is not appreciably affected.

Scientists generally use schematic diagrams to represent the essential parts of a circuit. Standard symbols are used to represent specific apparatus. Many of the standard wiring diagram symbols are shown in Appendix VI. The student will become acquainted with the more common ones as we proceed. Note the symbols for the ammeter and the voltmeter in Fig. 25.6*c*. The figure illustrates how the ammeter and the voltmeter are connected in a single circuit. Note that the ammeter is inserted directly *in the line* and the voltmeter is hooked *across* the portion of the circuit whose voltage drop is to be measured.

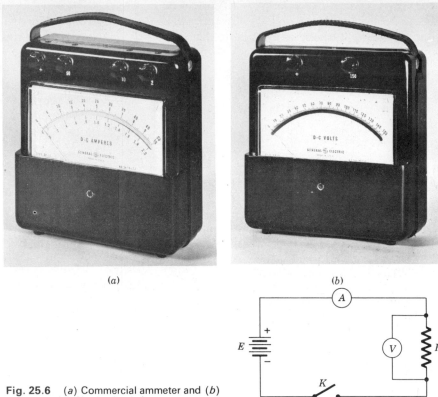

Fig. 25.6 (*a*) Commercial ammeter and (*b*) voltmeter. (*c*) Single circuit for measurement of amperes and volts.

In Fig. 25.6*c* the source of the emf is a battery *E*. Note its symbol. The ammeter *A* is connected *in* the circuit, and the voltmeter *V across* the load *R*. *K* is a single-pole switch. The ammeter is said to be connected in *series*, while the voltmeter is connected in *parallel* with resistor *R*. Thus if we read *A* and *V*, we will get the current *through* and the potential difference *across R*.

It should be pointed out that in a schematic diagram, the conductors connecting elements of a circuit are idealized to have no resistance. Such conductors are usually represented by straight lines. Thus, for example in Fig. 25.6*c*, all parts of the conductor connecting the battery and the ammeter to one side of the load *R*, are at the same potential. There is a potential drop in *R*, in the conventional direction of current. All parts of the conductor from the other end of the load *R*, through the switch and back to the battery, are at

the same potential also. The potential in this second part of the circuit is lower than that in the first by the magnitude of the potential drop measured by the voltmeter, *V*.

25.13   Resistors

In Tables 25.1 and 25.2 it can be noted that the resistivity of nichrome is approximately 60 times the resistivity of copper, and its temperature coefficient is about 10% that of copper. These properties make nichrome a good metal from which to make resistors. Spools wound with nichrome (or chromel) wire are used in an electric circuit to provide either *fixed* or *variable* amounts of resistance. Fixed nichrome resistors are used in electric heaters, electric ranges, and toasters. Such commercial resistors are manufactured in sizes ranging from a fraction of 1 Ω up to more than

Fig. 25.7 Slide-wire rheostat (Central Scientific Co.)

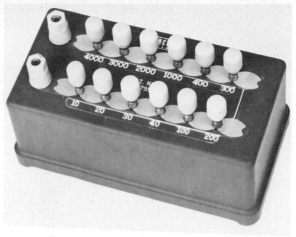

(a)

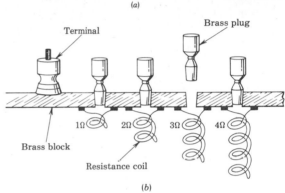

(b)

Fig. 25.8 (a) Resistance box. (b) Cutaway view of the inside of a resistance box.

100,000 Ω. The *slide wire rheostat* (Fig. 25.7) consists of a spiral coil of nichrome or chromel wire wound on a porcelain or enamel tube. Connections are made to one end and to a sliding contact. In Fig. 25.7, the current enters the coil at the terminal on the left, passes through the coil, and leaves it at the moving contact. From there it travels along a massive copper bar (assumed to have zero resistance) and leaves the rheostat at the right terminal. By moving the slider, the resistance can be varied within the range of the resistance of the entire length of the coil. In electronics, variable resistors are smaller. Some are made by winding the wire around a toroid (doughnut-shaped) form. The moving contact in this case moves along a circular path.

A resistance box used in the laboratory is shown in Fig. 25.8a. Resistance coils are soldered to solid brass blocks as shown in Fig. 25.8b. Tapered brass plugs are inserted between adjacent blocks. When all the plugs are in place, the electric current will flow through the bar which has negligible resistance. Removing any plug will cause the current to flow through one of the coils. Thus, if the 2-Ω and 3-Ω plugs are out, the resistance of the box will be 5 Ω.

Dial type variable resistance boxes are commonly used today in electrical laboratories. The principle of the dial type is the same as that for the plug type. However, the plugs are replaced by a set of three, four, or five dials which provide a mechanism to switch in various resistances.

Each switch has a multiple leaf phosphor bronze spring which provides a positive, wiping action against heavy brass plates to which are attached the various resistance coils (or high-grade carbon resistors). Each dial has ten positions, with each dial representing a multiple decade value of resistance. For example, in the four dial resistance box with a range of 0.1–999.9 Ω, the dials are marked from left to right: 100, 10, 1, 0, 1. The resistance is indicated by reading, from left to right across the box, numbers appearing next to the dials on the plastic box which houses the mechanism. Thus, if the position numbers of the four dials (left to right) are 7, 0, 5, and 3, a resistance of 705.3 Ω would be indicated.

Table 25.3 EIA resistor color code

Color	Band A (1st significant figure)	Band B (2nd significant figure)	Band C (multiplier)	Band D (tolerance, %)
Black	0	0	$\times 10$	
Brown	1	1	$\times 10^1$	
Red	2	2	$\times 10^2$	
Orange	3	3	$\times 10^3$	
Yellow	4	4	$\times 10^4$	
Green	5	5	$\times 10^5$	
Blue	6	6	$\times 10^6$	
Purple	7	7	$\times 10^7$	
Gray	8	8	$\times 10^8$	
White	9	9	$\times 10^9$	
Silver				10
Gold				5

Fig. 25.9 EIA color-coded bands of a carbon resistor.

Small commercial resistors have wide applications in electronics; one widely used type is the *carbon resistor*. A mixture of carbon (or graphite) granules and a binding material can be varied to yield a wide range of high resistances. The hardened mixture is enclosed in a ceramic housing with a wire connector attached to each end of the resistor. Since the carbon resistor is quite small in size, it is easier to use a color code to indicate the resistance value than to print the resistance on the housing. The carbon resistor has four color-code bands. The EIA (Electronics Industries Association) has adopted the color code described in Table 25.3 and Fig. 25.9. Band *A* is closest to one edge of the resistor. Band *B* is farther away from the other edge. Bands *A* and *B* indicate the first two significant figures of the resistance value. The first two significant numbers are multiplied by the multiplier indicated by band *C*. Band *D* indicates the accuracy of the indicated resistance. A missing fourth (*D*) band indicates that the resistance value is correct within $\pm 20\%$ of the coded value.

Illustrated examples 25.9

Color bands	Resistance, Ω
orange-green-red-silver	$35 \times 10^2 \pm 10\%$
brown-black-orange-gold	$10 \times 10^3 \pm 5\%$
purple-yellow-black-gold	$74 \pm 5\%$
purple-yellow-black	$74 \pm 20\%$

Composition resistors are generally less expensive than metal ones. They are commonly used in low-power situations. Wire-wound or metal-film resistors are usually more accurate than carbon resistors.

SERIES AND PARALLEL CIRCUITS

25.14 Series Circuits

Most electrical circuits consist not merely of a single source of emf and a single external resistor. They usually comprise a number of emfs, resistors, or other elements such as capacitors, interconnected in various ways. To begin with, we will discuss a few of the simpler ways resistors may be interconnected in a circuit.

Consider the circuit illustrated in Fig. 25.10. In it there is one source of emf, and the resistors provide a single path between the end points of

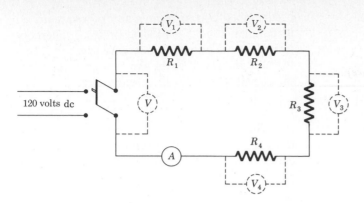

Fig. 25.10   A circuit with resistors in series.

the emf source. In such a circuit the resistors are said to be connected in *series*. Other circuit elements such as capacitors, cells, motors, etc., are also said to be connected in series with one another if they are similarly connected so as to provide a single path for the current.

> *A series combination of elements in a circuit is one in which the current is the same in every part of the circuit.*

An ammeter can be connected anywhere in a series circuit because the current is the same at every point in the circuit. Thus, in Fig. 25.10,

$$I = I_1 = I_2 = I_3 = I_4 \qquad (25.8)$$

where I, I_1, I_2, I_3, I_4 represent the current from the source, and the current in the resistors R_1, R_2, R_3, R_4, respectively.

The total *resistance R* of the circuit is equal to the *sum of the resistances of all the parts*. If we neglect the resistance of the wires connecting the loads, we get

$$R_T = R_1 + R_2 + R_3 + R_4 \qquad (25.9)$$

The emf, measured by the voltmeter V, must be equal to the sum of the potential drops measured by V_1, V_2, V_3, and V_4. If we express the voltages as V, V_1, V_2, V_3, V_4, respectively, we get

$$V = V_1 + V_2 + V_3 + V_4 \qquad (25.10)$$

Equation (25.9) can also be expressed in another manner. Since from Ohm's law, $V_1 = IR_1, V_2 = IR_2, V_3 = IR_3$, and $V_4 = IR_4$, we

can substitute these values in Eq. (25.9) and get

$$V = IR_1 + IR_2 + IR_3 + IR_4$$

or $$V = I(R_1 + R_2 + R_3 + R_4) \qquad (25.10')$$

To summarize, *for a series circuit* (1) *the current is the same in every part of the circuit,* (2) *the resistance of the combination of resistors is equal to the sum of the resistances of the individual resistors, and* (3) *the total voltage across the combination is equal to the sum of the voltage drops across the separate resistors.* The voltage drop across any individual resistor in the circuit is directly proportional to the resistance of that portion of the circuit. Such a circuit as in Fig. 25.10 is referred to as a *voltage divider network*.

Illustrative Problem 25.10   A series circuit consisting of three resistors having resistances of 40, 50, and 20 Ω, respectively, is connected across a 120-V line (Fig. 25.11). Find (*a*) the current in the circuit and (*b*) the voltage drop across each resistor.

Solution
(*a*) The total resistance R of the circuit is equal to the sum of the individual resistances.

$$R = R_1 + R_2 + R_3$$
$$= 40\ \Omega + 50\ \Omega + 20\ \Omega$$
$$= 110\ \Omega$$

The current is equal to the total voltage divided by the total resistance.

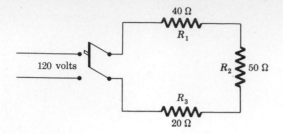

Fig. 25.11   Illustrative Problem 25.10.

$$I = \frac{\mathscr{E}}{R}$$

$$= \frac{120 \text{ V}}{110 \text{ }\Omega}$$

$$= 1.09 \text{ A} \qquad ans.$$

(b) The amount of current in each resistor is then equal to 1.09 A. Then

$$V_1 = IR_1 = 1.09 \text{ A} \times 40 \text{ }\Omega = 43.6 \text{ V}$$

$$V_2 = IR_2 = 1.09 \text{ A} \times 50 \text{ }\Omega = 54.5 \text{ V}$$

$$V_3 = IR_3 = 1.09 \text{ A} \times 20 \text{ }\Omega = 21.8 \text{ V} \quad ans.$$

The answers can be checked by showing that

$$\mathscr{E} = V_1 + V_2 + V_3$$

25.15   Parallel Circuits

When two or more resistors are joined in *parallel* (Fig. 25.12*a*), the effect is similar to that which occurs when two or more water pipes are joined in parallel (Fig. 25.12*b*). The pressure difference (water) or the potential difference (electricity) between *A* and *B* will be the same for each branch of the circuit. Hence, in the electric circuit,

$$\mathscr{E} = V_1 = V_2 = V_3 \qquad (25.11)$$

The total amount of current that will flow from *A* to *B* is equal to the sum of the separate currents passing through each channel. For a parallel circuit, then,

$$I = I_1 + I_2 + I_3 \qquad (25.12)$$

If we denote the total resistance of the combination of parallel resistors of Fig. 25.12*a* as R_T, we know from Eq. (25.7) that $R_T = \mathscr{E}/I$; that is, the total resistance of any system is equal to the quotient of the potential difference divided by the current for that system. Thus, if we invert Eq. (25.7),

$$\frac{1}{R_T} = \frac{I}{\mathscr{E}}$$

Replacing I by its equivalent $I_1 + I_2 + I_3$, gives

$$\frac{1}{R_T} = \frac{I_1 + I_2 + I_3}{\mathscr{E}}$$

$$= \frac{I_1}{\mathscr{E}} + \frac{I_2}{\mathscr{E}} + \frac{I_3}{\mathscr{E}}$$

$$= \frac{I_1}{V_1} + \frac{I_2}{V_2} + \frac{I_3}{V_3} \quad \text{from Eq. (25.11)}$$

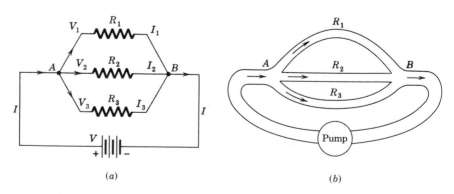

(*a*)   (*b*)

Fig. 25.12   Hydraulic analogy of resistors connected in parallel.

or $\quad \dfrac{1}{R_T} = \dfrac{1}{R_1} + \dfrac{1}{R_2} + \dfrac{1}{R_3} \qquad$ (25.13)

Therefore, *the reciprocal of the total resistance of a combination of resistors in parallel is equal to the sum of the reciprocals of the separate resistors.*

Illustrative Problem 25.11 A 100-W incandescent lamp, when hot, has a resistance of 144 Ω, while a hot 60-W lamp has a resistance of 240 Ω. Compute (*a*) the current in each lamp, (*b*) the total current in the circuit, and (*c*) the total circuit resistance when two 100-W lamps and four 60-W lamps, all connected in parallel, operate from a 120-V line.

Solution

(*a*) The potential difference across each lamp is 120 V. Hence, we can readily compute the current through each lamp by using Ohm's law.

For each 100-W lamp: $I_1 = \frac{120}{144} = 0.83$ A

For each 60-W lamp: $I_2 = \frac{120}{240} = 0.50$ A $\qquad$ *ans.*

(*b*) The total current I will equal the sum of the individual currents.

$$I = 2I_1 + 4I_2$$
$$= 2(0.83 \text{ A}) + 4(0.50 \text{ A})$$
$$= 3.66 \text{ A} \qquad\qquad\qquad ans.$$

(*c*) To determine the total resistance use Ohm's law as follows:

$$R_T = \dfrac{V}{I_T}$$
$$= \dfrac{120 \text{ V}}{3.66 \text{ A}}$$
$$= 32.8 \ \Omega \qquad\qquad\qquad ans.$$

25.16 Series-Parallel Circuits

Thus far we have dealt only with separate series and parallel circuits. Practical electric circuits, however, very often consist of combinations of the two basic types. There are a great number of combinations possible. Figure 25.13 illustrates two of the more elementary types. In type (*a*) R_1 and R_2 are in series, R_3 and R_4 are in series, and the two groups are connected in parallel. In type (*b*) R_1 and R_2 are in parallel, R_3, R_4, and R_5 are in parallel, and these two groups are connected in series. There is no definite procedure that can be followed in solving combination series-parallel circuits. The method of approach depends upon the arrangement. However, whenever possible, it is a good practice to replace each parallel branch by an equivalent series branch.

Illustrative Problem 25.12 Compute the total resistance of the circuit of Fig. 25.13*a* if $R_1 = 10 \ \Omega$,

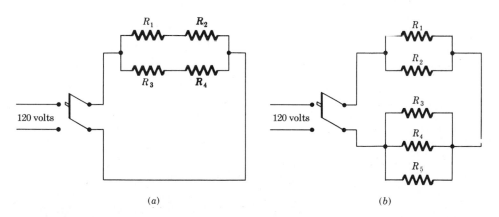

(*a*) $\qquad\qquad\qquad\qquad\qquad\qquad$ (*b*)

Fig. 25.13 Typical series-parallel circuits.

$R_2 = 30\ \Omega$, $R_3 = 20\ \Omega$, and $R_4 = 40\ \Omega$. What is the total current in the circuit?

Solution The resistors R_1 and R_2 can be replaced by a single resistor R_x, where

$$R_x = R_1 + R_2$$

$$= 10\ \Omega + 30\ \Omega$$

$$= 40\ \Omega$$

In like manner, R_3 and R_4 can be replaced by R_y, where

$$R_y = R_3 + R_4$$

$$= 20\ \Omega + 40\ \Omega$$

$$= 60\ \Omega$$

The total resistance of the system (Fig. 25.14) can then be found by applying Eq. (25.13):

$$\frac{1}{R_T} = \frac{1}{R_x} + \frac{1}{R_y}$$

$$= \frac{1}{40\ \Omega} + \frac{1}{60\ \Omega} = \frac{5}{120\ \Omega}$$

$$R_T = \tfrac{120}{5} = 24\ \Omega \qquad\qquad ans.$$

The current can then be found by using Ohm's law.

$$I = V/R_T$$

$$= 120\ \text{V}/24\ \Omega$$

$$= 5\ \text{A} \qquad\qquad ans.$$

Illustrative Problem 25.13 If, in Fig. 25.13b, $R_1 = 5\ \Omega$, $R_2 = 40\ \Omega$, $R_3 = 30\ \Omega$, $R_4 = 20\ \Omega$, and $R_5 = 60\ \Omega$, compute (a) the total resistance R_T and (b) the total current I in the circuit.

Solution Replace the parallel resistors R_1 and R_2 by one resistor R_x (Fig. 25.15) solved as follows:

(a) $\qquad \dfrac{1}{R_x} = \dfrac{1}{5\ \Omega} + \dfrac{1}{40\ \Omega} = \dfrac{9}{40\ \Omega}$

$\qquad\qquad R_x = \tfrac{40}{9} = 4.4\ \Omega$

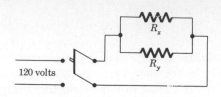

Fig. 25.14 Illustrative Problem 25.12.

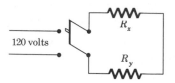

Fig. 25.15 Illustrative Problem 25.13.

In like manner replace R_3, R_4, and R_5 by one resistor R_y:

$$\frac{1}{R_y} = \frac{1}{30\ \Omega} + \frac{1}{20\ \Omega} + \frac{1}{60\ \Omega} = \frac{6}{60\ \Omega}$$

$$R_y = 10\ \Omega$$

Hence

$$R_T = R_x + R_y$$

$$= 4.4\ \Omega + 10\ \Omega = 14.4\ \Omega \qquad ans.$$

(b) Applying Ohm's law for I, yields

$$I = V/R_T$$

$$= 120\ \text{V}/14.4\ \Omega$$

$$= 8.3\ \text{A} \qquad\qquad ans.$$

25.17 Kirchhoff's Laws

Many circuits contain more than one source of emf and several devices which dissipate electric energy by converting it into heat. These more intricate networks are not easily solved by the method used in the preceding pages. In more complicated networks it is necessary to employ two laws given by Gustav Robert Kirchhoff (1824–1887) an eminent German physicist. Kirchhoff's laws refer to conditions that exist in a network of conductors where currents are either steady-state or alternating.

First, let us define two terms. A *junction* is a point of a circuit at which three or more conductors are joined. A *loop* is a closed conducting path in an electric circuit.

Kirchhoff's first law is essentially a restatement of the conservation of electric charge. It states that

At any junction in an electric circuit, the sum of the currents directed toward the junction equals the sum of the currents directed away from the junction.

Currents toward a junction are considered positive and those away from the junction negative. With this convention applied, the first law can be rephrased as follows:

The algebraic sum of the currents at a junction equals zero.

$$\Sigma I = 0 \qquad (25.14)$$

Kirchhoff's second law is essentially a restatement of the conservation of energy. It states that

In any closed loop of a circuit, the algebraic sum of the emfs equals the algebraic sum of the potential drops in the same loop.

$$\Sigma \mathscr{E} = \Sigma IR \qquad (25.15)$$

A rise in potential will be considered positive. A drop in potential will be negative. Remember that a current I in a resistance R corresponds to a potential difference IR. An alternative equation for the second law is

$$\Sigma \mathscr{E} - \Sigma IR = 0 \qquad (25.15')$$

The second law simply states that the total energy supplied by the sources of emf is equal to the total energy dissipated as the charges move around the loop.

In applying Kirchhoff's laws in circuit analysis it is convenient to carry out the following steps:

1. Arbitrarily assign a direction and a symbol to the current in each independent loop of the network. It is not essential that the selected direction be correct. If an incorrect direction is chosen, the current will simply show up as a negative number.

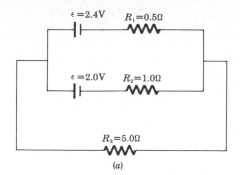

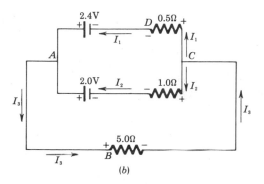

Fig. 25.16   Illustrative Problem 25.14.

2. Place appropriate positive and negative signs at the terminals of each source of emf and at every resistor for each loop. It must be remembered that the current external to the emf source is from the plus terminal to the negative terminal.

3. Use the two laws to develop as many independent equations as there are unknowns in the circuit. Be sure that no two of these equations have exactly the same unknowns in them.

4. Solve the equations for the desired unknowns.

Illustrative Problem 25.14   Two cells are connected to three resistances as shown in Fig. 25.16a. (a) Determine the current through each battery. (b) Determine the current through R_3.

Solution   Label points as shown in Fig. 25.16b. Arbitrarily select directions of the currents in the loops. At C, from Kirchhoff's first law,

$$I_3 = I_1 + I_2 \qquad (1)$$

For loop $ABCDA$, from Kirchoff's second law,

$$2.4 = 5.0I_3 + 0.5I_1 \qquad (2)$$

For loop $ABCFA$,

$$2.0 = 5.0I_3 + 1.0I_2 \qquad (3)$$

Subtract Eq. (3) from Eq. (2):

$$0.4 = 0.5I_1 - 1.0I_2 \qquad (4)$$

Substitute Eq. (1) in Eq. (3):

$$2.0 = 5.0(I_1 + I_2) + 1.0I_2$$

$$2.0 = 5.0I_1 + 6.0I_2 \qquad (5)$$

Multiply Eq. (4) by 6 and add to Eq. (5):

$$2.4 = 3.0I_1 - 6.0I_2 \qquad (4')$$

$$4.4 = 8.0I_1 \qquad (6)$$

Then

$$I_1 = \frac{4.4}{8.0} = \frac{11}{20} = 0.55 \text{ A} \qquad ans.$$

Substitute $I_1 = 0.55$ in Eq. (4) and solve for I_2:

$$0.4 = 0.5(0.55) - 1.0I_2$$

$$I_2 = 0.275 - 0.4 = -0.12 \text{ A} \qquad ans.$$

The negative sign indicates a wrong selection for current direction in loop $ABCFA$. The current is charging the 2.0-V battery.

THE MEASUREMENT OF RESISTANCE

25.18 Ammeter-Voltmeter Method of Measuring Resistance

When a high degree of accuracy is not required, an easy and simple method of measuring resistance is with an ammeter and a voltmeter. Figure 25.17 illustrates the circuit employed. A direct current is passed through the unknown resistance R. An ammeter A is connected in series with R, while a voltmeter V is connected across (parallel with) the resistance. Simultaneous readings are taken on the meters, and R is found by applying

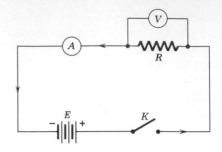

Fig. 25.17 Ammeter-voltmeter method of measuring resistance.

Ohm's law. This method is well adapted for general use in the electrical repair shop, where the electrician is interested in measuring field-coil resistances of generators and motors, resistances of armature windings, and resistances of various electric appliances while in operation.

Several precautions must be taken in using the ammeter-voltmeter method. Too much current should not be used, because if the resistance to be measured is heated, the computed results will be too high. (Of course, if it is the heated resistance which is desired, as in the case of lamps and heating appliances, the full rated current should be used.)

It will be noticed from Fig. 25.17 that A measures not only the current which passes through the unknown resistance R but also the current passing through V. However, it will be recalled that the voltmeter is an instrument of high internal resistance; hence relatively little

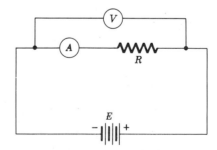

Fig. 25.18 When the resistance is high and the current small, the voltmeter is connected across both the unknown resistance and the ammeter.

current will pass through it, unless R is also of very high resistance. If R is a high resistance, the ammeter-voltmeter method does not provide very precise results.

When a very low resistance is to be measured, the potential drop across R will be small. Hence a sensitive voltmeter (*millivoltmeter*) is used for V. The millivoltmeter measures potential differences of thousandths of one volt. On the other hand, when the resistance is high, the amount of current passing through R may be small. In this case, a *milliammeter* can be used to good advantage. Generally, when the resistance is high and the current is small, the voltmeter is connected across both the unknown resistance and the ammeter, as shown in Fig. 25.18. In this case the ammeter will measure only the current going through R. Since the current is small, the potential drop across the ammeter (which has a very low resistance) will not be appreciable.

25.19 The Wheatstone Bridge

A more precise instrument for measuring resistances is the *Wheatstone bridge*. It consists essentially of a network of three known adjustable resistances R_1, R_2, R_3, and the unknown resistance R_x, joined by heavy connecting wires of very low resistance, as shown in Fig. 25.19a. The resistors are arranged in two parallel circuits. A sensitive current-measuring device, called a *galvanometer*, is bridged across the two circuits at points C and D. The construction and operation of the galvanometer will be discussed in Chap. 27.

In general, the current at A divides unequally into the two channels. As a rule, there will also be a potential difference between C and D as indicated by a deflection on the galvanometer G. However, by adjusting the values of the known resistances, it is possible to get the same potential at C and at D. When this occurs, no deflection will be shown by the galvanometer. This adjustment is known as *balancing the bridge*.

When the bridge is balanced, the potential difference between A and D will equal the poten-

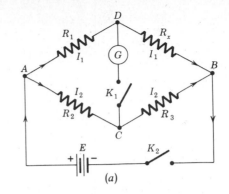

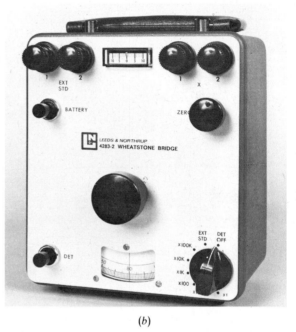

(b)

Fig. 25.19 (a) Wheatstone bridge. (b) Commercial Wheatstone bridge. (Leeds-Northrup Co.)

tial difference between A and C; also the potential difference between D and B will equal the potential difference between C and B. Let us call I_1 the current in the ADB circuit and I_2 the current in the ACB circuit when the bridge is balanced. Equating these potential differences, we write

$$I_1 R_1 = I_2 R_2$$

and

$$I_1 R_x = I_2 R_3$$

Basic Electric Circuits   **531**

Dividing the first equation by the second results in

$$\frac{R_1}{R_x} = \frac{R_2}{R_3}$$

or

$$R_x = R_1 \frac{R_3}{R_2} \qquad (25.16)$$

This is the basic equation for the Wheatstone bridge.

Illustrative Problem 25.15 The Wheatstone bridge diagramed in Fig. 25.19*a* is used to determine an unknown resistance R_x. It is balanced when $R_3 = 1000\ \Omega$, $R_2 = 10\ \Omega$, and $R_1 = 156\ \Omega$. Compute the value of R_x.

Solution

$$R_x = R_1 \frac{R_3}{R_2}$$

$$= 156\ \Omega \times \frac{1000\ \Omega}{10\ \Omega}$$

$$= 15{,}600\ \Omega \qquad\qquad \textit{ans.}$$

It will be observed from Eq. (25.16) that it is not necessary to know the values of R_2 and R_3 in order to solve for the unknown R_x. As long as the value of R_1 and the ratio of R_3 to R_2 are known, we can compute the value of R_x. Commercial bridges (Fig. 25.19*b*) are so constructed that the ratio of R_3 to R_2 is selected by means of a rotary switch. These ratios generally include multiples of 10 from 0.0001 to 1000. The values for R_1 can be adjusted by a system of four or more rotary switches, one of which varies the resistance in steps of 1 Ω from 1 to 10, the second varies in steps of 10 Ω from 10 to 100, the third varies in steps of 100 Ω from 100 to 1000, etc. Thus the value of R_1 can be read directly by summing up the resistances (in series) of these four or more rotary switches. R_x can then be readily computed by multiplying R_1 by the value indicated by the ratio switch.

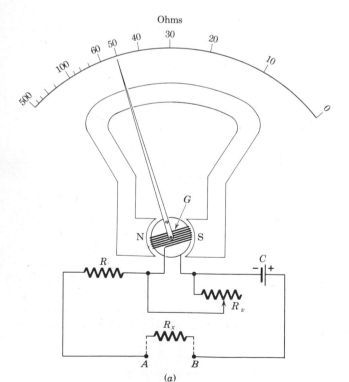

(a)

(b)

Fig. 25.20 (*a*) Single-range ohmmeter. (*b*) Commercial multirange ohmmeter. (Central Scientific Co.)

25.20 The Ohmmeter

The *ohmmeter* is a convenient, portable, direct-reading instrument for measuring resistances in ohms. A diagram of the essential parts of a simple single-range ohmmeter is shown in Fig. 25.20a. It consists of a series circuit which includes a dry cell C, a fixed known resistance R, and a specially calibrated milliammeter G. To this circuit is added the unknown resistance R_x across the terminals A and B. A variable resistor R_v is connected in parallel with the coil of the milliammeter. To operate the instrument, a conductor of extremely low resistance, called a *shunt*, is first connected to terminals A and B. The milliammeter will then reach its maximum deflection (in a clockwise direction). Next the resistance R_v is varied until the pointer of the meter reaches zero on the scale. This will indicate no external resistance between terminals A and B. The shunt is then replaced by the unknown resistance R_x. This added resistance in the series circuit will diminish the current flowing through the sensitive milliammeter. Since the current varies inversely as the resistance of the circuit, the meter can be calibrated to read directly in ohms. It should be observed that because of this inverse relationship, the scale readings increase as the scale is read to the left and also that the scale divisions are not uniform. Figure 25.20b shows one of the commercial types of multirange ohmmeters.

SUMMARY

The unit of electric charge is the *coulomb*; the unit of electric current is the *ampere*; the unit of electric resistance is the *ohm*; the unit of electric potential difference is the *volt*.

The direction of *electron flow* is from negative to positive.

The direction of conventional *current* is the same as the direction in which a positive charge will move.

The resistance of a wire conductor varies directly as the length and inversely as the cross-sectional area and depends upon the material (resistivity) and the temperature. Resistance can be calculated from the formula

$$R = \rho \frac{L}{A} \quad \text{in metric system}$$

$$R = \rho \frac{L}{d^2} \quad \text{in English system}$$

In the metric system L is measured in m (or cm) and A in m^2 (or cm^2).

In the English system wire diameters are measured in *mils*, while areas are measured in *circular mils*.

Resistivity is defined as the resistance in ohms of a wire 1 m long and 1 m^2 in cross-section (SI system) or as the resistance of a wire 1 ft long with a cross-sectional area of 1 cmil (English system).

The resistivity of most pure metals increases with a rise in temperature.

The resistivity of nonmetallic conductors decreases with a rise in temperature.

Wire sizes are standardized according to AWS gauge numbers in the English system.

Ohm's law: *amperes = volts/ohms*. This law applies to the whole circuit or to any part of it. As a formula,

$$\text{Current} = \frac{\text{potential difference}}{\text{resistance}}$$

The ammeter, which measures the current, is a low-resistance instrument and is connected in series in the circuit.

The voltmeter, which measures potential difference, is a high-resistance instrument and is connected in parallel across a circuit or a portion of a circuit.

For *series circuits:* (1) the current is the same throughout the circuit, (2) the total resistance of the circuit equals the sum of the individual resistances in the circuit, (3) the total voltage drop is equal to the sum of the individual drops along the whole circuit.

For *parallel circuits:* (1) the total current is equal to the sum of the separate currents, (2) the voltage drop is the same across each resistor, (3) the reciprocal of the total resistance is equal to the sum of the reciprocals of each resistance in the circuit.

Kirchhoff's laws for more complex electrical circuits state:

1. The sum of currents directed toward a junction equals the sum of the currents directed away from the junction.

2. In any closed loop the algebraic sum of the emfs and potential drops equals zero.

Resistances can be measured by (1) the ammeter-voltmeter method, (2) the Wheatstone bridge, and (3) the ohmmeter.

QUESTIONS AND EXERCISES

1. A Wheatstone bridge is commonly used to determine what electrical measurement?

2. What happens to the resistance of a copper wire if its temperature is raised?

3. If the diameter of a wire is d mils what is its area expressed in circular mils?

4. How is an ammeter connected in a circuit? a voltmeter? Show both in a circuit diagram.

5. What is meant by the term *resistivity*?

6. Name the factors that influence the resistance of a piece of wire and explain how each factor affects the resistance.

7. As a general rule, what kinds of materials make wires of low resistivity? of high resistivity?

8. Explain why such conductors as *chromel* and *nichrome* are desirable as heating elements.

9. Give as many reasons as you can for connecting house lights in parallel instead of series.

10. In what ways is the current in a conductor similar to the flow of water through a pipe? In what ways is it different?

11. Explain how the voltmeter-ammeter method is used for measuring resistance. What are some reasons why the method is not a particularly precise one to measure resistance?

12. When a Wheatstone bridge is balanced, will interchanging the galvanometer and the battery affect the balance? Give reasons for your answer.

PROBLEMS

1. What is the diameter in mils of a wire $\frac{3}{4}$ in. in diameter?

2. What is the area in circular mils of a wire whose diameter is $\frac{1}{4}$ in.?

3. How many coulombs per hour pass a point in a circuit that is carrying 5 A of current?

4. How much does 1 mi of AWG No. 16 copper wire weigh? Consult Appendix V.

5. What would be the likely AWG number of a copper wire if 10 ft of the wire had a resistance of 1.67 Ω? Consult Appendix V.

6. The resistance coil of an electric ironer is 12 ft long and has a resistance of 14.4 Ω. The coil is made of nichrome wire. What is the area of the wire in circular mils?

7. Manganin wire of diameter 0.051 cm is used to make a resistor of 15 Ω. What length of wire will be needed?

8. A wire having a cross-sectional area of 808 cmils has a resistance of 13.24 Ω. What is the resistance of a wire of the same length and the same material if its cross-sectional area is 101 cmils?

9. What is the resistance of 4500 ft of annealed copper wire at 20°C if the wire has a diameter of 0.064 in.?

10. What size wire on the AWG scale would have an area twice that of a No. 24 wire? What size wire would have an area four times that of a No. 24 wire?

11. How much current does a 50-Ω electric iron draw when operating from a 115-V line?

12. What is the electric resistance of 12 m of commercial copper wire at 20°C if it has a diameter of 1.5 mm?

13. A wire 500 ft long with a diameter of 102 mils has a resistance of 0.5 Ω. What would be the resistance of 500 ft of wire 51 mils in diameter and made of the same material?

14. A farmer has a generator which will deliver 120 V. He intends to run two wires from the generator to an appliance 300 ft from the generator. If the appliance is to be operated at 15.0 A and 110 V, what is the diameter of the smallest copper wire (in inches) which he can use? (Assume no drop in terminal potential difference at the generator when it is delivering current to the appliance.)

15. Four resistors having resistances of 1, 2, 4, and 5 Ω are connected in series with a battery of 24 V. (a) What is the total resistance of the circuit? (b) What is the current flow in the circuit?

16. Three resistors of 5, 10, and 20 Ω, respectively, are connected in parallel across a 12-V battery. (a) Find the combined resistance. (b) Find the current flowing through each resistor. (c) Find the total current flow.

17. Three coils R_1, R_2, and R_3 having resistances of 4, 8, and 16 Ω, respectively, are connected in parallel. These are connected in series with two other coils R_4

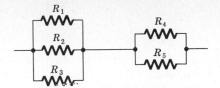

Fig. 25.21 Problem 17.

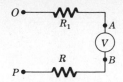

Fig. 25.24 Problem 21.

and R_5 of 12 and 20 Ω, respectively, as shown in Fig. 25.21. Find the total resistance of the combination.

18. If 2 A is flowing through R_2 of Prob. 17, find the amount of current passing through each of the other four coils.

21. The wiring diagram for a direct-reading voltmeter is shown in Fig. 25.24. When 0.612 V is placed across AB, the voltmeter registers a full-scale deflection calibrated to read 15 V. What will the voltmeter read when 20 V is applied across OP? The internal resistance of the meter is 50 Ω, $R_1 = 600$ Ω, and $R_2 = 1800$ Ω.

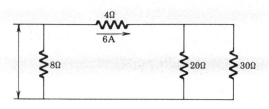

Fig. 25.22 Problem 19.

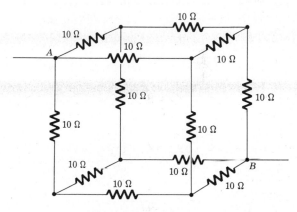

Fig. 25.25 Problem 22.

19. What voltage must be impressed on the circuit shown in Fig. 25.22 in order to have a current of 6 A in the 4-Ω resistor?

22. What is the resistance across AB of Fig. 25.25 if each edge of the cube has a resistance of 10 Ω?

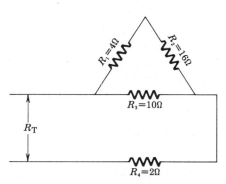

Fig. 25.23 Problem 20.

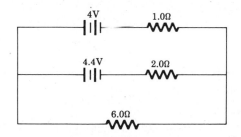

Fig. 25.26 Problem 23.

20. What is the magnitude of the total resistance R_T for the arrangement shown in Fig. 25.23?

23. Two dissimilar batteries are connected to three resistors as shown in Fig. 25.26. Determine the current through the three resistors.

Basic Electric Circuits **535**

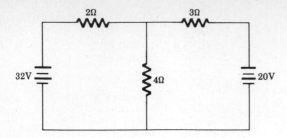

Fig. 25.27 Problem 24.

24. Use the data in Fig. 25.27 to determine the currents in each of the resistors.

25. Solve for the current in the 6-, 4-, 5-Ω resistors of Fig. 25.28.

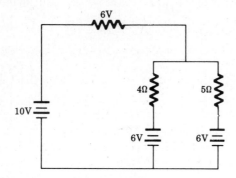

Fig. 25.28 Problem 25.

26 Sources and Effects of Electric Current

Electric charges in motion are referred to as an electric current. In order to maintain a steady current in a conductor, there must be a device for maintaining a steady difference of potential across the ends of the conductor. There are many devices capable of doing this, and in each device energy in some other form must first be changed into electrical energy. In batteries, chemical reactions take place, converting chemical energy into electrical energy. In the generators at power plants, mechanical energy is converted into electrical energy. Heat energy is converted to electrical energy in an instrument called the thermocouple. Solar energy is converted to electrical energy in the photoelectric cell and the solar cell.

Electric energy can, in turn, be converted into chemical, mechanical, heat and radiant energy. In this chapter we will be studying some of the more common sources of electric currents. We will then study how the processes can be reversed to convert electric energy into other forms of energy.

26.1 Electromotive Force and Potential Difference

A device that produces electric energy is called a *source of electromotive force*. It should be emphasized that such a device does not manufacture electrical charge. It simply moves the charge through a circuit. The use of the term "electromotive force" is unfortunate; "electromotive force" is not a force at all. It is the work per unit charge done as the charge is moved through the generating source. Hereafter we shall avoid the words "electromotive force" and use instead the abbreviation *emf* (see Sec. 25.4).

As the charge moves through a source of emf, work is done *by the source* in raising the electrical potential energy of the charge. This emf causes differences of potential energy to exist between points in the external circuit. As the charge moves through the external circuit, work is done *by* the electric field on the charge. The electric potential energy supplied by the emf source is dissipated as heat energy in a pure resistor. In an electric motor, the energy is dissipated as useful work plus heat energy. In some circuits, the emf results in chemical energy. In each of the above instances the energy provided by the source of the emf exactly equals the energy lost in the external circuit. A source of emf has an emf of 1 volt if 1 J of energy is gained for each coulomb of charge passing through it. We shall use the script letter $\mathscr{E}$ for the value of the emf. Thus, *the value of the emf* ($\mathscr{E}$), *in volts* is defined by

$$\mathscr{E} = \frac{W}{Q} \qquad (26.1)$$

where W is the amount of energy, in joules, gained by a charge Q, in coulombs, passing through the source.

Although the emf and potential difference are both measured in *volts*, there is a distinction

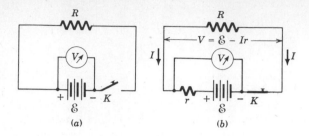

Fig. 26.1 The terminal voltage in (b) is always less than in (a) because of the potential drop *Ir* within the battery itself.

between them. In a source of emf, not only is some other form of energy transformed into electric energy, but the transformation is *reversible*. For example, a storage battery is charged by forcing electric current in a direction opposite to that in which it travels when the battery is discharging. By the same token, a generator can be converted into a motor. On the other hand, a pure resistor is not a source of emf. A potential difference across the resistor will establish a current in the circuit which will heat up the resistor. But building a fire under the resistor does not produce a current.

The potential difference across a battery, generator, or other *source* of electric energy *when it is not connected to any external circuit, is its emf*. When the source of electric energy is a part of a closed circuit, the circuit carries a current, and the potential difference across the terminals of the source is always less than the emf because of the *internal resistance* of the source. The circuit diagram of Fig. 26.1 will illustrate the situation.

A 6.0-V battery is connected across a resistance *R* of 3 Ω. When the switch *K* is open, the circuit will carry no current. A voltmeter connected across the terminals of the battery will register an emf of 6.0 V. When the switch is closed, about 2 A of current will be established through *R*, and the voltmeter will read a potential difference *V* across the battery terminals of about 5.7 V.

The drop in voltage from 6.0 on the open circuit to 5.7 V on the closed circuit is due to the internal resistance *r* of the battery. Hence the actual terminal voltage V across a source of emf

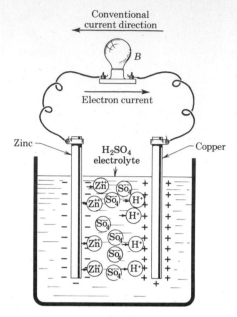

Fig. 26.2 Simple voltaic electric cell.

$\mathscr{E}$ and internal resistance *r* is

$$V = \mathscr{E} - Ir \qquad (26.2)$$

Terminal voltage = emf − potential drop within source. On an open circuit, no current exists, and $V = \mathscr{E}$; on a closed circuit the current that is established through the battery lowers the value of *V* by an amount equal to *Ir*.

SOURCES OF EMF

26.2 Electric Currents Can Be Produced by Chemical Means

About 1800, Alessandro Volta (1745–1827), an Italian scientist, discovered a chemical means of generating electricity. His first simple cell can be duplicated in the laboratory. Place a strip of zinc and a strip of copper in a glass vessel filled with dilute sulfuric acid, and connect these two strips by a copper wire to a flashlight bulb *B* (Fig. 26.2). An electric current in the conductor will be evident by the lighting of the bulb. It can be shown by means of a sensitive electroscope that the zinc plate has a negative charge, while the copper plate has a positive (less negative than the zinc) charge. Almost any two conductors could have been used for the strips (which are called *electrodes*), but the two strips must not be identical. Also, many other solutions could be used instead of the sulfuric acid; all that is necessary is that the solution, called the *electrolyte*, should chemically attack one of the metals. The copper strip shown in Fig. 26.2 is called the *positive* (+) *electrode*, and the zinc strip is called the *negative* (−) *electrode*.

The dry cells used in flashlights and the storage batteries used in automobiles, trucks, etc., are common examples of continuous sources of electric energy that result from chemical action.

Detailed study of what happens inside a "voltaic" cell, dry cell, or storage battery requires an understanding of chemical reactions. Only the simplest reactions will be discussed here.

Ions are atoms or groups of atoms that have a net positive or a net negative charge, due to an excess or deficiency of electrons. Ions containing opposite electrical charges attract each other. It is believed that the molecules of sulfuric-acid solution (chemical formula H_2SO_4) break up into ions. The "sulfate" combination of sulfur and oxygen (SO_4) seizes one electron from each of the hydrogen atoms, forming a "sulfate ion" (written SO_4^{--}). Each hydrogen atom, having lost an electron, becomes a positive ion (written H^+). The equation for this chemical reaction is written

$$H_2SO_4 \longrightarrow 2H^+ + SO_4^{--}$$

When substances in solution form ions, they are said to be *ionized*. Such a solution is called an *electrolyte*.

In simple terms this is what happens in the voltaic cell (Fig. 26.2). When the zinc is placed in the acid solution, the zinc begins to dissolve. Zinc atoms leave the strip and enter the solution. Each zinc atom that leaves the strip leaves behind two electrons. Thus the zinc atom when in solution becomes a positive zinc ion (Zn^{++}). At the same time, because of the electrons left behind, the zinc

strip becomes negatively charged. This reaction can be written

$$Zn \longrightarrow Zn^{++} + 2e^-$$

The positively charged zinc ions will repel the positive hydrogen ions toward the copper strip. As each positive hydrogen ion reaches the copper, it seizes an electron from a copper atom. The neutralized hydrogen atoms will then form hydrogen molecules and bubble off into the air. The copper strip, having lost electrons, becomes positively charged. Thus a potential difference (emf) is created between the zinc and copper strips.

26.3 Local Action

In the simple voltaic cell just described the commercial zinc used contains many impurities, such as small particles of carbon, iron, and lead. Even though no current is being drawn from the cell, the zinc may rapidly be eaten away. The small particles of impurities, when in contact with the zinc in the presence of an acid solution, will form small local cells with closed circuits (see Fig. 26.3). The result is a wasting away of the zinc even though no current is flowing in the external circuit. This process is called *local action*. Local action will continue as long as the impurity is in contact with the zinc and the acid. The effect of

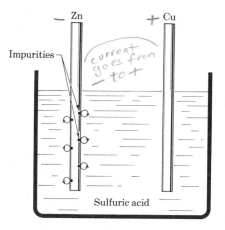

Fig. 26.3 Local action will eat away the negative electrode.

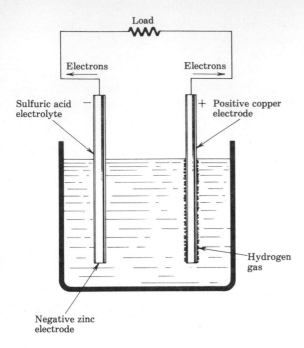

Load

Electrons   Electrons

Sulfuric acid —   + Positive copper
electrolyte   electrode

Hydrogen
gas

Negative zinc
electrode

Fig. 26.4   Polarization in a simple voltaic cell.

local action in a cell can be prevented by *amalgamating* the zinc, i.e., by polishing the zinc and then rubbing it with mercury until a thin coating of mercury surrounds the zinc. The mercury will dissolve some of the zinc but none of the impurities. In this way the impurities are prevented from making contact with the acid. The amalgamated zinc will then react with the acid only when the cell is in operation.

26.4   Polarization

When a heavy current is drawn from a simple voltaic cell, the current will quickly fall off in intensity because hydrogen bubbles, which result from the chemical action that takes place, collect on the positive electrode (Fig. 26.4). These bubbles not only reduce the effective surface area of the copper but alter the material of the electrode, thus reducing the effective emf of the cell. The effect of such action is called *polarization*. Polarization increases the *internal resistance* of a cell. In commercial cells where polarization may occur, another substance, called a *depolarizer*, is

introduced to combine with the hydrogen and remove it from the electrode. The oxygen of the depolarizer combines with the hydrogen to form water.

26.5   The Dry Cell

A dozen or more types of voltaic cells have been devised for commercial purposes, but the most widely used is the common *dry cell* (Fig. 26.5). Actually the cell is not dry. The negative electrode of the dry cell consists of a zinc cylinder which forms the walls of the cell. This cylinder contains the positive electrode (a carbon rod), the electrolyte (ammonium chloride), and the depolarizer (manganese dioxide). The cell also contains small amounts of zinc chloride to aid in depolarization. The electrolyte and the depolarizer are combined in the form of a paste. Powdered coke and graphite are added to the paste to reduce the internal resistance of the cell. The carbon electrode is placed in the center of the zinc cylinder, the space around it is filled with the paste, and the cell is then sealed with wax or pitch to make it watertight and airtight. Thus the cell can be easily transported and can be used in any position. Dry cells are used by the millions today in flashlights, toys, and portable radios.

A dry cell should be used only *intermittently* or with very low current drain, or it will polarize faster than the manganese dioxide and the zinc chloride can depolarize it. The emf of a new cell is about 1.5 V. The emf of a cell gradually declines even when not in use; hence the useful life of dry cells is ordinarily about 1 or 2 years.

26.6   Secondary or Storage Cells

The cells we have discussed thus far are termed *primary cells*. When a primary cell is "dead," it cannot be regenerated, because the active materials have been consumed. Primary cells are nonreversible cells. *Secondary cells* are reversible cells that may be recharged by reversing the direction of the electric current. The *storage cell* is one in which electric energy is *stored* in the form of chemical energy. In this cell two electrodes of dis-

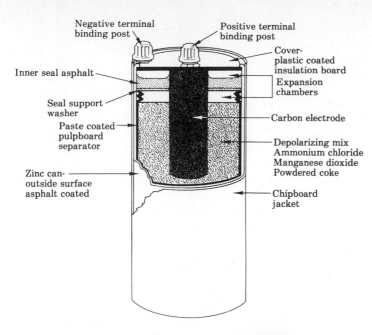

Negative terminal
binding post

Positive terminal
binding post

Inner seal asphalt

Cover-
plastic coated
insulation board

Expansion
chambers

Seal support
washer

Paste coated
pulpboard
separator

Carbon electrode

Zinc can-
outside surface
asphalt coated

Depolarizing mix
Ammonium chloride
Manganese dioxide
Powdered coke

Chipboard
jacket

Fig. 26.5 Cross-sectional view of common dry cell. (Union Carbide Inc.)

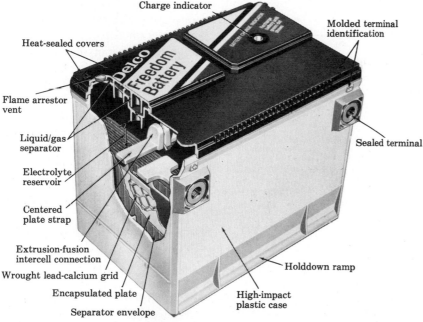

Charge indicator

Molded terminal
identification

Heat-sealed covers

Flame arrestor
vent

Sealed terminal

Liquid/gas
separator

Electrolyte
reservoir

Centered
plate strap

Extrusion-fusion
intercell connection

Wrought lead-calcium grid

Encapsulated plate

Separator envelope

Holddown ramp

High-impact
plastic case

Fig. 26.6 Cutaway showing internal construction of a lead storage battery. A built-in charge indicator replaces the hydrometer test used on other batteries. (Delco-Remy.)

similar materials are immersed in an electrolyte. After they have been supplying current to an external circuit, they can be restored to their original condition by sending an electric current from an external source through the cell in a direction opposite to that of current flow from the cell. Only those electrodes and electrolytes can be used in a storage cell in which the chemical action can be readily reversed by changing the direction of the current. When the cell is delivering a current to an external circuit, it is said to be *discharging*, and while energy is being restored, it is *charging*. A *storage battery* consists of two or more such cells connected in series-parallel combinations.

The most commonly used storage cell is the lead storage cell (Fig. 26.6), the type used in automobiles and for many other industrial purposes. The active material of the *positive* electrode of a lead storage cell is a chemical compound called lead oxide (PbO_2), while the active material of the *negative* electrode consists of finely divided, spongy, metallic lead (Pb). The electrolyte is dilute sulfuric acid (H_2SO_4). The positive and negative electrodes are held apart by separators of wood, rubber, or glass. The emf of such a cell is approximately 2.1 V. When the cell is discharging, the active materials of both plates change to lead sulfate ($PbSO_4$) while the electrolyte is converted to water (H_2O). As the cell continues to discharge, both plates eventually are covered with the same material ($PbSO_4$), and in this condition no emf is developed. The cell is said to be run down, dead, or discharged. Also the acid becomes more and more dilute as the battery discharges (see Sec. 11.18).

To charge a lead storage battery, a current is passed through it in a direction opposite to that of the normal flow of the battery. This reverses the chemical action until the electrolyte and the electrodes are restored to their original condition. The chemical reaction that takes place in the storage battery can be expressed by the reversible reaction:

$$\overset{\text{Discharging}}{\underset{\text{Charging}}{\xrightleftharpoons{}}}$$

$$PbO_2 + Pb + H_2SO_4 \rightleftharpoons 2\,PbSO_4$$
$$+ 2H_2O + \text{Electrical Energy} \quad (26.3)$$

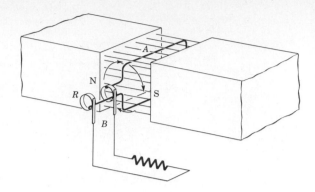

Fig. 26.7 A simple generator.

26.7 Electromagnetic Generators

For industrial purposes, the most important source of electric energy is the electromagnetic generator, in which mechanical energy is converted into electric energy. The essential parts of a simple generator are illustrated in Fig. 26.7. The simple form diagramed consists of a set of field magnets and a rotating loop of wire *A*. If the loop is rotated in the magnetic field, a flow of electric current will take place in the loop. This current can be taken off on brushes *B*, which touch a pair of collecting rings *R*, one attached to each end of the loop. A complete discussion of electric generators will be given in later chapters, after the principles of electromagnetism have been explained.

26.8 Thermoelectricity

A means of converting heat energy into electric energy was discussed in Sec. 13.9 and illustrated in Fig. 13.7, which shows a circuit of two different wires, copper and iron. They have been fused or brazed together at both ends to form a closed loop. If one of these junctions is kept at a temperature different from that of the other, a small electric current will flow around the circuit, passing from the iron to the copper at one junction and from the copper to the iron at the other. The current will continue to flow as long as the temperature at one junction is higher than at the

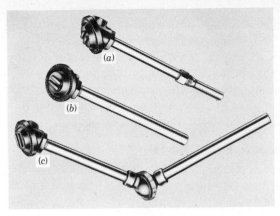

Fig. 26.8 Commercial thermocouple. (*a*) Pipe extension type. (*b*) Straight-tube protection type. (*c*) Angle type. (*d*) Cross section of straight-type assembled thermocouple. Length *C* of the couple can be varied over a considerable range. *W* is the tube used to protect the wires against mechanical damage and corrosive or contaminating atmospheres. *H* is the covered terminal head of the thermocouple which connects directly, or by means of extension wire contained in the conduit, to the calibrated instrument. (Bristol Co.)

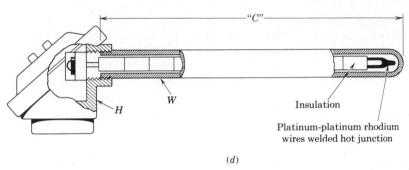

(*d*)

other, and the intensity of the current is a measure of the difference in the temperatures. Such a device is called a *thermocouple* (Fig. 26.8). Thermocouples are used for the measurement and control of temperature. A number of thermocouples may be joined in series, to form a *thermopile*. Thermopiles can produce voltages sufficient to operate radio receivers, relays, and other low-power devices.

26.9 Photoelectricity

Certain metals, such as selenium and cesium, have the ability to convert light energy into electric energy. When light falls on the surface of these metals, electrons are emitted from the surface. Such a phenomenon is called the *photoelectric effect*. An arrangement which permits the electric current emitted by the light-sensitive metal to be included in a closed circuit is called a

photoelectric cell. The photoelectric current is generally extremely feeble but can be amplified by means of other electric devices. The photoelectric cell has numerous applications in industry because (1) the photoelectric effect is almost instantaneous and (2) the intensity of the current is proportional to the intensity of the light beam. These properties have led to its application in sound motion pictures, in television, footcandle meters, exposure meters, automatic sorting machines, automatic burglar-alarm systems, door openers, drinking fountains, radio transmission of pictures, and many industrial devices for counting and control (see Chaps. 32 and 33).

26.10 The Solar Cell

The *solar cell* is a special type of photoelectric cell which converts light energy directly to electric energy. A typical solar cell consists of a silicon

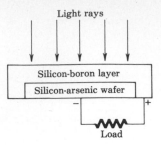

Light rays

Silicon-boron layer

Silicon-arsenic wafer

− +

Load

Fig. 26.9 The solar cell.

crystal "sandwich" (Fig. 26.9). One layer of the sandwich consists of a crystal of pure silicon, a semiconductor (see also Secs. 32.8 and 32.9), that has been "doped" with a slight amount of arsenic on the inside layer. Since arsenic is an electron donor, the wafer contains an excess of free electrons. The outer layer consists of a very thin layer of silicon to which has been added a slight amount of boron, an electron acceptor. When the cell is exposed to light, the silicon absorbs some of the light energy. This energy frees a quantity of electric charge carriers (both negative and positive). Some of these are collected and separated at the junction, giving the silicon-arsenic wafer a negative charge and the silicon-boron layer a positive charge. Thus an electric potential difference (emf) is produced between the wafer and the layer.

The solar cell is light in weight, maintenance free, long lasting and pollutant free. However, solar cells are quite expensive. The area of a single solar cell cannot be very large, since it is very difficult to produce large silicon crystals. A typical cell has an area of about 20 cm^2. One such cell can produce an output of about 0.6 V at about 150 mA. Solar cells can be connected in series to achieve higher voltages or in parallel for higher currents. In a typical installation, the solar batteries will be used to charge up storage batteries. Thus, power is available even when the solar cells are not exposed to the sun.

In the original Telstar satellite (see Sec. 32), 3600 solar cells were connected to form a series-parallel battery which produced about 15 W of power. Today solar batteries are being designed that produce peak-power outputs of about 250 W. Solar cells are used to power electronic equipment on space satellites. They can develop sufficient power to transmit radio messages from millions of miles in space. To decrease their price per energy output, different materials and manufacturing processes are being developed.

26.11   Fuel Cells

Considerable effort has been expended in the research laboratories of industry and government to develop a type of electrochemical cell known as the *fuel cell*. In the fuel cell, the energy of a conventional fuel is converted directly into electric energy. In the common battery, chemical energy is converted into electric energy until one or more of the active materials is used up, after which the battery is either discarded or recharged. In the fuel cell active materials are supplied, and the reaction products are removed continuously. Thus the fuel cell can operate as long as the required materials are supplied and the chemical by-products removed.

Many fuels have been experimented with in the research on fuel cells. The reactants consist of an oxidizing substance (one that combines with oxygen molecules) and a reducing material (one that removes oxygen molecules). Perhaps the simplest fuel cell to describe, construct, and to understand is one that uses hydrogen gas as the fuel and oxygen as the oxidizing agent. Fuel cells of this type which develop 2 kW of power have been used very effectively in spacecraft. A schematic diagram of the hydrogen-oxygen fuel cell is shown in Fig. 26.10. The cell consists of three basic parts: fuel and oxidizer, electrodes, and an electrolyte. The electrodes are made of inert conducting materials and are porous to permit the gases to come gradually in contact with the electrolyte. The electrolyte of the cell is a solution of potassium hydroxide (KOH). The solution breaks up into K$^+$ and OH$^-$ ions. (An *ion* is an atom or group of atoms that carries an electric charge.) At the negative electrode, hydrogen molecules combine with the hydroxide ions to

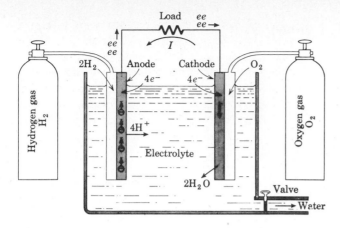

Fig. 26.10 Hydrogen-oxygen fuel cell.

form water. In the process electrons are released to the external current. The ionic reaction is

$$2H_2 + 4OH^- \longrightarrow 2H_2O + 4e^- \quad (26.4)$$

The electrons move through the external circuit to the positive electrode, where oxygen molecules combine with water molecules and the incoming electrons to produce new hydroxide ions:

$$O_2 + 2H_2O + 4e^- \longrightarrow 4OH^- \quad (26.5)$$

The hydroxide ions thus produced react with more hydrogen molecules at the anode. Thus for the entire cell the reaction is

$$2H_2 + O_2 \longrightarrow \begin{array}{l} 2H_2O + \text{the transfer of } 4e^- \\ \text{from one electrode to the} \\ \text{other.} \quad (26.6) \end{array}$$

The electrons can do useful work in passing from the anode to the cathode in the external circuit: 1 lb of hydrogen and 8 lb of oxygen in a fuel cell will produce 9 lb of water and enough electric power to light a 100 W bulb for nearly two weeks. The water produced must be removed from the electrolyte to maintain the proper electrolytic concentration.

An attractive feature of the fuel cells is their high efficiency. They are thermodynamically not heat engines and therefore are not subject to the (Carnot) cycle limitations of heat engines. Efficiencies close to 100 percent are theoretically possible with fuel cells. Another advantage of fuel cells is that the waste products of the conversion, if any, are not pollutants. In the hydrogen-oxygen cell, the by-product is water. Fuel cells have found applications in space vehicles, where efficiency and reliability are critically important.

Ideally, a fuel cell should use readily available materials such as natural gas, coal, air, or plant residues as fuel. The fact that hydrogen is neither cheap nor readily available limits the use of the fuel cell just described to applications where cost is not an important factor.

Research on cells fueled with natural gas could open the door to a new source of electricity. Fuel cells installed in homes and factories to convert piped-in natural gas into electricity in the future is a distinct possibility. It is five times cheaper to deliver energy as gas than as electricity. Researchers in fuel-cell technology are experimenting with cells which may produce power for such widely divergent uses as automobiles, artificial hearts, deep-water submergence vehicles, and portable battlefield communications systems.

26.12 Grouping of Cells

Often a single cell does not give enough voltage or enough current for a specific need. It then becomes necessary to group two or more cells in combinations (to form a battery) to gain the desired results. Let us consider three dry cells

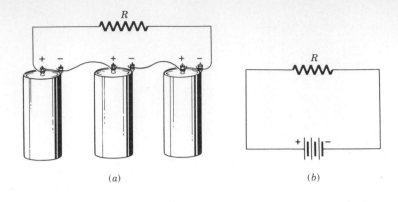

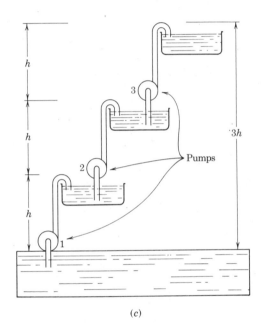

Fig. 26.11 (a) Three cells connected in series. (b) Wiring diagram of series circuit. (c) Hydraulic analogy of three cells in series.

connected in series (Fig. 26.11a). It will be noted that the zinc electrode of the first is connected to the carbon of the second; the zinc of the second is connected to the carbon of the third; and the zinc of the third is connected through the external resistance R to the carbon electrode of the first cell. If the emf of each cell is 1.5 V, the emf of the three cells will be $3 \times 1.5 = 4.5$ V.

For *cells arranged in series* the following statements apply:

1. *The emf of the combination is equal to the sum of the emfs of the cells.*

2. *The current in each cell and in the external circuit is the same.*

3. *The total internal resistance of the combination is equal to the sum of the internal resistances of the individual cells.*

Figure 26.11b is the wiring diagram, which illustrates the series arrangement of Fig. 26.11a. If we refer again to a hydraulic analogy (Fig. 26.11c), we would think of the first cell as pumping electricity up to a certain potential (level), and the second cell as pumping it to a still higher potential, and the third to a potential three

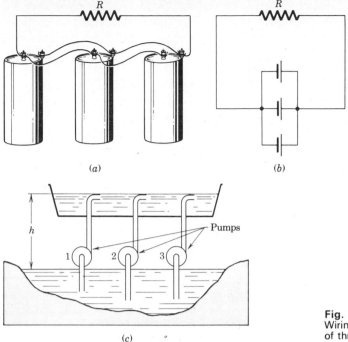

Fig. 26.12 (*a*) Three cells connected in parallel. (*b*) Wiring diagram of parallel circuit. (*c*) Hydraulic analogy of three cells in parallel.

times as high as the potential of the first cell, just as the water pumps shown in the figure raise water to successively higher levels.

When *cells are connected in parallel* (Fig. 26.12*a*), the positive (carbon) electrodes are joined together and so are all the negative (zinc) electrodes. With this arrangement (identical cells in parallel), the following statements apply:

1. *The emf of the combination is the same as that for each cell.*

2. *The total internal resistance is equal to the resistance of one cell divided by the number of cells.* (Assuming that the cells all have equal internal resistances).

3. *The current in the external circuit is the sum of the currents in each cell.*

Again, a hydraulic analogy (Fig. 26.12*c*) may help in understanding these relationships. Think of each cell as pumping the same amount of electricity to the same potential (level). The potential difference created by the battery of three cells is just the same as that contributed by one cell. The pumps of Fig. 26.12*c*, since they also act in paral-

lel, discharge the water into a common reservoir, at the same *level* (potential). Thus, the total amount of water pumped equals the sum of the amounts contributed by each pump.

EFFECTS OF ELECTRIC CURRENTS

Scientists conceive of current electricity as a motion of large numbers of electrons; yet the exact nature of the current flow is still not completely understood. However, the effects of electric current flow are in most cases quite clearly understood. It is because of these effects that electricity has become such a universally used source of energy. The electric current in itself has no value; it is useful only after the electric energy has been converted into some other form of energy, e.g., heat energy, chemical energy, mechanical energy. It is only because man has been able to understand these effects and develop basic laws for them that we now have the numerous applications of electricity. In our study of electricity, we shall not be so much interested in the electric current itself as in its effects. The individual effects are not many or difficult to understand. Even complicated electric devices are only combinations of several simple effects.

26.13   The Heating Effect of an Electric Current

Conductors carrying an electric current are heated by the current. In some cases the heating is desirable, while in many other cases, such as in electric motors, generators, household circuits, and transformers, it is highly undesirable. The amount of heating in an electric conductor depends on (1) the intensity of current flowing, (2) the time the current flows, (3) the kind and size of the conductor, and (4) the nature of the surrounding media. It is possible to regulate the controlling factors so that optimum heating effect can be obtained. Some devices in which the heating effects of an electric current are desirable are incandescent lamps, toasters, waffle irons, electric

stoves, percolators, irons, heating pads, and electric furnaces. The tungsten filament of an incandescent lamp operates at a temperature of about 2700°C. At this temperature, tungsten would evaporate rapidly and would gradually burn (oxidize) if the lamps were not filled with an inert gas. Here we see electric energy converted into both heat and light energy. The heating elements of percolators, flatirons, waffle irons, and radiant heaters are embedded in a refractory (heat-resistant) material which serves to keep the element in place and to retard oxidation. Special conductors such as nickel-chromium alloys must be used if the heating element is exposed to air, as in toasters. The electric arc used in welding develops temperatures of around 4000°C.

Heat energy developed in motors, generators, and lighting circuits not only may represent an expensive loss of energy but also may be dangerous should the temperature developed be too high. Protection from the danger of too high temperature is provided by *fuses* or *circuit breakers*

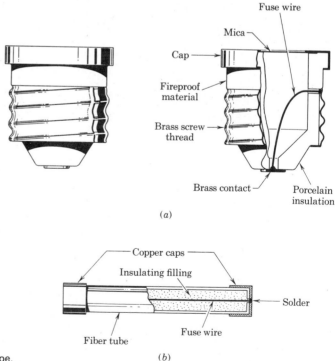

Fig. 26.13   Fuses. (*a*) Plug type. (*b*) Cartridge type.

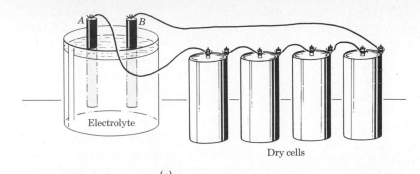

Electrolyte

Dry cells

(a)

(b)

Fig. 26.14 (*a*) Electrolytic cell. (*b*) Electrolytic refining of aluminum. Molten aluminum is tapped from a series of electrolytic smelters. Metal from this tap will be delivered in molten form to a nearby foundry.

which will interrupt the current when too "heavy" a load is carried by the circuit. The fuse is a strip of a low-resistance, low-melting-point alloy. The alloy melts when the current increases above a certain value, thus "breaking" the circuit. Fuses may be *plug type* (Fig. 26.13*a*) or *cartridge type* (Fig. 26.13*b*). After the cause of the overload is removed, it is a simple task to replace the burned-out fuse with a good one. Electromagnetic and thermal devices (called *circuit breakers*) that open the circuit when the current exceeds an established value and that can be reset when the

overload is removed have almost completely replaced low-melting-point fuses (see Sec. 27.17).

26.14 The Chemical Effects of Electric Current

We have seen that there is a close connection between electricity and chemical action. To illustrate the electrochemical action of a current, we use a device called an *electrolytic cell* (Fig. 26.14*a*). It consists essentially of two metallic strips *A* and *B*, called *electrodes*, immersed in a solution called the *electrolyte*, which is a solution of an acid, an alkali, or a salt. When a current of electricity from an external source is passed through such a cell, a chemical change takes place. Sometimes the solution is decomposed into its constituent parts and can be collected as gases at the electrodes. Oxygen, hydrogen, and chlorine are gases that are commercially obtained in this manner. In some cases, metal is removed from one electrode or from the solution and deposited on the other electrode. The refining of many metals, such as copper, gold, aluminum, magnesium, sodium, and potassium, is achieved by putting into solution (or melting to a liquid) the compound containing the desired element and then depositing it at the negative electrode (see Fig. 26.14*b*). Many important modern processes such as copper plating, silver plating, nickel plating, and chromium plating depend on electrochemical action for their success.

26.15 The Magnetic Effect of an Electric Current

For many years scientists knew of no relationship between electricity and magnetism. In the year 1820 Hans Christian Oersted (1777–1851), a Danish physicist, accidentally discovered that a flow of electricity produces magnetic effects. After one of his lectures at the University of Copenhagen he quite by accident placed a current-carrying wire *parallel to* and directly above a suspended magnetic compass needle. To his surprise, he saw the compass turn and assume a position perpendicular to the conductor carrying the current.

If we place a conductor *parallel to* and above a compass (Fig. 26.15), the compass needle will swing in a direction dependent on the direction of the current until it is perpendicular to the conductor. If we change the direction of the current, the compass will turn in the opposite direction. We could get the same effect by placing the conductor under the compass instead of changing the direction of the current. Hence it is evident that there is a magnetic field around a current-carrying conductor.

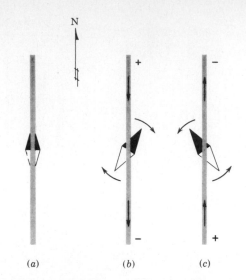

Fig. 26.15 Effect of current-carrying conductor on suspended compass needle.

If a current-carrying conductor is dipped into a pile of iron filings (Fig. 26.16), the filings cluster around and cling to the wire. When the current is cut off, the filings drop off. We shall learn in Chap. 27 how this *electromagnetic* effect can be greatly strengthened by using coils of wire instead of straight wires and by employing cores of iron. We shall also study the principles and laws which relate to the phenomenon of *electromagnetism*. These principles apply in the effective use of electromagnetic forces in such common devices and machines as electric bells, telephones, telegraphs, relays, solenoids, loudspeakers, motors, meters, and generators.

26.16 Electric Power

In Chap. 5 power was defined as the time rate of doing work

$$\text{power} = \frac{\text{work}}{\text{time}} \qquad (5.12)$$

The SI-metric system unit of power is the joule per second, called the *watt*. The time rate at which electrical energy is delivered or consumed is called *electric power*.

Fig. 26.16 Iron filings cling to a current-carrying conductor.

In earlier chapters the following relations were developed.

W (joules)

$$= V \text{ (volts)} \times Q \text{ (coulombs)} \quad (24.11')$$

Q (coulombs)

$$= I \text{ (amperes)} \times t \text{ (seconds)} \quad (25.1)$$

Substituting Eq. (25.1) in Eq. (24.11') gives

$$W = VIt \quad (26.7)$$

Therefore

$$P = \frac{W}{t} = VI$$

Electric power (watts) $= V$ (volts)

$$\times I \text{ (amperes)} \quad (26.8)$$

Since the watt is a quite small unit of power, a larger unit, the *kilowatt*, is often used. The kilowatt (kW) is equal to 1000 W.

$$kW = \frac{VI}{1000} \quad (26.9)$$

Most electrical appliances have labels which state the voltage at which the appliance is to operate and the power it consumes. Thus, a $\frac{1}{2}$-hp electric motor may be rated for 114 V at 670 W.

Illustrative Problem 26.1 What is the power consumption of a car radio if, when it is operating from a 12 V battery, it draws 1.5 A of current?

Solution Substituting in Eq. (26.8), we get

$$P = VI$$

$$= 12 \text{ V} \times 1.5 \text{ A}$$

$$= 18 \text{ W} \qquad ans.$$

Since $V = IR$, we can substitute for V in Eq. (26.8) and get an equation for power in terms of amperes and ohms. Thus

$$P = VI$$

$$= (IR)I$$

or

$$P = I^2R \quad (26.10)$$

Substituting V/R for I in Eq. (26.8), we get an equation for power in terms of volts and ohms.

$$P = VI$$

$$= V(V/R)$$

or

$$P = V^2/R \quad (26.11)$$

Because *electric* energy is frequently used to do *mechanical* work, it is important to relate the unit of electric power to the English-engineering unit of mechanical power. Experience (and calculation) has shown that 1 hp *is equivalent to* 746 W or that 1 kW *is equivalent to* 1.34 hp.

$$1 \text{ hp} = 746 \text{ W}$$

$$1 \text{ kW} = 1.34 \text{ hp}$$

Illustrative Problem 26.2   A dc electric motor operating from a 440 V line is used to hoist automobiles from a dock to the deck of a ship. The motor and winch which operate the hoist have a combined efficiency of 60 percent. (*a*) What is the power, in kilowatts, necessary to raise a car with a mass of 1600 kg a distance of 12 m in 90 sec? (*b*) What current will the motor draw from the line which is operating the hoist?

Solution   The work done to raise the car 12 m is equal to

$$mgh = 1600 \text{ kg} \times 9.81 \text{ m/sec}^2 \times 12 \text{ m}$$

$$= 188,000 \text{ J}$$

Since the efficiency is only 60 percent, the motor will expend $188,000/0.60$ or 313,000 **J**.
(*a*) The power of the motor will be

$$\text{Power} = \frac{\text{work}}{\text{time}}$$

$$= \frac{313,000 \text{ J}}{90 \text{ sec}}$$

$$= 3480 \text{ J/sec}$$

$$= 3480 \text{ W}$$

$$= 3.48 \text{ kW} \qquad \qquad ans.$$

(*b*) Substitute known values in Eq. (26.9) and solve for *I*.

$$\text{Power in kW} = \frac{VI}{1000}$$

$$3.48 = \frac{440 \, I}{1000}$$

$$I = \frac{3.48 \text{ kW} \times 1000 \text{ VA/kW}}{440 \text{ V}}$$

$$= 7.91 \text{ A}$$

$$ans.$$

26.17   Electric Power and EMF

In Sec. 26.1 it was explained that the terminal voltage *V* across a battery source of emf $\mathscr{E}$ and internal resistance *r*, which is also the voltage drop across the external circuit, is

$$V = \mathscr{E} - Ir \qquad (26.2)$$

Thus the power delivered by the battery is

$$P = VI$$

or $\qquad \qquad P = \mathscr{E}I - I^2 r \qquad (26.12)$

It will be noted that the first term on the right is the rate at which chemical energy is converted into electric energy and the second term is the rate of heating.

   In charging a storage battery the terminal voltage of the battery during the charging process is greater than the emf by the amount of the internal voltage drop within the battery. Thus, if the battery charger produces a terminal voltage *V* in sending a charging current *I* through the internal resistance *r*, we have

$$V = \mathscr{E} + Ir$$

The power *P* to charge is found by

$$P = VI$$

$$P = \mathscr{E}I + I^2 r \qquad (26.13)$$

The first term on the right side of the equation is the rate at which electric energy is being transformed and stored as chemical energy. The second term is the rate at which electric energy is being converted into unusable thermal energy in the battery electrolyte and plates.

Illustrative Problem 26.3   A battery charger sets up a potential difference of 13.7 V at the terminals of a storage battery with an emf of 12.6 V and internal resistance of 0.221 Ω. (*a*) What is the charging current? (*b*) What is the rate at which electric energy is being converted into chemical energy? (*c*) At what rate is heat produced?

Solution
(*a*) On charging,

$$V = \mathscr{E} + Ir$$

$$13.7 \text{ V} = 12.6 \text{ V} + IA \times 0.221 \, \Omega$$

$$I = 4.98 \text{ A} \qquad \qquad ans.$$

(b) $\qquad P = \mathscr{E}I + I^2 r$

$$\mathscr{E}I = 12.6 \text{ V} \times 4.98 \text{ A}$$

$$= 62.7 \text{ W} \qquad \qquad ans.$$

(c) $\qquad I^2 r = (4.98 \text{ A})^2 \times 0.221 \, \Omega$

$$= 5.48 \text{ W} \qquad \qquad ans.$$

26.18 Electric Energy

Energy was defined in Chap. 5 as the *capacity for doing work*. Since *power = work/time*, *energy (work) = power × time*. Electric-energy units are therefore the *watt-hour* and the *kilowatt-hour* (kW-hr). One watthour of energy is expended when 1 W of power is used for 1 hr. Electric bills are paid on the basis of the kilowatt-hours used.

In the laboratory we frequently use the smaller unit, the watt-second, equivalent to the *joule*.

$$\text{Joules} = \text{volts} \times \text{amperes} \times \text{seconds}$$

$$\mathbf{J} = VIt \qquad \qquad (26.14)$$

Illustrative Problem 26.4 How much will it cost to operate a $\frac{1}{2}$-hp swimming pool filter motor continuously for 30 days if the rate for electric energy is 5.86 cents per kilowatthour? (Assume 100 percent efficiency.)

Solution Converting $\frac{1}{2}$ hp to watts, we have

$$\tfrac{1}{2} \text{ hp} = \tfrac{746}{2}$$

$$= 373 \text{ W}$$

$$= 0.373 \text{ kW}$$

The energy consumed by the motor in 30 days will be

$$\text{Kilowatthours} = 0.373 \text{ kW} \times 30 \text{ days}$$

$$\times 24 \text{ hr/day}$$

$$= 268.56 \text{ kW-hr}$$

The total cost will then be

$$268.56 \text{ kW-hr} \times \$0.0586/\text{kW-hr} = \$15.74$$

$$ans.$$

26.19 Relationship between Electric Energy and Heat Energy

We have seen that one equation relating electric energy to mechanical energy is $\mathbf{J} = VIt = I^2 Rt$, where $\mathbf{J}$ is measured in wattseconds (joules), V in volts, I in amperes, R in ohms, and t in seconds. Many thorough and careful experiments have been conducted to relate electric energy to heat energy. The following relationship is the accepted outcome of these experiments: The *heat in calories* is equal to 0.24 times the joules (watt-seconds) of electrical energy expended. Let

$$H = \text{heat, cal} \quad V = \text{voltage}$$

$$I = \text{current} \quad t = \text{time, sec}$$

then $\qquad \qquad H = 0.24VIt \qquad (26.15)$

The value of V in Eq. (26.14) can be replaced by its equivalent IR to get the equation

$$H = 0.24I^2 Rt \qquad (26.16)$$

or I can be replaced by V/R to yield

$$H = 0.24 \frac{V^2}{R} t \qquad (26.17)$$

This second form of the Eq. (26.16) is more widely used, since, if the resistance of the load is known, it will be necessary only to measure the current with an ammeter and record the time. This equation can be used to measure the *heat energy* developed in heating devices and also the *heat losses* in transmission lines, dynamos, and other electrical instruments whose function it is to do work not related to heating. Such energy losses (in nonheating equipment) are often referred to as "$I^2 R$ losses." If H, V, I, and t are in Btu, volt, ampere, and hour units, the following equation can be derived:

$$H = 3.413 \times 10^6 VIT \qquad (26.15')$$

$$H = 3.413 \times 10^6 I^2 Rt \qquad (26.16')$$

$$H = 3.413 \times 10^6 \frac{V^2}{R} t \qquad (26.17')$$

Illustrative Problem 26.5 The heating element of an electric heater has a resistance of 10.6 Ω. It is operating from a 115-V outlet. How many calories of heat will it give off in 5 min?

Solution Substitute known values in Eq. (26.17) to get

$$H = 0.24 \frac{(115 \text{ V})^2}{10.6 \ \Omega} \times 300 \text{ sec}$$

$$= 89,800 \text{ cal} \qquad\qquad ans.$$

Illustrative Problem 26.6 A 5-hp 220 V dc motor having an efficiency of 90 percent is installed 210 ft from a power line with a two-wire service using AWG No. 12 copper wire. Calculate (*a*) the current used by the motor, (*b*) the resistance of the wires connecting the motor to the power line, (*c*) the I^2R loss.

Solution
(*a*) Changing 5 hp to watts, we get

$$\text{Power output} = 5 \text{ hp} \times 746 \text{ W/hp}$$

$$= 3730 \text{ W}$$

Since the motor is only 90 percent efficient, the power input will be

$$\text{Input} = \frac{3730 \text{ W}}{0.90} = 4144 \text{ W}$$

Using Eq. (26.8), solve for the current flowing through the motor,

$$P = VI$$

$$I = \frac{P}{V}$$

$$= \frac{4144 \text{ VA}}{220 \text{ volts}}$$

$$= 18.84 \text{ A} \qquad\qquad ans.$$

(*b*) From Appendix V, the resistance of No. 12 wire is 1.588 Ω/1000 ft. Then the resistance of the two lead-in wires will be

$$R = \frac{2(210) \text{ ft} \times 1.588 \ \Omega}{1000 \text{ ft}}$$

$$= 0.667 \ \Omega \qquad\qquad ans.$$

(*c*) The power loss will be

$$I^2R = (18.84 \text{ A})^2 \times 0.667 \ \Omega$$

$$= 237 \text{ W} \qquad\qquad ans.$$

Note that 18.84 A $\times$ 0.667 Ω > 12 V; so the power line must carry more than 232 V to get 220 V to the motor.

SUMMARY

The electromotive force (emf) is the energy per unit charge supplied by a source of electricity.

$$\mathscr{E} = \frac{W}{Q} \qquad\qquad (26.1)$$

The true direction of *electron flow* is from negative to positive. The conventional direction of *current* is from positive to negative.

Electric energy can be converted into chemical, mechanical, heat and radiant energy. Likewise, chemical, mechanical, heat, and radiant energy can be converted into electrical energy.

A simple voltaic cell consists of two electrodes of dissimilar substances in contact with an electrolyte. The substance which has the greater tendency to be chemically attacked by the electrolyte is generally the negative electrode.

Two or more cells connected in combination form a battery.

Primary cells are nonreversible; secondary cells are reversible.

For cells in series: (1) the total emf is equal to the sum of the emfs of the cells, (2) the current in each cell is the same as the current in the external circuit, (3) the total internal resistance is equal to the sum of the internal resistances of the cells.

For cells in parallel: (1) the total emf is equal to the emf of one cell, (2) the total internal resistance is equal to the resistance of one cell (if all cells have equal internal resistance) divided by the number of cells, (3) the total current from the combination equals the sum of the currents from the separate cells.

Cells are usually connected in series when the external resistance is high and in parallel when the external resistance is low.

The time rate at which electrical energy is delivered or consumed is called *electric power*.

$$P = VI \qquad (26.8)$$

$$P = I^2 R \qquad (26.10)$$

$$P = V^2/R \qquad (26.11)$$

The heat (in calories) generated in an electric circuit can be found from the equation:

$$H \text{ (cal)} = 0.24VIt \qquad (26.15)$$

$$H \text{ (cal)} = 0.24I^2Rt \qquad (26.16)$$

$$H \text{ (cal)} = 0.24 \frac{V^2}{R} t \qquad (26.17)$$

The electric energy lost in the form of heat in nonheating devices is termed the I^2R power loss.

$$\text{Watts} = \text{volts} \times \text{amperes}$$

$$1 \text{ hp} = 746 \text{ watts}$$

$$1 \text{ kW} = 1.34 \text{ hp}$$

Electric energy units are the *watthour*, the *kilowatthour* (kW-hr), and the *joule* (watt-sec).

QUESTIONS AND EXERCISES

1. How do primary and secondary cells differ?

2. Explain the difference between "potential difference" and "emf."

3. Which needs replacement more often in a voltaic cell, the negative or the positive electrode? Why?

4. Give the advantages of (*a*) the dry cell, (*b*) the storage cell, as a source of electrical energy.

5. Why should cells of different emfs *not* be connected in parallel?

6. Explain why the wires leading to the electric bulb in a reading lamp do not get as hot as the filament, even though they carry the same amount of current. Give a justification based on mathematical formulas.

7. What would happen if you tried to charge a run-down storage battery with the current from another battery of the same make and size. Explain.

8. Why should you not use a 25-A rated plug fuse in a circuit which can safely carry only 18 A?

9. The common 60-W incandescent lamp will draw about $\frac{1}{2}$ A when operating on the house lighting circuit. One new dry cell will deliver on short circuit a current of 30 A or more. Will this new cell light the lamp? Give (numerical) reasons for your answer.

10. A wire lying on a table top carries a constant current. When a pocket compass is slid slowly along the table top toward the wire, the current produces no magnetic effect on the compass. How is this possible? Why?

11. Which of the following is not equal to the watt? (*a*) Volt-ampere. (*b*) (Ohms)^2/Volt. (*c*) (Amperes)^2-ohm. (*d*) Joule/sec.

12. How are batteries rated as to their capacities?

PROBLEMS

1. Convert 30 hp to kilowatts.

2. A dc motor operating under full load draws 18 A at 240 V. What is the horsepower input to the motor?

3. A 32 V electric power plant has an output of 4.5 kW. What current (amperes) can be drawn from the generator?

4. What size plug fuse, rated in amperes, should be used in a house circuit that has a maximum load of 2100 W from a 120 V line?

5. If the cost of electric power is 6.5¢/kW-hr, how long can you operate a 250-W incandescent lamp for 1 ¢?

6. A soldering iron has a resistance (hot) of 350 Ω. What power does it draw from a 118-V line?

7. Determine the resistance of a 1650-W, 120-V electric heater.

8. A 120-V potential difference maintains 2.4 A in a resistor. Find (*a*) the resistance of the resistor, and (*b*) the potential difference required to maintain 3.6 A of current in the resistor.

9. A dc motor rated at 5 hp, operating at full load, was found to draw 23 A at 240 V. What is its efficiency?

10. Given 20 dry cells, each with emf of 1.5 V and internal resistance of 0.2 Ω. Find the amount of current that will flow through an external load of 0.03 Ω when the cells are connected (*a*) in series; (*b*) in parallel.

11. Given 20 dry cells, each with emf of 1.5 V and internal resistance of 0.2 Ω. Find the current that will flow through an external resistance of 30 Ω when the cells are connected (*a*) in series; (*b*) in parallel.

12. Determine the current through an external resistance R if it is connected to n cells in series, if each cell has an emf $\mathscr{E}$ and an internal resistance r.

13. Determine the current through an external resistance R if it is connected to n cells in parallel, if each cell has an emf $\mathscr{E}$ and an internal resistance r.

14. A storage battery has an open-circuit voltage of 13.2 V. When 100 A of current is delivered, the battery voltage drops to 10.6 V. Calculate the internal resistance of the battery.

15. How much will it cost per 40-hr week to run a motor having an average load of 50 hp and an average efficiency of 90 percent if the power rate is 6.50 ¢/kW-hr?

16. A 6-V storage battery with 0.6-Ω internal resistance is to be charged from a 15-V circuit. What resistance must be placed in series with the battery if the charging rate is set at 5.0 A? What is the total energy supplied to the battery in 8 hr?

17. A ½-hp (output) furnace fan motor operates for an average of 7 hr per day for a 30-day month. The motor is 92 percent efficient. If electric energy costs 4.8 ¢/kW-hr, what is the monthly bill?

18. A bank of 80 incandescent lamps (connected in parallel), each having a resistance of 240 Ω when hot, is connected to a 120-V power line. Find the cost of operating these lamps for an 8-hr day at 5.6 ¢/kW-hr.

19. A storage battery for farm lighting produces 96 V across its terminals in open circuit. The battery has 0.034 Ω internal resistance and is connected through a pair of copper wires, which have a resistance of 1.835 Ω/1000 ft, to a load 250 ft distant. The load draws 21 A under these conditions. (*a*) What is the voltage at the load, 250 ft from the battery? (*b*) What will be the voltage across the battery terminals when delivering this load?

20. A battery charger applies a potential difference of 13.5 V at the terminals of a 12.0-V battery. The internal resistance of the battery is 0.95 Ω. (*a*) What is the charging current? (*b*) What is the rate at which electric energy is being converted into useful chemical energy? (*c*) At what rate is nonusable heat produced? (*d*) How long will it take to fully charge a dead battery if it is rated at 70 A-hr?

21. A chicken brooder is to be built that will produce 220 cal of heat each second. Nichrome wire with a mean diameter of 0.180 in. is to be used as the heating element. The resistivity of nichrome is 660 Ω cmils/ft.

What length of nichrome wire is needed if the brooder is to operate off a 120-V line?

22. In the circuit of Fig. 26.17 determine (*a*) the supply voltage, (*b*) the power supplied, and (*c*) the power dissipated in each resistor.

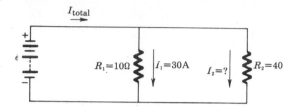

Fig. 26.17   Problem 22.

23. Two identical batteries are connected in series across a 5.0-Ω resistor, and the current is found to be 0.18 A. When the same batteries are connected in parallel with the same resistor, the current is 0.14 A. (*a*) What is the internal resistance of each battery? (*b*) What is the open circuit voltage of each battery?

24. Battery A has a no-load terminal voltage of 6.0 V and an internal resistance of 1.5 Ω. Battery B has a no-load terminal voltage of 8.5 V with an internal resistance of 2.1 Ω. The batteries are connected in parallel. A 3.5-Ω resistor is then connected to the terminals of the battery combination. What current will flow through each battery? [HINT: See Illus. Prob. 25.14.]

27 Magnetism and Electromagnetism

Magnetism and electricity have a great deal in common. They both have force fields and they both display forces of attraction or repulsion under proper conditions. Some phenomena of electrostatics and magnetism were known 2500 years ago.

As early as 600 B.C. certain rocks (later called *lode stones*) were known to exist. These rocks had the power of attracting iron and other pieces of the same rock. Unable to explain forces existing between two bodies separated at a distance, the early writers attributed the phenomenon to some magical or spiritual factor. Thales of Miletus (640?–546? B.C.), a Greek philosopher, regarded the magnet as having a "soul" which caused it to attract iron.

We still do not know the exact nature of magnetism, but we have learned a great deal about how it behaves and how to use its forces. The harnessing of magnetic forces has revolutionized man's life by opening up the age of electricity and electronics. In this and succeeding chapters we shall learn how the effects of magnetic forces are used in compasses, electrical measurements, generators, motors, telephones, radios, television equipment, and in modern electronic and nuclear devices.

27.1 Artificial Magnets

Lodestone ore is a *natural magnet*. Artificial magnets can be made in a variety of ways. When an iron or steel bar is stroked with a natural (or artificial) magnet, the iron or steel bar itself becomes a magnet. Artificial magnets can also be produced by placing a bar of iron or steel in a coil of wire in which a heavy direct current of electricity is flowing. A magnet produced in this manner is generally a much stronger magnet than the one produced by stroking. Soft iron is rather easily magnetized, but such magnets readily lose their magnetism. They are called *temporary magnets*. If the magnet is made from hardened steel, it will retain its magnetic properties much longer. Such magnets are termed *permanent magnets*. One of the strongest and most permanent of artificial magnets is made of an alloy called *alnico* containing aluminum, nickel, and cobalt. As pure metals, however, aluminum, nickel, and cobalt are only very weakly attracted by a magnet. Most substances such as copper, gold, silver, lead, wood, glass, etc., are not noticeably attracted by a magnet. A strong magnet attracts iron filings at a considerable distance, even though nonmagnetic substances such as copper, wood, glass, etc., are placed between the magnet and the filings.

Some substances are even repelled by magnets, although the repelling forces are very slight.

The following terms are used to describe magnetic materials.

1. *Paramagnetic.* Materials that are weakly attracted by a magnet. Examples of paramagnetic materials are: aluminum, chromium, sodium, platinum, wood.

2. *Ferromagnetic.* Materials that are strongly attracted by a magnet. Examples are: iron, cobalt, nickel,

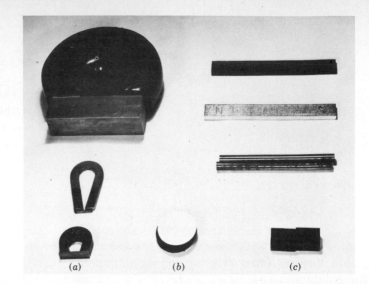

Fig. 27.1 Forms of magnets. (*a*) Horseshoe magnets. (*b*) Disk magnet. (*c*) Bar magnets. (Central Scientific Co.)

(*a*)   (*b*)   (*c*)

and alloys (such as permalloy and alnico) of ferromagnetic metals.

3. *Diamagnetic.* Materials that are weakly repelled by both poles of a magnet. Examples: bismuth, carbon, gold, zinc, salt.

A magnet in the form shown in Fig. 27.1*a* is called a *horseshoe magnet*, the one in (*b*) is a *disk magnet*, and the type in (*c*) is a *bar magnet*. The horseshoe magnet is the most common form, since it gives a stronger field of magnetic force.

27.2 Magnetic Poles

If a magnet is lifted out of a pile of iron filings or small nails, the filings or nails cling to the magnet near the ends but scarcely at all near the middle (Fig. 27.2). These regions near the ends are called *magnetic poles*. Magnetic poles always exist in pairs; it is impossible to have a magnet with only one pole. If magnets—whether natural or artificial, bar- or horseshoe-shaped—are suspended (on the earth) so as to turn freely about a vertical axis, they always come to rest in approximately a north-south position (Fig. 27.3). This fact was used by early mariners as they fashioned crude compasses by floating on water or otherwise suspending a lodestone magnet so that it

Fig. 27.2 Iron filings cling near the poles of the magnet.

might swing freely about a vertical axis. The pole of the suspended magnet that points toward the geographic north is called the *north pole*, while the other pole, which points toward the south, is called the *south pole*.

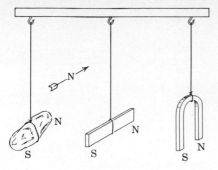

Fig. 27.3 Suspended magnets seek a north-south position when free to rotate.

The fact that a magnet placed on a piece of cork floating on water will take a north-south position without moving longitudinally in either a north or a south direction shows that *the two poles of a magnet are of equal strength.*

27.3 Laws of Attraction and Repulsion of Magnetic Poles

If the north pole of a magnet is brought near the north pole of a second magnet which is free to rotate (Fig. 27.4), the poles will repel each other. But if a south pole is brought near the north pole of the second magnet, the two will be attracted toward each other. A general law of magnetic attraction states:

Like poles repel each other, while unlike poles attract each other. These attractive and repulsive forces between magnetic poles vary directly as the

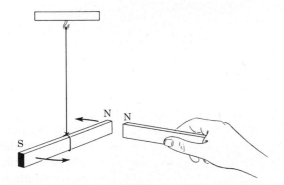

Fig. 27.4 Like poles repel each other.

product of the pole strengths and inversely as the square of the distance between them.

This law was developed as a result of carefully conducted experiments by the French physicist Coulomb and is known as *Coulomb's law.* Pole strengths in the SI-metric system are expressed in *ampere-meter* (A-m) units. *A magnetic pole has a pole strength of one ampere-meter if it repels a like pole of equal strength one meter distant with a force of 10^{-7} newtons.* Stated as a formula, Coulomb's law becomes

$$F = k \frac{m_1 \times m_2}{d^2} \qquad (27.1)$$

where

F = force in newtons
m_1, m_2 = pole strengths in ampere-meters
d = distance between the poles in meters
k = constant of the medium between poles in newtons/(ampere)^2
$k = 10^{-7}$ newtons/(ampere)^2 for a vacuum or air medium.

Equation (27.1) for magnetic forces is similar to Eq. (24.1), which was developed for electrostatic forces. The difference lies in the fact that Eq. (27.1) was developed on the assumption that magnetic poles are points. Since magnetic poles are not points, the equation can result only in approximations. This will not destroy the validity of the discussions that follow, however. Nonetheless, it should be emphasized that a unit magnetic pole does not actually exist. Since magnetic poles always exist in pairs, Eq. (27.1) can properly be applied only to long, thin magnets with well separated poles.

Illustrative Problem 27.1 A magnetic pole of 30 A-m strength exerts a force of 90×10^{-5} N upon a second pole placed 4 cm away. What is the strength of the second pole if they are in air?

Solution Solve Eq. (27.1) for m_2.

$$m_2 = \frac{Fd^2}{km_1}$$

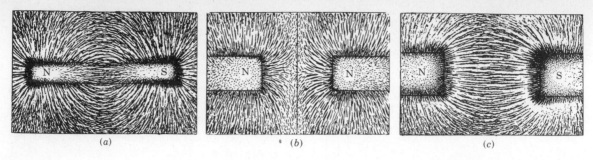

Fig. 27.5 Magnetic lines of force as shown by iron filings (*a*) about a bar magnet, (*b*) between like poles, and (*c*) between unlike poles.

Substitute given values to get

$$m_2 = \frac{90 \times 10^{-5} \text{ N} \times (0.04 \text{ m})^2}{10^{-7} \text{ N/A}^2 \times 30 \text{ A-m}}$$

$$= 0.48 \text{ A-m} \qquad \qquad ans.$$

27.4 Magnetic Lines of Force

A *magnetic field* is said to exist in the region around a magnetic pole where the influence of the pole can be detected. If we can conceive of an isolated unit north pole placed in a magnetic field, the *direction of the magnetic field will be the same as the direction of the force acting on the isolated north pole.* Michael Faraday (1791–1867) was perhaps the first to study the configuration of a magnetic field. Although the magnetic field cannot be seen, it can be demonstrated and mapped in several ways. Place a sheet of glass or cardboard on top of a magnet whose field is to be studied. Then sprinkle the sheet with iron filings. Upon tapping the sheet, you will see that the filings align themselves with the magnetic field and form "lines" or "strings" of definite lengths and shapes (Fig. 27.5*a*). Repeating the experiment using the field between two like poles, you will obtain the configuration illustrated in Fig. 27.5*b*. The field between two unlike poles is shown in Fig. 27.5*c*. Faraday drew what he called *lines of force* around a magnet to correspond to the lines formed by the iron filings.

Similar lines of force can be obtained by placing a small compass needle in the magnetic field and moving it always in the direction pointed to by its north pole while tracing its path. A magnetic line of force (called *line of flux*) can be more technically defined as a *line which indicates at every point along its length the direction in which a north pole would be urged.* The lines should have small arrowheads on them to indicate the direction of the field (see Fig. 27.6).

Note that we arbitrarily assumed that the force acts from the north pole to the south pole. It should also be noted that the pattern of lines of force represented in Fig. 27.6 represents the magnetic field only in the horizontal plane. To obtain a true picture, the lines of force should be drawn in three dimensions—showing the magnetic field over and under the magnet as well as on either side.

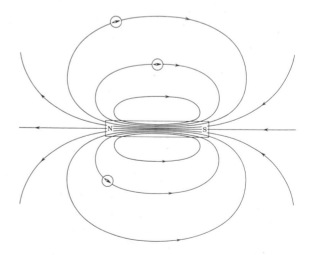

Fig. 27.6 Magnetic lines of force run from the north pole to the south pole.

The total number of magnetic lines of force (called magnetic flux) is used to represent the *direction* and *magnitude* of a magnetic field. The SI-metric unit of magnetic flux is one line of force and is called a *weber* (Wb). The number of webers per unit area, if the area is perpendicular to the field, is called *magnetic flux density* or *magnetic induction* (**B**). Magnetic flux and magnetic flux density will be discussed at greater length in Sec. 27.10. The lines of force are drawn close together where the magnetic field is strong and farther apart where the field becomes weaker. Remember that lines of force do not actually exist. They are just aids to the imagination.

The cause of magnetic force is not completely understood. Magnetic fields of force exist, just as gravitational and electrostatic fields of force exist. These fields of force are regions in which forces act on bodies without actual physical contact between the material bodies themselves. Gravitational and electrostatic forces act between *any two bodies* possessing mass or charge, respectively, whereas magnetic forces act *only* between certain types of materials known as *magnetic substances.*

Gravitational, electrostatic, and magnetic forces are known as *action-at-a-distance* forces, since they exist without any apparent contact between bodies. Actually it is more useful to think in terms of fields rather than specific forces. That is, it is the interaction between the magnetic field and the piece of iron that leads to the observed force of magnetic attraction or repulsion.

27.5   The Earth's Magnetic Field

The fact that a suspended magnet will seek a north-south direction if free to rotate is evidence that the earth itself is surrounded by a huge magnetic field. Thus the earth acts as a large magnet, with magnetic poles near the geographic poles.

The magnetic north, as indicated by a compass, in general differs from the true geographic north. This is due to the fact that the north magnetic pole of the earth does not coincide exactly with its geographic north pole. The angle between the magnetic north and the geographic north at a given point is called the *declination* at that point. In the United States the declination will vary roughly from 20°E in the northwestern section of the country to 20°W in the extreme northeastern section. The navigator and surveyor must continually make corrections for these declinations (see chart of Fig. 27.7), if magnetic compasses are being used.

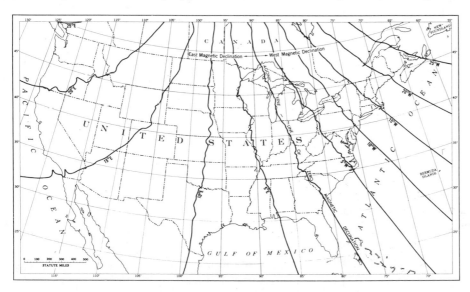

Fig. 27.7 Declination chart. (U.S. Coast and Geodetic Survey.)

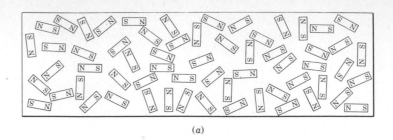

(a)

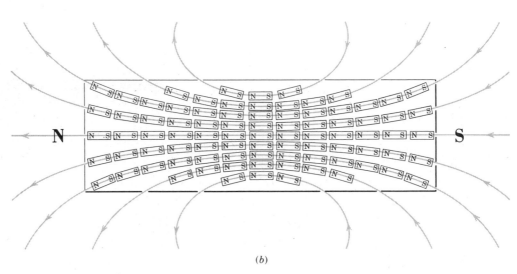

(b)

Fig. 27.8 Schematic diagram of Ewing's theory of magnetism. Elementary magnets or magnetic domains in (a) unmagnetized iron and (b) magnetized iron.

27.6   Theories of Magnetism

Although it was stated that the real cause of magnetism and magnetic fields is not clearly known, there are *theories* about the cause of magnetism, two of which are outlined here.

Ewing's Theory According to this theory of Sir Alfred Ewing (1855–1935), a piece of iron consists of millions of tiny elementary magnets. These submicroscopic magnets may be molecular in size, or they may possibly be made up of groups of molecules aligned to form tiny iron domains. In unmagnetized iron these elementary or molecular magnets are assumed to be pointing in all conceivable directions, with a random orientation which provides no external magnetic effect

(Fig. 27.8a). When this unmagnetized iron is placed in a magnetic field, the tiny molecular magnets are forced to align themselves with the field (Fig. 27.8b), the degree of alignment depending on the intensity of the field in which the iron is placed. Soft iron, this theory assumes, is more easily magnetized because its tiny molecular magnets can be more readily turned than those of hard steel. Conversely, the molecular magnets of soft iron revert more readily to their random orientations when the magnetizing field is removed, and soft iron is therefore not as suitable as hard steel for making permanent magnets.

Support for Ewing's molecular theory of magnetism comes from a simple experiment. A bar of soft iron *initially unmagnetized* can be converted into a weak magnet by holding it parallel

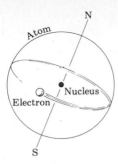

Fig. 27.9 Electrons revolving about the nucleus of an atom are thought to be a source of atomic magnetism.

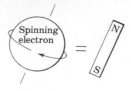

Fig. 27.10 The electron spinning on its axis is the major source of atomic magnetism.

to the earth's field and striking it repeated blows with a hammer. At each blow the molecular magnets are thought to be momentarily jarred out of their customary orientation and in their "unsettled" condition are more readily urged into alignment with the earth's field. The fact that a magnet will lose most of its magnetism if heated to a cherry-red heat also seems to be compatible with this molecular theory of magnetism.

The Modern Theory of Magnetism The present view attributes the phenomenon of magnetism to motion of electrons within the atoms of the molecules. Electron movements within the atom contribute to magnetic behavior in two ways. First, the electrons rotate in concentric shells around the nucleus (Fig. 27.9). We will learn in Sec. 27.9 that a current in the shape of a loop will show magnetic polarity. Currents are made up of electrons in motion. Hence, the electron movement around a nucleus imparts a magnetic property to the atoms.

Second, and the major source of atomic magnetic properties, the electron is conceived as resembling a charged sphere spinning around an axis. The spinning electron is equivalent to an extremely small current loop.* Hence every electron, due to its spin, is considered to have a mag-

netic field equivalent to that of a tiny bar magnet (Fig. 27.10). It is believed that in a magnet each atom has many more electrons *spinning* in one direction than in another. In a submicroscopic region called a *domain*, many of these atoms with electron spin in one direction create a magnetic field which effectively supplements the field engendered by the *revolving* electrons. Each *domain* thus becomes a tiny permanent magnet. When the domains are in random orientation, the substance as a whole is not a magnet, but the presence of an external field will reorient the domains or make the favorably oriented domains grow in size, and produce a magnet in the manner described above. When, in the presence of a very strong field, all the domains have been aligned, the condition of *magnetic saturation* occurs and any further increase in strength of the external field will not increase the magnetization of the iron.

The modern theory of magnetism is similar to that proposed by Ewing, except that now we talk of magnetic domains instead of tiny magnetic molecules.

ELECTROMAGNETISM

The classic discovery by Oersted relating electricity and magnetism (Sec. 26.15) opened a vast new era of scientific advance. The fact that an electric current sets up a magnetic field at right angles to the conductor through which it is flowing has led to some of the most fruitful achievements in the entire history of industrial and technical development. The electromagnetic effect of a current is employed in electric bells, meters, telephones, telegraph sets, radios and phonographs, electric

---

* Scientists believe that the phenomenon of magnetism is associated with the spin of the electrons within the third shell of the atoms of magnetic materials.

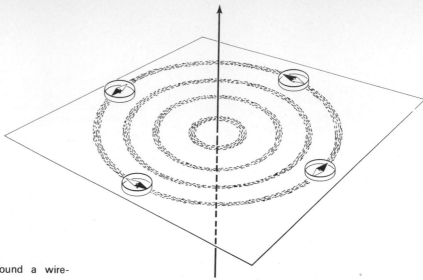

Fig. 27.11 Magnetic field around a wire-carrying current.

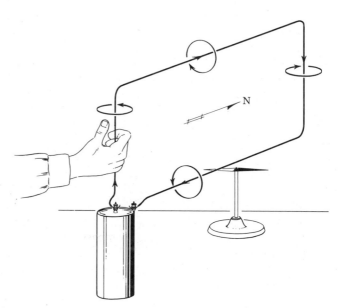

Fig. 27.12 Right-hand rule for determining the direction of a magnetic field.

motors, relays, electromagnetic control apparatus, radar, and nuclear devices. The relationship between electricity and magnetism is the basis for the development of most of our modern electric machines.

27.7 Magnetic Field around a Straight Conductor

If magnetism is attributed to motion of charge (the modern theory), a magnetic field should be

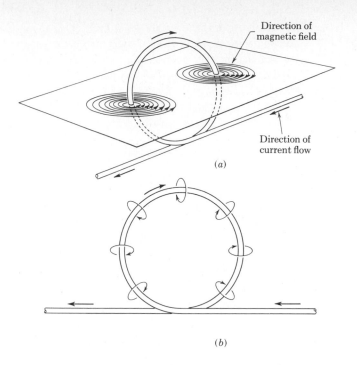

Direction of
magnetic field

Direction of
current flow

(a)

(b)

Fig. 27.13 Magnetic field about a single conducting loop.

expected to surround an electric current. Oersted discovered that a compass needle placed above and parallel to a current-bearing conductor will turn until the orientation of the magnetic N-S poles is perpendicular to the direction of the current. If the compass is then placed below the conductor (or the direction of the current is reversed), the compass will turn 180° to reverse the N-S poles' orientation (see Fig. 27.11).

The direction of the magnetic field around a straight current can readily be determined by noting the orientations of the compass in Oersted's experiment or by mounting a piece of cardboard horizontally and passing a heavy conductor through it (Fig. 27.12). When a strong current (say 30 A) is passed through the wire a moderate magnetic field will be established. Iron filings sprinkled on the cardboard will arrange themselves around the conductor in concentric rings, showing the pattern of the field. The density of the filings will be greater near the wire than farther out. If small compasses are placed at various positions on the cardboard, the needles align

themselves tangent to the rings of iron filings. When the current is flowing upward (as shown), the north poles of the compass needles will point in a counterclockwise direction. If the current is then reversed, the north poles will point in a clockwise direction around the wire. If the direction of a current is known, a convenient way to remember the direction of the circular magnetic field is the so-called *right-hand rule* (Fig. 27.11): *If one grasps a current-carrying wire with one's right hand and extends one's thumb so it points in the conventional direction of the current flow (positive to negative), then the fingers will indicate the direction of the magnetic field.*

27.8 Magnetic Field of a Solenoid

If a current-carrying wire is bent into a loop, the faces of the loop will show magnetic polarity. Two views of such a single loop of wire with some of the lines of force around the wire are illustrated in Fig. 27.13. It can be seen that all the magnetic lines enter the loop at one face and leave at the

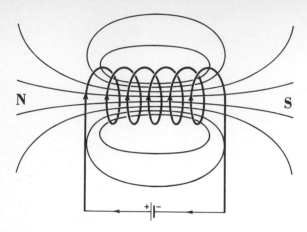

Fig. 27.14   Magnetic field about a solenoid.

other. In effect, the loop acts as a disk magnet. The polarity of the loop can be determined by applying the convention that magnetic lines leave a magnet from the north pole and enter at its south pole. Hence the left-hand faces of the illustrations would have north magnetic polarity and the right-hand faces south polarity. If such a loop of wire were free to turn, it would seek a north-south position when current flows through the wire.

The polarity becomes more pronounced and the magnetic field stronger if we wind a number of turns of wire side by side in a tight spiral. Such an arrangement is called a *solenoid* or *helix*. When current passes through a solenoid, the magnetic field developed resembles that of a bar magnet (Fig. 27.14). If the coils of wire are close together, most of the lines of flux around the loops merge and are in the same direction throughout the center of the solenoid (parallel to the axis). At the ends the lines of flux flare out, leaving the solenoid at one end and returning at the other. The solenoid has north polarity at the end where the magnetic lines of force leave and south polarity at the end where they return. The polarity of a solenoid can be determined by using a compass. That pole of the solenoid which repels the north pole of the compass is a north pole, and that pole which repels the south pole is a south pole. The polarity of a solenoid can be predicted

by applying another *right-hand rule: Grasp the solenoid with the right hand so that the fingers wrap around it in the conventional direction of the current; the outstretched thumb will then point to the north pole of the solenoid.* The converse of this rule can be applied to determine the direction of the current in the coil if the polarity is known.

27.9   Magnetic Induction

In Sec. 27.4 it was established that a magnetic field can be represented by (flux) lines of force and the total number of lines of force of the field is called the *magnetic flux.*

Magnetic flux is represented by the symbol ϕ (Greek phi). The unit of magnetic flux is the *weber.* The *magnetic induction* (also called *magnetic flux density*) **B** is the number of flux lines per unit area that cross the magnetic field at right angles.

$$\mathbf{B} = \frac{\phi}{A} \qquad (27.2)$$

Thus, the unit of magnetic induction in the SI system is the weber per square meter (also called the *tesla* T). An older unit which remains in use today is the *gauss* (G), where 1 tesla = 10^4 gauss

$$1 \text{ T} = 1 \text{ Wb/m}^2 = 10^4 \text{ G} \qquad (27.3)$$

The earth's magnetic flux density is about $\frac{1}{2}$ G. Lodestone magnets possess a magnetic density of approximately several hundred gauss. Permanent magnets made of modern alloys can produce fields with an intensity of 10,000 G.

Flux density is a vector quantity. The direction of **B** at any point of a magnetic field is the direction of the magnetic lines of force at that point. In Sec. 27.19 we shall see that a magnetic field exerts a force on a current-bearing wire placed in it in a direction perpendicular to its lines of force and perpendicular to the current direction. This phenomenon is used to define **B** in the SI system. The *magnetic induction* **B** is the force per unit length exerted by a magnetic field on a current-carrying conductor when the conductor is oriented so that it is perpendicular to the lines of force of the magnetic field. As a

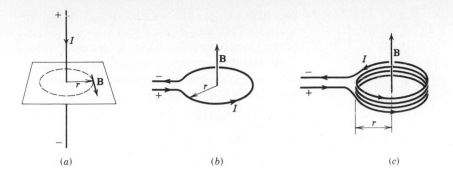

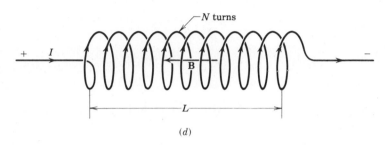

Fig. 27.15 Magnetic field intensity (*a*) near a single current-carrying wire, (*b*) at the center of a single loop, (*c*) at the center of a loop of *N* turns, (*d*) along the axis of a long solenoid.

defining equation

$$\mathbf{B} = \frac{\mathbf{F}}{IL} \qquad (27.4)$$

where

$\mathbf{F}$ = force, newtons
I = current, amperes
L = length of conductor, meters
$\mathbf{B}$ = magnetic induction, newtons/
   ampere-meter (abbreviated N/A-m) (or T)

In summary,

$$1 \text{ N/A-m} = 1 \text{ magnetic line/m}^2$$

$$= 1 \text{ Wb/m}^2 = 1 \text{ T}$$

By equating 1 Wb/m² = 1 N/A-m, it is easy to show that dimensionally 1 Wb is equivalent to 1 N-m/A.

The calculation of magnetic induction for circuits of various configurations is often quite difficult and involves the calculus. The following formulas can be developed for some common and practical circuits. Each formula involves μ (Greek mu), called the *permeability*. Permeability (μ) is a constant of the particular medium in which the magnetic induction is occurring. For air (or vacuum),

$$\mu = 4\pi \times 10^{-7} \text{ kg-m/C}^2$$

or $$\mu = 4\pi \times 10^{-7} \text{ Wb/A-m}$$

Around a Single Straight Wire   The *magnitude* of the magnetic induction B at a distance r from a straight wire carrying current I (Fig. 27.15*a*) is given by the expression

$$B = \frac{\mu I}{2\pi r} \qquad (27.5)$$

where B is in webers/square meter, I in amperes, and r in meters.

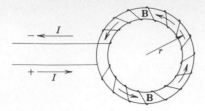

Fig. 27.16   Solenoid wound in the form of a toroid.

Inside a Single Loop of Wire   In Fig. 27.15*b*, if a circular loop of wire carries current *I* in amperes, the magnitude of the flux density *B* at the center of the loop of radius *r* (meters) is given by

$$B = \frac{\mu I}{2r} \qquad (27.6)$$

The magnetic induction is perpendicular to the plane of the loop (Sec. 27.9). Note that the induction *B* is not uniform at all points in the interior of the loop.

Inside a Multiple Loop   If a multiple loop of *N* turns is made, the magnitude of the flux density at the center is

$$B = \frac{\mu N I}{2r} \qquad (27.7)$$

At the Center of a Long Solenoid   Figure 27.15*d* is a schematic diagram of a long solenoid, of length *L* meters and total number of turns *N*. The magnitude of the flux density *B* along the axis within the solenoid is

$$B = \frac{\mu N I}{L} \qquad (27.8)$$

These formulas assume that the radius of the loops is small compared with the length of the solenoid.

One type of solenoid consists of a coil of wire which has been wound around a doughnut-shaped form. Such a closed solenoid is called a *toroid* (Fig. 27.16). Here *L* in Eq. (27.8) is the mean circumference of the toroid ($2\pi r$). The magnetic field is nearly uniform within the ring or

core material, and there is almost no field outside the core. Such toroids are frequently used for laboratory measurements where a source of a uniform field is needed or as magnetic-core memory devices in computers where the field in one core will not affect its neighbors.

Illustrative Problem 27.2   A solenoid 25 cm long has 840 turns of wire. What is the magnitude of the flux density of the magnetic field through the center of the solenoid when the current is 2.6 A?

Solution   Substituting the known values in Eq. (27.8) and solving, we get

$$B = \frac{\mu N I}{L}$$

$$= \frac{(4\pi \times 10^{-7}\ \text{Wb/A-m}) \times 840 \times 2.6\ \text{A}}{0.25\ \text{m}}$$

$$= 1.1 \times 10^{-2}\ \text{Wb/m}^2 \qquad \textit{ans.}$$

Illustrative Problem 27.3   A horizontal power line carries a current of 80 A in a south-to-north direction. (*a*) What is the magnitude of the magnetic induction, due to the current, at a point 1 m below the wire? (*b*) What is the direction of the field at that point?

Solution

(*a*) $\quad B = \dfrac{\mu I}{2\pi r}$

$$= \frac{(4\pi \times 10^{-7}\ \text{Wb/A-m}) \times 80\ \text{A}}{2\pi\ (1\ \text{m})}$$

$$= 1.6 \times 10^{-5}\ \text{Wb/m}^2 \qquad \textit{ans.}$$

(*b*) By the right-hand rule, the magnetic field is found to be directed westward.

Illustrative Problem 27.4   The core of a toroid has inner and outer diameters of 24 cm and 28 cm, respectively. It has 4000 turns. When the current in the wire is 5 A the magnetic induction inside the core is 2.5 T. Determine the permeability of the core.

Solution   The mean diameter is (24 cm + 28 cm)/2 = 26 cm. Solve Eq. (27.8) for μ and substitute known values.

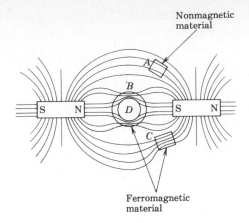

Nonmagnetic material

Ferromagnetic material

Fig. 27.17 Ferromagnetic materials become magnetized by induction. Ring B shields the area D against the magnetic field.

$$\mu = \frac{BL}{NI} = \frac{(2.5 \text{ T})(\pi \times 0.26 \text{ m})}{(4000 \text{ turns})(5 \text{ A})}$$

$$= 1.02 \times 10^{-4} \frac{\text{T-m}}{\text{A}} \qquad \textit{ans.}$$

or $\qquad \mu = 1.02 \times 10^{-4} \frac{\text{T-m}}{\text{A}} \left(\frac{\text{Wb/m}^2}{\text{T}}\right)$

$$= 1.02 \times 10^{-4} \text{ Wb/A-m} \qquad \textit{ans.}$$

27.10 Magnetic Shielding

If a sheet of soft iron (*C* in Fig. 27.17) is placed in a magnetic field, the lines of force are drawn away from the immediate region and concentrated so as to pass through the iron. The iron, in effect, draws the lines to it. The iron thus becomes magnetized by induction. Hence the term *magnetic induction* is used for **B**. On the other hand, a sheet of nonmagnetic material will have little effect on the flux distribution in a given region.

If a soft iron ring (*B* in Fig. 27.17) is placed in the magnetic field, the region *D* will be almost free of magnetic lines of force. The region *D* is said to be *shielded*. The property of magnetic shielding is used in many pieces of apparatus where magnetic fields are undesirable such as in color TV picture tubes. However, magnetic shielding is never as complete as electrostatic shielding (Sec. 24.8).

27.11 Magnetic Field Intensity or Magnetic Field Strength

A magnetic field accompanies any system of moving charges. We noted in Secs. 27.9 and 27.10 that magnetic flux density depends not only on the magnitude of the current and the geometric configuration of the conductor (e.g. loop, solenoid, toroid), but also on the material of the field (e.g. air, soft iron). For instance, the flux density produced by a current in a solenoid with a soft iron core will be considerably greater than that produced in a solenoid with an air core.

Therefore, it is convenient to define a new magnetic vector **H**, known as *magnetic field intensity* or *magnetic field strength*. Magnetic field intensity (field strength) at a point where the permeability of the medium is μ is defined as the ratio

$$\mathbf{H} = \frac{\mathbf{B}}{\mu} \qquad (27.9)$$

The SI unit for *H* is amperes per meter (A/m) or newtons per weber (**N**/Wb).

It should be apparent that when the medium is a vacuum or air, then $\mathbf{H} = \mathbf{B}/\mu_0$. If a piece of soft iron is placed in a magnetic field of intensity **H**, magnetic lines of force will be concentrated in the iron. Magnetism is said to be induced in the iron. The flux density (magnetic induction) *B* in the iron is found by the equation

$$\mathbf{B} = \mu \mathbf{H} \qquad (27.9')$$

The magnetic field intensity **H** might be thought of as the field intensity produced by the current, and the magnetic flux density **B** as the total field including the contributions made by the material in the field.

27.12 Magnetic Permeability

Iron and certain alloys of iron have the property of being able to multiply the strength of a magnetic field. In a highly permeable medium, the flux density may be thousands of times larger than the field intensity. Permeability may be thought of as a measure of how effective a ferromagnetic material is in multiplying a magnetic

Table 27.1 Electromagnetic concepts and their units

Electromagnetic concepts	SI units
Total magnetic flux, ϕ	Wb or N-m/A
Magnetic field intensity, **H** or magnetic field strength	N/Wb or A/m
Magnetic flux density, **B**, or magnetic induction	Wb/m² or T or N/A-m
Permeability, μ	Wb/A-m or T-m/A

field intensity. Permeability is not a constant for any given ferromagnetic material but varies markedly with the field intensity. From Eq. (27.9)

$$\mu = \frac{\mathbf{B}}{\mathbf{H}} = \frac{\text{magnetic flux density}}{\text{magnetic field intensity}} \quad (27.9'')$$

Using the units for **B** and **H**, we can show that the SI unit for μ is Wb/A-m or T-m/A.

Table 27.1 summarizes the electromagnetic units used in this chapter.

27.13 Magnetization Curves— Hysteresis

Magnetization curves are used to show how flux density B varies with field intensity H for an initially unmagnetized sample. Figure 27.18 shows a magnetization curve for a test sample and also shows how the permeability μ varies with the field intensity.

It will be noted from the magnetization curve that as the field intensity H is increased, the flux density B increases slowly at first (section AB of the curve). This may be explained by assuming that there is considerable difficulty in bringing magnetic domains into alignment at first. Then, as the magnetized field is further increased, the magnetization curve rises steeply, showing that the magnetic domains are now rapidly and easily being realigned in the magnetic field. Section BC of the curve illustrates this phase of the process. Finally, as *magnetic saturation* is approached, fur-

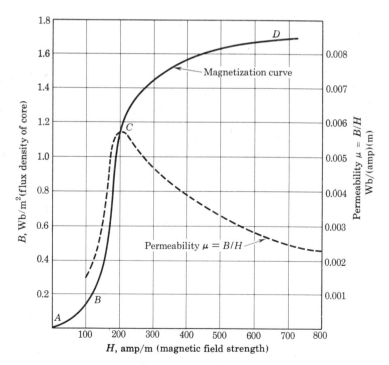

Fig. 27.18 Magnetism curve of a test sample. The variation of permeability with magnetic-field strength is also shown.

ther increase of the magnetizing field can realign fewer and fewer magnetic domains and the curve flattens out until finally (point *D*), no further increase in flux density can be obtained.

If a piece of iron is used as the core of a solenoid and the current applied is slowly increased, a magnetization curve like that of *ABCD* of Fig. 27.18 will be obtained. If the current to the solenoid is then gradually decreased, it will be found that the field intensity and flux density decrease but the demagnetization curve does not follow back along the same path as the magnetization curve.

In Fig. 27.19 let the curve *abcd* represent the same magnetization curve as that of Fig. 27.18. As the current to the solenoid is decreased to zero, the curve of decreasing magnetization takes some such path as *de*. When current to the solenoid, and therefore the magnetic field intensity, has been reduced to zero, the flux density may still be as much as 1.3 Wb/m². A considerable

amount of *residual flux density* (*residual inductance*) is thus indicated. This residual flux density corresponds to *permanent magnetization*.

A reversal of the current to the solenoid will reverse the direction of the magnetizing field, which will gradually bring the flux density in the test core back to zero at *f*. For the sample of Fig. 27.19, a reversed field of 120 A/m was required to reduce *B* to zero. If the current is further increased in the negative direction, the *fg* section of the curve will be obtained, the point *g* illustrating magnetic saturation with magnetic domains aligned in the opposite direction. Decreasing the *negative* current will result in the *gh* section of the curve, and finally, if the current is switched to its original (positive) direction and increased, the *hid* section of the curve is obtained.

It is seen that flux density *B* lags behind field intensity *H*. This lag is called *hysteresis*. The closed loop *dfgid* is called a *hysteresis loop*. As iron is magnetized, work must be done on the

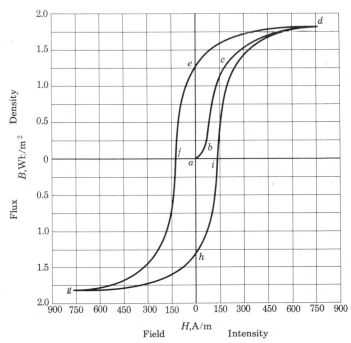

Fig. 27.19 The hysteresis loop.

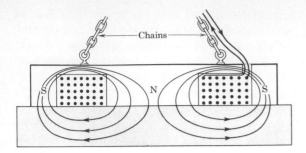

Fig. 27.20 Cross-sectional view of lifting magnet.

Fig. 27.21 Magnetic crane moving steel slab at steel plant. (Bethlehem Steel Corp.)

domains in order to align them, and this work shows up as heat. In alternating current apparatus (see Chap. 29), the magnetizing and demagnetizing process takes place many times a second, and the *hysteresis loss* (heat) may be considerable. For example, it may be shown by use of the calculus that the hysteresis loss (or heat loss) for one cycle in alternating current is proportional to the area enclosed within the hysteresis loop of a diagram like that just analyzed.

The magnitude of the area of the hysteresis loop is important in the design of some electrical machinery. This is particularly true for ac machinery where the direction of the current is continually changing at the rate of 120 times per second. Not only is the loss of energy in the form of heat energy expensive, but the accumulation of too much heat energy in the apparatus can be dangerous as well as damaging to equipment.

Since soft iron shows lower hysteresis losses than steel, it is used for the cores of rotors and stators in electric machinery. Soft iron also has high permeability, which is another factor in its favor for use where magnetic effects are desired.

27.14 Lifting Magnets

Lifting magnets are constructed in various shapes. One of the more common types is illustrated in Fig. 27.20. It consists of an energizing coil of copper wire placed around a soft-iron core. The soft-iron core is the center part of a larger iron shell which furnishes the permeable material to carry the magnetic lines of force. These powerful magnets are used in transferring scrap iron and steel from stockpiles to railroad cars or from railroad cars to river barges (Fig. 27.21). Such magnets are also used in manufacturing plants, where heavy iron or steel products have to be hoisted in orderly fashion, e.g., kegs of nails, iron sheets, pipes, and steel billets. Electromagnets have been made which will exert a force of 200 lb for every square inch of pole face. The electromagnet is effective so long as sufficient current flows through its coils. Thus it allows an operator to turn the magnetic field on or off, merely by the flick of a switch. Large electromagnets generally are equipped with portable dc generators that furnish the required current. Surgeons use a very small electromagnet for removing small splinters of steel from the eye or other parts of the body.

Electromagnets also are effective with alternating current. When alternating current is passed through the coils of an electromagnet, the polarity of the magnet will change periodically in step with the alternations of the current (60 cycles per second in the United States). However, it is usually not important whether it is a north or

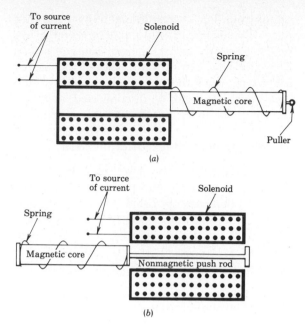

Fig. 27.22 (*a*) Puller-type solenoid. (*b*) Pusher-type solenoid.

the center of the solenoid. The field acts as though it sucked the plunger into the solenoid. Hence the coil and the plunger are termed a *sucking coil*. The sucking coil is also called *solenoid*, even though only the coil is the solenoid.

The sucking coil can be used either as a *puller-type* solenoid (Fig. 27.22*a*) or a *pusher-type* solenoid (Fig. 27.22*b*).

Thus the solenoid can be used to provide either a mechanical push or pull to operate various devices. When the current does not flow in the solenoid, a spring forces the plunger back to its original position. Pusher-type solenoids properly motivated by an electric current find frequent use in removing objects from a moving belt. Small solenoids are also used in operating such devices as door chimes, gas valves, water valves, and pipe organ stops. The schematic diagram of such a device is left as an exercise for the student.

south pole that is doing the attracting. Since the alternations occur so rapidly, the interruptions of the magnetic field are, for most purposes, negligible.

There are some problems inherent in the use of ac electromagnets. The changing magnetic field will cause an electric current to be induced in the core. These currents are called *eddy currents*.

Eddy currents in the electromagnet are undesirable for two reasons: the eddy currents represent power losses and the core tends to get quite hot. Eddy currents and the means of eliminating them will be discussed in Chap. 29.

27.15 Solenoids

The magnetic flux density along the axis of a solenoid was discussed in Sec. 27.10. If the solenoid is provided with a movable soft-iron core, called the *plunger*, and a current (either dc or ac) flows through the turns of the coil, the resultant magnetic field will tend to pull the plunger into

27.16 Magnetic Separators

Electromagnets are used in removing iron from many industrial processes when its presence is undesirable. For example, the coal industry uses separator magnets, which are quite similar to the lifting magnets discussed in Sec. 27.14, suspended over conveyor belts to remove "tramp" iron in the coal, which, if it were not removed, might break the rolls used to crush or pulverize the coal. Often *magnetic pulleys* replace the end pulleys on a belt-conveyor system. When material which contains iron reaches the magnetic pulley, all the tramp iron is attracted by the pulley and held against the belt on the bottom side of the magnetic pulley. When the tramp iron is carried past the magnetic field set up by the pulley, it falls harmlessly into a separate compartment or onto another conveyor belt.

Magnetic separators are used a great deal today in reclaiming iron and magnetic metal from foundry refuse or slag. They are also used to separate magnetic and nonmagnetic borings and turnings.

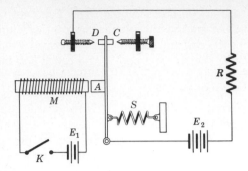

Fig. 27.23 Electric relay or magnetic switch.

27.17 Electric Relays

The *electric relay* is an electromagnetic device by means of which contacts in one electric circuit are operated by a change in current in the same circuit or in a different circuit. The essential parts of the relay are illustrated in Fig. 27.23. When the electromagnet M is energized by passing a current through it, the soft-iron armature A is attracted to it. The armature, which is a part of a second circuit, is provided with a tongue which can make contact with the fixed points C and D. As the armature is attracted to the electromagnet core, the tongue breaks contact with C and makes contact at D, thus *closing* the secondary circuit. When the current does not flow through the electromagnet, the attraction for A disappears and a spring S pulls the armature until the contact at D is broken.

Applications of electromagnetic relays are numerous. They are used extensively for local or remote control of circuits of all types. Electric motors at inaccessible locations, such as the motor of an elevator, can be operated by a system of relays. Thermostats and other control devices can actuate solenoids which close or open magnetic switches to start or stop electric motors. Small relay currents can be used to control large currents through relay controls. *Electric protective relays* are used to protect apparatus and entire electric systems from dangerous conditions. When these conditions arise, the relays disconnect the apparatus from the system or shut off the main-line current. Some of the protective relays include *circuit relays*, *voltage relays*, *power relays*, *frequency relays*, and *temperature relays*. They can be set to operate as either the *over* or *under* type, i.e., if the factor being guarded is too great, the circuit can be disconnected, or if it is too small, it can be closed.

27.18 Uses of Electromagnetism

We have discussed only a few uses of electromagnets and electromagnetic principles. There are hundreds that could be listed. The following partial list should serve to illustrate the great variety of uses to which electromagnetism has been applied.

1. Generators
2. Motors
3. Telephone receivers
4. Magnetic brakes
5. Magnetic switches
6. Circuit breakers
7. Magnetic clutches
8. Magnetic couplings
9. Ammeters
10. Voltmeters
11. Loudspeakers for radio, TV, and audio equipment
12. Electric chimes
13. Brakes for hoists, cranes, and elevators
14. Magnetic chucks
15. Electric clocks
16. Cyclotrons and other "atom smashers"
17. Holding coils for motor starters
18. Relays for automatic control of equipment
19. Track switches
20. Solenoid valves

ELECTRICAL MEASURING INSTRUMENTS

There are many and diverse types of instruments designed to measure electrical quantities. Practically all of them, either directly or indirectly, involve the measurement or detection of electric current. These instruments measure electric quantities by using one of the following effects of an electric current: (1) chemical effect, (2) heating effect, or (3) magnetic effect. The magnetic effect is the most universally used in these instruments. The following pages present a discussion of a few common electrical measuring instruments which employ the magnetic effect of an electric current.

27.19 Force of a Magnetic Field on a Current-Carrying Conductor

In Sec. 27.8, we studied the magnetic field around a wire carrying a current. Let us place such a wire in the field established between the opposite poles of a magnet. In Fig. 27.24a the current, represented by a dot, is directed toward the reader. The lines of force of its field are in a counter-clockwise direction (apply the right-hand thumb rule). In Fig. 27.24b, lines of force between the poles of a magnet are shown traveling from the north to the south pole. When the current-carrying wire is placed in the magnetic field (c), it can be seen that the lines of force (produced by the poles and by the current in the wire) below the conductor all move from left to right, while those above the conductor travel in opposite directions. The result is a strong magnetic field below the conductor and a weak field above. The effect of this distribution of lines on the conduc-

tor can best be visualized if we conceive of the magnetic lines of force as having elastic properties. As the lines contract to become as short as possible, the conductor will be pushed upward. An easy way to remember the direction of motion of the conductor is to apply what is commonly called the *right-hand* or *motor rule* (also called Ampère's rule). *Place the thumb and the first and second fingers of the right hand at right angles to each other* (as in Fig. 27.25). *Then hold the hand so that the first finger points in the direction of the conventional current, the second finger in the direction of the magnetic field; then the thumb will point in the direction in which the conductor will tend to move.*

When the wire and magnetic field are at right angles to each other, the force exerted upon the wire by the field is directly proportional to each of three factors: (1) *the magnetic induction B* of the field, (2) *the current I* flowing in the wire, and (3) *the length L* of the part of the wire that is in the field. Thus, the magnitude of the force is given by

$$F - ILB \qquad (27.10)$$

where F is expressed in newtons if I is in amperes, L in meters, and B in newtons per ampere-meter (see Eq. 27.4).

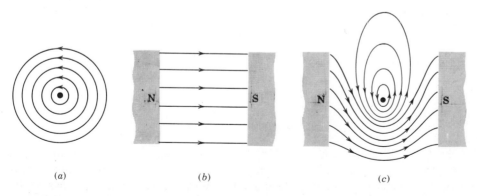

(a)   (b)   (c)

Fig. 27.24 Lines of force (a) around a conductor, (b) between magnetic poles, (c) when conductor is in the magnetic field.

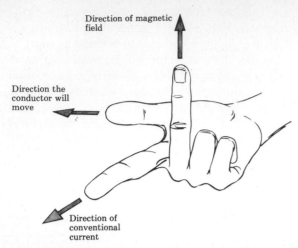

Direction of magnetic field

Direction the conductor will move

Direction of conventional current

Fig. 27.25 Right-hand rule for finding direction of force on a conductor.

In discussing the development of Eq. (27.10), we assumed that the wire must be perpendicular to the direction of the magnetic field. If the wire makes an angle θ with the field, the magnitude of the force is given by

$$F = ILB \sin \theta \qquad (27.11)$$

Illustrative Problem 27.5 A wire carrying a current of 40 A is at right angles to a uniform magnetic field in which the magnetic induction is 0.50 N/A-m. If the length of the wire is 15 cm, what is the force on the wire?

Solution

$I = 4.0$ A $L = 0.15$ m $B = 0.50$ N/A-m

$F = ILB$

$\quad = 4.0$ A $\times 0.15$ m $\times 0.50$ N/A-m

$\quad = 0.30$ N   *ans.*

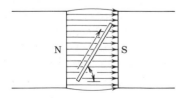

Fig. 27.26 Illustrative Problem 27.6

Illustrative Problem 27.6 A wire 40 cm long carries 25 A at an angle of 60° to a uniform magnetic field of flux density 12×10^{-4} N/A-m (see Fig. 27.26). What is the magnitude and direction of the force on the wire?

Solution

$I = 25$ A $L = 0.40$ m

$B = 12 \times 10^{-4}$ N/A-m $\theta = 60°$

$F = ILB \sin \theta$

$\quad = 25$ A $\times 0.40$ m $\times (12 \times 10^{-4}$ N/A-m)

$\qquad \times 0.866$

$\quad = 0.010$ N   *ans.*

Using the motor rule, we find that the wire will be pushed away from the reader (into the page).

27.20   Force between Two Parallel Conductors

When two long parallel current-carrying conductors are adjacent, each gives rise to a magnetic field and the interaction of the fields gives rise to mechanical forces which act on both wires. The two circular fields combine as shown in Fig. 27.27. Note that two long straight parallel wires carrying current in the same direction attract each other and the same two wires carrying current in opposite directions repel each other.

We can use Eq. (27.5) to determine the field caused by I_1:

$$B = \mu \frac{I_1}{2\pi d}$$

From Eq. (27.10), the force on current I_2 by B will be

$$F = I_2 LB = I_2 L\mu \frac{I_1}{2\pi d}$$

or $F = \mu \dfrac{I_1 I_2 L}{2\pi d}$ (27.12)

Definition of the Ampere (A) and the Coulomb (C)

The force of attraction between long current-carrying parallel conductors, derived from

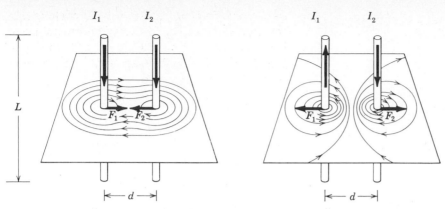

Fig. 27.27 (*a*) Parallel wires carrying current in the same direction attract each other. (*b*) Parallel wires carrying current in opposite directions repel each other.

Eq. (27.12), was officially adopted in 1947 by the U.S. National Bureau of Standards in defining the ampere unit. The *ampere* is defined as the current in each of two long parallel conductors 1 m apart in free space which causes them to exert a force on each other of 2×10^{-7} N for each meter of length. Having defined the ampere in this way we can now define the *coulomb* as the charge transferred through any cross section of a conductor in 1 sec by a current of 1 A.

Illustrative Problem 27.7 Two long straight wires are 6 cm apart. What force per meter length will the wires exert upon each other if one wire carries 2 A of current while the other carries 3 A?

Solution

$$\mu = 4\pi \times 10^{-7} \text{ kg-m/C}^2$$

$$I_1 = 2 \text{ A} \qquad I_2 = 3 \text{ A}$$

$$d = 0.06 \text{ m} \qquad F = \mu \frac{I_1 I_2 L}{2\pi d}$$

$$\frac{F}{L} = \frac{(4\pi \times 10^{-7} \text{ kg-m/C}^2) \times 2 \text{ A} \times 3 \text{ A}}{2\pi(0.06 \text{ m})}$$

$$= 2 \times 10^{-5} \frac{\text{kg-A}^2}{\text{C}^2} \left(\frac{1 \text{ C}}{\text{A-sec}}\right)^2 \left(\frac{1 \text{ N}}{1 \text{ kg-m/sec}^2}\right)$$

$$= 2 \times 10^{-5} \text{ N} \qquad\qquad ans.$$

27.21 Galvanometers

The basic current-measuring or current-detection instrument is the *galvanometer*. There are many kinds of galvanometers, of which we shall consider only two. The essential parts of the *d'Arsonval galvanometer* are shown in Fig. 27.28. It consists of a flat coil of wire C suspended by means of a light metallic ribbon between the poles of a permanent U-shaped magnet. Current is conducted to and from this coil by means of the light ribbon by which it is suspended from the top, and a helix of similar material below the coil. When a current flows through the coil, it produces a magnetic field with opposite poles on either side of the coils. In the figure (*top view*) the direction of the current in the coil is indicated by dots when it is toward the reader and by +'s when it is away from the reader. Thus it is seen that a north polarity is produced above the coil and a south polarity below. The interaction between this field and that of the permanent magnet will establish a clockwise torque which is proportional to the amount of current passing through the coil. The coil is provided with a soft-iron core B to form a uniform magnetic field. The torque is opposed by the restraining force of the suspension ribbon. The movable system carries a mirror M which reflects the numbers from a distant scale (see Fig. 27.28*b*). The farther the scale is from the galvanometer, the greater the sensitivity of the instrument. A more rugged and portable (but less sensitive) instrument is obtained by replacing the optical (mirror) system of deflection by a pointer, which moves over a graduated scale, and by replacing the metallic-ribbon suspension by

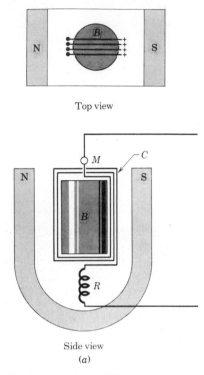

Top view

Side view
(a)

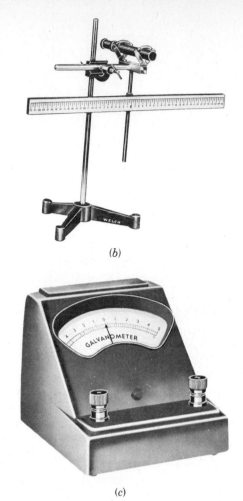

(b)

(c)

Fig. 27.28 (*a*) Moving system of d'Arsonval galvano-meter, (*b*) D'Arsonval galvanometer telescope scale, (*c*) Portable galvanometer. (Welch Scientific Co.)

two spiral springs. Such an instrument is shown in Fig. 27.28c. The springs, besides balancing the magnetic deflection torque exerted on the coil, provide the external leads to the instrument. The coil is mounted in jeweled bearings.

27.22   The Ammeter

It was emphasized in Sec. 25.12 that the ammeter is connected in *series* with the line in which the current is to be measured. The ammeter is simply a *low-resistance galvanometer* whose scale is calibrated to read current in amperes. Since the moving coil and the springs used in the galva-nometer are designed to carry only very small cur-rents (about 0.05 A), the instrument, if it is to

serve as an ammeter, must be equipped with a *shunt of* such low resistance that only a small part of the total current flows through the moving coil. The greater part of the current in the main circuit bypasses the coil and travels through the shunt (see Fig. 27.29). Many ammeters are fitted with several shunts of different resistances in order to provide an instrument which will mea-sure currents over several different ranges. These instruments require the use of several sets of figures on the scale (see Fig. 25.6a). Judicious selection of the proper range by the technician is required in using the instrument.

Several precautions must be observed in connecting an ammeter in a circuit. The movable coil is quite light in weight and is mounted on

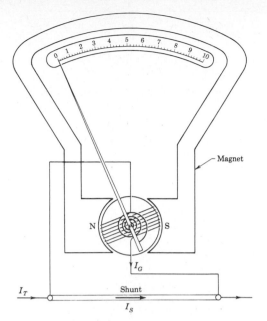

Fig. 27.29 Most of the current passes through the shunt in an ammeter. Only a small part passes through the moving parts of the instrument.

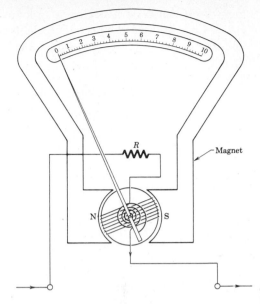

Fig. 27.30 Circuit of voltmeter.

jeweled bearings. This increases its sensitivity but makes it more susceptible to damage from jarring. Since it is a *low*-resistance instrument, care must be taken never to connect it in a circuit which will cause too much current to flow through the moving coil. The instrument can easily *burn out* if subjected to too much current. Also, since most ammeters permit only clockwise deflection of the needle, it is imperative that the instrument terminal marked $(+)$ be connected to the positive side of the current source, when used in dc circuits.

Illustrative Problem 27.8 The moving coil of an ammeter has a resistance of 5 Ω and the shunt a resistance of 0.05 Ω. Full-scale deflection of the meter results when 10 A flows through it. What is the actual current in the moving coil?

Solution Let

I_T = total current through ammeter

I_c = current in coil $\Big\}$ in parallel
I_s = current in shunt

Then $\qquad I_T = I_c + I_s = 10$ A

But, from Sec. 25.15

$$V_s = V_c$$

From Ohm's law

$$I_s R_s = I_c R_c$$

$$\frac{I_s}{I_c} = \frac{R_c}{R_s} = \frac{5\ \Omega}{0.05\ \Omega}$$

or $\qquad I_s = \dfrac{5}{0.05} I_c = 100 I_c$

Substituting yields

$$I_c + 100 I_c = 10 \text{ A}$$
$$I_c = 0.099 \text{ A} \qquad \qquad ans.$$

27.23 The Voltmeter

The galvanometer can also be used satisfactorily to measure voltages. For a given resistance, according to Ohm's law, the voltage is proportional to the current. The voltmeter is connected in *parallel*. Hence to measure voltage, we need only use a galvanometer of such high resistance that the instrument itself will draw a negligible amount of the current, not enough to affect appreciably the voltage drop between the points in the circuit to which it is connected (see Fig. 27.30). Multiple-range voltmeters are produced by providing the instrument with several coils of different resistances in series with the moving coil. The proper range is obtained by connecting across the correct terminals. With the voltmeter, as with the ammeter, it is important to avoid using a (voltage) range on the instrument below that which is to be measured. Also the positive (+) terminal of the voltmeter must be connected to the positive side of the line when used with dc circuits. The voltmeter must be connected across the load, not in series in the circuit.

27.24 Alternating-Current Instruments

Although we do not discuss *alternating currents* in detail until Chap. 29, it should be made clear at this point that the instruments discussed thus far are used only in dc circuits (circuits in which the charges flow always in only one direction). For ac circuits the measuring instruments must be of different design.

Alternating-current instruments for measurement of currents and voltages sometimes make use of the repulsion between two pieces of iron of like magnetic polarity. A simple demonstration will explain the principle involved. Support two pieces of soft iron side by side within a coil of wire (Fig. 27.31a). When current is allowed to pass through the coil, as shown in Fig. 27.31b, the two pieces of iron will acquire magnetic properties with north polarity at the tops and south polarity at the bottoms of the pieces. As a result the pieces will repel each other. The same result would be obtained if the direction of the current were reversed, even though the polarity, too, would be reversed. So, even though alternating current reverses its direction 120 times every second (60-cycle current), repulsion always takes place.

Thus if we use two soft-iron vanes (Fig. 27.32), a stationary one A, fixed at one side of the coil, and the other B, next to it and at-

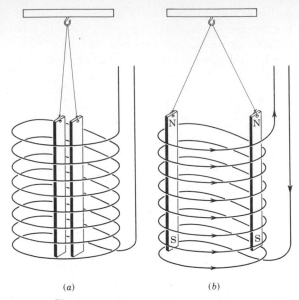

Fig. 27.31 When current flows in any direction in the coil, the two pieces of iron will repel each other as shown in (b).

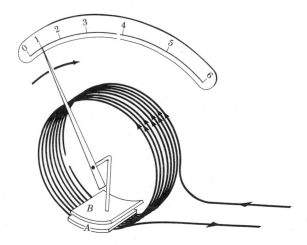

Fig. 27.32 Principle of moving-iron meter.

tached to a shaft which is mounted to turn at the center of the coil, a mutual repulsion between the two will result in a rotation of the center shaft. When current flows in the coil, the two vanes will repel each other, causing the moving element to turn in a clockwise direction. The force of repulsion is proportional to the strength of the current and is restrained by a spiral spring at the top of the moving element. By proper use of shunts in parallel or resistances in series, the instrument can be calibrated to read directly in amperes or in volts.

SUMMARY

Magnets are classified as natural or artificial, and as permanent or temporary.

Ferromagnetic materials are strongly attracted by a magnet.

Paramagnetic materials are weakly attracted by a magnet.

Diamagnetic materials are weakly repelled by a magnet.

Magnetic poles are those regions at which the magnetic properties are most evident; they usually exist at the ends of the magnet.

Scientists associate magnetism with the movement of electrons about the positive nuclei of the atoms and with electron spins about parallel axes.

The earth acts as a huge magnet with one pole in the Northern Hemisphere and one in the Southern Hemisphere.

When a magnet is freely suspended in the earth's magnetic field, the end which turns toward the north is called the *north pole* and the end which turns toward the south is called the *south pole*.

Like poles repel each other; unlike poles attract each other.

Coulomb's law for the force between magnetic poles in air is

$$F = k \frac{(\text{pole strength})_1 \times (\text{pole strength})_2}{(\text{distance between poles})^2} = k \frac{m_1 m_2}{d^2}$$

Lines of force are used in visualizing a magnetic field. A magnetic line of force in the SI system is called a *weber*, (Wb).

A conductor carrying a current is surrounded by a magnetic field.

The right-hand rule for a magnetic field around a current-carrying wire: *Grasp conductor with right hand so that the thumb points in the direction of the current; then the fingers will show the direction of the magnetic lines of force around the conductor.*

The magnetic field within a solenoid carrying an electric current resembles that of a bar magnet.

The strength of an electromagnet depends on (1) ampere-turns, (2) the quality or grade of core material, and (3) design and construction.

The north pole of a solenoid is shown by another right-hand rule: *Wrap the fingers around the coil in the direction of current (from + to −). The thumb then points to the north pole of the solenoid.*

Magnetic induction (or flux density) **B** is the magnetic flux per unit area:

$$\mathbf{B} = \frac{\phi}{A} \qquad (27.2)$$

It is also the force per unit length per unit current when a current-carrying conductor is placed in the magnetic field:

$$\mathbf{B} = \frac{F}{IL \sin \theta}$$

$1 \text{ N/(A-m)} = 1$ magnetic line/m$^2 = 1$ Wb/m$^2 = 1$ T.

The flux density in empty space may be found from the following formulas:

Near a long, straight current-carrying conductor:
$$B = \mu I / 2\pi r$$

At center of a single loop of a conductor: $B = \mu I / 2r$

At center of a multiple loop: $B = \mu N I / 2r$

At center of a long solenoid: $B = \mu N I / L$

The magnetic field **H** is the magnetic field produced by the flow of current in wires and the magnetic field **B** is the total magnetic field including also the contribution made by the various magnetic properties of the materials in the field.

Permeability μ is the measure of the ability of a substance to concentrate the lines of force when placed in a magnetic field:

$$\mu = B/H$$

In the SI system of units the permeability of a vacuum (air, approximately) has a value of

$$4\pi \times 10^{-7} \text{ Wb/(A-m)}, 4\pi \times 10^{-7} \text{ T-m/A}$$

or

$$4\pi \times 10^{-7} \text{ kg-m/C}^2$$

Ferromagnetic materials may have relative permeabilities considerably greater than 1 (taking $\mu_0 = 1$, for a vacuum). Paramagnetic materials have relative permeabilities only slightly greater than 1; diamagnetic materials slightly less than 1.

The force of a wire carrying current I_1 on a parallel wire carrying current I_2 is given by

$$F = \mu \frac{I_1 I_2 L}{2\pi d}$$

The magnetic behavior of iron can be determined from *magnetization curves.*

The lag of magnetic flux density behind magnetic field intensity is called *hysteresis.*

Electromagnets have numerous applications in industry and in everyday modern conveniences.

Most electrical measuring devices make use of the magnetic effect of an electric current.

QUESTIONS

1. Distinguish between the magnetic field near a stationary electrostatic charge and the field near a charge that is moving.

2. One end of a surveyor's compass needle is made slightly heavier than the other. Why? Which should be made the heavier in the Northern Hemisphere? [HINT: Look in your library references for material on *magnetic dip.*]

3. List similarities and differences between magnetism and electrostatics.

4. Distinguish between a diamagnetic and paramagnetic substance.

5. The current in a conductor is directed westward. What is the direction of the lines of force of the magnetic field (a) above the conductor, (b) below the conductor?

6. Which of the following will serve best to shield a watch from a magnetic field: gold, lead, iron? Give reasons for your answer.

7. A method for magnetizing a bar of soft iron by tapping when it is placed parallel to the earth's magnetic field was discussed in Sec. 27.6. In your locality at what angle from true north, clockwise or counterclockwise, should the rod be held while hitting it with the hammer? [HINT: Consult Fig. 27.7.]

8. Use the (modern) theory of magnetism to give an explanation of how a magnet will attract an unmagnetized piece of iron.

9. You are given a bar magnet. List at least two methods for determining which is the north pole of the magnet.

10. How can a galvanometer be made into an ammeter of a stipulated full-scale reading? into a voltmeter of a stipulated full-scale reading?

11. What precautions should be observed in using (a) an ammeter and (b) a voltmeter?

12. A stream of electrons is moving in a westerly direction. The stream passes through a uniform magnetic field directed downward. In what direction will the electrons be deflected?

13. Why is it impossible to magnetize an iron bar beyond a certain maximum flux density?

14. Show by a diagram how an ammeter can be wired to provide for two ranges of current.

15. Make a diagram to show how a voltmeter can be provided with two ranges.

16. What is the significance of the area of the hysteresis loop for a given ferromagnetic material?

PROBLEMS

1. Two unlike magnetic poles 10 and 20 A-m units strong are placed 5 cm apart. Find the force of attraction between the two poles. (Assume $k = 10^{-7}$ N/A².)

2. What is the flux through an area of 575 cm² if a magnetic flux density of 0.25 Wb/m² is directed normal to the area?

3. The coil of an electromagnet has 2400 turns. If the resistance of the coil is 13.6 Ω, how many ampere-turns will it develop when it is connected to a 36-V dc source? (An ampere-turn is the product of the amperes and the number of turns in the coil.)

4. What is the flux density of the magnetic field at a distance of 10 cm from a straight wire carrying a current of 6 A?

5. Calculate the magnetic induction at a distance of 15 cm from a long straight wire carrying a current of 25 A.

6. A magnetic flux density of 12×10^{-3} Wb/m² is desired at the center of a solenoid 1.5 m long carrying a current of 2.5 A. How many turns should the solenoid have?

7. A solenoid is 20 in. long. It is wound with 5000 turns of copper wire. What current must flow through the windings to produce a magnetic field strength of 40 A/cm?

8. A current of 3 A is flowing through a long solenoid with 12 turns/cm. If the cross-sectional area of the solenoid is 4 cm^2, what is the magnetic flux density inside the solenoid? What is the total flux inside the solenoid?

9. A U-core electromagnet has two similar exciting coils. Each coil has 300 turns and a resistance of 8 Ω. What ampere-turns (the product of the amperage and the number of turns in the coil) are produced by the electromagnet if it is connected to a 32-V dc source and the coils are connected (*a*) in series; (*b*) in parallel?

10. What is the flux density of the magnetic field at the center of a circular coil of 75 turns having a radius of 8 cm if the current in the coil is 0.5 A?

11. A wire 50 cm long carries a current of 25 A at an angle of 30° with a magnetic field whose flux density is 7.5×10^{-4} Wb/m^2. What is the magnitude of the force on the wire?

12. Find the magnetic flux density inside a solenoid 38 cm long with 300 turns when it is carrying a current of 2 A. What would be the flux density in a core of soft iron ($\mu = 1.6 \times 10^{-3}$ Wb/A-m) inserted in the solenoid?

13. The cables connecting an auto battery with the starter motor carry 250 A. What would be the force between the cables if they are 50 cm long and 1 cm apart?

14. How many turns should a solenoid 1.5 m long have for a current of 2 A to produce a magnetic flux density of 8×10^{-3} Wb/m^2 at its center?

15. What force per meter do two parallel conductors 15 cm apart exert upon each other if one carries 2.4 A and the other carries 3.6 A of current?

16. The resistance of a milliammeter (range is 0–1 mA) is 18 Ω. What resistance must be added to convert it into a voltmeter with a 0 to 1.0 V range?

17. The resistance of a milliammeter is 20 Ω. What resistance must be shunted across the instrument to convert it to an ammeter, range 0 to 1.0 A?

18. Two identical magnets are 8 cm long and have pole strengths of 75 A-m. One magnet is placed on a table and the other is placed so that its north and south poles are directly above the north and south poles of the first magnet. The second magnet will then "float"

6 cm above the first magnet. What is the weight of each magnet?

19. The moving coil of a galvanometer has a resistance of 50 Ω and requires 0.0080 A to produce a full-scale deflection. What resistance must be added in series with the coil to convert the galvanometer into a voltmeter capable of reading 10 V at full-scale deflection? Find the resistance to be added in series to give the meter a 15-V range.

20. The moving coil of a galvanometer requires 0.0045 A to produce a full-scale deflection. The coil has a resistance of 50 Ω. What must be the resistance shunted across the moving coil to convert the galvanometer into an ammeter reading 1 A at full-scale deflection? Find the resistance to give the meter a 5-A range.

28 Electromagnetic Induction

We have seen how an electric current gives rise to a magnetic field. It is only natural, then, to question the existence of the reverse phenomenon. Can magnetic fields produce electric currents? Michael Faraday in England and Joseph Henry (1797–1878), an American scientist, working independently, discovered in 1831 that an emf is set up in a conductor that moves through a magnetic field or in a conductor that is cut by a moving magnetic field.

The findings of Oersted, Faraday, and Henry form the basis for the development of the entire electrical industry. The limitations inherent in the development of electric power by chemical means were overcome. With the discoveries of Faraday and Henry it became possible to convert mechanical energy from such sources as falling water, steam, nuclear reactions, petroleum, and fossil fuels to produce mechanical energy which in turn can be converted into electric energy. Electric energy today is transmitted with little loss over large distances to create light and heat, to drive motors, and to accomplish many functions in the home as well as in business and industry.

In the next six chapters we will study how induced electromotive and electromagnetic forces are used in electric generators, motors, transformers, and numerous other devices in industrial electricity, electronics, and electronic communication.

28.1 Electromagnetic Induction

The experiments which led to Faraday's and Henry's discoveries can readily be repeated by the student in the laboratory. Consider the equipment arrangement depicted in Fig. 28.1. The ends of a wire loop are connected to a sensitive galvanometer. When a strong horseshoe magnet is moved down quickly over the wire (Fig. 28.1a) there will be a momentary deflection of the galvanometer, indicating the flow of an electric current in the conductor. The current is said to be an *induced current* and the process by which the current is produced is called *electromagnetic induction.* The galvanometer needle will show no deflection when there is no relative motion between the magnet and the conductor. Thus, it can be established that *the induced current is related not merely to the presence of the magnet, but to the motion of the magnet.*

If the magnet is next moved up from the wire (Fig. 28.1b), there will be another momentary deflection of the galvanometer. However, this time the needle will be deflected in the opposite direction, indicating a current whose direction is reversed with respect to that of the first current.

If the experiment is repeated, but with the polarity reversed (Fig. 28.1c, d), the directions of the current will also be reversed.

Thus, if the magnet is moved up and down past the wire, the current will alternate in direction with each change in direction of motion. Such a current is called an *alternating current* (ac).

The effects just described are also obtained if the magnet is held stationary and the wire is moved up and down so that it cuts across the flux of the magnet. *Relative motion* between magnetic

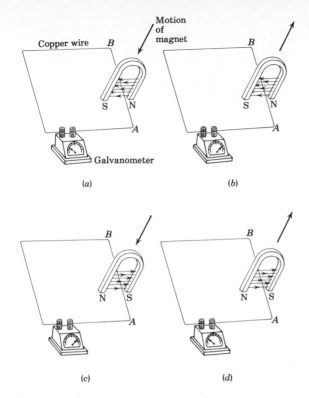

Fig. 28.1 Electromagnetic induction. A current is induced in the conducting loop whenever it cuts through the flux of a magnetic field.

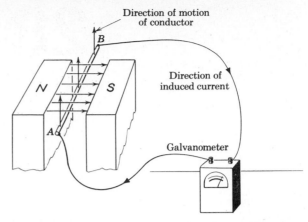

Fig. 28.2 Induced current in a conductor moving through a magnetic field.

field and conductor is necessary to initiate and sustain an induced current.

28.2 Induced EMF

We know that to make a current flow in a circuit there must be a source of energy or an emf. In Chap. 26, we studied how emfs are produced by chemical means in voltaic cells, storage batteries, and electrochemical cells. Attention was also given to thermoelectric and photoelectric sources of emf. With Faraday's and Henry's discoveries yet another, and perhaps the most important, means for producing emfs has become available. By introducing a relative motion between a conductor and a magnetic field, which results in a change of magnetic flux linked by the conductor, an *induced emf* is produced in the conductor. The greater the change of magnetic flux or the faster the *relative motion of the wire and the magnet*, the

greater the induced emf. In Sec. 28.5 we will discuss why emfs are induced in the circuit.

28.3 Direction of Induced Current —Fleming's Rule

In Fig. 28.2, let a single straight wire *AB* be moved as shown in a magnetic field. A complete *electric circuit* is provided by connecting the ends of the moving wire to a sensitive galvanometer. If the wire is moved upward through the field, the current (in accordance with the positive-to-negative convention) will be from *A* to *B* and around the circuit in the direction shown by the arrows. If the wire is thrust downward through the field, the galvanometer will show a reversal of the current direction. Or, if the poles of the magnet are reversed, the current direction will be reversed.

If the moving wire segment *AB* is not part of a closed circuit, a negative charge will pile up at one of its ends. An equal positive charge will appear at the other end. The difference of potential $\mathscr{E}$ between the two ends will be similar to that between the terminals of a battery on open circuit.

The direction of the induced emf, then, depends on (1) the direction of motion of the conductor and (2) the direction of the magnetic field. *Fleming's right-hand rule* relates these three

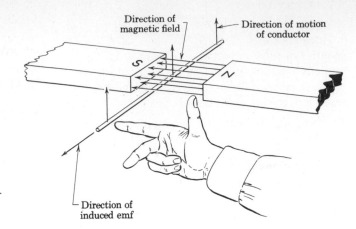

Direction of magnetic field

Direction of motion of conductor

Direction of induced emf

Fig. 28.3 Fleming's right-hand rule for determining the direction of induced emf.

directions, and is used to predict the direction of magnetically induced electric currents.

Fleming's Rule Position the thumb, forefinger, and middle finger of the *right hand* so that they are all at right angles to each other (Fig. 28.3). Let the thumb point in the direction of motion of the conductor, and the forefinger in the direction of the magnetic field (N to S). Then the middle finger *will indicate the direction of the current induced in the conductor.*

28.4 Strength of Induced EMF

Faraday found experimentally that the magnitude of the emf $\mathscr{E}$ induced by a conductor moving at right angles to the lines of force of a magnetic field (or by the lines of force moving at right angles to the conductor) is proportional to the number of lines of force cut per second. In other words, $\mathscr{E}$ depends on the rate of change of magnetic flux. The average emf $\bar{\mathscr{E}}$ is given by

$$\bar{\mathscr{E}} = -k\frac{\Delta\phi}{\Delta t} \qquad (28.1)$$

In the SI system $\bar{\mathscr{E}}$ is measured in volts when $\Delta\phi$ is expressed in webers, Δt in seconds, and $k = 1$. Thus, a flux change of 1 Wb/sec will induce an emf of 1 V in the conductor. Equation (28.1) expresses what is known as *Faraday's law.* The reason for the negative sign will be explained in Sec. 28.5.

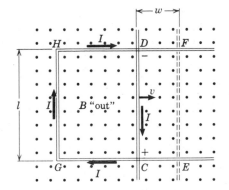

Fig. 28.4 Emf is induced in conductor *CD* moving across a magnetic field **B** (directed out of page).

Consider a U-shaped conductor *EGHF* with a second straight conductor *CD* moving with velocity v across a uniform magnetic field **B**, as shown in Fig. 28.4. When the straight conductor is in position *CD*, the total flux ϕ_1 linking the closed circuit *GHDC* is BA_1, where A_1 is the cross-sectional area of the magnetic field enclosed by *CGHD*. Assume that in time Δt the conductor reaches position *EF* and the total flux ϕ_2 linking circuit *EGHF* has increased to BA_2 enclosed by *EGHF*. The change in flux $\Delta\phi$ is equal to $\phi_2 - \phi_1$ or $B(A_2 - A_1)$. But $A_2 - A_1 = lw$. Hence, for the single loop,

$$\bar{\mathscr{E}} = -k\frac{\Delta\phi}{\Delta t} = -k\frac{Blw}{\Delta t} = -kBl\frac{w}{\Delta t}$$

$$\bar{\mathscr{E}} = -Blv \qquad (28.2)$$

where $k = 1$, B is in webers per square meter, l is in meters, and v is in meters per second. If a length l of the conductor is perpendicular to the velocity v and to the magnetic field, and the velocity is inclined at an angle θ (Greek theta) with respect to the magnetic field B, the induced emf $\bar{\mathscr{E}}$ is found by the equation

$$\bar{\mathscr{E}} = -Blv \sin \theta \qquad (28.2')$$

Equations (28.1) and (28.2) give the average emf induced by the relative motion of the conductor and the magnetic flux. Frequently, scientists are more concerned with the instantaneous value for the induced emf, which can be obtained by taking the rate of change of the magnetic flux over a very small time interval. As a formula, the instantaneous emf e is determined by

$$e = -\lim_{\Delta t \to 0} \frac{\Delta\phi}{\Delta t} \qquad (28.3)$$

Equation (28.3) is read "the instantaneous emf is equal to the limiting value of $\Delta\phi/\Delta t$ as Δt approaches zero." Hereafter the lower-case letters e and i will be used to represent the values of *instantaneous emf* and *instantaneous current*.

The induced emf depends upon the number of lines cut per second. Thus, to increase the emf induced to an outside circuit, a coil of many loops of wire can be used, the magnetic field can be made stronger, or the coil can be turned faster. An equal change of flux occurs in each loop of the coil, inducing the same emf in each loop of the coil. Since the loops are in series with each other, the total emf between the ends of the coil is the sum of the emfs for each loop. Thus, if there are N turns in the circuit and if $k = 1$, Eqs. (28.1) and (28.2) become

$$\bar{\mathscr{E}} = -N\frac{\Delta\phi}{\Delta t} \qquad (28.4)$$

and

$$\bar{\mathscr{E}} = -NBlv \qquad (28.5)$$

Illustrative Problem 28.1 A wire 60 cm long moves across a uniform magnetic field in a direction that makes an angle of 30° with respect to

the field. The magnetic induction (magnetic flux density, B) of the field is 2.4×10^{-2} Wb/m². The velocity of the conductor is 5 m/sec. (*a*) What is the average emf induced in the wire? (*b*) If the wire is connected to a circuit whose resistance is 0.3 Ω, what current flows through it?

Solution

$$L = 0.60 \text{ m} \qquad B = 2.4 \times 10^{-2} \text{ Wb/m}^2$$

$$v = 5 \text{ m/sec} \qquad \theta = 30°$$

(*a*)

$$\bar{\mathscr{E}} = Blv \sin \theta$$

$$= 2.4 \times 10^{-2} \text{ Wb/m}^2 \times 0.60 \text{ m} \times 5 \text{ m/sec} \times \tfrac{1}{2}$$

$$= 3.6 \times 10^{-2} \text{ Wb/sec}$$

$$= 36 \text{ mV} \qquad\qquad ans.$$

(*b*) $\qquad V = 0.036 \text{ V} \qquad R = 0.3 \text{ Ω}$

$$I = \frac{V}{R} = \frac{0.036 \text{ V}}{0.3 \text{ Ω}}$$

$$= 0.12 \text{ A} \qquad\qquad ans.$$

Illustrative Problem 28.2 A small test coil has 200 turns, each with an area of 5 cm². It is placed in a magnetic field of unknown flux density with its plane perpendicular to the lines of induction. It is then quickly jerked out of the field in 0.08 sec. The average emf developed during the process is 0.4 V. Find the flux density B of the magnetic field.

Solution The flux through each turn of the coil when in the magnetic field is $\phi = BA$. It drops to zero in 0.08 sec. Hence the *change* in flux $\Delta\phi$ is also equal to BA. If we solve Eq. (28.4) for $\Delta\phi$, we get

$$\Delta\phi = \frac{\bar{\mathscr{E}} \, \Delta t}{N}$$

Then

$$BA = \frac{\bar{\mathscr{E}} \, \Delta t}{N}$$

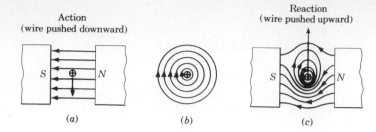

Action
(wire pushed downward)

Reaction
(wire pushed upward)

Fig. 28.5 Reaction opposes action when wire is moved across a magnetic field.

(a)   (b)   (c)

and

$$B = \frac{\bar{\mathscr{E}} \, \Delta t}{NA}$$

$$= \frac{0.4 \text{ V} \times 0.08 \text{ sec}}{200(5 \times 10^{-4} \text{ m}^2)}$$

$$= 0.32 \text{ V-sec/m}^2 \times \frac{1 \text{ Wb/sec}}{1 \text{ V}}$$

$$= 0.32 \text{ Wb/m}^2 \qquad \textit{ans.}$$

28.5 Lenz's Law and the Direction of Induced EMF

The German physicist Heinrich Lenz (1804–1865) conducted extensive experiments on induced electromotive forces which finally led to certain generalizations now ascribed to him.

Lenz's Law

Electromagnetically induced currents always have such a direction that the magnetic field set up by the induced currents tends to oppose the motion which produces them.

This law can be illustrated by using a simple diagram (Fig. 28.5). Figure 28.5a shows a wire located in a magnetic field. When the wire is pushed down through the field, the current induced in the conductor (see Sec. 27.19) will be directed into the page (represented by the tail of an arrow). The magnetic field set up by this current is shown in Fig. 28.5b. We see the two fields superimposed and interacting in Fig. 28.5c; the effect is the production of a stronger field below the wire than above it. We can conceive of the lines of force acting as taut rubber bands tending to straighten out with a resultant push upward on the wire. Work must be done against these forces to push the conductor downward through the magnetic field.

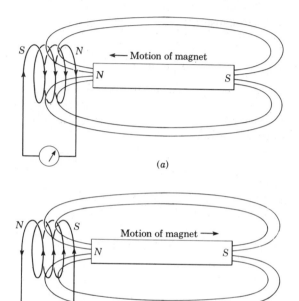

(a)

(b)

Fig. 28.6 Induced emf. (a) Magnet moves toward coil. (b) Magnet moves away from coil.

If a permanent bar magnet is thrust into a coil of wire, an emf is induced in the coil. If the coil is connected across a closed circuit, current will flow. According to Lenz's law, the direction of the current in the coil will be such as to establish a polarity which will oppose the direction of motion of the bar magnet. Thus in Fig. 28.6, if the north pole of a magnet is moved toward the coil, the induced current in the coil will produce north polarity at the face nearest the magnet, resulting in a force of repulsion between magnet and coil. If the magnet is moved away from the coil, the polarity of the magnetic field of the coil will be

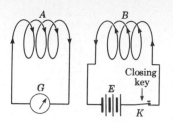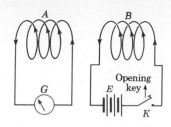

reversed and attraction between magnet and coil results. In either case, external work must be done to cause a current to flow in the coil. The negative signs of Eqs. (28.2) and (28.4) express the fact that the induced emf is of such direction as to oppose the change which produced it.

The example just cited is another case of the law of *conservation of energy. Energy must be expended* to produce electric current by induction. The mechanical energy expended in moving the magnet against the forces of repulsion and attraction set up by the interacting magnetic fields of the permanent magnet and of the induced current is converted to the electric energy of the current. No force is required to move the magnet near the coil if the circuit is open, for no magnetic field is established to oppose the motion when no current flows. Lenz's law can be looked upon, then, as stating the law of conservation of energy for electromagnetic induction.

In many electric generators (Chap. 30) emf is induced by moving conductors through a magnetic field at high speeds. In all electric generators, the direction of the induced emf is such that the resulting current through the conductors sets up a magnetic field which opposes the movement of the conductors across the original magnetic field. This opposition is overcome by the mechanical energy supplied the hydraulic or steam turbine or other driving engine.

28.6 Mutual Induction

We have seen how an emf can be induced either by moving magnetic lines of force past a conductor or by moving the conductor so it cuts lines of force. Electromotive forces can also be induced in a conductor, without moving either the magnet or the conductor, by *varying the strength or direc-*tion of a magnetic field in which a conductor is placed. For example, we might replace the bar magnet of Fig. 28.6 by a second coil as shown in Fig. 28.7. When the key *K* is closed, a current is built up in coil *B*, called the *primary coil*. The increasing current will produce an increasing magnetic flux density around the coil. When no current is flowing through the primary coil, there will be no lines of force around it, but as the current builds up, flux lines originate in the coil and continue to grow outward by expanding into larger and larger loops. The extent and intensity of this magnetic flux depend upon the current and the number of turns (*ampere-turns*) of the coil. While the current is building up, the expanding magnetic lines cut across the nearby loops of the coil *A* called the *secondary coil*, thereby inducing an emf in the secondary which is indicated by the deflection of the galvanometer *G*.

After the current in the primary has reached its maximum value, the flux in it will be steady; hence no emf will then be induced in the secondary. However, when the key is opened, the magnetic field of the primary will collapse until once more no lines are present in the coil. As these lines shrink inward, they cut the coils of the secondary again but in an opposite direction, inducing a reverse emf in the secondary. This will be evident by a deflection of the galvanometer opposite to that which occurred when the key was closed. The development of an induced emf in one circuit by the change of current in another is called *mutual induction*.

The instantaneous emf induced in the secondary coil is directly proportional to the rate of change of current in the primary coil

$$e_s = -M \frac{\Delta i_p}{\Delta t} \tag{28.6}$$

where M is a constant called the *mutual induction* of the system, and e and i represent *instantaneous values*.

The mutual induction M is defined as the ratio of the emf induced in the secondary to the time rate of change of the current in the primary.

$$M = \frac{-e_s}{\Delta i_p/\Delta t} \qquad (28.7)$$

Here, again, the negative sign is used to show that the induced emf is in such a direction as to oppose the change of current in the primary.

Mutual induction is expressed in henrys (H) when e_s is in volts and $\Delta i_p/\Delta t$ is in amperes per second. Thus a pair of adjacent circuits is said to have a mutual inductance of *one henry* if a time rate of change of current of one ampere per second in the primary induces an emf of one volt in the secondary. Where the henry is too large a unit, the millihenry (mH) or the microhenry (μH) is used: 1 mH is equal to 10^{-3} H; 1 μH is equal to 10^{-6} H.

Illustrative Problem 28.3 A magnetic flux density of 0.3 T flows through a coil having 400 turns and an area of 20 cm^2. If in $\frac{1}{100}$ sec the flux density is reduced to 0.2 T, what is the average emf induced in the coil?

Solution

$$N = 400 \qquad \Delta B = 0.1 \text{ T}$$

$$A = 20 \text{ cm}^2 \qquad \Delta t = \tfrac{1}{100} \text{ sec}$$

$$\overline{\mathscr{E}} = N\frac{\Delta\phi}{\Delta t} = N\frac{\Delta B \times A}{\Delta t}$$

$$= 400 \times 0.1 \text{ T} \times 20 \text{ cm}^2$$

$$\times \frac{1 \text{ m}^2}{10,000 \text{ cm}^2} \times \frac{1}{\frac{1}{100} \text{ sec}}$$

$$= 8 \text{ Wb/sec} = 8 \text{ V} \qquad ans.$$

Illustrative Problem 28.4 Determine the mutual inductance of a pair of adjacent coils if an average emf of 115 V is induced in the secondary by a current in the primary that changes at the rate of 25.0 A/sec.

Solution

$$e_s = 115 \text{ V} \qquad \frac{\Delta i_p}{\Delta t} = 25.0 \frac{\text{A}}{\text{sec}}$$

Substitute the known values in Eq. (28.7):

$$M = \frac{e_s}{\Delta i_p/\Delta t} = \frac{115 \text{ V}}{25.0 \text{ A/sec}} = 4.6 \text{ H} \qquad ans.$$

Illustrative Problem 28.5 The mutual inductance of a pair of adjacent coils is 2.5 H. The current in the primary coil changes from 0 to 30 A in 0.045 sec. (*a*) What is the average emf induced in the secondary coil? (*b*) What is the change of flux in the secondary coil if it has 750 turns?

Solution

$$M = 2.5 \text{ H} \qquad \frac{\Delta i_p}{\Delta t} = \frac{30 \text{ A}}{0.045 \text{ sec}} \qquad N_s = 750$$

(*a*)

Use Eq. (28.6) to find e_s.

$$\overline{\mathscr{E}}_s = M\frac{\Delta i_p}{\Delta t} = 2.5 \text{ H}\left(\frac{30 \text{ A}}{0.045 \text{ sec}}\right) = 1670 \text{ V}$$

(*b*)

Now substitute this value for e_s in Eq. (28.4)

$$\overline{\mathscr{E}}_s = N_s\frac{\Delta\phi}{\Delta t} = 750\left(\frac{\Delta\phi}{0.045 \text{ sec}}\right) = 1670 \text{ V}$$

Then

$$\Delta\phi = \frac{1670 \text{ V}(0.045 \text{ sec})}{750} = 0.100 \text{ Wb} \qquad ans.$$

28.7 The Induction Coil

Frequently a high voltage is desired when only a source of low voltage is available. One device for obtaining high voltages is the *induction coil* (see Fig. 28.8). It consists of a primary coil of a relatively few turns of heavy insulated wire and a secondary coil of many thousands of turns of insulated fine copper wire, both coils being wound around a cylindrical core of a bundle of soft-iron wires. The primary is connected through an automatic interrupter, similar to that of an electric bell or buzzer. When the switch to the *primary*

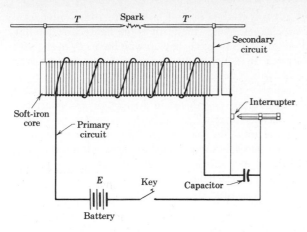

Fig. 28.8 Wiring diagram of an induction coil.

circuit is closed, current alternately builds up and diminishes in the coil, with a frequency controlled by the interrupter. This establishes a changing flux through and about the core. This rapidly changing magnetic flux cuts the many turns of the secondary, inducing a high emf across the terminals T and T'. This emf is high enough to cause a spark of an inch or more to jump across the terminals of the secondary, even in a small coil.

When the current in the primary is interrupted, the current tends to continue to flow just as if it possessed inertia. The result is a *jumping* of the current across the interrupter gap even after it has opened slightly. This, in time will pit and burn the contact points of the interrupter. A capacitor is usually connected across the interrupter gap to diminish the arcing at make and break. Then, as the contact points are separated, the electrical energy between them will charge up the capacitor, rather than produce a spark across the interrupter.

It should be made clear that the induction coil provides a high *voltage* at the sacrifice of *current*. Actually, the power (watts) which can be taken from the secondary is always a little less than that supplied to the primary, owing to losses within the coils.

28.8 Induction Coils in Automobile Ignition Systems

An important practical application of the induction coil is the production of sparks in automobile-engine cylinders for ignition of the gasoline vapors. A potential difference of several

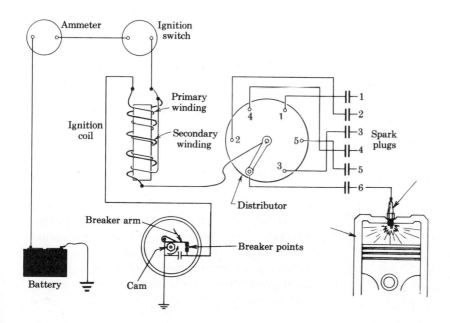

Fig. 28.9 Ignition system of a six-cylinder gasoline engine. (General Motors Corp.)

thousand volts is required between the terminals of the *spark plug* to ignite the vapors; the higher the cylinder pressure, the greater the voltage required. Figure 28.9 illustrates one type of ignition system for a six-cylinder engine. The primary (low-voltage) circuit includes the storage battery, the primary of the induction coil, the circuit-breaker mechanism (points), and the capacitor. It is completed through the frame of the automobile (ground). The secondary (high-voltage) circuit includes the secondary of the induction coil, the distributor, and the spark plugs of the engine. A cam in the circuit-breaker mechanism causes the making and breaking of the circuit in the primary of the induction coil, which in turn causes high induced emfs in the secondary circuit. The breaker is timed to break the circuit at just the instant a spark is needed in a cylinder. The distributor makes connection between the secondary of the induction coil and the proper cylinder. It is important that the ignition system be adjusted to give a spark at just the proper instant. Each automobile ignition system is provided with some means for advancing or retarding the spark to afford the most effective engine performance. Various automatic-control mechanisms are used to give the desired spark advance for varying engine speeds.

Electronic ignition Electronic ignition systems (EIS) are now being used on most American passenger cars and many trucks. The function of the electronic system is the same as that of the conventional system illustrated in Fig. 28.9—to produce a high-voltage spark and to distribute it at the proper time to the proper spark plug to fire the plug. But most of the newer ignition systems eliminate the conventional breaker points and the capacitor (condenser). These components are replaced by a magnetic pick-up coil and pulse-distributor assembly inside the distributor. The magnetic-pulse distributor eliminates the problem of pitting and burning of the contact points in the conventional interrupter.

The newer ignition system is a magnetic-pulse triggered, transistor-controlled, inductive discharge system. The magnetic pick-up assembly located inside the distributor contains a permanent magnet, a pole piece with internal teeth, and a pick-up coil. There are

as many teeth in the pole piece as there are cylinders in the engine. When the teeth of the timer core, rotating inside the pole piece, line up with the teeth of the pole piece, an induced voltage in the pick-up coils signals an electronic device, called a module, to trigger the coil primary circuit. As the primary current decreases, a high voltage is induced in the ignition secondary coil which is directed through the rotor and secondary leads to fire the spark plugs. There is a capacitor in the distributor, but it is not a part of the ignition system. It is used to suppress static for car radio reception.

All of the ignition components for EIS are housed within the distributor (see Fig. 28.9).

28.9 Self-Induction

We have observed how a changing electric current in one circuit will induce emfs in neighboring circuits. Moreover, when the current varies in a single coil, the changing flux within the coil induces an emf in the coil itself. The direction of the emf, according to Lenz's law, is such that it opposes the changing current that causes it. Such a process is called *self-induction* and the emf produced is called *counter emf.*

Consider a coil with a soft-iron core and many turns of wire that is connected to a storage battery. When the switch in the circuit is closed the current does not instantly reach its full value as determined by Ohm's law ($I = V/R$). Nor does it instantly decrease to zero when the switch is opened. The rate of increase and decrease can be shown by a graph (Fig. 28.10). The explanation of this phenomenon can be illustrated by another diagram (Fig. 28.11). The figure illustrates the growth of magnetic flux in a coil as the current increases from left to right in the turns. As the current moves from left to right in the coil, a flux of circular lines of force radiates from each turn of the coil. These lines, as they expand outward, cut the turns of wire which produce them as well as all the other turns in the coil. There is thus induced in the coil a voltage in such a direction that it opposes the growth of the current. When the switch connecting the coil to the source of emf is opened, the current decreases, but as the current's own magnetic field dies away, the lines

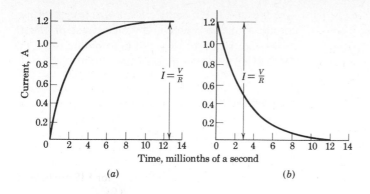

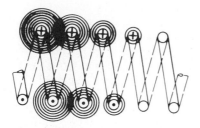

Fig. 28.10 (*a*) Increase and (*b*) decrease of current in an inductive circuit, such as a coil of wire.

Fig. 28.11 Cross-sectional view of the growth of flux lines in a solenoid. (Front half of solenoid is represented by dotted lines.)

of force again cut the turns of wire—this time in a reverse direction. An emf is then induced which tends to *maintain* the current in the coil. When a coil has this property of opposing any change of current in a circuit, it is said to possess *self-inductance* or simply *inductance*. The self-inductance of an electric circuit might be thought of as an *electromagnetic inertia* because it behaves very much like the inertia of material bodies.

The instantaneous induced counter emf(e_L) is directly proportional to the time rate of change of current $\Delta i/\Delta t$ in the coil.

$$e_L = -L\frac{\Delta i}{\Delta t} \qquad (28.8)$$

or

$$L = -\frac{e_L}{\Delta i/\Delta t} \qquad (28.9)$$

where L is called the *self-inductance* of the circuit and e and i represent *instantaneous values*. The unit of self-inductance is the henry (H). Self-inductance is defined (from Eq. 28.9) as the ratio of the emf of self-induction to the rate of change of the current in the coil. Thus, a circuit has a self-inductance of one henry if a time rate of change of current equal to one ampere per second induces a counter emf of one volt.

Illustrative Problem 28.6 It takes 75 μsec for a current of 35 μA in a circuit to fall to zero. An average back emf of 20 V is induced in the circuit. Calculate the self-inductance of the circuit.

Solution

$$L = \frac{\overline{\mathscr{E}}_L}{\Delta i/\Delta t}$$

$$= \frac{20\ V}{75\ mA/35\ m\text{-}sec}$$

$$= 9.3\ H \qquad\qquad ans.$$

The effect of self-induction can be demonstrated by connecting a coil C which has an iron core, with a lamp L in parallel, and this combination in series with a dc source G, a suitable resistance R, and a switch K, as shown in Fig. 28.12. When the switch is closed, the lamp will flash for a moment and then become quite dim. This is because the back emf built up in the coil temporarily retards the passage of current through the coil and hence most of the current flashed through the lamp. Then, since the actual ohmic resistance of the coil is considerably less than that of the lamp, as the self-induced emf diminishes, most of the current flows through the coil. When the switch is opened, the lamp once more lights up brightly. This is due to the large

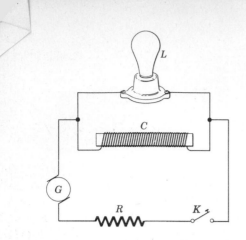

Fig. 28.12 Self-inductance in a coil.

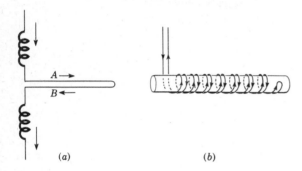

(a)   *(b)*

Fig. 28.13 Elimination of self-inductance in coils.

self-inductance set up in the coil by the rapidly decaying magnetic flux, causing the current to flow back through the lamp.

In some instruments it is necessary to have coils of wire that have little or no self-inductance. In order to eliminate self-inductance, some means must be devised to eliminate the magnetic flux around the wires. This is generally done by looping the wire back on itself so that two conductors side by side will have current going in opposite directions, thereby having the magnetic fields around them nullify each other (see Fig. 28.13*a*). The closer the wires *A* and *B* in the figure are together, the smaller the self-induction. Coils wound with such looped wires (Fig. 28.13*b*), even though they have soft-iron cores, have little self-induction.

Noninductive resistance coils are used in commercial ammeters and voltmeters, in most other electrical measuring instruments, and in some electronic communications circuits.

SUMMARY

Most commercial sources of electric energy involve the converting of mechanical energy to electric energy by means of electromagnetic induction.

Induced emfs are generated only when lines of force cut an electric circuit or whenever the flux through the circuit varies with time.

Lenz's law states that *an electromagnetically induced current always has such a direction that the magnetic field set up by it tends to oppose the cause which produces it.* The cause can be relative motion of a conductor in a magnetic field or a change of flux through a closed circuit.

The value of an induced emf depends upon the rate of change of the magnetic flux which cuts the conductor. This is called *Faraday's law.*

Fleming's rule is a right-hand rule for predicting the direction of an induced current.

The average emf $\overline{\mathscr{E}}$ in a coil of N turns produced by a magnetic flux change of $\Delta\phi$ webers in time Δt sec is found by

$$\overline{\mathscr{E}} = -N\frac{\Delta\phi}{\Delta t} \qquad (28.4)$$

Mutual inductance is that property of two circuits by which a change of flux in one circuit will cause an induced emf in the other.

In the case of mutual inductance, the instantaneous emf induced in the secondary coil is directly proportional to the rate of change of current in the primary coil,

$$e_s = -M\frac{\Delta i_p}{\Delta t} \qquad (28.6)$$

where e_s is in volts, M in henrys, i_p in amperes, and t in seconds.

Induction coils are used to provide high voltage emfs for the ignition systems of automobile engines.

When a current changes in a coil, a *self-induced emf* exists in the coil. The induced emf opposes the change in the current (Lenz's law). The instantaneous self-induced counter emf is directly proportional to the time rate of change of current in the coil,

$$e_L = -L \frac{\Delta i}{\Delta t} \qquad (28.8)$$

where e_L is in volts, L in henrys, i in amperes, and t in seconds.

QUESTIONS

1. Explain why the current in a secondary coil is zero when a steady (dc) current flows in a nearby primary coil.

2. What is the purpose of the older type capacitor in the ignition system of a gasoline engine?

3. List several ways in which an emf can be induced in a coil of wire.

4. Why is the self-inductance of an iron-core coil greater than a similar air-core coil?

5. What are the factors upon which an induced emf depend?

6. Why is the secondary of the induction coil wound with many more turns than the primary?

7. Why must the wire in the primary of an induction coil be heavier than that of the secondary?

8. Discuss the difference in the flow of a charge from a battery and that from the secondary of an induction coil.

9. If the north pole of a bar magnet is dropped through a closed-loop coil whose plane is horizontal, what is the direction of the induced current as the north pole approaches the loop?

10. What physical quantities are measured in (*a*) webers; (*b*) webers per second; (*c*) webers per square meter; (*d*) webers per ampere-meter; (*e*) joules per coulomb; (*f*) newtons per ampere-meter?

11. Which of the following is true for Fig. 28.14? A current will be induced in *B* (*a*) only when the key is being closed, (*b*) only when the key is being opened, (*c*) whenever the key is either being closed or opened, (*d*) only when the key is kept closed, (*e*) none of these.

12. Which of the following is not true for Fig. 28.14? Whenever a current is induced in the right-hand circuit (*a*) Lenz's law applies to the circuit, (*b*) the current in coil *B* will have the same value as that in coil *A*, (*c*) the voltage of coil *B* can be more than that of coil *A*, (*d*) the voltage of coil *B* can be less than that of coil *A*, (*e*) the direction of the current in *B* will alternate on closing and opening the key.

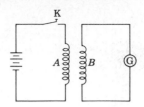

Fig. 28.14 Questions 11 and 12.

PROBLEMS

1. A straight wire 25 cm long moves perpendicular to and through a field of magnetic flux with a velocity of 75 cm/sec. What emf is induced in the wire if the flux density is 0.065 Wb/m^2?

2. A flux of 1.5×10^{-3} Wb emanates from the north pole of a bar magnet. What emf will be induced at the ends of a coil of 400 turns when it takes 0.75 sec to thrust the magnetic pole through the coil? Assume each turn of the coil is cut by the flux in that time.

3. A single conductor passes through a flux of 8.2 Wb. How long did it take the conductor to cut the magnetic flux to produce an emf of 18 V?

4. A coil has 450 turns of wire and an area of 25 cm^2. What is the average emf induced in the coil if the magnetic flux density B increases from 0.10 to 0.65 Wb/m^2 in $\frac{1}{60}$ sec?

5. Find the emf induced in a coil of 840 turns and 12 cm diameter if the magnetic flux density B inside the coil is changed from 0.15 to 0.55 Wb/m^2 in 1 sec.

6. With what velocity must a straight wire 50 cm long cut across a magnetic field of flux density 0.5 Wb/m^2 in order to induce across the ends of the wire a potential difference of 0.2 V?

7. A coil having 150 turns and an area of 7.5 cm^2 is inserted between the poles of a magnet having a uniform flux density of 2.5 Wb/m^2 in 0.025 sec. (*a*) What voltage is induced across the coil? (*b*) If the coil has 45 Ω resistance, what current will flow in it?

8. The secondary of an induction coil has 30,000 turns. If the total flux in the coil changes from 4.5×10^{-4} to 9.0×10^{-5} Wb in 1.8×10^{-4} sec, what is the magnitude of the induced emf?

9. Find the potential difference between the tips of the 20-m wing of a jet plane that is traveling 1600 km/hr east if the vertical component of the earth's magnetic field has a flux density of 7.5×10^{-5} Wb/m^2.

10. An induction coil has 2.5×10^4 turns. The number of flux lines ϕ through it changes from 5.0×10^{-3} to 0 in $\frac{1}{100}$ sec. What is the average induced emf?

11. The current in a coil of wire decreased from 6.2 A to 0.5 A in 0.015 sec. What is the self-inductance of the coil if the average induced emf was 1200 V?

12. The self-inductance of a coil of wire is 3.5 H. When a switch is opened, the current through the coils decreases from 8.2 A to 0 in 0.15 sec. What is the average induced emf which opposes the decrease in the current?

13. Two separate coils are wound on the same iron core. When the current in one coil drops from 6.4 A to 2.6 A in 0.015 sec, the potential difference between the terminals of the second coil is 64 V. What is the value of the mutual inductance?

14. Two circuits have a mutual inductance M of 85 mH. What is the average emf induced in the secondary by a change from 0 to 25 A in 46 μsec in the primary?

15. Two coils placed side by side have a mutual inductance of 0.35 H. The secondary coil has 55 turns. The current in the primary increases from 1.2 A to 6.2 A in 15 μsec. Calculate (a) the average induced emf in the secondary, and (b) the total change in the magnetic flux $\Delta\phi$ in the secondary during the time interval of 15 μsec.

29 Alternating Current

The kind of electric current we have dealt with mostly up to this point is direct current (dc). This means the direction of the current does not change. In contrast to this, *alternating* current (ac) is constantly changing in both direction and magnitude.

For many years dc electricity was the only electric energy in use. Although dc electricity still has many applications in industry, more than 99 percent of all the generating stations in this country generate electricity as alternating current. Even those industries which require huge amounts of direct current, such as those involved in electroplating and electrolytic refining of metals, use energy which is originally generated as alternating current. The alternating current is rectified into direct current by some machine or electronic device.

The chief reason for the widespread use of alternating current has been the fact that it can be transmitted over long distances much more efficiently and economically than if the energy were in dc form. Before methods were perfected for producing and transferring electric energy as an alternating current, the costs of electrochemical processes were economically prohibitive. Cheaper electrical power in ac form which is then converted to direct current is what has made the processes practical. However, over very long distances it is cheaper to rectify the high-voltage alternating current to direct current (for transmission) and then convert back to alternating current.

The fundamental element in a dc circuit is the *resistor*. In an ac circuit, in addition to the resistor, there are two other important fundamental elements. They are the *inductor* and *capacitor* (sometimes called condenser). In this chapter, we study the basic characteristics of alternating current, and how resistors, inductors, and capacitors affect the characteristics. We will also see why electric energy in ac form is much more economical to produce and transport.

29.1 Nature of Alternating Current

An *alternating current* periodically reverses its direction. We saw in Sec. 28.5 how an alternating current can be produced by alternately thrusting a bar magnet into a coil of wire and then withdrawing it. The same effect can be obtained by rotating a coil of wire in a uniform magnetic field. In order to investigate how the emf varies as a coil turns with uniform angular velocity in a magnetic field, let us consider the motion of the coil from moment to moment. Figure 29.1a shows 12 positions of the rotating coil, 30° apart. The instantaneous emfs in the coil corresponding to these 12 positions are shown plotted in Fig. 29.1b. The horizontal axis is marked off in units of degrees of rotation (which also may be considered as units of time, since we assume uniform angular velocity) and the vertical axis shows the emfs produced.

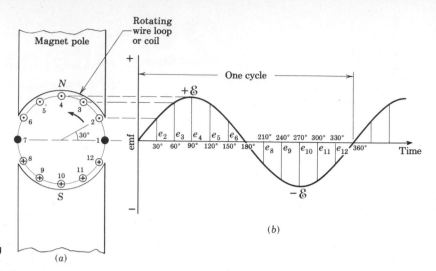

Magnet pole

One cycle

$+\mathcal{E}$

e_2 e_3 e_4 e_5 e_6 210° 240° 270° 300° 330°

30° 60° 90° 120° 150° 180° e_8 e_9 e_{10} e_{11} e_{12} 360°

Time

emf

$-\mathcal{E}$

(b)

Fig. 29.1 Graph of alternating emf.

(a)

When the rotating coil is in position 1, the coil motion is parallel to the lines of force; hence no lines are cut and no emf is induced in the coil. When the coil has reached position 2, lines of force are being cut and an emf is produced, giving rise to a current in a direction which is *out* from the page (indicated by a dot). The vertical line e_2 in part (b) represents the instantaneous emf in the coil when it is at position 2. In like manner, e_3, e_4, ... represent the instantaneous emfs as the coil occupies positions 3, 4, ... From Fig. 29.1a it is seen that as the coil changes from position 1 to position 4, the number of lines of force cut in a given period of time increases from zero at position 1 to a maximum at position 4. Then as the coil moves on from 4, the number of lines cut in that same period of time diminishes from that maximum to zero when the coil reaches position 7.

As soon as the coil passes position 7, it cuts the lines of force in an opposite direction. An emf will now be induced in the coil which results in a current directed toward the page (indicated by a cross). This induced emf will continue to rise in a negative direction (assuming the original direction as positive) until a maximum value at position 10 is reached. As the coil continues to rotate from position 10, the emf gradually decreases

until it once more is zero at position 1. Thus we see that the emf of an alternating current is continually changing as the coil rotates.

To investigate exactly how the emf varies from moment to moment, consider the diagram of Fig. 29.2. Let A and B represent the cross sections of the two segments of the conductor which makes up the loop turning with angular velocity ω in a magnetic field. With the loop in the AB position no emf is produced in the loop, since the conductor segments are at that instant moving parallel to the magnetic lines of force.

When the coil is in the $A'B'$ position one-quarter turn (90°) later, lines of force are being cut at the greatest rate and the emf produced will be a maximum. Let this maximum emf be designated e_{max}. At some intermediate coil position such as ab the coil will have turned through an angle $\theta = \omega t$. The emf produced in the coil depends on the rate at which magnetic force lines are being cut, and this rate, in turn, is proportional to that component of the linear velocity v which is perpendicular to the flux. If we let v be numerically equal to e_{max}, then the instantaneous electromotive force e at any time t is equal to

$$e = v_x = v \sin \theta = e_{max} \sin \theta$$

or $\qquad e = e_{max} \sin \omega t \qquad\qquad (29.1)$

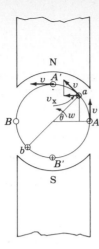

Fig. 29.2 Variation of alternating emf and current throughout a complete revolution.

The current produced (in a pure resistance circuit) will follow the same pattern of variation as the emf, and the instantaneous current

$$i = i_{max} \sin \omega t \qquad (29.2)$$

The curve of Fig. 29.1b is called a *sine curve*, because the same curve would result in plotting the sines of angles from 0 to 360°. Although our graph has been used to illustrate emfs, a curve of the same general shape would illustrate the changing instantaneous values of *current* established in the coil when it is connected to a closed circuit.

In terms of electrons we can define an alternating current as one in which the transfer of electrons in the circuit is first in one direction and then in the opposite direction. Each electron travels only a very short distance along the circuit before it is urged the opposite way. In the simple ac cycle just described, the number of electrons that pass through a given cross section of the circuit varies from moment to moment.

29.2 The AC Cycle–Frequency

When the curve which represents the alternating emf (or current) rises from zero to a maximum, falls to zero, increases to a maximum in the reverse direction, and then falls back to zero again, it is said to complete a *cycle*. A complete cycle would occur when the coil of Fig. 29.1a makes a complete revolution of 360°. The number of complete *cycles per second* of alternating current is called the *frequency*. Often the word *cycles*, with the words *per second* being omitted but understood, is used to express frequency. Thus, we speak of a 60-cycle current.

Today the word *hertz* (Hz) is more commonly used in place of cycles or cycles per second. The most prevalent ac frequency generated for commercial power in the United States is 60 Hz. In Europe, Asia, and Africa 50 Hz is a common frequency. Lower frequencies, such as 25 Hz and $16\frac{2}{3}$ Hz, are used for electric railways in many countries. In aircraft and guided missiles 400-Hz generators are widely used. Radiobroadcast frequencies are of the order of a million hertz; and some super-high frequency transmitters operate at several billion hertz (see Fig. 21.14 and Table 32.1).

If we use $\omega = 2\pi f$, where f is the frequency in hertz, Eqs. (29.1) and (29.2) become

$$e = e_{max} \sin 2\pi f t \qquad (29.1')$$

and

$$i = i_{max} \sin 2\pi f t \qquad (29.2')$$

29.3 Effective Current and Effective Voltage

A glance at the curve of Fig. 29.1 will reveal that the mathematical average of the values for emf (and, hence, for current also) for one complete cycle is zero. As much of the curve lies above emf = 0 as lies below it. We defined one ampere of current in terms of the force of mutual attraction or repulsion between two parallel conductors carrying direct current. This definition cannot be used for alternating current because the forces alternate 120 times a second. We know that alternating currents do possess energy. Hence, the ampere for alternating current must be defined in terms of some property which is independent of the direction of the current.

The heat produced by an electric current is independent of the direction of the current. Thus *an alternating current of one ampere is defined as that current which produces the same average* heating effect as a dc ampere under the same conditions. The ampere thus defined is the unit of *effective value* of the alternating current.

The heat (power) dissipated at any instant in a metallic resistor which has a resistance R and which carries an instantaneous current i is (from Eqs. 26.10 and 29.2')

$$P = i^2 R = i_{max}^2 R \sin^2 (2\pi f t)$$

Squaring i will have the effect of eliminating the negative sign for i. Figure 29.3 shows how i, which is equal to $i_{max} \sin 2\pi f t$, and i^2 vary with angle θ (where $\theta = 2\pi f t$). The dashed line represents the value of $\frac{1}{2}(i \, max)^2$. It should be evident from the graph for i^2 that as much of the curve lies above as below the dashed line. Thus, the average (mean) power for a complete cycle of alternating current is

$$P = \frac{(i_{max})^2}{2} R \qquad (29.3)$$

The effective value I_{eff} of the alternating current is equivalent to the direct current that produces the same power dissipation in R.

$$P = I_{eff}^2 R = \frac{(I_{max})^2}{2} R$$

or

$$I_{eff}^2 = \frac{(I_{max})^2}{2}$$

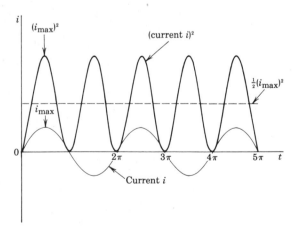

Fig 29.3 For alternating current $i_{eff} = \sqrt{i^2} = \frac{1}{2}\sqrt{i_{max}^2}$.

and

$$I_{eff} = \frac{I_{max}}{\sqrt{2}} = 0.707 I_{max} \qquad (29.4)$$

Thus, a direct current with a value of 0.707 of the maximum value of an alternating current has the same power output or heating effect as the alternating current.

The electric power for direct current can also be expressed in terms of $\mathscr{E}$ and R as $P = \mathscr{E}^2/R$. Thus we can follow the same reasoning as above by drawing graphs of e and e_{max} as a function of t to determine that the average power

$$P = \frac{(e_{max})^2/2}{R} \qquad (29.5)$$

from which it can be shown that

$$E_{eff} = \frac{e_{max}}{\sqrt{2}} = 0.707 e_{max} \qquad (29.6)$$

Equations (29.4) and (29.6) can also be written

$$I_{max} = \sqrt{2} I_{eff} = 1.414 I_{eff} \qquad (29.7)$$

and

$$E_{max} = \sqrt{2} E_{eff} = 1.414 E_{eff} \qquad (29.8)$$

In measuring alternating currents we are interested in measuring effective values. Hence, unless otherwise noted, we will assume the terms *ampere* and *volt* in ac circuits to mean effective values. We will also drop the subscript *eff.* I and $\mathscr{E}$ for ac circuits will refer to effective current and electromotive force.

Most ac meters are so made that they read directly the values of $\sqrt{(i^2)_{av}}$ and $\sqrt{(e^2)_{av}}$ which we will label I and $\mathscr{E}$ (or V).

NOTE: $\sqrt{(i^2)_{av}}$ is the square *root* of the *mean* (average) of the *square* of i. It is called the *root mean square* (abbreviated "*rms*") of i.

Illustrative Problem 29.1 An ac voltmeter is used to determine that the potential difference of the house line is 115 V. Write the equation for the voltage (sine) wave if it is known that its frequency is 60 Hz.

Solution The meter reads effective voltage. Using Eq. (29.8), we can find the maximum voltage.

$$e_{max} = 1.414(115 \text{ V}) = 163 \text{ V}$$

Substitute $e_{max} = 163$ V and $f = 60$ Hz in Eq. (29.1) to get

$$e = 163 \sin(120\pi t) \qquad ans.$$

for the equation of the voltage wave.

Illustrative Problem 29.2 When a purely resistive radiant heater is connected to a standard 120-V, 60-Hz line, it dissipates power at the rate of 1320 W. Calculate (*a*) the maximum voltage, (*b*) the effective current, and (*c*) the maximum current.

Solution

(*a*) $V_{max} = 1.414V = 1.414(120$ V$) = 170$ V *ans.*

(*b*) Since average power $\bar{P} = IV$, we know that

$$I = \frac{\bar{P}}{V} = \frac{1320 \text{ W}}{120 \text{ V}}\left(\frac{1 \text{ VA}}{1 \text{ W}}\right)$$

$$= 11.0 \text{ A} \qquad ans.$$

(*c*) $I_{max} = 1.414I = 1.414(11) = 15.6$ A *ans.*

PROPERTIES OF AC CIRCUITS

29.4 Pure-Resistance Circuit

A pure-resistance circuit is one in which there is no inductance or capacitance. In such a circuit the amperage and voltage curves would both be sine curves, so related that when the voltage is zero, the current is zero; when the voltage is a positive maximum, the current is a positive maximum; and when the voltage is a negative maximum, the current is a negative maximum. When this condition exists, the current is said to be *in phase* with the voltage. Figure 29.4 illustrates this condition graphically.

Pure-resistance circuits can be treated as if they were dc circuits. Thus the following laws apply, if we use *effective values* for the ac units:

$$I(\text{amperes}) = \frac{V(\text{volts})}{R(\text{ohms})} \qquad (29.9)$$

and

$$P(\text{watts}) = V(\text{volts}) \times I(\text{amperes}) \qquad (29.10)$$

$$= I^2(\text{amperes}^2) \times R(\text{ohms}) \qquad (29.11)$$

Circuits whose sole function it is to provide heat are generally pure-resistance circuits and are used in such devices as incandescent lamps, electric irons, electric blankets, water heaters, radiant heaters, and the like.

29.5 Circuits Containing Inductance

Consider that an ac voltage source is connected across an inductor L as shown in Fig. 29.5. The sinusoidal applied voltage will cause a sinusoidal current in the inductor. Note the symbol for an inductor. But the inductor will induce a counter emf which will oppose the change in the current.

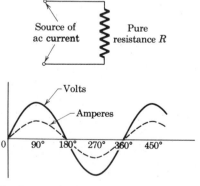

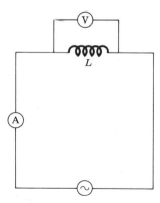

Fig. 29.4 Pure-resistance circuit, voltage and current in phase.

29.5 Setup for measuring voltage across and current through an inductor.

that property of the circuit which opposes the change in value of the current is termed *self-induction* or *inductance* (see Sec. 28.9). The electrical property of inductance is analogous to the mechanical property of inertia. It will be recalled (Chap. 28) that inductance is measured in henrys or millihenrys. A circuit has an inductance of one henry if, when the current changes at the rate of one ampere per second, there is induced in it a counter emf of one volt. The inductance L is found by the formula

$$L = \frac{-e}{\Delta i / \Delta t} \qquad (28.9)$$

An inductive circuit usually occurs in the form of a coil. The windings of motors, generators, and transformers are examples of circuits that have inductance. The inductance of a circuit is determined by (1) the number of turns in the coil, (2) the cross-sectional area, (3) the magnetic permeability of the core, (4) the transverse length of the coil, and (5) the shape of the coil and its core. The property of inductance can be illustrated by connecting a lamp in series with a "choke" coil first to a 120-V dc line and then to a 120-V ac line, as shown in Fig. 29.6. When the lamp and coil are connected to the dc line, the lamp will burn brightly. If we then insert a soft-iron core in the coil, there will be no change in intensity of light coming from the lamp. If we next remove the iron core and connect the lamp and coil to the ac line, the lamp lights only dimly. Inserting the iron core in the coil will dim the lamp a great deal more.

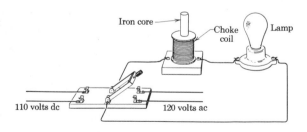

Fig. 29.6 Choke coil and lamp connected in series to a dc or ac circuit.

29.6 Pure-Inductance Circuit

If we could have a circuit which includes only a coil of many turns of wire of negligible resistance wrapped around a good quality of magnetic steel, we would have what is known as a *pure-inductance circuit*. We know that such a circuit is theoretically impossible, because every wire (except superconductors) has some resistance; but a zero-resistance circuit can be approximated in practice. In such a circuit the counter emf, or self-inductance, will offer continual resistance to any change in current. Thus if an *alternating current* is sent through such a circuit, since the current is continually changing, there will be constant opposition to the flow of current. This opposition, called *inductive reactance*, is a resistance to current flow and is measured in ohms. It is generally represented by X_L. It can be proved both experimentally and mathematically that

$$X_L = 2\pi f L \qquad (29.12)$$

where X_L = inductive reactance, ohms
f = frequency, hertz
L = inductance, henrys

Illustrative Problem 29.3 The "choke" coil (illustrated in Fig. 29.6) has an inductance of 400 mH and a negligible resistance. If it is connected across a 120-V, 60-Hz line, find (*a*) the inductive reactance of the coil and (*b*) the current which flows through the coil.

Solution

(*a*) $L = 400 \text{ mH} = 0.4 \text{ H}$

$$X_L = 2\pi f L$$
$$= 2(3.14) \times 60 \text{ Hz} \times 0.4 \text{ H}$$
$$= 151 \ \Omega \qquad\qquad ans.$$

(*b*)
$$I = \frac{V}{X_L}$$
$$= \frac{120 \text{ V}}{151 \ \Omega}$$
$$= 0.79 \text{ A} \qquad\qquad ans.$$

A graph relating the current and voltage in a pure-inductance circuit is shown in Fig. 29.7.

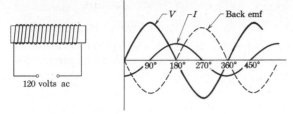

29.7 Variation of current, applied voltage, and self-induced emf in a pure inductance circuit.

When the current is increasing, the induced emf tends to oppose the increase, and when the current is decreasing, the induced emf opposes the decrease and tends to keep the current flowing. Hence the rise in the current takes place later than does the rise in the source voltage. The fall of the current will take place later than does the fall of the source voltage. Thus the current lags behind the voltage throughout the entire cycle. From the graph it will be seen that the current lags 90° behind the voltage. The current and the voltage are said to be 90° *out of phase*. Thus a pure-inductance circuit will not only offer opposition (inductive reactance) to a flow of current (in ohms) but also cause the current to lag behind the voltage by exactly 90°. The self-induced emf will be opposite to the applied voltage at all times. *Current lags behind voltage in an inductive circuit.**

At 0° the current is momentarily neither increasing nor decreasing, and so the induced voltage is zero. At 90° the current is increasing at its greatest rate and so the back emf is at its most negative to oppose this increase. The voltage drop *V* across the coil has the opposite sign of the back emf at its maximum value. At 180° the current is again momentarily neither increasing nor decreasing, and so the induced voltage is zero. At 270° the current is decreasing at its greatest rate and so the back emf has its most positive value to oppose this decrease; *V* then is most negative. At 360° the cycle starts over.

---

* Some people use the "code letters" *ELI* to remember that "*E* (voltage) in an inductive circuit (*L*) leads *I*."

29.7 Resistance and Inductance in Series–Impedance

Usually we have to consider circuits in which both resistance and inductive reactance are present. In such cases, the current will lag behind the voltage by an angle greater than 0° but less than 90°. If, for example, the ohmic resistance equals the reactive resistance, the current will lag behind the voltage by 45° (see Fig. 29.8). It should be made clear that although the ohmic resistance and the reactance are shown as separate entities in the figure, they are actually properties of the same circuit.

The combined effect of a resistance and a reactance is known as *apparent resistance* or *impedance*. The letter *Z* is generally used to indicate impedance. Thus, in an ac circuit, we can write Ohm's law as

$$I(\text{amperes}) = \frac{V(\text{volts})}{Z(\text{ohms})} \qquad (29.13)$$

Impedance can be represented *vectorially* as the hypotenuse of the right triangle whose two sides are the ohmic resistance and the reactance (see Fig. 29.9). Thus

$$Z = \sqrt{R^2 + X_L^2} \qquad (29.14)$$

And from Eq. (29.12) and Eq. (29.13) Ohm's law can be written

$$I = \frac{V}{Z} = \frac{V}{\sqrt{R^2 + (2\pi f L)^2}} \qquad (29.15)$$

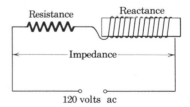

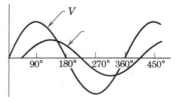

Fig. 29.8 Current and voltage curves when $R = X_L$.

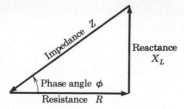

Fig. 29.9 Vector triangle relating resistance, reactance, and impedance.

The angle ϕ between R and Z is called the *phase angle* and is equal to the lag of the current behind the voltage, in electrical degrees.

Illustrative Problem 29.4   A coil has a resistance of 2.4 Ω and an inductance of 5.8 mH. If it is connected to a 120-V, 60-Hz source, find (*a*) the reactance, (*b*) the impedance, and (*c*) the current in the coil. (*d*) What will be the current in the coil when it is connected to a 120-V, 6400-Hz source?

Solution

(*a*) $X_L = 2\pi f L$

$\qquad = 2(3.14) \times 60 \text{ cycles/sec} \times 0.0058 \text{ H}$

$\qquad = 2.19 \ \Omega$ *ans.*

(*b*) $Z = \sqrt{R^2 + X_L^2}$

$\qquad = \sqrt{(2.4 \ \Omega)^2 + (2.19 \ \Omega)^2}$

$\qquad = 3.25 \ \Omega$ *ans.*

(*c*) $I = \dfrac{V}{Z}$

$\qquad = \dfrac{120 \text{ V}}{3.25 \ \Omega}$

$\qquad = 36.9 \text{ A}$ *ans.*

(*d*) $I = \dfrac{120 \text{ V}}{\sqrt{(2.4 \ \Omega)^2 + (2\pi \times 6400 \times 0.0058 \ \Omega)^2}}$

$\qquad = 0.51 \text{ A}$ *ans.*

Note that the resistance effect is negligible in comparison to the effect of the inductance. This illustrates how an inductor can be used to keep currents low at high frequencies.

Illustrative Problem 29.5   When a coil whose resistance is 15 Ω is connected to a 120-V, 60-Hz source, it draws 5 A. Find (*a*) the impedance and (*b*) the inductance of the coil.

Solution
(*a*) We first solve for the impedance of the circuit by using Eq. (29.13)

$$Z = \frac{V}{I}$$

$$= \frac{120 \text{ V}}{5 \text{ A}}$$

$$= 24 \ \Omega \qquad\qquad ans.$$

(*b*) Knowing values for R and Z, we can solve for X_L, using Eq. (29.14):

$$Z^2 = R^2 + X_L^2$$

$$X_L = \sqrt{Z^2 - R^2}$$

$$= \sqrt{(24 \ \Omega)^2 - (15 \ \Omega)^2}$$

$$= 19 \ \Omega$$

But since $X_L = 2\pi f L$, or $L = X_L/2\pi f$,

$$L = \frac{19 \ \Omega}{6.28 \times 60 \text{ cycles/sec}} = 0.0504 \text{ H}$$

$$= 50.4 \text{ mH} \qquad\qquad ans.$$

Such problems can also be solved by vector methods, with vectors drawn to scale on graph paper (Sec. 29.10).

29.8   Capacitive Reactance

The effect of a capacitor in a dc circuit is to stop the flow of current entirely. In an ac circuit the effect requires some study. If we connect an incandescent lamp L in series with a capacitor C (about 10 μF capacity), then connect the two across a 110-V dc line (Fig. 29.10*a*), the lamp will not light because the circuit is *open* at the capacitor plates. When the lamp and capacitor are connected to an ac line, the lamp will light despite the fact that the circuit is open between the plates of the capacitor. If the capacity of the capacitor is varied, it will be noted that as the capacity is made smaller the lamp will grow dimmer.

The action of the capacitor can be illustrated by again referring to a hydraulic analogy

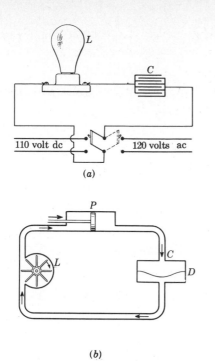

Fig. 29.10 (*a*) Effect of capacitor in dc and ac circuits. (*b*) Hydraulic analogy of a capacitor in an ac circuit.

(Fig. 29.10*b*). The alternating current is analogous to the back-and-forth surging of the water through the line as the reciprocating pump *P* moves the water first in one direction and then in the other. The box *C* corresponds to the capacitor; the diaphragm stretched across the box would have the same effect as the dielectric of the capacitor. The elastic strength of the diaphragm and the size of the box determine the "capacity" of the capacitor. The water wheel *L* on which the surging current does work corresponds to the incandescent lamp in the previous example.

Applying these principles learned from the water circuit to an electric circuit, it should be clear that no electrons actually flow *through* the capacitor, but electricity does flow *into* and *out of* the capacitor. As the current moves in one direction in the circuit, one set of plates is charged positively while the other is charged negatively. When the current changes direction, the charges on the plates reverse. Hence there is an oscillation of electron flow in the circuit which depends on the frequency of the ac source. A 60-Hz source

will cause 120 changes per second in the direction of current into and out of the capacitor.

Every capacitor placed in a circuit will offer some resistance to the motion of an alternating current in the circuit. The effect may be regarded as similar to that offered by resistance and by inductive reactance. The opposition to electron flow offered by a capacitor is termed *capacitive reactance* to distinguish it from inductive reactance. It, too, is measured in ohms. The symbol for capacitive reactance is X_C. Its value is given by the formula

$$X_C = \frac{1}{2\pi f C} \qquad (29.16)$$

where X_C = capacitive reactance, ohms
f = frequency, cycles/sec
C = capacitance, farads

Illustrative Problem 29.6 What is the capacitive reactance of a capacitor of 50 μF capacitance when an alternating current of 60 Hz is impressed on it?

Solution 1 F = 1,000,000 μF

Hence 50 μF = 50 × 10⁻⁶ F

$$X_C = \frac{1}{2\pi f C}$$

$$= \frac{1}{2(3.14) \times 60 \text{ cycles/sec} \times (50 \times 10^{-6} \text{ F})}$$

$$= 53 \ \Omega \qquad\qquad\qquad ans.$$

29.9 Comparison of Inductive Reactance and Capacitive Reactance

The *lagging* of current behind voltage in a *pure-inductive* circuit was illustrated in Fig. 29.7. It can be shown that in a *pure-capacitive* circuit the *current will lead the voltage* by 90°* (see Fig. 29.11). Thus the effect of capacitance is just the reverse of that of inductance. When a resistor and a capacitor are in series in an ac circuit, the current will

---

* The code letters *ICE* are used by some individuals to recall that "current leads voltage in a capacitive circuit." (It should be noted that the voltage in this case should be represented by *V* rather than *E*.)

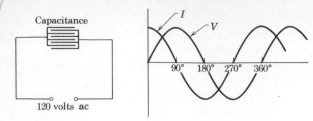

Capacitance

90° 180° 270° 360°

120 volts ac

Fig. 29.11 The current leads the voltage by 90° in a pure-capacitance circuit.

lead the voltage by amounts varying from 0 to 90°. The combined effect Z (*impedance*) of resistance R and capacitive reactance X_C is found by the equation

$$Z = \sqrt{R^2 + X_C^2} \qquad (29.17)$$

This impedance can be represented vectorially as the hypotenuse of a right triangle whose sides are R and X_C (Fig. 29.12). The angle ϕ between R and Z is the *phase angle* and is equal to the "lead" of the current over the voltage. The current I in such resistance–capacitive-reactance circuits can be expressed by the equation

$$I = \frac{V}{Z} = \frac{V}{\sqrt{R^2 + X_C^2}} \qquad (29.18)$$

Illustrative Problem 29.7 What current will flow in a 60-Hz ac circuit in which an emf of 120 V is impressed upon a capacitor of 80 μF capacitance in series with 24 Ω resistance? Replace the 60-Hz ac source with a 6400-Hz source and determine I.

Solution X_C is found by using Eq. (29.12), as follows:

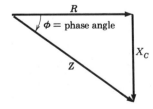

R

ϕ = phase angle

X_C

Z

Fig. 29.12 Vector diagram of relations between impedance, resistance, and capacitance.

$$X_C = \frac{1}{2\pi f C}$$

$$= \frac{1}{2(3.14) \times 60 \text{ Hz} \times (80 \times 10^{-6} \text{ F})}$$

$$= 33.2 \ \Omega$$

Then $\quad I = \dfrac{V}{\sqrt{R^2 + X_C^2}}$

$$= \frac{120 \text{ V}}{\sqrt{(24 \ \Omega)^2 + (33.2 \ \Omega)^2}}$$

$$= 2.93 \text{ A} \qquad\qquad ans.$$

For a 6400-Hz alternating source,

$$I = \frac{120 \text{ V}}{\sqrt{(24 \ \Omega)^2 + (2\pi \times 6400 \text{ Hz} \times 80 \times 10^{-6} \text{ F})^2}}$$

$$= 4.96 \text{ A} \qquad\qquad ans.$$

Note how the *current increases with increasing frequency*. In this latter case the capacitive reactance is almost negligible.

When both inductance and capacitance are present in a (zero-resistance) series circuit, one tends to neutralize the other since their effects are opposite. The net reactance in such a circuit is the difference between them. If we let X equal the net reactance, X_L the inductive reactance, and X_C the capacitive reactance, then

$$X = X_L - X_C$$

or $\qquad\qquad X = X_C - X_L \qquad (29.19)$

depending on which is greater, X_L or X_C.

29.10 Resistance, Inductance, and Capacitance in Series—RCL Circuits

It is impossible to have a pure-inductance circuit, a pure-capacitance circuit, or simply a combination of the two, since any circuit must have some resistance (except for superconductors). Hence we must consider all three in practical computations. Often, however, one of the effects may be negligible. We can represent the relationship between the three geometrically by the vector dia-

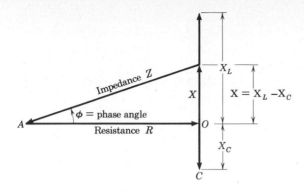

Fig. 29.13 Vector diagram relating impedance, resistance, inductance, and capacitance.

gram of Fig. 29.13. From the diagram we can deduce the relation

$$Z = \sqrt{R^2 + X^2}$$
$$= \sqrt{R^2 + (X_L - X_C)^2} \qquad (29.20)$$

Illustrative Problem 29.8 What is the impedance of a circuit which contains an inductance of 0.7 H in series with a capacitance of 40 μF if the resistance of the circuit is 50 Ω and the frequency of the current is 60 Hz?

Solution

$$X_L = 2\pi f L$$
$$= 2(3.14) \times 60 \text{ cycles/sec} \times 0.7 \text{ H}$$
$$= 264 \ \Omega$$

$$X_C = \frac{1}{2(3.14) \times 60 \text{ cycles/sec} \times (40 \times 10^{-6} \text{ F})}$$
$$= 66.3 \ \Omega$$

$$Z = \sqrt{(50 \ \Omega)^2 + (264 - 66.3 \ \Omega)^2}$$
$$= 204 \ \Omega \qquad \qquad ans.$$

We may now write Ohm's law for ac circuits containing resistance, inductance, and capacitance in series:

$$I = \frac{V}{Z} = \frac{V}{\sqrt{R^2 + (2\pi f L - 1/2\pi f C)^2}} \qquad (29.21)$$

29.11 Series Resonance

It can be seen from Eq. (29.21) and from Fig. 29.13 that when the inductive reactance equals the capacitive reactance, they will nullify each other and Z will equal R. A circuit in which such a condition exists is said to be *in resonance* with the applied voltage, and the current will then have its maximum value equal to V/R. The effect of resonance can be clearly demonstrated by using a circuit like that illustrated in Fig. 29.14. Let L be a choke coil of many turns, C a fixed capacitor of about 20 μF capacity, B an incandescent lamp, and K a double-pole double-throw switch connected to both a 120-V dc source and a 120-V ac source of current. The circuit is provided with switches to bypass either the choke coil or the capacitor. If the capacitor is left out of the circuit, the lamp will glow brightly when the switch is thrown to the dc line but only faintly when connected to the ac line. (Why?) If the choke coil is now omitted but the capacitor included, the lamp will burn brightly when the switch is thrown to the ac line and not at all when connected with the dc line. (Why?)

It is seen that the coil partially blocks the passage of alternating current, while the capacitor stops the direct current completely. If we now include both the coil and capacitor in series in the circuit and pass alternating current through the circuit, the lamp will glow brightly. We can adjust the inductive reactance of the coil by inserting a moveable soft-iron core in the coil (or we could

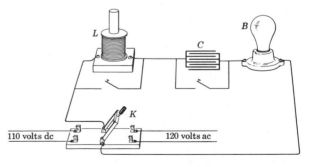

Fig. 29.14 Series resonance demonstration. The lamp will glow brightly when $X_L = X_C$.

vary the capacitance) until the lamp gives the normal illumination that it would give if only the resistance of the circuit were considered. We would then have obtained the condition of *series resonance*, when $X_L = X_C$.

Resonant circuits have great usefulness in the field of electronics. A particular frequency is *tuned in* by establishing a resonant circuit for that frequency. This is often done by the use of a variable condenser (capacitor) (see Fig. 24.31) to change the capacitive reactance of the circuit until resonance is established. We can find the frequency which will give resonance in a series circuit by recalling that resonance occurs when $X_L = X_C$. Set

$$X_L = X_C$$

Then

$$2\pi f L = \frac{1}{2\pi f C}$$

and

$$f_{\text{(resonance)}} = \frac{1}{2\pi\sqrt{LC}} \qquad (29.22)$$

Thus, by altering the value of C (or L), the circuit can be made resonant for any fixed frequency.

Illustrative Problem 29.9 Find the resonant frequency of a circuit in which a 0.03-mH inductance is in series with a 0.005-μF capacitor.

Solution From Eq. (29.22) for series resonance

$$f = \frac{1}{2\pi\sqrt{LC}}$$

$$= \frac{1}{6.28\sqrt{3 \times 10^{-5} \text{ F} \times 5 \times 10^{-9} \text{ H}}}$$

$$= \frac{1}{6.28 \times 3.87 \times 10^{-7}}$$

$$= 4.11 \times 10^5 \quad \text{or} \quad 411 \text{ kilohertz (kHz)} \quad ans.$$

29.12 Power in AC Circuits— Power Factor

Power in a dc circuit is equal to the product of volts and amperes. In a pure-resistance ac circuit, i.e., one in which the current and voltage are in phase, the average power is equal to the product of the effective voltage and the effective current. However, this does not hold true when reactance is present in the circuit. If one connects a voltmeter and an ammeter in a circuit which has inductance and/or capacitance, the product of the readings of the two instruments will be the *apparent power* in *volt-amperes*, not the *true power* (*watts*) in the circuit. This is due to the fact that (except at resonance) the current and voltage will be out of phase. Energy is stored in E and B fields of the capacitors and inductors. Thus, energy is not dissipated. A check of Figs. 29.7 and 29.11 will reveal that at times during each cycle the current is negative while the voltage is positive, and vice versa. This results in a value of true power in the circuit which is less than the product of V and I.

The ratio of true power to apparent power is called the *power factor*. Thus if P is the true power, V the voltage, and I the amperage in a single-phase ac circuit, we get the equation

$$\text{Power factor} = \frac{\text{true power (wattmeter reading)}}{\text{volts} \times \text{amperes}}$$

$$= \frac{P}{VI} \qquad (29.23)$$

or $\quad P(\text{watts}) = VI \times \text{power factor} \qquad (29.23')$

It can be shown that the power factor is equal to the ratio of the resistance to the impedance. This ratio is equivalent to the cosine of the phase angle ϕ (see Fig. 29.9). Thus we have the following equations for ac power in watts:

$$P = VI \frac{R}{Z} \qquad (29.24)$$

$$= VI \frac{R}{\sqrt{R^2 + (R_L - R_C)^2}}$$

$$= VI \cos \phi \qquad (29.25)$$

where ϕ is the phase angle, and (again see Fig. 29.9)

$$\cos \phi = \frac{R}{Z} \qquad (29.26)$$

Thus we see that the power factor depends upon how much the voltage and current are out of phase. In a pure-resistance circuit the power factor is unity and power equals VI. In a pure-inductance or in a pure-capacitance circuit the current and voltage are 90° out of phase. In such case the power factor is zero, *resulting in a zero value for the true power.* In circuits which contain both resistance and reactance, the power factor will have a value which lies somewhere between 1 and 0, depending on the relative values of the resistance and the reactance of the circuit. Power factors may be expressed as decimals or as percentages.

Illustrative Problem 29.10 A 240-V, 60-Hz emf is applied to a circuit containing a resistance of 15 Ω, an inductance of 0.08 H, and an 80-μF capacitor, all in series. Find (*a*) the circuit impedance, (*b*) the current flow, and (*c*) the true power.

Solution

(*a*)

$$X_L = 2\pi f L$$

$$= 2(3.14) \times 60 \text{ cycles/sec} \times 0.08 \text{ H}$$

$$= 30.1 \ \Omega$$

$$X_C = \frac{1}{2\pi f C}$$

$$= \frac{1}{2(3.14) \times 60 \text{ cycles/sec} \times (80 \times 10^{-6} \text{ F})}$$

$$= 33.2 \ \Omega$$

$$Z = \sqrt{(15 \ \Omega)^2 + (30.1 - 33.2 \ \Omega)^2}$$

$$= 15.3 \ \Omega \qquad\qquad ans.$$

(*b*)

$$I = \frac{V}{Z}$$

$$= \frac{240 \text{ V}}{15.3 \ \Omega}$$

$$= 15.7 \text{ A} \qquad\qquad ans.$$

(*c*) Power factor = R/Z

$$= \frac{15 \ \Omega}{15.3 \ \Omega}$$

$$= 0.98$$

$$P = VI \times \text{power factor}$$

$$= 240 \text{ V} \times 15.7 \text{ A} \times 0.98$$

$$= 3690 \text{ W} \qquad\qquad ans.$$

TRANSFORMERS

We have noted that alternating currents are used more widely than direct currents because of the ease with which the voltage can be stepped up or down. Voltage changes for alternating currents are accomplished by means of a device called a *transformer*. Its construction and principles of operation will now be explained.

29.13   Action of the Transformer

A simple transformer usually consists of two coils of conducting wire wound on a closed iron core (Fig. 29.15). Electric energy is supplied to one of the coils, called the *primary coil*, and is delivered to the load from the other coil, called the *secondary coil*. If the primary coil P is connected to a source of alternating potential difference, the current in the coil will cause an alternating magnetic flux ϕ in the iron core which will surge first in one direction and then in the other. This alternating magnetic flux will also cut the secondary coil S,

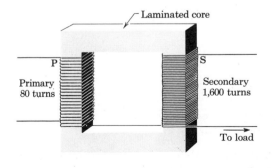

Fig. 29.15   Step-up transformer.

inducing an alternating emf in it which usually differs in magnitude from that applied to the primary coil, depending on the relative number of turns in the two coils.

If the number of turns in the primary coil is N_p, then the self-induced emf in the primary is

$$(\text{emf})_p = -N_p \frac{\Delta\phi}{\Delta t}$$

If we assume that there is *no flux leakage* in the iron core, then the same flux passes through the secondary coil as through the primary coil. If we let N_s equal the number of turns in the secondary coil, according to Faraday's law, the emf in the secondary is

$$(\text{emf})_s = -N_s \frac{\Delta\phi}{\Delta t}$$

Since $\Delta\phi/\Delta t$ is the same for both the secondary and primary coils, we can divide the first equation by the second one. Then

$$\frac{V_s}{V_p} \frac{(\text{emf})_s}{(\text{emf})_p} = \frac{N_s}{N_p} \qquad (29.27)$$

or $\quad \dfrac{\text{output voltage}}{\text{input voltage}} = \dfrac{\text{total secondary turns}}{\text{total primary turns}}$

Equation (29.27) is strictly true only when no current is flowing in the secondary coil. When current flows in the secondary, counter emfs are set up in the primary (and the secondary) which will make V_p somewhat less than the voltage impressed on the primary from the outside source. The voltage at the terminals of the secondary will therefore be slightly less than V_s, the value predicted from Eq. (29.27).

Thus we see how an input voltage at the primary can be converted to a higher or lower output voltage at the secondary. In a *step-up transformer*, N_s/N_p is greater than 1. In a *step-down transformer*, N_s/N_p is less than 1.

Illustrative Problem 29.11 Voltage from a power transmission line is reduced from 2400 V to 220 V by a transformer. The primary winding of the transformer has 4800 turns. (a) How many turns are there in the secondary? (b) If the power output of the transformer is 12 kW, what is the current in each of the two coils? Assume 100% efficiency and a power factor of 1.

Solution Use Eq. (29.27) to solve for N_s.

(a) $$\frac{\mathcal{E}_s}{\mathcal{E}_p} = \frac{N_s}{N_p}$$

or $\quad N_s = N_p\left(\dfrac{\mathcal{E}_s}{\mathcal{E}_p}\right) = 4800\left(\dfrac{220 \text{ V}}{2400 \text{ V}}\right)$

$$= 440 \text{ turns} \qquad ans.$$

(b) The power input = the power output because we assume 100% efficiency. Then

$$P = \mathcal{E}_p I_p = \mathcal{E}_s I_s = 12{,}000 \text{ W}$$

$$I_p = \frac{12{,}000 \text{ W}}{2400 \text{ V}} = 5.0 \text{ A} \qquad ans.$$

$$I_s = \frac{12{,}000 \text{ W}}{220 \text{ V}} = \frac{600}{11} \text{ A} = 55 \text{ A} \qquad ans.$$

If we assume 100% efficiency, and a power factor of 1, we can equate the power input to power output; or $\mathcal{E}_p I_p = \mathcal{E}_s I_s$. Thus, we can write

$$\frac{\mathcal{E}_s}{\mathcal{E}_p} = \frac{I_p}{I_s} = \frac{N_s}{N_p} \qquad (29.28)$$

In other words, a stepped-up voltage means a consequent reduction in current, and vice versa. This can be verified by checking the answers in Illus. Prob. 29.11.

29.14   Uses of Transformers

One may wonder what advantages are to be gained by stepping up or stepping down voltages. Let us consider a few examples. When an electric current flows through a conductor, a certain amount of heat is developed, depending on the current and the resistance of the circuit. We have determined its value in Chap. 25 as I^2R (watts). For electric power transmission purposes it is desirable to keep the heat developed at a minimum, because whatever heat is developed in the transmission lines will be just that much energy loss. The rate of electric energy produced by a generator is fixed and equal to the product of emf generated times the current furnished. Let us consider a generator which furnishes 10 A of current at 550 V. The power output of the generator is $P = VI = 5500$ W. Let us assume further that this energy is to be transmitted for a considerable distance over wires that have a resistance of 20 Ω. The energy loss in transmission will equal

$$I^2R = 10^2 \times 20 = 2000 \text{ W}$$

which is 36 percent of the original power. By stepping up the voltage to 5500 V before transmission, the current will then be only 1 A and the I^2R loss will equal $1^2 \times 20$, or 20 W, which is only 0.36 percent of the original power. Thus by stepping up the voltage 10 times, the efficiency of transmission is increased from 64 to 99.64 percent.

By stepping up the voltage, it is possible to furnish electric power over long distances with little loss in energy. A step-down transformer (or series of transformers) at the other end of the line can be used to deliver the electric power at any desired voltage. A more detailed discussion of electric-power transmission will be found in Chap. 31.

Where large currents are desired, as in electric welding, step-down transformers are used. The action of such a transformer in electric welding can be illustrated by winding the primary of a transformer with many turns of light wire and the secondary with only a turn or two of very large copper wire. If the primary is connected to a 120-V ac source and a pair of nails are attached to the ends of the secondary coil (Fig. 29.16a), the current flowing through the secondary when the nail tips are brought together will be sufficient to melt the nails at the tips and weld them together.

Small step-down transformers are also used when low voltages are needed. These transformers are used for operating electric bells (see Fig. 29.16b), radios, thermostatic controls, toy electric trains, and the like. Picture tubes in TV sets generally require potential differences of 15,000 to 30,000 V. A step-up transformer is used for that purpose.

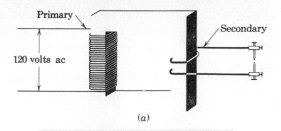

(a)

(b)

Fig. 29.16 (a) Circuit for a step-down transformer. (b) Power transformer for radio use made with laminated closed core, with steel covers protecting the windings. Marked binding posts on the cover provide connections for the input and output currents. (Central Scientific Co.)

29.15 Operation of a Transformer Under Load

When no current is drawn from the secondary (secondary circuit "open"), the impedance set up by the counter emf in the primary circuit is such that *practically no current will flow in the primary*. However, when the secondary circuit is closed and current is drawn from the secondary, a magnetic field is set up around it which, according to Lenz's law, tends to neutralize some of the magnetic field around the primary winding. This reduces the counter emf force established by the primary current. As a result, more current will flow through the primary coil. The primary current will build up until its counter emf will once more exactly balance the impressed emf due to the secondary current. If more current is drawn from the secondary, the counter emf of the primary is further reduced and more current is drawn from the source through the primary circuit. Conversely, if the load in the secondary is decreased, less current will be drawn from the primary circuit. Thus the transformer will adjust itself to changes in load in the secondary circuit. However, if the load in the secondary becomes too great, the resultant heavy current in the primary could burn out the primary windings.

29.16　Eddy Currents

It is desirable that the core of an induction coil, the core of a transformer, and the armatures of ac motors and generators be made of material of high magnetic permeability, in order to increase their magnetic flux densities when current flows in the coils surrounding them. However, these highly permeable substances are also good conductors of electricity. Hence when there is any change in flux density in them, induced currents will be sent through portions of their mass. Such induced currents, set up in cores and armatures, are referred to as *eddy currents*. Eddy currents produce heat which may be harmful to the electric apparatus as well as a source of energy loss.

To eliminate this wasted energy, cores and armatures of electric equipment are *laminated*; that is, they are made up of many very thin sheets of iron, with insulation between them in a direction perpendicular to that in which the eddy currents tend to flow.

Eddy currents are desirable in induction heating and in shielding from electromagnetic radiation.

Ferromagnetic materials, such as iron oxide (Fe_3O_4) are good insulators. As insulators, virtually no eddy currents are induced in them. They are used for coil cores, as well as memory and microwave devices.

For the sake of clarity, in Figs. 29.15 and 29.16 we have shown the primary and secondary windings on separate legs of the core. Commercial transformers are not constructed in this manner because with this arrangement a considerable amount of the magnetic flux produced by the primary current does not cut the secondary winding and the transformer is said to have a large leakage flux. In order to keep this leakage flux to a minimum, the two coils are wound around a common core (or cores). The low-voltage coil is placed next to the core, with the high-voltage coil placed around the low-voltage coil.

Most transformers have relatively high efficiencies. Even with the energy losses due to flux leakage, resistance of the windings, hysteresis of the iron core, and eddy currents, transformers usually have efficiencies in excess of 97%.

SUMMARY

Alternating current has been more generally used than direct current because of its economy of transmission over great distances.

Frequency denotes the number of complete cycles per second, called hertz (Hz), of the alternating current.

The common frequency of ac systems in the United States is 60 Hz.

Alternating currents and voltages are always expressed in terms of their *effective values*. For single-phase (sinusoidal) currents these are equal to 70.7 percent of their maximum values. However, ac voltmeters and ammeters read effective values directly.

In ac circuits (1) the current and voltage are in phase for a pure-resistance circuit; (2) the current lags 90° behind the voltage in a pure-inductance circuit; and (3) the current leads the voltage by 90° in a pure-capacitance circuit.

The amount the current lags or leads the voltage is called the *phase angle*. The phase angle varies from 0 to 90° in an *RCL* circuit, which includes *resistance, capacitance, and inductance.*

The combined opposition of a circuit to an alternating emf, because of the circuit's inductance and capacitance, is called *reactance*. The reactance X is equal to the difference between the inductive reactance X_L and the capacitive reactance X_C:

$$X = X_L - X_C \text{ (or } X_C - X_L) \qquad (29.19)$$

The inductive reactance in ohms is found by the equation

$$X_L = 2\pi f L \qquad (29.12)$$

The capacitive reactance in ohms is found by the equation

$$X_C = \frac{1}{2\pi f C} \qquad (29.16)$$

Impedance Z in a series circuit is the combined effect of resistance and reactance and is computed from the equation

$$Z = \sqrt{R^2 + (X_L - X_C)^2} \qquad (29.20)$$

Ohm's law for ac circuits which contain resistance, inductance, and capacitance can be expressed by the equation

$$I = \frac{V}{\sqrt{R^2 + (2\pi f L - 1/2\pi f C)^2}} \qquad (29.21)$$

Maximum current occurs in a series ac circuit when the inductive reactance is just equal and opposite in sense to the capacitive reactance. When $X_L = X_C$, the circuit is said to be *in resonance*.

The *resonant frequency* for a series ac circuit is

$$f = \frac{1}{2\pi\sqrt{LC}} \qquad (29.22)$$

The actual power in an ac circuit is not the product of voltage times current but is the product of these two quantities multiplied by a percentage called the *power factor*. The power factor of a single-phase circuit is equal to the cosine of the angle by which the current lags behind or leads the applied voltage.

$$P(\text{watts}) = VI \times \text{power factor} \qquad (29.23')$$

$$P = VI \frac{R}{\sqrt{R^2 + (X_L - X_C)^2}}$$

$$= VI \cos \phi \qquad (29.25)$$

where ϕ is the phase angle, and

$$\cos \phi = \frac{R}{Z} \qquad (29.26)$$

The basic equation for an ideal transformer is

$$\frac{\mathscr{E}_s}{\mathscr{E}_p} = \frac{I_p}{I_s} = \frac{N_s}{N_p} \qquad (29.28)$$

QUESTIONS

1. List as many applications as you can (a) where direct current is preferable to alternating current and (b) where alternating current is preferable to direct current.

2. A single-phase circuit contains only resistance, no reactance. What effect would doubling the resistance have on the power factor?

3. What is the power factor of a single-phase ac circuit containing only (a) resistance, (b) inductance, (c) capacitance?

4. A circuit contains a capacitor. Describe the difference between the currents in the circuit when (a) dc voltages and (b) ac voltages are impressed across the capacitor.

5. What is meant by the effective value of an alternating current?

6. What is meant by (a) inductance, (b) capacitance, (c) reactance, and (d) impedance? (e) Give the unit in which each of the above is measured.

7. What is meant by resonant frequency? How can it be varied in an *RCL* circuit?

8. In an ac circuit that has only a pure inductance load, the average power delivered to the load is zero. Does this mean that no energy is transferred from the source to the load? Explain.

9. A common demonstration for the physics laboratory is illustrated in Fig. 29.17. An aluminum ring is slipped over an extra long iron core set in a solenoid. At the instant an alternating emf is applied to the coil, the ring will be flipped violently into the air (see Fig. 29.17). Explain.

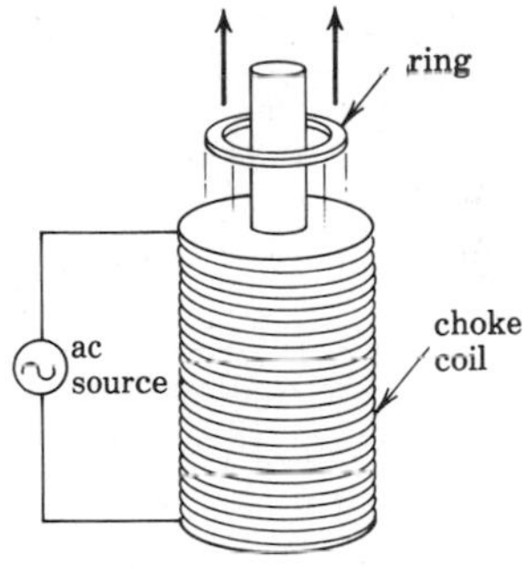

Fig. 29.17 Figure for Question 9.

10. A transformer used for operating an electric buzzer is designed for 120-V, 60-Hz current. Explain why the transformer is likely to burn out if it is connected to a 120-V dc source of current.

11. Check definitions for the henry (unit of inductance for L) and the farad (unit of capacitance for C). Show how $X_L = 2\pi f L$ and $X_C = 1/2\pi f C$ will give ohms of resistance, even though the L appears in the numerator while C appears in the denominator.

12. Under what conditions will the total power of an ac circuit equal the product VI?

13. Explain why a small step-down transformer connected to a door bell will take only very little, if any, electric energy from the supply line when the bell is not ringing, even though the primary is connected directly across the line.

14. Plot a rough graph of X_L as a function of frequency f for an inductance L of (*a*) 1.5 H and (*b*) 3.0 H. (*c*) What is the significance of the fact that $X_L = 0$ when $f = 0$ regardless of the value for L?

15. Plot a rough graph of X_C as a function of frequency f for a capacitance C of (*a*) 1.5 f and (*b*) 3.0 f. (*c*) What is the significance of the fact that when $f = 0$, X_C is extremely high regardless of the value for C?

PROBLEMS

1. If the effective voltage of a circuit, as shown by an ac voltmeter, is 115 V, what is the maximum voltage?

2. A 60-Hz alternating emf has a maximum value of 156 V. What is the value of the emf $\frac{1}{720}$ sec after it passes through its zero value?

3. A 60-Hz alternating current has an effective value of 2.4 A. What is the instantaneous value of the current at an instant $\frac{1}{360}$ sec after it passes the zero value?

4. A door chimes transformer is connected to a 120-V house circuit. The primary coil has 720 turns. The secondary coil has 180 turns. What is the voltage delivered to the door chimes?

5. An ac generator with emf of 550 V supplies electric energy to an 11,000-V line. If the primary of the step-up transformer has 80 turns, how many turns are there in the secondary?

6. A spot welder operates on a current of 300 A. The step-down transformer has a secondary of a single loop. If the transformer draws 1.5 A from the power line, how many turns must there be in the primary coil?

7. What is the capacitive reactance of a 4.0-μF capacitor when connected to an alternating current of 60 Hz?

• **8.** What is the reactance of a capacitor of 12 μF capacitance when it is connected to an alternating emf which has a frequency of 1800 Hz?

9. A single-phase 60-Hz motor draws 5.4 A at 115 V and has an inductive power factor of 0.88 at this load. How much power in watts does the motor use?

10. An impedance coil has a negligible resistance and an inductance of 0.0036 H. What is the frequency of the emf which will maintain a current of 0.15 A through the coil at 5.5 V?

11. A 50-kW ac generator delivers power at 12.5 kV to a step-up transformer which has a primary coil with 60 turns and a secondary coil with 900 turns. Find (*a*) the emf induced in the secondary and (*b*) the current in the secondary. Assume 100% efficiency.

12. A resistance of 15 Ω, a coil with a reactance of 25 Ω, and a capacitor with a reactance of 45 Ω are connected in series to an alternating 60-Hz voltage. What is the voltage required to maintain a current of 3.5 A through the circuit?

13. What inductance must be placed in series with a 32-μF capacitor to produce a circuit with a resonant frequency of 4000 Hz?

14. What is the resonant frequency of an ac series antenna circuit which has a 5.7-Ω resistance, an inductance of 25 μH, and a capacitance of 2.8×10^{-10} F?

15. A capacitor, a 20-Ω resistor, and a 0.20 H coil are connected in series to a 60-Hz applied voltage of 120 V. What capacitance will produce resonance in the circuit?

16. A series circuit consists of a 120-V, 60-Hz source of emf, a resistor of 15 Ω, an inductor of 0.25 H, and a capacitor of 35 μF. (*a*) What is the impedance of the circuit? (*b*) How much current will flow in the circuit?

17. The transmitter of a television station transmits at a frequency of 78 megahertz (MHz). What inductance is needed with a capacitance of 16 pF to tune in the station? (Remember that 1 pF $= 10^{-12}$ F.)

18. Radio waves from the transmitter of a radio station are broadcast with a frequency of 1200 kHz. What capacitance is needed with an inductance of 4×10^{-8} H to form a circuit resonant to this frequency?

19. A coil having a resistance of 2 Ω and an inductance of 0.15 H is connected to a 30-V, 60-Hz source. Calculate the current that will flow through the coil.

20. A coil with a resistance of 15 Ω and an inductance of 0.25 H is connected to a 220-V, 60-Hz line. Compute (*a*) the impedance of the coil, (*b*) the current through it, (*c*) the phase angle, and (*d*) the power factor.

21. A 4-Ω resistor, a 6-μF capacitor, and an unknown inductor are connected in series across a 120-V, 60-Hz alternating source of emf. (*a*) What must be the magnitude of the inductor (in henrys) in order for the maximum current to flow through the resistor, capacitor, and inductor? (*b*) What is the maximum current? (*c*) What is the voltage across the capacitor?

22. A series ac circuit consisting of a resistor of 24 Ω, an inductance of 0.30 H, and a capacitor of 36 μF has a 120-V, 60-Hz voltage impressed upon it. (*a*) What are the inductive and capacitive reactances in the circuit? (*b*) What is the impedance of the circuit? (*c*) Show by a vector diagram the relation between the resistance, the inductive and capacitive reactances, and the impedance. What is the value of the phase angle? (*d*) What is the power factor of the circuit? (*e*) How much current will flow in the circuit? (*f*) How much power is consumed in the circuit?

30 Generators and Motors

Forced motion of a conductor in a magnetic field produces a potential difference between the ends of the conductor. When the ends of the conductor are connected to a closed circuit, electrons will flow in the circuit and an electric current is produced. If a way could be provided for a continuous motion of the conductor in the field, then a continuous emf would be generated. This can be accomplished by rotating a coil of wire in the magnetic field. The mechanical energy to turn the coil can be provided by a *prime mover* such as a water turbine, steam turbine, or a diesel engine.

A machine that converts mechanical energy into electric energy is called a *generator*. It has already been pointed out that a generator does not generate electricity. It develops an emf that produces a current (electron flow) in a closed circuit external to it. The study of electric generators is a highly technical field in itself. In this chapter, we deal only with some of the more basic principles of operation, design and use of dc and ac generators and motors.

In an electric *motor*, electric energy is converted into mechanical energy. An electric motor is then just a generator in reverse. The word *dynamo* is applied to machines that convert either mechanical energy into electric energy or electric energy into mechanical energy. Thus, the dynamo is a reversible machine capable of operation as either a generator or motor.

30.1   The Simple AC Generator

The principles involved in the electric generator can probably best be explained in terms of the simplest possible generator made up of one loop of wire turning in a uniform magnetic field, as shown in Figs. 30.1 and 30.2. In Chap. 28 we discussed the generating of an alternating current in a closed coil which is rotated in a magnetic field. We discussed further in Chap. 29 the fluctuation (in sinusoidal fashion) of the emf or current established in the coil as it rotates in the magnetic field. The current generated *in the coil* is always an *alternating* one. However, the current

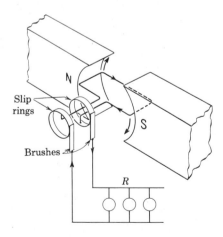

Fig. 30.1   Generator with slip rings.

delivered *from the coil* to an external circuit can be either alternating or direct, depending on how it is picked up from the coil. If the current that is established in the coil is picked up by means of *brushes* on *slip rings* attached to the ends of the coil (Fig. 30.1), an alternating current will also flow through the external circuit R. The current will reverse direction twice in each complete revolution. The value of the emf, and hence that of the current, will not be constant as the coil turns with constant velocity but will follow the conventional sine curve (see Fig. 29.1).

DC GENERATORS

30.2 The DC Generator

Even though the current that flows in the rotating coil of a generator is always alternating, it is quite possible to obtain a unidirectional current through an external circuit. A current is said to be *rectified* if it is changed from alternating to direct. Figure 30.2 shows how an ac generator can be modified by means of a split ring, called a *commutator*, and two brushes, so that it will deliver a pulsating direct current to an external circuit.

In the dc generator the terminals of the coil are connected to the two halves of a *split-ring commutator*. The segments of the ring are insulated from each other. The brushes are placed on opposite sides of the ring so that when the emf changes direction (at zero emf), the brushes pass over the division between the ring segments.

When the conductor AE in Fig. 30.2 is moving upward, the induced emf will cause a current to flow in the direction from A to E. In like manner, as conductor CD moves downward, the induced emf will cause a current to flow in the direction from C to D. This will make the semiring a of the split-ring commutator negative and the semiring b positive. During the half-revolution when AE moves downward and CD moves upward, the signs of the charges on semirings a and b will be reversed. But the brushes are so adjusted that at the instant the signs of the charges on the semirings change, the segments will

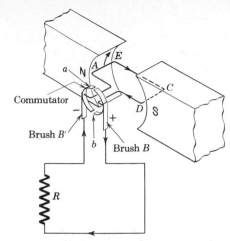

Fig. 30.2 Generator with commutator, for dc output.

make contacts with the other brushes. Hence brush B will always have a positive charge and brush B' will always be charged negatively, with a resultant unidirectional flow through the external load R. Thus, although the emf still fluctuates, it is always in the same direction in the *external* circuit.

It will be noticed that the direction of current in the circuit *external* to the generator flows from a higher or positive potential to a lower or negative potential, while in the *internal circuit* of the generator the current flows from a lower potential to a higher potential. Using the hydraulic analogy, work is done *on* the coil (in this case the rotating part) by a prime mover, and this will force current to flow "uphill" just as water is pumped from a lower to a higher level. It must be remembered that the real direction of flow of electricity in a circuit is the direction of flow of electrons, but in this discussion we are assuming the conventional *direction of current flow* opposite to the direction of electron flow.

Figure 30.3 illustrates graphically the pulsating nature of the rectified emf from a single-loop dc generator. The curve is similar to that in Fig. 29.1, except that the second half of the cycle is inverted.

A pulsating direct current such as the one just described is not suitable for commercial users

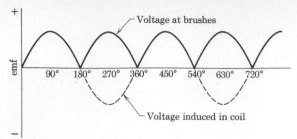

Fig. 30.3 Curve showing unidirectional pulsating emf at the brushes of a single loop with commutator.

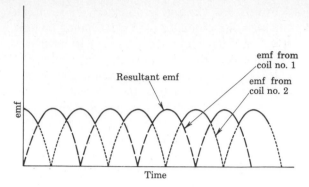

Fig. 30.5 Emf fluctuations from a two-coil armature with coils at right angles to each other. The coils are insulated from each other.

of direct current. A *steady* unidirectional current is necessary, and this is obtained in the commercial dc generator by using many armature coils distributed evenly around a soft-iron laminated drum. Each coil has a pair of commutator bars (segments) that are insulated from the other bars and from the shaft.

Figure 30.4 illustrates how two coils at right angles can be used to produce a pulsating direct current. Note that the commutator is divided into four segments or bars. Each coil terminates in two opposite commutator bars. The emf fluctuations in the two coils are shown with the resultant emf in Fig. 30.5. We see that when either coil has a maximum emf, the other has an emf of zero.

The arrangement shown in Fig. 30.4 is not a practical one. Note that at any given time only

the armature coil that connects the bars in contact with the brushes produces current. The induced voltages in the other coil do not produce a current because it is not connected to the outside circuit. Thus only half the induced voltages are used to produce a current.

However, if the emfs coming from the two coils are joined in series, the instantaneous value of the combined emfs of the two coils will equal the sum of the individual emfs of each coil at that instant. This resultant is shown in Fig. 30.6. Thus the resultant emf is seen to fluctuate considerably less than the emfs of the individual coils. The resultant emf never falls to zero in this arrangement.

Most generators, today, use *drum-type armatures*, that is armatures in the form of a cylinder or drum. These are wound with a large number of coils of wire so that the current supplied to the external circuit will be almost free from fluctuation. The core of the armature is made of sheet-steel laminations that are keyed to a *spider*, which in turn is keyed to the shaft. The armature winding consists of coils of wire or bars connected in series and lying in slots on the surface of the core. Each end of a coil is soldered to a commutator segment. Figure 30.7a shows a drum-type armature with 8 coils and segments. Figure 30.7b illustrates the commutator segments and armature coils laid on a flat surface to show how they are connected. Note that each slot contains conduc-

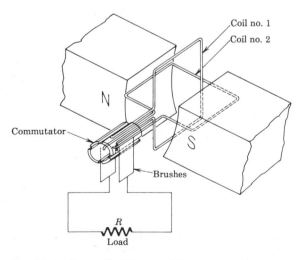

Fig. 30.4 Two-coil armature with commutator.

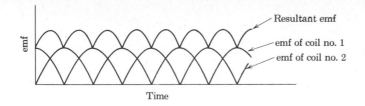

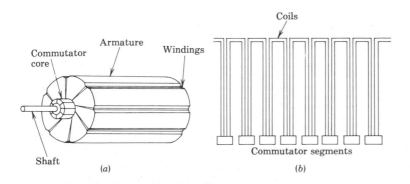

Fig. 30.6 Emf fluctuations from a two-coil armature with coils at right angles to each other and joined in series.

Fig. 30.7 Drum-type armature with 8 armature coils and 8 commutator bars.

Fig. 30.8 Partly wound armature of an 830-kW, 800-V dc generator. The formed coils can be seen joined to the commutator segments. (Fairbanks, Morse Co.)

tors from two adjacent coils. These coils are inserted in the slots and kept in place by retaining wedges. A partly wound armature of a dc generator is shown in Fig. 30.8.

The use of formed coils provides for superior ventilation and permits speedier repair of the generator in case one of the coils needs to be

removed. On small generators, the coils are generally not form-wound but are wound by hand or by machine directly into the slots.

30.3 Field Excitation

The magnetic field of a dc generator is usually produced by using a portion of the induced current to energize the field magnets. Such generators are said to be *self-exciting*. If a permanent magnet is used for the field, the generator is commonly called a *magneto* (Fig. 30.9a).

There are three common ways of self-exciting the field magnets. When the field magnets are connected in series with the armature loops so that all the generator current passes through the field coil windings (Fig. 30.9b), the generator is called a *series-wound* generator. Since the entire current furnished by the generator flows through the field coils, the coils must be wound with heavy wire.

When no load is connected to the generator, there is no current in the armature and the field windings. When a load is connected, the small emf induced because of the residual magnetism of the pole pieces will cause a current in the arma-

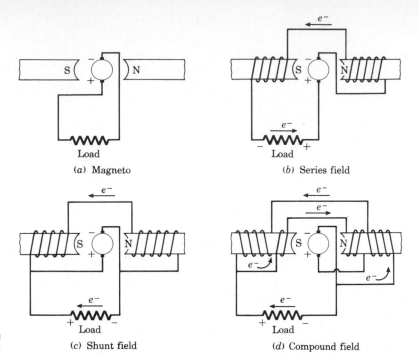

Fig. 30.9 Methods of connecting field and armature in dc generators.

(a) Magneto

(b) Series field

(c) Shunt field

(d) Compound field

ture winding, the load, and the field winding. As the current flows, the magnetic flux in the field increases, causing an increase in the terminal emf. This, in turn, will increase the output current.

As the terminal voltage and output current build up, there is an accompanying increase in the back emf developing in the armature and field windings. This tends to reduce the terminal voltage. Soon a point will be reached where the two processes balance and the terminal voltage ceases to rise.

Since the terminal voltage varies so much with the load, series-wound generators are seldom used. They are used, however, in arc-welding because once the generator reaches its peak terminal voltage, the output remains fairly constant for large variations in the output current.

In the *shunt-wound* dc generator, the field magnets are connected in parallel with the armature so that only a portion of the generated current is used to excite the field (Fig. 30.9c). Usually

less than 5% of the armature current is fed to the field coils. To keep the current low, the coils have a high resistance. In order to gain an adequate magnetic flux in the field coils, many turns of fine wire are used.

The principles involved in the shunt-wound generator are similar to those applying to the series-wound type. As the armature is rotated, lines of residual magnetic flux are cut to produce an emf at the brushes. A current flows in the armature windings and field windings, causing an increase in the magnetic flux of the field windings which, in turn, increases the current. The cumulative process continues until equilibrium is reached for a given load.

For light loads the terminal voltage remains quite constant for variations in load currents. However, if the load is too heavy, the terminal voltage drops sharply.

The series-wound generator and the shunt-wound generator have generally been replaced by the compound-wound generator. In the *compound-wound* generator both series and shunt field windings are employed. The series field windings consist of a few turns of heavy wire over a shunt winding of many turns of fine wire. The two windings are connected in series with the armature winding (Fig. 30.9*d*). In such a generator the potential difference across an external circuit may be kept fairly constant, since an increase in the load will cause an increase in the current in the series windings and a decrease in the shunt windings. By using the proper number of turns of each type of winding, a constant flux density can be maintained under varying loads, so that the generator voltage will be constant regardless of the load in the external circuit.

30.4 Multipolar DC Generators

Generators such as the ones discussed so far, having one north pole and one south pole, are called *bipolar generators.* Commercial generators, especially large ones, commonly have four, six, eight, or more poles (an even number). The greater number of poles, the slower the rotation necessary to produce a given emf. A four-pole machine rotates half as fast as a two-pole machine to develop the same voltage. The poles of a four-pole generator (Fig. 30.10) are magnetized alternately north and south; alternate brushes are positive and connected to the same generator

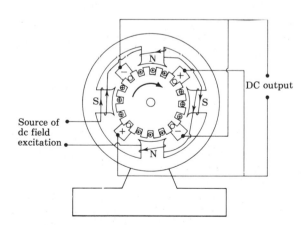

Fig. 30.10   Schematic diagram of a four-pole generator.

terminal. Each of the four flux paths leaves the north pole, passes through part of the armature, enters the south pole of an adjacent field pole, and completes the path through the pole cores and the yoke.

30.5 EMF Field of DC Generators

In Sec. 28.4 we discussed the *average* or *mean* voltage generated in a single wire moving at right angles to a magnetic field. The emf of any dc generator can be computed by finding the average emf induced in each conductor in the armature and multiplying this value by the number of conductors wound in series around the armature. The average emf of the generator can be calculated from the formula

$$\mathscr{E} = \frac{\phi P n Z}{60b} \qquad (30.1)$$

where

$\phi =$ flux density (webers) per pole
$P =$ number of poles
$n =$ angular speed of armature, rev/min
$Z =$ total number of conductors on armature
$b =$ number of paths in parallel through armature

For any given generator, the quantities P, Z, and b are fixed. The average emf for a given generator then depends only on the flux per pole ϕ and the speed n. If we substitute K for all the fixed values or *constants* of a given generator, Eq. (30.1) can be simplified to

$$\mathscr{E} = K\phi n \qquad (30.2)$$

Thus it can be seen that generated emf can be raised or lowered by increasing or decreasing the speed of rotation of the armature and/or increasing or decreasing the flux per unit pole.

Illustrative Problem 30.1   A four-pole generator rotates at 1500 rev/min. It has 240 armature conductors arranged in four parallel paths. A flux of 3.6×10^{-2} weber extends from each magnet pole. Compute the average emf of the generator.

Solution

$$\phi = 3.6 \times 10^{-2} \text{ Wb}$$

$$P = 4 \qquad n = 1500 \text{ rev/min}$$

$$Z = 240 \quad b = 4$$

$$\mathcal{E} = \frac{\phi PnZ}{60b}$$

$$= \frac{3.6 \times 10^{-2} \text{ Wb} \times 4 \times 1500 \text{ rev/min} \times 240}{60 \text{ sec/min} \times 4}$$

$$= 216 \text{ Wb/sec}$$
$$= 216 \text{ V} \qquad\qquad\qquad ans.$$

DC MOTORS

30.6 Generators and Motors Compared

We have seen that a generator converts mechanical energy to electric energy. An electric motor does just the reverse; i.e., it converts electric energy into mechanical energy. In fact, any dc generator will run as an electric motor, and conversely the dc motor will operate as a generator.

Structurally, both the dc generator and the dc motor consist essentially of field magnets, armature, commutator, and brushes. Both are constructed in bipolar or multipolar form. One significant structural difference between motors and generators lies in the housing of the two machines. Since generators ordinarily operate in more sheltered locations, their construction is of the *open* type. Motors often operate in locations that are exposed to moisture, dust, foreign particles, and corrosive gases. The housings of motors are designed to protect the machine against these elements. The motors are generally either semienclosed or totally enclosed by the housing. Ventilating openings are provided in the semienclosed type of motor but are so placed as to protect the motor against falling foreign particles or liquids. Totally enclosed motors (Fig. 30.11) are completely sealed in the housing. The heat developed in these motors must be entirely dissipated by radiation from the housing or by circulating cooling fluids.

(a)

(b)

Fig. 30.11 (*a*) Totally enclosed fan-cooled dc motor. (*b*) Cutaway view showing the high capacity system of controlled ventilation in which blowers and heat exchanger are mounted at the shaft end.

The load of a generator consists of devices which convert electric energy into other forms of energy, while the load of a motor consists of countertorques that tend to oppose the rotation of the moving part of the motor. When the load of a generator changes, there is a tendency for the emf of the generator to change; in the case of the motor, a change in load tends to result in a change in speed. The emf of a generator can be changed by adjusting the strength of the magnetic field or by changing the speed of the prime mover; the speed of a dc motor can be altered by changing the voltage impressed across the armature or by changing the field strength of the motor.

Generators are always started with no electric load in the external circuit; motors may be started either with or without a mechanical load.

30.7   Counter EMF of a Motor

The armature of a dc motor produces its own magnetic field, which reacts with the flux of the field coils to cause a rotary motion. However, the conductors on the rotating armature cut the lines of force of their own field just as if the machine were a generator driven by some prime mover. Consequently, there is induced a counter emf in the conductors of the armature which is opposite in direction to the flow of current in the conductors of the armature (Lenz's law). Thus every machine that is operating as a motor is at the same time a generator.

The effect of counter emf on the amount of current used by a motor can be demonstrated by a simple experiment. Connect an ammeter and an incandescent lamp in series with a small dc motor (Fig. 30.12). If we forcibly hold the armature stationary while the line current is turned on, the lamp will glow with its normal brilliance; but when the armature is allowed to turn freely, the lamp grows dim and the ammeter reading drops considerably. Thus it is evident that the motor passes more current when the rotation is stopped or retarded than when it is allowed to run freely. Since the voltage of the line source has not changed in the experiment, the current must be diminished by the development of a counter emf

which opposes that of the driving emf. The faster the motor turns, the greater the counter emf, and therefore the smaller the difference between the impressed emf and the counter emf. It should be clear that the counter emf can never be equal to, but must always be less than, the impressed voltage on the armature terminals. The difference in the two will be equal to the potential drop in the motor armature. The current flowing through the armature, then, depends upon and is controlled by the counter emf. Using Ohm's law, we find the dc armature current to be

$$I_a = \frac{V_a - \mathscr{E}_c}{R_a} \tag{30.3}$$

where

I_a = armature current
V_a = impressed voltage on armature
$\mathscr{E}_c$ = counter emf generated in armature
R_a = armature-circuit resistance

The counter emf of a dc motor operating at any speed will be equal to the emf developed by the machine if it is operated as a generator at the same speed, provided the field strength of the machine remains the same in both cases. Hence the counter emf of a motor at any speed can be determined by operating the motor as a generator at that speed and measuring the emf generated.

Illustrative Problem 30.2 A small dc motor operates at 120 V and has an armature resistance of 2.8 Ω. When the motor is running at 3600 rev/min the armature current is 3.5 A. (*a*) What counter emf is developed in the armature? (*b*) What current would flow through the armature if it were held fast and not permitted to turn?

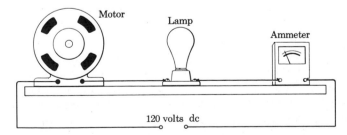

Motor   Lamp   Ammeter

120 volts dc

Fig. 30.12   Effect of counter emf. More current is required to start the motor than to keep it running.

Solution

(a)
$$I_a = \frac{V_a - \mathscr{E}_c}{R_a}$$

$$3.5 = \frac{120 - \mathscr{E}_c}{2.8}$$

$$120 - \mathscr{E}_c = 2.8(3.5)$$

$$\mathscr{E}_c = 120 \text{ V} - (2.8 \text{ }\Omega)(3.5 \text{ A})$$

$$= 110 \text{ V} \qquad\qquad ans.$$

(b) If the armature is not permitted to rotate, the value of $\mathscr{E}$ will be 0.

$$I_a = \frac{V_a - \mathscr{E}_c}{R_a}$$

$$= \frac{120 \text{ V} - 0 \text{ V}}{2.8 \text{ }\Omega}$$

$$= 42.9 \text{ A} \qquad\qquad ans.$$

Such a large current would burn out the windings of a small motor.

The counter emf regulates the amount of current drawn by a motor; hence a motor is, to some extent, a self-regulating device. No counter emf exists in the motor until it begins to turn. When the motor first starts to turn, the current passing through the armature is very large. In order to avoid burning out the motor with these high starting currents, adjustable starting resistors to take the place of counter emfs are frequently inserted in series with the motor. As the speed increases, the resistance may be cut out gradually because of the increasing counter emf. When the motor attains its normal speed, all the adjustable resistance can be cut out. When the motor turns at constant speed, the impressed voltage will be balanced by the counter emf and by the drop in potential due to resistance. If the load upon the motor is increased, the torque being developed no longer is sufficient to overcome the larger load. Hence the speed of the motor falls. This decreases the counter emf, permitting a greater amount of current to flow through the armature, which in turn produces the larger torque required.

AC GENERATORS

30.8   The AC Generator or Alternator

In dc generators the armature windings are always placed on the rotating part, called the *rotor*, while the field poles are placed on the stationary part, called the *stator*. This is necessary in order to provide a means of converting the ac voltage generated in the armature (the rotating coils) windings to dc voltage at the terminals of the windings by the use of a commutator. Almost any dc generator can be converted to an ac generator, commonly called *alternator*, by replacing the commutator by a pair of slip rings properly connected to the armature windings.

Large ac generating plants are usually of two kinds. In hydroelectric generating plants, slowly rotating water turbines furnish the mechanical energy to turn the alternator rotor. The alternators for hydroelectric generators use four or more poles to produce the 60-Hz current. In steam-generating plants, heat from burning oil, gas, or coal, or from nuclear reactors is used to produce steam. The steam drives a rapidly rotating steam turbine, which, in turn, rotates the rotor. The alternator in these generators uses only two or four poles.

The larger generators generate emfs of thousands of volts and may produce thousands of amperes of current in the armature. Consequently the armature coil must be very heavy and well insulated. If the armature is the rotor, the contact between the slip rings and brushes tends to produce serious losses of energy at these high voltages and currents. For this reason it is better to have the armature remain stationary and to rotate the field coils within it.

You will remember that it is the relative motion between the armature windings and field magnets that is essential to any generator; it does not matter which of the two revolves. Alternators can be classified into two types: (1) *field stator*, in which the field windings are stationary and the armature revolves in the field; and (2) *field rotor*, in which the armature windings are stationary and the field poles revolve. Slip rings are used in either type to conduct current to and from the rotor.

All but the small low-voltage alternators are built with a *stationary armature* and a *revolving field*. The construction of this type of alternator is more economical and simplifies the problems of insulation of the alternator. Perhaps the chief advantage of this type of construction lies in the fact that relatively high voltages can be drawn from the *stationary* armature to the switchboard without the use of sliding-contact slip rings and brushes. Also the problem of centrifugal forces acting on the armature windings is eliminated by making them stationary. Every alternator must have its field coils energized with direct current from some kind of separate dc generator, called an *exciter* (see also Chap. 31).

30.9 Revolving-Field Alternator

The field structure of a revolving-field alternator may be of two distinct types—the *salient-pole type* and the *cylindrical type*. In the salient-pole type the field poles are laminated pole pieces which, with their field coils, are mounted on the outer rim of a steel *spider*, which in turn is keyed to the shaft. This type of construction is illustrated in Fig. 30.13. The field windings are generally form-wound, and the completed coils are then fitted around the pole pieces and attached rigidly in place. The coils are connected in series in such a way as to give alternate polarity to the poles. The ends of these field windings are then connected to the slip rings. The salient-type alternator generally is a slow-speed generator driven by diesel engines or water turbines and may have 30 or more poles in the rotor.

The *stationary* armature of the alternator consists of carefully insulated form-wound coils placed in iron core slots supported by a circular cast-iron or fabricated steel frame. The core of the armature is formed of sheet-steel laminations placed together and held rigidly in place by suitable wedges. A partly wound stationary armature with some of the coils placed in their core slots is shown in Fig. 30.14.

Fig. 30.13 Rotor of salient-pole alternator. The rotor, containing 18 poles, is driven by a diesel engine at 333 rev/min and delivers three-phase 60-Hz current rated at 6300 and 3640 V. (General Electric Co.)

High-speed alternators have cylindrical rotors whose diameters are considerably shorter than their axial lengths. The diameters must be relatively small to withstand the large centrifugal stresses that are set up at high speed. The prime mover of this type of alternator is usually a steam turbine. The rotor must be an extremely well-constructed and rigid structure, and is generally forged with the shaft in a single piece to the proper dimensions. The rotor is then slotted longitudinally by a milling machine to a depth sufficient to house the field windings. The windings for the rotor are form-wound and fitted in the slots and held in place by steel wedges placed at the outer ends of the slots. The ends of these windings are further held in place by a steel retaining ring attached to the rotor. These ends are then attached to slip rings from which the direct current for excitation is drawn through brushes

Fig. 30.14 Coils being placed in slots of stationary armature for a steam-driven generator. (General Electric Co.)

riding on the surface of the rings. The rotor of a turbo-alternator of this type is shown in Fig. 30.15.

30.10 Frequency of Alternators

The frequency of an ac generator depends upon the number of field poles and the speed of the rotor. A complete cycle is generated in a given armature coil when a pair of poles (one north and one south) is moved past the coil. Thus a complete cycle is developed in a bipolar alternator when the rotor makes one complete revolution. In a four-pole alternator two cycles will be developed per revolution.

Let

$$N = \text{rotor speed, rev/min}$$
$$p = \text{number of poles}$$
$$f = \text{frequency, Hz}$$

Then the frequency can be found by the equation

$$f = pN/120 \qquad (30.4)$$

Fig. 30.15 The rotor of a steam-turbine generator is being machine-slotted on a huge horizontal slot miller. (General Electric Co.)

Illustrative Problem 30.3 Find the frequency of an alternator of 30 poles that is revolving at 250 rev/min.

Solution This alternator will have 30 poles, so $p = 30$.

$$f = \frac{pN}{120}$$

$$= \frac{30 \text{ poles} \times 240 \text{ rev/min}}{120}$$

$$= 60 \text{ Hz} \qquad\qquad ans.$$

Illustrative Problem 30.4 How many revolutions per minute must an eight-pole alternator make to generate a 60-Hz current?

Solution The alternator has eight poles. So $p = 8$. Then

$$N = \frac{120f}{p} = \frac{120 \times 60}{8} = 900 \text{ rev/min} \quad ans.$$

As previously noted, the most common frequency in the United States is 60 Hz. In some 50-Hz currents are in use.

30.11 Polyphase Alternators

Alternators may be grouped not only according to their design and construction but also as to the type of connections of their armature windings. When all the armature windings are connected in series so that their individual emfs will add together, the alternator is a *single-phase* machine. This is distinguished from the *polyphase* alternator, which generates two or more alternating emfs at the same time.

The *two-phase alternator* has two separate and independent armature windings which are mounted 90° apart. The alternating emfs developed in the two windings are said to be 90° apart in phase. Figure 30.16a illustrates the emf curves of a two-phase alternator. It can be seen from the figure that at no time during a cycle will the emf be zero if a two-phase alternator is used.

Engineers have found the operating characteristics of three-phase machinery to be far superior to those of either the single-phase or two-phase equivalent. The three-phase alternator has three separate sets of windings symmetrically placed on the armature so that three alternating emfs of the same frequency, differing by 120 electrical degrees, are obtained. The emf curves for such an alternator are shown in Fig. 30.16b. The three sets of coils that are mounted 120° apart may each be brought out from the alternator to form three separate single-phase circuits. However, these coils are generally interconnected in such a way as to bring out only three, or sometimes four, terminals from the alternator. Three-phase alternator windings will be discussed in Chap. 31.

The power output of a dc generator is usually rated in watts or kilowatts. However, in

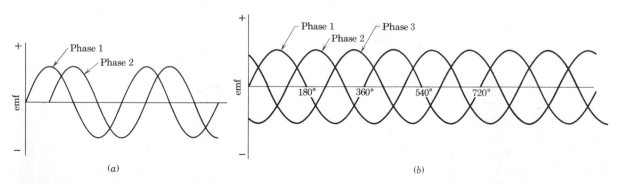

Fig. 30.16 Emf curves of (*a*) two-phase alternator, (*b*) three-phase alternator.

an ac circuit, the inductance and capacitance affect the relationship between the current and voltage (Chap. 29). Hence the power is affected. Inductance and capacitance are variable factors that cannot accurately be determined in advance. For this reason, the power output of an ac generator is normally rated in volt-amperes (VA) or kilovolt-amperes (kVA).

AC MOTORS

30.12 Alternating-Current Motors vs. DC Motors

Alternating-current motors are used much more widely in industry than dc motors. There are several reasons for this. Alternating current can be transmitted over long distances with little loss of power; hence most utility companies offer only ac service to the consumer. The commutators of dc motors are subject to sources of trouble and must be frequently checked and serviced. Sparks passing between the commutator and brushes of dc motors not only present a source of power loss but may also be a serious hazard in areas filled with combustible gases or dust particles. Most ac motors do not require brushes and a commutator.

Although the ac motor is generally less expensive to build, is more rugged, and can stand more abuse, dc motors are used in a large number of applications where large torques at slow speed are required. Speed in the dc motor can be regulated with greater accuracy than in the ac motor.

30.13 Universal (Series) Motors

The series motor is essentially a dc motor (with commutator segments) in which the field coils and the armature are connected in series. Such a motor will operate effectively on either an alternating or a direct current because the current in the field and that in the armature reverse at the same instant. Since both the field polarity and the direction of the current through the armature change together, the armature and the field poles repel and attract each other in such a way as to give a continuous rotary motion to the armature. Since the windings of the stator and the rotor are connected in series, the speed of the motor increases with decrease in load and the starting torque is high. This type of motor is called a *universal motor*. It is used when it is desirable to use a motor that operates satisfactorily on either ac or dc circuits. It has application where high speeds are required. Small universal motors with load speeds of 10,000 rev/min are not uncommon. It also is used when a motor is required which automatically adjusts itself to the magnitude of the load—high speed for light loads and low speeds for heavy loads.

Universal motors find wide application in vacuum cleaners, sewing machines, electric shavers, electric fans, hair dryers, kitchen food mixers, portable drills, power saws, pipe threaders, small grinders, and certain woodworking lathes.

30.14 Induction Motors

The simplest and most common type of ac motor is the *induction motor*, invented in 1888 by Nicola Tesla (1856–1943). Like any other motor, it consists of a stationary part and a rotating part. However, the principle upon which the induction motor operates is quite different from those on which other motors are based. In all the motors discussed thus far the current has been "fed" into the *rotating* part of the machine through brushes and a commutator. That is, the electric energy has been conducted directly to that part of the machine where it is converted into mechanical energy. In the induction motor the *stator* is connected to the ac supply. The rotor is not connected to any supply but has current *induced* in it by inductive action from the stator. The induction motor will operate satisfactorily only when it is connected to an ac source of the frequency for which it was designed.

The stator core is built of a stack of circular sheet-steel laminations with slots on the inner surface and parallel to the axis (Fig. 30.17). This core is attached to a cast-iron or cast-steel yoke

Fig. 30.17 Typical small-horsepower induction motor disassembled, showing main components. (Siemens-Allis, Inc.)

for support. Stator windings are placed in these slots according to the number of phases and the required speed of the motor. The stator windings are very similar to those of the alternator from which it will draw its current.

The rotor of an induction motor is built up of a laminated-steel core attached to a cast spider that is then pressed onto a shaft. The rotor windings are placed in the rotor slots with the conductors parallel or approximately parallel to the shaft. These conductors are not insulated from the core since the current will flow more readily through the copper conductors than through the iron. The ends of the rotor conductors are joined (short-circuited) together by a cast ring. Thus the rotor of an induction motor is essentially a number of parallel copper bars cast into slots and joined together by a ring. Because the appearance of such a rotor is similar to a revolving squirrel cage, it is commonly called a *squirrel-cage rotor*. Figure 30.18 shows a typical squirrel-cage rotor with the copper windings cast into the steel rotor core.

The operating principle of the induction motor is based on a *rotating magnetic field*. This

Fig. 30.18 Squirrel-cage rotor assembly. The rotor fans are cast as an integral part of the squirrel-cage winding and are used to provide air turbulence and better heat transfer. (Robbins & Myers, Inc.)

concept can be illustrated by suspending, with a piece of string, a horizontal bar magnet over a compass (Fig. 30.19). The compass needle will align itself parallel to the bar magnet. If the magnet is now rotated, its rotating magnetic field will cause the needle of the compass to follow. In the induction motor, the magnetic field revolves even though the windings which produce it do

Fig. 30.19 Illustration of a rotating magnetic field.

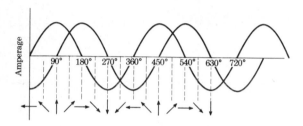

Fig. 30.20 Curves of two currents 90° out of phase.

not move. The production of a rotating magnetic field in an induction motor can be illustrated by considering a two-phase alternator from which we may have two alternating currents of the same frequency but differing in phase by 90 electrical degrees. Curves of the two currents which differ in phase by 90° are shown in Fig. 30.20. Let us connect these currents to coils of a four-pole circular ring, as shown in Fig. 30.21.

When the current of phase 1 has maximum value, the current of phase 2 will be zero. As a result, poles B and B_1 will be fully magnetized and poles A and A_1 will be unmagnetized. The flux will then be directed from north to south, as shown by the arrow in Fig. 30.21a. One-eighth of a cycle later (45 electrical degrees) the current in phase 1 will have diminished by as much as the current in phase 2 will have increased, and the magnetic flux will take the direction indicated in Fig. 30.21b. Another eighth of a cycle later the current in phase 1 will be zero and the current in phase 2 will be at a maximum. At this stage, poles A and A_1 will be fully magnetized, poles B and B_1 will be unmagnetized, and the flux will have the direction indicated by the arrow in Fig. 30.21c. As the two-phase current continues to flow through the coils, the magnetic flux will continue to rotate, as shown by the arrows drawn in Fig. 30.20.

When the induction-motor rotor is placed in this rotating field, there will be emfs induced in the rotor conductors. Since the rotor windings form a series of closed paths, large currents will flow in the conductors. This current will produce a magnetic field about the rotor which will react with the magnetic field of the stator and cause rotation. It should be clear that the rotor can never spin as fast as the magnetic field. If it did, no lines of force would be cut by the rotor conductors, no current would be induced, and hence no torque would be exerted by the conductor.

Commercial induction motors have multipolar stators. These poles are not projecting poles, as shown in Fig. 30.21, but are embedded

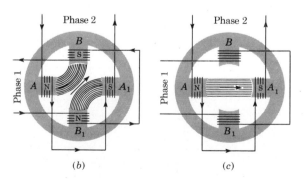

Fig. 30.21 Fields produced in the stator of an induction motor by a two-phase alternator.

(a) (b) (c)

Fig. 30.22 Wound stator for a polyphase induction motor. (General Electric Co.)

in slots in steel-laminated plates. Figure 30.22 shows a typical commercial stator of an induction motor. Most large ac motors in use today are induction-type motors. They have fair starting torques and fairly constant speeds with varying loads.

30.15 Synchronous Speed and Slip

The speed at which the *magnetic field* revolves is called the synchronous speed of the motor. It corresponds to the value of N in Eq. (30.4) and is determined by the frequency of the source of alternating current and by the number of poles in the stator. Thus, solving Eq. (30.4) for N yields

$$\text{Synchronous speed } N_s = \frac{120 \times \text{frequency } f}{\text{number of poles}}$$
(30.5)

or

$$N_s = 120f/p$$

where N_s is expressed in revolutions per minute. We have pointed out that an induction motor will not function unless the speed of the rotor is less than that of the revolving field. The difference between the rotor speed and the synchronous speed is known as the *slip* of the motor. It can be

expressed in revolutions per minute but is usually expressed as a percentage of the synchronous speed.

Percent slip S

$$= \frac{\text{synchronous speed} - \text{rotor speed}}{\text{synchronous speed}} \times 100$$

Or, written as an equation,

$$S = \frac{N_s - N_r}{N_s} \times 100$$
(30.6)

The equation for the speed of the induction motor then becomes

$$N_r = \frac{120f(1 - S)}{p}$$
(30.7)

where N_r is the rotor speed expressed in revolutions per minute.

Illustrative Problem 30.5 Calculate the speed of (*a*) a two-pole, (*b*) a four-pole, and (*c*) a six-pole induction motor connected to a 60-Hz source of current, assuming a 2.8 percent slippage in each case.

Solution

(*a*) Two-pole motor

$$N_r = \frac{120 \times 60}{2}(1 - 0.028)$$

$$= 3500 \text{ rev/min} \qquad \textit{ans.}$$

(*b*) Four-pole motor

$$N_r = \frac{120 \times 60}{4}(1 - 0.028)$$

$$= 1750 \text{ rev/min} \qquad \textit{ans.}$$

(*c*) Six-pole motor

$$N_r = \frac{120 \times 60}{6}(1 - 0.028)$$

$$= 1170 \text{ rev/min} \qquad \textit{ans.}$$

The most common speeds for large commercial motors are 1750 and 1150 rev/min.

30.16 Synchronous Motors

In the induction motor the current necessary to produce torque is induced in the rotor only when the rotor rotates at a speed slower than the rotating magnetic field. The speed of most electric motors changes as the load changes; the greater the load the less the speed, and vice versa. But if we made the rotor an electromagnet by a means other than the induced current, there would be no need for the rotor to slip behind the rotating field. One way this could be done would be to use a dc source to magnetize the rotor. Then, if by some external means the rotor is brought up to the speed of the rotating field, its poles would lock in step with the rotating field. The rotor would then be *synchronized* with the rotating field. Such a motor is called a *synchronous motor*. It is the only truly constant-speed ac motor. Let us consider two alternators connected in parallel and supplying current with the same frequency and voltage and in phase. If the source of mechanical energy (the prime mover) of one of the generators is removed, that generator will continue to turn, but as a motor, and will draw its

electric energy from the second alternator. The speed of the motor will be exactly the same as when operating as a generator but will slip back a few degrees from the position it held when operating as a generator. Such a synchronous motor is essentially a generator operating as a motor.

If the synchronous motor has the same number of poles as the alternator from which it is drawing its electric energy, it will rotate with exactly the same speed as the alternator. If the motor has twice as many poles as the generator, it will turn exactly half as fast as the generator; if it has half as many poles, it will turn twice as fast.

Modern synchronous motors do not need external starting motors. Their rotors are constructed as a combination squirrel cage induction motor and a wound-type motor with slip rings. The motor is started as an induction motor. The induction motor brings the rotor speed to about 95 percent of the speed of rotation of the stator field. The direct current is then applied to the wound portion of the rotor through the slip rings. This sets up north and south magnetic poles in the rotor which lock in step with the poles of the rotating field. Synchronization results, with the

Fig. 30.23 A 4,150 horsepower synchronous motor (right foreground) powers a grinding mill in a copper mining project in Arizona. (Siemens-Allis, Inc.)

rotor revolving at the same speed as the rotating field.

Once synchronization is established the squirrel-cage portion of the rotor will not cut any lines of force of the rotating field. Hence, in effect, it is removed from the operation of the motor.

The speed of rotation of a synchronous motor depends upon (1) the frequency of the alternating current supplied to it and (2) the number of poles on the stator. As a formula $N = 120f/p$ (see Eq. 30.4). The number of poles in commercial synchronous motors generally ranges from four to 100 poles.

Synchronous motors are used where constant speed is essential. Some of the common applications of this type of motor are driving extremely large air and refrigerant compressors, large blowers, fans, pulverizers, and dc generators which need a large source of direct current. These motors are usually larger than 100 hp (Fig. 30.23). Small, nonexcited single-phase synchronous motors are used to drive electric clocks, phonograph turntables, and other devices where constant speed is essential.

30.17 Measurement of AC Energy Consumption

Power companies charge the consumer according to the amount of electric energy used. This electric energy is measured by means of a watt-hour meter. *Energy* is the *product of power and time*, and consequently an instrument must be used which takes into account both the power and the time. Electric power is measured in terms of watts, which is the product of amperes and volts. The common commercial *ac watt-hour meter* (Fig. 30.24a) is essentially a small induction motor in which the speed of rotation is a measure of the power being consumed at that instant. The total number of revolutions of the meter rotor in a given time interval is then a measure of the total energy used during that interval. Figure 30.24b illustrates the essential parts of the meter: the *electromagnetic field*, a *rotor*, a *damping system*, and a *register*. The field is set up by a voltage coil and two current coils mounted on an iron core.

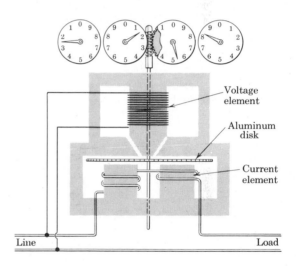

Fig. 30.24 (a) Single-phase watt-hour meter. (b) Essential parts of the meter. (General Electric Co.)

The voltage coil, wound with many turns of fine wire in order to have a very high inductance, is connected directly across the line. The current in this coil is proportional to the line voltage V and lags behind the voltage by nearly 90°. In order to make the voltage-coil flux lag exactly 90° behind the line voltage, a small "lagging coil" is used.

The current coils consist of only a few turns of heavy wire and are connected in series with the line. The flux produced by the current coils is proportional to, and is in phase with, the line current I.

The rotating element (rotor) of the meter is an aluminum disk mounted on a vertical shaft. The shaft rotates on jeweled bearings to reduce friction to a minimum. Since the fluxes produced by the voltage and current coils are out of phase, *eddy currents* will be induced in the disk. These eddy currents will set up their own field, which will react with that of the stator to produce a motor action. The speed of the rotating element is proportional to the product of V and I. The motor would run too fast with a given load or, because of its inertia, would continue to run for some time when the load is removed, unless some damping or braking device were provided. This is done by placing permanent magnets above and below the aluminum disk so that their fluxes are cut by the disk. This arrangement develops eddy currents in the disk which are proportional to the speed and provides a braking torque proportional to that speed.

SUMMARY

A *dynamo* is a machine that will convert mechanical energy to electric energy or vice versa. Dynamos are classified as *generators or motors*.

The *generator* transforms mechanical energy to electric energy.

The *motor* transforms electric energy to mechanical energy.

The fundamental principle of a generator involves magnetic flux being cut by conductors or conductors cut by magnetic flux. *Relative motion* between the flux and the conductors is essential to the generation of emf.

The rotating part of a dynamo is called the *rotor* (or *armature*). The stationary part is the *stator*.

The current in the armature of a generator is always an alternating one. A generator with *slip rings* delivers an alternating current, while the generator with a *commutator* delivers a pulsating direct current to an external circuit.

When a generator operates without load, the average emf developed is proportional to the flux per pole ϕ and the speed n of rotation.

$$\mathscr{E} = K\phi n$$

Every motor when running is also at the same time acting as a generator. The emf of this *generator action* (termed *counter emf*) is in such a direction as to oppose the flow of the current which drives the motor (Lenz's law).

In dc *generators* the field magnets are stationary and the armature revolves. In most *ac generators*, the armature is stationary and the field magnets revolve.

Alternating-current generators (called *alternators*) must have their field poles excited by an auxiliary dc source.

Alternators may be classed as *single-phase* and *polyphase*. Most large commercial alternators produce *three-phase* current. Alternators of this type have three separate sets of windings symmetrically placed on the armature so that alternating emfs of the same frequency, differing in phase by 120 electrical degrees, are developed.

Some of the more common type ac motors include the *universal* (series), *induction*, and *synchronous* types. Only the synchronous motor will operate at a constant speed, regardless of load. Induction motors may slow down as much as 5 percent under their rated speed if loads are heavy.

QUESTIONS AND EXERCISES

1. What is a dynamo?

2. Name the essential parts of a dc generator. What are the essential parts of an alternator?

3. Upon what factors does the emf of a generator depend? Which factors are constant and which are variable?

4. Why does a generator require more power to turn it when it is delivering current to a load than when there is no load in the external circuit?

5. Why does a motor take more current as it starts from rest than when it is turning at its rated speed?

6. List the advantages and disadvantages of dc series and dc shunt motors. List instances where each type could be used.

7. Sketch the windings for a dc compound motor. What are its advantages?

8. What determines the speed of an induction motor?

9. What is a universal motor? How is it constructed?

10. Name three *prime movers* used to drive commercial alternators.

11. What is meant by the *generator action* of a motor?

12. What is meant by *synchronous speed* and *slip*?

13. What conditions must be fulfilled before a synchronous motor will function properly?

PROBLEMS

1. A 25-hp motor draws 70 A at 440 V. What is its efficiency?

2. What is the frequency of an alternator that has 24 poles if it is rotating at 250 rev/min?

3. How fast must the rotor of a two-pole turbo-alternator rotate to produce current with 60-cycle frequency?

4. How many poles does a 60-Hz generator have if it rotates at 300 rev/min?

5. What should be the speed (revolutions per minute) of an ac generator with four pairs of poles to develop 60 Hz?

6. What is the speed of a synchronous motor operating from a 60-Hz dc supply if it has 24 poles?

7. A six-pole generator has 300 armature conductors arranged in four parallel paths. Compute the average emf generated if the armature turns at the rate of 1750 rev/min in a field flux of 4.2×10^{-2} Wb.

8. The armature of a motor has a resistance of 1.5 Ω and operates on a 120-V line. In starting the motor, a rheostat is connected in series with the armature. What resistance should the rheostat add to the circuit to limit the starting current to not more than 6 A?

9. Compute the slip of an eight-pole induction motor operating from a 60-cycle ac supply if the motor turns at 875 rev/min.

10. What is the synchronous speed of a six-pole 60-cycle induction motor? If the motor has a slip of 5 percent, what is the speed of the motor in revolutions per minute?

31 Production and Distribution of Electric Energy

It has been emphasized repeatedly that physics is the science of matter and energy. Energy is available in many forms, and we have studied in detail mechanical energy, heat energy, and light energy. Solar energy is becoming a significant factor in many parts of the nation, and nuclear energy is already providing a tenth of our stationary power plant production. For the present, electricity is the energy source preferred in industry, commerce, and in most homes. There is a definite trend to "all-electric" homes, commercial buildings, and industrial plants. In the not-too-distant future it may be necessary to turn to electric energy to power surface transportation. Electric railways may become the principal means of heavy transport within a few decades. Streetcars ("urban people movers") may make a comeback. Electric automobiles and buses, now almost curiosities in an era dominated by gasoline and diesel engines, could become the principal mode of personal transportation by the turn of the century. If these developments come about, the demand for electric energy, produced in a vast network of central-station generating plants, will increase steadily in the years ahead.

Fuel oil and natural gas now generate most of the electric energy in the United States, but supplies of these fuels are dwindling. Coal, nuclear energy, and eventually solar energy will be called upon to produce much of the electricity for the future.

The purpose of this chapter is to describe the commercial production of electric energy and the methods by which it is transmitted and distributed to consumers.

Electricity is a *commodity*. It is produced, transported, and sold at wholesale and retail to both jobbers and ultimate consumers. However, except in the case of batteries and fuel cells (which have been treated in Chap. 26), electricity cannot be stored awaiting sale, nor can it be produced during off-season for consumption later when load demands are higher. When a consumer closes a switch, he wants electricity immediately, not later in the day or the following week. Electricity must ordinarily be used as it is produced, and there must be enough production to satisfy instantaneous demand.

31.1 Review and Development of Terminology

Electric current has been previously defined as *the flow of electric charge*. We have defined unit current as the *ampere*, which is a *rate of one coulomb of charge per second*. The *coulomb* itself has been evaluated as the equivalent of about 6.24 billion billion (6.24×10^{18}) electron charges (e). An ampere, then, is a flow of 6.24 billion billion electron charges past a given point in a wire per second.

Staggering as these figures are, the user of electricity is not interested in purchasing electrons or electric charges. What he wants to buy is *energy*, the *ability to do work*. Elsewhere we have defined the unit of electric power—the *watt*—and the larger, practical units, the *kilowatt* and the *megawatt*. The student should also recall that 1 hp is equivalent to 746 W, or that 1 kW equals 1.34 hp. Since power = work/time, work (i.e., energy) = power × time. Electric energy, therefore, is expressed in *kilowatts × hours*, or *kilowatthours* (kW-hr), and it is *kilowatthours of energy consumed* for which the customer pays.

The nature of certain ac circuits and loads is such that the *actual power that can be consumed* from the circuit is not as great as the power *supplied to and available in* the circuit. Consequently, it is common practice to measure *electric power consumed* in *kilowatts* (kW) and *electric power produced and distributed* in *kilovolt-amperes* (kVA). The reason for this discrepancy is the fact that current and voltage are frequently not in phase with each other on ac lines (see Sec. 29.12). It will be recalled that both *inductance* and *capacitance* affect the phase relations in ac circuits. The amount by which current and voltage are out of phase with each other is expressed in electrical degrees. The angle of *lead or lag* is called the *phase angle* ϕ (see Sec. 29.12), and the power in watts that can be drawn from an ac circuit is given by the expression

$$P = VI \cos \phi \qquad (29.25)$$

Watts = volts × amperes × cos (phase angle)

The term *power factor* is applied to the expression $\cos \phi$, and both V and I must be *effective values*. It is apparent, then, that

$$\text{Power factor} = \frac{\text{watts}}{\text{volt-amperes}} = \frac{\text{kW}}{\text{kVA}} \qquad (31.1)$$

The *production* of ac power is rated in kilovolt-amperes, and the capacity of generators, transformers, transmission lines, switches, and associated equipment is also so rated. *At the point of use* of the power the power rating is in kilowatts or megawatts. Alternating current voltmeters and ammeters read *effective values* of voltage and current, and wattmeters read *true power*.

With this brief review of fundamental relations and terminology, we now turn our attention to the production of ac electric power.

PRODUCTION OF ELECTRIC POWER

Generating electricity involves the conversion of mechanical energy to electric energy through the medium of electromagnetism. Commercial production of electric power is currently carried on in four different types of plants—*steam plants, internal-combustion-engine plants, hydroelectric plants*, and *nuclear-reactor plants*. The first three types will be briefly described here, and nuclear plants will be discussed in Chap. 34.

31.2 Steam Plants

At one time reciprocating steam engines were widely used for generating electricity, but they have been almost entirely replaced by steam-turbine installations in the United States. The present discussion will be limited to steam-turbine plants. Steam turbines (see Chap. 16) are now in use for generating electricity in capacities from 500 to more than 200,000 hp.

High steam pressures are the rule with turbine plants, and a high degree of superheat is also used. Pressures of 2000 lb/in² and temperatures of 1050°F are not at all uncommon. All such turbines discharge into condensers which are designed to create a high degree of vacuum, usually 1 to 2 mm Hg absolute.

The turbines themselves have been previously discussed at some length in Chap. 16. Small sizes (up to 50,000-kVA capacity) normally operate at 3600 rev/min, while the larger units usually run at 1800 rev/min, though with modern improvements in metallurgy some 100,000-kVA units now operate at 3600 rev/min.

There is, of course, a maze of complex equipment for the control and automatic operation of a steam-generating plant. We cannot describe in an elementary book of this kind the full details of the interconnected series of operations required

Fig. 31.1 Control room of a steam-electric generating plant. Combustion and turbine-generator operation are monitored and controlled by the switches, relays, gauges, and other equipment shown. Computers are also used to control certain operations. The control center shown governs a large plant which produces 3,580,000 lb of steam per hour at 1000°F and 2400 psi (gauge) pressure. Electric power produced is 530 MW. (Detroit Edison Company.)

at a large plant. The photo of Fig. 31.1 and the diagram of Fig. 31.2 will, however, give some idea of the scope of operations in such a plant.

31.3 Internal-Combustion-Engine Plants

The use of internal-combustion engines for generating electric power is generally limited to two types of plants: (1) very small plants for intermittent operation, as for mountain resorts, or for portable industrial plants and arc-welding machines; and (2) larger plants where the peculiar local circumstances, e.g., abundance of natural gas or fuel oil combined with a scarcity of water suitable for boilers, indicate the relative economy of diesel or natural-gas engines over a steam-turbine installation.

Small generators may be V-belt driven, but larger installations are either direct drive (750 to 1800 rev/min) or gear-driven to obtain the proper generator speed. Several engines may be connected to a single generator through a torque-converter drive. For larger installations 16-cylinder diesel engines are often used as the prime mover. Hot-gas turbines are also in common use (see Sec. 16.20).

31.4 Hydroelectric Plants

Where a steady flow of water is available reasonably close to the consumer load, and if reservoir development costs are not excessive, hydroelectric plants are an economical method of generating electricity. Many more hydropower sites will have to be developed in the future, as fossil fuels get scarcer.

Classification of Hydroelectric Plants Hydroelectric plants are of three types:

1. *Low-head plants*, heads up to about 30 ft of water

2. *Medium-head plants*, 30 to 300 ft

3. *High-head plants*, 300 to 2000 ft or more.

Classification of Water Turbines Water turbines are of two general types: *impulse turbines* and *reaction turbines*.

Impulse turbines develop power from the impact of a stream of high-velocity water against the blades or buckets of the turbine. For greatest efficiency, the water emerging from the *outlet* of the turbine should have a velocity as near zero as possible, since this means that all the kinetic

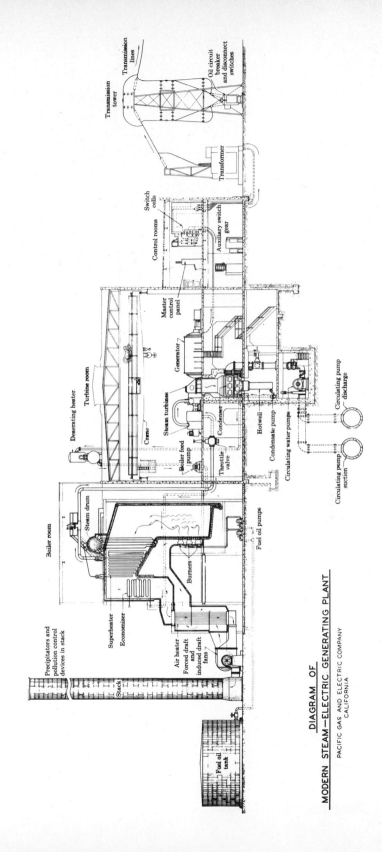

DIAGRAM OF
MODERN STEAM—ELECTRIC GENERATING PLANT

PACIFIC GAS AND ELECTRIC COMPANY
CALIFORNIA

Fig. 31.2 Diagram of the physical layout of a steam-powered electric generating plant, using fuel oil as the energy source. (Pacific Gas and Electric Co.)

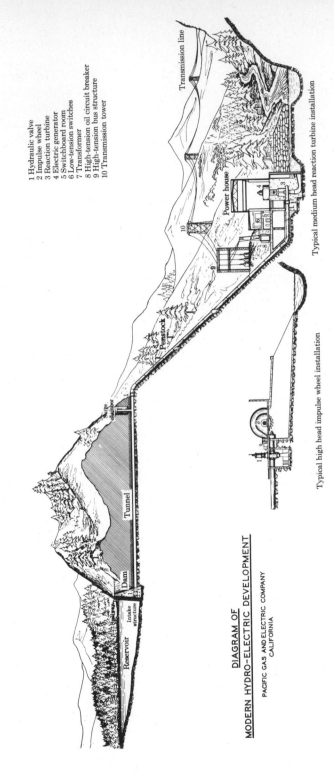

1 Hydraulic valve
2 Impulse wheel
3 Reaction turbine
4 Electric generator
5 Switchboard room
6 Low-tension switches
7 Transformer
8 High-tension oil circuit breaker
9 High-tension bus structure
10 Transmission tower

Transmission line

Power house

Penstock

Surge chamber

Tunnel

Dam

Intake structure

Reservoir

DIAGRAM OF
MODERN HYDRO-ELECTRIC DEVELOPMENT
PACIFIC GAS AND ELECTRIC COMPANY
CALIFORNIA

Typical medium head reaction turbine installation

Typical high head impulse wheel installation

Fig. 31.3 Diagram of a hydroelectric development in a mountain setting. The insert at the bottom left shows the modification of design that would be required for a high-head impulse turbine installation. (Pacific Gas and Electric Co.)

energy of the water has been absorbed by the turbine. Impulse turbines are best adapted to high pressure heads with relatively small quantities of water. Their efficiencies may run as high as 80 percent. They find frequent application on the mountain streams of the West, where extremely high heads are available and water flow, especially in the late summer and fall, may be quite limited.

Reaction turbines are commonly employed with medium- and low-head plants where there is an ample water flow the year around. They are also used for high-head installations where there is a large volume of water flow. The rivers of the Northwest and of the Mississippi Valley have many reaction-turbine power-plant installations. Reaction turbines develop their power in accordance with the third law of motion—action and reaction. The turbine blades are curved in such a way that as the water leaves the turbine, its momentum gives a reactive kick to the turbine rotor. Large reaction turbines may have

Fig. 31.4 Two 75,000-kVA, 60-cycle three-phase generators installed in the Haas Powerhouse on the Kings River in California. The powerhouse is 500 ft underground in solid rock. The high-head turbines operate under a pressure head of 2444 ft of water. (Pacific Gas and Electric Co.)

efficiencies ranging from 85 to 92 percent. The turbine and generator are usually mounted vertically on the same shaft, with the generator above the turbine. The weight of the rotating parts and the force of the downward thrust of the water are carried by thrust bearings in the generator. Figure 31.3 shows, in diagram form, a hydroelectric development in a mountain setting. Figure 31.4 illustrates the arrangement of turbo-generators in a powerhouse.

31.5 Power from Water

If the water pressure head and the quantity of water flow are known, it is easy to compute the theoretical power available at a hydroelectric plant. Let h be the head in feet and Q the rate of water flow in cubic feet per second. Since the weight density of water $D = 62.4$ lb/ft^3, the horsepower available may be obtained by starting with Eq. (4.13), $P = Fv$, from which we obtain

$$P_{hp} = \frac{F(\text{lb}) \times v(\text{ft/sec})}{550 \dfrac{\text{ft-lb}}{\text{sec-hp}}}$$

But force F equals pressure times area, and pressure equals hD. Making these substitutions, we have

$$P_{hp} = \frac{hDAv}{550}$$

The quantity Av (area × velocity) equals the water flow Q in cubic feet per second. Therefore, in terms of the units above,

$$P_{hp} = \frac{62.4 \text{ lb/ft}^3 \times Q \text{ ft}^3/\text{sec} \times h \text{ ft}}{(550 \text{ ft-lb/sec})/\text{hp}} \quad (31.2)$$

Since 1 hp is equivalent to 0.746 kW, the electric power theoretically available at any site will be

$$P_{kw} = \frac{46.55Qh}{550} \quad (31.3)$$

To these theoretical values the overall water turbine-generator efficiency must be applied to obtain the actual power output. The value of this overall efficiency is usually in the range 65 to 80 percent.

DESIGN AND OPERATION OF ALTERNATORS

The principles and underlying theory of operation of ac generators have been described in Chap. 30. That discussion should be reviewed at this time. In this section some fundamentals of design and construction of large alternators will be discussed, such as speeds of rotation, number of poles, types of armature and field windings, and problems of synchronizing two or more alternators.

31.6 Synchronous AC Generators

Generators for the production of commercial electric power are, for the most part, *three-phase alternators*, because (1) alternating current is desired in most commercial applications, and (2) three-phase power is more economical to produce than two-phase or single-phase power. If two-phase or single-phase power is desired by the consumer, it is easily obtained by merely tapping the appropriate wires of the three-phase distribution system. More capacity can be obtained from a given alternator when it is wound for three phases.

Three-phase generators are usually Y (*wye*) or *star-connected*, as shown in the diagram of Fig. 31.5. With this type of connection the *current* in any line L_1, L_2, L_3 is the same as that in the corresponding phase winding 1, 2, or 3; but the *voltage* V_L between any two lines is $\sqrt{3}$ times the voltage V_C in the phase windings. Thus a higher line voltage is obtained than that actually induced in the armature windings. Mathematically,

$$V_L = \sqrt{3}V_C \tag{31.4}$$

This type of connection makes a common point N available to all three phase windings. This point, called the *neutral*, is ordinarily grounded as shown in the diagram.

31.7 Standard Generating Voltages and Frequencies

The voltage at which a plant is designed to operate depends on the nature of the connected load and the length of the distribution system. A small

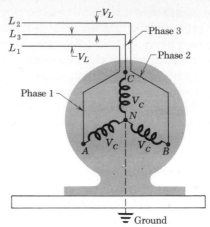

Fig. 31.5 The wiring diagram of a wye-connected three-phase alternator. The armature windings of such alternators are on the stator.

domestic plant designed for local lighting only might generate at 115 to 120 V. For industrial plants a local generating system might operate at 440 to 480 V and distribute on a three- or four-wire system. Motors, lights, and heating equipment are all well served by such a system. Small city plants where the distribution lines are only a few miles long may generate at 2400 V. In large plants, the standard generating voltage is 13,200 V, and if the distribution system is short, the same voltage may be used for transmission. If, as is usually the case, longer transmission lines are required to reach the consumer load, the voltage is stepped up before sending the power out on the line (see Fig. 31.11).

Frequency It will be recalled that the frequency of an ac generator is a function of the number of poles on the armature and the speed of rotation of the rotor. In general, if f is the frequency, p the total number of poles, and N the number of revolutions per minute, then

Frequency, cycles/sec (Hz),

$$f = \frac{pN}{120} \tag{30.4}$$

Frequency is one of the rigid standards of operation of a power plant for two reasons: (1) electric clocks of thousands of consumers must keep accurate time; and (2) when generators are connected in parallel to the same distribution system, their outputs must be *in phase* and *remain at the same frequency*. Automatic controls govern the revolutions per minute of all three types of prime movers to give constant generator speed.

Engine-driven alternators may turn at 300 to 500 rev/min. Hydroplant alternators vary from 50 to 750 rev/min, while steam-turbine-driven units range from 1500 to 3600 rev/min, with four-pole 1800 rev/min machines being most common for 60-cycle operation. We have already seen that 60 cycles is the standard frequency in the United States. In Europe 50-cycle frequency is standard in many countries.

Although the approved unit for frequency is now the *hertz* (Hz), many engineers and technicians in the electric power industry still prefer the older designation, *cycles per second*, or just *cycles*, as in "60-cycle current." We will use both designations interchangeably in this chapter.

Details of alternator construction were given in Chap. 30. Figures 30.14, 30.15, and 30.16 show some of these construction features for engine- and steam-turbine-driven alternators. Figure 31.6 shows the rotor for the world's largest hydroelectric generator.

31.8   Connecting Alternators to the Line

Many different methods are used to connect ac generators to the line. On a small city system where no voltage step-up is required the alternators are connected directly to the line through the necessary control and protection equipment. Larger systems, where voltage step-up is required for long-distance transmission and where several alternators may be required to serve the load, must have a very flexible connection system. One or several alternators may be operating, according to load demand, and provision for *synchronizing* them, i.e., keeping them in phase with each other and paralleling them onto the line must be made. Power may be sent to several substations, some near, some far away. This situation may require feeding a portion of the power being generated to local consumers at 13,200 V (the generator voltage); another portion to transformers for step-up to, say, 44,000 V for transmission to a consumer area 30 to 50 mi away; and possibly another portion to transformers for step-up to 300,000 V or more for a long-distance transmission line (see Fig. 31.11).

Fig. 31.6 Rotor for the world's largest hydroalternator being lowered by giant crane into position in the stator. This hydroelectric generator is rated at 600,000 kW. Three identical units are being installed at Grand Coulee Dam on the Columbia River in Washington. (Westinghouse Electric Corp.)

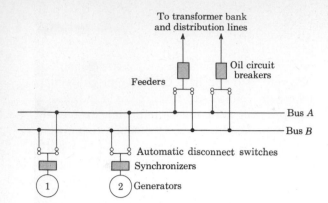

Fig. 31.7 One possible system for connecting a small power plant to its distribution lines.

The alternators are connected in parallel to heavy copper bars called *busses*. Provision for synchronizing each alternator with those already in operation is made at the master switchboard. The main busses run the entire length of the switchboard behind the panel, but they are separated from the switchboard, to lessen the danger of injury to personnel. Each generator has a main switch to connect it to the bus. This switch opens and closes in oil, to prevent arcing.

Wires or cables leading out from the busses to the powerhouse distributing station are called *feeders*. Each feeder has its associated meters, switchboard, and overload circuit breaker. The feeders supply the step-up transformers for the transmission lines. *Lightning arresters* are installed just beyond the transformers. Figure 31.7 gives a simplified schematic diagram of a system for two alternators. Figure 31.8 shows equipment installed at a large generating station.

TRANSMISSION OF ELECTRIC POWER

Most of the electric power produced in the United States today is 60-Hz, three-phase, alternating current. Our discussion of transmission and distribution systems will therefore concentrate on this form of electric power.

The reason why alternating current is preferred is that its voltage may be stepped up or down, i.e., *transformed*, with very little loss at any stage in the distribution system. Three-phase power is preferred over two-phase or single-phase because it is far more economical of copper, both in the alternator construction and in the transmission and distribution system. A three-phase three-wire system may require only 75 percent as much copper to transmit a given amount of power as a single-phase or two-phase system; and a three-phase, four-wire system only one-third as much copper.

31.9 Transmission Lines and Equipment

Not all the power sent out on the transmission system reaches its destination. The principal losses are classified as follows:

1. *Line losses*, due to actual ohmic resistance of the lines themselves. Line losses are a major factor in transmission losses and may amount to 10 percent or more of the power put on the line at the generating station.

2. *Leakage losses*—to the air, to ground, to trees and buildings, and from wire to wire. This factor is small at low voltages, but at high voltages (200,000 V and up) it becomes a serious factor. Weather conditions appreciably affect this loss.

3. *Transformer losses*. With well-designed transformers this loss should not exceed 2 to 4 percent.

The first two of these losses will be discussed in detail in this section. Transformer losses take the form of heat produced in the windings from eddy currents and hysteresis losses.

Line Losses From *Ohm's law*, the voltage drop along a line is given by $V = IR$. Now, electric power $P = VI$, in watts where V and I are *effective values*. If we substitute in the power equation from Ohm's law, we obtain $P = I^2R$, which the student will recognize as the power loss in the line (see Sec. 26.16). Since the power loss varies as the square of the current, it is imperative that the current be kept at a low value. This is, of course, equivalent to saying that the voltage should be at a high level. A numerical example will serve to illustrate the principle.

Suppose a manufacturing plant is to be

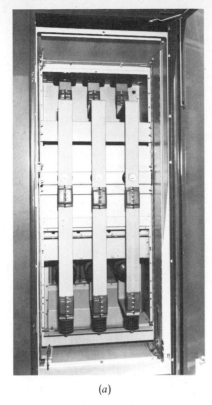

(b)

(c)

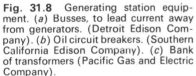

(a)

Fig. 31.8 Generating station equipment. (*a*) Busses, to lead current away from generators. (Detroit Edison Company). (*b*) Oil circuit breakers. (Southern California Edison Company). (*c*) Bank of transformers (Pacific Gas and Electric Company).

supplied electric power from a hydroelectric plant in the mountains nearby. Let the average demand be 500 kW and the voltage desired at the plant 440 V. The line is 20 mi of No. 2 copper, whose resistance is about 0.85 ohm/mi (see Fig. 31.9). For simplicity a two-wire line is shown. A power factor of unity is assumed, and other simplifications have been made to allow for easy computation.

In diagram (*a*) the 500-kW load is shown drawing power from the line through a 4400-440-volt step-down transformer. The line current is therefore

$$I = \frac{P}{V} = \frac{500,000 \text{ W}}{4400 \text{ V}} = 114 \text{ A}$$

The voltage drop over the 20 mi of line is then, by Ohm's law,

$$V = IR = 114 \text{ A} \times 20 \text{ mi} \times 0.85 \text{ ohm/mi}$$
$$= 1930 \text{ V}$$

The step-up transformer at the generating station must supply the power at 6330 V (4400 + 1930 = 6330).

The line power loss is

$$P = I^2R = (114)^2 \times 20 \times 0.85$$
$$= 220,000 \text{ W} = 220 \text{ kW}$$

The generator must therefore produce 720 kW to supply a 500-kW load. This represents a very inefficient operation (30.5 percent line loss), one

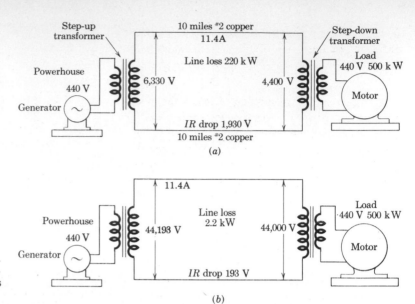

Fig. 31.9 High-voltage transmission minimizes line losses. The diagram is schematic only and assumes unity power factor.

which can be materially improved by using a higher voltage for transmission, as will be shown.

In diagram (b) the same generator, line, and load are shown. In this case, however, the load draws power from the line through a 44,000–440-V step-down transformer. The line current in this case will be

$$I = \frac{500,000}{44,000} = 11.4 \text{ A}$$

The voltage drop over the 20-mi line is

$$V = 11.4 \times 20 \times 0.85 = 193 \text{ V}$$

The generating-station transformer must supply the power at 44,193 V to compensate for the line-voltage drop. The line power loss in this case is

$$P = (11.4)^2 \times 20 \times 0.85$$

$$= 2220 \text{ W} = 2.22 \text{ kW} \qquad \text{only 0.4 percent}$$

This example illustrates the economic necessity for high-voltage transmission.

Leakage Losses The higher the transmission voltage, the greater the tendency for electrons to leak off the wires. Alternating current tends to flow in the outer layers of the wire (this is called *skin effect*), and this crowding of the surface of the wire, together with the very large potential difference with respect to ground, ionizes the surrounding air so that a conducting path to ground, trees, nearby buildings, or other wires is encouraged. This type of leakage is accentuated during stormy weather because moist air is a better conductor than dry air and also because the air may already be ionized due to atmospheric electric disturbances.

Insulators several feet long support high-voltage wires, holding them away from the towers and away from the other wires of the system, to minimize leakage losses within the system itself.

With extremely high voltages (300,000 and up), atmospheric conditions sometimes set up an ionization of the surrounding air which permits an extremely large line leakage known as *corona loss*. The discharge from the wire, in such cases, can actually be seen as a bluish-purple haze surrounding the wire. It can also be heard as a persistent 60-cycle tone, accompanied by spitting and crackling noises. It is this loss which sets a practical upper limit on transmission voltages.

This practical upper limit is generally considered to be in the region of 500,000 V, although a few lines have been designed for and are operating at one million volts.

Direct-current Transmission Lines With the availability of high-capacity electronic rectifiers and oscillators, and rotary converters, it is feasible to first step up the ac voltage to 500,000 V or more with transformers, and then rectify it to direct current. It is transmitted as direct current at the high voltage (one power company now has a 750,000 V dc line), and then converted back to alternating current at substations where local distribution begins. Direct-current (dc) transmission materially reduces line losses due to corona problems, and also those resulting from inductive and capacitive reactance. Appreciable operating economies have been reported from dc transmission systems, especially for large loads over distances of from 40 to 400 mi. Although still not very common, dc transmission will probably increase in the future.

31.10 Transformers for Power Transmission

The basic theory of transformer construction and operation has been explained in a previous chapter.

For small stations transmitting at medium and low voltages to several load areas, the best practice is to provide a transformer bank for each generator at the power station. For large plants, especially hydroelectric plants at remote locations, where all the power generated is to be sent for a long distance at high voltage, the entire bank of transformers is connected to the main busses through automatic-overload circuit breakers.

Transformer Windings Power-station transformers are usually three-phase and may be of either the core type or the shell type. In the core type the copper windings encircle the iron core, and in the shell type the iron circuit is placed around the copper windings. Provision must be made for removing the heat generated in the transformer.

31.11 Lightning Arresters A high-voltage transmission line is very likely to transmit to the generating station, or to substations, extremely high-voltage high-frequency electric discharges if the line is struck by lightning. Even if the line is not actually struck, passing clouds may induce large charges in the wires, and when the lightning bolt discharges the cloud, high-frequency oscillations are set up in the transmission line and a heavy surge travels to the stations along the line. These surges may start arcing of generators and similar equipment if the surges are not channeled off to ground. Recent disastrous "blackouts" (power outages) in New York and Atlantic Coast states were initiated by lightning-induced surges, with consequent arcing and equipment breakdown. The devices which provide lightning protection are called *lightning arresters* (see Fig. 31.10).

Fig. 31.10 Thyrite (patented) lightning arresters mounted on a large three-phase transformer bank. (General Electric Co.)

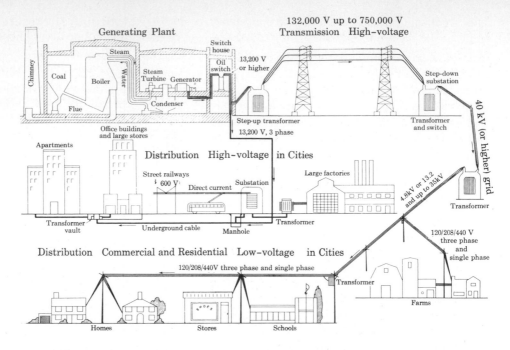

Fig. 31.11 Schematic arrangement of a power production and distribution system. Coal is the energy source for this system. (Bureau of Labor Statistics, U.S. Department of Labor.)

31.12 Overall Arrangement of Power-station Equipment

One possible arrangement of a coal-fired power station, showing the disposition of the units and equipment discussed in the preceding paragraphs, is shown in Fig. 31.11. A three-phase three-wire system is shown, and a medium-length (20 to 60 mi) transmission line is assumed. The diagram also shows a typical distribution system for industrial, commercial, and residential consumers.

SUBSTATIONS

The function of a substation in an electric-energy distribution system is to modify certain properties of the transmission current to conform to the needs of the consumer load which the substation serves. There are substations designed to accomplish a variety of purposes, and we shall discuss only a few of the more common types. A single station may, of course, accomplish two or more of the functions discussed.

31.13 Substations for Reducing Voltage

This type of substation is located near the city or load area to be served and steps down the voltage from the high transmission voltage to an intermediate voltage (say 40,000, or 13,200 or 4800 V) for local distribution. Such substations are sometimes known as step-down substations, or *transformer substations*, since transformers are the key equipment involved.

Feeders lead out from the substation, and they branch into *subfeeders*, *mains*, *submains*, and finally into *services*, which are the final connections to the premises of the consumer.

In general, the lines from a generating plant to a step-down substation are called the *transmission system*, and the network from the substation to the ultimate consumers is called the *distribution system* (see Fig. 31.12).

31.14 Frequency-changer Substations

In cases where a different frequency is required by an industrial consumer, a substation may be built primarily for frequency changing. The principal components of such a substation are synchronous machines in pairs, one of which runs as a motor on the applied frequency and the other as a generator delivering the desired frequency. If the frequency-changer substation operates off a high-voltage transmission line, the high voltage must first be stepped down to a value suitable for synchronous machine operation—usually 13,200 V or less.

It is common practice to tie in the facilities of one power company with those of another so that electric power can be shared between two localities as load demands require. If the two systems are of different frequencies, frequency-changer substations are required to tie in the two systems. The United States and Canada have extensive facilities for tying in electric power networks.

31.15 Substations for Supplying Direct Current

Many industrial operations require large quantities of dc power. Substations for rectifying alternating to direct current may use either *rotary converters* or huge *electronic rectifiers*. The rotary converter is merely a synchronous ac motor, direct-connected to a dc generator. Electronic equipment is also used for rectification. *Ignitrons* and *multianode tanks* (see Chap. 32) are used for this purpose. Street-railway systems, trolley-bus systems, metal refining companies, metal rolling mills, and paper and textile mills are heavy users of dc power.

31.16 Distribution Substations

Substations whose principal function is to act as the distributing center for electric power in a given area are called *distribution substations*. They may incorporate one or all of the above functions. In addition, they have a variety of complex associated equipment for switching, phasing, protection, voltage regulation, power-factor control, and emergency service in case of casualty.

DISTRIBUTION SYSTEMS

As previously defined, the *distribution system* is the network which has to do with local distribution of electric power. Distribution systems are of many types, but we shall discuss here, in an elementary way, a simple one which has the following specifications:

Current: alternating
Connection system: multiple, feeder, and main
Phases: three-phase and single-phase
Voltages: 40 kV, 13.2 kV, 120/240 volts
Frequency: 60 Hz (cycles/sec)

31.17 Feeder-and-Main Multiple Distribution System

In this type of distribution system, four-wire three-phase *feeders* from the central substation supply other smaller transformer stations at relatively high voltages, perhaps 40,000 V. From these stations, *primary mains* carry current to other central points at 13,200 V. Local distribution to commercial and residential consumers at 240 and 120 V is effected from these central points. Industrial consumers are supplied with 440-V three-phase service from individual transformers off 13.2-kV lines. Ordinarily 440 V is not used as a distribution voltage. Figure 31.12 shows in diagram form how such a multiple distribution system might be organized. For 13.2- and 4.8kV feeders and mains the trend is to underground installation. In many states the law now requires that all residential, commercial, and light industrial distribution networks be underground in specified urban areas.

Load Factors In planning the distribution network a great deal of attention must be given to balancing out the load served from a substation. The greatest diversification possible (in type of consumer load) is desired, in order to keep the power factor as high as possible. If the load is largely of an *inductive* nature (induction heating, induction motors, etc.), *the current will lag the voltage*, while loads with elements of *capacitance* predominant will cause the *current to lead the*

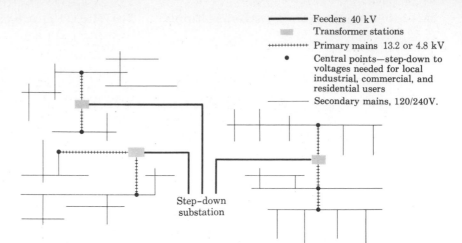

- ——————— Feeders 40 kV
- ▨ Transformer stations
- ++++++++++ Primary mains 13.2 or 4.8 kV
- • Central points—step-down to voltages needed for local industrial, commercial, and residential users
- ——————— Secondary mains, 120/240V.

Step-down substation

Fig. 31.12 A possible arrangement of a feeder-and-main multiple distribution system.

voltage. In either case, the power factor (cos ϕ, where ϕ is the angle of lead or lag, the so-called *phase angle*) will be reduced, and the distribution of the power becomes an uneconomical operation (see Sec. 29.12). Residential-type loads tend to be largely resistance loads, and these circuits tend to have a comparatively high power factor. Industrial-type loads, made up as they usually are of a predominant motor load, are highly inductive in nature, and it requires continuous attention and planning by power-company engineers to keep power factors in industrial areas as high as possible. In general, a residential network may have a power factor as high as 90 to 95 percent; industrial loads, one of 60 to 75 percent; and a combined residential and industrial network, one of 75 to 90 percent. The increasing use of electric motors in homes (air conditioners, workshops, etc.) is tending to cause lower power factors on these lines also. Banks of capacitors may be installed to correct this condition (see Fig. 31.13). Capacitor banks are ordinarily installed with time switches which cut them out of the line after about 11 p.m. Otherwise, as the inductive load falls, current may lead voltage, with a resultant low power factor from too much capacitive reactance.

Illustrative Problem 31.1 An industrial-plant service line has an impressed power of 100 kVA on a 60-cycle line at 440 V. An installed motor

Fig. 31.13 Bank of capacitors installed on a rural electrification line to balance the inductive load of electric motors. The capacitors will bring current and voltage more nearly in phase and this will result in a better power factor. (Westinghouse Electric Corp.)

offers a total ohmic resistance of 350 ohms and an inductance of 1 henry. Find (*a*) the inductive reactance X_L of the circuit, (*b*) the impedance Z, (*c*) the current to the motor, (*d*) the power factor, (*e*) the power (in kilowatts) being used by the motor, and (*f*) the size capacitor (microfarads) which would have to be installed to increase the power factor to 95 percent. (Review Secs. 29.4 through 29.12).

Solution

(a) $X_L = 2\pi f L = 6.28 \times 60 \times 1 = 377\ \Omega$ *ans.*

(b) $Z = \sqrt{R^2 + X_L^2} = \sqrt{(350)^2 + (377)^2} = 514\ \Omega$ *ans.*

(c) $\qquad I = \dfrac{V}{Z} = \dfrac{440\ \text{V}}{514\ \Omega} = 0.856\ \text{A}$ *ans.*

(d) Power factor $= \dfrac{\text{kW}}{\text{kVA}} = \cos\phi = \dfrac{R}{Z}$

$\qquad = \dfrac{350\ \Omega}{514\ \Omega} = 0.681$ or 68 percent *ans.*

(e) Power used $= 100\ \text{kVA} \times 0.68$

$\qquad\qquad = 68\ \text{kW}$ *ans.*

(f) For the power factor to be 95 percent, the phase angle must be

$$\phi = \cos^{-1} 0.95 = 18°$$

Then $\quad \dfrac{X}{R} = \tan 18°$, where X is the

net reactance

or $\qquad X = 350 \tan 18°$

$\qquad\qquad = 113.7\ \Omega$

Now $\quad X = X_L - X_C = 113.7\ \Omega$

or $\qquad X_C = X_L - 113.7\ \Omega$

$\qquad\qquad = 377\ \Omega - 113.7\ \Omega$

$\qquad\qquad = 263.3\ \Omega$, the *capacitive reactance* required in the circuit

But

$$X_C = \frac{1}{2\pi f C} \qquad \text{and} \qquad C = \frac{1}{2\pi f X_C}$$

Substituting gives

$$C = \frac{1}{2\pi \times 60 \times 263.3}$$
$$= 10.1 \times 10^{-6}\ \text{farad (F)} = 10.1\ \mu\text{F} \quad \textit{ans.}$$

ELECTRICITY AND THE FUTURE

The foregoing paragraphs have presented an elementary introduction to the present "state of the art" in the electric power industry. The future use of electric energy will probably be far beyond presently extrapolated trends. For example, most factories, commercial and public buildings, and homes in the United States are now heated by fossil fuels (coal, oil, or natural gas). Shifting the greater part of this energy burden to electricity will be an endeavor of great magnitude, and it may have to be accomplished in less than 30 years, as we shift to coal-, nuclear-, and solar-powered generating stations.

Further, consider the fact that all the automobiles and trucks in the nation, and most of the railways, are powered by petroleum-base fuels *carried by the vehicles themselves*. If it can be assumed that perhaps half of this transportation energy requirement will have to be shifted to electricity by the year 2000, the increase in electric energy output needed presents a challenge difficult for the industry to meet.

Moreover, merely generating the increased gigawatts of power (with new coal-fired and nuclear plants) would be only the first step. Building and electrifying a national network of electric railways and the power transmission networks for urban and interurban passenger buses and freight delivery vehicles would be a far more difficult task than providing the central-station generating plants. And, if these developments appear difficult, what about the replacement of the millions of gasoline-powered personal automobiles? Will we have "two cars in every garage" in the year 2000? How will they be powered? Will new, heavy-duty, long-lived (and lightweight) batteries be developed which will operate all day long and recharge overnight from standard house current? Will there be coin-operated "quick-charge" stations all over town and at frequent intervals on highways between towns? Will major expressways be electrified, so that motorists can "hook on" to the expressway power system, switch off the car's propulsion batteries, and head for Dallas, or Seattle, or San Francisco, or Philadelphia? How will such a system be made safe, dependable, everywhere available, and economically sound for both user and power company?

These are only questions for the present, since the answers belong to the future. Today's students of engineering and technology will be involved all their working lives with these questions. With persistence and skill and "a little bit of luck" they will find the answers, one by one.

SUMMARY

Production of electric power is carried on in four different types of plants: *steam plants, hydroelectric plants, internal-combustion-engine plants,* and *nuclear-reactor plants.*

The *steam turbine* is the prime mover in steam plants. Fossil fuels or nuclear energy may furnish the heat. *Coal-fired plants* will become more numerous as fuel-oil and natural-gas supplies are further depleted.

Diesel engines, natural-gas engines, and *hot-gas turbines* are the prime movers in internal-combustion-engine plants.

Hydroelectric plants utilize *impulse turbines* (for high heads only) or *reaction turbines* (for medium, low, and high heads).

Alternators for the commercial production of electric power are usually three-phase machines, *Y*-connected; and a common practice is to generate at 13,200 V, 60-cycle frequency.

Large alternators are of the *field-rotor, armature-stator* type.

Transformers are used to step up the voltage for long-distance transmission of ac power.

Losses in the transmission of electric power are classified as *line losses, leakage losses,* and *transformer losses.*

Line losses due to ohmic resistance can be minimized by high-transmission voltages, but exceedingly high voltages may result in appreciable leakage losses.

Leakage losses can be reduced by converting to direct current (dc) for long-line transmission.

Substations accomplish one or more of the following functions: *voltage changing, frequency changing, rectification,* and *distribution.*

If economical distribution of electric power is to be attained, the *power factor* must be kept high. This is accomplished by keeping *inductive reactance* and *capacitive reactance* as nearly equal as possible.

It is probable that much of today's motor vehicle traffic (propelled by fuels carried with the vehicle) will, in a few decades, have to be shifted to electric power generated at central locations from coal, nuclear, or solar energy.

QUESTIONS AND EXERCISES

1. Distinguish clearly between the terms *kilowatt* and *kilovolt-ampere.* Under what conditions would the kilowatts used by a customer be equal to the kilovolt-amperes available on the line?

2. From library research find the per capita consumption of *total energy* in the United States (or your own country) for a recent year. What percent of this was electric energy? If electric energy were to replace one-half of the energy now supplied by petroleum fuels and natural gas, by what factor would electric-energy output have to be increased?

3. Recalling that 1 W equals 1 J/sec, show that the kilowatthour is a unit of energy.

4. Discuss the factors which often make steam plants a more economical installation than hydroelectric plants, even though the "fuel" for the latter is "free," once the installation is completed.

5. Contrast the basic principle of operation of the hydro-plant impulse turbine with that of the reaction turbine. What laws of motion are involved? What conditions at the power-plant site would influence the decision as to which of the two types to use?

6. Summarize the reasons for winding most commercial alternators as three-phase *Y*-connected machines.

7. Enumerate the kinds of leakage losses suffered by electric-power transmission lines. Discuss the causes of each loss, and some of the measures taken to minimize the loss.

8. Name and describe the principal function of four types of substations. Visit (get permission from the power company) a nearby substation and get an explanation of the operations there.

9. Why is it of critical importance to balance inductive loads with capacitors in electrical distribution and service systems?

10. Explain why cos ϕ (where ϕ is the *phase angle*) is equal to the ratio kW/kVA.

11. Why is it necessary to synchronize two or more alternators before connecting them to the busses and feeders? Use a sketch in your explanation.

12. What is the basic physical principle on which transformers operate? Why is alternating current necessary?

13. Write a short research paper on the details of construction, size and material, and the methods of support of the conductors (wires and cables) used in long-line power-transmission systems for loads up to 500 MW at 250,000 V and higher.

14. Make arrangements to visit your local electric power generating station, and then write a short paper describing the production and distribution equipment and facilities.

PROBLEMS

1. Your breakfast waffle requires 2.5 min in a waffle iron drawing current at the rate of 7 A. How many electron charges are involved in cooking the waffle?

2. The power company delivers 580 kVA to the line serving an industrial plant, during a period when the power *used* by the plant is metered at 500 kW. Find the power factor and the phase angle on the factory's circuit.

3. A small stream runs through your mountain ranch with an average flow of 15 sec-ft (ft^3/sec). It has a total drop in elevation of more than 100 ft as it runs through your property. What pressure head (feet of water) must you provide for a small turbogenerator (efficiency 70 percent) if it is to generate 35 kW?

4. A hydroelectric-plant site has a steady flow of 1200 ft^3/sec under a head of 1600 ft. Calculate (*a*) the theoretical horsepower available and (*b*) the theoretical kilowatt output of the plant.

5. An alternator has 20 field poles and is to be used with a reaction-type water turbine. At what speed in revolutions per minute must it turn to generate at 60 Hz?

6. A steam turbine-generator turns at 1500 rev/min and has four poles. What is the frequency of the alternating current generated?

7. A steam generator produces one million pounds of saturated steam per hour (pressure 500 lb/in.² abs) from boiler feedwater whose temperature is 80°F. (Steam turbines are actually operated on *superheated*

steam. This example uses saturated steam to avoid undue complexity of the problem.) (*a*) What is the input in Btu/hour to this boiler? (*b*) What horsepower turbine would it operate at 35 percent efficiency? (*c*) What is the fuel-consumption in gallons per hour if burner-boiler efficiency is 60 percent? Take fuel-oil density as 6.25 lb/gal. (HINT: See Table 14.6 for properties of saturated steam.)

8. A transmission line is solid aluminum, diameter 0.875 in., and spans a distance of 22.5 mi. Assume a two-wire line and unity power factor for simplicity. A 10,000-kVA generator is generating at 13,800 V, which is the voltage impressed on the line. (*a*) What is the *IR* drop along the line? (*b*) What is the line loss in kilowatts? (*c*) If the voltage is stepped up to 85,000 V before transmission, what is the line loss?

9. A steam plant produces ac power at 13,200 V and has been transmitting its energy to nearby towns at that voltage. A new industrial plant and town site have located 20 mi away, and the average demand for power at the new site will be 40,000 kW. The new town wants to purchase electric energy delivered to its distribution substation. Assume (for simplicity) a two-wire line whose total length is 40 mi and whose ohmic resistance is 0.15 ohm/mi. Assume unity power factor. (*a*) Compare the ohmic line losses under two possibilities: (1) transmitting at a voltage which provides 100,000 V at the town's substation; or (2) transmitting at a voltage which provides 320,000 V at the town's substation. (*b*) If electric energy costs $0.0125 per kVA to produce, and the wholesale price delivered to the town's substation is $0.025 per kVA, how many dollars per day would the power company gain from transmitting at the higher voltage?

10. The power factor on a 60-cycle, 440-V three-phase industrial service line is unsatisfactory, and it is desired to build it up to 90 percent. The plant load is found to have a total inductance of 2.5 henrys and an ohmic resistance of 450 ohms. What should be the capacitance installed on this line?

11. After several years of operation, the power factor on a 230-V residential distribution circuit has fallen below acceptable levels. When the load impressed on the line is 350 kVA (60-cycle), the power *actually being consumed is* 200 kW. An estimate of the total ohmic resistance of the line and its load is 500 ohms. Find (*a*) the power factor, (*b*) the impedance, (*c*) the inductive reactance, and (*d*) the amount of capacitance in microfarads which must be installed to bring the power factor up to 85 percent.

PART 8 Modern Physics

The internal vacuum vessel for the Poloidal Divertor Experiment *(PDX)* —a magnetic-containment nuclear fusion experiment utilizing hot plasmas at temperatures in the range of 100 million degrees. The PDX version is another in the "family" of *tokomak* machines being developed in this country and in Russia. (Princeton Plasma Physics Laboratory.)

32 Electronics

Much of the physics studied up to this point depends on observations made and theories formulated before 1900. We turn now to a brief look at electronics, a phenomenon of the 20th century. Electronics is partly based in electrical theory well established nearly a century ago, but for the most part it is rooted in the theories and discoveries of "modern physics." In a sense, electronics can be regarded as bridging the gap between classical physics and modern physics—between the physics of force, mass, and motion on the one hand, and the physics of atoms, particles, and waves on the other.

Electronics in the Real World In just the last 30 years, electronics has become a major industry worldwide. In advanced industrial nations the impact of electronics on the daily lives of people has reached proportions unimagined in the 1950s. Here are just a few selected features of that impact. Dozens more could easily be identified.

. . . Electronics in communications and entertainment affects the lives of millions—radio, telephone, worldwide color television networks, CB radio and mobile transmitters, stereo hifi sound systems, motion pictures, and home videotape recorder/players.

. . . Electronic devices and circuits now control and automate design, manufacturing, sorting, and inspection processes in thousands of industries.

. . . The electronic computer has moved from the research laboratory into the business offices, homes, and pockets of millions of people around the world.

. . . Electronics has fostered "The Learning Society," providing tapes, films, programmed lessons, and student-operated teaching machines, allowing information retrieval and encouraging study on a scale not dreamed of a generation ago.

. . . The aerospace industry, with manned moon explorations, space vehicle voyages to the planets, and (perhaps soon) space-shuttle commuter traffic between the earth and orbiting space stations, is completely dependent on electronics.

. . . Current work on energy resource development in all fields—solar, wind and tides, thermonuclear, and geothermal—is heavily involved with electronics.

. . . Semiconductor chips and printed circuits have resulted in the miniaturization and simplification of hundreds of common devices, from the telephone to the ignition systems of automobiles.

. . . Lasers have made available incredibly powerful sources of coherent light used in surgery, industrial tools, and communications systems.

. . . Electronics controls the operation of orbiting satellites as they serve worldwide communications networks, chart the earth's natural resources, and

Fig. 32.1 Electronics in communications. This electronic switching system in a telephone toll station incorporates transistors, microprocessor chips, integrated circuits, computers, relays, plug-in units, coils, and many other components. It serves circuits that make use of many electronic devices, including amplifiers, modulators, microwave transmitters and receivers, antennas, and communications satellites. (American Telephone and Telegraph Co.)

collect and disseminate data on weather and climate.

. . . And finally, in the business world, banking, billing, purchasing, accounting, shipping and receiving, inventories, payrolls, taxes and deductions, insurance and pensions, market analyses and economic forecasts—all are controlled or coordinated by electronic computers and by electronic networks and devices.

We shall divide this very elementary introduction to the field of electronics into four parts: first, an introduction to electronic theory; second, a brief survey of industrial electronics; third, some applications of electronics to communications; and fourth, electronics in special applications such as the military, aerospace, and medicine.

BASIC PRINCIPLES OF ELECTRONICS

What is Electronics? The word *electronics* is now commonplace, and like many such technical words which have been appropriated by the public at large, it is as often misused as it is

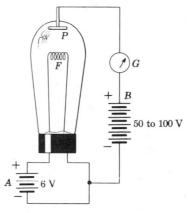

Fig. 32.2 Edison's discovery of thermionic emission, or the *Edison effect*. The filament F is heated to incandescence by battery A. When a positive charge is applied to plate P by battery B, a current is noted in the *PBF* circuit. If P is made sufficiently negative, however, there is no flow of electrons.

properly used. Technically, electronics is *the science of freeing electrons from atoms of matter and controlling their flow through vacuum or gas-filled tubes and in conductors and semiconductors.*

The control of the electron-stream movement is ordinarily effected by electric and magnetic fields. Much of the science of electronics is based on the electron tube, the fundamental principle of which was discovered by Thomas A. Edison (1847–1931). In 1883, while working to improve his early electric light bulb, Edison noted that the glowing filament emitted a small flow of electric charges (see Fig. 32.2). Associating this current with the heat of the filament, Edison referred to the phenomenon as *thermionic emission*. It is also known as the *Edison effect* in honor of its discoverer.

32.1 Thermionic Emission

The science of electronics involves *free electrons*, i.e., electrons which have been separated from their parent atoms and are thus free to move at the urging of electric or magnetic fields. The mass of an electron is so small (9.11×10^{-31} kg) that its inertia is almost negligible. Consequently its response to electric and magnetic forces is almost instantaneous.

One method of causing electron emission is to raise a metallic surface to a high temperature. Free electrons in a metal are thought to be in continual random motion, somewhat like the molecules of a gas. The rate of motion of free electrons increases with increase in temperature, and their increased velocity results in increased kinetic energy. This increased kinetic energy allows some of them to escape from the metallic surface in much the same way that high-energy molecules escape from the surface of a liquid which is being heated. This analogy holds true in actual practice, for even as the surface layer of a liquid offers resistance to the escape of molecules by *surface tension*, there is a retarding electric field at the surface of metals which opposes the escape of free electrons. This retarding field is called a *potential barrier*, and only electrons with relatively high energies cross through this barrier and escape from the metal surface (Fig. 32.3). Since the process bears some resemblance to boiling, we speak of electrons "boiling off" the heated metal. This process of "evaporation" of electrons from heated metallic surfaces is called *thermionic emission*. Other forms of electron emission include: *field emission* (stripping electrons from a metal surface by a very high-voltage elec-

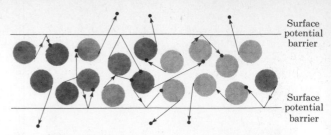

Fig. 32.3 Diagrammatic representation of thermionic emission. The large circles represent the outer orbits of atoms; the black dots, electrons. Only the higher-energy electrons possess the ability to escape through the retarding field of the surface potential barrier. Heating the metal increases the energy of the electrons, and many more of them can escape.

trostatic field); *photoelectric emission* (light impinging on certain metals causes electron emission); and *secondary emission* (electrons emitted as a result of bombarding the metal surface with high-speed atomic particles).

Space Charge Just as molecules "boiling off" from a liquid surface create vapor pressure above the liquid which retards further evaporation, electrons being emitted from metallic surfaces form a kind of "cloud" around the emitter. This cloud has a high negative charge, and repels other electrons being emitted. As the *space charge* builds up, an equilibrium condition is finally reached in which the number of electrons being repelled or forced back to the metal surface equals the number being emitted per unit time.

Herein lies the function of the *plate*, or *anode*, in a vacuum tube. A positive charge applied to the anode will draw off the space charge so that electron emission from the heated filament (*cathode*) can continue. When the positive charge on the anode is high enough to pull away the space-charge electrons as fast as electrons are being emitted from the cathode, the current through the tube will be a maximum.

Types of Emitters The rate of emission of electrons from a given area of metallic surface depends on the kind of metal and upon the temperature. Many different metals have been used in attempts to find combinations of high emission rate and high melting point. Three have proved especially suitable with respect to both of these criteria—pure tungsten, thoriated tungsten,

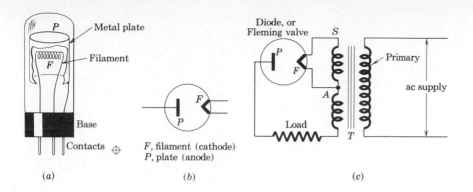

Fig. 32.4 The Fleming valve, or diode. (*a*) Construction. (*b*) Diagrammatic representation of a diode, as used in electronic circuit diagrams. (*c*) Circuit diagram showing how a diode can be used to rectify alternating current.

and oxide-coated materials. *Pure tungsten* has the highest melting point of any common substance (about 3300°C), and in order to get relatively good electron emission it must be operated at about 2130°C.

Thoriated tungsten actually contains a very small amount of thorium as an impurity, but as the filament (or cathode) is heated, the thorium covers the surface with a layer only about one atom thick and the rate of electron emission may be many hundreds of times greater than that for a pure tungsten surface. Thoriated-tungsten filaments operate at about 1700°C.

The third type of filament is obtained by coating tungsten or a suitable alloy with certain oxides, notably barium oxide, calcium oxide, or strontium oxide. *Oxide-coated filaments* emit large quantities of electrons and can be operated successfully at a low red heat.

At the high operating temperatures noted, filaments would rapidly burn up if oxygen were present, so the emitter (whether it is the filament or an indirectly heated cathode) is operated either in a high vacuum or in the presence of an inert gas. High-vacuum tubes (common in communications equipment) are called " hard " tubes. Gas-filled tubes (called " soft " tubes) find more frequent use as heavy-duty rectifiers and as control devices in industrial electronics work.

VACUUM TUBES

32.2 The Diode

A few years after Edison's discovery, Sir John A. Fleming (1849–1945) devised a two-element tube

with an associated circuit which acted as a one-way valve, or *rectifier*. Figure 32.4*a* shows a diagram of the construction of the early Fleming "valve." In vacuum-tube circuit diagrams a two-element tube, or *diode*, is indicated by the symbol shown in Fig. 32.4*b*. The diagram of Fig. 32.4*c* shows how a diode can be used to change alternating current to pulsating direct current. This process is called *rectifying*. The ac line voltage is applied to the primary of a transformer *T*. The secondary *S* is tapped at *A* to supply the current to heat the filament. During the half cycle when the plate *P* is positive, electrons flow from the heated filament to the plate and on through the load. When *P* is negative during the other half cycle, no current flows. The tube thus acts as a one-way valve and rectifies or changes alternating current to pulsating direct current, as indicated in the diagram of Fig. 32.5.

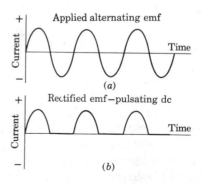

Fig. 32.5 Rectifying action of a diode, diagramed. (*a*) The alternating current in the transformer-filament circuit. (*b*) The pulsating direct current in the plate-load circuit.

Fig. 32.6 The triode. (*a*) Diagrammatic representation of a triode. (*b*) Circuit diagram showing how the grid of a triode can control large currents in the plate-load filament circuit.

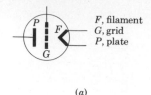

F, filament
G, grid
P, plate

(*a*)

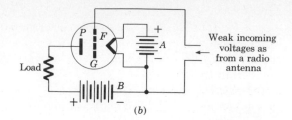

Load

Weak incoming voltages as from a radio antenna

(*b*)

32.3 The Triode

In 1907, while experimenting with wireless telegraphy, Lee DeForest (1873–1961), an American inventor, devised a method of controlling the action of a vacuum tube by adding a third element, which he called the *grid*. This grid, ordinarily a metal screen, is inserted between the filament (cathode) and the plate (anode) and acts as a shutter to control the passage of electrons from filament to plate. Figure 32.6*a* shows how such a tube is diagramed in electronic circuits. If the grid is made negative (−), it repels the space-charge electrons and throttles down the filament-to-plate current. If made positive (+), it attracts electrons and strengthens the plate current. A circuit like that of Fig. 32.6*b* shows the action of the triode. Weak voltages may be applied to the grid *G*. These very weak *voltages* on the grid effectively control the flow of rela-

tively large *currents* in the plate-load circuit. The *A* battery heats the cathode, and the *B* battery keeps a high positive potential on the plate *P*. This tube was originally called an *audion* because it made *audible* (in earphones) signals which otherwise would have been too weak to detect. This function of controlling large plate *currents* with weak *voltages* applied to the grid is called *amplification.*

NOTE: Batteries are shown in circuits to simplify circuit diagrams. Electronic equipment may be battery-operated or ac-operated, from transformers and/or vacuum-tube or solid-state rectifiers.

32.4 Characteristic Curves of Triodes

The action of a triode can be shown by a curve which results from plotting plate current I_p against grid voltage V_g for a given set of values of plate voltage. Figure 32.7 shows a typical set of

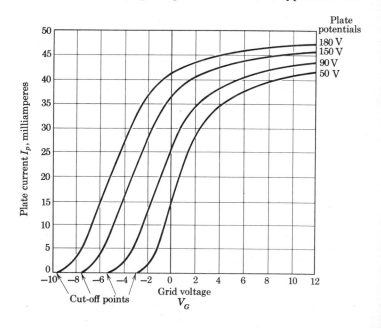

Fig. 32.7 A family of characteristic curves for a single triode operated at four different plate potentials.

four such *characteristic curves* obtained from a triode for four different values of plate (anode) voltage. The filament heater current is the same for all the curves. Note the following interpretations from these curves: (1) As the grid is made more positive, the plate current rises. (2) Small changes in grid voltage (called *grid bias*) cause large changes in plate current. (3) The higher the plate voltage (positive), the more current will flow. (4) When the grid is highly positive, the curves flatten out, indicating that nearly all the electrons leaving the filament are reaching the plate. (5) For each plate potential there is a certain value of the grid bias (negative) which will stop all flow of electrons. This value of V_g is called the *cutoff point*.

32.5 Uses of Characteristic Curves—Amplification

The characteristic curve of a triode is an indicator of its performance both as an amplifier and as a rectifier. Consider for example, the characteristic curve of a triode, as diagrammed in Fig. 32.8. For the triode tested, the plate potential was held at 200 V (positive), and values of the plate current I_p were determined for values of grid voltage V_g from -10 to $+8$ V. The cutoff point for this tube is at -10 V (V_g). If we wanted this tube to amplify some weak ac frequencies without distortion, we could apply them to the grid along the -4-V V_g line, for this marks the *center of the linear portion* of the characteristic curve. The incoming signals would be applied *through* a battery set to hold a steady voltage of -4 V on the grid. The weak ac voltage is shown coming in at the bottom of the figure along a vertical axis at -4 V. By projecting each point of the incoming sine wave up to the curve and then horizontally to the right, we obtain the undistorted but greatly amplified sine wave shown, indicating a strong current in the plate circuit of the same frequency and general characteristics as the weak signal impressed on the grid.

Voltage Amplification Factor Since the grid is so much closer to the filament (cathode) than the plate is, a very small change in grid potential ΔV_g will produce a greater change in plate current than would be produced by the same change in

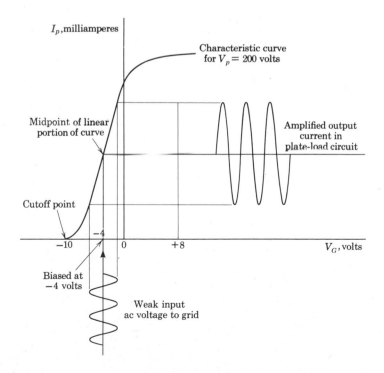

Fig. 32.8 Operation of a triode at the midpoint of its characteristic curve, to secure undistorted amplification. This is commonly called "Class A" operation.

plate potential ΔV_p. Stated another way, if the plate current I_p is to be held constant during a small *increase* of grid voltage, the plate voltage will have to be *decreased* by a much larger amount.

The *voltage amplification factor μ* of a triode is defined as the ratio of the change in plate voltage to the change in grid voltage (in the opposite direction) such that the plate current remains constant. As a formula

$$\mu = \frac{\Delta V_p}{\Delta V_g} \qquad (32.1)$$

for constant I_p.

Referring to the family of characteristic curves for a typical triode in Fig. 32.7, consider the line of constant $I_p = 20$ mA cutting across the linear portion of all four of the curves. From the 180-V curve to the 50-V curve, $\Delta V_p = 180 - 50 = 130$ V. Between these same two curves (for I_p constant at 20 mA) $\Delta V_g = 5.8$ V (from -5.1 to $+0.7$ V), and $\mu = 130/5.8 = 22.4$, the voltage amplification factor of this tube.

Amplification with Rectification The triode can be used to accomplish both amplification and

rectification with the same tube by operating the grid at the proper point. Let us take the same tube and operate it at $V_g = -8$ V, the point where the lower portion of the curve breaks into the linear portion (Fig. 32.9). Projecting the sine wave representing the incoming signal up to the curve as before, we now find that amplification and (partial) rectification (also called *detection*) take place. It is this ability of the triode to amplify and rectify which makes it effective as a detector tube in radio receivers.

In radio equipment, triodes are used to produce high-frequency oscillations, as detectors or rectifiers, and as amplifiers. However, solid-state devices (transistors) have almost completely replaced vacuum tubes in most radio communications equipment and in sound systems (see Secs. 32.8 through 32.11).

32.6 Classification of Electronic Tubes

In the preceding paragraphs we have traced the history of the development of the electronic tube, and in so doing we have discussed in a prelimi-

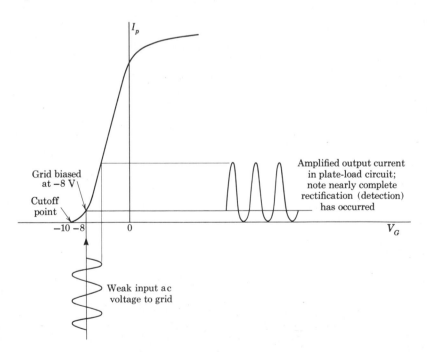

Fig. 32.9 Operation of a triode at the lower end of the linear portion of the characteristic curve. Both amplification and rectification (detection) take place.

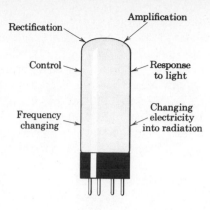

Fig. 32.10 The basic functions of electronic tubes.

nary way two of the functions of electronic tubes—*rectification* and *amplification*. There are four other primary functions of electronic tubes—*control, frequency changing, electrical response to light,* and *converting electricity into radiation* (Fig. 32.10). We shall discuss theory and industrial applications of each of these classifications by function in the pages to follow.

Electronic tubes are also classified by type, specifically by the number of electrodes sealed within. *Diodes* and *triodes* have already been discussed; *tetrodes, pentodes,* and tubes with even more than five elements are frequently used in complex electronic circuits. It should be emphasized that electronic tubes by themselves can accomplish little. They must be operated in electric circuits together with familiar electric devices such as resistors, rheostats, capacitors, inductances, transformers, relays, contactors, switches and the like. In an introductory physics text we can indicate only a few of the many complex circuits and components of electronic equipment now in everyday use.

32.7 Construction and Nomenclature of Electronic Tubes

The simplest vacuum tube is the diode, and its component parts are the *envelope* (glass or metal), the *cathode* or heater (sometimes the cathode is the *filament*), the *anode* or *plate*, and the *terminals*, which are the outside connectors to the inner components. If the cathode is actually the filament (dc-operated tubes), the diode diagram is as shown in Fig. 32.4b. If an indirectly heated cathode (ac operation) is used, the diagram is as shown in Fig. 32.11. In any event, *the cathode is always the source of electron emission and is negative*; and *the anode, or plate, is always positive and attracts electrons.* Electrons flow only one way in electronic tubes—*from cathode to plate.* Cathodes are made of tungsten, thoriated tungsten, and oxide-coated nickel alloys. *Pure tungsten cathodes* are operated at very high temperatures (2500 K) and require considerable power. They are used in large transmitting tubes and are rugged, dependable, and long-lived. *Thoriated tungsten* contains a small amount of thorium oxide which results in electron emission at lower temperatures (around 1900 K) than for pure tungsten. *Nickel-alloy cathodes* coated with barium oxide or strontium oxide operate at even lower temperatures (1150 K). The bulk of small electron tubes for radio/TV equipment and sound amplification have cathodes of this type.

In *vacuum tubes,* the space within the envelope is evacuated as completely as possible to make the flow of electrons a maximum. In *gas-filled tubes,* either the envelope is filled with the gas itself (helium, neon, etc.) or an arrangement for heating and vaporizing the working substance (sodium-vapor and mercury-vapor tubes) is in the envelope. The gases or vapors used easily become ionized and offer a good conducting path for electrons from cathode to plate. (When the atoms of a gas have gained or lost electrons, the *charged* atoms are called *ions* and the gas is said to be *ionized.*) Figure 32.12 illustrates a number of different types of electronic tubes, and shows some construction details.

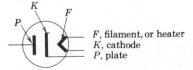

Fig. 32.11 Circuit diagram of a heater-type diode.

Fig. 32.12 Representative electronic tubes for light and heavy duty. (General Electric Company).

THE SOLID STATE AND TRANSISTORS

One of the important fields of scientific study and research today is solid-state physics. Historically, it was much easier to study the properties of gases and liquids than those of solids, since it is impossible to "get inside" a solid to examine its molecular or crystalline structure. It is only comparatively recently, with the aid of such modern devices and techniques as the electron microscope, the oscilloscope, x-rays, and cryogenics, that researchers have been able to learn some of the secrets of the solid state. We now know (or can hypothesize, in the absence of "hard" knowledge) a great deal about such matters as the electric conductivity of pure vs. impure substances; the arrangement of atoms and molecules in solids; the behavior of electrons in a crystal lattice; and the magnetic and electrical properties of solids at very high and very low temperatures.

Good electric *conductors*, like silver, copper, gold, and aluminum, have a crystal structure such that the outermost, or *valence*, electrons in uncompleted shells (see Sec. 33.5) are shared by other atoms. These valence electrons are therefore free to wander throughout the substance, and as they move in response to a potential difference, a *current* exists in the conductor.

Poor conductors, or *insulators*, on the other hand, have very few free electrons, even when a high potential difference is applied, and since there are very few electrons free to move, very little current results. Glass, ceramic, quartz, and rubber are examples of insulators.

32.8 Semiconductors

In the class of materials called semiconductors the electric conductivity lies between that of the insulators like glass and mica and the good metallic conductors like copper. Silicon and germanium are two such substances. Figure 32.13 shows how the electric resistance of semiconductors compares with that of certain conductors and insulators.

Studies of the properties of semiconductors began early in the present century. These studies are a branch of *solid-state physics*, and research in these fields has been extremely productive in the past 30 years. (See also the discussions of the *thermoelectric effect* and the *Peltier effect* in Chaps. 13 and 17, respectively.)

The *crystal detectors* of the early days of radio were actually *semiconductor diodes*. They acted as one-way valves and served to rectify or detect the incoming radio-frequency signals. They were replaced by the vacuum tube, since tubes could handle more current and were more stable in their operation. As electronic components became smaller, however, and as the number

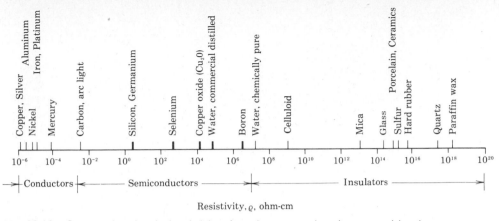

of tubes multiplied, both space and heat from the cathodes became serious problems. Designers renewed their interest in semiconductors and found that germanium and silicon gave the most promise of being useful in electronic circuits.

The theory of semiconductors is extremely complicated, and the following explanation deals only in very general concepts.

32.9   Electrons and Holes

Semiconductors are crystalline materials whose atoms are arranged in a rather rigid structure by the forces between the outer, or valence, electrons. Pure silicon is a very poor conductor because all four of the outer (valence) electrons of each atom are locked fast with those of another atom, thus binding each atom to its neighbors. The sketch of Fig. 32.14 illustrates such a crystal lattice in schematic two-dimensional form. Since there are no really *free electrons*, the material is a poor conductor.

***N*-type Silicon**   If pure silicon is contaminated (" doped ") with a mere trace (say one part in a million) of some element whose atoms have excess electrons in the outer orbit, e.g., *arsenic* (As), which has five valence electrons, these *extra electrons* are available within the silicon crystal structure for carrying an electric current (see Fig. 32.15*a*). Since there is an excess of (negative) electrons in this material, it is called *N*-type silicon. Germanium is also used as a semiconductor.

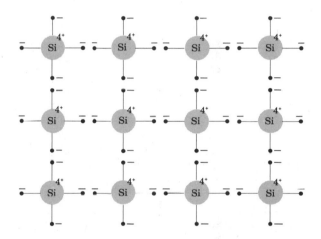

Fig. 32.14   Two-dimensional diagrammatic representation of the arrangement of atoms and outer (valence) electrons in a crystal of pure silicon. Actually, the bonds exist in three dimensions.

***P*-type Silicon**   If pure silicon is doped with a trace of some element which *borrows* electrons (such as boron or gallium), the crystal structure is also out of balance electronically. Each atom of boron or gallium has three valence electrons, and is therefore short one electron needed to fit it into the crystalline structure of silicon. This leaves an electronic " hole " in the crystal structure (see Fig. 32.15*b*). These " holes " behave like positive charges in that they attract electrons from neighboring atoms, causing a condition of disturbed electronic equilibrium in the crystalline structure. *Holes* can move from place to place in the crystal

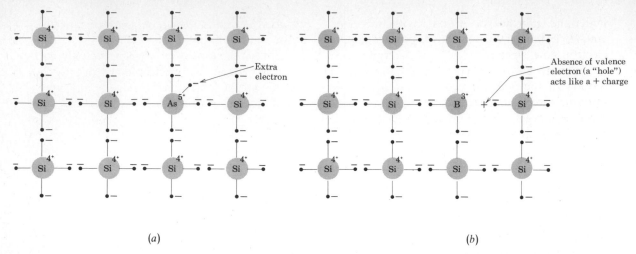

(a)

(b)

Fig. 32.15 Two-dimensional diagrams of two types of silicon. (a) Crystalline structure of N-type silicon. The contaminant is arsenic (As). (b) P-type silicon crystal. The contaminant shown is boron (B), but gallium, with the same valence as boron, is often used.

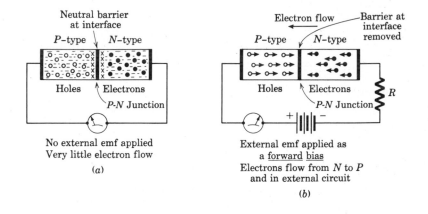

Fig. 32.16 Diagrams explaining semiconductor diode operation. (a) P-N junction with no externally applied emf sets up neutral barrier and the flow of charge is very small. (b) External emf applied as a *forward bias* removes neutral barrier and encourages the flow of charge (electrons and holes) across junction and in external circuit. (c) External emf applied as a *reverse bias* creates wide neutral barrier and flow of charge is completely blocked.

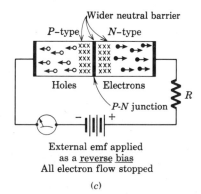

under the influence of an externally impressed emf. *Holes* flow in the opposite direction from the electron flow, since motion of an electron in one direction to fill a *hole* produces the effect of motion of the *hole* (or positive charge) in the opposite direction. *Holes* behave as if they were positive charges, since their effective motion is toward a negative electrode.

32.10 Semiconductor Diodes

If a block of *P*-type silicon and a block of *N*-type silicon are formed or compressed together, holes from the *P*-silicon combine with the free electrons of the *N*-silicon at the interface and effectively deplete the junction region of charge carriers. The neutral barrier thus formed (Fig. 32.16) inhibits further movement of holes or electrons, and electric conduction is effectively prevented (Fig. 32.16*a*). However, if an *external* emf is applied, as shown in Fig. 32.16*b*, positive to the *P* material and negative to the *N* material, new holes will be supplied to the *P* material and new electrons to the *N* material. An impressed emf of about 0.5 V is sufficient to overcome the neutral-barrier effect for *P*- and *N*-silicon and cause the diode to be a good conductor. Applying an external emf in the manner described is called *forward biasing*.

In Fig. 32.16*c reverse biasing* is shown. The neutral barrier is now markedly widened, and the diode becomes a poor conductor. A semiconductor diode is, then, a one-way valve for electron flow. When it is connected to an external *alternating* emf, current will readily flow for the half cycle during which the junction is *forward-biased* and will hardly flow at all during the half cycle of *reverse bias*. We have then, a simple rectifier. Semiconductor diodes are widely used in electronic circuits as simple one-way "gates" or valves, or as rectifiers.

32.11 Transistors

The transistor is a semiconductor device originally developed by the Bell Telephone Laboratories in 1948. Since 1960 the transistor has assumed a position of commanding importance in the electronics industry.

Although transistors are not the only components that go into electronic computers, they are the essential components. Computers were a $50 billion a year industry in a recent year. Computer-controlled games depend on transistors, as do also radios, television equipment, telephones, and stereo hifi sound systems. Auto ignition systems are transistorized. Most of the complex "hardware" for military and aerospace vehicles and equipment is controlled or operated by transistorized circuits.

Transistors, as noted above, can serve as "Go-No Go" gates, and also as rectifiers. They will also amplify weak signals when connected in suitable circuits and properly *biased*. Some elementary examples of transistor circuits will be given below.

Junction transistors are made by fabricating a "wafer" composed of a thin section of one type of silicon or germanium between two sections of the opposite type. The resulting wafer may be either *N-P-N* type or *P-N-P* type. The middle section of this sandwich acts to control the flow of electrons (or *holes*) across it much as a grid controls electron flow in a vacuum tube. The outer sections act as *emitter* and *collector* respectively.

Figure 32.17*a* is a diagrammatic representation of a typical *N-P-N*-type junction transistor, and Fig. 32.17*b* shows how such a transistor is represented in a circuit. Figure 32.17*c* is the symbol for a *P-N-P* transistor in circuit diagrams.

In an *N-P-N* transistor, *P*-type semiconductor material is used for the base wafer, which is ordinarily made quite thin. The two *N*-type blocks serve as emitter and collector. With forward bias in the emitter-base circuit the emitter has an excess of electrons and continually supplies them to the base. In this sense the *emitter* acts like the *cathode* of a triode vacuum tube.

Since the base is positive, electrons are attracted into the base from the emitter. If the base were at a *high* positive potential, many holes

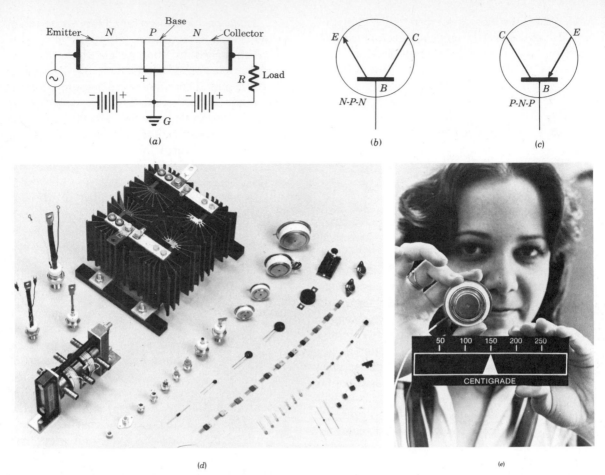

Fig. 32.17 Junction transistors. (*a*) Diagrammatic representation of *N-P-N*-type junction transistor, showing forward bias on emitter to base and reverse bias on collector to base. (*b*) Symbol used to indicate an *N-P-N* transistor in circuit diagrams. (*c*) Symbol for *P-N-P* transistor. (*d*) A collection of semiconductor devices ranging from heavy-duty rectifiers to junction transistors no larger than a pin head. (General Electric Co.) (*e*) A high-junction-temperature, silicon-controlled rectifier (SCR) rated up to 1200 V and 400 A. SCRs can replace heavy-duty vacuum tubes like thyratrons, and can serve as relays and fuses. They provide excellent power control with little power loss. (Westinghouse Electric Corp.)

would be produced, all (or almost all) the electrons emitted from the emitter would recombine with these holes in the base, and very little or no electron flow would move on to the collector and load. Keeping the positive potential on the base at a *low value* with respect to the emitter still attracts electrons from the emitter in numbers far in excess of the number of *holes*, and these surplus electrons can then continue across the base to the collector so that a current flows in the collector-load circuit. Consequently, the relative emfs of the two batteries are chosen so that the base posi-

tive charge is kept low. Minor fluctuations in base potential can thus control major flows of current from emitter to collector. In this sense the *base* of a transistor is analogous to the *grid* of an electronic tube.

The *collector* is biased so that no current will flow from the collector to the base (both are charged positively). It is said to have *reverse bias*. The collector receives the electron flow through the transistor and sends it along to the external load. The *collector* is then the counterpart of the *plate* (or anode) in a vacuum tube.

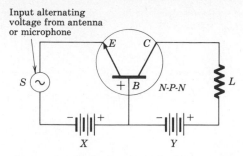

Input alternating
voltage from antenna
or microphone

Fig. 32.18 A simplified transistor amplifier circuit. *S* is any source of weak ac pulses, as from a microphone. *L* is a resistance load, as a loudspeaker. Note that the emitter-base circuit is forward-biased, and the collector-load circuit is reverse-biased.

When used as an *audio amplifier* an *N-P-N* transistor could be placed in a circuit like that diagramed in Fig. 32.18. Note that battery *X* and battery *Y* are in series and that the net effect of the two batteries (in a practical circuit a voltage-divider device would be used) is to put a high negative potential on the emitter, a high positive potential on the collector, and a relatively low positive potential on the base. (The emitter bias voltage is usually 1 V or less.) When the emitter-base circuit is properly forward-biased, the weak, fluctuating voltages from *S* cause very minor changes in the potential on the base, but these minor *voltage* changes control large *current* fluctuations in the emitter-collector-base circuit.

P-N-P transistors act like *N-P-N* transistors, but with electrons and *holes* interchanged and applied voltages reversed.

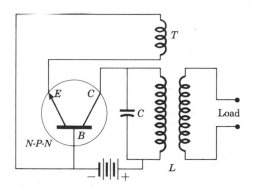

Fig. 32.19 A basic transistor-oscillator circuit, in diagram form.

Transistor oscillators are common components of modern electronic equipment. Figure 32.19 shows how a junction transistor (*N-P-N*) might be used in an oscillator circuit. The inductance *L* and the capacitance *C* determine the frequency of the oscillation. Feedback of a weak signal to the emitter is from a coil *T*, in which alternating emfs are induced by virtue of its proximity to *L*.

Advantages and Uses of Transistors Since there is no filament to be heated in transistors, energy consumption, as compared to that of vacuum tubes, is very small. (A typical small transistor in a hand-size radio receiver may operate on less than 0.05 W.) The extremely small size of transistors is a definite advantage in designing electronic equipment. Without the development of the transistor and integrated circuits much of the personalized electronic equipment on the market

Fig. 32.20 "Growing" crystals artificially. These quartz crystals are called "*R*-face," and plates made from them can operate at frequencies above 10^6 Hz. They will be used in the manufacture of filters and oscillators for telephone equipment. (Western Electric Co.)

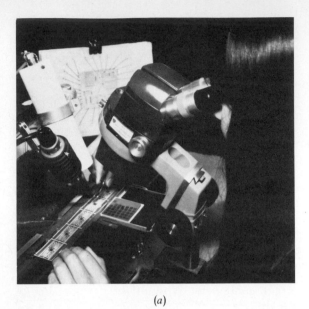

(a)

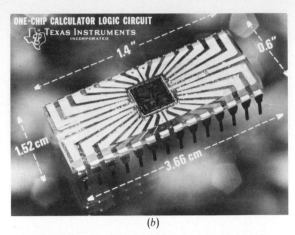

(b)

Fig. 32.21 A computer on a chip (a) Technician engaged in final assembly of single-chip pocket calculator logic circuits. Five complete circuits are seen in the microscope tray. (b) Close-up of a single circuit, with size indicated. The chip itself (dark square in center) incorporates all the logic and memory circuits needed to perform 8-digit, 3-register calculator functions, including the four fundamental operations of adding, subtracting, multiplying and dividing. (Texas Instruments, Inc.)

today would not have been possible. Semiconductor diodes may be as small as the head of a pin, and junction transistors of many types are no larger than an eraser on the tip of a pencil (see Fig. 32.17d).

The main disadvantage of transistors is that they cannot handle large amounts of current at high voltages and therefore they do not lend themselves well to heavy-duty industrial electronics applications. Also their conductivity tends to increase as they heat up, so they must be protected from drawing a heavy current.

32.12 Microprocessor Chips

In recent years there has been large-scale combination of semiconductor diodes and junctions into *integrated circuits* (ICs) and *microprocessors* on silicon chips. Thousands of microminiaturized transistors can now be placed on a sliver or "chip" of silicon only a fraction of a centimeter on a side. A master drawing showing the details of the desired circuitry is first made, perhaps 500 times as large as the final chip. By means of a process similar to silkscreening or photolithography, and then through photographic reduction, the circuitry is created in and on the silicon

wafer, one layer at a time. The "doping" of the wafers is accomplished by baking them in an atmosphere of hot gases containing the doping agents.

Single-chip circuits (see Fig. 32.21) are used in many kinds of electronic equipment, including radios, television, telephones, and microwave communication equipment. So-called "miracle chips" (silicon wafers) find increasing application in microcomputers, industrial controllers, supermarket checkout stands, microwave ovens, microprocessors for auto ignition and caburetion systems, and in medical electronics.

The extreme miniaturization of the single-chip microprocessor can be appreciated by comparing it to the first fully electronic computer (ENIAC), which ushered in the computer age in 1946. It weighed 30 tons and utilized 18,000 vacuum tubes. A 5-hp air conditioner was needed just to carry away the heat it generated. Modern microcomputers crowd more than 5000 microminiaturized transistors onto a chip only 0.5 cm square. They also contain the circuitry which

Fig. 32.22 Microprocessor chip used in telephone circuits. This tiny MAC-8 chip (held by the tweezers) is only one-tenth the size of a postage stamp, but it contains the equivalent of 7,000 transistors. (American Telephone and Telegraph Co.)

enables them to exceed the computational power of ENIAC (see Fig. 32.22).

INDUSTRIAL ELECTRONICS

Industry uses electronics in nearly every step of design, manufacturing, testing, packaging, and shipping. Only a few of these applications can be described here. Any standard text on industrial electronics would give ample coverage of this topic.

32.13 AC-to-DC Conversion in Industry—Full-wave Rectification

It will be recalled that the diode rectifies only in the sense that it passes, by its valvelike action, one-half of each ac wave (see Figs. 32.4 and 32.5). By sealing two filaments and two plates in the same envelope, *full-wave rectification* can be obtained. Such a tube is diagramed in Fig. 32.23*a*, and the full-wave rectification is shown in Fig. 32.23*b* and *c*. A circuit for operating such a *duodiode* for full-wave rectification is shown in Fig. 32.23*d*. The transformer T has separately wound secondaries S_1 and S_2. S_1 is a low-voltage winding and supplies the heater current to

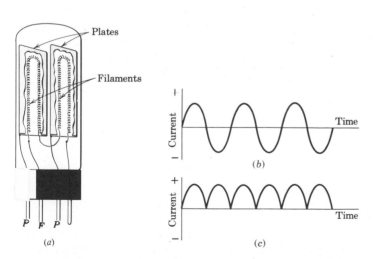

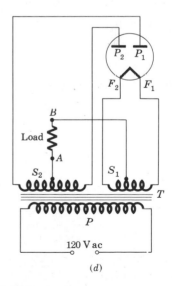

Fig. 32.23 (*a*) Schematic diagram of the construction of a duodiode. (*b*) Alternating-current wave in filament circuit. (*c*) Pulsating direct current in plate-load circuit of tube. (*d*) Circuit for full-wave rectification of alternating current by a duodiode.

filaments F_1 and F_2, which are connected in series. S_2 is a high-voltage winding and supplies an alternating potential to plates P_1 and P_2, so that when P_1 is positive and P_2 is negative, the electron emission from the filaments goes to P_1, and, on the next half cycle, when P_1 is negative and P_2 is positive, the electrons flow to P_2. The current will go through the load AB in the same direction in either case and is pulsating direct current, as shown in Fig. 32.23c. This pulsating character of the rectified current can be smoothed out almost perfectly by the introduction of a properly designed *filter circuit.*

Both diodes and duodiodes are sometimes termed *kenotrons* in the electronics industry. They vary in size from tubes hardly larger than a peanut to tubes several feet long for heavy-duty applications. In industry, they find use in electrostatic precipitators for air cleaning, in the recovery of valuable byproducts from the smokestacks of industrial plants by electrostatic

precipitation, and still, to some extent, in amplifiers for public-address and sound-motion-picture equipment.

Gas-filled Electronic Tubes for Rectification
Filling a tube with a gas or vapor which, when charged (*ionized*), will assist the flow of electrons across the space permits much higher currents to flow. Such tubes may be filled with argon, zenon, or mercury vapor. They serve in many industrial rectifier equipments where medium-to-heavy currents at low voltages are required. Examples of their use are the supplying of current for magnetic separators, for magnetic chucks, and for battery chargers.

Heavy-duty Electronic Rectifiers Where very heavy direct currents are required, as for operating electric railways, mine equipment, steel-mill rolling operations, and electrolytic refining equipment, *mercury-pool tubes* are used. The

(a)

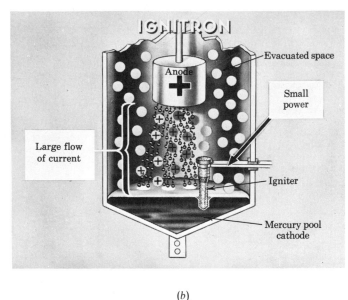

(b)

Fig. 32.24 (a) An industrial ignitron tube. (b) Interior arrangement of an ignitron tube. (General Electric Co.)

cathode is actually the surface of a pool of mercury, which, when heated, vaporizes and furnishes the ionized gas for high-current operation. The *envelope* of these heavy-duty tubes is metal, and special "igniting" apparatus must be provided to start the operation of the tube. The name *ignitron* is applied to one such heavy-duty tube. Figure 32.24*a* shows an ignitron, and Fig. 32.24*b* diagrams the interior arrangement of an ignitron tube. To start the operation, a strong positive voltage is applied to the igniter temporarily, which causes sparks around the igniter and heat which vaporizes some of the mercury to start the action of the tube. The ac voltage is then applied to the mercury-pool cathode, and electron flow to the anode begins. Ignitrons are made in a size range from a sealed tube about 6 in. long to huge tanks the size of oil drums. Where very large direct currents are required, as in electrolytic refining processes, the ignitron principle is employed, but several anodes are placed in the same tank and the mercury pool is made large enough to act as the cathode for all the anodes. Such huge rectifiers, called *multianode tanks*, are capable of carrying currents of 3000 A at voltages of 1000 to 2500 V. They are finding increased use in rectifying high-voltage alternating current to direct current for long-distance transmission lines.

32.14 Amplifying with Electronic Tubes

Amplifiers are used for many industrial applications such as public-address and intercommunication systems, theater sound systems, photoelectric control and counting systems, and for inputs to the control systems of motors and drives for heavy industrial equipment. The basic principle of amplification by the three-element vacuum tube was explained in Sec. 32.5.

Audio-frequency Amplifiers Amplifiers designed for supplying music and the spoken word to groups of people are classed as *audio amplifiers*. Good-quality audio amplifiers are designed to give nondistorted amplification for frequencies from 30 Hz up to about 15,000 Hz. In multistage amplifiers, the early stages usually effect *voltage amplification* at small power outputs. The grid circuits of the succeeding tubes require very little power, so the entire amplifying operation can be carried out up to the final stage by voltage amplification. Since the output of the final stage will ordinarily operate one or more loudspeakers, considerable power is required and this final stage is a *power amplifier*.

Many factory and institutional amplifier systems are designed to operate from microphone, phonograph, and radio inputs. Plant officials may want to broadcast music at certain hours, make general announcements, or pipe in a radio broadcast or news release at other times. The one basic amplifier cannot handle all these types of inputs equally well, so *preamplifiers*, *mixers*, *volume indicators*, *monitors*, and associated equipment may be supplied. The block diagram of Fig. 32.25 shows the arrangement of such components as they might be used in a large system. Smaller systems would have some but not

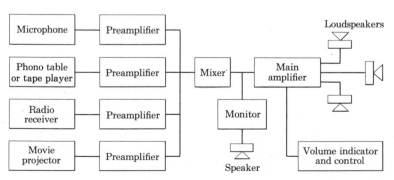

Fig. 32.25 Block diagram of a public-address system for a theater, ball park, factory, or convention hall.

all of these components. Large public-address (PA) systems operating several speakers will have a power output of from 20 to 200 W per channel and may require a technician in constant attendance during operating periods.

32.15 Automated Industrial Processes

Many industrial processes move at high speeds and require controls which operate with split-second timing and great precision. Wiredrawing, hosiery-mill operations, papermaking, steel and aluminum sheet rolling, textile weaving, precision grinding, and sequential machinery operations are a few examples of such processes. Electronic amplification of very weak voltages is used to operate electromagnetic relays, which in turn control the main switches to the driving machinery. For example, in wiredrawing, the wire itself as it is pulled from the die may be used as a conductor of a very weak current, a fraction of a milliampere. A sliding contact feeds this weak current to an amplifier, which, operating through a relay and magnetic switch, keeps the power on to the driving machinery. The slightest flaw in the drawn wire will cause a potential drop across the flaw to the contact. This voltage variation is amplified, and the amplified voltage is sufficient to operate the relay, stopping the machinery for an analysis and correction of the cause of the difficulty. Similar controls prevent thread breakage in textile and hosiery mills. Sensors at one stage of a manufacturing process may *feed back* a weak signal to the grid of an electronic tube and thus control a current to a motor which will initiate corrective action in the process. This self-correcting action through electronics is often called *automation*.

The operations controlled by automation need not be "on-off" or "all-or-none" processes. Utilizing *feedback* signals and electronic tubes, combined with *synchros* or *servomotors* (motor-like devices which synchronize themselves with one another, in pairs), it is possible to make continuous, very fine adjustments on heavy machinery *while it is operating*, from signals fed back

Fig. 32.26 Control room of a rolling mill for cold-rolled steel sheets. Automated electronic controls are monitored by the operator, but feedback signals from sensors on the rolling mill itself actually govern the rolling process. (United States Steel Corp.)

from a sensor which is testing the end product. As an example, suppose it is desired to control the thickness of sheet steel emerging from the final rolls of a rolling mill, at 0.015 ± 0.001 in. (see Fig. 32.26). A sensor (optical, electric-resistance, or radioisotope) continually monitors the thickness of the sheet as it comes from the final set of rolls (Fig. 32.27). If thickness varies from the specified 0.015 in., either *plus* or *minus*, the sensor detects the variation, converts it into an electronic signal, which is fed back through the necessary *comparer* and amplifying equipment to the *transmitter* half of a synchro or servo unit. This transmitter is the "master" for its "slave," the *receiver* synchro or servo unit. The necessary amount of correction (as indicated by the signal from the original sensor) is finally made by the receiver synchro or servo. In a steel-rolling mill, the final element of the process would be a heavy-duty dc servomotor which would either increase or decrease the pressure on

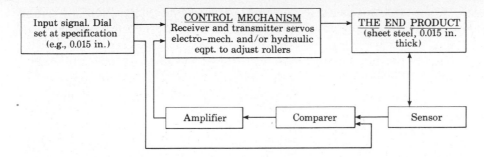

Fig. 32.27 Block diagram of a closed-loop control system for automatic control of a sheet-steel rolling mill. The comparer "compares" the input signal (specified thickness) with the actual thickness determined by the sensor. The discrepancy (plus or minus) is sent as a signal to the control unit.

the rolls through a mechanical or hydraulic coupling.

Such self-correcting control devices are used by the thousands in industry, in business, and in the home. Called *closed-loop control systems*, they make use of such electronic devices as photoelectric tubes, vacuum tube and transistor amplifiers, microcomputers, thermocouples, thermistors, synchros, and servomotors, as well as a wide variety of sensors such as thermometers, pressure gauges, optical devices, radioisotopes, strain gauges, and similar devices.

32.16 Frequency Control by Electronic Devices

Most commercial electric power in the United States is produced and distributed as alternating current at a frequency of 60 Hz. However, there are many specialized applications where a different frequency is required. Radio utilizes very high frequencies (10,000 Hz up to many millions of cycles—*megahertz*) to broadcast the carrier wave. Induction furnaces and dielectric heating of sub-

stances which are nonconductors of electricity require high-frequency currents. Diathermy machines for hospitals and clinics need high frequencies, as do microwave ovens. Low frequencies of high power are required for certain steel-mill operations.

For frequency conversions up to 10,000 Hz, motor-generator sets are often used. The motor operates off the standard 60-cycle power, and the generator is wound to deliver the required frequency.

For delivering power in the higher frequency ranges, vacuum tubes with grid control are ordinarily used. Below 2000 Hz gas-filled grid-controlled tubes (*thyratrons*) may also be used.

The first step in frequency conversion by electronic tubes is to obtain the proper voltage to apply to the tube. A transformer does this job. This ac voltage is then applied to a rectifier tube to produce direct current at the same voltage, which, in turn, is supplied to an oscillator circuit which produces the required ac frequency. The block diagram of Fig. 32.28 illustrates the steps

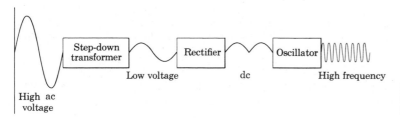

Fig. 32.28 Block diagram of the frequency-conversion process as performed with electronic tubes.

just described. Transistor oscillators now replace vacuum tube oscillators, except in circuits requiring heavy current.

32.17 Photoelectric Cells in Industry

Not many years ago the photoelectric cell was so rare as to be classed as a scientific curiosity. Today photoelectric cells or *phototubes*, are performing an imposing list of jobs in many varied industrial applications. They count, sort, and measure in the mass production of small parts and serve in safety devices to protect the operators of production machinery. They turn lighting systems on and off, stand guard against illegal entry to buildings, and open doors automatically for those who are welcome. They detect smoke, control traffic signals, indicate liquid levels in boilers, and control burner operation for furnaces and boilers. They control sound reproduction for the motion-picture industry, and many camera fans use a photocell-operated exposure meter, either built in to the camera or as a separate instrument.

Construction and Operation of Phototubes Certain metallic oxides, notably *cesium oxide*, *selenium oxide*, and *rubidium oxide*, emit electrons when exposed to light, just as thoriated tungsten emits electrons when heated. The electron emission is, within limits, proportional to the intensity of the light striking the cathode. Emission of electrons as a result of light is called the *photoelectric effect*. (See also Secs. 26.9 and 33.7.) The diagram

of Fig. 32.29 shows the construction of a typical phototube. The cathode is coated with the light-sensitive substance, and the anode is usually just a slender rod or wire along the axis of the cathode. A positive potential is applied to the anode, and when light strikes the cathode, electrons are emitted and attracted to the anode. Anode currents are usually very small, of the order of 10 μA or less. Such a current will not operate anything but the most delicate and sensitive relays, so an amplifier must be used to obtain currents of useful magnitude.

32.18 The Electronic Computer

Modern computers are of two general types, the *digital computer* and the *analog computer*. The former is the type universally used in business and government and in the solution of problems where data can be reduced to some form of binary arithmetic. Analog computers lend themselves to the solution of problems of a scientific and engineering nature. Descended from the original "giant brains" of two decades ago, the "fourth and fifth generations" of electronic computers have been getting smaller as they become more complex and sophisticated (see Fig. 32.30).

Fig. 32.30 A high-performance digital computer undergoing final test at the factory. The technician and engineer are running a set of computer programs designed to check on the performance of every circuit and function of the computer. (International Business Machines Corp.)

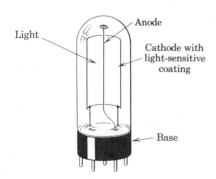

Fig. 32.29 A typical phototube in diagram form.

Vacuum tubes, with their disadvantages of size and heat dissipation, have been largely replaced by transistors and solid-state devices of many kinds. The arithmetic and logic circuits of modern computers operate at speeds measured in billionths of a second. *Communications networks* are now operating which, with electronic computers as integral components, provide speedy information retrieval on almost any desired subject, by "interrogating" a nearby computer center from suitable equipment installed in an office, research laboratory, or school. If the information desired is not available in a data bank at the nearby computer center, a computer there "knows" where it is available and automatically makes the necessary connections to a location where the requested information is stored, all in a matter of seconds. The information desired may arrive at the interrogator's station by television (video), voice (audio), or computer printout sheet, depending on the equipment available and the nature of the requested information. Though such services are currently limited by cost considerations, the use of computers in such networks is increasing at a rapid rate.

In the short span of 30 years, the electronic computer has developed from a rather crude laboratory curiosity into an essential tool of modern societies. Industry, business, agriculture, medicine, government, even recreation and games are all so dependent on computers that one can hardly imagine industrialized societies functioning without them. The extreme miniaturization made possible by transistors, microprocessor chips, and integrated circuits (ICs) has brought the computer out of the laboratory and the industrial plant into almost every facet of our daily lives. Figures 32.30 and 32.31 show two examples.

ELECTRONICS IN COMMUNICATIONS

Electronic communications equipment has been revolutionized during the past 25 years thanks to the transistor, solid-state physics, integrated circuits, and miniaturization. Many of the equipment improvements have been brought about by

Fig. 32.31 A minicomputer for hospital laboratories. This model is designed to automate laboratory data analysis during actual diagnosis and treatment of cardiac disorders. (Hewlett Packard.)

the need to perfect systems for space exploration and satellite operations. The further development of electronic computers, new discoveries in optical electronics, e.g., the laser and fiber-optics, the application of electronics to medicine, and the efforts to harness thermonuclear fusion—all have drawn upon or contributed to developments in electronic communications. It is the purpose of this section to explain some of the fundamental principles of electronic communications and to discuss briefly such devices and systems as the telephone, radio, television, radar, and microwave transmission and reception.

The discussions to follow are, of course, elementary, and serve merely as an introduction to the subject. Students desiring details of theory and operation should consult a standard text on communications electronics.

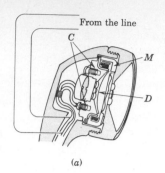

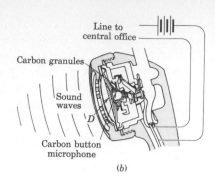

Fig. 32.32 Schematic diagrams of (*a*) a telephone receiver, and (*b*) a telephone transmitter.

(*a*)

(*b*)

THE TELEPHONE

Electromagnetic induction is one of the physical principles of telephone communication. The telephone transmits and receives sounds of diverse character and considerable range of frequency. It is readily apparent, then, that a simple electric-current pulse is not sufficient to carry voice signals; that fluctuating currents capable of reproducing sounds with a considerable degree of fidelity are required. Let us see how this is accomplished.

32.19 Telephone Receivers and Transmitters

The basic components of a telephone circuit are the *receiver*, the *transmitter*, and the *line*. Figure 32.32*a* shows a telephone receiver in diagram form. (The diagrams are schematic but show the principles of operation more clearly than photographs of compact modern instruments would.) The receiver magnet *M* has coils *C* of many turns of fine insulated copper wire wound on each pole. These coils are connected to the transmitter at the other station, through the local telephone exchange, and with a battery or other current source in the line. The thin elastic diaphragm *D* is of iron so that it will respond to the variable magnetic field strength in the instrument. A microphone-type telephone transmitter is shown in diagram form in Fig. 32.32*b*. The mouthpiece directs the sound waves to the thin diaphragm *D*, which reproduces mechanically the waveforms of the speaker's voice. As the diaphragm vibrates, the variable pressure on the carbon granules is

continually varying their resistance to electric-current flow. A loud sound causes a strong push on the diaphragm, and this presses the carbon granules closer together, lowering the resistance, and a large current flows in the line. The line current is at all times a more or less faithful electrical counterpart of the sound waves entering the mouthpiece.

At the receiver end of the line, these variable currents flow in the coils *C* and modify the magnetic field existing between the poles of the magnet. This varying magnetic field causes the receiver diaphragm to duplicate the movements of the transmitter diaphragm, and the original sound is thus reproduced with a considerable degree of fidelity.

The first telephone was invented in 1876 by Alexander Graham Bell, and crude though it was, judged by modern standards, it marked the beginning of the era of electrical communication by the human voice.

32.20 Modern Telephone Sets

Both receiver and transmitter are combined in a single mounting called the *handset*. Figure 32.33 shows a handset in cutaway view. Handsets are subject to rough handling and are held in any position the user may desire. Consequently the transmitter's output must be independent of position. Carbon granules must not become more loosely or more tightly packed merely because of the position in which the transmitter is held. Diaphragms have been ribbed for stiffening, which makes them more rugged; and the present designs have the inner surface of the diaphragm

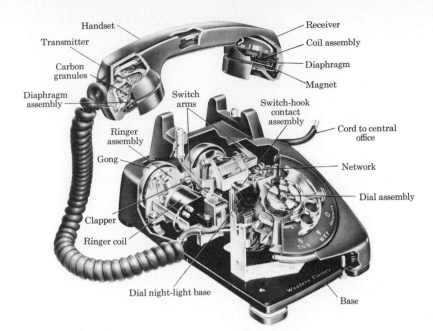

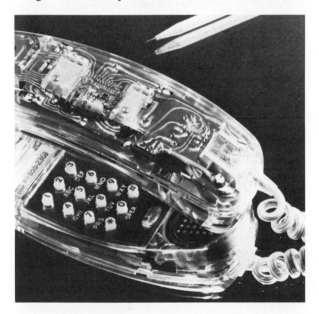

Fig. 32.33 Phantom view of a telephone handset with standard dialing system. (Western Electric Co.)

Fig. 32.34 Transparent model of a telephone handset with "touch-tone" or digital "dialing". The pencil points to the tone-generating integrated circuit. (American Telephone and Telegraph Co.)

in direct contact with the carbon granules. Receiver magnets have been improved and acoustic fidelity is good throughout the frequency range of human speech.

Figure 32.34 shows a recent handset which combines the transmitter, receiver, and push-button dialing system all in one instrument. Note the integrated circuitry seen through the clear plastic. These units make considerable use of transistors.

32.21 Telephone Exchanges—Central Office Operations

Special inquiries and some long-distance calls are still handled by switchboard operators, but nearly all telephone traffic today is handled by automatic electronic equipment in local and central exchanges.

If a dial telephone is used, the calling party inserts a finger in the appropriate hole of the dial and turns until a stop is reached. A spring returns the dial to the rest position, and as it does so, a small cam periodically interrupts the direct current to the central office. These current interruptions actuate relays in the central office, the number of pulses (interruptions) being equal to the digit dialed. For example, if the number 764-9472 is to be dialed, the first dialing sends seven current pulses to the central office; the second six; the third, four; the fourth, nine; and so on. A push-button (or "touch-tone") handset sends signals which are interpreted as digits "dialed" by the calling party.

Central-office Equipment The central office, which may serve thousands of subscribers, is a maze of complex equipment consisting of trunk lines, selector switches, amplifiers, relays, contactors, and similar devices. It also contains the source of electricity to operate the system. The essential equipment is the switching apparatus. From the original manual switchboard which required an operator to complete every call, the telephone industry has gone through four "generations" of switch gear. The first was called *step-by-step* switching, because the dialed pulses activated a series of stepping switches in the central office to connect one telephone line to another. "Step-by-step" is still used in small communities where there are not more than a few thousand lines.

Panel switching was introduced about 40 years ago, and it was soon followed by a system called *crossbar switching*, the first system to incorporate a *storage* or *memory* capability. Crossbar switching can accommodate up to 10 simultaneous connections, compared to only one for each step-by-step switch. Crossbar equipment is still in use in many central telephone exchanges.

Currently, telephone companies are installing *electronic switching systems* (ESS), which incorporate computer programming, microprocessor chips, and the very latest developments in electronic memory and logic circuits.

The details of construction and operation of these switching systems are beyond the scope of a physics text. Figures 32.35 and 32.36 give some indication of their complexity.

Fig. 32.35 Crossbar switching equipment in a large toll center switching office. The selecting and control equipment in the crossbar system can perform simultaneously for a number of talking circuits. It makes a connection in a fraction of a second and then becomes immediately available for the next call. (Western Electric Co.)

Fig. 32.36 An electronics technician checks out a circuit in a "plug-in" unit of an electronic switching system. Electronic switching systems (ESS) represent the very latest in sophisticated equipment for telephone communications. (American Telephone and Telegraph Co.)

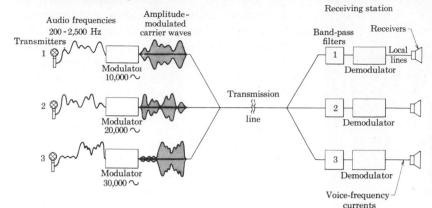

Fig. 32.37 Diagram illustrating carrier telephony. The carrier frequencies of 10,000, 20,000, and 30,000 Hz are illustrative only. In practice, a complex array of frequencies for two directions and for different channels have been standardized.

Long-distance telephony would be impossible without *amplification*, for losses along the line soon reduce signal strength below acceptable levels. Voice-frequency currents for transmission over distances greater than about 50 mi are passed through amplifying units called *repeaters* at key points along the way.

32.22 Simultaneous Messages on the Same Line

Since it takes two wires to make a telephone line, the number of separate wires which would have to be strung between large cities to carry the present volume of traffic would be prohibitive if a method of sending simultaneous messages over the same line had not been devised. This technique is called *carrier telephony*, which had its beginnings as early as 1918. Only the general simplified theory of carrier telephony will be explained here.

Suppose the voice-frequency currents of several simultaneous conversations can be limited to the range 200 to 2500 Hz without objectionable fidelity loss (see Fig. 32.37). Now let these several voice-frequency transmissions each be superimposed on (i.e., *modulate*) a separate but much higher frequency signal known as the *carrier*. These higher *carrier frequencies* are produced by electronic-tube or transistor oscillators and are transmitted to the other end of the line, carrying the voice frequency with them. At

the other end, *bandpass filters* separate the several carrier frequencies, each still associated with its superimposed voice frequency. *Demodulators* then pick the voice-frequency currents off the several carrier frequencies, and the ordinary voice-frequency currents are amplified and fed out on local lines.

In Fig. 32.37, let 1, 2, and 3 be separate voice-frequency currents from three different telephone transmitters. These signals arrive at the local exchange and are fed into three separate modulators, one being supplied a carrier frequency of 10,000 Hz from its oscillator, the second a carrier frequency of 20,000 Hz, and the third a carrier frequency of 30,000 Hz. These modulated carrier waves are sent over the same line (with carrier-frequency amplification as needed) to the distant exchange where the bandpass filters pick them off. (A bandpass filter is a circuit containing inductance and capacitance in such an arrangement that it will allow to pass only a relatively narrow "band" of frequencies, from 18,000 to 22,000 Hz, for example.) The demodulators strip the voice frequencies from the carrier waves and send them along the local lines to the three parties being called.

Variations of carrier telephony permit many two-way conversations to take place simultaneously over a single pair of wires. One method makes use of *coaxial cable* (Fig. 32.38*a*) which provides for thousands of telephone, radio, and

Fig. 32.38 Simultaneous messages on telephone lines. (*a*) Coaxial cable. This 22-conductor cable can carry up to 90,000 telephone calls simultaneously. (*b*) Telephone cable made of glass fibers. Message information is communicated by pulses of light along the glass fibers in the cable ("fiber-optics"). Total internal reflection within each fiber keeps the energy of the light pulses from dissipating. (American Telephone and Telegraph Co.)

television channels, all simultaneously traversing the continent in either direction at the speed of light. Another system makes use of *fiber optics.* Thousands of individual glass fibers, each only about 0.005 mm in diameter, are assembled into a *bundle*. The bundle, if properly put together, will act as a "light pipe," provided the curves along the bundle are not too sharp. Laser light is then used for message transmission. Fiber-optics

cables reportedly can carry several hundred times as many channels as can be carried by large coaxial cables. Presently, a 30-mi optical-fiber telephone cable is one such system under development. It is to link telephone switching centers in and near Calgary, Alberta, Canada (see Fig. 32.38*b*).

RADIO

We have studied radiant energy in two forms previously—heat energy and light energy. We found that radiant energy travels in waves through space at the prodigious speed of 3×10^8 m/sec or 186,000 mi/sec. By the work of Young, Huygens, and others *wavelengths* of light were measured, and it was found that visible light varied in wavelength from 0.00004 to 0.00007 cm (4000 to 7000 Å).

In 1856 the brilliant theoretical physicist James Clerk Maxwell proposed the theory, on purely mathematical grounds, that light and heat were alike in every respect except in wavelength and further proposed that there would be discovered many other types of such waves, differing from light and heat only in wavelength and frequency. He suggested that all such radiations were *electromagnetic* in nature and that all would travel with the speed of light.

32.23 Electromagnetic Waves

It was some 30 years later (1888) that Heinrich Hertz (1857–1894) devised an apparatus that was to prove Maxwell's electromagnetic-wave theory. He worked with a spark-gap apparatus like that diagramed in Fig. 32.39. An induction coil I supplies high voltage to the spark-gap or transmitting circuit $AB_2 SB_1 B$. B_1 and B_2 are large metal balls of considerable electric capacity. They can be moved along the rods A and B, which will vary the capacitance and inductance of the $AB_2 SB_1 B$ system. Now, the natural (resonant) frequency of oscillation of such a circuit is given by

$$f = \frac{1}{2\pi\sqrt{LC}} \qquad (29.22)$$

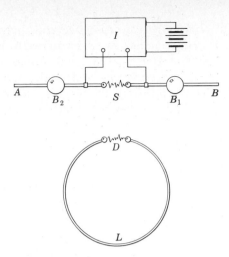

Fig. 32.39 Hertz's electromagnetic-wave apparatus.

where

f = frequency, cycles/sec (Hz)
L = inductance, henrys
C = capacitance, farads

By moving B_1 and B_2, and by adjusting the length of the air gap at S, the capacitance and inductance of the circuit can be varied. It is therefore possible to vary, within limits, the frequency of oscillation of the sparking circuit.

Hertz found that if he brought a loop L with another spark gap D within a few meters of the spark gap S, sparks could also be observed jumping the gap D when the proper adjustment of the two circuits was obtained. The DL circuit can be considered as a receiver and the $AB_2\,SB_1\,B$ circuit as a transmitter. Later careful measurements revealed that the maximum transfer of energy from one circuit to the other was effected when the receiver's natural oscillation frequency as given by Eq. (29.22) above was exactly equal to the transmitter circuit's natural frequency. When this is the case, a condition of *electric resonance* is said to occur.

By stepping up the power of his transmitting circuit and using larger loops as receivers, Hertz was able to send and receive signals by an apparatus of this type over distances of more than 100 meters. Later he was able to measure the wavelength of these radiations quite accurately, and, in every case, multiplying the wavelength by the frequency gave a velocity equal to the speed of light. Hertz's experiments, then, definitely supported Maxwell's original proposals. These new waves of energy were called *electromagnetic waves*.

NOTE: The velocity of any wave motion is given by the expression $v = f\lambda$ where f is the frequency of the motion and λ is the wavelength. By international agreement the term *hertz* (Hz) is now used as the unit of frequency; 1 Hz equals 1 cycle/sec. The terms *kilohertz* (kHz), *megahertz* (MHz), and *gigahertz* (GHz) are used also, for 1000 cycles, 1 million cycles, and 1 billion cycles, respectively.

32.24   Early Radio Messages

In 1895 Guglielmo Marconi (1874–1937), a young Italian inventor, became interested in Hertz's work and its possibilities for wireless communication. He found that he could greatly increase the distance of reception of electromagnetic radiations by increasing transmitter power, by using an antenna high in the air, and by improved detection methods. By 1900 he had sent and received signals at distances of 300 km, and wireless telegraphy was born. Early transmitters used spark coils to actuate the oscillatory circuit, but with the development of the vacuum tube and its application to oscillatory circuits, the science of radio began its rapid rise. At first mere signals, as in telegraphy, could be sent, but in 1906 Reginald A. Fessenden (1866–1932) of the University of Pittsburgh was successful in *modulating* a radio-frequency carrier wave with audio-frequency currents from a microphone, and the first transmission of human speech was carried out. We shall explain here only the basic principles of radiotelephony, not the design and operation of actual radio transmitters and receivers.

32.25   The Vacuum-tube Oscillator

Voice transmission by electromagnetic waves requires a carrier wave of very high frequency and constant amplitude. The three-element vacuum

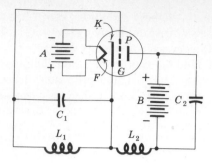

Fig. 32.40 A vacuum-tube oscillator circuit for producing radio-frequency (RF) waves of constant amplitude, known as a continuous-wave (CW) oscillator.

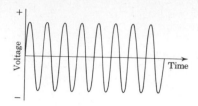

Fig. 32.41 Graph of CW oscillations—sine waveform.

tube (*triode*) can be employed to produce these oscillations. A circuit like that shown in Fig. 32.40 may be used. Application of a high positive potential to the plate P from the B battery accelerates the electrons emitted from the cathode K, heated by filament F. A current begins to flow in the plate circuit $PBL_2 K$. This current, building up in L_2, creates a magnetic field expanding from L_2. This growing field cuts the coil L_1 and induces in it a current whose direction is such as to apply a negative potential to the grid G. This negative grid causes the plate current to decrease, resulting in a collapsing of the field about L_2. As this L_2 field collapses, the current induced in L_1 is reversed, applying a positive potential to the grid. This positive grid potential builds up the plate current again, and the process is repeated over and over many times per second. Condenser C_2 passes these high-frequency pulses around the B battery to the plate.

The actual frequency of oscillation is determined by the $L_1 C_1$ circuit, since it controls the grid potential. The natural frequency of such a vacuum-tube oscillator is given by the equation

$$f = \frac{1}{2\pi\sqrt{L_1 C_1}}$$

Illustrative Problem 32.1 An oscillator circuit like that of Fig. 32.40 has $C_1 = 0.0200 \ \mu f$ and $L_1 = 0.0010$ mh. What is its natural frequency?

Solution Substitute in Eq. (29.22):

$$f = \frac{1}{2\pi\sqrt{0.02 \times 10^{-6} \times 0.001 \times 10^{-3}}}$$

$$= \frac{1}{2\pi\sqrt{2 \times 10^{-14}}}$$

$$= \frac{1}{6.28 \times 1.414 \times 10^{-7}} = 1,125,000 \ \text{Hz}$$

$$= 1.125 \ \text{MHz} \qquad\qquad ans.$$

By choosing the values of C_1 and L_1 properly, it is possible for the vacuum tube to produce constant-amplitude oscillatory frequencies of more than 100 MHz. If voltage is plotted against time, the continuous oscillations would result in a graph like that of Fig. 32.41. Specialized equipment, involving the use of *klystrons* and *magnetrons* can produce ultrahigh frequencies (UHF) and superhigh frequencies (SHF) of 10,000 MHz and up. Transistors are also used in many oscillator circuits. Oscillators can also generate "square wave" pulses and "sawtooth" waves (see Fig. 32.42).

32.26 Transmitting Radio Waves

Continuous-wave (CW) oscillations used as a source of electromagnetic waves for carrying audio frequencies can be produced with a circuit like that diagramed in Fig. 32.43. The high-frequency, constant-amplitude oscillations of the $L_1 C_1$ circuit are transferred to an antenna-ground circuit which is *inductively coupled* by means of L_3 to the $L_1 C_1$ circuit as shown. The high-frequency oscillations, which will constitute the carrier wave, are modulated with voice-frequency currents from the MP circuit. M is a

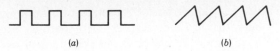

(a)   (b)

Fig. 32.42 Two additional forms of waves produced by oscillators. (a) Square waves. (b) Sawtooth waves.

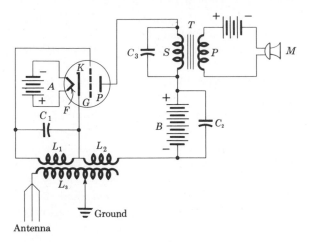

Fig. 32.43 Circuit diagram of a radio transmitter, amplitude-modulation type (greatly simplified).

microphone, and P is the primary winding of a transformer T. S is the secondary winding of the transformer, and C_3 is a capacitor to bypass the high-frequency oscillations around S. The rest of the transmitter circuit is merely the oscillator circuit discussed in the preceding section.

The voice-frequency currents, ranging from 20 to about 10,000 cycles, are superimposed on the radio-frequency oscillations, and these modulated voltages are transferred to the antenna and broadcast as electromagnetic waves with the speed of light. Figure 32.44 illustrates graphically the process of *amplitude modulation*.

A transmitter built on the circuit shown in Fig. 32.43 would leave much to be desired in the way of performance. Its frequency control would be poor, and such transmitters drift away from their assigned frequency band markedly. Exact frequency control can be effected by the introduction in the grid circuit of certain crystals. A quartz crystal, for example, will pass frequencies in inverse proportion to its thickness. A 1-mm thick quartz crystal will give a resulting frequency of about 3 MHz. If a crystal of the proper thickness (already prepared and calibrated) is inserted in the grid circuit of the transmitter oscillator and the values of C_1 and L_1 adjusted to the crystal's rated frequency, fairly exact frequency control of the transmitter can be obtained. Quartz-crystal-controlled oscillators are also used to regulate the action of fine watches and clocks.

32.27   Radio Broadcasting

Radio waves travel through space with the speed of light ($c = 3 \times 10^8$ m/sec). This value is a constant, so it is a relatively easy matter to determine either the wavelength λ or the frequency f if one of the two is known.

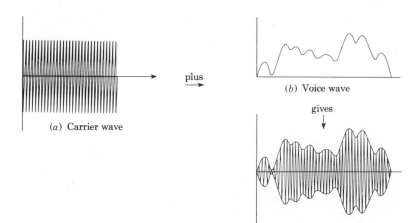

(a) Carrier wave

plus →

(b) Voice wave

gives ↓

(c) Amplitude-modulated wave

Fig. 32.44 Graphic representation of amplitude modulation of a carrier wave for radio broadcasting. Carrier wave (a), plus voice wave (b) combine to produce amplitude-modulated wave (c).

Illustrative Problem 32.2 What is the wavelength in meters of the radio waves from a transmitter broadcasting on a frequency of 970 kHz?

Solution Use the equation

$$c = f\lambda$$

Solve for λ:

$$\lambda = \frac{c}{f}$$

Substitute given values:

$$\lambda = \frac{3 \times 10^8 \text{ m/sec}}{970 \times 10^3 \text{ cycles/sec}}$$

$$= 310 \text{ m} \qquad\qquad ans.$$

The complete frequency range of radio waves (10 kHz to 30,000 MHz and up) offers a wide selection of frequencies for communication purposes. Since radio communication affects so vitally the life and security of a nation, most governments exercise considerable control over radio transmissions. All transmitters in the United States must be licensed and must operate on the frequency band assigned by the Federal Communications Commission (FCC). The usual classification of frequency bands is listed in Table 32.1, along with the typical type of assignment for each.

Commercial (AM) radio-broadcasting stations (550 to 1600 kHz) are assigned frequencies and licensed for their power output by the FCC. Their frequency control must be quite exact, so that stations in a common area will not "wander" into one another's assigned frequency. Power-output assignments must be rigidly observed also so that signals from a local station will not interfere with those of a station in

Table 32.1 Designation and assignment of radio frequencies

Frequency f, kHz except as noted	Wavelength λ, m, except as noted	Designation and Usual Assignment
15–100	20,000–3000	Low radio frequency (LRF) (government and commercial, point-to-point)
100–400	3000–750	Medium RF (telegraph and government, ship to shore, navigation)
500	600	Medium RF (international SOS)
550–1600	550–187.5	Medium RF, standard radio broadcast band (AM)
1600–6000	187.5–50	Medium-high frequency (short-wave, amateur, police)
6–30 MHz	50–10	High frequency [police, amateur, citizens' band (CB), government, aviation]
30–80 MHz	10–5	Very high frequency (VHF) (government, military, police, forestry, television)
80–890 MHz	5–1	Ultrahigh frequency (UHF) (military and television; FM radio; aeronautics, amateur radio)
800–30,000 MHz	1 m–1 cm	Superhigh frequency (SHF) (also called *microwaves* and *centimeter waves*) (radar, microwave telephony, satellite and space travel communications)
30,000–300,000 MHz (30–300 GHz)	1 cm–1 mm	Extremely high frequencies (EHF) (experimental, government)

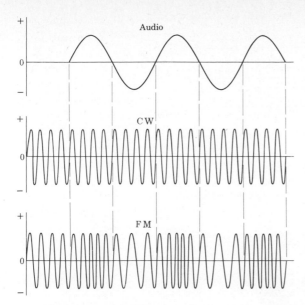

Fig. 32.45 Graphic representation of frequency modulation. An audio sine waveform is shown modulating the frequency of the carrier wave.

another locality which may be assigned nearly the same frequency. Power outputs of commercial radio stations may vary from 250 W for small stations to 50,000 W for large stations.

32.28 Frequency Modulation

Voice-frequency currents can be made to modulate the carrier *frequency* rather than the *amplitude* of the carrier wave. With this system the amplitude of the carrier wave remains constant and depends on the power output of the station. *Frequency modulation* (FM) stations operate in the VHF and UHF bands, say 58 to 108 MHz. At these frequencies, the static from electrical storms, power lines, electric machinery, and the like can more easily be eliminated by so-called "limiter circuits." Also, frequency modulation is more adaptable than amplitude modulation to the higher audio frequencies (8000 and up), and therefore frequency modulation is considerably better than amplitude modulation for the high-fidelity broadcasting of programs of fine music. Figure 32.45 shows graphically how frequency modulation functions.

Circuits and equipment for transmitting and receiving radio broadcasts (both AM and FM) are very complex, well beyond the scope of an introductory physics book.

Radio *receivers* are of so many different designs and types that their inclusion in an introductory physics book is not feasible. Interested students may consult any standard text on radio communications.

THE CATHODE-RAY TUBE (CRT)

So far we have discussed the uses of electronic tubes and transistors for *rectification, amplification, control, frequency changing,* and *electrical response to light.* There remains a sixth and final function, *converting electricity into radiation.* The most common examples of this function are found in the incandescent and gas-filled (neon and fluorescent) electric lights and signs. These have been briefly treated in an earlier chapter and will not be further discussed here. We shall, however, explain the basic principles of two specialized electronic tubes, the *cathode-ray tube* and the *x-ray tube,* in which electricity is converted into useful radiations.

32.29 The Cathode-ray Tube and the Oscilloscope

The cathode-ray tube, diagramed in Fig. 32.46 and shown in phantom view in Fig. 32.47, is an electronic tube of almost unlimited potentialities. Its use in television receivers, computer consoles, and radar and sonar receivers is widely known. No less important is its application to Loran (Long-Range Navigation) systems, to analysis of electronic system performance, to sound-wave analysis, to the measurement of energy from nuclear reactions, and to medical electronics.

The cathode K (Fig. 32.46) is a heater type and is hooded to permit electrons to emerge only through a small hole O in the grid G. The slender beam of electrons is accelerated by an *electrostatic plate system* (A_1 and A_2). The potentials applied to these anodes are such as to give great acceleration to the electron beam and narrow it

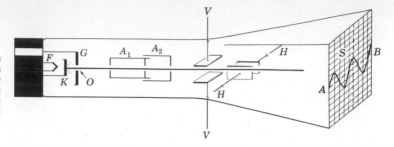

Fig. 32.46 Simplified diagram of a cathode-ray tube. The $FKGA_1A_2$ system is called an *electron gun*. V and H are pairs of electrostatic plates (or magnetic coils in some models) which can cause the beam of electrons to be diverted in the vertical and horizontal planes, respectively.

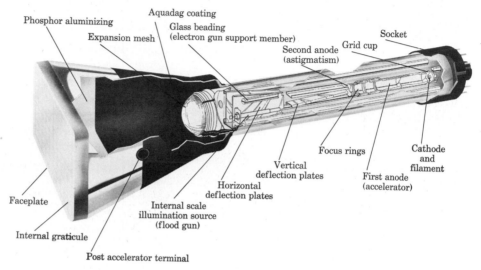

Fig. 32.47 Phantom view, with parts labeled, of cathode-ray tube (CRT) as used in an oscilloscope for science research. (Hewlett Packard Co.)

down to a slender "pencil" of electrons heading straight down the tube. The portion of the tube so far discussed is often called an *electron gun*.

The beam now passes through two pairs of deflector plates (or coils) V and H to which alternating potentials (or alternating magnetic fields) are applied by various equipments for the specific purposes in mind. For example, as used in a cathode-ray oscilloscope (Fig. 32.48), a high "sweep-frequency" ac potential is applied to the H (horizontal) plates, which causes the electron beam to sweep horizontally across the fluorescent screen S (Fig. 32.46) so rapidly that the visual result is merely a straight line. It sweeps from A to B and then jumps back to A to repeat the sweep. Now, if a variable potential of some sort is to be studied, it is applied across the V (vertical) plates, and the net result of the two motions will

be a pattern on the screen which is a composite of the two inputs. *Standing waves* can be obtained by adjusting the sweep frequency of the instrument to a submultiple of the frequency of the alternating voltage applied to the V plates.

Cathode-ray oscilloscopes constitute an important tool for electronics technicians in testing radio, television, sound, and radar equipment. *Amplification* (*gain*), *fidelity* (or its opposite—*distortion*), power output, frequency measurements, and many other features of electronic-equipment performance can be accurately determined from oscilloscope tests.

The cathode-ray tube, modified to suit specific requirements, is the basic unit in many modern devices, including television and radar, computer consoles, electrocardiographs (EKG) (Fig. 32.49*a*) and other medical-electronics de-

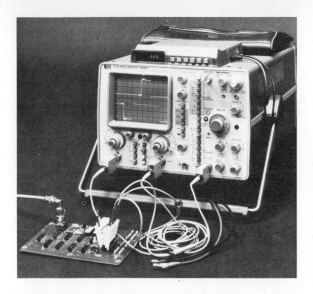

Fig. 32.48 A 200-MHz cathode-ray oscilloscope with direct readout for time interval measurements. Precision instruments of this kind are used in scientific and engineering research. (Hewlett Packard Co.)

vices, electronic games and teaching machines. It is also an essential tool in diagnosing malfunctions in auto-engine performance in the "tune-up" operation, (Fig. 32.49*b*) and in many industrial operations involving precision measurements. Modern scientific research would be severely handicapped without the cathode-ray oscilloscope. The energies liberated from atomic and nuclear processes, as well as other characteristics of subatomic particles (see Chap. 33), are often determinable only from oscilloscope observations and measurements.

TELEVISION

Worldwide communication is really a phenomenon of only the past 15 years. Satellite communications networks and the cathode-ray tube, modified for television, are the key factors.

32.30 Television Components

The tube of a television *camera*, called the *iconoscope*, is a kind of cathode-ray tube. The image of the scene being televised is formed by optical methods on a light-sensitive screen inside the

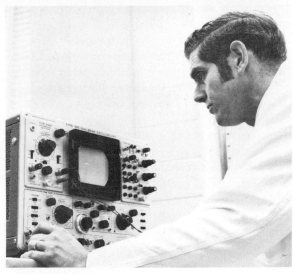

(*a*)

(*b*)

Fig. 32.49 (*a*) Monitoring the electric action of the heart with a dual-beam oscilloscope. (The University of Michigan Medical School.) (*b*) Cathode ray tubes (CRTs) and a computer are combined in this modern auto-engine diagnostic unit. Note the traces on the CRT screen at left, portraying actual performance conditions within the engine cylinders; and the design specifications for the test, displayed on the CRT screen in center. (Sun Electric Corporation.)

tube. The sensitive surface is composed of several hundred thousand tiny photosensitive elements, each of which emits electrons in an amount proportional to the intensity of the light falling on it at that instant. The problem is then to pick up from this screen the varying electronic signals which represent the *video information*.

This pickup is effected by a fast-moving electron beam which *scans* the image screen in a series of horizontal sweeps across it. In the United States a complete scan contains 525 lines, and, in order to present smooth motion to the viewer, 30 complete scans per second are made. The *horizontal scanning rate* is therefore, 30×525, or 15,750 total lines per second. The electron beam is thus seen to be an exceedingly fast-moving physical phenomenon. Its scanning and sweeping action is controlled by a magnetic deflection system. Millions of signal variations per second are thus picked up and sent out from the iconoscope, and this information must be "sent" by electromagnetic waves through space to the receiving antenna and set. The carrier wave, in order to be modulated by the video signal, must be of extremely high frequency, 54 to 216 MHz for VHF transmitters and 470 to 890 MHz for UHF transmitters.

The television *receiver* also has as its principal element a kind of cathode-ray tube, called the *kinescope* or *picture tube*. It has a magnetic deflection system similar to that in the iconoscope, and the electron beam sweeps and scans in exact synchronism with the beam in the iconoscope. Electronic information for this synchronization is transmitted with the carrier wave from the transmitter, as is also the *audio information*.

The incoming signals have to be amplified, the audio portion routed to the audio circuits, the sweep and scan information to the magnetic deflector plates, and the video information to the electron gun, all in proper sequence and timing, with time intervals which may be as short as a few hundredths of a microsecond.

The surface of the picture tube has a phosphor coating which emits light as the beam from the electron gun hits it. The successive sweeps in the scan and the number of complete scans per second reproduce a "picture" of the original scene which has satisfactory optical definition when viewed from several feet away. Both the persistence of human vision and the persistence of the fluorescing process on the picture-tube surface help make for satisfactory viewing.

The problems of scanning, transmitting, and receiving the video information are complex enough in "black and white" TV, and the complexity is greatly multiplied for color TV. Such "spectaculars" as live telecasts from the moon or from space vehicles, and the daily programs from overseas via satellite, are examples of the state of the art in communications electronics today.

ELECTRONICS IN SPECIALIZED OPERATIONS

32.31 Radar and Sonar

In *radar* and *sonar* receivers a cathode-ray tube is modified to give a visual indication of surrounding land masses (navigation), other ships on the surface of the sea (or under the surface—sonar), and aircraft in the surrounding skies.

A complete radar system (the word *radar* stands for *ra*dio *d*etection *a*nd *r*anging) has a *transmitter*, a *receiver*, and various *indicating devices*. The high-frequency radar waves (600 MHz and up) are sent out from the transmitter in a narrow cone-shaped beam. They travel outward with the velocity of light ($c = 3 \times 10^8$ m/sec) and are reflected when they strike a distant object. The very small part of the reflected wave energy which returns and strikes the receiving antenna is amplified in the radar receiver, and these strengthened signals are fed to the desired indicating devices.

The inverse-square law governs the attenuation of the radar signal strength. The transmitted energy falls off as the square of the distance from the transmitter; and the reflected energy arriving at the receiving antenna falls off as the square of the distance from the reflecting target. Consequently, the actual signal strength received from a target is inversely proportional to the fourth power of the distance to the target (see Fig. 32.50).

Signal strength is affected markedly by the size and nature of the target. Metallic objects and land masses give relatively high reflected signal strengths. The short wavelengths used necessitate almost a line-of-sight tracking situation, since these extremely short waves act much like light in that they do not bend over the horizon to any

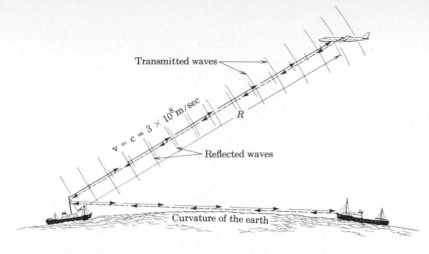

Transmitted waves

$v = c = 3 \times 10^8$ m/sec

R

Reflected waves

Curvature of the earth

Fig. 32.50 Sketch showing the basic principles of radar. Radar waves are electromagnetic waves which travel at the speed of light (3×10^8 m/sec). They travel in straight lines like light and are therefore not very effective beyond the visual horizon. They obey the inverse-square law as concerns intensity of reflected energy. If R is the range to a target, the reflected signal strength back at the ship is inversely proportional to R^4.

great extent. Unlike light, however, they do penetrate fog, haze, smoke, and clouds, which makes radar very useful for air and surface navigation, as well as for military purposes, and for tracking satellites and space vehicles. Other very important uses of radar include air-traffic control, weather observations and storm warnings, and highway traffic (speeding) control.

Methods of Radar Ranging Ordinarily the transmitter and the receiver use the same antenna system, so that the receiving device will have exactly the same bearing and elevation as the transmitting device. This scheme aids in target acquisition. The use of the same antenna, and the proximity of the receiver to the powerful transmitter necessitates a means of "blocking out" the transmitter from the receiver while transmitting, and turning off the transmitter completely during a small time interval while the returning signal is received. This is accomplished by an intermittent transmitting system called *pulsing*. [*Pulse radar* is only one of several types. Two other commonly used systems are *Doppler radar* (frequency shift) and *FM radar*.]

The transmitter is on for a few microseconds, during which time the receiver is deadened; then the transmitter is shut off while the receiver is switched fully on to pick up the returning signal. This on-again, off-again pulsing is repeated many hundreds of times per second.

This switching is effected by a component of the radar sometimes called the *T-R* box (Fig. 32.51).

One type of indicator is a special cathode-ray tube. A horizontal sweep maintains a continuous line across the face of the tube, and special circuits control its steady movement back and forth. A small amount of each radiated pulse is fed to the vertical deflection plates and the "pip" *T* of Fig. 32.51*a* is thus formed. As the echo signal returns it is also applied to the vertical deflection plates, and the "pip" *R* is formed. The time for the signal to reach the target and return determines the distance between *T* and *R*. The operator can position a pointer at *R*, and a dial then reads range to the target directly in miles or yards. Plan position indicator (PPI) screens with concentric circles to mark distances in miles are used to track the target as it approaches or recedes from the transmitter.

Range Rate In the case of an aircraft, ship, space vehicle, or speeding car, which may be coming toward or going away from the transmitter station, its *relative velocity* will cause an apparent frequency difference between the outgoing waves and the returning echo waves. This is a manifestation of the *Doppler effect* (see Sec. 20.3).

By accurately measuring this frequency change, and applying the proper Doppler formula, the velocity of approach or escape can be determined. This velocity of approach or

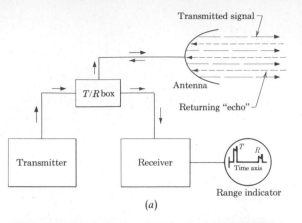

(a)

Fig. 32.52 Giant antenna for air route surveillance radar. The detection range of the system served by this antenna is 234 nautical miles (433 km), with a coverage altitude of 100,000 ft (30.5 km). (Westinghouse Electric Corp.)

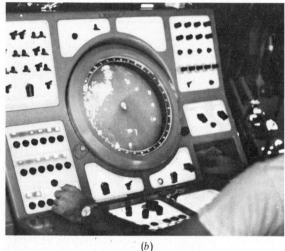

(b)

Fig. 32.51 (a) Block diagram of a radar system showing transmitter, receiver, *T-R* (switching) box, antenna, and range indicator. The distance (on the range indicator) between the initial pulse "pip" (*T*) and the echo pulse pip (*R*) is a measure of the elapsed time between transmitting and receiving a return signal and therefore a direct measure of the distance to the target. It takes 12.4 sec for a radar wave to travel 1 nautical mile to a target, be reflected, and return to the sending radar antenna. (b) A radarman monitoring the scope aboard a U.S. Navy ship. (Official U.S. Navy photo).

escape is known as the *range rate*, and it is a necessary factor for both navigation and gunfire control purposes. The calculation of the velocity or range rate is of course performed electronically within a computer section of the radar equipment itself, and the operator reads the answer directly; or it may be fed directly to other computing equipment in navigation or fire-control systems,

where the signal may be used to position weapons for firing or to indicate needed changes in course and speed to intercept or avoid the target.

One of the most important uses of radars is in the monitoring and control of air traffic. Without these electronic ranging devices, safe and efficient control of air traffic would be impossible. At night, or under conditions of reduced visibility, aircraft could not operate at all without radars. Electronic computers are tied in with radars, and air traffic controllers can "see" the present position and predicted (short-time) future position of any aircraft in range of the "tracking" radar (see Fig. 32.52).

Sonar is a method of locating (and tracking, if desired) underwater objects such as submarines, sunken vessels, the sea floor itself, or schools of fish. The energy waves employed are sound waves, not electromagnetic waves. The same general idea, however, is involved—sending out the signal and receiving the reflected return wave. The Doppler effect is again used to determine range rate; and the elapsed time between a leaving pulse and its return is the base for distance determination. The returning information is processed by a computer and displayed on a cathode-ray tube (sonar screen). The returning signal can also be made audible, when desired by the operator.

32.32   Communications by Satellite

Brief mention was made of communications satellites in Chap. 7, in connection with an explanation of the dynamics of orbital motion. Since Telstar I was launched in 1962 and operated for nearly a year, a series of communications satellites has been giving more or less continuous intercontinental communications service.

The term *synchronous* means that a satellite's motion is synchronized with the earth's rotation so that the satellite will "hover" over the same location on the earth's equator, the altitude required being about 22,250 mi. Modern communications satellites (see Fig. 32.53) are often powered by solar cells which convert the sun's radiant energy into electricity.

Design of modern satellites provides for the reception and retransmission of thousands of telephone conversations simultaneously. They are also designed for television transmissions, and live television programs between the United States and other countries by communications satellites are now a daily occurrence. Satellites are also used for navigation control, military surveillance, and weather monitoring; and for gathering information on crops, minerals, forests, ocean currents, and other earth data.

Nearly all of the electronic equipment and circuitry (vacuum tubes, transistors, integrated circuits, telephony, television, radar, microwave equipment, etc.) described in earlier sections is involved in communications by satellite. Both solar energy and nuclear energy (see Chap. 34) are utilized to provide the power to operate communications satellites.

32.33   Production and Uses of X-Rays

A discovery made in 1895 has had tremendous applications to medical science and industrial technology in the past 60 years. Wilhelm Röntgen (1845–1923), while experimenting with high-

Fig. 32.53 A communications satellite (COMSTAR), undergoing tests during manufacture. This unit is a part of a satellite communications network. Note the panels of solar cells, which furnish electric power to operate the satellite's equipment. (American Telephone and Telegraph Co.)

voltage discharges through gas-filled tubes, noticed a bright fluorescence in some crystals which happened to be close by (see also Sec. 33.8). In follow-up experiments Röntgen soon discovered that radiation of some sort seemed to be emanating from the tube and that these rays had extraordinary penetrating power, for they would pass readily through paper, cloth, the experimenter's hand, sheets of wood, and other like materials. Since the rays were of unknown origin and had unknown potentialities,

Röntgen called them x-rays. Within a few weeks he discovered that x-ray penetration was inversely proportional to the density of the material and, using photographic plates, took the first "x-ray picture" of a human hand. In less than a year, x-rays were being used throughout Europe as an aid to surgeons in setting broken bones. The industrial use of x-rays has developed only during the last 60 years and depended on the development of extremely high voltage sources (500,000 to 1 million volts and higher) to operate the tubes.

The Modern X-ray Tube In 1913 Dr. William Coolidge developed the forerunner of the modern x-ray tube, in which he used a tungsten target embedded in a copper anode, a heater-type cathode from which the electron flow could be controlled by the filament temperature, and a high vacuum tube. Figure 32.54 shows a picture and a diagram of the Coolidge x-ray tube, with hot cathode. As the cathode is heated under careful control, electrons boil off and are drawn to the anode, the speed of impact being dependent upon the potential difference between anode and cathode, which for medical diagnostic uses may be from 20,000 to 100,000 V, but for industrial purposes or medical therapy, may be 1 MV or higher. These high-speed electrons bombard the atoms of the tungsten target, resulting in actual changes in atomic structure, one result of which is the giving off of x-radiation, which, like all other electromagnetic radiations, travels with the speed of light. Like light, x-rays have been shown to be waves, but of very short wavelength—of the order of 2.8×10^{-8} cm or 2.8 Å.

X-rays in Industry Industrial use of x-rays in the United States began in 1922, with the installation at the Watertown Arsenal, of a 200,000-V x-ray machine. Its principal use was in the examination of castings for cracks and blowholes. Since this first installation, and particularly in the past 40 years, the use of industrial x-ray systems has caught on with amazing speed. Portable units have now been developed in sizes up to 2 MV and are used for inspecting tires, welds, castings,

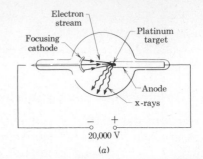

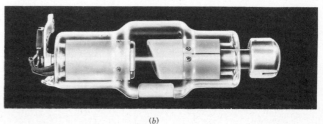

(b)

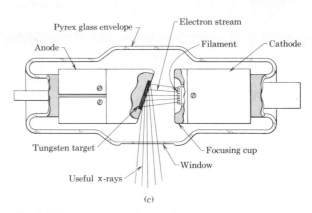

(c)

Fig. 32.54 The x-ray tube. (a) Diagram of an early x-ray tube. It is basically an electron gun with a platinum target. The high-speed electrons hit platinum atoms, creating atomic changes which result in the emission of x-radiation (see Chapter 33). (b) The modern Coolidge x-ray tube, as used in diagnostic medicine. (c) Diagram of Coolidge x-ray tube with parts labeled. (X-ray Department, General Electric Co.)

forgings, crankshafts, boilers, tanks, and highly stressed parts for aircraft and military equipment. The welded joints on the Alaska oil pipeline were all x-rayed by portable units of this kind. The penetrating power of 2-MV x-rays is tremendous. An 8-in. steel casting can be penetrated and a satisfactory radiograph ("x-ray picture") obtained in about 3-min exposure time.

Fig. 32.55 X-rays in industry. A 2-MV x-ray unit being prepared to obtain a radiograph of a steam turbine housing. (General Electric Co.)

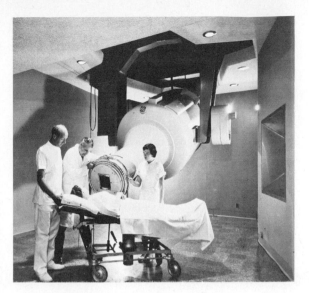

Fig. 32.56 X-rays in medicine. A 2-MV x-ray therapy unit, as used in patient treatment. (General Electric Co.)

Savings to industry are, of course, twofold: (1) use of x-rays prevents faulty parts from being used, with costly breakdowns and product deterioration later; and (2) x-ray analysis makes it unnecessary to scrap parts which have certain obvious surface flaws but which may actually be perfectly sound throughout (see Fig. 32.55).

32.34   Electronics in Medicine

In addition to the intensive use of x-rays in medicine, both for radiographs (x-ray "pictures") and for radiation therapy (see Fig. 32.56), recent years have seen the adoption by the medical profession of a multitude of electronic devices for the diagnosis, treatment, and simulation of human ills. The oscilloscope, in modified form, serves to monitor the vital signs of patients under intensive care; and another modification of the oscilloscope is the principal component of the electrocardiograph machine. Kidney machines, heart-lung machines, respirators, and other modern operating-room equipment depend on electronic circuits and components for their operation. Radioisotope and ultrasonic "scanners" used to check for abnormalities of human organs, including tumors, are dependent upon cathode-ray tubes and electronic circuits. The advent of the transistor and microcomputers on chips made the *heart pacer* possible, and the combination of a computer with sophisticated equipment for chemical analysis has made available to hospitals technological aids like the multichannel analyzer for making a number of different but simultaneous laboratory analyses from a patient's blood sample. Biomedical electronics technicians and engineers are now valued members of the staff of many modern hospitals, and the equipment they operate and service is in demand on all the wards.

SUMMARY

Electronics is the science of controlling free electrons as they pass through vacuum or gas-filled tubes or through semiconductors.

Electronic tubes are classifed by the number of electrodes as *diodes, triodes, tetrodes,* and *pentodes.*

Electronic tubes can be used to accomplish the following six functions: *rectification, amplification, control, frequency changing, electrical response to light*, and *converting electricity into radiation.*

The *characteristic curve* of a triode is obtained by plotting *plate current* I_p against *grid voltage* V_g for a constant value of *plate voltage* V_p. The curve thus obtained gives the rectification and amplification characteristics of the tube.

Transistors, which have been developed using new knowledge about solid-state physics, are devices for the control of free electrons. They are made of semiconductors and can rectify and amplify in electronic circuits.

Microprocessors on *silicon chips* are leading the way to extreme miniaturization in the electronics industry.

Electronic rectifiers include (1) the *duodiode*, for full-wave rectification for radio, public-address, and communications systems, (2) *gas-filled rectifiers* for medium-to-heavy current flow; and (3) *mercury-pool* tubes (*ignitrons* and *multianode tanks*) for extremely heavy-duty operation.

Amplifier circuits using electronic tubes include (1) radio-frequency amplifiers, (2) audio-frequency amplifiers, (3) power amplifiers, and (4) voltage amplifiers. Power amplifiers and voltage amplifiers are used in industrial circuits and use gas-filled grid control tubes. Electronic tubes are the essential components of *high-frequency* industrial equipment for *induction heating.*

Phototubes are sensitive to light, and their electron emission is an example of the *photoelectric effect.* They are installed in relay and amplifier circuits for the control of processes where light is a factor.

Electronic components, combined with sensors and gauges, are used in industry to control production operations by *feedback* from the end product. This is called *closed-loop control*, or more commonly, *automation.*

Computers have been revolutionized by *solid-state physics.* By means of *transistors*, *integrated circuits* (ICs), and *silicon chips* the pocket computers of today can be programmed to do the work of the "giant brains" of the 1950s.

The *telephone* depends on the principles of electromagnetic induction and makes use of many electronic components.

The basic elements of telephony are the *transmitter*, the *receiver*, and the *line.*

In *carrier telephony*, it is possible to send many messages over a single line simultaneously, by *modulating* different high-frequency carrier voltages with audio-frequency voice currents.

Coaxial-cable, *fiber-optics*, and *microwave telephony* make possible thousands of simultaneous messages over the same line or channel.

Radio communication is carried by *electromagnetic waves* traveling at the speed of light (3×10^8 m/sec or 186,000 mi/sec).

Spark gaps were used in early radio transmitters, but crystal-controlled *vacuum-tube* or *transistor oscillators* are now used for producing the electromagnetic waves.

A *radio transmitter* modulates the radio-frequency carrier wave with audio-frequency waves, either with amplitude modulation (AM) or frequency modulation (FM).

The *cathode-ray tube* (*CRT*) and the *oscilloscope* are among the most important developments of this century, with applications in science, engineering, business, industry, education, and medicine.

The *cathode-ray tube* is the integral component in oscilloscopes; in television, radar, and sonar receivers; and in many devices in medical laboratories. It is essentially a high-vacuum *electron gun* whose electron beam is controlled by voltages (or magnetic fields) applied to *deflector plates* (or *coils*). The resulting path of the electron beam is observed visually on a fluorescent screen.

Both *television* and *radar* make use of high-frequency electromagnetic radiation and cathode-ray tubes.

Communications satellites involve nearly all the basic principles and devices of electronics—photoelectricity, electromagnetic waves, oscillators, amplifiers, transistors, integrated circuits, silicon chips, receivers, and transmitters being a few examples.

X-rays are extremely short-wavelength (2.8 Å) radiations given off by certain metals (notably tungsten) when bombarded by high-speed electrons. Modern x-ray equipments use the *Coolidge tube* and operate at voltages from 50,000 V to 2 MV. In industry, x-rays are used to detect hidden flaws in castings, forgings, and highly stressed machine components.

A completely new field for electronics is developing as biologists, physicians, and electronic engineers work together in the field of biomedical engineering.

QUESTIONS AND EXERCISES

1. List all the devices and/or appliances in your home that have any kind of electronic component. Do not include devices that merely use electricity in ordinary wired circuits. Many of these may have electronic components, however, so mention the components.

2. Explain the process of *thermionic emission*. What is meant by *space charge?*

3. Describe how a simple diode can act as a one-way valve for the passage of electric current. Use a circuit diagram, and also show how alternating current becomes pulsating direct current.

4. What is the function of the grid in a triode? Show how the input to the grid can control the action of the tube.

5. Of what value are characteristic curves of triodes? (*a*) How are they obtained? (*b*) What does *cutoff point* mean?

6. Where (on the characteristic curve) would you operate a triode if amplification is the sole desired result? If rectification is the principal result desired? If both are desired?

7. List the six major functions of electronic tubes, and opposite each function give one industrial example.

8. Explain the operation of the ignitron rectifier. Why is the mercury-pool type of construction capable of such heavy-duty operation?

9. On the basis of additional reading, diagram the basic circuits and explain the operation of two different industrial applications of phototubes and photoelectric relays.

10. Explain what is meant by the terms: *semiconductor, valence electrons, holes, collector, base-emitter circuit, collector-load circuit, forward bias, reverse bias.*

11. Explain the operation of a cathode-ray oscilloscope as it might be used in analyzing a musical sound, such as a note from a trumpet.

12. Compare the functions of the elements in a triode vacuum tube with those in an *N-P-N* transistor.

13. Steam must enter the first stage of a turbine at 950°F and 2000 lb/in.^{2} from a natural-gas fired boiler. By means of a block diagram show how a closed-loop control system could be applied to the turbine-boiler system to utilize feedback and self-correcting mechanisms to maintain the proper pressure and temperature. Indicate the general nature of the electronic and other devices which could be used.

14. Explain in detail how it is possible to carry several simultaneous conversations, in both directions, over a single telephone line.

15. Describe Hertz's experiment with electromagnetic waves and show how it illustrates the phenomenon of electric resonance.

16. How is modulation of a radio-frequency carrier wave accomplished in a radio transmitter? Explain with diagrams.

17. In using the triode as an amplifier, why is it operated near the center of the linear portion of the characteristic curve? (Use a diagram.)

PROBLEMS

1. An oscillator for an induction heating machine has a resonance circuit consisting of an inductance L of 0.3 mH and a capacitance C of 0.04 μF. What is the frequency of oscillation?

2. Microwave telephony uses waves as short as 1 cm. What would the frequency (megahertz) of such waves be?

3. A broadcasting station transmits on a frequency of 1410 kHz. What is the wavelength (meters) of its radio waves?

4. A television station broadcasts on a carrier frequency of 500 MHz. What is the wavelength (meters) of these waves?

5. A radar receiver shows an elapsed time of 10^{-3} sec for the pulse to travel to an airborne target and return. How far away (miles) is the target?

6. A vacuum-tube oscillator has in the grid-cathode circuit a condenser of 0.0025 μF and an inductance of 0.06 mH. What is the natural frequency of the oscillator?

7. A transistor oscillator circuit has a frequency of 150 MHz. What is the period (time for one complete oscillation) for this oscillator?

8. How many times must an input signal be amplified (voltage amplification) to have a gain of 40 dB?

9. One of the citizens' band (CB) channels is on a frequency of 27 MHz. How long (meters) are these waves?

10. A radio signal is sent from "Houston Control" to a space vehicle in a low-altitude orbit around Mars. How long does it take for the signal to reach the vehicle if Mars is 85 million miles away at the time?

11. A sonar operator notes an elapsed time of 1.8 sec for his ship's sonar wave to travel to a submerged object and return. If the velocity of sound in water at that location is 1480 m/sec, how far away is the object?

33 Atomic Structure —Quantum Physics —Particles and Waves

We continue, in this chapter, with a look at some elementary principles and applications of "modern physics." As the nineteenth century drew to a close, most of the unifying theories of "classical" or "Newtonian" physics were well established. Almost all the basic principles had many practical applications, first in the industrial revolution, and from 1880 on, in the age of electricity. By about 1890 the entire body of knowledge known as physics seemed to be internally consistent and remarkably complete. Most intelligent persons, including many scientists of that period, probably felt that physics was a well-rounded body of knowledge and that little, if anything, remained to be discovered.

Hardly had such ideas taken root however, before a series of discoveries revealed that they were premature and ill-considered. In 1895 Röntgen discovered x-rays. In 1896, Henri Becquerel (1852–1908), in experimenting with uranium sulfate, discovered natural radioactivity; and in 1898, Marie Curie (1867–1934) separated two new elements, *polonium* and *radium*, from uranium ores. In 1897, J. J. Thomson (1856–1940), experimenting with electric discharges in evacuated tubes (cathode rays), identified the electron. And, shortly after the turn of the century, Albert Einstein (1879–1955), proposed an explanation of the *photoelectric effect*, and in 1905, suggested the now famous equation for *mass-energy equivalence*, $E = mc^2$. None of these discoveries was explainable on the basis of classical physics, and indeed some of them were actually incompatible with Newtonian physics.

The present chapter deals with atomic structure, and with the behavior and energy of the electron and other atomic particles. We will also look further into the question of how energy is transmitted through space. Is it carried by *waves*, or do infinitely small *particles* of mass travel with the speed of light and give up their kinetic energies on impact? Or do both phenomena occur, under different sets of conditions?

First, a brief and elementary review of atomic structure.

33.1 Early Concepts of Atomic Structure

The original idea of the atom goes back to the philosophers of ancient Greece. Democritus, about 450 B.C., conceived of all matter as being composed of tiny, indivisible spheres. The Greek word for *indivisible* is *atomos*, and from this root word comes the modern word *atom*. For 22 centuries the idea of the atom lay undeveloped until, in 1805, John Dalton (1766–1844), an English chemist, proposed the theory that all matter was made up of just a few basic *elements* and suggested that the unit building block of each element be called the *atom*. Dalton conceived of atoms as tiny, indestructible spheres, all atoms of the same element being alike but the atoms of one element differing from those of the other elements. Chemical reactions, said Dalton, were reactions between atoms, and *compounds* were built up from elements by chemical combination of the atoms of the elements.

Scientists have since identified and named 92 elements which occur naturally, ranging from *hydrogen*, the lightest element, to *uranium*, the heaviest. In recent years additional elements, heavier than uranium, have been produced in laboratories, but as far as we know they do not occur naturally. These are the so-called *transuranic elements* of modern physics.

33.2 Nineteenth-Century Discoveries —The Electron

Dalton's atomic theory aided the development of the science of chemistry through the entire nineteenth century. About 1865, however, Sir William Crookes (1832–1919), an English scientist, while experimenting with high-voltage (about 5000 V) electric discharges through evacuated tubes, discovered a visible discharge from the negative electrode, or *cathode*, of the tube. He called this visible stream of energy *cathode rays*. This discovery gave rise to speculation that the atom was not indivisible after all.

J. J. Thomson in 1896 discovered that cathode rays consist of particles which are electric in nature and that they possess a negative charge. He found that the rays can be bent by magnetic and electrostatic fields (see Fig. 33.1). He measured the velocity of the particles and found

it to be thousands of miles per second and discovered that the velocity depends upon the voltage applied across the terminals of the tube. And—most significant of all—he found that regardless of the material used for the cathode, the particles which "boil off" and shoot toward the anode always have the same properties. Thomson concluded that atoms therefore are *not indivisible*, that these cathode-ray particles are present in all atoms, and that they, not atoms, are a basic unit of matter. Since they possessed the properties of both *matter* and *electricity*, he called them *electrons*. (In modern physics the ending "-on" implies *particlelike* behavior, as distinguished from wavelike behavior.

33.3 Late Nineteenth- and Early Twentieth-Century Discoveries

Robert A. Millikan (1868–1953), an American, measured the mass of the electron in 1913 and found it to be about $\frac{1}{1835}$ the mass of the lightest atom, hydrogen. This makes the mass of the electron (at rest) 9.109×10^{-31} kg. He also measured the charge on the electron and found it to be so small that, for a current of 1 A about *six billion billion electrons* (6×10^{18}) per second would have to pass a given point in a wire! The charge on the electron was determined as $e = 1.602 \times 10^{-19}$ coulomb.

Other research physicists and chemists were active in the field of atomic physics during these years. Since the electron was a negative particle and whole atoms were thought to be *neutral* electrically, search for the positive parts of the atom

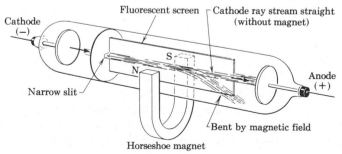

Fluorescent screen — Cathode ray stream straight (without magnet)

Cathode (−)

S

N

Anode (+)

Narrow slit

Bent by magnetic field

Horseshoe magnet

Fig. 33.1 Diagram showing the bending of cathode rays in a magnetic field. If the polarity of the magnet is reversed, the stream of rays will bend up instead of down.

was going on. Also the question of the remainder of the mass of the atom was explored. Some careful calculations revealed that atoms were exceedingly small (of the order of 10^{-8} cm in diameter)—so small that more than a billion of them, side by side, would not be equal to the width of a man's hand. Small as atoms were, the electron was found to be only a very tiny fraction of the atom. What filled the rest of the space? Or was it empty? Where were the positive charges? What and where was the greater part of the mass of an atom?

33.4   The Proton

In 1886, Eugen Goldstein (1850–1930), while experimenting with cathode rays, designed a special type of electric-discharge tube with the cathode in the center, as diagrammed in Fig. 33.2. A small hole was provided in the center of the cathode, as shown. Despite the fact that the cathode rays traveled from cathode to anode, Goldstein observed that a fluorescent screen at the other end of the tube would glow as the tube was operated. A stream of invisible particles seemed to be flowing *counter to the cathode-ray stream* through the small hole in the cathode, to strike the fluorescent screen and cause it to glow. Goldstein called these new "rays" *canal rays*, in reference to the streamlike character of the discharge.

Later (in 1896) Wilhelm Wien (1864–1928) was successful in deflecting canal rays in a magnetic field and proved them to be not "rays" but *positively charged particles*. Subsequent workers found their charge to be exactly equal to that of the electron, but of opposite sign. It was found that *any material* used as an anode would give off these particles; and so they, too, came to be recognized as one of the basic units of matter. The name *proton* was given to this particle, and later research has revealed that protons reside in the central core, or *nucleus*, of the atom. The proton has a mass (at rest) about 1835 times that of an electron, or approximately 1.673×10^{-27} kg.

It is important to note that these early workers in the field of modern physics were almost immediately confronted with these questions: Is it a wave? Is it a ray? Is it a particle? Or does it exhibit attributes of both waves and particles?

33.5   The Bohr-Rutherford Atom Model

On the basis of experimental evidence available up to 1910, Lord Rutherford (1871–1937) proposed a theory of atomic structure which imagined the atom to be like a miniature solar system with a heavy center which he called the *nucleus*, surrounded by electrons revolving in various orbits, much as the planets of our solar system revolve around the sun. Niels Bohr (1885–1962), a Danish physicist and a student of Rutherford's, was also instrumental in the development of this *planetary* concept or Bohr-Rutherford model of the atom (see Fig. 33.3).

As developed during the decade 1910 to 1920, the planetary-atom idea included the following hypotheses:

1. The atom is mostly empty space; e.g., if a uranium atom were a mile in diameter, the nucleus would be only about the size of a baseball!

2. Electrons (negative particles or charges) revolve around the positive nucleus in definite *orbits* or *shells*.

3. Almost all the mass of the atom resides in the central nucleus.

4. The nucleus is made up of protons, which are the unit positive particles.

5. Every normal atom has as many protons in the nucleus as there are electrons in the orbits: the atom is electrically neutral.

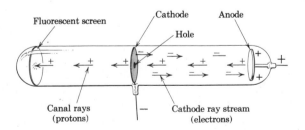

Fig. 33.2   Diagram of Goldstein's apparatus with which he observed "canal rays," later identified as protons.

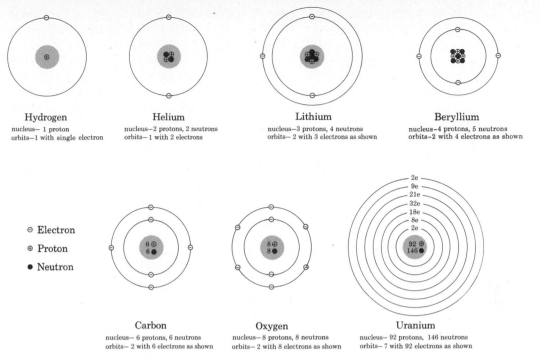

Hydrogen
nucleus— 1 proton
orbits—1 with single electron

Helium
nucleus—2 protons, 2 neutrons
orbits—1 with 2 electrons

Lithium
nucleus—3 protons, 4 neutrons
orbits— 2 with 3 electrons as shown

Beryllium
nucleus—4 protons, 5 neutrons
orbits–2 with 4 electrons as shown

⊖ Electron
⊕ Proton
● Neutron

Carbon
nucleus— 6 protons, 6 neutrons
orbits— 2 with 6 electrons as shown

Oxygen
nucleus— 8 protons, 8 neutrons
orbits— 2 with 8 electrons as shown

Uranium
nucleus— 92 protons, 146 neutrons
orbits– 7 with 92 electrons as shown

Fig. 33.3 Two-dimensional atom diagrams according to the Bohr-Rutherford, or planetary model of the atom.

This theory seemed to fit all the observed facts but one—the observation that most atoms had a greater mass than the number of protons and electrons would seem to indicate. There arose a belief among some atom scientists that there must be another type of particle in the nucleus, a particle with no charge, *almost equal in mass to the proton*. In 1932, Sir James Chadwick, an English physicist, discovered this particle. It was called the *neutron*, because of its neutral electric state. The neutron will be given further attention in Chap. 34.

Atoms, as pointed out above, are extremely small. No one has ever seen an atom. Only by scattered bits of evidence gathered over half a century of experimentation can scientists even attempt to "picture" an atom, and indeed an actual visual picture is not possible at all. It is only by diagrams (and these are not to scale) that we attempt to portray some of the known facts and hypotheses about atoms. The diagrams of Fig. 33.3 show in a schematic way the supposed structure of several atoms, according to the planetary-atom (Bohr-Rutherford) hypothesis.

The smallest and lightest atom is *hydrogen*. Its nucleus is a single proton, and it has only one electron revolving around the nucleus. *Helium*, the next heavier, has a nucleus composed of 2 protons and 2 neutrons, with 2 electrons revolving about the nucleus. *Oxygen* has 8 neutrons and 8 protons in the nucleus, with 8 electrons in its orbits. *Uranium*, the heaviest of the *natural* elements, has 92 electrons in its orbits and 92 protons and 146 neutrons in the nucleus.

33.6   The Quantum Idea

Energy can be carried by both particles and waves. Energy transfer with particles involves the movement of a concentrated mass of matter, that is, a mass in motion. Wave motion involves energy moving from one point in space to another, by means of the oscillatory motion of particles, or as a result of disturbances in electric and magnetic fields. Particles and waves would seem to be quite different things, from the vantage point of classical physics.

However, when atomic phenomena are studied, two surprising findings emerge: (1) both particles and waves possess only steplike amounts of energy, and (2) particles sometimes act like waves in other ways as well (recall the discussion of the nature of light in Chap. 21). Based on classical physics, neither finding seems reasonable. However, both findings are essential to explain the observed characteristics of such now common devices as the photocell, the electron microscope, x-ray tubes, and the laser, for example.

Quantization Quantum physics deals with the study of phenomena in which only discrete or *quantized* amounts of energy are evidenced. *Discrete variables* are those that can take on only definite values within a given range. The number of people in a room is a quantized variable. It could be 7, 8, 22, or 27, but not 7.4 or 22.9. *Continuous variables*, on the other hand, can assume any numerical value between given limits. Water flow in a pipe or flume is an example of a continuous variable. The flow is not just in exact whole units of cubic feet or even cubic centimeters, but

is continuous. Any increment specified could be infinitely small, and even then that small increment could, in theory, be further divided.

Our everyday experience suggests that energy should be a continuous variable. The kinetic energy of an automobile, $\frac{1}{2}mv^2$, could have almost any value within specified limits, since velocity is a continuous variable. Any speed between 20 mi/hr and 90 mi/hr is quite possible (although perhaps not accurately measurable). Similarly, the potential energy associated with a pile driver, *mgh*, is a continuous variable, since the height of the falling mass can assume any value between reasonable limits.

In the light of twentieth-century discoveries however, it turns out that the energies of electromagnetic radiation and atomic particles, for example, are not continuous quantities. Their energies are limited to values with steplike or incremental (discrete) differences from one value to the next. We say their energies are *quantized*. The word "quantum" means "a small increment of energy."

A number of quantities you have already studied are quantized variables. Some examples are: (1) the electric charge on a particle or object, which can have only those values that are multiples of 1.602×10^{-19} C; (2) the atomic number

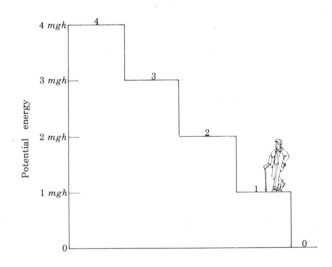

Fig. 33.4 A "quantized" man. True if gravitational potential energy (GPE) could exist only in discrete increments, or "quanta," of 1 mgh. Steps 1, 2, 3, 4 . . . could be regarded as energy levels, or energy states.

(Z) of an element, which can only be a positive integer; and (3) the harmonic frequencies of a vibrating string or air column, which are whole number multiples of the first harmonic. Figure 33.4 shows a "quantized man," that is, what it would be like if gravitational potential energy were always quantized, the way the energy of atomic particles is quantized.

Stair steps are not foreign to our experience, but suppose that steps of a specified (discrete) height were the only way to change gravitational potential energy level. No ramps, no rolling hills, and no slides would be possible.

Planck's Constant An early suggestion of quantization came in 1900 when Max Planck (1858–1947), a German physicist, used the idea to explain the distribution of radiant energy that is emitted by a hot object (see Sec. 15.6). Classical wave theory held that the radiation emitted from an "ideal" (blackbody) radiator was due to oscillations of the heated (excited) atoms in the body. It was known that electrical discharges (sparks) produced oscillations called *radio waves* (see Sec. 32.23), and it was assumed that all forms of radiation would behave in a similar manner. The amount of energy associated with atomic oscillations was therefore considered to be continuous by early investigators of the period 1885 to 1900.

As laboratory equipment became capable of greater precision, it was noted that with regard to blackbody radiation, the distribution of frequencies predicted from classical physics theory was at variance with the observed actual distribution. To account for this discrepancy, Planck assumed that the energy of atomic oscillators was not continuous, but instead *quantized*. Stated another way, Planck's laboratory observations and detailed mathematical calculations made it essential to assume that the energy of each atomic oscillator must vary by discrete increments or "jumps," and that therefore the energy emitted by the blackbody radiator must also vary by increments or jumps. By theoretical work and by "fitting" mathematical constants to observed radiation curves, Planck derived a simple equa-

tion for the minimum possible energy change or *quantum*. If E is the quantum of energy, and f is the frequency of oscillation, then

$$E = hf \qquad (33.1)$$
(Planck's quantum condition)

The proportionality constant, h, is known as *Planck's constant*.

The accepted value of h is 6.626×10^{-34} J-sec. What Planck's equation says is that the bundles of energy emitted by an oscillator are limited to those values that are integral multiples of one quantum. For any given source of radiation, the magnitude of the quantum is proportional to the frequency in cycles per second (Hz). The proportionality constant h was determined from experimentally observed radiation curves. *Planck's constant* is a fundamental constant of nature and now ranks in importance with such other well-known constants as g, G, c, e, and J. It will appear repeatedly in the discussions to follow.

Quantum theory indicates that an oscillator of frequency f may not emit any and all energies, but only *discrete* energy values which are whole-number multiples of hf, that is, $1hf$, $2hf$, $3hf$, ... Planck's quantum equation can therefore be written in the form

$$E = nhf \qquad (33.1')$$

where n is restricted to integral values, 1, 2, 3, ... n is called a *quantum number*.

Illustrative Problem 33.1 An atomic oscillator has a frequency of 3.28×10^{14} Hz. What are its three lowest possible levels of energy, assuming Planck's quantum condition?

Solution Use Planck's quantum equation (33.1'), where

$$n = 1, 2, \text{ and } 3$$
$$E_1 = 1 \times 6.626 \times 10^{-34} \text{ J-sec}$$
$$\times 3.28 \times 10^{14} \text{ Hz}$$
$$= 2.17 \times 10^{-19} \text{ J} \qquad ans.$$
$$E_2 = 2E_1 = 4.34 \times 10^{-19} \text{ J} \qquad ans.$$
$$E_3 = 3E_1 = 6.51 \times 10^{-19} \text{ J} \qquad ans.$$

33.7 The Photoelectric Effect—Einstein's Hypothesis

The ability of a beam of light to cause electrons to be ejected from certain metallic surfaces has been mentioned in Chaps. 26 and 32 (see Fig. 32.29). This conversion of light energy to electric energy is called the *photoelectric effect*. The ejected electrons are called *photoelectrons*.

The wave theory of light (see Chap. 21) suggests that light waves might transmit energy to electrons in the same manner that a water wave transmits motion to a floating body. With water, the more energy there is in the water wave, the more violently a floating body will be agitated. Reasoning along this line, the energy imparted to a single electron by photoelectric action should be related to the intensity of the light. But this was not found to be the case. The following rather unexpected observations were confirmed by many investigators of photoelectricity at about the turn of the century:

1. For each metallic substance, there is a certain minimum frequency of the light required, before any electron emission occurs, *no matter how great the intensity of illumination.* This frequency is referred to as the *threshold frequency* of that metal.

2. When the threshold frequency is reached, emission of electrons begins immediately, *no matter how weak the intensity of illumination.*

3. For radiation of a given frequency (above the threshold), the rate of emission of electrons (measured electric current) is directly proportional to the intensity of the illumination.

4. The light intensity has no effect on the maximum kinetic energy of the emitted electrons. The kinetic energy is related instead to the *frequency* of the light.

These findings, consistently reported by the most able researchers of the period, were for the most part incompatible with a wave "model" theory of light. According to the wave theory, the energy in a light beam is proportional to the intensity of illumination, not the frequency, and ejected electrons should therefore have kinetic energies related to light intensity, not frequency. Also, at low light intensities the wave theory would predict either no electron ejection at all, or a significant time delay before each electron would absorb enough energy to eject. A new explanation was needed to account for these experimental findings.

Light as a Stream of Photons Albert Einstein (1879–1955) in 1905 proposed a theoretical explanation for the photoelectric effect. He revived the old particle (or "corpuscular") theory of light and suggested that:

1. Light travels as a stream of individual packets of energy, which he termed *photons*.

2. The amount of energy carried by a *single photon* is

$$E_{photon} = hf \qquad (33.2)$$

a form of Planck's quantum equation, where

f = the frequency of the light

h = Planck's constant (6.626×10^{-34} J-sec)

3. Each electron emitted acquires its energy from a single photon, never from two or more. In other words, the photon-to-electron interaction is a one-to-one proposition.

4. As a photon gives its total energy hf to one electron, a part of the energy is required to do the work of removing the electron from the metal, and the remainder will show up as kinetic energy of the emitted photoelectron. Stated another way, max. electron KE = photon energy − work to remove the electron

$$\tfrac{1}{2}mv_{max}^2 = hf - \phi \qquad \text{or}$$

$$E = hf = \phi + \tfrac{1}{2}mv_{max}^2 \qquad (33.3)$$

(the photoelectric equation)

where $E = hf$ is the energy of a single photon, $\tfrac{1}{2}mv_{max}^2$ = the KE$_{max}$ of the photoelectron produced, and ϕ is the *work function*—the minimum amount of work done to remove the electron from the metal, m is the mass of the photo electron, and v is its velocity.

In terms of the wavelength of (monochromatic) light,

$$E = h\left(\frac{c}{\lambda}\right) = \phi + \tfrac{1}{2}mv^2 \qquad (33.4)$$

The work function ϕ is an attribute of each cathode material.

The photoelectric effect is, then, an energy-conversion process. The light photon with energy $E = hf$ ceases to exist after it transfers its energy to an electron. The electron uses its newly acquired energy (a) to accomplish the work of separation through the surface potential barrier, and (b) for residual kinetic energy of motion after it leaves. If the photon's total energy is less than the work function for that metal ($hf < \phi$), no electron emission will occur. The *threshold frequency* is thus found when

$$hf = \phi \tag{33.5}$$

High-intensity light means that more photons are striking the metal surface, causing more electrons to be emitted. The kinetic energy of the photoelectrons is not increased by greater light intensity as would be expected from the wave theory. The one-photon/one-electron interaction governs the process and unless $hf > \phi$, an electron will not be emitted at all. The entire process is not continuous on a scale of energy, but *quantized*.

The work function of most metals is between 1 and 5 electron-volts (eV). The *electron-volt* is a unit of energy equal to that acquired or lost by an electron when it is accelerated through a potential difference of 1 V. It is a very useful energy unit in atomic and nuclear physics.

$$1 \text{ eV} = 1.602 \times 10^{-19} \text{ C} \times 1 \text{ V}$$

$$= 1.602 \times 10^{-19} \text{ J} \tag{33.6}$$

The photoelectric effect constitutes one of the major pieces of evidence that light must be considered as having particlelike characteristics. A photon, like a particle, carries a concentrated amount of energy. *Photon* means a "particle of light." Einstein received the Nobel Prize in physics in 1921 for his explanation of the photoelectric effect, in which he extended Planck's quantum idea to include the emitted radiation itself, not just the oscillator energy that produced it.

Illustrative Problem 33.2 Light of wavelength 6000 Å (1 Å = 10^{-10} m) shines on a metal whose work function is 1.3 eV. What is the velocity of the emitted photons of maximum kinetic energy? (NOTE: Some electrons deeper in the metal may also escape, but the work necessary to separate them will be greater than the work function of 1.3 eV and their KE will be less than maximum. We are concerned here only with the electrons of *maximum* KE.)

Solution Solve Eq. (33.4) for KE_{max}:

$$KE_{max} = \tfrac{1}{2}mv^2 = h\frac{c}{\lambda} - \phi$$

$$= 6.626 \times 10^{-34} \text{ J-sec} \frac{3 \times 10^8 \text{ m/sec}}{6 \times 10^{-7} \text{ m}}$$

$$- 1.3 \text{ eV} \times 1.602 \times 10^{-19} \text{ J/eV}$$

$$= 1.23 \times 10^{-19} \text{ J}$$

Rearranging algebraically yields

$$v^2 = \frac{2 \times 1.23 \times 10^{-19} \text{ J}}{m}$$

But (see tables inside back cover), m = rest mass of the electron = 9.11×10^{-31} kg. Substituting,

$$v^2 = \frac{2.46 \times 10^{-19} \text{ kg-m}^2/\text{sec}^2}{9.11 \times 10^{-31} \text{ kg}}$$

from which,

$$v = 5.2 \times 10^5 \text{ m/sec} \qquad ans.$$

Illustrative Problem 33.3 Zinc has a work function of 4.23 eV. What threshold wavelength is needed to produce electrons?

Solution Use Eq. (33.5) and recall that $f = c/\lambda$. Solving for the threshold wavelength, we get

$$\lambda = \frac{hc}{\phi}$$

$$= \frac{6.626 \times 10^{-34} \text{ J-sec} \times 3 \times 10^8 \text{ m/sec}}{4.23 \text{ eV} \times 1.602 \times 10^{-19} \text{ J/eV}}$$

$$= 2.93 \times 10^{-7} \text{ m}$$

$$= 2930 \text{ Å} \qquad ans.$$

This frequency is in the ultraviolet region. (See Sec. 21.14 and Fig. 21.14.) *Visible light* is not energetic enough to cause photoelectrons to be emitted from zinc.

Confirmation of Planck's Constant Planck's constant h was originally determined from measurements connected with blackbody (cavity) radiation. Since it is more or less the basis of quantum theory, one would expect that it could be verified from observations of other physical phenomena. Einstein's work on the photoelectric effect in 1905, and the work of Robert A. Millikan in 1916 provided just such verification.

From Einstein's photoelectric equation (33.3), $\frac{1}{2}mv^2$ represents the maximum kinetic energy of the photoelectrons emitted. This KE_{max} can be determined by the use of a *stopping voltage* in a circuit like that shown in Fig. 33.5. The photocell consists of a clean metal surface called the cathode, and a "collecting" anode, both housed in an evacuated quartz tube. The galvanometer detects the presence or absence of electron flow (current). The battery can be connected either to aid or to oppose the flow of photoelectrons from the cathode to the collecting anode. Provision must be made for continuous variation in the battery voltage. Phototubes with different metals as cathodes are used, and provision is made for using incident light of several different and known frequencies. The exact opposing voltage needed to reduce the photoelectron current to zero is determined for each frequency. At this *stopping voltage*, the photoelectrons possessing maximum KE have been stopped. Those possessing less than maximum KE would have been stopped before the stopping voltage was reached.

From Eq. (33.3), equating photon energy to the work function plus the maximum kinetic energy of emitted electrons,

$$hf = \phi + KE$$

But to stop the flow of photoelectrons, a repelling energy field must be used such that

$$Ve = KE = \tfrac{1}{2}mv^2 \tag{33.7}$$

where V is the applied negative voltage (*stopping voltage*), and e is the electronic charge (charge on the electron).

Substituting in Eq. (33.3), and rearranging yields

$$Ve = hf - \phi \tag{33.8}$$

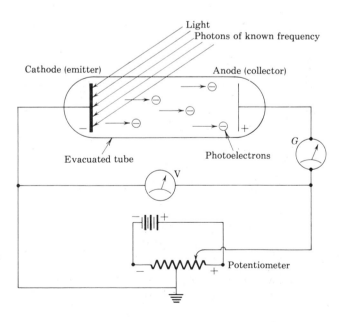

Fig. 33.5 Diagram of an apparatus for measuring photoelectric current and for detecting the maximum kinetic energy (KE_{max}) of the photoelectrons. With the battery and potentiometer arranged as shown, photoelectrons are emitted from the cathode and are attracted to the anode on which the battery holds a positive charge. The galvanometer, G, measures the photoelectric current in the external circuit. The experimenter can determine the exact negative voltage V needed on the anode to stop the flow of photoelectrons, by reversing the battery current and adjusting V by means of the potentiometer until the galvanometer reads zero.

When experimental data are collected, and values of Ve are plotted against values of the light frequencies applied, a graph like that of Fig. 33.6 is obtained. Note that regardless of the metal used for the cathode, the plots are straight lines, and they all have the same inclination to the horizontal axis (*slope*). From basic algebra, any equation of the form $y = mx + b$ plots to a straight line, with m being the slope of the line and b being the point of intercept on the y axis. In the case of Eq. (33.8), values of Ve are plotted vertically (y values), and values of f are plotted horizontally (x values). When the slope m is measured from such plots, it is found to be equal to Planck's constant h. Thus, the numerical value of Planck's constant and the quantum idea itself are confirmed by experiment in a field different from that of Planck's original investigations.

The negative y intercept is identified as the work function, ϕ, of the particular metal. The intercept on the x axis (frequency) is the minimum (*threshold*) frequency f_0 that will liberate photoelectrons from that metal. At these points, $Ve = 0$, and Eq. (33.8) reduces to Eq. (33.5), for *the threshold frequency*:

$$hf_0 = \phi \qquad \text{or} \qquad f_0 = \frac{\phi}{h} \qquad (33.9)$$

Illustrative Problem 33.4 The threshold frequency f_0 for gold is 1.16×10^{15} Hz. If ultraviolet light of frequency 1.60×10^{15} Hz is incident on the gold cathode of a phototube, what is the maximum kinetic energy of the emitted photoelectrons, in electron-volts?

Solution From Eq. (33.3), recall that

$$\tfrac{1}{2}mv^2 = KE_{max}$$

From Eq. (33.8), $\phi = hf_0$. Substituting these terms in Eq. (33.3) gives

$$KE_{max} = hf - hf_0 = h(f - f_0)$$

Substituting given values results in

$$KE_{max} = 6.626 \times 10^{-34} \text{ J-sec} \times (1.60 - 1.16)$$
$$\times 10^{15} \text{ cycles/sec (Hz)}$$
$$= 2.92 \times 10^{-19} \text{ J}$$
$$= \frac{2.92 \times 10^{-19} \text{ J}}{1.602 \times 10^{-19} \text{ J/eV}} = 1.82 \text{ eV} \qquad ans.$$

33.8 The Production of X-Rays

In a prior chapter we discussed the x-ray tube and some of the industrial and medical applications of x-rays. In this section the emphasis will

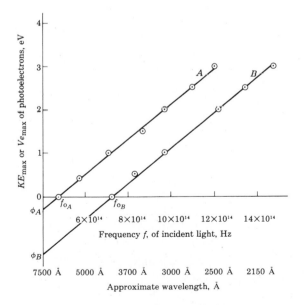

Fig. 33.6 A plot of photoelectron KE_{max} as a function of the frequency of incident light, for two different metals, A and B. The x-intercepts f_{oA} and f_{oB} mark the threshold frequencies for A and B. The y-intercepts mark the work functions for the two metals. The mathematical slope of such lines is the same for all metals, and it is numerically equal to h, Planck's constant.

be on the energy processes in atoms that produce x-rays and on the characteristics of the x-ray spectrum.

X-rays are short-wavelength radiations (photons), with wavelengths on the order of 10 Å to 0.01 Å. They are highly penetrating but can be absorbed by dense matter such as lead. In being absorbed they transfer all or part of their energy to electrons in the material, and thereby *ionize* the material. X-rays ionize atoms in biological tissue, and therefore are harmful in large doses and, with continuing exposure, even in small doses, since effects are cumulative.

Braking Radiation (Bremsstrahlung) X-rays can be produced in two ways. The first process is called *bremsstrahlung* (from the German, meaning *braking radiation*). It is depicted in Fig. 33.7. As an electron at *A* moves at high speed in the immediate vicinity of a heavy nucleus (i.e. impacting on a solid metal target), it has rapid negative acceleration, loses energy, and in terms of the wave theory, radiation is emitted. In quantum terms, the energy emitted is in the form of photons, each having a definite frequency. The negative acceleration (braking) of the electron can occur either as a result of its actually being stopped by collision with nuclei of the target

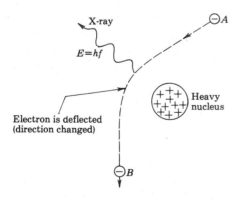

Fig. 33.7 Schematic diagram illustrating braking radiation (*bremsstrahlung*). A high-velocity electron at *A* is slowed and redirected as it passes near an atomic nucleus. The change in speed *and direction* represents a large negative acceleration. The electron's loss in energy shows up as an x-ray photon whose energy is *E = hf*.

material, or merely by having its direction changed (velocity is a vector quantity) as shown in Fig. 33.7.

The target in an x-ray tube (see Fig. 32.54) must be a metal with heavy nuclei, atomic numbers of $Z \geq 40$ being typical for x-ray production. A great deal of heat is generated in the collision (or braking) process, so materials with high melting points must be used. Tungsten is a commonly used target material. Provision for cooling the tube is ordinarily made.

For the emission of a single photon, as an electron is "braked" from velocity v to velocity v', the conservation-of-energy equation is

$$E = hf = \tfrac{1}{2}mv^2 - \tfrac{1}{2}mv'^2 \qquad (33.10)$$

The *maximum* photon frequency is produced when an electron is completely stopped and v' becomes zero. In that case the frequency can be found from

$$E = hf_{max} = \tfrac{1}{2}mv^2 \qquad (the\ x\text{-}ray\ equation) \qquad (33.11)$$

In order to provide electrons of high kinetic energy, they are accelerated in the x-ray tube by a high potential difference, ordinarily of 10 kV or more. The electron potential energy-to-photon kinetic energy conversion can be expressed by Eq. (33.7) as

$$Ve = \tfrac{1}{2}mv^2$$

Combining Eqs. (33.11) and (33.7), and using $f = c/\lambda$, we obtain

$$Ve = h\frac{c}{\lambda} \quad \text{from which} \quad \lambda = \frac{hc}{Ve}$$

Substituting known physical constants the *minimum* x-ray wavelength is given by

$$\lambda_{min} = \frac{(6.626 \times 10^{-34}\ \text{J-sec})(3 \times 10^8\ \text{m/sec})}{(1.602 \times 10^{-19}\ \text{C}) \times \text{V}}$$

Since coulombs = joules/volts, or C = J/V, and 1 Å = 10^{-10} m, it follows that

$$\lambda_{min} = \frac{12,400\ \text{volts}}{V}\ \text{Å} \qquad (33.12)$$

Equation (33.12) says that an applied voltage V of 12,400 volts produces x-ray photons with a minimum wavelength of 1 Å. To obtain the wavelength in Å of the x-rays from an x-ray tube, divide 12,400 volts by the voltage applied across the tube. Typical accelerating voltages range from 10 kV up to 2 MV or more.

Braking-radiation x-ray production is often referred to as the inverse photoelectric effect. In the photoelectric effect, photon energy is converted to photoelectron energy. In x-ray production, electron energy is converted to the energy of x-ray photons. The basic equation for x-ray production, $hf_{max} = \frac{1}{2}mv^2$, is just the photoelectric equation, $hf = \frac{1}{2}mv^2 + \phi$, with the work function ϕ = zero.

Illustrative Problem 33.5 An x-ray machine uses an accelerating voltage V of 25,000 V. Find (*a*) the maximum electron kinetic energy produced, and (*b*) the (minimum) x-ray wavelength.

Solution (*a*) Using Eq. (33.7) gives

$$\frac{1}{2}mv^2 = Ve = (1.602 \times 10^{-19} \text{ C})(25 \times 10^3 \text{ V})$$

$$= 4 \times 10^{15} \text{ J} \qquad \qquad ans.$$

(*b*) Use Eq. (33.12) to solve for wavelength:

$$\lambda_{min} = \frac{1.24 \times 10^4 \text{ V}}{2.5 \times 10^4 \text{ V}} \text{ Å}$$

$$= 0.5 \text{ Å} \qquad \qquad ans.$$

Characteristic X-rays The second mechanism for creating x-rays involves electron transitions between energy levels, as an electron "jumps" from one orbit to another orbit of an atom. In heavy atoms (i.e. with heavy nuclei), *energy shells* have been identified and labeled K, L, M, N, etc. When a high-speed electron collides with a tightly bound electron in the innermost (K) shell, it may transfer enough energy to knock that electron out of the atom. A vacancy is thus created, and an electron from a higher-energy shell may move to fill the vacancy, thereby creating a new vacancy (Fig. 33.8). Other electrons change shell locations until the vacancy moves to the outside of the atom. A number of photons are emitted,

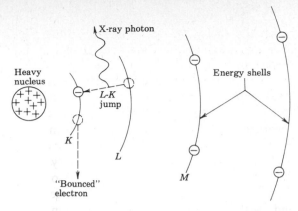

Fig. 33.8 Schematic diagram of the production of characteristic x-rays. Only those electron "jumps" that occur in the innermost shells or orbits involve enough energy to produce x-ray photons.

but only the innermost electron changes involve enough energy to produce x-ray photons. These x-rays are called *characteristic x-rays*, because they are characteristic of the energy differences between shells involved in the electron jumps in a particular atom.

33.9 The Compton Effect

We have described Planck's theoretical deductions from his work with blackbody (cavity) radiation, Einstein's theory of photoelectricity, and the production and behavior of x-rays, as supportive of the idea that "light waves" or electromagnetic waves are particlelike in nature. Other experiments and theoretical work in modern physics support the quantum or photon concept, and we will briefly describe one more—the *Compton effect*. Arthur H. Compton (1892–1962) was investigating the properties of x-rays in 1922. He found that when a beam of x-rays strikes a target of solid carbon, the x-rays are "scattered" and their wavelength is slightly *increased*. Classical wave theory would predict that when an electron is "struck" by an electromagnetic wave, the electron would oscillate with the same frequency as the (x-ray) wave, and would radiate energy of that same frequency (wavelength).

Since classical wave theory could not account for the *increased* wavelength, Compton

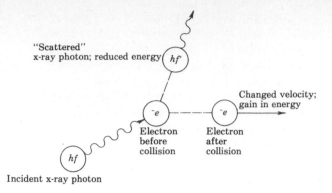

Fig. 33.9 Diagram illustrating the *Compton effect.* A photon of energy $E = hf$ collides with a free electron, charge $-e$. The collision is assumed to be elastic. The electron acquires energy from the incident photon. The "scattered" photon has a reduced energy, $E' = hf'$. Since frequency has decreased, wavelength must increase. This mathematical approach is confirmed by Compton's experimental results.

"Scattered" x-ray photon; reduced energy hf'

Changed velocity; gain in energy

$-e$
Electron before collision

$-e$
Electron after collision

hf
Incident x-ray photon

turned to the photon idea. He assumed that the x-rays used in the bombardment were x-ray photons (i.e. particlelike rather than wavelike). Each photon is assumed to possess kinetic energy and momentum, and to have the capability of having elastic collisions with other particles of matter. In Fig. (33.9) suppose hf is an x-ray photon approaching collision with a free electron ^-e, in a carbon block. This incident photon transfers some of its energy to the electron, which is accelerated and moves off in the direction shown. The energy gained by the electron from the incident x-ray photon shows up in a change of velocity (and also direction, since velocity is a vector quantity). The "scattered" x-ray photon is shown moving upward and to the right, in the diagram. For the photon, since $E = hf$ is less than it was before the collision, and since h is a constant, f must decrease to some value f'. A decreased frequency means an increased wavelength, which is precisely the experimental finding. The Compton effect, then, is one more in a series of twentieth-century experimental findings that support the quantum idea and the photon theory of light.

33.10 Wave-Particle Duality

Einstein's explanation of the photoelectric effect shows that light does not always behave as if it were a wave. In forcing the emission of electrons from a metal, light acts like a particle with a quantized energy, $E = hf$. Einstein's "model" is supported by actual laboratory observations, in

the finding that the predicted relationship among energy, frequency, and the constant h (Planck's constant) is confirmed in the laboratory.

For x-rays also, the quantum-theory model calls for the equation $E = hf = \frac{1}{2}mv^2 = Ve$. Careful laboratory investigations again confirm the particle or photon theory. And likewise with the Compton effect. But equally competent experimenters and theorists had earlier developed the wave theory of light, and the entire science of optics is based on light as a wave motion. Also, electromagnetic wave theory is the basis for radio, television, radar, and microwave devices. Since these devices—optical equipment and communications equipment—are conceived, designed, manufactured, and successfully operated on a wave-theory model, it is unlikely that the wave theory is wrong.

It would appear therefore that there are two mutually incompatible theories on the nature of light and electromagnetic radiations. And, rather than ask which theory is right and which is wrong, we are forced by circumstances to admit that they are probably both right. It seems that light is both wavelike and particlelike, with irrefutable evidence supporting either view. However, wave characteristics and particle characteristics do not seem to occur simultaneously or in the same process. For example, a beam of light can be refracted and undergo dispersion (color separation) through optical equipment, obeying well-known wave-theory patterns of behavior, and immediately thereafter shine on a photocell and act like a stream of photons. But the two behaviors do not occur simultaneously in the same process. Two generalizations are possible: (1) Concerning the *propagation* of light (and electromagnetic radiations in general), the wave theory seems to be obeyed. (2) When light (or electromagnetic radiation) strikes or *interacts* with atoms, molecules, electrons, neutrons, or other subatomic units, it acts like particles or photons.

In summary, the two theories complement each other; both are "correct" in some situations, though neither is "correct" in all situations. Neither gives us a "picture" of the true nature of light, but both help us understand the behavior of light and electromagnetic radiations. Pending further discoveries which might provide a basis for a single unifying theory, it seems that we must accept both the wave theory and the photon (quantum) theory of light.

33.11 Matter Waves—The de Broglie Hypothesis

If light can exhibit particlelike behavior, can particles (i.e. small masses) act like waves? Louis de Broglie, a French physicist, in 1924 suggested an equation which assumes that all particles of the atomic and subatomic realms (electrons, protons, neutrons, etc.) have associated with them waves of a specific wavelength. He reasoned more or less as follows: If a photon is assumed to be an elastic particle of energy $E = hf$, then, according to Einstein's mass-energy equation $E = mc^2$, it would have a mass of

$$m = E/c^2 = hf/c^2 \quad \text{(photon mass)} \quad (33.13)$$

Since the velocity of a photon is c by definition (a photon is a "particle" of light), its momentum is

$$p = mc = hfc/c^2 = hf/c \quad \text{(photon momentum)} \quad (33.14)$$

Since $f/c = 1/\lambda$, then $p = h/\lambda$ and

$$\lambda = \frac{h}{p} \quad \text{(photon wavelength)} \quad (33.15)$$

It was suggested by de Broglie that these *photon* relationships, based on Einstein's mass-energy equivalence, be applied to all atomic particles, even though their velocities are at some value v much less than c.

The momentum of a particle of mass m and velocity v is $p = mv$. Suppose that these small particles of matter behave like photons, and that each has its own associated *matter wave* whose wavelength is given by Eq. (33.15). Substituting $p = mv$ in that equation gives an equation for the wavelength of *matter waves*:

$$\lambda = h/mv \quad (33.16)$$

This is the well-known *de Broglie wave equation*. It predicts that the faster a particle moves (e.g. a high-speed electron), the shorter the wavelength associated with it. We will illustrate the de Broglie hypothesis (matter-wave theory) with one example from the macrorealm, and another from the atomic realm.

Illustrative Problem 33.6 A bullet of mass 10 gm has a muzzle velocity of 800 m/sec. What is its associated wavelength?

Solution Change mass in grams to kilograms and substitute in Eq. (33.16):

$$\lambda = \frac{h}{mv} = \frac{6.626 \times 10^{-34} \text{ J-sec}}{(10 \times 10^{-3} \text{ kg})(800 \text{ m/sec})}$$
$$= 8.28 \times 10^{-33} \text{ m} \quad \text{ans.}$$

This wavelength is much too short to be detected, a result characteristic of masses in the macrorealm. It should be noted that de Broglie did not intend to suggest that his hypothesis and the matter-wave theory would give meaningful results with anything but particles of very small mass. With particles in the microrealm, measurable wavelengths do result, as the following example shows.

Illustrative Problem 33.7 An electron whose mass is 9.1×10^{-31} kg has a velocity of 2.65×10^7 m/sec. Find its associated wavelength.

Solution   Using Eq. (33.16),

$$\lambda = \frac{h}{mv} = \frac{6.626 \times 10^{-34} \text{ J-sec}}{9.1 \times 10^{-31} \text{ kg} \times 2.65 \times 10^{7} \text{ m/sec}}$$

$$= 2.75 \times 10^{-11} \text{ m} = 0.275 \text{ Å} \qquad ans.$$

The concept of "matter waves" was a radical idea—almost a figment of the imagination—when de Broglie first proposed it and derived his wave equation. Shortly, however, experimental findings began to confirm the de Broglie hypothesis, and out of this "radical" assumption, there has developed over the past 40 years, the entire field of electron optics. The electron microscope (see Fig. 33.11), is a good example of this development. In 1929, de Broglie was awarded the Nobel Prize in physics for his work on matter waves.

33.12   The Hydrogen Atom —Energy Levels

Once the planetary atom model (Fig. 33.3) had gained some acceptance as to its ability to explain observed phenomena, Bohr and other workers turned their attention to the related questions of the structure of the atom and how energy is distributed in the atom. Since ordinary hydrogen is the simplest atom—one (negative) electron revolving around a nucleus of one (positive) proton, according to the model, we will use it for a brief discussion of energy levels in the atom. Only the barest introduction to what is a highly complex subject is possible here.

The first hint of quantized energy states in atoms came from observations of spectral lines. Consequently, we begin with some comment on the spectrum of hydrogen. In Chapter 23, bright-line and absorption spectra were mentioned briefly. Fig. 23.9 portrays the solar spectrum and the bright-line spectra of the elements hydrogen, helium, mercury, and uranium. Line spectra are the equivalent of "fingerprints" for different elements. Each element, when incandescent in a gaseous state, exhibits a unique line spectrum.

Since hydrogen is the simplest atom, we might expect it to have a simple, orderly spectral pattern. Indeed, this is the case, as Fig. 23.9 shows. J. J. Balmer (1825–1898) derived the following equation in 1884 for calculating the wavelength of each visible bright line in the spectrum of hydrogen (only four were then known—see Fig. 33.10).

$$\frac{1}{\lambda} = R\left(\frac{1}{2^2} - \frac{1}{n^2}\right) \text{ where } n = 3, 4, 5 \cdots$$

(Balmer series equation)   (33.17)

R is called *Rydberg's constant* and is equal to $1.097 \times 10^{7} \text{ m}^{-1}$, and λ is the wavelength in meters of the bright lines in the *Balmer series* of spectral lines in the hydrogen spectrum (often called the *visible series*).

Illustrative Problem 33.8   Find the wavelength of the third line in the visible (Balmer) series for hydrogen.

Solution   In the Balmer series equation (33.17), substitute $n = 5$, the third permitted value, obtaining

$$\frac{1}{\lambda} = (1.097 \times 10^{7} \text{ m}^{-1})\left(\frac{1}{2^2} - \frac{1}{5^2}\right)$$

$$= 2.30 \times 10^{6} \text{ m}^{-1}$$

and $$\lambda = \frac{1 \text{ m}}{2.30 \times 10^{-6}} = 4.34 \times 10^{-7} \text{ m}$$

$$= 4350 \text{ Å} \qquad ans.$$

Other series, outside the visible spectrum, were soon discovered for hydrogen, by Lyman, Paschen, and others. The Lyman series lies in the ultraviolet region and the Paschen series in the infrared (see Fig. 33.10). Each has an equation, similar to the Balmer equation, and involving the Rydberg constant, for predicting the wavelength of the series lines.

Electrons in Orbits   Chronologically, these findings from analyses and interpretation of spectral lines preceded by some 30 years the inquiry into atomic structure and energy levels in the atom begun about 1910 by Rutherford and Bohr. One of the puzzling questions about the plan-

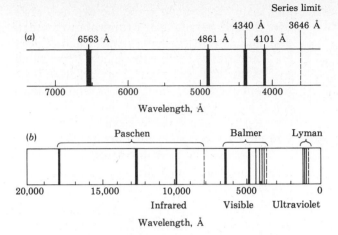

(a)

Series limit

4340 Å   3646 Å

6563 Å   4861 Å   4101 Å

7000   6000   5000   4000

Wavelength, Å

(b)   Paschen   Balmer   Lyman

20,000   15,000   10,000   5000   0

Infrared   Visible   Ultraviolet

Wavelength, Å

Fig. 33.10 Diagrams of three series of spectral lines of hydrogen. (*a*) Four lines of the Balmer series. (*b*) The Lyman, Balmer, and Paschen series (stronger lines of each only) on the same scale of wavelengths, showing their relative positions.

etary theory of the atom was the question of orbital stability. An electron circling a nucleus is in many ways similar to a satellite circling the earth. A centripetal force of some kind is needed to keep both in their orbits, since they are undergoing continual acceleration. For the satellite, gravity supplies the force; for the electron the force is one of attraction between the positively charged nucleus (^+e) and the negatively charged electron (^-e). In a circular orbit (assumed) around the nucleus, an electron would be continually changing direction, and would have an acceleration directed toward the center, of $a_c = v^2/r$ (Eq. 7.5). Classical electromagnetic theory requires that accelerating charges continually radiate energy. But if this were the case with electrons, their orbits would "decay," and they would spiral into the nucleus, just as an earth satellite crashes after its orbit decays. According to classical physics analysis, the atom must collapse if it is giving off radiations. But it does not. Why? In 1913, Bohr proposed a radical (at the time) explanation, which embodied the principle that classical (Newtonian) physics does not apply to particles in the atomic and subatomic realm. Quantization of electron energy was a key element in this statement. Again we see the influence of Planck.

Only a sketchy outline of Bohr's complete development can be given here. Not only is the

theoretical treatment beyond the scope of an elementary text, but the intervening decades have seen major modifications in atomic theory.

Three assumptions were made, using the Bohr-Rutherford atom as the operating model. These were:

1. Certain orbits exist where electrons can continue to orbit *without radiating energy*. These are *stable orbits*, and they later came to be referred to as *energy levels*.

2. As an electron changes energy levels (orbits), it either loses energy accompanied by photon emission or gains energy by absorbing a photon. If E_A and E_B are the energies of an electron in adjacent energy levels A and B, the frequency of the radiation emitted when the electron "jumps" from A to B is given by the quantum condition

$$E_A - E_B = hf = h\frac{c}{\lambda} \qquad (33.18)$$

3. The condition that determines the radii of the possible orbits or energy states—the quantized condition—is

$$mvr_n = \frac{nh}{2\pi} \qquad (33.19)$$

where
$n = 1, 2, 3 \cdots$
m = the electron mass
v = the orbital speed
h = Planck's constant (as before)

The quantity mvr_n is the *angular momentum* of an electron as it orbits the nucleus in one of the "allowed" levels.

The assumption that angular momentum must be a multiple of $h/2\pi$ is referred to as *Bohr's quantum hypothesis*. In effect, it says that angular momentum of electrons cannot vary in a continuous fashion, but only in incremental or quantized fashion. There was probably no basis in either experimental or theoretical physics for Bohr's hypothesis or "hunch" at the time (1913). It was an assumption introduced because classical methods failed entirely in explaining observed results. Later, however, (1924), with de Broglie's hypothesis and his wave equation, the Bohr assumption could be checked mathematically, as indicated here. Equation (33.19) can be rewritten as

$$2\pi r_n = n\frac{h}{mv}$$

where $2\pi r_n$ is the circumference of the "n" orbit, and h/mv is the de Broglie electron wavelength (see Eq. 33.16). The equation

$$2\pi r_n = n\frac{h}{mv}$$

states that a whole number (n) of electron waves must be fitted into the circumference of an electron orbit for the orbit to be stable, again confirming the idea of discrete, or quantized elements.

From his quantum hypothesis and by further mathematical calculations, Bohr was later able to predict from theoretical considerations the exact frequencies of the four bright lines in the Balmer series, as previously determined by laboratory observations. Only by adopting the quantum hypothesis could experiment and theory be brought into agreement.

With the passage of time, the "Bohr atom" concept has proved to be far too restricted in its scope. As a "picture" of an actual atom, it has become outdated. Since 1913, physicists have learned that the atom, and especially the atomic nucleus, is vastly more complicated and uncertain than was thought to be the case in Bohr's time. Indeed, in talking or writing about present-day particle physics, the key words and phrases are "uncertainty," "indeterminacy," "probability," "statistical mechanics," and "models." Even so, Bohr's ideas of *energy levels* and *quantum states* have stood the test of time, and are still essential elements in modern extensions of quantum theory and atomic and nuclear "modeling."

We close this brief theoretical treatment of quantum physics with just one example of the modern view, clothed as it is in probability, statistics, and indeterminacy. Then the final section of this chapter will discuss a few applications of particle physics to devices with which technicians often work.

33.13 Heisenberg's Principle of Indeterminacy

This principle is often called "the uncertainty principle." It was first enunciated in 1927 by Werner Heisenberg (1901–1976), and it is a sweeping generalization about the observation and precise measurement of microevents. Heisenberg pointed out that in every scientific observation or measurement there has to be some sort of physical interaction between the observer or his "agent" (a beam of light, a photon, or a ray), and the process or particle being observed or measured. For example, in "observing" an electron by reflecting a photon (or a quantum of radiant energy) from it, the interaction affects the electron so that what we observe is not really the way it was before the interaction. If the agent photon (sent as our contact device) has a wavelength small enough to pinpoint the location and perhaps the size of the electron, then its frequency is very high ($f = c/\lambda$), and its energy will be very high also ($E = hf$). When our agent photon strikes the electron, a "Compton effect" collision (see Fig. 33.9) takes place and the electron that we wanted to observe immediately acquires a different momentum. If we decided to avoid this by using a photon "agent" of low energy, it would have low frequency, a longer wavelength, and would therefore not be able to "tell us" with any precision where the electron was at the time we made the observation. Heisenberg's *principle of indeterminacy* states that both the position and the momentum of a particle cannot *simultaneously* be known with any reasonable degree of precision.

If a photon (or radiation) of wavelength λ is to be used for a measurement of location or length, the smallest distance or length Δx that can be measured is of the order of the magnitude of λ. Since $\lambda = c/f$, $\Delta x \geq c/f$. Δx cannot be less than c/f. Moreover, in the attempt to measure the *momentum* of the electron, the indeterminacy of the measurement is of the same order of magnitude as the momentum that it could acquire from the "agent" photon. Momentum of radiation in free space is given by the relation $p = hf/c$ (Eq. 33.14), and therefore $\Delta p = hf/c$. Combining these limiting equations, we get

$$\Delta x\, \Delta p \geq \frac{chf}{fc}$$

or
$$\Delta x\, \Delta p \geq h \qquad (33.20)$$

Since the product of Δx and Δp cannot be less than h, the more precise the *position* determination is (indicated by Δx), the less precise the *momentum* determination (indicated by Δp) can be, and conversely.

Applied to the macroworld (objects of finite size and mass) the uncertainty principle poses no real limitations on the accuracy of measurements, as the following problem shows.

Illustrative Problem 33.9 A softball of mass exactly 230 gm is travelling at an unknown speed v. Its position can be determined with an uncertainty of 0.0001 m by electro-optical methods. What is the indeterminacy of the velocity, v?

Solution Since the mass is known "perfectly," $\Delta p = m\,\Delta v$. But, from Eq. (33.20),

$$\Delta p = \frac{h}{\Delta x}$$

Subst. for Δp gives

$$m\,\Delta v = \frac{h}{\Delta x}, \quad \text{and} \quad \Delta v = \frac{h}{m\,\Delta x}$$

Subst. numerical values,

$$\Delta v = \frac{6.626 \times 10^{-34} \text{ J-sec}}{0.23 \text{ kg} \times 10^{-4} \text{ m}}$$

$$= 28.81 \times 10^{-30} \text{ m/sec} \qquad \textit{ans.}$$

This result represents the absolute limit of precision allowed by nature, according to the principle of indeterminacy. But, for the macrorealm, it really is not much of a limitation, since we hardly need to know the velocity of baseballs in flight (or the velocity of projectiles of any kind, for that matter), with any such degree of precision.

For measurements in the atomic realm however, the implications are vastly different. In measuring the velocity of an electron, we need to know the time taken to travel a given distance. But we must use at least one photon of light (or radiation of some kind) to interact with the electron in order to know when it passes the beginning and end points in our distance interval. The photon interaction has a significant effect on the energy, momentum, and the speed of the electron. The electron's position at the end point of the observation is thus affected by the attempt to measure its position at the beginning.

The point made by Heisenberg, and the difficulty faced by experimenters in the field of particle physics, is that the status (energy, position, momentum, spin, etc.) of any atomic particle is disturbed or distorted by whatever observation is being made. It has been said that experimenters in particle physics cannot be merely "spectators"—they must be "participants" as well.

Illustrative Problem 33.10 An electron in the first Bohr orbit of hydrogen, $r = 0.51$ Å, has a position uncertainty of 10^{-11} m. Assuming that the mass of the electron is known "perfectly" as 9.11×10^{-31} kg, what is the uncertainty in the velocity?

Solution With mass known perfectly,

$$\Delta p = m\,\Delta v$$

Then from Eq. (33.20),

$$\min \Delta p = h/\Delta x$$

Substituting, $$m\,\Delta v = \frac{h}{\Delta x}$$

and

$$\Delta v = \frac{h}{m\,\Delta x} = \frac{6.626 \times 10^{-34}\ \text{J-sec}}{(9.11 \times 10^{-31}\ \text{kg})(10^{-11}\ \text{m})}$$

$$= 7.27 \times 10^7\ \text{m/sec} \qquad ans.$$

The position uncertainty of 10^{-11} m is about one-fifth the $n = 1$ orbit radius of 0.51 Å. Even with this rather large uncertainty in position, the uncertainty in velocity is nearly one-fourth the speed of light! And (recall Eq. 33.20), if the position determination could be made more accurately, the uncertainty principle says that the velocity determined from such an observation would be even more indeterminate, since the product of Δx and Δp *cannot be less than h.*

The principle of indeterminacy provides some clarification of the wave-particle duality problem. When we restrict position, as in passing a narrow light beam through a slit or a stream of electrons between charged plates, we get a momentum spread. If, on the other hand, we restrict momentum to a narrow range, we lose information about position, and what were photons (single particles with exact positions) now appear to be waves. Our present state of knowledge does not permit an exact description of how photons, electrons, and waves vary in their many attributes. One might conclude that with more time and better laboratory observations, the present controversies will be resolved and a unifying theory capable of accounting for all experimental findings will emerge. On the other hand, the principle of indeterminacy predicts that, as the accuracy of any one measurement improves, the accuracy with which some other equally important measurement can be made, decreases. The mere fact of observation changes the result. The situation is crudely analogous to the old question: Why does the light go off when you close the refrigerator door? If you hold the door open a crack to see if the light goes off, it is still on. If you close the door, you cannot observe whether it goes off or not. If you crawl inside (Don't do it!) to observe the event firsthand, as the door is closed the light goes off, but in the dark you cannot see why it went off.

SOME APPLICATIONS OF PARTICLE PHYSICS

A number of applications of modern or particle physics have been described in previous chapters. Vacuum tubes, transistors, lasers, cathode-ray tubes, photocells, x-ray machines, television, and radar are a few examples. Solar cells, which turn the sun's rays (photons) into electricity, are another example. Energy from nuclear particles will be discussed in the next and final chapter. This chapter closes with a few more applications.

33.14 The Electron Microscope

Optical microscopes are useful for viewing and magnifying only those objects whose dimensions are greater than about 10^{-6} m (10,000 Å). Below this object size, the wavelength of visible light begins to become comparable to the size of the object being viewed. (The visible spectrum ranges from about 4000 Å to about 7000 Å). The problem of indeterminacy (*diffraction* in this case) then enters in, and the image becomes so fuzzy as to be useless. In optical terms we say that the *resolution* is unsatisfactory.

Many objects that we need to "see" and magnify for the purposes of modern technology, are much too small to be viewed and magnified by light. Some examples, beginning at about the limit of satisfactory resolution of an optical microscope and getting smaller, are listed in Table 33.1.

Since light is useless for viewing very small objects, another source, of shorter wavelength, has to be used. The electron itself can be this source.

The de Broglie hypothesis predicts that the electron has an associated wavelength which depends on its velocity—the higher the velocity, the shorter the wavelength.

$$\lambda = h/mv \qquad (33.16)$$

Table 33.1 Sizes of selected micro-objects

Chromosome, inside a cell	10^{-6} m
Small bacterium	2×10^{-7}
Virus	10^{-8}
Atom diameter	1 to 2×10^{-10}
1 Å	10^{-10}
Bohr radius, hydrogen atom	5.292×10^{-11}
de Broglie wavelength of	
a 1-MeV electron	1.23×10^{-12}
Diameter, atomic nucleus	10^{-14}

Using a high voltage to accelerate electrons, one can obtain an electron beam which will give good resolution (sharp definition) of objects of molecular and atomic size. The electron microscope makes use of the *wave nature* of the electron, another example of the *wave-particle duality* concept discussed earlier. An example of accelerating voltage and associated wavelength follows.

Illustrative Problem 33.11 An electron beam is given an accelerating voltage of 10,000 V. What is the associated wavelength of the electrons?

Solution Use Eq. (33.7), relating KE to the accelerating voltage.

$$Ve = \tfrac{1}{2}mv^2 \quad \text{and} \quad v = \sqrt{\frac{2Ve}{m}}$$

Substituting,

$$v = \sqrt{\frac{2 \times 1.602 \times 10^{-19} \text{ C} \times 10^4 \text{ V}}{9.11 \times 10^{-31} \text{ kg}}}$$

$$= 5.93 \times 10^7 \text{ m/sec}$$

Now use the de Broglie wave equation (33.16):

$$\lambda = \frac{h}{mv} = \frac{6.626 \times 10^{-34} \text{ J-sec}}{9.11 \times 10^{-31} \text{ kg} \times 5.93 \times 10^7 \text{ m/sec}}$$

$$= 0.123 \times 10^{-10} \text{ m} = 0.123 \text{ Å} \qquad ans.$$

Electron beams can be focused by electrostatic and/or magnetic "lenses" (*electron optics*) in a manner analogous to the use of lenses to refract (bend) light rays and bring them to a focus. The

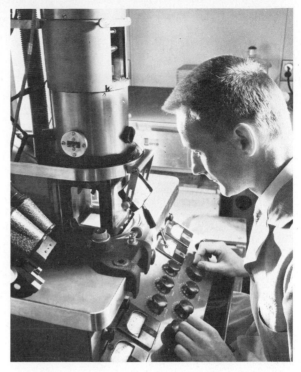

Fig. 33.11 An electron microscope in use in a research laboratory. Combining an electron gun with electrostatic or magnetic lenses, these instruments readily attain magnifying powers of 100,000—100 times the power of microscopes using ordinary light. Potentials of up to 100,000 V enable details as small as 1 Å to be observed.

electron microscope brings its "image" to a focus on a fluorescent screen, where it can be viewed or photographed (Fig. 33.11). The electrostatic or magnetic "lenses" are charged plates or coils whose field lines interact with the charged electrons, producing the bending needed to bring them to a focus. The electron stream itself is supplied by an electron gun, like that used in a cathode-ray tube.

In a *transmission* electron microscope, the "object" must be in the form of a thin section, so the electron beam will pass through it. Tissues are often embedded in plastic with the slice being less than 200 Å in thickness.

In a *scanning* electron microscope or SEM, a sharply focused electron beam scans back and forth across the sample (object) to be examined, much like a television-tube "sweep" or scan. This type of instrument was first developed in the late 1960s and is used to study surface details of objects such as blood cells. The great detail and the depth of field possible with the SEM gives an almost three-dimensional quality to the image.

The electron microscope has contributed important new knowledge in such fields as molecular biology, studies of the human cell, research on viruses, cancer, and leukemia over the past three decades. Precautions must be taken by the operators of these instruments. High-speed electrons often produce x-rays, and operating personnel should be adequately protected.

33.15   The Laser

A laser is a device that emits a concentrated, narrow beam of light. Lasers come in several varieties—ruby lasers, helium-neon lasers, etc. They can generally be classified as either a *solid* laser or a *gas* laser, depending on the nature of the material used for the production of light. An introductory discussion of lasers was presented in Chap. 21. That section (p. 21.24) should be reviewed at this time. A more detailed and quantum-related explanation of laser operation follows. It should be recalled that the word *laser* is an acronym made up of the first letters of the words, *l*ight *a*mplification by *s*timulated *e*mission of *r*adiation.

Solid lasers are usually *pulsed* lasers. They are operated to emit bursts of energy with a pulse time of (typically) one nanosecond (10^{-9} sec) and may have power ratings of up to one trillion (10^{12}) watts. Gas lasers, on the other hand, emit a *continuous* beam of light energy and have a much lower power rating, in the milliwatt (10^{-3} W) range.

Properties of Laser Light   A laser produces a *monochromatic*, or single frequency (wavelength), source of light. Ordinary light sources, on the other hand, emit photons of many different wavelengths. Laser light is *coherent* light. That is, all of the light waves are in a fixed phase relationship everywhere in the beam, and the phase re-

mains steady over time. By way of contrast, an incandescent lamp produces random-phase *incoherent* light. Beams of ordinary light diverge and dissipate their energy. Coherent light results in a collimated narrow beam of great intensity, capable of traveling to the moon and beyond without undue divergence.

Because a laser beam is highly concentrated, safety precautions must be taken. *Never look down a laser beam!* A low-power, one milliwatt laser can cause a retinal burn in a few seconds.

Quantum Explanation of Stimulated Emission
Ordinary atoms have their electrons in stable orbits in "ground" energy states. An atom that has been "excited" for any reason has electrons in higher energy states. It may be about ready to emit light (photons). Electrons in higher energy states are not very stable and usually return to the $n = 1$ ground state, the most stable state, in a very short time, of the order of 10^{-8} sec. No prodding is needed; they spontaneously jump back to the ground state, losing energy and giving off a photon in the process. Since many excited states are possible, many wavelengths result from electron jumps. The jumps do not ordinarily occur at the same time, so *spontaneous* emission results in multifrequency, noncoherent light emitted in all directions.

In *stimulated* emission, electrons change energy states due to "stimulation" from photons passing nearby. This can occur in two ways, as illustrated in Fig. 33.12. Consider a photon of energy hf (equal to the difference in energy between states A and B), passing close by the electron in Fig. 33.12*a*. If the electron is in the lower energy level A, *absorption* of the photon's energy results, and the electron moves to the higher energy state, B. If the electron is in the higher energy state B, the photon can stimulate it to emit a photon (Fig. 33.12*b*) and return to the lower state, A. This stimulation process has the unique characteristic of creating an additional photon, identical with the stimulating photon, and "in step" with it. The process is roughly analogous to a bystander "falling in," in-step with a marching

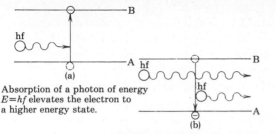

Absorption of a photon of energy $E=hf$ elevates the electron to a higher energy state.

Stimulated emission. Photon of energy $E=hf$ causes electron in higher energy state to emit a photon of the same frequency and phase, and "jump" to a lower energy state.

Fig. 33.12 Schematic diagram of electron "stimulation" by a passing photon. (*a*) Photon is absorbed and electron moves to higher-energy state. (*b*) Photon stimulates electron in higher-energy state to emit photon and "jump" to ground state.

band as it passes by. When the two photons in Fig. 33.12*b* encounter two more electrons in energy state *B*, four photons will become available. This chain reaction process quickly produces great amplification of light and a very intense beam.

Population Inversion and Metastable States Two conditions are necessary for the stimulated emission reaction to continue. First, there must be more electrons in level *B* than in level *A*, otherwise the photons will stimulate more absorptions than emissions. It happens, however, that normally there are more electrons in the lower energy state *A* than in state *B*, so a *population inversion* is essential. The second condition is that level *B* be semistable, or *metastable*. Photons in a metastable condition do not immediately emit radiation spontaneously, but remain stable long

enough (10^{-3} sec or longer) for the stimulated emission process to build up the number of monochromatic (i.e. same wavelength) coherent (i.e. same phase) photons.

Population inversion in a ruby laser is accomplished by "optical pumping" with a bright flash of 5500 Å yellow-green light. Chromium atoms in the ruby absorb the 5500 Å photons and go to an "excited" state. From there they quickly move to a metastable state, in which condition, photons of 6943 Å (red) will stimulate them to emit photons and return to ground state. Stimulated emission of photons is possible in the ruby laser because the chromium in the ruby has a well-defined metastable state.

In a helium-neon gas laser, helium is first excited by an electrical discharge to a metastable energy state 20.61 eV above its ground state. Neon happens to have a metastable energy state of 20.66 eV, very nearly equal to that of helium. The helium atoms transfer energy to neon atoms by collision, moving neon atoms to metastable state *B* (Fig. 33.13). Neon emits 6328 Å photons as stimulated emission, causing electron jumps from *B* to *A*. The population inversion is maintained between levels *B* and *A*, because state *A* contains few electrons to begin with (it is not the ground state), and spontaneous emission between *A* and ground occurs rapidly.

A schematic diagram of a laser is shown in Fig. 21.20. It should be reviewed at this time.

Despite the high-energy intensity of a laser beam, the efficiency of lasers is rather low. Only about 5 percent of input energy into a laser finds its way into the coherent beam. Consequently, in

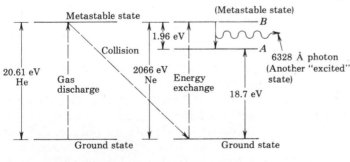

Fig. 33.13 Diagram illustrating metastable states and population inversion in a helium-neon gas laser.

Table 33.2   Some uses of lasers

Use	Comment
Welding of tissues in certain types of surgery, notably "retinal welding"	First used in 1964. Increasingly successful
Measuring distances in space	Earth-to-moon accuracy on an order of a meter or less
Guidance systems for weapons	So-called "smart bombs" and other ordnance
Alignment of large structural components—a "chalk line"	Lining up tunnels, particle accelerators, turbine shafts, etc.
Information carrier	"Laser optics"—through space, or in optical fibers
Three-dimensional photography	Holograms created and viewed with a laser
Detection of pollutants	Laser beam absorbed at the resonant frequency of the pollutant molecule

order to accomplish the tasks for which lasers are suited (welding, cutting, initiating nuclear fusion, etc.) large amounts of energy must be supplied to the laser apparatus.

The use of lasers for initiating nuclear fusion will be explained in Chap. 34. Some other uses are given in Table 33.2.

SOME ADDITIONAL APPLICATIONS

The practical applications of quantum (or particle) physics increase as the months go by, and we make no attempt to catalog them here. Three brief comments on other applications close the chapter.

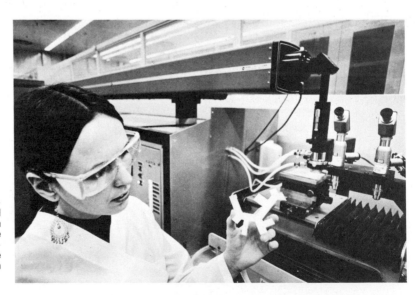

Fig. 33.14 A high-power carbon-dioxide laser (top of photo) is used to cut the very hard ceramic base in the irregular shape shown. The ceramic shape is the base for a thin-film electronic circuit in telephone microwave equipment. (Western Electric Co.)

Fig. 33.15 Experimental underground cable for power transmission at extremely low (cryogenic) temperatures. At temperatures approaching absolute zero, molecules, atoms, and electrons behave in strange ways. One of these characteristics is *superconductivity* in metals. Cables like the one shown are undergoing tests to see whether their increased conductivity at cryogenic temperatures is sufficient to offset the cost of keeping them chilled by liquefied gases and associated cryogenic apparatus. (General Electric Research and Development Center.)

Low-Temperature Physics At absolute zero, according to the original kinetic-molecular hypothesis, molecular motion ceases. But according to quantum theory, some energy is left in the particles making up the atoms of the molecule. Evidently the amount remaining is less than that discrete amount which can be radiated, the quantum itself. Physicists and technicians in low-temperature research are discovering many strange and important facts about molecules, atoms, and atomic particles at or near this zero-energy boundary.

Quantum Tunneling From probability theory and quantum mechanics theory, a bound particle needing energy E_0 to escape from a system through or over a boundary has a small but finite probability of escaping even though its energy is less than E_0. This effect is called *tunneling* (it "tunnels" through the boundary). Quantum tun-

neling explains alpha-particle emission in radioactive decay. It also explains how a crystal diode with a thin layer of insulation at the junction (a tunnel diode) is able to conduct electrons.

Nuclear Physics A general principle of quantum physics is that any bound system contains quantized energy states. The atomic nucleus is a bound system and therefore has such energy levels. These are of much greater magnitude than the energy levels or states that are involved with electrons in their orbits. *Gamma rays* of very high energy and penetrating power result when energy-level changes occur in *nuclei* (see Chap. 34), whereas ultraviolet, infrared, and visible light result when an *atom* changes energy states.

The practical applications of nuclear energy and some of the many nuclear devices now in use in industry and medicine—gamma-radiation devices, radiation counters, radioisotope devices, and neutron scanners, for example—will be discussed in the next and final chapter.

SUMMARY

The Rutherford-Bohr planetary atom, although it has been superseded by later "models," was instrumental in introducing the concept of *quantized energy levels* in atoms.

The atom as conceived by Bohr featured *electrons* (negative charges or particles) revolving in orbits around a central heavy *nucleus*. The nucleus is made up of *protons*, (the unit positive charge) and neutrons. Protons have a mass about 1835 times that of the electron.

A third important atomic particle—*the neutron*—was discovered in 1932 by James Chadwick. It is neutral in charge, and has a mass about equal to that of the proton. Neutrons, like protons, are in the atomic nucleus.

Planck's quantum hypothesis is one of the most pervasive generalizations in all of physical science. It marks the point at which *modern physics* departs from *classical* or *Newtonian* physics.

Einstein's hypothesis to account for the *photoelectric effect* conceived of light as a stream of individual energy packets called *photons*. Each photon has an amount of energy given by $E = hf$, where f is the frequency of the light source and h is Planck's constant.

When light (i.e. photons) shines on certain metal surfaces, electric current (*photoelectrons*) is produced. A certain minimum photon energy is needed for photoelectrons to be emitted. This minimum value for each metal is called the *work function*, ϕ, of that metal.

X-rays are produced either by (*a*) deceleration or "braking" of high-speed electrons by a heavy nucleus, or (*b*) by electron jumps from one orbit to another in the inner energy levels of heavy atoms. The first method is termed *braking radiation*, and the second produces *characteristic x-rays*.

X-ray photons have higher energies than (visible) light photons. Both travel with the speed of light, c, but x-rays have higher frequencies (shorter wavelengths) than visible light.

Many light phenomena support the *wave theory of light*—for example, refraction, interference, dispersion into colors, image formation, etc. On the other hand, the photoelectric effect, x-rays, the Compton effect, etc.,

support the particle, or *quantum theory of light*. At present we are forced to accept the idea of *wave-particle duality*.

The concept of *matter waves* is associated with the *de Broglie hypothesis*. The de Broglie wave equation is $\lambda = h/mv$, where λ is the de Broglie *wavelength* of a particle of matter, h is Planck's constant, and mv is the momentum of a moving particle of mass m and velocity v.

The *Balmer series* is one of several groups of lines in the spectrum of hydrogen. These lines are caused by electrons "jumping" from one orbit or energy level to another, emitting or absorbing energy as they do so.

The wavelength of light associated with a particular electron shift from energy level A to adjacent energy level B is given by

$$E_A - E_B = hc/\lambda \qquad (33.18)$$

Heisenberg's *principle of indeterminacy* (often called the *uncertainty principle*) states that it is impossible to study atomic or subatomic phenomena experimentally, without significantly changing the state of the unit or system being studied. In the microrealm, the observer is automatically a participator as well.

Among the applications of particle physics are the following: cathode-ray tubes, photocells, fluorescent lamps, television, radar, solar cells, x-ray machines, lasers, electron microscopes, electron and neutron scanners, and tunnel diodes, to name only a few.

QUESTIONS AND EXERCISES

1. Express Planck's hypothesis in your own words and then show how the SI-metric units of Planck's constant are obtained.

2. Describe J. J. Thomson's discovery and identification of the electron. How was his technique related to the cathode-ray tube?

3. Explain how Goldstein discovered and identified the proton.

4. Use a simple diagram and explain the *photoelectric effect*. What was Einstein's hypothesis with respect to the photoelectric effect? What is the physical meaning of the *work function* ϕ?

5. What determines the wavelength of the shortest waves produced in an x-ray tube?

6. How are braking-radiation x-rays different from characteristic x-rays?

7. Explain the *Compton effect* and show how it supports the quantum theory.

8. Summarize the half-dozen or so essential findings of both classical and modern physics that make it necessary, at least for the present, to accept the concept of wave-particle duality in explaining electromagnetic energy in general and light in particular.

9. Interpret de Broglie's hypothesis that high-speed particles are associated with *matter waves*. What were his principal assumptions?

10. If, in observing the spectrum of an incandescent gaseous element, you detected *bright lines*, how would you explain this in terms of electron behavior in the energy levels or orbits of the atom?

11. Explain Heisenberg's indeterminacy(uncertainty) principle as it applies to the observation and measurement of atomic and subatomic particles and events. Then think of and describe several ways in which the general principle might apply in everyday life in the macroworld of instruments, machines, and people.

PROBLEMS

1. How much energy does an electron acquire if it is accelerated through a potential difference of 20,000 V? Answer in joules.

2. An atomic oscillator has a frequency of 2.5×10^{15} Hz. What is the lowest energy it can have according to Planck's quantum hypothesis? If it emits radiation of the same frequency, what wavelength would this radiation have? Answer in angstrom units.

3. Find the energy possessed by a photon whose frequency is 5.7×10^{14} Hz.

4. The earth's surface receives solar radiation at the rate of about 1.4 kW/m² on a sunny day if the sun is directly overhead. If the average wavelength of this radiation is 6700 Å, compute the number of photons per second in a beam of sunlight 1 cm² in cross section.

5. Light of wavelength 4200 Å shines on a metal with a work function of 3.8 eV. No electrons are emitted. Show mathematically why this is the case.

6. The threshold wavelength for emission of photoelectrons in tungsten is 2.73×10^{-7} m. (*a*) What is tungsten's work function? (*b*) What is the maximum KE of the electrons ejected from tungsten by monochromatic light of wavelength 1.6×10^{-7} m? Answer in electron-volts.

7. Light of wavelength 5000 Å shines on a metal whose work function is 1.0 eV. Calculate the maximum velocity of ejected photoelectrons.

8. The work function of aluminum is 4.2 eV. If monochromatic light of wavelength 2500 Å illuminates the surface, what "stopping voltage" will be required to reduce the photoelectron current to zero?

9. Monochromatic light of wavelength 4100 Å strikes the surface of a metal in a photoelectric experiment. The "stopping potential" is found to be 0.85 V. What is the value of the work function ϕ?

10. A photoelectron has been accelerated by an applied voltage until it has an energy of 20,000 eV. What is its wavelength in angstroms?

11. What is the minimum x-ray wavelength of x-rays produced in a 1.5 MV x-ray machine?

12. From the Balmer-series formula determine the wavelengths of the two longest bright lines of the four lines shown in the visible series for hydrogen (Fig. 33.10).

13. Calculate the associated de Broglie wavelength for an electron traveling at 2.5×10^6 m/sec. For a proton at the same speed.

14. A helium nucleus (alpha-particle) is accelerated to a velocity of 2.4×10^7 m/sec. Calculate its de Broglie wavelength. The mass of an α-particle is approximately 6.64×10^{-27} kg.

15. What is the momentum of a high-speed proton whose matter wave has a wavelength of 0.65 Å?

16. An x-ray tube operates at 1.0 MV. Electrons in the tube are stopped in a distance of 4.2×10^{-6} m when they strike the metal target. Calculate their (negative) acceleration as they are brought to rest.

17. A neutron, mass assumed known "perfectly" as 1.675×10^{-27} kg, has a position uncertainty of 0.75×10^{-14} m. What is the uncertainty in the velocity expressed as a decimal fraction of c?

34 Nuclear Energy

In the previous chapter we introduced and explained some ideas about the atom, with especial emphasis on the electron and its energy relationships. It was noted that sometimes electrons are in orbits, sometimes they change orbits and radiate or absorb energy, and sometimes they are ejected or emitted from the atom. In all cases, an exchange of energy in quantum steps is involved. The energy given off or absorbed was found to be in the form of visible and invisible radiations, sometimes acting like waves and sometimes like particles, referred to as *photons* and also as *rays*.

Now, we look inward to the nucleus of the atom where most of the mass of the atom resides, and where most of the energy is. Two nuclear particles have already been mentioned—the *proton*, a particle with unit positive charge and a mass about 1835 times that of the electron; and the *neutron*, a particle with no charge, almost equal in mass to the proton.

BACKGROUND THEORY AND EARLY DEVELOPMENTS

Since 1932, when Chadwick isolated and identified the neutron, several other particles have been discovered, and quite a few more have been hypothesized. We will mention some of these in paragraphs to follow. As a general operating principle, however, the presently held concept of the nature of matter is that it is composed of atoms, which are in turn made up of three kinds of elementary particles— *electrons*, *protons*, and *neutrons*. Electrons are in orbits or energy states circling the nucleus (except for "free electrons"), and protons and neutrons constitute the nucleus.

34.1 Nuclear Structure—Atomic Number and Atomic Mass Number

Until fairly recently, scientists referred to the relative masses of atoms by the term *atomic weight*. A scale of relative atomic weights, originally based on hydrogen as 1.000, and later on oxygen as 16.000, was painstakingly determined from thousands of chemical analyses over a period of two hundred years. In 1960, an international commission established a new base for a scale of *atomic masses*, taking the mass of the "natural" carbon atom as exactly 12. The unit of atomic mass is defined below.

The nucleus is very small, compared to the diameter of the atom's outer electron shell. The atom is mostly empty space. A typical atom may have a diameter in the range of 10^{-10} m (1 Å), with the diameter of its nucleus being of the order of 10^{-14} m (0.0001 Å), only one ten-thousandth as large.

As is the case with the atom as a whole, there is no "picture" or conceptual reality of an atomic nucleus. Figure 34.1 shows one method of depict-

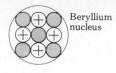

Beryllium nucleus

$\oplus$ Proton

◯ Neutron

Fig. 34.1 Schematic diagram of the beryllium nucleus, containing four protons and five neutrons. The illustration is conceptual only. We do not know what an atom or an atomic nucleus actually "looks like."

ing nuclei in diagrams, but it is not intended to give an idea of what a nucleus really "looks like." We do not know what one looks like.

The Nuclear Force Scientists were very early concerned with what holds the atomic nucleus together. Protons, for example, all being positively charged, repel each other. Moreover, they would not be attracted to neutrons *electrically* (the *Coulomb force*), since neutrons have no charge. For the same reason, neutrons would not attract each other. *Gravitational* attractive forces could not possibly hold the nucleus together, since they are much smaller than the electrical forces of repulsion tending to force the protons apart. In order to account for stable nuclei, it has been necessary to resort once again to an hypothesis: Within the atomic nucleus there is a *nuclear*

force, the exact nature of which is still unknown. It is extremely great at short distances, and it is markedly different from electric and gravitational forces.

Some Definitions

1. Both neutrons and protons are called *nucleons*.

2. The *atomic number Z* of an atom is given by the number of protons in the nucleus. Since each proton has a unit charge of ^+e, the charge on the nucleus is ^+Ze. Also, for a *neutral* atom, Z indicates the total number of electrons in all the orbits of that atom. Their total charge is ^-Ze.

3. The *neutron number N*, of an atom is given by the number of neutrons in the nucleus.

4. The *mass number A* of an atom is equal to the number of *nucleons*, that is to the sum of the numbers of protons and neutrons. The mass number A is the nearest whole number to the actual *atomic mass*, expressed in atomic mass units (see Table 34.1). As an equation,

$$A = Z + N \qquad (34.1)$$

34.2 Isotopes

Most elements have been found to exist in nature with atoms of more than one atomic mass. The *atomic number Z* of all atoms of the same element is the same, but the *listed atomic mass* of most

Table 34.1 Atomic data on isotopes of some light elements

Isotope	Atomic Mass, amu	Mass Number A	Atomic Number Z	Neutron Number N
$_1H^1$	1.007825	1	1	0
$_1H^2$	2.014102	2	1	1
$_2He^3$	3.016030	3	2	1
$_2He^4$	4.002604	4	2	2
$_3Li^6$	6.015126	6	3	3
$_5B^{10}$	10.012939	10	5	5
$_6C^{12}$	12.00000 (defined as the standard reference)	12	6	6
$_7N^{14}$	14.003074	14	7	7
$_8O^{16}$	15.994915 (formerly the standard, at 16.000000)	16	8	8

elements is merely the average of the masses of the several types of atoms possessing the same atomic number. Since the atomic number specifies the number of protons in the nucleus (and also the number of electrons in the orbits), it must be the number of *neutrons* in the nucleus which varies, in order to allow for atoms of the same element to have different atomic masses. Such atoms, called *isotopes*, have the same *chemical* properties as the "ordinary" atom, but they differ in mass and in nuclear structure. "Ordinary" uranium, for example, has a mass number of 238 and is designated U^{238}. One isotope of uranium is U^{235}

In order to distinguish between isotopes of the same element, it is conventional to write the symbol of the element, as C (carbon), with a subscript in front which is its *atomic number Z* and a superscript following, which is its mass number. The notation $_6C^{12}$, for example, designates "natural" carbon whose nucleus contains six protons ($Z = 6$) and six neutrons for a mass number of $12(6 + 6 = 12)$. The notation $_6C^{14}$ indicates an isotope of carbon with eight neutrons in the nucleus for an atomic mass of 14. In like manner, $_{92}U^{238}$ is "natural" uranium, and $_{92}U^{235}$ is an isotope.

34.3   Atomic Mass Unit

As mentioned above, the "natural" carbon atom, $_6C^{12}$, is now the reference standard for atomic mass units. One-twelfth of the mass of the natural carbon atom is defined as 1.0000 *atomic mass unit* (amu). The *actual mass* of 1 amu is

$$1 \text{ amu} = 1.6605 \times 10^{-27} \text{ kg}$$

Table 34.1 gives some atomic data on a few of the lighter elements.

The three major components of the atom, the *electron*, the *proton*, and the *neutron*, have the following rest masses and charges:

34.4   The Energy in the Atom

On the basis of experimental evidence and much theoretical and mathematical development, Albert Einstein (1879–1955) suggested in 1905 that *mass* and *energy* are equivalent, and that the equivalence can be expressed by the equation

$$E = mc^2 \qquad (34.2)$$

where E is in joules when m is in kilograms and c, the *velocity of light*, is 3×10^8 m/sec. In accordance with this equation (which modern experimental evidence has verified), 1 kg of mass, if converted *completely* to energy in 1 sec, would result in the staggering power production of 40,000 billion kW! From the Einsteinian concept of mass-energy, the law of conservation of energy and the law of conservation of matter have been combined into the newer idea of a *law of conservation of mass-energy.*

Most atomic nuclei are extremely stable. The protons and neutrons of the nucleus are bound together with tremendous force—the *nuclear force*. So far we have discussed *forces* in the nucleus. It is more useful, however, to consider the *energies* related to these forces. The so-called *binding energy* of the atomic nucleus is the source of atomic energy from nuclear processes. Suggested by Einstein's mass-energy equation, $E = mc^2$, many experiments were performed to try to show that adding energy could increase mass, or that mass could be made to decrease, liberating energy. Sir John Cockcroft and Ernest T. Walton, two English scientists, were able to show in 1932 that atoms of the light metal *lithium* could be broken up, forming helium nuclei and at the same time liberating radiant energy. A small amount of mass disappeared in the process. High-speed protons were used to "bombard" the lithium nuclei. The diagram of Fig. 34.2 shows the results of this early "atom-smashing" experiment. The amount of energy released was

	Rest mass	Electric charge
Electron	$m_e = 0.000549$ amu $= 9.109 \times 10^{-31}$ kg	$^-e = 1.602 \times 10^{19}$ C
Proton	$m_p = 1.007277$ amu $= 1.6725 \times 10^{-27}$ kg	$^+e = 1.602 \times 10^{19}$C
Neutron	$m_n = 1.008665$ amu $= 1.6748 \times 10^{-27}$ kg	Zero

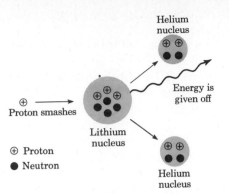

Proton smashes

Lithium nucleus

Helium nucleus

Energy is given off

Helium nucleus

⊕ Proton
● Neutron

Fig. 34.2 Diagram of Cockcroft and Walton's experiment. Lithium is converted to helium, and energy is released as mass decreases.

precisely the amount that Einstein's equation predicted.

Another classic experiment (or series of experiments) showed that mass is associated with the *binding energy* of the atomic nucleus. After countless careful experiments, scientists found that the nuclei of certain lighter elements have a mass which is less than that obtained by adding up the masses of their constituent parts. As an example, helium ($_2He^4$) has 2 protons and 2 neutrons in its nucleus. The mass of a proton is, to five decimal places, 1.00728 amu and that of a neutron 1.00867 amu.

For the helium ($_2He^4$) nucleus:

> 2 protons, each 1.00728 = 2.01456 amu
> 2 neutrons, each 1.00867 = 2.01734 amu

Calculated mass of "unbound" helium nucleus
(adding masses of its constituent parts)

$\qquad\qquad\qquad\qquad\qquad = 4.03190$ amu

Actual mass of "bound" helium nucleus from
hundreds of careful experiments $\quad = 4.00151$ amu

Discrepancy, or mass defect $\qquad\quad = 0.03039$ amu

This *mass defect*, or apparent loss in mass was assumed to represent the energy which binds the helium nucleus together. Calculations revealed that if methods for liberating these "binding energies" could be perfected, staggering amounts of energy could be released. Using helium as an example again, Einstein's equation predicts the production of almost 200,000 kW-hr of energy in the *formation* of 1 gm of helium from

protons and neutrons. This is enough to run a 10-hp electric motor continuously for more than a year and a half! The problem is, of course, how to release this binding energy or mass-energy, under controlled conditions.

NOTE: In 1947, the English physicist C. F. Powell and his coworkers discovered the π *meson*, or *pion*, an elementary particle which, it is believed, supplies the nuclear force to bind protons and neutrons together within the nucleus. Mesons are facetiously referred to as "nuclear glue."

Illustrative Problem 34.1 Calculate the energy equivalent of 1 amu of mass as it "disappears" in the process of formation of an atomic nucleus. Answer (*a*) in joules; (*b*) in electron-volts.

Solution Given: 1 amu = 1.6605×10^{-27} kg.
(*a*) From Einstein's mass-energy equation, $E = mc^2$, the energy equivalent of 1 amu is

$$E_{amu} = 1.6605 \times 10^{-27} \text{ kg}$$
$$\times (2.998 \times 10^8 \text{ m/sec})^2$$
$$= 1.492 \times 10^{-10} \text{ J} \qquad ans.$$

(*b*) Since 1 J = 6.242×10^{18} eV, it follows that

$$E_{amu} = 1.492 \times 10^{-10} \text{ J} \times 6.242 \times 10^{18} \text{ eV/J}$$
$$= 9.31 \times 10^8 \text{ eV} = 931 \text{ MeV} \qquad ans.$$

34.5 Atomic Disintegration—Natural Radioactivity

Early in the development of nuclear physics, it was discovered that some extremely heavy atoms, e.g., *radium* and *uranium*, have relatively *unstable* nuclei and that certain of these atoms undergo natural disintegration. The disintegration results in lighter nuclei and energy liberated in the form of radiations. The total (at rest) mass of the products is less than the mass of the original nucleus. The *mass defect* × c^2 equals the sum total of the energies liberated.

Natural atomic disintegration with consequent energy liberation had been observed by Antoine Henri Becquerel (1852–1908) and Pierre (1859–1906) and Marie (1867–1934) Curie, near the close of the nineteenth century. They had studied and observed ores of uranium for years and had noted that three types of radiations were emitted spontaneously from materials containing uranium. These radiations were studied, their energies were determined, and they were named alpha (α), beta (β), and gamma (γ), for the first three letters of the Greek alphabet. Subsequently, alpha "rays" were found to be not rays but particles and were identified by Rutherford as *helium nuclei*. Beta "rays" proved to be *electrons*. Gamma rays were shown to be true energy radiations, electromagnetic in character and possessing extreme penetrability.

The behavior of alpha (α), beta (β), and gamma (γ) rays was studied by Rutherford and his associates by passing the rays through a strong magnetic field. A radioactive sample (e.g. uranium ore) was placed in a cavity in a lead block (Fig. 34.3) in such a manner that the emerging stream or pencil of radiations would pass through the magnetic field. If the direction of the magnetic field is in toward the page in the diagram, the alpha "rays" are deflected to the left and are thus shown to be not rays at all, but particles with a positive charge. The beta "radiations" are deflected to the right, and are therefore particles also, with a negative charge. Only the gamma rays, undeflected by the magnetic field, are true energy radiations.

Alpha particles are emitted at velocities of about 10^7 m/sec. Each carries a double positive charge. Their energies are of the order of several MeV and they have relatively low penetrating power, being stopped by very thin metal foil.

Their double charge gives them high ionizing power, however.

Beta particles are electrons with speeds very nearly equal to that of light. Being of such small mass, their KE is ordinarily less than that of alpha particles. They can penetrate about 1 mm of lead. Since it is thought that there are no electrons in the nucleus, the question as to the source of beta radiation is a puzzling one. At present it is believed that a neutron initiates beta radiation by changing to a proton. The proton stays in the nucleus, and an electron (beta particle) and a new particle called a *neutrino* (see page 751) are simultaneously ejected.

Gamma rays are highly penetrating electromagnetic waves usually of wavelength shorter than 10^{-10} m (1 Å). They can penetrate several centimeters of lead, and are very dangerous to humans in uncontrolled situations. Like x-rays however, they are used for both diagnosis and therapeutic treatment in medicine. They are also used in industry.

Half-life A characteristic of radioactive elements and isotopes is that each has a particular period of time during which half of any given initial amount will undergo radioactive decay. The *half-life* of a radioactive element or isotope is the time required for the number of radioactive nuclei in the sample to decrease to one-half the original number. Half-lives can be determined by measuring the *rate of decay* with a suitable radiation counter, such as a *Geiger counter*. The half-life of $_{92}U^{238}$ (natural uranium) is about 4.5 billion years. That of $_{88}Ra^{224}$ (an isotope of radium), 3.64 days; of $_{27}Co^{60}$ (Cobalt-60, commonly used in industry as a gamma-radiation source), 5.25 years; and $_6C^{14}$ (Carbon-14), about 5700 years. Half-lives of isotopes are often used for dating events of the past. Carbon-14 dating is very commonly used with objects or artifacts containing carbon (all matter that was once plant or animal), to establish the time of historic or prehistoric events.

These phenomena led to the discovery of *radium, polonium, thorium, actinium, plutonium,*

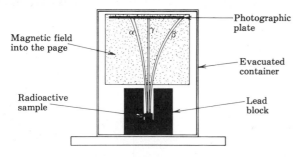

Fig. 34.3 Diagram illustrating the behavior of alpha and beta particles and gamma rays in the presence of a magnetic field.

and other radioactive elements. In these elements we find nuclei which are *not* stable, which, in fact, are undergoing constant natural disintegration at a slow rate, with a consequent liberation of energy and disappearance of mass. It was to these heavy elements, then, that atomic scientists turned in their attempts to release the energies of atomic nuclei.

34.6 Atomic Bombardment—Artificial Radioactivity

Early work by Rutherford and later experiments by Cockcroft and Walton had shown that one way to break down atomic nuclei was to bombard them with certain atomic particles. Rutherford had changed nitrogen into oxygen by bombarding nitrogen nuclei with alpha particles (helium nuclei). The result of such a process is shown in Fig. 34.4. Cockcroft and Walton had shot protons at lithium nuclei and were able to show that the lithium nucleus (after "capturing" a proton) then split into two alpha particles (helium nuclei), as shown in Fig. 34.2.

Similar lines of investigation were carried on by many researchers in the United States, England, and Germany during the 1930s, and more than 60 elements were bombarded, with various radioactive products resulting. Also, there was some small energy liberation. Neutrons proved to be the most successful "bullets" since they carried no electric charge and were not repelled by the nuclei which they were supposed to hit. It was assumed that the higher the energy of the bombarding particle, the better the chance of pene-

Fig. 34.5 The 184-in. cyclotron at the Lawrence Radiation Laboratory. This machine was first completed in 1946 with an energy capability for deuterons of 200 MeV. Subsequent modifications have enabled it to operate in the 900-MeV range for alpha particles. This range is identified with "medium-energy physics." Professor Lawrence's first cyclotron, built at the physics laboratory of the University of California in the 1930s, was not much larger than a good-sized frying pan. (University of California.)

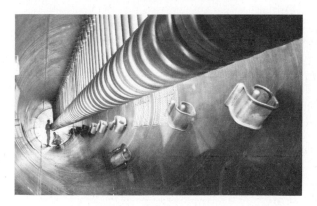

Fig. 34.6 The heavy ion linear accelerator (HILAC) at the Lawrence Radiation Laboratory. The power requirements of such machines are tremendous, up to 8 megawatts for this modification. Particle accelerators are essential for research in high-energy physics. (University of California, Berkeley.)

trating into the nucleus to demolish it. Particle accelerators were developed for the purpose of accelerating the various particles used for nuclear bombardment. The *cyclotron* (Fig. 34.5), the *heavy ion linear accelerator* (HILAC) (Fig. 34.6),

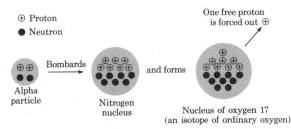

Fig. 34.4 Rutherford's famous experiment, shown in a two-dimensional diagram. Nitrogen was changed to oxygen by alpha-particle bombardment.

Fig. 34.7 A portion of the Cockcroft-Walton accelerator at the Fermi National Accelerator Laboratory in Batavia, Illinois. Standing over 20 feet high, this unit is operated to provide initial acceleration for protons. (Fermi National Accelerator Laboratory.)

and the *Cockcroft-Walton accelerator* (Fig. 34.7) are three of the machines devised for this purpose. Despite some successes in transforming minute quantities of certain elements and in building bigger and more powerful equipment for atomic bombardment, it became increasingly apparent that the answer to the problem of energy from the atomic nucleus lay not necessarily in higher bombardment energies but in the search for a particular type of reaction which would be self-sustaining.

34.7 Uranium Fission

Enrico Fermi (1901–1954) in the United States and Otto Hahn, F. Strassmann, and Lise Meitner in Germany were concentrating, in the 1930s, on the bombardment of uranium with *extremely high-speed* neutrons. These investigations were not very productive. Later they reversed the trend and *slowed* their neutron "bullets" by passing them through paraffin moderators. They found that these *slow neutrons* seemed to be more effec-

tive in terms of nuclear changes than high-speed neutrons.

In 1939, Hahn and Strassmann announced that in bombarding one of the isotopes of uranium (U^{235}) they had observed an unusually high energy liberation and had verified that one of the products formed was an isotope of *barium*. Lise Meitner repeated these experiments and came to the conclusion that the U^{235} nucleus had nearly split in two. She coined the term *nuclear fission* for this reaction, verified the presence of barium, and estimated the energy liberated from each fission as about 200 MeV.

It should be recalled that the *electron-volt* (eV) is the amount of energy acquired by an electron as it is accelerated through a potential difference of 1 V. It is a *very small* unit of energy. The *MeV* (million electron-volts) and *GeV* (billion electron-volts) are more practical units. $1\ GeV = 1.6 \times 10^{-10}$ **J**, and $1\ eV = 1.6 \times 10^{-19}$ **J**.

Other experiments in 1939 and 1940 soon showed the complete nature of the *slow-neutron–U^{235}* reaction, observed by Meitner, and, as is often the case, the most significant item was one of the last noted—the fact that when all the products of the reaction (including the energy liberated) were accounted for, there were from 1 to 3 neutrons left over! One possible reaction is diagrammed in Fig. 34.8.

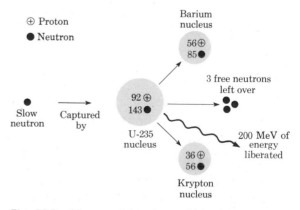

Fig. 34.8 Diagram of the process of U^{235} fission.

Another way of depicting such a U^{235} fission is

$$_{92}U^{235} + _0n^1 \rightarrow _{56}Ba^{141} + _{36}Kr^{92} + 3_0n^1$$

The theoretical energy release *per fission* can be calculated from mass discrepancy considerations as follows:

Masses before capture	Masses after fission
U^{235} 235.044 amu	Ba^{141} 140.914 amu
n 1.009 amu	Kr^{92} 91.897 amu
	$3n$ 3.026 amu
236.053 amu	235.837 amu

Mass defect = 0.216 amu

Since (see Sec. 34.4, Illus. Prob. 34.1) 1 amu is the energy equivalent of 1.492×10^{-10} J, and 1 eV $= 1.602 \times 10^{-19}$ J, the *energy equivalent* of 1 amu is

$$\frac{amu}{eV} = \frac{1.492 \times 10^{-10} \text{ J/amu}}{1.602 \times 10^{-19} \text{ J/amu}}$$

$$= 9.31 \times 10^8$$

or

1 amu (mass) = 931 MeV (energy)

From the above example of U^{235} fission, then, the energy to be expected from one fissioning nucleus is the product of the mass defect in atomic mass units (amu) and the energy equivalent of 1 amu.

$$E = 0.216 \text{ amu} \times 931 \text{ Mev/amu} = 201 \text{ MeV}$$

which checks with the laboratory measurements made by Meitner in her early experiments.

The three "extra" neutrons were the key to a self-sustaining process. They were liberated *within the body* of the mass of U^{235} and would, under proper control conditions, set up a self-sustaining or *chain reaction*. Uranium fission had been attained, the *possibility* of a self-sustaining reaction proved, and energy from nuclear reactions was only a few years away. Figure 34.9 diagrams the beginning of the chain reaction, assuming two spare and effective neutrons from each fission of a U^{235} nucleus.

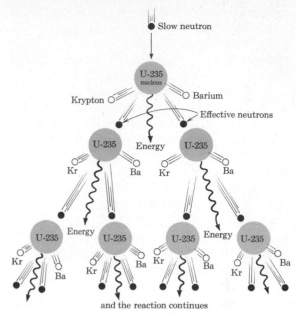

Fig. 34.9 U^{235} fission chain reaction. Two effective neutrons from each split nucleus are assumed. Such a reaction is self-sustaining.

Such a chain reaction must be controlled or it would rapidly accelerate into an explosion of cataclysmic proportions as in the so-called "atomic" bomb. The idea is to produce a controlled number of *effective* neutrons (i.e. those which will blast apart another nucleus) from each nuclear fission. At a rate of one effective neutron per fission the reaction will sustain itself, and a constant level of power output can be attained.

The amount of energy theoretically available from U^{235} fission is remarkable. When 1 kg of U^{235} undergoes controlled fission, the potential energy output is equivalent to that from 5000 tons of coal. As another comparison, a 1-MW nuclear reactor would use only about 1 gm of fissionable U^{235} per day, compared to the use of perhaps 6000 gal of fuel oil per day for the same power output from a thermal combustion plant.

34.8 Early Nuclear Reactors

Nuclear-energy research was put on a "crash" basis just before and during World War II, both in the United States and in Germany. The emphasis was on the development of the "atom bomb." In 1942, Fermi and his colleagues, working in a makeshift laboratory under the football stadium at the University of Chicago, operated the first successful *nuclear reactor*. It contained more than six tons of uranium and used graphite as the moderator for "slowing" the neutrons. Cadmium, an effective *absorber* of neutrons, was used for an array of control rods, which could be inserted in or drawn out of the reactor pile, until just the right rate of nuclear fission and energy output was achieved. Nearly all nuclear reactor activity of the early 1940s was for military purposes, but the possibilities for using the tremendous amounts of heat generated for the commercial production of electrical energy was immediately recognized.

The normal form of uranium, $_{92}U^{238}$, does not undergo fission except when bombarded with extremely high-speed neutrons. Instead, it tends to capture a neutron and become U^{239}. The fissionable isotope of uranium is $_{92}U^{235}$. Natural uranium contains only about 0.7 percent of U^{235}, and commercial reactor operation with such low-grade "fuel" is not feasible. United States government-owned installations, operating originally under the Atomic Energy Commission and currently under the Department of Energy, have produced (and controlled the supply and distribution of) "enriched" uranium since the late 1940s.

Figure 34.10 is a cutaway diagram of an enriched-uranium fission reactor typical of the period of the 1950s. The graphite serves as the *moderator* to produce slow neutrons, which, it will be recalled, are most effective in causing fission of another U^{235} atom. Neutrons liberated by the U^{235} atoms enter into two reactions: (1) some of them hit other U^{235} nuclei and keep the chain reaction going; (2) others are *captured* by U^{238} nuclei, resulting in U^{239} and eventually in plutonium. The rods of natural uranium, together with the graphite moderator block, form a structure known as a *pile* or a *reactor*. An atomic pile, or reactor, is merely an arrangement of fissionable uranium and a suitable moderator, which is above the *critical size*. A sustained reaction is therefore possible, but, unlike the bomb, the rate

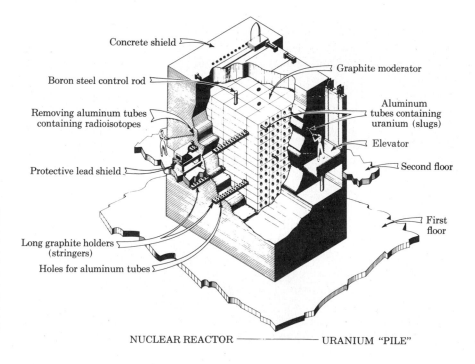

Fig. 34.10 Schematic diagram of an enriched uranium pile, or atomic-fission reactor of an early (1950s) design. Boron-steel control rods, tubes for preparation of radioisotopes, and graphite moderator blocks for slowing neutrons are all shown. (Atomic Energy Commission, now the Department of Energy.)

Concrete shield
Boron steel control rod
Removing aluminum tubes containing radioisotopes
Protective lead shield
Long graphite holders (stringers)
Holes for aluminum tubes
Graphite moderator
Aluminum tubes containing uranium (slugs)
Elevator
Second floor
First floor

NUCLEAR REACTOR ———————— URANIUM "PILE"

of the reaction can be controlled, in a manner to be described.

Multiplication Factor Nuclear physicists use the term *multiplication factor* to express the likelihood of a sustained reaction. If, for example, 100 U^{235} nuclei, upon fission, produce 100 neutrons which blast apart other nuclei, the factor is said to be 1.00. If 100 original neutrons produce 125 *effective* neutrons, the factor is 1.25, while if only 75 effective neutrons are produced from 100 original neutron "bullets," the multiplication factor is only 0.75 and the reaction will die out. At the *critical size*, the multiplication factor is 1.00. Above the critical size, the nuclear reaction goes faster and faster, and if it were uncontrolled, a "meltdown" or other nuclear disaster would occur. Control of the pile is effected by inserting in the graphite some rods of cadmium, boron-steel, or other neutron-absorbing material. Sliding the cadmium rods in or out will vary the multiplication factor of the pile. The pile is kept operating at a multiplication factor slightly above 1.00, but under control due to neutron absorption by the cadmium or boron-steel rods.

As U^{235} nuclei undergo fission in the pile, much energy in the form of heat and radioactive materials is given off. Workers must be protected from the harmful radioactive rays, so the reactor is shielded by thick walls of steel and concrete. The entire reactor must be enclosed within a radiation-tight and airtight compartment, usually concrete, 3 ft or more thick. The rods themselves, after removal from the reactor, are intensely radioactive, and the entire chemical process of plutonium separation must be carried out by remote control from behind heavy shields. Any material, whether it is solid, liquid, or gas, can become dangerously radioactive if exposed to the reactor for a brief period of time. Figure 34.11 shows the kind of precautions which must be observed in handling radioactive materials.

34.9 Plutonium Is Also Fissionable

During the early work on U^{238} and U^{235}, it had been discovered by Fermi and others that when U^{238} is bombarded with a neutron, it captures the neutron and becomes U^{239}. This isotope is very unstable, and it decays by emitting a beta

Fig. 34.11 Technician examines reactor fuel unit after depletion in the reactor. The operator is protected by thick layers of concrete and leaded glass. (Oak Ridge National Laboratory.)

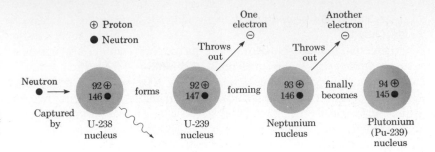

Fig. 34.12 Diagram illustrating the process of formation of plutonium $_{94}Pu^{239}$ from neutron capture by U^{238}.

particle (high-speed electron), leaving a nucleus of *neptunium*, $_{93}Np^{239}$. Neptunium is unstable also, and another beta particle is emitted, creating another element, *plutonium* Pu^{239}, $Z = 94$. This process is diagramed in Fig. 34.12. Plutonium is fissionable, like U^{235}.

Plutonium proved to be easier to produce in large quantities than was U^{235}, and extensive facilities for Pu^{239} production were established in the 1940s at Hanford, Washington, where the Columbia River could be used to absorb the large amounts of heat generated by fission reactions. Most nuclear weapons ("nukes" as they are called in the military) contain plutonium. A new concept for nuclear reactors (fast-breeder reactors) designed to use plutonium is now ready for prototype construction and evaluation. This reactor will be briefly described in the next section, along with some of the controversy over whether or not to go ahead with full-scale development.

MODERN NUCLEAR-FISSION REACTORS

34.10 Power From Nuclear Fission

The heart of a nuclear-power plant is the reactor core. As diagramed in Figs. 34.8, 34.9, and 34.10, U^{235} fission can result in a self-sustaining chain reaction regulated by the *moderator* (now usually water rather than graphite) and the *control rods*, which are usually of boron steel. At each U^{235} fission, about 200 MeV of energy is liberated, and two or three "left-over" neutrons are ejected. If each fission produces just one *effective* neutron,

the *multiplication factor* is 1.00, and the reactor is said to be "critical." The actual rate of U^{235} fission and of the resulting heat energy output is determined by the control rod adjustment. Fully inserted, the boron-steel rods absorb the extra neutrons and slow the rate of fission to near zero. When the rods are withdrawn, the chain reaction can resume.

In modern nuclear plants there are some 400 to 500 variable factors that have to be measured, adjusted, or controlled. Needless to say, most of these processes are automated and computer-controlled. Back-up systems are everywhere, and the engineers and technicians who operate the plants must be highly trained, and safety conscious to the extreme (see Fig. 34.13).

Light-Water Reactors A common type of nuclear fission reactor now operating at many locations is the *light-water reactor* (LWR). The term "light-water" means that ordinary water whose hydrogen is $_1H^1$ is used as the moderator and coolant. *Heavy water* is water whose hydrogen is $_1H^2$, which is called *deuterium*. We shall have more to say about deuterium in a later section.

One form of LWR is the boiling-water reactor (BWR), illustrated in the diagram of Fig. 34.14. In this reactor, water circulates around the reactor core where it is heated by the energy given off from U^{235} fission in the fuel rods. At a pressure of about 1000 psi, its boiling point is in the range of 580°F. The steam produced is used to drive a turbine generator for the production of ac electric power. The spent steam from

Fig. 34.13 The control room of the Dresden nuclear power plant near Chicago. This plant uses a boiling-water reactor to produce about 200,000 kVA. Note the hundreds of meters, dials, warning lights, and timing devices. (Commonwealth Edison Co.)

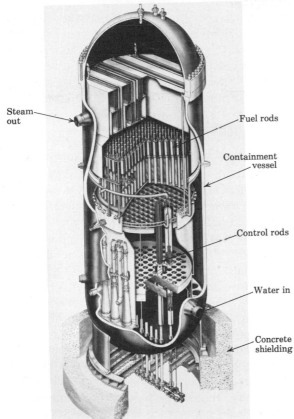

Steam out

Fuel rods

Containment vessel

Control rods

Water in

Concrete shielding

Fig. 34.14 Cutaway view of a boiling-water reactor (BWR). In this early design (1960s), water is boiled in the reactor vessel, producing saturated steam (about 580°F, 1000 psia) which passes directly to the turbine. The fuel rods (with enriched U^{235} as the fuel) and control rods are shown, along with associated equipment within the containment vessel. The heavy concrete shielding is indicated. The many protective devices, gauges, meters, valves, pumps, etc., are not shown in this illustration. (General Electric Co.)

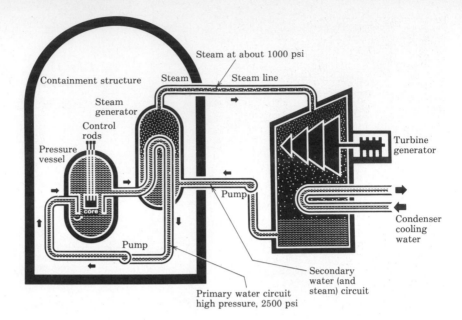

Fig. 34.15 Cross-sectional diagram (schematic) of a pressurized water reactor (PWR). This system is representative of the "present generation" of nuclear electric power plants. (Westinghouse Electric Corp.)

Labels on figure: Containment structure — Steam — Steam at about 1000 psi — Steam line — Steam generator — Control rods — Pressure vessel — Core — Pump — Pump — Primary water circuit high pressure, 2500 psi — Secondary water (and steam) circuit — Pump — Turbine generator — Condenser cooling water

the turbine is then condensed in the condenser and pumped back into the reactor for another cycle. The condenser water, in turn, is cooled by an atmospheric cooling tower.

A second kind of light-water reactor is the pressurized-water reactor (PWR). This reactor has two circulating water systems, one entirely within the reactor containment vessel, operating at high pressure, called the *primary system*; and one in which steam is formed and supplied to the turbine (see Fig. 34.15). The water circulated in the primary system is under a pressure of 2000 to 2500 psi. Consequently, it does not boil at the reactor temperature of around 600°F. In the heat exchanger (steam generator) its heat boils the water in the secondary system and produces steam at about 1000 psi. After driving the turbine, the steam is condensed and the condensate is pumped back to the heat exchanger again.

Gas-Cooled Reactors Steam turbines are more efficient as the steam temperature rises. The present 600°F practical limit for steam temperature from LWRs has led to the development of still another type of U^{235} fission reactor, called the high-temperature, gas-cooled reactor (HTGR). This reactor is still in the development stage, although a few have been in actual operation.

The gas which circulates around the core is usually helium, and it is heated to a temperature of 1300 to 1500°F by fission energy. The hot gas in its closed cycle is passed through a heat exchanger (steam generator) where steam at 1000°F or more can be produced. This very hot superheated steam increases turbine efficiency appreciably.

Fast-Breeder Reactors It was mentioned above that only about 0.7 percent of uranium is the readily fissionable U^{235} isotope. Even with its fantastic energy yield per kilogram, our resources of U^{235} could be exhausted by the year 2010 if presently planned U^{235} reactors go "on line" as scheduled. If nuclear-fission reactors are to hold any real promise as an energy source for the long-term future, some nuclear "fuel" other than U^{235} will have to be exploited. The *fast-breeder reactor* (FBR) is a proposed solution to this problem; it uses plutonium ($_{94}Pu^{239}$) as the active ingredient in its fuel rods. Plutonium does not occur naturally. It is one of the *transuranic* (i.e. beyond uranium in the periodic table) elements produced artificially. Plutonium is readily manufactured, and is under more or less continuous production at several plants in the United States. It is relatively pure as produced, in contrast to "en-

Fig. 34.16 Enriched U^{235} fuel rods for a conventional PWR undergo final inspection by a technician. The principal metal in the rods is zirconium. Each rod contains an amount of uranium-dioxide fuel pellets which, when used by a nuclear reactor, can provide as much energy as three railroad cars of high-quality coal. (Westinghouse Electric Corp.)

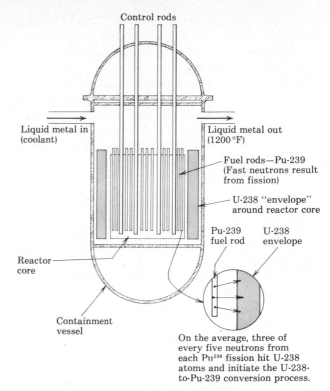

Fig. 34.17 Cross-sectional diagram showing how "fast neutrons" from fissioning Pu^{239} in a liquid-metal fast-breeder reactor (LMFBR) convert ordinary U^{238} in the uranium "envelope" surrounding the reactor core, into plutonium. The actual conversion process was diagramed in Fig. 34.12.

riched" uranium fuel, which is only about 5 percent U^{235}. The process of plutonium manufacture was diagramed in Fig. 34.12; it can be written as follows:

$$_{92}U^{238} + _{0}n^{1} \rightarrow _{92}U^{239} \qquad (34.3)$$

The isotope $_{92}U^{239}$ is unstable: it undergoes two beta emissions in rapid succession (remember that beta particles are electrons) and ends up as plutonium.

$$_{92}U^{239} - e^{-} \rightarrow _{93}Np^{239} - e^{-} \rightarrow _{94}Pu^{239}$$
$$(34.4)$$

Plutonium-239 is readily fissionable, like U^{235}. It liberates energy as heat, and several neutrons in the process. The heat goes into liquid metal (usually sodium) or gas (helium) which surrounds the reactor core. In the liquid metal

version, hot sodium at 1200°F circulates in the primary system and surrenders its heat to generate steam in a steam generator or boiler, as before. The turbine, condenser, and steam generator constitute the secondary system.

The big difference between a breeder reactor and a conventional U^{235} reactor is in the reactor core itself. In the U^{235} core the (conventional) U^{235} (about 5 percent of the enriched uranium fuel rod) is used up, and the rods must be replaced periodically with new ones (Fig. 34.16). In the fast-breeder (plutonium) reactor, the fissioning plutonium emits *fast neutrons* (no moderator is used since slow neutrons are not needed), which bombard an adjacent "envelope" of ordinary uranium, U^{238}. Figure 34.17 shows one possible arrangement in schematic form.

It will be recalled (Fig. 34.12 and Eqs. 34.3

and 34.4) that neutron bombardment of U^{238} produces plutonium 239. Here, then, is the key to FBR operation. A plutonium FBR, in addition to supplying vast amounts of heat energy, "breeds" more of its own fuel from ordinary uranium. Hence its name, "breeder reactor." The "fast" part of its name comes from the fact that fast neutrons are used, and no moderator is needed.

If FBRs can be brought to successful operation in large numbers, the nuclear-fuel limitations of U^{235} will be effectively bypassed. On the average, three out of every five fast neutrons emitted from the fissioning Pu^{239} strike U^{238} atoms in the uranium envelope and initiate the process of conversion from U^{238} to Pu^{239}. After a time, the uranium envelope is nearly all converted to plutonium and it can be processed into fuel pellets or rods to be used in the same or another FBR.

Fast-breeder reactors would operate at 1000° to 1200°F and permit a much higher steam turbine efficiency than light-water reactors do. Also, instead of using only 3 to 5 percent of its fuel rods, an FBR uses almost 90 percent of its fuel rods and breeds new fuel from ordinary uranium at the same time.

FBR development in the United States is a highly controversial issue (see below), and it is difficult to predict the future of this reactor.

34.11 Current Status of Fission Reactors

Light-water reactors constitute a proven energy resource. About seventy of them are currently operating, competing successfully on a daily basis with fossil-fueled plants, and generating energy equivalent to about 1.4 million barrels of oil per day. Their output is only about 14 percent of the total electric energy production in the United States, but many more plants are under construction with the expectation of being "on line" in the 1980s. Others are in the design and contract-negotiation stage. The (former) Atomic Energy Commission predicted in the early 1970s that by 1980 the United States would have about 140 nuclear plants, averaging 1000 MW each in output, and supplying 20 percent of the nation's

electrical energy. By the year 2000, according to the same AEC study, nearly 1000 large fission plants should be in operation, supplying 60 percent of the nation's electricity. Many of these would be FBRs, a type which would be emphasized after 1985, according to the predictions.

Currently however, it appears that these predictions were much too ambitious, and it seems unlikely that the goals set then will be met. Progress has been disappointingly slow, even as the energy crisis deepens year by year. There are a variety of reasons for lack of progress on nuclear power, many of them sociopolitical in nature. We shall try to avoid discussing those, and merely mention a few that have at least some basis in scientific and technological concerns. First, we cite four problems related to *feasibility and safety*.

1. There is a concern that a long-term commitment to a U^{235}-based energy program may be ill-advised, since proven reserves of that "fuel" will last for perhaps only 25 to 30 years. Critics of nuclear power have demanded that the hundreds of millions of dollars being spent in research and development in that field be switched to projects on solar, wind, tidal, or geothermal energy, because they believe that the latter have better long-term potential as "renewable" resources.

2. There are reservations about the safety of all nuclear plants not only for the plant personnel, but for the residents in the plant locality. FBRs are a special concern since they operate at very high pressures—2000 to 2500 psi—and use molten metals as the heat-transfer medium. These substances are known to be highly toxic and corrosive.

3. LWRs require the perfect synchronization of some 500 operating variables, and FBRs 1500 to 2000 variables. Even with modern automated controls, back-up systems, and "fail-safe" devices, there cannot be 100 percent assurance that nothing will go wrong. Critics contend that 99.99 percent assurance is not good enough. A "melt-down" with release of substantial quantities of radioactive materials could poison an entire region for a considerable period of time and perhaps result in many deaths.

4. There is concern about the security of the plutonium itself in FBRs. The Pu^{239} which fuels them is essentially the same material that goes into nuclear bombs and artillery warheads. The possibility of theft of Pu^{239} by organized gangs of terrorists for the pur-

pose of nuclear blackmail and perhaps even nuclear attack cannot be entirely dismissed.

Other concerns relate to the real or perceived *impact on the environment* of nuclear power plants. Several of these are:

1. There is the immediate problem of thermal pollution (heat dissipation) involving rivers, the seacoast, or the atmosphere in the vicinity of nuclear plants.

2. The problem of safe disposal of nuclear wastes is a major concern. Burying them under mountains or under the polar ice caps has been proposed. Dumping them at sea and even shooting them into space to orbit around the earth forever are suggestions. All such proposals have serious drawbacks.

3. The power plants themselves are regarded by many as unsightly, and as having a negative effect on the visual environment and on property values. "We need the power, but put the plant somewhere else," is a common attitude.

Undoubtedly many of these concerns are justifiable. But so is the concern and the reality that the nation is running out of energy. The technology to exploit nuclear-fission power to the point of meeting a major share of the nation's needs for electricity is already developed. The question is not: Can we do it?, but rather, Should we do it? Time is running out while we debate the question.

FUSION POWER—PROMISE FOR THE FUTURE

34.12 Thermonuclear Fusion

We have noted that the nuclei of certain lighter elements have a mass which is less than that obtained by adding up the masses of their constituent parts. This discrepancy, or mass defect, for the helium nucleus (alpha particle) has been calculated to be about 0.03 amu (see Sec. 34.4).

According to Einstein's mass-energy conservation equation, if mass decreases, an amount of energy $E = mc^2$ will be produced. If two protons and two neutrons come together to form a helium nucleus, the process should yield energy, since there is a mass decrease. Consider the following problem.

Illustrative Problem 34.2 What amount of energy is released as two protons and two neutrons fuse to form a helium nucleus?

Solution The mass decrease in this fusion process (see Sec. 34.4) is 0.03039 amu. Expressed in kilograms, this becomes

$$m = 0.03039 \text{ amu} \times 1.6605 \times 10^{-27} \text{ kg/amu}$$
$$= 5.05 \times 10^{-29} \text{ kg}$$

Now, use Einstein's equation, $E = mc^2$. Substituting,

$$E = 5.05 \times 10^{-29} \text{ kg} \times (2.998 \times 10^8 \text{ m/sec})^2$$
$$= 4.54 \times 10^{-12} \text{ J} \qquad \qquad ans.$$

or, expressed in electron-volts,

$$E = 4.54 \times 10^{-12} \text{ J} \times 6.242 \times 10^{18} \text{ eV/J}$$
$$= 28.3 \text{ MeV} \qquad \qquad ans.$$

The same result is obtained by recalling that the energy equivalent of 1 amu is 931 MeV. $E = 0.03039 \text{ amu} \times 931 \text{ MeV/amu} = 28.3 \text{ MeV}$, as before.

Nuclear fission and nuclear fusion are, in a way, opposite processes. In the former, a heavy nucleus under neutron bombardment splits into two or more different particles, liberating energy in the process; in the latter, nucleons combine to form light nuclei, also releasing energy. A single stray neutron can start a chain reaction in fissionable material whose mass is of critical size, but in order to bring light nuclei in sufficiently close proximity to initiate fusion, they must be given prodigiously high speeds to overcome the forces of repulsion (*Coulomb force*) between nuclear particles. These speeds can be imparted only by fantastically high temperatures, of the order of 10 million kelvins and above. Such temperatures are present in the interior of the stars and on the sun. It is the necessity for an extremely high temperature that gives the process the name *thermonuclear fusion.*

Two isotopes of hydrogen—$_1\text{H}^2$(*deuterium*) and $_1\text{H}^3$(*tritium*)— and *lithium* have been used in research and development on thermonuclear fusion, including the thermonuclear bomb. One of these is quite "plentiful" in nature, since deuterium makes up about 1 part in 600 of ordinary hydrogen, and hydrogen is an essential element in water. Deuterium can be separated from ordinary hydrogen without great difficulty, but it is a costly process. Lithium is present in the earth's crust, but in amounts not quite equivalent to the remaining energy of fossil fuel reserves. Tritium must be manufactured at considerable cost. The following reactions, which occur only at the extremely high temperatures noted above, are typical:

$$_1H^2 + {}_1H^2 \rightarrow {}_2He^3 + {}_0n^1 + 3.2 \text{ MeV} \quad (34.5)$$

$$_1H^2 + {}_1He^3 \rightarrow {}_2He^4 + {}_0n^1 + 18 \text{ MeV} \quad (34.6)$$

These reactions are actually simplifications of complex reactions that include more steps than are shown here.

Once a sufficiently high temperature is reached (perhaps 100 million kelvins) and a thermonuclear reaction is started, enormous amounts of energy can be liberated. The problem is to keep the process going and keep it under control. *Uncontrolled* thermonuclear fusion was achieved when the first hydrogen bomb was detonated in 1952. To date, although there has been intensive research in laboratories all over the world, *controlled* fusion as a sustained reaction delivering more output energy than the input energy consumed, has not been attained. The research continues at an accelerating pace because of the world energy shortage, and because of the almost unlimited energy which would become available were fusion reactors in operation. Considering only one of the possible "fuels" as a source of protons, there is enough deuterium (heavy hydrogen) present in 1 ft³ of sea water to provide nearly 300 million Btu of heat energy. At this rate, a body of water only a tenth the size of Chesapeake Bay would provide the equivalent of all the energy used by mankind since fire was first kindled in prehistoric times.

Such comparisons are dramatic but they are also deceiving. It must be kept in mind that if there were a simple solution to the problem of controlled thermonuclear fusion, we would have the reactors running now. On the contrary, the problem is one of almost indescribable difficulty. Only a few beginning "successes" are being reported. Two lines of investigation are currently favored. Both are concerned first with attaining the high temperatures required and then with containing, controlling, and sustaining the reaction. Both will be described very briefly, and with much oversimplification.

Toroidal Fusion Research—The "Magnetic Bottle" If a gas is heated to a high-enough temperature that its atoms lose most of their electrons, it is called a *plasma* and it is said to be highly ionized. An entire new branch of physics is now called *plasma physics*, and some scientists have proposed plasmas as the *fourth state of matter*, after solids, liquids, and gases.

The "matter" of all hot stars and of the sun is thought to be in the plasma state. Their almost incalculable energies result from thermonuclear fusion. In order to attain nuclear fusion on earth, plasmas at high temperatures must first be produced. Then the temperature must be maintained long enough for the reaction to take place. The plasmas must be contained in some sort of vessel, and the rate of fusion must be controlled lest disastrous consequences result.

Some laboratories have already attained temperatures in the 30-million degree range. Something of the order of 100-million degrees is believed to be necessary to attain reactions which provide a net energy yield. The problem of containment has been the focus of years of research and billions of dollars. The hot plasma cannot be allowed to touch its container, since all known solid matter vaporizes at temperatures above about 3000°C. Scientists and technicians are attempting to "contain" the stream of hot plasma within a strong magnetic field whose lines are generally parallel to the stream of plasma in the "bottle." Thus the hot plasma is confined to a limited region in space and does not touch the container at all. The heating is done by a high-voltage electric discharge through the bottle after

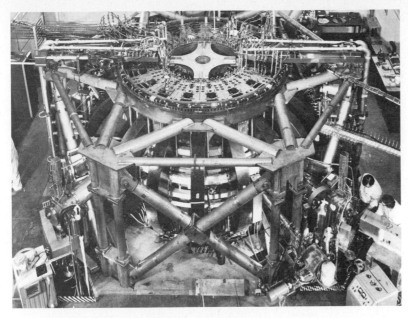

(a)

(b)

Fig. 34.18 The Princeton Large Torus (PLT) nuclear fusion research machine. Hot ionized gas (plasma) is confined by a stainless steel, dough-nut-shaped vessel. Very powerful magnetic fields hold the plasma away from the vessel's walls. During 1977 the PLT attained a temperature of 27 million degrees, the record for plasma machines up to that time. (a) Overall view of the PLT, with engineers and technicians running tests. (b) View during construction, showing positioning of huge toroidal magnets. (Princeton University Plasma Physics Laboratory.)

the deuterium gas (and/or other "fuel") has been introduced. The natural "pinch effect" of electric discharge in the plasma assists in the heating process.

Recent versions of the magnetic bottle have taken the form of a toroidal (doughnut-shaped) ring (see Fig. 34.18). Russian scientists have favored this approach in their "Tokamak" ma-

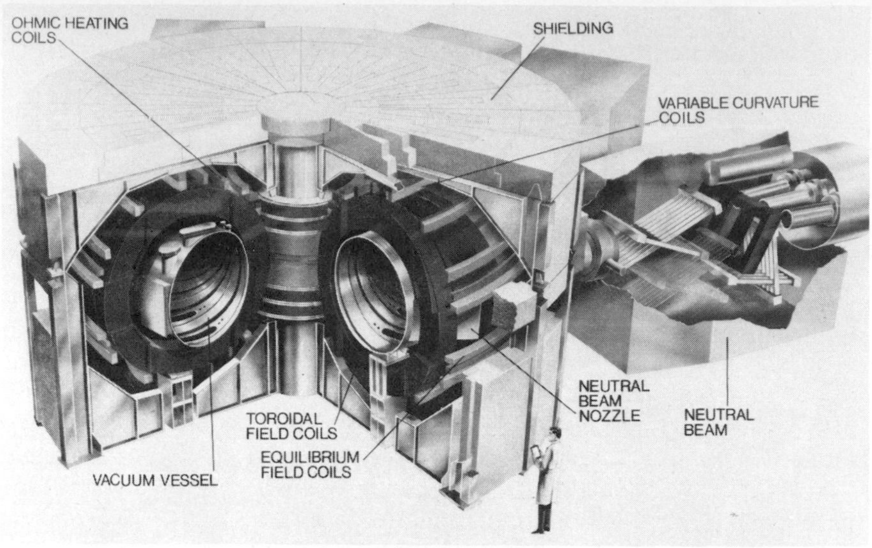

Fig. 34.19 Artist's rendition of the Tokamak Fusion Test Reactor (TFTR). The TFTR will use a two-component plasma (deuterium and tritium). The TFTR is expected to advance fusion research to the break-even point, that is to the point where fusion power output will be comparable to the power input. It will operate throughout the decade of the 1980s. (Princeton University Plasma Physics Laboratory.)

OHMIC HEATING COILS

SHIELDING

VARIABLE CURVATURE COILS

TOROIDAL FIELD COILS

VACUUM VESSEL

EQUILIBRIUM FIELD COILS

NEUTRAL BEAM NOZZLE

NEUTRAL BEAM

Fig. 34.20 The LLL 2XIIB Magnetic Fusion Research machine. Using preheated bursts of plasma "fuel" from the plasma guns in the left foreground, this machine is expected to be able to attain temperatures of 250 million degrees during its research "lifetime". Such a temperature is considered to be more than sufficient to initiate a self-sustaining reaction. (Lawrence Livermore Laboratory of the University of California.)

chine. One United States version uses a deuterium-tritium fusion "fuel" in a toroidal "bottle" surrounded by a huge solenoid capable of producing extremely strong magnetic fields. Among the laboratories in the United States working on magnetic containment fusion are the Princeton University Plasma Physics Laboratory with its *Tokamak Fusion Test Reactor* (*TFTR*); the Lawrence Livermore Radiation Laboratory of the University of California with its open-ended (nontoroidal) MFE/2XII-B machine; and the Los Alamos Scientific Laboratory, with its huge 15-meter machine, called *Scyllac*. Figures 34.19 and 34.20 give some idea of the equipment involved.

Temperature, containment, and control are the three problems to be solved. Furthermore, they must combine for a steady-state condition, for if any one of the three fails, fusion would cease or, even worse, disaster might result. Reactor disaster, if it should occur, would be a "melt-down," not an explosion.

Laser-Induced Fusion Research In this approach, powerful lasers are used to heat the "fuels" to the fusion ignition temperature. Fusion fuels (either deuterium–deuterium or deuterium–tritium—see Eq. 34.6) are first formed into small pellets of much less than pinhead size. A pellet is injected into a vacuum chamber where it becomes the target for several synchronized laser beams of tremendous power, all of which hit the pellet simultaneously from different sides (Fig. 34.21). The laser is "pulsed," concentrating its energy in the beams for about 10^{-9} sec (1 nanosecond). This "implosion" energy shock compresses the pellet by a factor of 10,000 times its normal density and raises its temperature to about 100 million degrees K within the time frame of 1 billionth of a second. Fusion ignition then occurs. As the fusion energy from one pellet falls off, another pellet is injected, the laser pulses, and fusion begins again (Fig. 34.22). At

least one scientific laboratory has already reported "significant" production of high-energy neutrons from laser implosion (Fig. 34.23).

All of this sounds as if thermonuclear fusion is only a few years away—as if progress were being made in large steps. Exactly the opposite is true. Progress is made in small, hard-won steps. Years and probably decades will be needed to solve the difficult problems that remain. A tentative timetable recently suggested by the U.S. Department of Energy announced these goals:

1980	Significant "fusion burn" in laser machines
1982–85	"Substantial" quantities of thermal energy in toroidal fusion machines. "Break-even" point in 1982 and "net energy gain" by 1985 in laser machines
1990–95	Demonstration model of a commercial power plant from both toroidal and laser approaches
2000–2010	Commercial fusion reactors of both types in operation, *if* the above goals are met on time

In almost every category of comparison, fusion power is to be preferred over any presently operating energy system and over any form of fission power. Fusion "fuel" is abundant and relatively cheap. It would appear to be nearly inexhaustible. Once in full operation, fusion-produced electricity should be much cheaper than fossil-fuel electricity is today. (However, visions of "dirt-cheap" electricity will probably not be realized, because a large share of the cost of electricity to the consumer is in *distribution and delivery*, rather than in production.) Nuclear fusion energy has only minor radiation problems, in contrast to the very serious radioactive waste problems of nuclear fission energy. As to plant safety itself, there would be no danger of a bomb-type explosion, since only small quantities of the fuels would be reacting (or even near the reactor) at any given time. The casualties that conceivably could occur would be minor (in-plant) explosions or ruptures of high-pressure steam lines, electrical casualties in the generator room, or at the

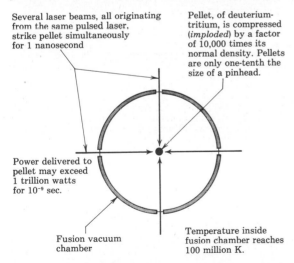

Several laser beams, all originating from the same pulsed laser, strike pellet simultaneously for 1 nanosecond

Pellet, of deuterium-tritium, is compressed (*imploded*) by a factor of 10,000 times its normal density. Pellets are only one-tenth the size of a pinhead.

Power delivered to pellet may exceed 1 trillion watts for 10^{-9} sec.

Fusion vacuum chamber

Temperature inside fusion chamber reaches 100 million K.

Fig. 34.21 Schematic diagram of combustion chamber for a laser-induced "implosion," resulting in thermonuclear fusion of a deuterium-tritium pellet. Experimental only at the present time. Temperatures of 100 million kelvin have been attained, and actual fusion has been achieved, but so far the energy output has been less than the input energy to operate the equipment.

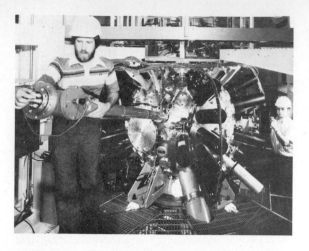

(a)

(c)

(b)

Fig. 34.22 Laser fusion research equipment currently in use. (*a*) Technicians prepare the SHIVA laser system's target chamber for a laser implosion of the deuterium-tritium target. (*b*) Inside the SHIVA combustion (target) chamber. Note the needle-like target positioner, on the tip of which is mounted the deuterium-tritium target, no larger than a grain of sand. (*c*) The SHIVA laser-amplifier chains. This laser system delivers 30 trillion watts of optical power to the target in less than one-billionth of a second. (Lawrence Livermore Laboratory of the University of California.)

extreme, a casualty in the plasma "bottle" or laser implosion chamber. The problem of theft and unauthorized use of nuclear materials for terrorism, which is a real danger with plutonium, is not applicable to fusion materials, since they will react only if brought to fusion-ignition temperature under reactor conditions.

In summary, the *promise* of fusion power is tantalizing in the extreme. But it is still too early to predict that the promise will be fulfilled. The difficulties yet to be overcome are awesome in their immensity. The cost, over the next quarter century, will be many billions of dollars. The *chances* of ultimate success are good, but success is by no means certain. As a consequence, all other energy options—solar, hydroelectric, coal, wind, tides, geothermal, and certainly nuclear fission and the fast-breeder reactor—should be kept open, and research should go forward in all these fields.

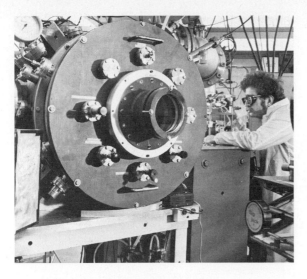

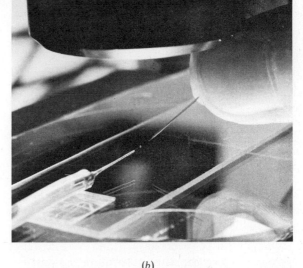

Fig. 34.23 Laser-fusion research. (*a*) The experimenter is observing the positioning of a deuterium-tritium pellet in the heavy-wall evacuatable target chamber. Powerful laser beams enter the chamber, one through the port shown (center foreground), and the other on the back side of the chamber. These beams are pulsed to provide enormous energy as they implode the pellet from opposite sides. The experimenter can observe and measure what takes place during a pellet implosion. (*b*) Pellets for laser-fusion research are so small (about the diameter of a human hair) that technicians must use a microscope when attaching the tiny bubble-like targets to a slender support fiber which is used to position them in the target chamber for laser irradiation. (Both photos courtesy of KMS Fusion, Inc., Ann Arbor).

34.13 Mobile Power from the Atomic Nucleus

So far our discussion of nuclear power has been limited to stationary power plants. The tremendous weight of a nuclear reactor and its associated shielding would seem to make atomic power unsuitable for use except by such plants. However, the extremely small amount of "fuel" required, compared to conventional fuels, makes a reanalysis of the possibilities interesting. Fissionable materials such as U^{235} or plutonium have an energy content of about 10 million kW-hr/lb. Nearly 300,000 gal of fuel oil would be required for the same energy content. So far we have not perfected the processes and machines for actually getting all this energy out of fissionable material (efficiencies run less than 1 percent), but even with only a small fraction of it available, there still results an almost weightless fuel. Considering the fuel load now carried by ships, submarines, aircraft, and spacecraft, the potential of nuclear energy should be fully exploited.

To date the greatest use of mobile nuclear power has been in submarines. Nuclear power plants enable submarines to cruise submerged for weeks at a time. The USS Triton cruised around the world underwater without once surfacing, in the 1960s, and made home port with more than half her fuel capacity unused. Several aircraft carriers of the U.S. Navy are also nuclear-powered, with these advantages: greatly extended cruising range, higher speed, and far more space for aircraft, jet fuel, spare parts, and crew accommodations, since fuel oil for their own propulsion does not have to be carried. Some cruisers, frigates, and other Navy surface ships are also nuclear-powered.

The prospect for nuclear propulsion of space vehicles while they are out in space seems good, but the tremendous thrust required to blast off from earth and that required by retrorockets to effect a safe return to earth, seems to be available—for the present, at least—only in conventional chemical reactions.

RADIOACTIVITY IN INDUSTRY, MEDICINE, AND AGRICULTURE

34.14 Uses of Radioactive Isotopes

An *isotope* of an element has been previously defined as a *form of the element which has an atomic mass different from the standard or most frequently occurring value.* This difference in atomic mass, it was pointed out, is due to a difference in the number of neutrons in the nucleus. The number of protons in the nucleus and electrons in the orbits remains the same as for the standard atom of the element. Consequently, isotopes of an element possess the same *chemical* properties as the standard element. Industrial and medical research have found countless uses for various radioactive isotopes in the past few years.

Making Radioisotopes To make a radioisotope in a nuclear reactor, a small amount of the highly purified element to be used is placed in an aluminum canister about the size of a rifle cartridge. Many such canisters are placed in the reactor and allowed to remain for a period of from one to several weeks, depending on the particular isotope being produced. When the required time has passed, the canister is removed. The originally pure element has been largely converted to a radioactive isotope of the same element by neutron bombardment in the reactor. For example, the element *phosphorous* (at. mass 31) may become radioactive phosphorus (at. mass 32); iodine (at. mass 127) may become radioiodine (at. mass 131), and cobalt (at. mass 59) becomes Cobalt-60. Several hundred radioisotopes are now produced in the United States by neutron bombardment.

The value of these isotopes lies in the fact that, being radioactive, they give off at varying but predictable rates a measurable amount of radiation. Some are *alpha-particle* emitters, others emit *beta particles*, and some give off *gamma* radiation (see Sec. 34.5).

Neutrons constitute another type of radiation. They are heavy, uncharged particles and may travel great distances in air. They are produced during fission and fusion processes. In themselves they have no ionizing power, being neutral in charge, but they create large numbers of ions by collision with other atoms.

Units of Radiation From Radioactive Sources The *quantity of radioactive material present* in a source is determined by the rate at which atoms are disintegrating. The more atoms that disintegrate per unit time, the greater the activity of the source. The amount of radioactive material that undergoes 3.7×10^{10} disintegrations per sec (dps) is called the *curie* (Ci).

$$1 \text{ curie} = 3.7 \times 10^{10} \text{ dps}$$

$$= 1000 \text{ millicuries (mCi)}$$

$$= 1,000,000 \text{ microcuries } (\mu\text{Ci})$$

The curie is an extremely large unit of radiation, approximately equivalent to the radiation emanating from 1 gm of pure radium. Only in industrial radiography and in cancer therapy are sources of multicurie strength used. Most sources used as tracers in industry and medicine are in the millicurie or microcurie range.

Units of Absorbed Radiation—Damaging Effects The *amount of radiation absorbed* is measured in *rads.* It is only the radiation which is *absorbed* by living cells which damages them. Radiations which *pass through* cells do no damage. Absorbed radiation results in *ionization*, and this ionization in turn promotes cell damage and bone marrow damage and initiates changes in body chemistry which all too frequently are irreversible and do not respond to treatment.

The *rad* is the absorbed *dose* when 1 gm of matter (e.g. human flesh and bone) absorbs 10^{-5} **J** of radiant energy.

You may encounter another unit, the *roentgen,* a unit of *radiation exposure.* One *roentgen* (R) is that exposure which will produce 2.08×10^9 ionizations in 1 cm³ of air at standard conditions (0°C, 760 mm Hg pressure).

The safe limit for persons exposed to low-level radiation of x-rays or γ-rays over the entire body is thought to be about 0.05 rad/week. The

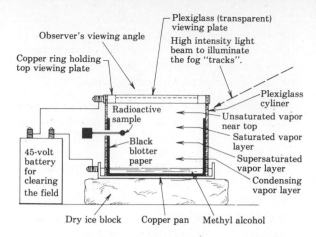

Fig. 34.24 The Wilson cloud chamber. (a) Diagram of essential components. Ions produced along the path of a ray or particle provide nuclei on which vapor from the supersaturated atmosphere condenses to form droplets of fog. The line of fog droplets delineates the path of the ray or particle.

body can repair and keep ahead of cell damage which occurs from doses below this rate. Acute radiation exposure for short intervals of time is another and far more serious matter. Exposure over the whole body of a 20- to 50-rad dose will cause blood-cell damage; 100 to 200 rad will cause severe illness from which most persons might recover; 400 rad is fatal to 50 percent of persons exposed; and 500 rad is fatal to nearly all persons. The effects of radiation are ordinarily cumulative, so small doses are also dangerous if they are repeated daily or weekly.

Since various kinds of radiation (alpha and beta particles and gamma rays, protons and neutrons) are absorbed differently, and create different amounts of biological damage, a third unit of radiation dosage, called the *rem* (*roentgen equivalent, man*) is used to measure the degree of harm to human beings. A factor called *relative biological effectiveness* (RBE) is rated as 1 (unity), for x-rays and gamma rays of 1 MeV. The *rad* and the *rem* are equivalent for these conditions. Other rays and high-energy particles do more biological damage and have RBEs approximately as follows: fast neutrons (and protons), 10; 1-MeV

alpha particle, 20; beta particle, 1.0+ . The *rem* and the *rad* are related as follows:

$$\text{rem} = \text{RBE} \times \text{rad} \qquad (34.7)$$

Radiation and Particle Detecting Instruments There are several types of radiation and particle detection and measuring instruments. The *Wilson cloud chamber* (Fig. 34.24) is one such instrument, and so is the *bubble chamber* (Fig. 34.25). The *Geiger counter*, described below, is another. *Scintillation counters* make use of materials which fluoresce when struck by particles or radiations. A scintillator is usually combined with a photomultiplier tube which converts light flashes from the fluorescing material into electric impulses that can be recorded on an oscilloscope or that will produce both a video and an audio response. Radiation-sensitive emulsions are used in the popular *film badge* type of *dosimeter* used by personnel working in the vicinity of radioactive materials.

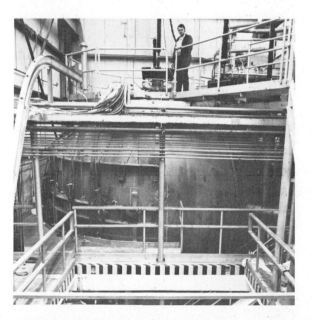

Fig. 34.25 The 15-foot Bubble Chamber at the Fermi National Accelerator Laboratory. High-energy particles enter the big spherical chamber from the right. The interactions of these particles with the liquid hydrogen in the chamber form "tracks" (see Fig. 34.30) which are photographed by seven cameras positioned in the top of the chamber. (Fermi National Accelerator Laboratory.)

Nuclear Energy **747**

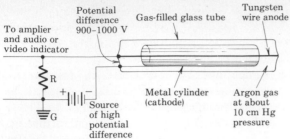

Fig. 34.26 Diagram of the operation of a Geiger counter.
The high potential difference between the metal-cylinder
cathode and the tungsten-wire anode puts an electrical
stress on the argon-gas atmosphere in the tube, just
slightly less than that which would be required to initiate
electrical discharge. When a ray or particle traverses the
tube, it ionizes the gas and "triggers" a momentary
electrical discharge, which flows in the external circuit. An
amplifier "boosts" the signal to operate audio or video
equipment.

Uses of Radioisotopes   There is a continually increasing use of radioisotopes in the treatment of
tumors and cancer, much as radium has been
used for years, but the largest use of radioisotopes
is as *tracers*. The radiations given off from any
radioactive material can be picked up and
recorded by a Geiger counter (Fig. 34.26). This
instrument is based on a small electronic tube
which is sensitive to the passage of a particle or
ray. The passage of any type of radiation through
the tube initiates a transient electric current in the
tube. This current is amplified and used to cause
a "ping" in a loudspeaker, to operate a mechanical counter, or to cause a light to flash. Fig. 34.27
shows two forms of radiation detection instruments in practical use.

Plant scientists are using radioisotopes to
test the absorption and use by plants of various
fertilizers applied to the soil. Veterinarians make
use of radiocobalt in studying the metabolism
rates of poultry, cattle, and sheep.

In the field of human medicine, radioactive
tracer materials can be followed through the body
by suitable detection equipment. Radiosodium 24
is used, for example, to check the effectiveness
of the heart in pumping blood. Iodine 131 is used
in the location of brain tumors and in the detection of tumors of the thyroid gland. Radioisotopes
are also used in tumor and cancer therapy in a
manner similar to x-rays. Phosphorus 32 is used

(a)

(b)

Fig. 34.27   (a) A Geiger counter being used to determine
the effectiveness of a fertilizer which has been "tagged"
with a small amount of radioactive phosphorus as a
"tracer." (b) Another form of radiation detector known as
a "cutie pie," being used to monitor the level of radiation
on top of a reactor core in a nuclear power plant.

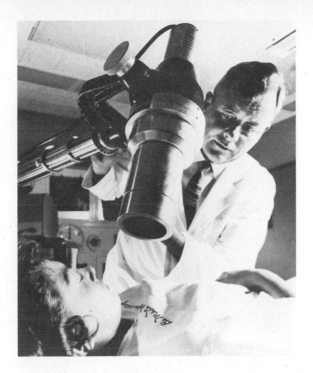

Fig. 34.28 Measurement of the uptake of radioactive iodine in the thyroid gland of a patient. This is a diagnostic procedure. In many diseases radiation therapy is also used. (Brookhaven National Laboratory.)

in the treatment of chronic leukemia. Neutron scanners record the size, position, and probable condition (normal or pathological) of bodily organs and systems.

In industry, radioactive tracers are being used more and more. The flow of chemicals, gas, or oil through pipes is checked by introducing a radioactive isotope in the fluid and checking the flow with a Geiger counter. Gamma radiation has been used in petroleum exploration to locate oil-bearing strata deep under the surface of the earth. Cobalt-60 is extensively used for making industrial radiographs.

Radioisotopes are used in many automated devices for the control of industrial processes, such as thickness control, tests of compaction, and density control (see Fig. 34.29).

One oil company, in developing new motor oils, has used radioisotopes extensively to test wear on auto-engine piston rings. The rings themselves are first made radioactive, then installed in the test engine. The engine is operated for the desired period, and a sample of the oil is drained off for Geiger counter analysis. The amount of radioactivity in the oil is a measure of

Fig. 34.29 Radioisotopes in industry. A radioisotope thickness gauge measuring the thickness of sheet brass as it emerges from the final rollers. The gauge itself is one element of a closed-loop automated system that uses feedback for self-correction of the process. (Revere Copper and Brass, Inc.)

piston-ring wear, since the radioactive metal worn off the rings circulates with the oil. A significant test of an oil's lubricating qualities can be obtained after only about 50 miles of driving, according to the research reports.

Only a few of the many applications of atomic and nuclear energy to industry and modern life have been described. Hardly a week passes without announcements of new uses in the press or in scientific, medical, and industrial journals.

PARTICLE PHYSICS— MATTER AND ENERGY

The emphasis in this book has been on matter and energy and on the applications of physics to industry, medicine, agriculture, commerce, and everyday life. In the two final chapters we have discussed the equivalence between mass and energy, the duality of particles and waves, and the energy of the atomic nucleus. The so-called *basic building blocks* of matter have been identified as the *electron*, the *proton*, and the *neutron*. All of this seemed well-established theory and was amply confirmed by experimental observation and by applications to engineering and science by the middle of the 1930s.

34.15   High Energy Physics— New Particles

In recent decades, as particle-physics research has grown almost to the status of an industry, higher and higher energies have been used in such machines as bevatrons, linear accelerators, synchrotrons, plasma physics machines, and laser-implosion devices. Some of these machines can produce particle energies of more than 400 billion electron-volts (400 GeV). With the aid of these machines, augmented by cloud chambers, bubble chambers, scintillation counters, cathode-ray oscilloscopes, and of course, computers, particle-physics research (now called *high-energy physics*) has identified a large number of new subnuclear particles (Fig. 34.30). Besides those actually *observed*, others have been hypothesized, in order to

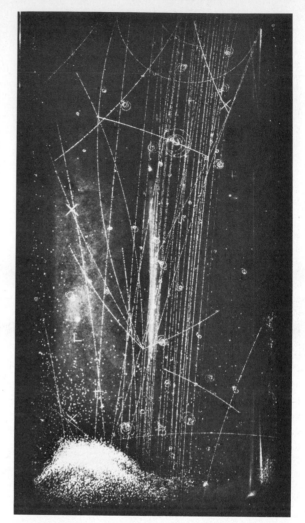

Fig. 34.30   Ray (particle) tracks showing proton-antiproton annihilation from a 3.3 GeV antiproton experiment in a 20-in. bubble chamber. (Brookhaven National Laboratory.)

explain observed events. A list of a few of these new particles, with brief comment about each, follows as a conclusion to the book and as a view of the future.

The *positron*, discovered in 1932. It was predicted earlier from considerations of quantum mechanics. It has the same mass and properties as the *electron*, but its charge is positive (^+e) instead of negative (^-e).

The *μ-meson* (muon), discovered in cosmic rays (radiation from outer space) in 1936. It has a mass approximately 200 times that of the electron, but its charge is equal to that of the electron.

The *π-meson* (pion), discovered in 1947. It, too, was first discovered in cosmic rays. Its mass is about 270 times that of the electron. There is evidence to suggest that π-mesons are the agents that supply the *nuclear force* that holds protons and neutrons together in the nucleus and that they are responsible for the *binding energy* of the atomic nucleus.

The *antiproton* and the *antineutron* were discovered in the 1950s (Fig. 34.30). Their existence had been hypothesized after the discovery of the positron, which is, in effect, an *antielectron*. Principles of *symmetry* seemed to demand these particle-antiparticle pairs.

The *neutrino* ("little neutron"), proposed in 1930, and actually observed experimentally in 1956. It is a high-speed neutral particle, thought to have zero rest mass, no electric charge, and no magnetic properties. It could be described as "nothing," except for evidence that it possesses energy, linear momentum, and angular momentum. Neutrinos have very little interaction with other particles and are therefore thought to be able to penetrate "solid" matter for vast distances. Detection equipment installed deep down in mines has given evidence of occasional neutrino interaction with matter. The fact that some muons formed from these interactions are observed moving *upward* toward the earth's surface, suggests that the neutrino that caused the muon formation must have come all the way through from the other side of the earth. "Natural" neutrinos originate in cosmic rays from outer space, but they have also been observed in the vicinity of operating nuclear reactors and high-energy particle accelerators.

Strange particles, so called because their mass, or their charge, or their behavior did not "fit" into formerly held theories. One group of these is called *hyperons*. As an example, when a proton is bombarded by a very high-energy "projectile"—say another proton—it undergoes a short period of internal excitation, and as it returns to the normal state, it may emit various unstable particles (*strange particles*), some of which are heavier than the proton itself! These are evidently *created* from the high energy of the projectile in accordance with the Einstein equation, $E = mc^2$. In high-energy physics, then, we not only *discover* particles—we probably *create* new ones as well.

Quarks are hypothetical particles, postulated in 1973. Their actual existence has not been verified. The idea was to try to bring some order to the field of high-energy physics, which some scientists were beginning to call the "particle zoo." Nearly 50 new particles had been discovered, postulated or suggested in a 20-year period, and a new concept of an *elementary* particle was needed. The initial *quark* proposal envisioned all nuclear particles as being made up of three kinds of quarks. Very recent observations in the extremely high-energy machines of the late 1970s predict a fourth quark; and theoretical and mathematical analyses in the last two or three years are suggesting that a recognition of six quarks may be just around the corner.

The list above contains only a few of the most well-known of the new particles. The descriptive comment with each is highly simplified and can at best lead to a surface familiarity, not to understanding. Interested students should consult any modern textbook on high-energy physics or particle physics, for an extended discussion.

The sentiment of the typical lay citizen might well be: So what? Will these particles ever be of significance to ordinary people, or are the billions of dollars expended on high-energy physics research merely populating a particle zoo where subnuclear particles are made to "jump through hoops" by physicists cracking 400-GeV whips? Will these findings ever mean anything to the societies that provide the universities, the laboratories, and the funds?

No one knows for sure the answer to this question. However, the experience of history tells us that knowledge sought for its own sake is always valuable. Has Planck's constant had any meaning? Did the Edison effect remain a laboratory curiosity? Was Bohr's concept of the neutron-and-orbiting-electron atom a useless theory? Was de Broglie dreaming when he likened particles to waves? Is Einstein's mass-energy equivalence equation just a mathematical puzzle for the history books?

These discoveries, hypotheses, and theories of a recent yesterday provided the basis for the entire electronics industry, for computers, and for aerospace exploration and satellite communications. They now also hold the key to solar and nuclear energy development for the future. These theoretical and experimental triumphs of two or three generations ago literally remade our world, but it is entirely possible that they may not have had any more impact on the twentieth century than high-energy physics will have on the twenty-first.

SUMMARY

The atomic nucleus is composed of *protons* and *neutrons*. The *nucleus* contains almost all of the mass of the atom.

Isotopes have the same chemical properties as the natural atom, and also the same atomic number, Z. They differ in *atomic mass* and in nuclear structure.

The Einstein equation, $E = mc^2$ is a statement of *mass-energy equivalance*.

The *binding energy* of the nucleus is the result of the *nuclear force*. The binding energy shows up as a *mass defect* in nuclear fusion.

The principal radiations (particles or rays) resulting from nuclear instability are *alpha particles*, *beta particles*, and *gamma rays*.

The *half-life* of a radioactive element or isotope is that period of time in which half of any given amount will undergo radioactive decay.

Fission is a process whereby energy is released as a heavy nucleus breaks apart, while *fusion* represents an energy release as protons and neutrons come together (fuse) to form light nuclei.

Nuclear fission reactors are of two general classifications, based on the "fuel" used: (1) *light water reactors* (LWRs) using enriched uranium, with U^{235} fission; and (2) *fast-breeder reactors* (FBRs), using plutonium (Pu^{239}) as the fuel.

Thermonuclear fusion releases energy from the formation of nuclei by the nucleons of light elements. *Deuterium*, *tritium*, and *lithium* are the substances currently being used in research and development.

The two most promising areas of present fusion research are: *plasma physics*, with the use of a magnetic-toroid containment "bottle"; and *laser-induced fusion*.

High-energy physics is the newest branch of physics. It is an outgrowth of *particle physics*. Two of its characteristics are: (1) the extremely high-energy accelerators being used; and (2) indications that new particles may be *created* as well as discovered.

QUESTIONS AND EXERCISES

1. Why is a neutron such a satisfactory particle for nuclear bombardment?

2. Define the following: alpha particle, beta particle, gamma ray.

3. What are isotopes? How can isotopes of the same element be separated?

4. Why is it unlikely that Newtonian gravitational forces hold nuclear particles together? Why is it even more unlikely that electrical attraction (the *Coulomb force*) holds them together?

5. Define the atomic mass unit, amu. How is its energy equivalent determined?

6. Explain the relationship between *mass defect* and the energy liberated when nucleons from light elements come together to form a new and heavier nucleus.

7. What is meant by the term "*transuranic elements*"?

8. Based on some library reading, explain how Carbon-14 dating could be used to establish the date of a prehistoric event.

9. List and explain the advantages and disadvantages of the plutonium fast-breeder reactor.

10. Take either side of the controversy you prefer, and write a three-page paper on why we should (should not) proceed with nuclear fission reactors to meet the nation's energy crisis.

11. Evaluate the current status of either magnetic-toroid fusion research or laser-induced fusion research (your choice). Ask your librarian to assist you in finding recent articles in suitable journals.

12. Do you see any connection between Heisenberg's indeterminacy (uncertainty) principle and the problems of high-energy physics? If not, explain why it does not apply. If so, list and explain a few instances where it may apply.

PROBLEMS

1. An element has an atomic number of 82. It has 125 neutrons in the nucleus. What is its atomic mass, to the nearest whole number?

2. Calculate the energy equivalent of the mass of a proton. Answer in (a) joules and (b) electron-volts.

3. If 1 gm of matter could be completely converted to energy in 5 msec, what would be the power produced in kilowatts? in horsepower?

4. Fifty atomic mass units of nucleons from light elements "disappears" in the fusion of heavier nuclei. Calculate the energy equivalent of this mass decrease, in joules; in electron-volts.

5. The half-life of cobalt 60 is about 5.3 years. How much of any initial amount would exist after 16 years?

6. How many rads of gamma radiation are absorbed if a person whose mass is 65 kg absorbs 1 J of radiant energy?

7. A worker is exposed to 40 rads of 1-MeV alpha radiation. How many rems is this?

8. The total energy radiated by the sun in 1 year is estimated to be 1.2×10^{34} J, based on measurements taken on the earth's surface. (The earth receives only a small fraction of this total.) If this total energy is the result of a decrease in the sun's mass in accordance with Einstein's equation, in how many years will the sun be consumed, assuming the rate stays the same? (See Table 7.1 for sun's mass.)

9. A plutonium reactor is operating at 250 MW of thermal power. Find the daily decrease in mass of the Pu^{239} fuel, if 100 percent efficiency is assumed.

10. One pound of U^{235} can be converted into available heat energy at 0.05 percent efficiency by the nuclear-reactor power plant in a submarine. What amount of fuel oil would the submarine have to carry to provide an equal amount of heat energy?

11. A nuclear-power plant generates at 200 MW of electric power from a U^{235} reactor-steam turbine-alternator system. If the turbine-alternator combination has an efficiency of 32 percent and the nuclear fission process an efficiency of 0.04 percent, calculate the annual mass decrease of the enriched uranium "fuel."

12. In a plutonium reactor, if the initial decrease in mass by energy conversion is 0.1 percent per day for a fuel load of 1000 kg of Pu^{239}, how many pounds of saturated steam at 500 psia can be produced per minute from boiler feedwater whose initial temperature is 62°F? Assume 0.3 percent efficiency in the reactor-heat exchanger-boiler system. (See Table 14.6 for properties of saturated steam.)

Appendixes

APPENDIX I Fundamental and Derived Units in the SI-Metric System

Quantity	Unit	Symbol	Dimensions
Fundamental Units			
Length	meter	m	
Mass	kilogram	kg	
Time	second	sec	
Electric current	ampere	A	
Temperature	kelvins or degrees centigrade	K or °C	
Luminous intensity	candela	cd	
Plane angle	radian	rad	
Solid angle	steradian	sr	
Some Derived Units			
Acceleration	meter/sec/sec	a	m/sec^2
Angular acceleration	radian/sec/sec	α	rad/sec^2
Angular velocity	radian/sec	ω	rad/sec
Area	square meter	A	m^2
Capacitance	farad	F	A-sec/volt
Charge	coulomb	C	A-sec
Density	kilogram/cubic meter	ρ	kg/m^3
Energy, heat	kilocalorie	kcal	kg-°C
Energy	joule	**J**	N-m
Entropy	joule/kelvin	ΔS	**J**/K
Force	newton	N	kg-m/sec^2
Frequency	hertz	Hz	sec^{-1}
Illumination	lux	lx	lm/m^2
Inductance	henry	H	V-sec/A
Luminance	candela/sq. meter		cd/m^2
Luminous flux	lumen	lm	cd-sr
Magnetic flux	weber	Wb	V-sec
Magnetic flux density	tesla	T	Wb/m^2
Potential difference	volt	V	watts/amp (also joules/coul)
Power	watt	W	**J**/sec
Pressure	newton/sq. meter (pascal)	Pa	N/m^2
Resistance, elect.	ohm	Ω	V/A
Stress	newton/sq. meter		N/m^2
Velocity, speed	meter/second	v	m/sec
Volume	cubic meter	V	m^3
Work	joule	**J**	N-m

Note that it is necessary in a few instances to use the same letter to symbolize two or more units.

APPENDIX II Numerical Prefixes and Their Symbols

Factors	Prefixes	Symbols
10^{12}	tera	T
10^{9}	giga	G
10^{6}	mega	M
10^{3}	kilo	k
10^{2}	hecto	h
10	deka	da
10^{-1}	deci	d
10^{-2}	centi	c
10^{-3}	milli	m
10^{-6}	micro	μ
10^{-9}	nano	n
10^{-12}	pico	p
10^{-15}	femto	f
10^{-18}	atto	a

APPENDIX III   Table of Natural Trigonometric Functions

Degrees	Sine	Cosine	Tangent	Degrees	Sine	Cosine	Tangent
0	0.000	1.000	0.000	46	0.719	0.695	1.03
1	0.017	1.000	0.017	47	0.731	0.682	1.07
2	0.035	0.999	0.035	48	0.743	0.669	1.11
3	0.052	0.999	0.052	49	0.755	0.656	1.15
4	0.070	0.998	0.070	50	0.766	0.643	1.19
5	0.087	0.996	0.087	51	0.777	0.629	1.23
6	0.105	0.995	0.105	52	0.788	0.616	1.28
7	0.122	0.993	0.123	53	0.799	0.602	1.33
8	0.139	0.990	0.141	54	0.809	0.588	1.38
9	0.156	0.988	0.158	55	0.819	0.574	1.43
10	0.174	0.985	0.176	56	0.829	0.559	1.48
11	0.191	0.982	0.194	57	0.839	0.545	1.54
12	0.208	0.978	0.213	58	0.848	0.530	1.60
13	0.225	0.974	0.231	59	0.857	0.515	1.66
14	0.242	0.970	0.249	60	0.866	0.500	1.73
15	0.259	0.966	0.268	61	0.875	0.485	1.80
16	0.276	0.961	0.287	62	0.883	0.470	1.88
17	0.292	0.956	0.306	63	0.891	0.454	1.96
18	0.309	0.951	0.325	64	0.899	0.438	2.05
19	0.326	0.946	0.344	65	0.906	0.423	2.14
20	0.342	0.940	0.364	66	0.914	0.407	2.25
21	0.358	0.934	0.384	67	0.920	0.391	2.36
22	0.375	0.927	0.404	68	0.927	0.375	2.48
23	0.391	0.920	0.424	69	0.934	0.358	2.61
24	0.407	0.914	0.445	70	0.940	0.342	2.75
25	0.423	0.906	0.466	71	0.946	0.326	2.90
26	0.438	0.899	0.488	72	0.951	0.309	3.08
27	0.454	0.891	0.510	73	0.956	0.292	3.27
28	0.470	0.883	0.532	74	0.961	0.276	3.49
29	0.485	0.875	0.554	75	0.966	0.259	3.73
30	0.500	0.866	0.577	76	0.970	0.242	4.01
31	0.515	0.857	0.601	77	0.974	0.225	4.33
32	0.530	0.848	0.625	78	0.978	0.208	4.70
33	0.545	0.839	0.649	79	0.982	0.191	5.14
34	0.559	0.829	0.674	80	0.985	0.174	5.67
35	0.574	0.819	0.700	81	0.988	0.156	6.31
36	0.588	0.809	0.726	82	0.990	0.139	7.11
37	0.602	0.799	0.754	83	0.993	0.122	8.14
38	0.616	0.788	0.781	84	0.995	0.105	9.51
39	0.629	0.777	0.810	85	0.996	0.087	11.4
40	0.643	0.766	0.839	86	0.998	0.070	14.3
41	0.656	0.755	0.869	87	0.999	0.052	19.1
42	0.669	0.743	0.900	88	0.999	0.035	28.6
43	0.682	0.731	0.932	89	1.000	0.017	57.3
44	0.695	0.719	0.966	90	1.000	0.000	∞
45	0.707	0.707	1.000				

APPENDIX IV Table of Common Logarithms of Numbers

No.	0	1	2	3	4	5	6	7	8	9
10	0000	0043	0086	0128	0170	0212	0253	0294	0334	0374
11	0414	0453	0492	0531	0569	0607	0645	0682	0719	0755
12	0792	0828	0864	0899	0934	0969	1004	1038	1072	1106
13	1139	1173	1206	1239	1271	1303	1335	1367	1399	1430
14	1461	1492	1523	1553	1584	1614	1644	1673	1703	1732
15	1761	1790	1818	1847	1875	1903	1931	1959	1987	2014
16	2041	2068	2095	2122	2148	2175	2201	2227	2253	2279
17	2304	2330	2355	2380	2405	2430	2455	2480	2504	2529
18	2553	2577	2601	2625	2648	2672	2695	2718	2742	2765
19	2788	2810	2833	2856	2878	2900	2923	2945	2967	2989
20	3010	3032	3054	3075	3096	3118	3139	3160	3181	3201
21	3222	3243	3263	3284	3304	3324	3345	3365	3385	3404
22	3424	3444	3464	3483	3502	3522	3541	3560	3579	3598
23	3617	3636	3655	3674	3692	3711	3729	3747	3766	3784
24	3802	3820	3838	3856	3874	3892	3909	3927	3945	3962
25	3979	3997	4014	4031	4048	4065	4082	4099	4116	4133
26	4150	4166	4183	4200	4216	4232	4249	4265	4281	4298
27	4314	4330	4346	4362	4378	4393	4409	4425	4440	4456
28	4472	4487	4502	4518	4533	4548	4564	4579	4594	4609
29	4624	4639	4654	4669	4683	4698	4713	4728	4742	4757
30	4771	4786	4800	4814	4829	4843	4857	4871	4886	4900
31	4914	4928	4942	4955	4969	4983	4997	5011	5024	5038
32	5051	5065	5079	5092	5105	5119	5132	5145	5159	5172
33	5185	5198	5211	5224	5237	5250	5263	5276	5289	5302
34	5315	5328	5340	5353	5366	5378	5391	5403	5416	5248
35	5441	5453	5465	5478	5490	5502	5514	5527	5539	5551
36	5563	5575	5587	5599	5611	5623	5635	5647	5658	5670
37	5682	5694	5705	5717	5729	5740	5752	5763	5775	5786
38	5798	5809	5821	5832	5843	5855	5866	5877	5888	5899
39	5911	5922	5933	5944	5955	5966	5977	5988	5999	6010
40	6021	6031	6042	6053	6064	6075	6085	6096	6107	6117
41	6128	6138	6149	6160	6170	6180	6191	6201	6212	6222
42	6232	6243	6253	6263	6274	6284	6294	6304	6314	6325
43	6335	6345	6355	6365	6375	6385	6395	6405	6415	6425
44	6435	6444	6454	6464	6474	6484	6493	6503	6513	6522
45	6532	6542	6551	6561	6571	6580	6590	6599	6609	6618
46	6628	6637	6646	6656	6665	6675	6684	6693	6702	6712
47	6721	6730	6739	6749	6758	6767	6776	6785	6794	6803
48	6812	6821	6830	6839	6848	6857	6866	6875	6884	6893
49	6902	6911	6920	6928	6937	6946	6955	6964	6972	6981
50	6990	6998	7007	7016	7024	7033	7042	7050	7059	7067
51	7076	7084	7093	7101	7110	7118	7126	7135	7143	7152
52	7160	7168	7177	7185	7193	7202	7210	7218	7226	7235
53	7243	7251	7259	7267	7275	7284	7292	7300	7308	7316
54	7324	7332	7340	7348	7356	7364	7372	7380	7388	7396
No.	0	1	2	3	4	5	6	7	8	9

No.	0	1	2	3	4	5	6	7	8	9
55	7404	7412	7419	7427	7435	7443	7451	7459	7466	7474
56	7482	7490	7497	7505	7513	7520	7528	7536	7543	7551
57	7559	7566	7574	7582	7589	7597	7604	7612	7619	7627
58	7643	7642	7649	7657	7664	7672	7697	7686	7694	7701
59	7709	7716	7723	7731	7738	7745	7752	7760	7767	7774
60	7782	7789	7796	7803	7810	7818	7825	7832	7839	7846
61	7853	7860	7868	7875	7882	7889	7896	7903	7910	7917
62	7924	7931	7938	7945	7952	7959	7966	7973	7980	7987
63	7993	8000	8007	8014	8021	8028	8035	8041	8048	8055
64	8062	8069	8075	8082	8089	8096	8102	8109	8116	8122
65	8129	8136	8142	8149	8156	8162	8169	8176	8182	8189
66	8195	8202	8209	8215	8222	8228	8235	8241	8248	8254
67	8261	8267	8274	8280	8287	8293	8299	8306	8312	8319
68	8325	8331	8338	8344	8351	9357	8363	8370	8376	8382
69	8388	8395	8401	8407	8414	8420	8426	8432	8439	8445
70	8451	8457	8463	8470	8476	8482	8488	8494	8500	8506
71	8513	8519	8525	8531	8537	8543	8549	8555	8561	8567
72	8573	8579	8585	8591	8597	8603	8609	8615	8621	8627
73	8633	8639	8645	8651	8657	8663	8669	8675	8681	8686
74	8692	8698	8704	8710	8716	8722	8727	8733	8739	8745
75	8751	8756	8762	8768	8774	8779	8785	8791	8797	8802
76	8808	8814	8820	8825	8831	8837	8842	8848	8854	8859
77	8865	8871	8876	8882	8887	8893	8899	8904	8910	8915
78	8921	8927	8932	8938	8943	8949	8954	8960	8965	8971
79	8976	8982	8987	8993	8998	9004	9009	9015	9020	9025
80	9031	9036	9042	9047	9053	9058	9063	9069	9074	9079
81	9085	9090	9096	9101	9106	9112	9117	9122	9128	9133
82	9138	9143	9149	9154	9159	9165	9170	9175	9180	9186
83	9191	9196	9201	9206	9212	9217	9222	9227	9232	9238
84	9243	9248	9253	9258	9263	9269	9274	9279	9284	9289
85	9294	9299	9304	9309	9315	9320	9325	9330	9335	9340
86	9345	9350	9355	9360	9365	9370	9375	9380	9385	9390
87	9395	9400	9405	9410	9415	9420	9425	9430	9435	9440
88	9445	9450	9455	9460	9465	9469	9474	9479	9484	9489
89	9494	9499	9504	9509	9513	9518	9523	9528	9533	9538
90	9542	9547	9552	9557	9562	9566	9571	9576	9581	9586
91	9590	9595	9600	9605	9609	9614	9619	9624	9628	9633
92	9638	9643	9647	9652	9657	9661	9666	9671	9675	9680
93	9685	9689	9694	9699	9703	9708	9713	9717	9722	9727
94	9731	9736	9741	9745	9750	9754	9759	9763	9768	9773
95	9777	9782	9786	9791	9795	9800	9805	9809	9814	9818
96	9823	9827	9832	9836	9841	9845	9850	9854	9859	9863
97	9868	9872	9877	9881	9886	9890	9894	9899	9903	9908
98	9912	9917	9921	9926	9930	9934	9939	9943	9948	9952
99	9956	9961	9965	9969	9974	9978	9983	9987	9991	9996
No.	0	1	2	3	4	5	6	7	8	9

APPENDIX V Selected Electrical Tables
Wire Table for Standard Annealed Copper at 20°C

AWG Gauge	Diameter d, Mils	Area d², Cir Mils	Ohms/ 1000 ft	Pounds/ 1000 ft
0000	460.0	211,600	0.04901	640.5
000	409.6	167,800	0.06180	508.0
00	364.8	133,100	0.07793	402.8
0	324.9	105,500	0.09827	319.5
1	289.3	83,690	0.1239	253.3
2	257.6	66,370	0.1563	200.9
3	229.4	52,640	0.1970	159.3
4	204.3	41,740	0.2485	126.4
5	181.9	33,100	0.3133	100.2
6	162.0	26,250	0.3951	79.46
7	144.3	20,820	0.4982	63.02
8	128.5	16,510	0.6282	49.98
9	114.4	13,090	0.7921	39.63
10	101.9	10,380	0.9989	31.43
11	90.74	8,234	1.260	24.93
12	80.81	6,530	1.588	19.77
13	71.96	5,178	2.003	15.68
14	64.08	4,107	2.525	12.43
15	57.07	3,257	3.184	9.858
16	50.82	2,583	4.016	7.818
17	45.26	2,048	5.064	6.200
18	40.30	1,624	6.385	4.917
19	35.89	1,288	8.051	3.899
20	31.96	1,022	10.15	3.092
21	28.45	810.1	12.80	2.452
22	25.35	642.4	16.14	1.945
23	22.57	509.5	20.36	1.542
24	20.10	404.0	25.67	1.223
25	17.90	320.4	32.37	0.9699
26	15.94	254.1	40.81	0.7692
27	14.20	201.5	51.47	0.6100
28	12.64	159.8	64.90	0.4837
29	11.26	126.7	81.83	0.3836
30	10.03	100.5	103.2	0.3042
31	8.928	79.70	130.1	0.2413
32	7.950	62.31	164.1	0.1913
33	7.080	50.13	206.9	0.1517
34	6.305	39.75	260.9	0.1203
35	5.615	31.52	329.0	0.0954
36	5.000	25.00	414.8	0.0757
37	4.453	19.83	523.1	0.0600
38	3.965	15.72	659.6	0.0476
39	3.531	12.47	831.8	0.0377
40	3.145	9.888	1,049	0.0299

APPENDIX VI   Wiring Diagram Symbols

Ammeter	—(A)—	Inductor, air core	⌇⌇⌇⌇
Anode or plate	—⊣	Lamp	—⊃◯
Antenna	▽	Resistance, fixed	—⋀⋀⋀—
Battery	—⊣||||⊢—	Rheostat (Variable resistance)	⋀⋀⋀
Capacitor	—|(—	Single-pole switch	—∘ ∘—
Cell	—+|-—	Transformer	⌇⌇⌇⌇⌇ ⌇⌇⌇⌇⌇
Inductor, iron core	⌇⌇⌇⌇⌇⌇	Triode	G F ⊳|| P
Directly heated filament	▷	Variable capacitor	—⊁|⊢
Double-pole switch		Voltmeter	—(V)—
Grid	— - - -	Wire crossing, connection	—┼—
Ground connection	⏚	Wire crossing, no connection	—⌒—
Source of alternating emf	—(∿)—	Junction transistor (NPN)	E C B

APPENDIX VII Letters of the Greek Alphabet

Capital	Small		English Equivalent
A	α	Alpha	a
B	β	Beta	b
Γ	γ	Gamma	g
Δ	δ	Delta	d
E	ε	Epsilon	e
Z	ζ	Zeta	z
H	η	Eta	$\bar{e}$
Θ	θ	Theta	th
I	ι	Iota	i
K	κ	Kappa	k
Λ	λ	Lambda	l
M	μ	Mu	m
N	ν	Nu	n
Ξ	ξ	Xi	x
O	o	Omicron	o
Π	π	Pi	p
P	ρ	Rho	r
Σ	σ	Sigma	s
T	τ	Tau	t
Υ	υ	Upsilon	u
Φ	ϕ	Phi	ph
X	χ	Chi	ch
Ψ	ψ	Psi	ps
Ω	ω	Omega	$\bar{o}$

Answers to Odd-Numbered Problems

Chapter One

1. (a) $6 \times 10^{-4} = 0.0006$
   (b) $3.5 \times 10^{-3} = 0.0035$
   (c) $5 \times 10^{-6} = 0.000005$
   (d) $75 \times 10^{-3} = 0.075$

3. (a) 2.225×10^4
   (b) $2.585 = 10^7$
   (c) 6.29×10^2
   (d) 6.29×10^{-3}
   (e) 7.85×10^{-5}
   (f) 8.4×10^{-2}

5. (a) $I = V/R$
   (b) $W = Fh/L$
   (c) $t = \sqrt{2s/g}$
   (d) $m = Fr/v^2$

7. 7.75 percent

9. 10.19 acres

11. 7.64 in.

13. 1260 ft^2

15. 143,900 gal

17. 268 ft^3

19. $27.2°$

21. 115.5 ft

23. 47 percent

25. 5.04 gal

27. 2510 ft^2

Chapter Two

1. (a) 0.787 in.
   (b) 17.7 in.
   (c) 1.99 cm
   (d) $0.131 \ \mu\text{m}$
   (e) 111.2 km
   (f) 64.75 hectares

3. 88.6 km/hr

5. 7.26 kg (equiv.)

7. (a) 324 mg
   (b) 15.4 (multiple)

9. 2.49 in.^3

11. 12.82 mm

13. 13.2 gal

15. 26 kilos

17. 316 in.^3

19. 1.283 sec

Chapter Three

1. 19.8 km at 340° T

3. 1825 mi

5. $v_N = 115$ m/sec;
   $v_E = 164$ m/sec

7. 3400 lb

9. $R = 84.5$ lb at 20.2° with horiz.

11. $N = 3.2$ lb; $P = 3.9$ lb

13. $AB = 405$ lb; $\theta = 37°$; $CD = 356$ lb

15. $F = 1200$ lb; $x = 4.92$ ft

17. Ball w/r auto, $\overline{BA} = 20$ m/sec;
    Ball w/r ground, $\overline{BG} = 29.9$ m/sec at 41° with auto's path

19. 28.4 kts at 93° with ship's heading

21. Compass heading 065°; Time 4.6 hr

Chapter Four

1. 1.57 hr

3. 61.6 ft/sec^2

5. 44.1 m

7. 280 m/sec

9. 39.4 ft

11. 1.15 m/sec^2

13. 67.9 N

15. 2730 N

17. 794 ft/sec

19. 61/6 m

21. 2.76 m/sec^2

23. 3.91 sec

25. $R = 52,000$ ft; $\theta = 42.7°$

Chapter Five

1. 17,170 N-m (joules)

3. 19,620 N-m (joules)

5. −4.76 m/sec (minus sign shows recoil in oppos. direction from projectile motion)

7. 1.88 m/sec

9. 110 lb/min

11. 88.3 kW

13. 3.17 hp

15. 47.1 sec

17. 95.9 percent

19. 22,500 lb (force)

21. -8.0×10^5 **N** (sign indicates direction of force)

23. 12 kg-m/sec

25. 58.4 MW

27. 10,000 lb

Chapter Six

1. 8.7 ft

3. $R = 10.6$ kg-f; $L = 30.6$ kg-f

5. 3750 **N**

7. 27.8 kg-f

9. 18.2 percent

11. Power = 21.4 kW; thrust at each rear wheel = 1925 **N**

13. (*a*) 38.78 ft/sec; (*b*) 38.74 ft/sec

15. 202 lb

17. 0.054

19. 54,000 lb

21. 122 lb

23. 27.4

25. 86 rpm

Chapter Seven

1. (*a*) 7.272×10^{-5} rad/sec; (*b*) 1667 km/hr

3. 183 rad/sec

5. 3.142 m/sec

7. 7.88 ft/sec

9. 1185 lb

11. 0.20

13. $4.57\,g$

15. 75.1°

17. 17,050 mi/hr Note that mass of satellite does not matter.

19. 27.5 kilos

21. (*a*) 12,950 km/hr; (*b*) 1.69 hr

23. $2.729 = 10^{-3}$ m/sec^2

25. 71.6 lb

27. 25,000 mi/hr Note mass is not a factor.

29. Show identity: 0.734 ft/sec^2 = 0.734 ft/sec^2

Chapter Eight

1. 25 **J**

3. 22.9 hp

5. 2270 rev/min (rpm)

7. (*a*) −5.24 rad/sec^2 (minus sign shows deceleration) (*b*) 1000 revs.

9. (*a*) −0.0524 rad/sec^2; (*b*) 10.46 rad

11. 5600 lb-ft^2

13. (*a*) 96.9×10^{36} kg-m^2; (*b*) 1.28×10^{22} **N**

15. 183 lb-ft

17. 3.75 rev/sec (rps)

19. KE$_{\text{trans}}$ = 158 kg-m^2/sec^2 (**J**); KE$_{\text{rot}}$ = 63.4 **J** KE$_{\text{tot}}$ = 221.4 **J**

21. 9.88×10^8 kg-m^2/sec^2 (**J**); Power (ave. over 4 hrs) = 68.6 kW

Chapter Nine

No numerical answers

Chapter Ten

1. 3 in.

3. 44.3 lb

5. 18,100 lb/in.^{2}

7. 6870 lb

9. 8.8

11. 0.663 cm

13. 0.029 cm

15. 17.1×10^{10} **Nm^2**

17. 0.0215 in.

19. 2.24 mm

21. Three cables

23. 2.1×10^{-3} m^3

25. 200 rad ~ 32 rev

Chapter Eleven

1. The forces will be the same.

3. 1800 kg/m^3

5. 52 lb/in.^{2}

7. 53.3 lb

9. 3 cm

11. 18,700 lb

13. 27.3 ft/sec

15. 26.7 m

17. 51.2 lb/ft^3

19. (*a*) 0.019 ft^3; (*b*) 439 lb/ft^3

21. 64,000 ft^3/hr

23. (*a*) v_B = 3.42 m/sec v_C = 0.855 m/sec (*b*) h_A = 0.118 m; h_B = 0.596 m; h_C = 0.149 m

25. 15.4 lb/in.^{2}

Chapter Twelve

1. 2.18 lb

3. 25 ft

5. 2.2 lb/in.^{2} abs (psia)

7. (a) 1.53 m/sec;
    (b) 1.3×10^{-2} m^3/min

9. (a) 2.42 lb/in.^{2};
    (b) 1.67 **N**/m^2

11. 177 cm^2

13. 1.62 ft^3

15. 647/1397

17. (a) 36.4 lb/in.^{2};
    (b) 147/364 V_1;
    (c) 0.203 lb/ft^3

19. 871 in.^{3}

Chapter Thirteen

1. 20°C

3. 185°C

5. (a) Not numerical;
    (b) +10.4°F

7. 350 K

9. 3.57 ft

11. 0.146 mm

13. 6.85 ft too long

15. 4.06 gal

17. 13.36 gm/cm^3

19. 336°C

21. 2.42×10^5 Pa

23. 25.7 lb/in.^{2} gauge

25. 12.56×10^5 **N**/m^2 (Pa)

Chapter Fourteen

1. 1501 Btu

3. 69,270 kcal

5. 66.0°F

7. 1190 gal

9. 206 liters/day

11. 9440 Btu

13. 44.7 kcal

15. 28 lb

17. 13°C

19. 1031 Btu/lb

21. 284,000 Btu/hr

23. 187 gal/hr

Chapter Fifteen

1. Pine wall thickness is 12.44 in.
    Brick wall thickness is 74 in.

3. 30,800 Btu/hr

5. 38,500 Btu/hr

7. 32,140 Btu/hr

9. 19 lb/day

11. 2.53 kcal/min

13. 7.10 lb/hr

15. 0.429 cm/hr

Chapter Sixteen

1. 22.1 percent

3. 30.6°F

5. 58.7 percent (No real turbine
    is this efficient. In actual
    practice, 38 to 40 percent is
    upper limit).

7. (a) 324 Btu;
    (b) 2.52×10^5 ft

9. 0.79 kcal

11. 43.1 percent

13. (a) 1.155×10^{12} ft-lb;
    (b) 34.3 percent

15. 1785 hp

17. 31.9 percent

19. 819 m/sec

21. 50.7 percent

Chapter Seventeen

1. 0.15 tons of refrigeration

3. 130.5 tons

5. 20.9 tons

7. c.c.p. = 5.17

9. 35.8 percent

11. 36,000 lb/hr

13. 19.4 gal/hr

15. 1060 ft^3/min (cfm)

17. 67.3 lb/min

19. 301,000 Btu/hr

21. 13.7 kW

Chapter Eighteen

1. 2.5×10^{-3} sec

3. 360 m

5. Multiplier is $\sqrt{2}$

7. (a) 9.42 in./sec;
    (b) 29.6 in./sec^2

9. (a) 5.7 cm;
    (b) 350 cm/sec^2

11. (a) 0.25 sec^{-1};
    (b) 4.7 cm/sec

13. (a) $\frac{1}{2}$ sec
    (b) 32 cm

15. 0.14 cm/sec

17. 3/1

19. 66 lb

21. 9.58 in.

23. 1.737 sec

25. (a) $v = 0$ cm/sec;
    $a = -26.30$ m/sec^2
    (b) $v = 109$ cm/sec;
    $a = -13.20$ m/sec^2
    (c) $v = 126$ cm/sec;
    $a = 0$ cm/sec^2

Chapter Nineteen

1. 0:1 (approx.)

3. 8.58 sec

5. 77°F

7. (a) 1150 m;
    (b) 3.74 cm

9. 1060 m

11. (a) 3220 ft;
    (b) 16,100 ft/sec

13. 1400 m/sec, or 4700 ft/sec

15. $v = 1480$ m/sec; $s = 3150$ m

17. I = 10^{-7} W/m^2

19. 950 m/sec

21. 2.6 ft

23. 13,400 ft

25. 133 db

Chapter Twenty

7. 8190 ft
9. 17 m
11. 450 ft
13. 361 Hz
15. (a) 4240 yd;
    (b) 28.4 kts
17. 833 ft^2

Chapter Twenty-One

1. 24.5 sec
3. 2.8 lux (ft-c)
5. 71 cd
7. 0.34 m
9. 600 km
11. 11.5 lux
13. 581.25 rev/sec (rps)

Chapter Twenty-Two

1. −4 diopters
3. 37.1°
5. 0.9 m
7. −0.30 m
9. 28.6 in.
11. 20 in
13. 3
15. (a) 2.5 cm
    (b) 5.0 cm
17. (a) 1 cm;
    (b) 0.5 cm
19. (a) 30°;
    (b) 19°
21. 26.7 in.
23. 26,400 ft

Chapter Twenty-Three

1. 4.1
3. 1.8 in.
5. 5.625

7. (a) 4.0 cm;
   (b) 4.03 cm
9. $M = 8; D_o = 5.25$
11. 85.8 cm
13. (a) −1.25 diopters;
    (b) 1.54 m
15. (a) −1.33 diopters;
    (b) 12 m
17. $M = -5.34$; image is 53.33 cm
    from object
19. Dispersion is 2.5°

Chapter Twenty-Four

1. 1.1×10^2 **N** (attraction)
3. 8.40×10^{-15} **N**
5. 1.5×10^5 V
7. 180 **J**
9. 20,000 **N/C**
11. 2.75×10^6 N/C (toward the 40
    μC charge)
13. (a) 8.0×10^{-15} **N**;
    (b) 8.8×10^{15} m/sec^2
15. 21.8 μF
17. 4.4 μF
19. 38,000 V
21. (a) 1230 m/sec;
    (b) 3.89×10^{-5} sec
23. 0.234 **N**

Chapter Twenty-Five

1. 750 mils
3. 18,000 C
5. AWG 32 (probably)
7. 7.13 m
9. 11.4 Ω
11. 2.3 A
13. 2.0 Ω
15. (a) 12 Ω;
    (b) 2 A
17. 9.79 Ω
19. 96 V

21. 10.0 V
23. $I_2 = 0.34$ A;
    $I_1 = 0.28$ A;
    $I_3 = 0.62$ A
25. $I_1 = 46/37$ A;
    $I_2 = 32/37$ A;
    $I_3 = 78/37$ A

Chapter Twenty-Six

1. 22.4 kW
3. 140 A
5. 37 min
7. 8.73 Ω
9. 67.6 percent
11. (a) 0.882 A;
    (b) 0.050 A
13. $L_T = \dfrac{\varepsilon}{r/n + R}$
15. $107.75
17. $4.09
19. (a) 76 V;   (b) 95.3 V
21. 770 ft
23. (a) 2.3 Ω;
    (b) 0.86 V

Chapter Twenty-Seven

1. 8×10^{-3} **N**
3. 6360 A-turns
5. 3.3×10^{-5} Wb/m^2 (T)
7. 0.406 A
9. (a) 1200 A-turns
   (b) 4800 A-turns
11. 4.7×10^{-3} **N**
13. 0.625 **N**
15. 1.15 **N/m**
17. 0.02 Ω
19. (a) 1200 Ω
    (b) 1825 Ω

Chapter Twenty-Eight

1. 1.2×10^{-2} V
3. 0.46 sec
5. 3.8 V
7. 11.25 V
9. 0.67 V
11. 3.2 H
13. 0.25 H
15. (a) 1.2×10^5 V;
    (b) 3.3×10^{-2} Wb

Chapter Twenty-Nine

1. 163 V
3. 2.9 A
5. $N_s = 1600$
7. 660 Ω
9. 550 W
11. (a) 187.5 kV;
    (b) 0.267 A
13. 49 μH
15. 35 μF
17. 2.60×10^{-7} H
19. 56.5 Ω
21. (a) 1.17 H;
    (b) 30 A;
    (c) 13,300 V

Chapter Thirty

1. 60.6 percent
3. 3600 rev/min (rpm)
5. 900 rev/min (rpm)
7. 550 V
9. 2.8 percent

Chapter Thirty-One

1. 6.55×10^{21} electron charges
3. 39.4 ft
5. 360 rev/min (rpm)
7. (a) 1.156×10^9 Btu/hr;
   (b) 159,000 hp;
   (c) 16,200 gal/hr
9. (a) 960 kW (Case 1);
   (b) 93.8 kW (Case 2);
   (c) $519.72/day
11. (a) 0.57;
    (b) 877 Ω;
    (c) 720 Ω;
    (d) 6.50 μF

Chapter Thirty-Two

1. 46 kHz (kilocycles, or kc)
3. 212.8 m
5. 80.7 naut. mi
7. 6.67 nanosec
9. 11.1 m

Chapter Thirty-Three

1. 32.04×10^{-16} J
3. 3.777×10^{-19} J
5. 2.954 eV (this is less than given value, 3.8 eV)
7. 7.225×10^5 m/sec
9. 2.176 eV
11. 8.267×10^{-3} Å
13. Electron $\lambda = 2.91$ Å;
    Proton $\lambda = 0.00158$ Å
15. 10.19×10^{-24} kg-m
17. $0.176\,c$ (the speed of light)

Chapter Thirty-Four

1. 207 amu (to the nearest whole number)
3. Power = 1.8×10^{13} kW, or 2.41×10^{13} hp
5. Of an initial number N, in 16 years 0.125 N will remain
7. 40 rems
9. 2.392×10^{-4} kg/day
11. 548.1 kg

Index

Lux, 425
Lux meter, 426

Mach number, 392
Machines:
  basic, 107–112
  compound, 128, 129
  defined, 5, 107
  efficiency of, 117–128
  mechanical advantage of, 108 ff.
  and the principle of work, 117
"Magnetic bottle" fusion research, 740–743
Magnetic field, 560–68
  around current-carrying wire, 564, 567
  direction of, 565
  around the earth, 561
  intensity of, 569–572
  of solenoid, 565, 566
Magnetic flux density, 561, 566
Magnetic permeability, 569
Magnetic poles, 558–561
Magnetic separators, 573
Magnetic shielding, 569
Magnetism:
  induced, 566–572
  terrestrial, 561
  theories of, 562, 563
Magneto, 619
Magnetohydrodynamics (MHD), 331, 332
Magnets, artificial, 557
  bar, 558
  types of, 558
Magnification, by lenses, 474, 476
  by mirrors, 446
Magnifier, 474
Malleability, 196
Manometer, 238, 239
Marconi, Guglielmo, 683
Maser, 435
Mass:
  of the earth, 145

defined, 25
and force, 74–78
and inertia, 73, 74
measurement of, 31, 32
standards and units of, 25, 26
and weight, contrasted, 25, 73–76
Mass defect of atomic nucleus, 727
Mass density, 181
Mass-energy equivalence, 728 ff.
Matrix, 414
Matter:
  definition, 2
  states of, 175, 270, 271
  structure of, 177, 698 ff.
  waves, 711
Maxwell, James Clerk, 283, 419
Maxwellian distribution curve, 283, 284
Measurement:
  accuracy and precision of, 32–34
  English system, 24–26
  SI-metric system, 23–25
  units of, 23 ff.
Measuring instruments, 27–31
Mechanical advantage, defined, 108
  of simple machines, 111–127
Mechanical equivalent of heat (Joule's constant, J), 309 ff.
Melting (fusion), 277–280
Melting points, table of, 279
Mercurial barometer, 233
Meson, 727, 751
Metal fatigue, 189
Meter, unit of length, 23
Metric units, 23–32
Michelson, Albert A., 421
Micrometer caliper, 27, 28
Microprocessor chips, 670
Microscope, electron, 716–718
  optical, 475, 476

Microwaves, 421
Mil, unit of wire diameter, 517
Millibar, 205
Millikan, Robert A., 726
Mirage, 449
Mirror equation, 445
Mirrors:
  aperture, 442
  concave, 443–447
  convex, 443–447
  focus of, 442
  magnification by, 446
  parabolic, 447
  plane, 440
  reflection in, 440
  spherical, 441
  spherical aberration of, 446
Missiles and rockets, 98, 99, 328–330
Mixtures, 178
Modulation, amplitude (AM), 685
  frequency (FM), 687
Modulus of elasticity, Young's, 186
Mohs' scale of hardness, 194
Molecular attraction, 175
Molecular kinetic energy, 254
Molecular motion:
  and heat, 253–255, 282–284
  and pressure of a gas, 177
Molecular repulsion, 178
Molecular theory, 175–179, 253–255, 282–284
Moment of a force, 47, 48, 157, 158
Moment of inertia, 163–170
  and torque, 163
  values for some common bodies, 165
Momentum:
  angular, 166–170
  and collision, 97
  conservation of, 96, 97, 166–169
  defined, 93